ELECTRONIC COMMUNICATIONS SYSTEMS

Fundamentals Through Advanced

Wayne Tomasi
Mesa Community College

 PRENTICE HALL, Englewood Cliffs, New Jersey 07632

Library of Congress Cataloging-in-Publication Data

TOMASI, WAYNE.
  Electronic communications systems.

  Includes index.
  1. Telecommunication systems. I. Title.
TK5101.T625 1988   621.38′0413   87–7205
ISBN 0-13-250804-4

Chapters 1 through 12 are published as
Fundamentals of Electronic Communications Systems
by Wayne Tomasi (© 1988);
chapters 13 through 22 are published as
Advanced Electronic Communications Systems
by Wayne Tomasi (© 1987).

Editorial/production supervision and
interior design: **Kathryn Pavelec**
Cover design: **Diane Saxe**
Cover photo: Courtesy of **Sperry Corporation**
Manufacturing buyer: **Margaret Rizzi/Lorraine Fumoso**

 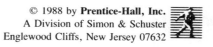
Printed in the United States of America

10 9 8 7 6 5 4 3 2 1

ISBN 0-13-250804-4

PRENTICE-HALL INTERNATIONAL (UK) LIMITED, *London*
PRENTICE-HALL OF AUSTRALIA PTY. LIMITED, *Sydney*
PRENTICE-HALL CANADA INC., *Toronto*
PRENTICE-HALL HISPANOAMERICANA, S.A., *Mexico*
PRENTICE-HALL OF INDIA PRIVATE LIMITED, *New Delhi*
PRENTICE-HALL OF JAPAN, INC., *Tokyo*
SIMON & SCHUSTER ASIA PTE. LTD., *Singapore*
EDITORA PRENTICE-HALL DO BRASIL, LTDA., *Rio de Janeiro*

*To my loving
and very patient wife,
Cheryl*

CONTENTS

wk 1

wk 2

wk 3

wk 4

wk 5 & 6

Chapter 11   **ANTENNAS**   423

Chapter 12   **BASIC TELEVISION PRINCIPLES**   452

Chapter 13   **DIGITAL COMMUNICATIONS**   487

Chapter 14   DATA COMMUNICATIONS   536

Chapter 15   DATA COMMUNICATIONS PROTOCOLS   584

wk 8

out

wk 9

WK 10

WK 11

WK 12

PREFACE

During the past two decades, the electronic communications industry has undergone some remarkable technological changes, primarily in the form of miniaturization. In the late 1950s and early 1960s, vacuum tubes were replaced with transistors. More recently, transistors are being replaced with multipurpose large-scale integrated circuits. The development of large-scale digital and linear integrated circuits has paved the way for many new and innovative approaches to electronic communications. Also, digital electronics principles are being implemented into electronic communications circuits and systems more and more each year. The need for these changes is attrributed to the continuing increase in the number of digital and data communications systems.

The purpose of this book is to introduce basic electronic communications fundamentals and to expand the knowledge of the reader to more modern digital and data communications systems. The book was written so that a reader with previous knowledge in basic electronic principles and an understanding of mathematics through trigonometry will have little trouble grasping the concepts presented. An understanding of calculus principles (i.e., differentiation and integration) would be helpful but is not a prerequisite. Within the text, there are numerous examples that emphasize the important concepts, and questions and problems are included at the end of each chapter. Also, answers to the odd-numbered problems are given at the end of the book.

Chapter 1 is an introduction to electronic communications. Fundamental communications terms and concepts such as modulation, demodulation, bandwidth, and information capacity are explained. Signal analysis using both frequency and time domain is discussed. Nonsinusoidal periodic waves are examined using fourier analysis, and the effects of bandlimiting on signals is discussed. Electrical noise, signal-to-noise ratio, and noise figure are explained and discussed. The basic principles of frequency and time division multiplexing are also explained in Chapter 1. Chapter 2 covers frequency generation. The basic requirements for oscillations to occur are outlined and discussed. Standard LC, crystal, and negative resistance oscillator configurations are explained. The basic concepts of frequency multiplication are discussed. Linear and nonlinear mixing with single and multiple frequency input signals are analyzed. Chapter 3 defines and explains the basic concepts of amplitude modulation transmission. A detailed analysis is presented on the voltage, power, and bandwidth considerations of amplitude modulation in both the frequency and time domain. Circuits for generating amplitude modulation are explained. High and low level transmitters are discussed. Chapter 4 introduces the basic concepts of radio receivers including a detailed analysis of tuned radio frequency and superheterodyne receivers. The primary functions of each receiver stage are explained. Amplitude modulation receivers and the detection of amplitude modulated signals are explained. The basic concepts of automatic gain control and squelch are discussed and double conversion receivers are introduced. Chapter 5 is dedicated to the analysis of phase-locked loops and frequency synthesizers. Basic phase-locked loop concepts are introduced and a detailed explanation of the operation of a phase-locked loop is given. Both direct and indirect frequency synthesizers with single and multiple crystals are discussed. Chapter 6 extends the coverage of amplitude modulation given in Chapters 3 and 4 to single sideband transmission. The various types of sideband transmission are explained and contrasted. Circuits for generating single sideband waveforms are explained. Several types of single sideband receivers are discussed in Chapter 6. Chapter 7 introduces the basic concepts of angle modulation. Frequency and phase modulation are explained and contrasted. The amplitude, power, and frequency characteristic of an angle modulated wave are explained in detail. Both direct and indirect frequency and phase modulation transmitters are shown and discussed in detail. Chapter 8 extends the coverage of angle modulation to receivers. Several types of angle modulation demodulators are introduced and discussed. Frequency modulation stereo transmission, two-way frequency modulation communications, and mobile telephone communications including cellular radio are discussed. Chapter 9 explains the characteristics of an electromagnetic wave and wave propagation on a metallic transmission line. Several basic transmission line configurations are discussed and contrasted. Incident and reflected energy and the concept of standing waves are discussed in Chapter 9. Transmission line characteristic impedance and input impedance are also introduced. The concept of a matched line is explained and the consequences of a mismatched line are discussed. Chapter 10 extends the coverage of wave propagation to free space. Electromagnetic radiation concepts are explained. Spherical wavefronts are analyzed and the inverse square law is derived. Wave attenuation and absorption are covered. Chapter 10 gives a complete explanation of the optical properties of electromagnetic waves: refraction, reflection, diffraction, and interference. Ground, space, and sky wave propagation are discussed, and the fundamental limits for free space wave propagation are defined and explained. Chapter 11 introduces the antenna and describes basic antenna operation. Fundamental antenna terms are defined and explained. Radiation patterns are explained. The most basic antenna, the elementary doublet, is explained. The basic half- and quarter-wave antenna are explained along with the effects of the ground on the wave, and several antenna-loading techniques are described. Some of the more common antenna arrays and special-purpose configurations are explained including the following: folded dipole, log-periodic and loop antennas. Chapter 12 introduces the basic concepts of television broadcasting. Both monochrome and color television transmission and reception are explained. Generation of the composite video signal is explained. Basic transmitter and receiver circuits are explained. The basic concepts of scanning, blanking, and synchronization are discussed.

Chapter 13 introduces the concepts of digital transmission and digital modulation. In this chapter the most common modulation schemes used in modern digital radio systems—FSK, PSK, and QAM—are described. The concepts of information capacity and bandwidth efficiency are explained. Chapter 14 introduces the field of data communications. Detailed explanations are given for numerous data communications concepts, including transmission methods, circuit configurations, topologies, character codes, error control mechanisms, data formats, and data modems. Chapter 15 describes data communications protocols. Synchronous and asynchronous data protocols are first defined, then explicit examples are given for each. The most popular character- and bit-oriented protocols are described. In Chapter 15 the basic concepts of a public data network and a local area network are outlined, and the international user-to-network packet switching protocol, X.25, is explained. Chapter 16 introduces digital transmission techniques. This includes a detailed explanation of pulse code modulation. The concepts of sampling, encoding, and companding (both analog and digital) are explained. Chapter 16 also includes descriptions of two lesser-known digital transmission techniques: adaptive delta modulation PCM and differential PCM. Chapter 17 explains the multiplexing of digital signals. Time-division multiplexing is discussed in detail and the operation of a modern LSI combo chip is explained. The North American Digital Hierarchy for digital transmission is outlined, including explanations of line encoding schemes, error detection/correction methods, and synchronization techniques. In Chapter 18 analog multiplexing is explained and AT&T's North American frequency-division-multiplexing hierarchy is described. Several methods are explained in which digital information can be transmitted with analog signals over the same communications medium. Chapter 19 introduces microwave radio communications and the concept of system gain. A block diagram approach to the operation of a microwave radio system is presented and numerous examples are included. In Chapter 20 satellite communications is introduced and the basic concepts of orbital patterns, radiation patterns, geosynchronous, and nonsynchronous systems are covered. System parameters and link equations are discussed and a detailed explanation of a satellite link budget is given. Chapter 21 extends the coverage of satellite systems to methods of multiple accessing. The three predominant methods for multiple accessing—frequency-division, time-division, and code-division multiple accessing—are explained. Chapter 22 covers the basic concepts of a fiber optic communications system. A detailed explanation is given for light-wave propagation through a guided fiber. Also, several light sources and light detectors are discussed, contrasting their advantages and disadvantages. Appendix A describes the Smith chart and how it is used for transmission line calculations. Examples are given for calculating input impedance, quarter-wave transformer matching, and shorted stub matching.

WAYNE TOMASI

ACKNOWLEDGMENTS

I would like to acknowledge the following individuals for their contributions to this book: Gregory Burnell, Executive Editor and Assistant Vice President, Electronic Technology, for giving me the opportunity to write this book; Kathryn Pavelec, Production Editor, for deciphering my manuscript, and providing a pleasant and professional working environment; and the two reviewers of my manuscript who corrected several of my mathematical errors and provided invaluable constructive criticism—Robert E. Greenwood, Ryerson Polytechnic Institute; and James W. Stewart, DeVry, Woodbridge.

Chapter 1

INTRODUCTION TO ELECTRONIC COMMUNICATIONS

INTRODUCTION

In essence, *electronic communications* is the transmission, reception, and processing of information with the use of electronic circuits. The basic concepts involved in electronic communications have not changed much since their inception, although the methods by which these concepts are implemented have undergone dramatic changes.

Figure 1-1 shows a communications system in its simplest form, which comprises three primary sections: a *source* (transmitter), a *destination* (receiver), and a *transmission medium* (wire pair, coaxial cable, fiber link, or free space).

The *information* that is propagated through a communications system can be *analog* (proportional), such as the human voice, video picture information, or music; or it can be *digital pulses*, such as binary-coded numbers, alpha/numeric codes, graphic symbols, microprocessor op-codes, or data base information. However, very often the source information is unsuitable for transmission in its original form and therefore must be converted to a more suitable form prior to transmission. For example, with digital communications systems, analog information is converted to digital form prior to transmission, and with analog communications systems, digital data are converted to analog signals prior to transmission. This book is concerned with the basic concepts of conventional analog radio communications and the concepts of digital radio communications.

Modulation and Demodulation

For reasons that are explained in Chapter 10, it is impractical to propagate low-frequency electromagnetic energy through the earth's atmosphere. Therefore, with *radio communi-*

1

FIGURE 1-1 Block diagram of a communications system in its simplest form.

cations, it is necessary to superimpose a relatively low-frequency intelligence signal onto a relatively high-frequency signal for transmission. In electronic communications systems, the source information (intelligence signal) acts upon or modulates a single-frequency sinusoidal signal. *Modulate* simply means to vary or change. Therefore, the source information is called the *modulating signal*, the signal that is acted upon (modulated) is called the *carrier*, and the resultant signal is called the *modulated wave*. In essence, the source information is transported through the system on the carrier.

With analog communications systems, *modulation* is the process of changing some property of an analog carrier in accordance with the original source information and then transmitting the modulated carrier. Conversely, *demodulation* is the process of converting the changes in the analog carrier back to the original source information. The total or *composite information signal* that modulates the main carrier is called *baseband*. The baseband is converted from its original frequency band to a band more suitable for transmission through the communications system. Baseband signals are *up-converted* at the transmitter and *down-converted* at the receiver. *Frequency translation* is the process of converting a frequency or band of frequencies to another location in the total frequency spectrum.

Equation 1-1 is the general expression for a time varying sine wave of voltage such as an analog carrier. There are three properties of a sine wave that can be varied: the *amplitude* (*V*), the *frequency* (*F*), or the *phase* (θ). If the amplitude of the carrier is varied proportional to the source information, *amplitude modulation* (AM) results. If the frequency of the carrier is varied proportional to the source information, *frequency modulation* (FM) results. If the phase of the carrier is varied proportional to the source information, *phase modulation* (PM) results.

$$v = V \sin (2\pi Ft + \theta) \tag{1-1}$$

where

v = time-varying sine wave of voltage
V = peak amplitude (V)

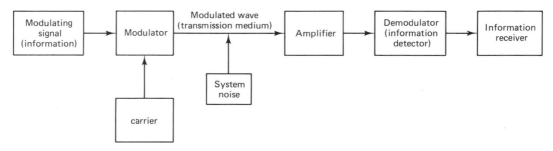

FIGURE 1-2 Communications system block diagram.

F = frequency (Hz)

θ = phase (deg)

Figure 1-2 is a simplified block diagram for a communications system showing the relationships among the modulating signal (information), the modulated signal (carrier), the modulated wave (resultant), and the system noise. Again, the frequency of the carrier is relatively high as compared to the frequency of the intelligence signal.

Transmission Frequencies

In the United States, frequency assignments for *free-space radio propagation* are assigned by the *Federal Communications Commission* (FCC). The exact frequencies assigned specific transmitters operating in the various classes of service are constantly being updated and altered to meet the nation's communications needs. However, the general division of the total usable frequency spectrum is decided at the *International Telecommunications Conventions*, which are held approximately once every 10 years.

The usable *radio-frequency* (RF) spectrum is divided into narrower frequency bands which are given descriptive names and band numbers. *The International Radio Consultative Committee's* (CCIR) designations are listed in Table 1-1. Several of these bands are further broken down into various services, which include shipboard search, microwave, satellite, mobile land-based search, shipboard navigation, aircraft approach, airport surface detection, airborne weather, mobile telephone, and many more.

TABLE 1-1   CCIR BAND DESIGNATIONS

Band number	Frequency range[a]	Designation
2	30–300 Hz	ELF (extremely low frequencies)
3	0.3–3 kHz	VF (voice frequencies)
4	3–30 kHz	VLF (very low frequencies)
5	30–300 kHz	LF (low frequencies)
6	0.3–3 MHz	MF (medium frequencies)
7	3–30 MHz	HF (high frequencies)
8	30–300 MHz	VHF (very high frequencies)
9	0.3–3 GHz	UHF (ultra high frequencies)
10	3–30 GHz	SHF (super high frequencies)
11	30–300 GHz	EHF (extremely high frequencies)
12	0.3–3 THz	Infrared light
13	3–30 THz	Unassigned
14	30–300 THz	Visible-light spectrum
15	0.3–3 PHz	Ultraviolet light
16	3–30 PHz	X-ray
17	30–300 PHz	Unassigned
18	0.3–3 EHz	Gamma rays
19	3–30 EHz	Cosmic rays

[a] 10^0, hertz (Hz); 10^3, kilohertz (kHz); 10^6, megahertz (MHz); 10^9, gigahertz (GHz); 10^{12}, terahertz (THz); 10^{15}, petahertz (PHz); 10^{18}, exahertz (EHz).

TABLE 1-2   EMISSION CLASSIFICATIONS

Type of modulation or emission	Type of information	Supplementary characters
A Amplitude	0 Carrier on only	None Double sideband, full carrier
F Frequency	1 Carrier on–off	
P Pulse	2 Carrier on, keyed tone—on–off	a Single sideband
	3 Telephony, voice, or music	b Two independent sidebands
	4 Facsimile, nonmoving or slow-scan TV	c Vestigal sideband
		d Pulse amplitude modulation (PAM)
	5 Vestigal sideband, commercial TV	e Pulse width modulation (PWM)
	6 Four-frequency diplex telephony	f Pulse position modulation
	7 Multiple sidebands	g Digital video
	8 Unassigned	h Single sideband, full carrier
	9 General, all others	i Single sideband, no carrier

Classification of Transmitters

For licensing purposes in the United States, radio transmitters are classified according to their bandwidth, type of modulation, and type of intelligence information. The emission classification is identified by combinations of letters and numbers as shown in Table 1-2. Emission designations include an uppercase letter which identifies the type of modulation, a number which identifies the type of emission, and a lowercase subscript which further defines the emission characteristics. For example, the designation A3a describes a single-sideband, reduced-carrier, amplitude-modulated signal carrying voice or music information.

Bandwidth and Information Capacity

The two most significant limitations on system performance are *noise* and *bandwidth*. The significance of noise is discussed later in this chapter. The bandwidth of a communications system is the minimum passband required to propagate the source information through the system. The bandwidth of a communications system must be sufficiently large to pass all significant information frequencies.

The *information capacity* of a communications system is a measure of how much source information can be carried through the system in a given period of time. The amount of information that can be propagated through a transmission system is proportional to the product of the system bandwidth and the time of transmission. The relationship among bandwidth, transmission time, and information capacity was developed by R. Hartley of Bell Telephone Laboratories in 1928. Simply stated, Hartley's law is

$$C \propto B \times T \qquad (1-2)$$

where

> C = information capacity
> B = bandwidth (Hz)
> T = transmission time (s)

Equation 1-2 shows that information capacity is a linear function and is directly proportional to both the system bandwidth and the transmission time. If either the bandwidth or the transmission time changes, the information capacity changes by the same proportion.

Approximately 3 kHz of bandwidth is required to transmit basic voice-quality telephone signals. Approximately 200 kHz of bandwidth is required for FM transmission of high-fidelity music, and almost 6 MHz of bandwidth is required for broadcast-quality television signals (i.e., the more information per unit time, the more bandwidth required).

SIGNAL ANALYSIS

When designing electronic communications' circuits, it is often necessary to analyze and predict the performance of the circuit based on the power distribution and frequency composition of the information signal. This is done with mathematical *signal analysis*. Although all signals in electronic communications are not single-frequency sine or cosine waves, many of them are, and the signals that are not can be represented by a series of sine and cosine functions.

Sinusoidal Signals

In essence, signal analysis is the mathematical analysis of the frequency, bandwidth, and voltage level of a signal. Electrical signals are voltage- or current-time variations that can be represented by a series of sine or cosine waves. Mathematically, a single-frequency voltage or current waveform is

$$v = V \sin (2\pi F t + \theta) \qquad \text{or} \qquad v = V \cos (2\pi F t + \theta)$$
$$i = I \sin (2\pi F t + \theta) \qquad \text{or} \qquad i = I \cos (2\pi F t + \theta)$$

where

> v = time-varying voltage wave
> V = peak voltage (V)
> i = time-varying current wave
> I = peak current (A)
> F = frequency (Hz)
> θ = phase (deg)

Whether a sine or a cosine function is used to represent a signal is purely arbitrary and depends on which is chosen as the reference (however, it should be noted that $\sin \theta = \cos \theta - 90°$). Therefore, the following relationships hold true:

$$v = V \sin (2\pi Ft + \theta) = V \cos (2\pi Ft + \theta - 90°)$$

$$v = V \cos (2\pi Ft + \theta) = V \sin (2\pi Ft + \theta + 90°)$$

The preceding formulas are for a single-frequency, repetitive waveform. Such a waveform is called a periodic wave because it repeats at a uniform rate (i.e., each successive cycle of the signal takes exactly the same length of time and has exactly the same amplitude variations as every other cycle—each cycle has exactly the same shape). A series of sine waves is an example of a periodic wave. Periodic waves can be analyzed in either the *time* or the *frequency domain*. In fact, it is often necessary when analyzing system performance to switch from the time domain to the frequency domain, and vice versa.

Time domain.　A standard oscilloscope is a time-domain instrument. The display on the CRT is an amplitude-versus-time representation of the input signal and is commonly called a signal waveform. Essentially, a signal waveform shows the shape and the instantaneous value of the signal but is not necessarily indicative of its frequency content. With an oscilloscope, the vertical deflection is proportional to the instantaneous amplitude of the total input signal, and the horizontal deflection is a function of time (sweep rate). Figure 1-3 shows the signal waveform for a single-frequency sinusoidal signal with a peak amplitude of V volts and a frequency of F hertz.

Frequency domain.　A spectrum analyzer is a frequency-domain instrument. Essentially, there is no waveform displayed on the CRT. Instead, an amplitude-versus-frequency plot is shown (this is called a frequency spectrum). With a spectrum analyzer,

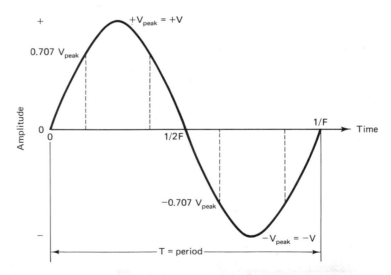

FIGURE 1-3　Time-domain representation (signal waveform) for a single-frequency sinusoidal wave.

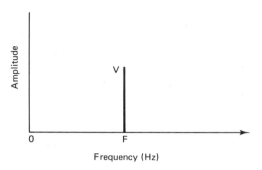

Amplitude

V

0   F

Frequency (Hz)

FIGURE 1-4   **Frequency-domain representation (spectrum) for a single-frequency sinusoidal wave.**

the horizontal axis represents frequency and the vertical axis amplitude. Therefore, there is a vertical deflection for each input frequency. Effectively, the input waveform is swept with a variable-frequency, high-Q bandpass filter whose center frequency is synchronized to the horizontal sweep rate of the CRT. Each frequency present in the input waveform produces a vertical line on the CRT (these are called spectral components). The height of each line is proportional to the amplitude of that frequency. A frequency-domain representation of a wave shows the frequency content but is not necessarily indicative of the shape of the waveform or the instantaneous amplitude. Figure 1-4 shows the spectrum for a single-frequency sinusoidal signal with a peak amplitude of V volts and a frequency of F hertz.

Nonsinusoidal Periodic Waves (Complex Waves)

Essentially, any repetitive waveform that comprises more than one sine or cosine wave is a *nonsinusoidal* or *complex periodic wave*. To analyze a complex periodic waveform, it is necessary to use a mathematical series developed in 1826 by the French physicist and mathematician Baron Jean Fourier. This series is appropriately called the *Fourier series*.

The Fourier series. The Fourier series is used in signal analysis to represent the sinusoidal components of a nonsinusoidal periodic waveform. In general, a Fourier series can be written for any series of terms that include trigonometric functions with the following mathematical expression:

$$f(t) = A_0 + A_1 \cos \alpha + A_2 \cos 2\alpha + A_3 \cos 3\alpha + \cdots + A_N \cos N\alpha$$
$$+ B_1 \sin \beta + B_2 \sin 2\beta + B_3 \sin 3\beta + \cdots + B_N \sin N\beta \qquad (1\text{-}3)$$

Equation 1-3 states that the waveform $f(t)$ comprises an average value (A_0), a series of cosine functions in which each successive term has a frequency that is an integer multiple of the frequency of the first cosine term in the series, and a series of sine functions in which each successive term has a frequency that is an integer multiple of the frequency of the first sine term in the series. There are no restrictions on the values or relative values of the amplitudes for the sine or cosine terms. Equation 1-3 is stated in words as follows: Any *periodic waveform* comprises an average component and a series of harmonically related sine and cosine waves. A *harmonic* is an integral multiple of the fundamental frequency. The *fundamental frequency* is the first harmonic, the second multiple of the fundamental is called the second harmonic, the third multiple is called the third harmonic, and so on. The fundamental frequency is the minimum frequency necessary to represent a waveform and is also the frequency of the waveform (i.e., the repetition rate). Therefore, Equation 1-3 can be rewritten as

$$f(t) = \text{dc} + \text{fundamental} + \text{2nd harmonic}$$

$$+ \text{3rd harmonic} + \cdot \cdot \cdot + n\text{th harmonic}$$

Wave symmetry

Even symmetry. If a periodic voltage waveform is symmetric about the vertical (amplitude) axis, it is said to have *axes* or *mirror symmetry* and is called an *even function*. For all even functions, the B coefficients in Equation 1-3 are zero. Therefore, the signal simply contains a dc component and the cosine terms (note that a cosine wave is itself an even function). The sum of a series of even functions is an even function. Even functions satisfy the condition

$$f(t) = f(-t) \tag{1-4}$$

Equation 1-4 states that the magnitude of the function at $+t$ is equal to the magnitude at $-t$. A waveform that contains only the even functions is shown in Figure 1-5a.

Odd symmetry. If a periodic voltage waveform is symmetric about a line midway between the vertical and horizontal axes and passing through the coordinate origin, it is said to have *point* or *skew symmetry* and is called an *odd function*. For all odd functions, the A coefficients in Equation 1-3 are zero. Therefore, the signal simply contains a dc component and the sine terms (note that a sine wave is itself an odd function). The sum of a series of odd functions is an odd function. This form must be mirrored first in the Y axis, then in the X axis for superposition. Thus

$$f(t) = -f(-t) \tag{1-5}$$

Equation 1-5 states that the magnitude of the function at $+t$ is equal to the negative of the magnitude at $-t$. A periodic waveform that contains only the odd functions is shown in Figure 1-5b.

Half-wave symmetry. If a periodic voltage waveform is such that the waveform for the first half cycle ($t = 0$ to $T/2$) repeats itself except with the opposite sign for the second half cycle ($t = T/2$ to T), it is said to have *half-wave symmetry*. For all waveforms with half-wave symmetry, the even harmonics in the series for both the sine and cosine terms are zero. Therefore, half-wave functions satisfy the condition

$$f(t) = -f(T/2 + t) \tag{1-6}$$

A periodic waveform that exhibits half-wave symmetry is shown in Figure 1-5c.

The coefficients A_0, B_1 to B_N, and A_1 to A_N can be evaluated using the following integral formulas:

$$A_0 = \frac{1}{T} \int_0^T f(t)\, dt \tag{1-7}$$

$$A_N = \frac{2}{T} \int_0^T f(t) \cos N\omega t\, dt \tag{1-8}$$

$$B_N = \frac{2}{T} \int_0^T f(t) \sin N\omega t\, dt \tag{1-9}$$

Solving Equations 1-7, 1-8, and 1-9 requires integral calculus, which is beyond the intent of this book. Therefore, in subsequent discussions, the appropriate solutions are given.

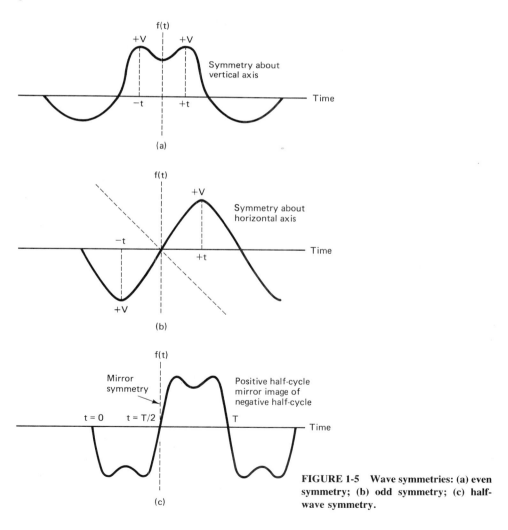

FIGURE 1-5   Wave symmetries: (a) even symmetry; (b) odd symmetry; (c) half-wave symmetry.

EXAMPLE 1-1

For the train of square waves shown in Figure 1-6:

(a) Determine the coefficients for the first 9 harmonics.

(b) Draw the frequency spectrum.

(c) Sketch the time-domain signal for frequency components up to the ninth harmonic.

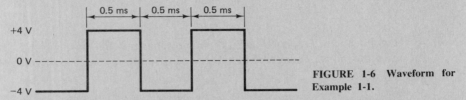

FIGURE 1-6 Waveform for Example 1-1.

Solution   (a) From inspection of the waveform, it can be seen that the average dc component is 0 V and the waveform has half-wave symmetry. Evaluating Equations 1-7, 1-8, and 1-9 yields the following Fourier series for a square wave:

$$v = \frac{4V}{N\pi}(\cos \omega t - \frac{1}{3}\cos 3\omega t + \frac{1}{5}\cos 5\omega t - \frac{1}{7}\cos 7\omega t + \cdot \cdot \cdot) \qquad (1\text{-}10)$$

From Equation 1-10 the following frequencies and coefficients are derived:

$$V_N = \frac{4V}{N\pi} \text{ (odd)}$$

where

$N = N$th harmonic (odd harmonics only)

V = peak amplitude of the complex waveform

N	Harmonic	Frequency (kHz)	Voltage (V)
0	0	dc	0
1	1st	1	5.09
2	2nd	2	0
3	3rd	3	1.69
4	4th	4	0
5	5th	5	1.02
6	6th	6	0
7	7th	7	0.728
8	8th	8	0
9	9th	9	0.566

(b) The spectrum is shown in Figure 1-7. (Note that although both + and − components are included in the waveform, all magnitudes are shown in the + direction on a waveform spectrum.)

(c) The time-domain signal for the first nine harmonics is shown in Figure 1-8. Although the waveform shown is not an exact square wave, it does closely resemble one.

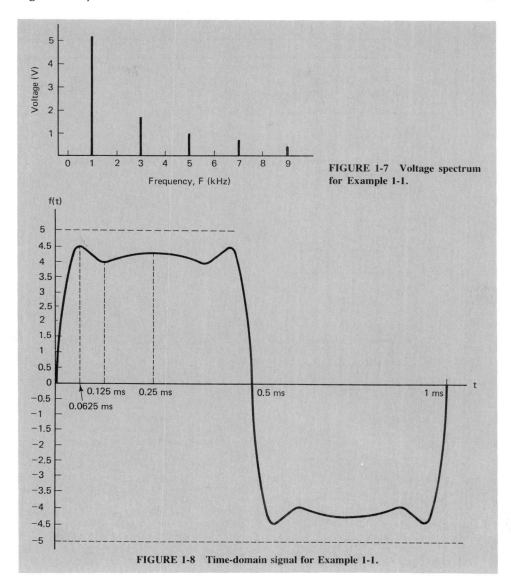

FIGURE 1-7 Voltage spectrum for Example 1-1.

FIGURE 1-8 Time-domain signal for Example 1-1.

Table 1-3 is a summary of the Fourier series for several of the more common nonsinusoidal periodic waveforms.

Fourier Series for a Rectangular Waveform

When analyzing electronic communications circuits, it is often necessary to use *rectangular pulses*. A waveform for a string of rectangular pulses is shown in Figure 1-9. The

TABLE 1-3   FOURIER SERIES SUMMARY

Waveform	Fourier series
	$v(t) = \dfrac{V}{\pi} + \dfrac{V}{2} \sin \omega t - \dfrac{2V}{3\pi} \cos 2\omega t - \dfrac{2V}{15\pi} \cos 4\omega t + \cdots$ $v(t) = \dfrac{V}{\pi} + \dfrac{V}{2} \sin \omega t + \displaystyle\sum_{N=2}^{\infty} \dfrac{V[1 + (-1)^N]}{\pi(1 - N^2)} \cos N\omega t$
	$v(t) = \dfrac{2V}{\pi} + \dfrac{4V}{3\pi} \cos \omega t - \dfrac{4V}{15\pi} \cos 2\omega t + \cdots$ $v(t) = \dfrac{2V}{\pi} + \displaystyle\sum_{N=1}^{\infty} \dfrac{4V(-1)^N}{\pi[1 - (2N)^2]} \cos N\omega t$
	$v(t) = \dfrac{2V}{\pi} \sin \omega t + \dfrac{2V}{3\pi} \sin 3\omega t + \cdots$ $v(t) = \displaystyle\sum_{N=\text{odd}}^{\infty} \dfrac{2V}{N\pi} \sin N\omega t$
	$v(t) = \dfrac{2V}{\pi} \cos \omega t - \dfrac{2V}{3\pi} \cos 3\omega t + \dfrac{2V}{5\pi} \cos 5\omega t + \cdots$ $v(t) = \displaystyle\sum_{N=1}^{\infty} \dfrac{V \sin N\pi/2}{N\pi/2} \cos N\omega t$
	$v(t) = \dfrac{Vt}{T} + \displaystyle\sum_{N=1}^{\infty} \left(\dfrac{2Vt}{T} \dfrac{\sin N\pi t/T}{N\pi t/T} \right) \cos N\omega t$
	$v(t) = \dfrac{4V}{\pi^2} \cos \omega t + \dfrac{4V}{(3\pi)^2} \cos 3\omega t + \dfrac{4V}{(5\pi)^2} \cos 5\omega t + \cdots$ $v(t) = \displaystyle\sum_{N=\text{odd}}^{\infty} \dfrac{4V}{(N\pi)^2} \cos N\omega t$

duty cycle (DS) for the waveform is the ratio of the active time of the pulse (*t*) to the period of the waveform (*T*). Mathematically, the duty cycle is

$$DS = \frac{t}{T} \tag{1-11a}$$

$$DS\ (\%) = \frac{t}{T} \times 100 \tag{1-11b}$$

FIGURE 1-9   Rectangular pulse
waveform.

Regardless of the duty cycle, a rectangular waveform is made up of a series of harmonically related sine waves. However, the amplitude of the spectrum components are dependent on the duty cycle. The Fourier series for a rectangular voltage waveform is

$$v = \frac{Vt}{T} + \frac{2Vt}{T}\left[\frac{\sin x}{x}(\cos \omega t) + \frac{\sin 2x}{2x}(\cos \omega t)\right.$$
$$\left. + \frac{\sin 3x}{3x}(\cos \omega t) + \cdot \cdot \cdot + \frac{\sin Nx}{Nx}(\cos \omega t)\right] \qquad (1\text{-}12)$$

where

$$x = \pi t/T$$
$$N = N\text{th harmonic and can be any whole integer}$$

From Equation 1-12 it can be seen that a rectangular waveform has a 0-Hz (dc) component equal to

$$V \times \frac{t}{T} \quad \text{or} \quad V \times \text{ duty cycle} \qquad (1\text{-}13)$$

The narrower the pulse width, the smaller the dc component. Also, from Equation 1-12, the amplitude of the Nth harmonic is

$$V_N = \frac{2Vt}{T}\frac{\sin Nx}{Nx} \qquad (1\text{-}14)$$

The (sin x)/x function is used to describe repetitive pulse waveforms. Sin x is simply a sinusoidal waveform whose instantaneous amplitude depends on x. With only x in the denominator, the denominator increases with x. Therefore, a (sin x)/x function is simply a damped sine wave. A (sin x)/x function is shown in Figure 1-10.

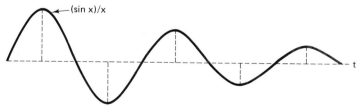

FIGURE 1-10   (sin x)/x function.

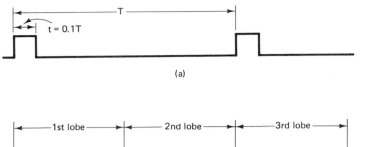

(a)

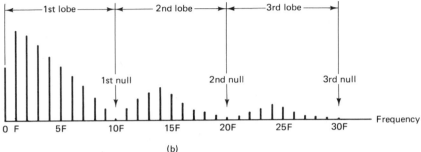

(b)

FIGURE 1-11   (sin *x*)/*x* function: (a) rectangular pulse waveform; (b) frequency spectrum.

Figure 1-11 shows the *frequency spectrum* for a rectangular pulse with a pulse width-to-period ratio of 0.1. It can be seen that the amplitudes of the harmonics follow a damped sinusoidal shape. The frequency whose period equals $1/t$ (i.e., at frequency 10*F* Hz), there is a 0-V component. A second null occurs at 20*F* Hz (period = 2/*t*), a third at 30*F* Hz (period = 3/*t*), and so on. All harmonics between 0 Hz and the first null frequency are considered in the first *lobe* of the frequency spectrum. All spectrum components between the first and second null frequencies are in the second lobe, frequencies between the second and third nulls are in the third lobe, and so on.

The following characteristics are true for all repetitive rectangular waveforms:

1. The dc component is equal to the pulse amplitude times the duty cycle.
2. There are 0-V components at frequency $1/t$ Hz and all integer multiples of that frequency.
3. The amplitude-versus-frequency time envelope of the spectrum components take on the shape of a damped sine wave.

EXAMPLE 1-2.

For the pulse waveform shown in Figure 1-12:
    (a) Determine the dc component.
    (b) Determine the peak amplitudes of the first 10 harmonics.
    (c) Plot the (sin *x*)/*x* function.
    (d) Sketch the frequency spectrum.

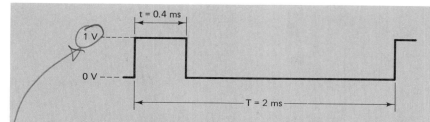

FIGURE 1-12 Pulse waveform for Example 1-2.

Solution (a) From Equation 1-13 the dc component is

$$V(0 \text{ Hz}) = \frac{1 \ (0.4 \text{ ms})}{2 \text{ ms}} = 0.2 \text{ V}$$

(b) The amplitudes of the first 10 harmonics are determined from Equation 1-14:

$$V_N = 2(1)\left(\frac{0.4 \text{ ms}}{2 \text{ ms}}\right)\left[\frac{\sin N\,180(0.4 \text{ ms}/2 \text{ ms})}{N3.14 \ (0.4 \text{ ms}/2 \text{ ms})}\right] \qquad x = \pi t / T$$

N	Frequency (Hz)	Amplitude (V)
0	0	0.2
1	500	0.374
2	1000	0.303
3	1500	0.202
4	2000	0.094
5	2500	0
6	3000	−0.063
7	3500	−0.087
8	4000	−0.076
9	4500	−0.042
10	5000	0

(c) The $(\sin x)/x$ function is shown in Figure 1-13.

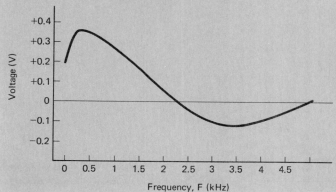

FIGURE 1-13 $(\sin x)/x$ function for Example 1-2.

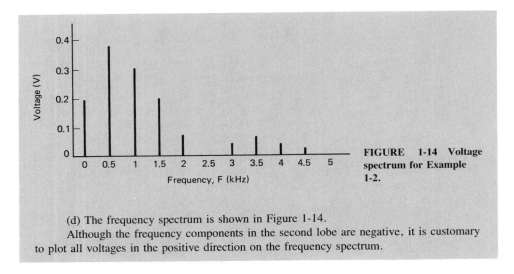

FIGURE 1-14 Voltage spectrum for Example 1-2.

(d) The frequency spectrum is shown in Figure 1-14.
Although the frequency components in the second lobe are negative, it is customary to plot all voltages in the positive direction on the frequency spectrum.

Figure 1-15 shows the effect that reducing the duty cycle (i.e., reducing the t/T ratio) has on the frequency spectrum for a nonsinusoidal waveform. It can be seen that narrowing the pulse width produces a frequency spectrum with a more uniform amplitude. In fact, for infinitely narrow pulses, the frequency spectrum comprises an infinite number of frequencies of equal amplitude. Increasing the period of a rectangular waveform while keeping the pulse width constant has the same effect on the frequency spectrum.

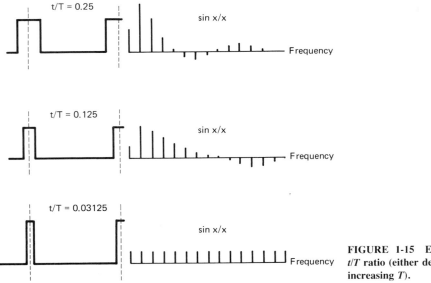

FIGURE 1-15 Effects of reducing the t/T ratio (either decreasing t or increasing T).

Effects of Bandlimiting on Signals

Every communications channel has a limited bandwidth and therefore has a limiting effect on signals that are propagated through them. We can consider a communications channel to be equivalent to an ideal linear phase filter with a finite bandwidth. If a

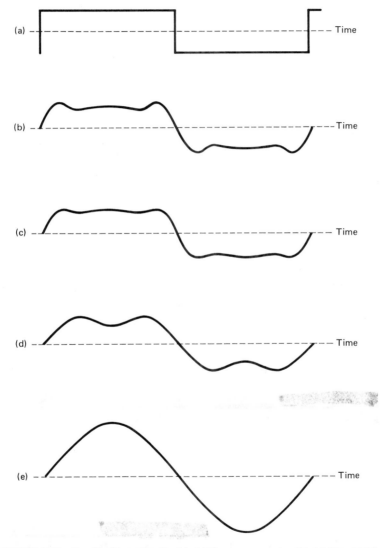

FIGURE 1-16 Bandlimiting signals: (a) 1-kHz square wave; (b) 1-kHz square wave bandlimited to 8 kHz; (c) 1-kHz square wave bandlimited to 6 kHz; (d) 1-kHz square wave bandlimited to 4 kHz; (e) 1-kHz square wave bandlimited to 2 kHz.

nonsinusoidal repetitive waveform passes through an ideal low-pass filter, the harmonic frequency components that are higher in frequency than the upper cutoff frequency for the filter are removed. Consequently, the shape of the waveform is changed. Figure 1-16a shows the time-domain waveform for the square wave used in Example 1-1. If this waveform is passed through a low-pass filter with an upper cutoff frequency of 8 kHz, frequencies above the eighth harmonic (9 kHz and above) are cut off and the waveform shown in Figure 1-16b results. Figures 1-16c, d, and e show the waveforms produced when low-pass filters with upper cutoff frequencies of 6, 4, and 2 kHz are used, respectively.

It can be seen from Figure 1-16 that *bandlimiting* a signal changes the frequency content and thus the shape of its waveform and, if sufficient bandlimiting is imposed, the waveform eventually comprises only the fundamental frequency. In a communications system, bandlimiting reduces the information capacity of the system and, if excessive bandlimiting is imposed, a portion of the information signal can be removed from the composite waveform.

ELECTRICAL NOISE

In general terms, *electrical noise* is defined as any unwanted electrical energy present in the usable passband of a communications circuit. For instance, in audio recording any undesired signals that fall into the band 0 to 15 kHz are audible and will interfere with the audio information. Consequently, for audio circuits, any unwanted electrical energy in the band 0 to 15 kHz is considered noise.

Essentially, noise can be divided into two general categories: correlated and uncorrelated. Correlation implies a relationship between the signal and the noise. Uncorrelated noise is noise that is present in the absence of any signal.

Correlated Noise

Correlated noise is unwanted electrical energy that is present as a direct result of a signal such as harmonic and intermodulation distortion. Harmonic and intermodulation distortion are both forms of nonlinear distortion; they are produced from nonlinear amplification. Correlated noise can not be present in a circuit unless there is an input signal. Simply stated, no signal, no noise! Both harmonic and intermodulation distortion change the shape of the wave in the time domain and the spectral content in the frequency domain.

Harmonic distortion. *Harmonic distortion* is the generation of unwanted multiples of a single-frequency sine wave when the sine wave is amplified in a nonlinear device such as a large-signal amplifier. *Amplitude distortion* is another name for harmonic distortion. Generally, the term "amplitude distortion" is used for analyzing a waveform in the time domain, and the term "harmonic distortion" is used for analyzing a waveform in the frequency domain. The original input frequency is the first harmonic and, as stated previously, is called the fundamental frequency.

There are various degrees or orders of harmonic distortion. Second-order harmonic distortion is the ratio of the amplitude of the second harmonic to the amplitude of the fundamental frequency. Third-order harmonic distortion is the ratio of the amplitude of the third harmonic to the amplitude of the fundamental frequency, and so on. The ratio of the combined amplitudes of the higher harmonics to the amplitude of the fundamental frequency is called *total harmonic distortion* (THD). Mathematically, total harmonic distortion is

$$\% \text{ THD} = \frac{V_{\text{higher}}}{V_{\text{fundamental}}} \times 100 \qquad (1\text{-}15)$$

where

$\% \text{ THD}$ = percent total harmonic distortion

V_{higher} = quadratic sum of the root-mean-square (rms) voltages of the higher harmonics

$V_{\text{fundamental}}$ = rms voltage of the fundamental frequency

EXAMPLE 1-3

Determine the percent second-order, third-order, and total harmonic distortion for the output spectrum shown in Figure 1-17.

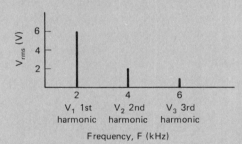

FIGURE 1-17   Harmonic distortion for Example 1-3.

Solution

$$\% \text{ 2nd-order harmonic distortion} = \frac{V_2}{V_1} \times 100 = \frac{2}{6} \times 100 = 33\%$$

$$\% \text{ 3rd-order harmonic distortion} = \frac{V_3}{V_1} \times 100 = \frac{1}{6} \times 100 = 16.7\%$$

$$\% \text{ THD} = \frac{\sqrt{V_2^2 + V_3^2}}{V_1} \times 100 = \frac{\sqrt{2^2 + 1^2}}{6} \times 100 = 37.3\%$$

Intermodulation distortion.   *Intermodulation distortion* is the generation of unwanted *cross products* (sum and difference frequencies) created when two or more frequencies are amplified in a nonlinear device such as a large-signal amplifier. As with harmonic distortion, there are various degrees of intermodulation distortion. It would be impossible to measure all of the intermodulation components produced when two or more frequencies mix in a nonlinear device. Therefore, for comparison purposes, a common method used to measure intermodulation distortion is percent *second-order* intermodulation distortion. Second-order intermodulation distortion is the ratio of the total amplitude of the second-order cross products to the combined amplitude of the original input frequencies. Generally, to measure second-order intermodulation distortion, four test frequencies are used; two designated the A band (FA1 and FA2) and two designated the B band (FB1 and FB2). The second-order cross products (2A-B) are: 2FA1 − FB1, 2FA1 − FB2, 2FA2 − FB1, 2FA2 − FB2, (FA1 + FA2) − FB1, and (FA1 + FA2) − FB2. Mathematically, percent second-order intermodulation distortion (IMD) is

$$\% \text{ 2nd-order IMD} = \frac{V_{\text{2nd-order cross products}}}{V_{\text{original}}} \times 100 \qquad (1\text{-}16)$$

where

$V_{\text{2nd order}}$ = quadratic sum of the amplitudes of the 2nd-order cross products
$V_{original}$ = quadratic sum of the amplitudes of the input frequencies

---

EXAMPLE 1-4

Determine the percent second-order intermodulation distortion for the A-band, B-band, and second-order intermodulation components shown in Figure 1-18.

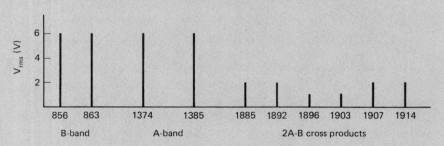

FIGURE 1-18   Intermodulation distortion for Example 1-4.

Solution

$$\% \text{ 2nd-order IMD} = \frac{V_{\text{2nd-order cross products}}}{V_{\text{original}}}$$

$$\% \text{ 2nd-order IMD} = \frac{\sqrt{2^2 + 2^2 + 2^2 + 2^2 + 1^2 + 1^2}}{\sqrt{6^2 + 6^2 + 6^2 + 6^2}} \times 100$$

$$\% \text{ IMD} = 35.4\%$$

Harmonic and intermodulation distortion are caused by the same thing, *nonlinear* amplification. Essentially, the only difference between the two is that harmonic distortion can occur when there is a single input frequency, and intermodulation distortion can occur only when there are two or more input frequencies. The generation of harmonic and intermodulation components is explained in Chapter 2.

Uncorrelated Noise

Uncorrelated noise is noise that is present regardless of whether or not there is a signal present. Uncorrelated noise is divided into two general categories: external and internal.

External noise. *External noise* is noise generated external to a circuit and allowed to enter into the circuit only if the frequency of the noise falls into the passband of the input filter. There are three primary types of external noise: atmospheric noise, extraterrestrial noise, and man-made noise.

Atmospheric noise. *Atmospheric noise* is naturally occurring electrical energy that originates within the earth's atmosphere. Atmospheric noise is commonly called static electricity. The source of most static electricity is natural electrical disturbances such as lightning. Static electricity generally comes in the form of impulses which spread its energy throughout a wide range of radio frequencies. The magnitude of the impulses measured from naturally occurring events has been observed to be inversely proportional to frequency. Consequently, at frequencies above 30 MHz, atmospheric noise is insignificant. Also, frequencies above 30 MHz are limited predominantly to line-of-sight propagation, which limits their interfering range to approximately 80 km.

Atmospheric noise is the summation of the energy from all sources both local and distant. Atmospheric noise propagates through the earth's atmosphere in the same manner as radio waves. Therefore, the magnitude of the static noise received depends on the propagation conditions at the time and is dependent, in part, on diurnal and seasonal variations. Atmospheric noise is the familiar sputtering, cracking, and so on, heard on a radio receiver predominantly in the absence of a received signal and is relatively insignificant compared to the other sources of noise.

Extraterrestrial noise. *Extraterrestrial noise* is noise that originates outside the earth's atmosphere and is, therefore, sometimes called *deep-space noise.* Extraterrestrial noise originates from the Milky Way, other galaxies, and the sun. Extraterrestrial noise is divided into two categories: solar and cosmic (galactic).

Solar noise is noise generated directly from the sun's heat. There are two components of solar noise: a "quiet" condition when a relatively constant radiation intensity exists and high-intensity sporadic disturbances caused by sun spot activity and solar flare-ups. The sporadic disturbances come from specific locations on the sun's surface. The magnitude of the disturbances caused from sun spot activity follows a cyclic pattern that repeats every 11 years. Also, these 11-year periods follow a supercycle pattern where approximately every 100 years a new maximum intensity is realized.

Cosmic noise sources are continuously distributed throughout our galaxy and other galaxies. Distant stars are also suns and have high temperatures associated with them. Consequently, they radiate noise in the same manner as our sun. Because the sources of galactic noise are located much farther away than our sun, their noise intensity is relatively small. Cosmic noise is often called *black body noise* and is distributed fairly evenly throughout the sky. Extraterrestrial noise contains frequencies from approximately 8 MHz to 1.5 GHz, although frequencies below 20 MHz seldom penetrate the earth's atmosphere and are, therefore, insignificant.

Man-made noise. Man-made noise is simply noise that can be attributed to man. The sources of man-made noise include spark-producing mechanisms such as commutators in electric motors, automobile ignition systems, power switching equipment, and fluorescent lights. Man-made noise is also impulsive in nature and therefore contains a wide range of frequencies that are propagated through space in the same manner as radio waves. Man-made noise is most intense in more populated metropolitan and industrial areas and is sometimes called *industrial noise*.

Internal noise. *Internal noise* is electrical interference generated within a device. There are three primary kinds of internally generated noise: thermal, shot, and transit-time.

Thermal noise. Thermal noise is a phenomenon associated with *Brownian movement* of electrons within a conductor. In accordance with the kinetic theory of matter, electrons within a conductor are in thermal equilibrium with the molecules and in constant random motion. This random movement is accepted as being a confirmation of the kinetic theory of matter and was first noted by the English botanist, Robert Brown (hence the name "Brownian noise"). Brown first observed evidence for the *kinetic* (moving-particle) nature of matter while observing pollen grains under a microscope. Brown noted an extraordinary agitation of the pollen grains that made them extremely difficult to examine. He later noted that this same phenomenon existed for smoke particles in the air. Brownian movement of electrons was first recognized in 1927 by J. B. Johnson of Bell Telephone Laboratories. In 1928, a quantitative theoretical treatment was furnished by H. Nyquist (also of Bell Telephone Laboratories). Electrons within a conductor carry a unit negative charge, and the mean-square velocity of an electron is proportional to the absolute temperature. Consequently, each flight of an electron between collisions with molecules constitutes a short pulse of current. Because the electron movement is totally random and in all directions, the average voltage produced by their movement is 0 V dc. However, such a random movement gives rise to an ac component. This ac component has several names, which include *thermal noise* (because it is temperature dependent), *Brownian noise* (after its discoverer), *Johnson noise* (after the person who related Brownian particle movement to electron movement), *random noise* (because the direction of electron movement is totally random), *resistance noise* (because the magnitude of its voltage is dependent upon resistance), and *white noise* (because it contains all frequencies). Hence thermal noise is the random motion of free electrons within a conductor caused by thermal agitation.

The *equipartition law* of Boltzmann and Maxwell combined with the works of

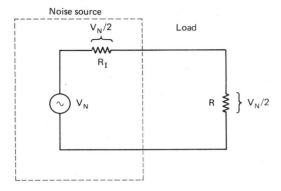

FIGURE 1-19 Noise source equivalent circuit.

Johnson and Nyquist states that the thermal noise power generated within a source for a 1-Hz bandwidth is

$$N_o = KT \qquad (1\text{-}17)$$

where

N_o = noise power density (W/Hz)
K = Boltzmann's constant
$\quad$ = 1.38×10^{-23} J/K
T = absolute temperature (K) (room temperature = 17°C or 290 K)[*]

Thus at room temperature, the available noise power density is

$$N_o = 1.38 \times 10^{-23} \text{ J/K} \times 290 \text{ K} = 4 \times 10^{-21} \text{ W/Hz}$$

The total noise power is equal to the product of the bandwidth and the noise density. Therefore, the total noise power present in bandwidth (B) is

$$N = KTB \qquad (1\text{-}18)$$

where

N = total noise power in bandwidth B (W)
$KT = N_o$ = noise power density (W/Hz)
B = bandwidth of the device or system (Hz)

Figure 1-19 shows the equivalent circuit for an electrical noise source. The internal resistance of the noise source (R_I) is in series with the rms noise voltage (V_N). For the worst case condition (maximum transfer of noise power), the load resistance (R) is made equal to R_I. Therefore, the noise voltage dropped across R is equal to $V_N/2$, and the noise power (N) developed across the load resistor is equal to KTB. Therefore, V_N is determined as follows:

[*] 0 K = −273°C.

$$N = KTB = \frac{(V_N/2)^2}{R} = \frac{V_N^2}{4R} \tag{1-19}$$

and

$$V_N^2 = 4RKTB$$

$$V_N = \sqrt{4RKTB}$$

Thermal noise is equally distributed throughout the frequency spectrum. Because of this property, a thermal noise source is called a "white noise source" (this is an analogy to white light, which contains all visible-light frequencies). Therefore, the noise power measured at any frequency is equal to the noise power measured at any other frequency. Similarly, the noise measured in any given bandwidth is equal to the noise measured in any other equal bandwidth regardless of the center frequency. In other words, the thermal noise power present in the band 1000 to 2000 Hz is equal to the thermal noise power present in the band 1,001,000 to 1,002,000 Hz.

EXAMPLE 1-5

For a device operating at a temperature of 17°C with a bandwidth of 10 kHz, determine:
(a) The noise power density (N_o).
(b) The total noise power (N).
(c) The rms noise voltage (V_N) for a 100-Ω internal resistance and a 100-Ω load resistor.

Solution   (a) By definition, the noise density is the noise measured in any 1-Hz bandwidth with constant temperature. Therefore, substituting into Equation 1-15, noise density equals

$$N_o = KT = 1.38 \times 10^{-23} \times 290 = 4 \times 10^{-21} \text{ W/Hz}$$

(b) Substituting into Equation 1-17, the total thermal noise is found to be

$$N = KTB$$

$$= 1.38 \times 10^{-23} \times 290 \times 10^4 = 4 \times 10^{-17} \text{ W}$$

(c) Substituting into Equation 1-19, we find that the rms noise voltage is

$$V_N = \sqrt{4RKTB}$$

$$V_n = \sqrt{4 \times 100 \times 1.38 \times 10^{-23} \times 290 \times 10^4} = 0.1265 \text{ } \mu\text{V}$$

For equal-value load and internal resistances, the noise voltage dropped across the load resistor is equal to one-half the noise voltage, or 0.06325 μV. Therefore, the total noise power is

$$N = \frac{E^2}{R} = \frac{(V_N/2)^2}{R} = \frac{(0.1265/2)^2}{100} = \frac{(0.06325)^2}{100} \text{ } \mu\text{V} = 4 \times 10^{-17} \text{ W}$$

Thermal noise is random, continuous, and occurs at all frequencies. Also, thermal noise is present in all devices, is predictable, and is additive. This is why thermal noise is generally the most significant of all the noise sources. Also, as previously stated, there are several names for thermal noise. In subsequent discussions in this book it will be referred to as simply thermal or random white noise.

Shot noise. *Shot noise* is caused by the random arrival of carriers (electrons and holes) at the output element of an active device such as a diode, field-effect transistor (FET), bipolar transistor (BJT) or tube. Shot noise is randomly varying and is superimposed on top of any signal present. Shot noise, when amplified, sounds like a shower of pellets falling on a tin roof.

Shot noise is proportional to the charge of an electron (1.6×10^{-19}), direct current, and system bandwidth. Also, shot noise is additive with thermal noise.

Transit-time noise. Any modification to a stream of carriers as they pass from the input to the output of a device (such as from the emitter to the collector of a transistor) produces an irregular random variation categorized as *transit noise*.

When the time it takes a carrier to propagate through a device is an appreciable part of the time of one cycle of the signal, the noise becomes noticeable. Transit-time noise in transistors is determined by ion mobility, the bias voltages, and the actual transistor construction. Carriers traveling from the emitter to the collector suffer from emitter delay times, base transit-time delays, and collector recombination and propagation delay times. At high frequencies and if transit delays are excessive, the device may add more noise than amplification to the signal.

Miscellaneous types of noise

Excess noise. *Excess noise* is a form of uncorrelated internal noise that is not totally understood. It is found in transistors and is directly proportional to emitter current and junction temperature and inversely proportional to frequency. Excess noise is also called *low-frequency* or *flicker noise*, *1/F noise*, and *modulation noise*. It is believed to be caused by or at least associated with *carrier traps* in the emitter depletion layer. These traps capture and release holes and electrons at different rates but with energy levels that vary inversely with frequency. Excess noise is insignificant above approximately 1 kHz.

Resistance noise. *Resistance noise* is a form of thermal noise that is associated with the internal resistances of the base, emitter, and collector of a transistor. Resistance noise is fairly constant from about 500 Hz up and therefore may be *lumped* into an equivalent resistance with thermal and shot noise.

Precipitation noise. *Precipitation noise* is a type of static noise caused when an airplane passes through snow or rain. The airplane becomes electrically charged to a potential difference high enough in respect to the surrounding space that a *corona* discharge occurs at a sharp point on the airplane. The interference from precipitation static is most annoying at shortwave frequencies and lower.

Signal-to-Noise Ratio

Signal-to-noise ratio (*S/N*) is a simple mathematical relationship of the signal level in respect to the noise level at a given point in a circuit, amplifier, or system. Signal-to-noise can be expressed as a voltage ratio or as a power ratio. Mathematically, *S/N* is:

As a voltage ratio:

$$\frac{S}{N} = \frac{\text{signal voltage}}{\text{noise voltage}} = \frac{V_s}{V_n} \tag{1-20}$$

As a power ratio:

$$\frac{S}{N} = \frac{\text{signal power}}{\text{noise power}} = \frac{P_s}{P_n} \tag{1-21}$$

The signal-to-noise ratio is often expressed as a logarithmic function with the unit dB (decibel):

For power ratios:

$$\frac{S}{N}(\text{dB}) = 10 \log \frac{P_s}{P_n} \tag{1-22}$$

For voltage ratios:

$$\frac{S}{N}(\text{dB}) = 20 \log \frac{V_s}{V_n} \tag{1-23}$$

Signal-to-noise is probably the most important and most often used parameter for evaluating the performance of a radio communications system. The higher the signal-to-noise ratio, the better the system performance. From the signal-to-noise ratio, the general quality of a system can be determined.

Noise Figure

Noise figure (F) (sometimes called *noise factor*) is a figure of merit that indicates the degradation of the signal-to-noise ratio as a signal passes through an amplifier. Noise figure is the ratio of the input signal-to-noise ratio (S_i/N_i) to the output signal-to-noise ratio (S_o/N_o). Therefore, noise figure is a ratio of ratios. Mathematically, noise figure is

$$F = \frac{S_i/N_i}{S_o/N_o} \tag{1-24}$$

$$F(\text{db}) = 10 \log \left(\frac{S_i/N_i}{S_o/N_o} \right) \tag{1-25}$$

An amplifier will amplify equally all signals and noise present at its input that fall within its passband. Therefore, if the amplifier is ideal and noiseless, the signal and noise are amplified by the same amount and the signal-to-noise ratio at the output is equal to the signal-to-noise ratio at the input. However, in reality, amplifiers are not ideal noiseless devices. Therefore, although the input signal and noise are amplified equally, the device adds internally generated noise to the waveform, which reduces the

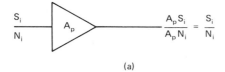

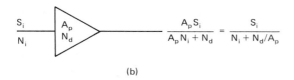

FIGURE 1-20 Noise figure: (a) ideal noiseless amplifier; (b) amplifier with internal noise.

overall signal-to-noise ratio. As stated previously, the most predominant form of electrical noise is thermal noise.

Figure 1-20a shows an ideal noiseless amplifier with a power gain (A_p), an input signal level (S_i), and an input noise level (N_i). It can be seen that the output S/N ratio is the same as the input S/N ratio. In Figure 1-20b, the same amplifier is shown except that instead of being ideal and noiseless, the amplifier adds an internally generated noise, N_d.

From Figure 1-20b, it can be seen that the output S/N ratio is less than the input S/N ratio by an amount proportional to N_d. Also, it can be seen that the noise figure of a perfect noiseless device is 1 or 0 dB.

EXAMPLE 1-6

For the amplifier shown in Figure 1-20b and the following parameters, determine:
(a) Input S/N.
(b) Output S/N.
(c) Noise figure.

$$\text{input signal voltage } (S_i) = 0.1 \times 10^{-3} \text{ V}$$

$$\text{input signal power } (P_i) = 2 \times 10^{-10} \text{ W}$$

$$\text{input noise voltage } (N_i) = 0.01 \times 10^{-6} \text{ V}$$

$$\text{input noise power } (P_n) = 2 \times 10^{-18} \text{ W}$$

$$\text{voltage gain } (A_v) = 1000$$

$$\text{power gain } (A_p) = 1,000,000$$

$$\text{amplifier noise voltage } (N_d) = 1 \times 10^{-6} \text{ V}$$

$$\text{amplifier noise power } (N_p) = 6 \times 10^{-12} \text{ W}$$

Solution (a) For the input signal and noise voltages given and substituting into Equation 1-20, we obtain

$$\frac{S_i}{N_i} = \frac{0.1 \text{ mV}}{0.01 \text{ }\mu\text{V}} = 10,000$$

Substituting into Equation 1-23 yields

$$\frac{S_i}{N_i} = 20 \log 10{,}000 = 80 \text{ dB}$$

For the input signal and noise powers given and substituting into Equation 1-21, we have

$$\frac{P_i}{P_n} = \frac{2 \times 10^{-10} \text{ W}}{2 \times 10^{-18} \text{ W}} = 100{,}000{,}000$$

Substituting into Equation 1-22 gives us

$$\frac{P_i}{P_n} = 10 \log 100{,}000{,}000 = 80 \text{ dB}$$

(b) For the calculated output signal and noise voltages and substituting into Equation 1-20, we obtain

$$\frac{S_o}{N_o} = \frac{A_v S_i}{A_v N_i + N_d} = \frac{(1000)\,(0.1 \text{ mV})}{(1000)\,(0.01 \text{ } \mu\text{V}) + 10\mu\text{V}} = 5000$$

Substituting into Equation 1-23 yields

$$\frac{S_o}{N_o} = 20 \log 5000 = 74 \text{ dB}$$

For the calculated output signal and noise powers and substituting into Equation 1-21, we have

$$\frac{S_o}{N_o} = \frac{A_p P_i}{A_p P_n + N_p} = \frac{(1 \times 10^6)(2 \times 10^{-10} \text{ W})}{(1 \times 10^6)(2 \times 10^{-18} \text{ W}) + 6 \times 10^{-12} \text{ W}}$$

$$= 25 \times 10^6$$

Substituting into Equation 1-22 gives us

$$\frac{S_o}{N_o} = 10 \log 25 \times 10^6 = 74 \text{ dB}$$

(c) From the results of parts (a) and (b) and using Equations 1-24 and 1-25, we obtain

$$\text{voltage noise figure} = \frac{10{,}000}{5000} = 2$$

$$F \text{ (db)} = 20 \log 2 \quad = 6 \text{ dB}$$

$$\text{power noise figure} = \frac{100{,}000{,}000}{25{,}000{,}000} = 4$$

$$F \text{ (dB)} = 10 \log 4 \quad = 6 \text{ dB}$$

A noise figure of 6 dB indicates that the power signal-to-noise ratio decreased by a factor of 4 and the voltage signal-to-noise ratio decreased by a factor of 2 as the signal propagated from the input to the output of the amplifier.

When two or more amplifiers or devices are cascaded together, the total noise figure (NF) is the accumulation of the individual noise figures. Mathematically, the total noise figure is[*]

$$NF = F_1 + \frac{F_2 - 1}{A_1} + \frac{F_3 - 1}{A_1 A_2} + \frac{F_4 - 1}{A_1 A_2 A_3} + \cdots \qquad (1\text{-}26)$$

where

NF = total noise figure
F_1 = noise figure of amplifier 1
F_2 = noise figure of amplifier 2
F_3 = noise figure of amplifier 3
A_1 = gain of amplifier 1
A_2 = gain of amplifier 2
A_3 = gain of amplifier 3

It can be seen that the noise figure of the first amplifier (F_1) contributes the most toward the overall noise figure. The noise introduced in the first stage is amplified by each of the succeeding amplifiers. Therefore, when compared to the noise introduced in the first stage, the noise added by each succeeding amplifier is effectively reduced by a factor equal to the product of the gains of the preceding amplifiers.

EXAMPLE 1-7

For three cascaded amplifiers each with noise figures of 3 dB and gains of 10 dB, determine the total noise figure.

Solution   Substituting into Equation 1-26 gives us

$$NF = F_1 + \frac{F_2 - 1}{A_1} + \frac{F_3 - 1}{A_1 A_2}$$

$$= 2 + \frac{2 - 1}{10} + \frac{2 - 1}{100}$$

$$= 2.11$$

and

$$10 \log 2.11 = 3.24 \text{dB}$$

An overall noise figure of 3.24 dB indicates that the *S/N* ratio at the output of A_3 is 3.24 dB less than the *S/N* ratio at the input to A_1.

MULTIPLEXING

Multiplexing is the transmission of information (either voice or data) from more than one source to more than one destination on the same transmission medium. Transmissions occur on the same medium but not necessarily at the same time. The transmission

---

[*] Noise figures and gains are given in absolute ratios rather than in decibels.

medium may be a metallic wire pair, a coaxial cable, a microwave radio, a satellite radio, or a fiber optic link. There are several ways in which multiplexing can be achieved, although the two most common methods are frequency-division multiplexing and time-division multiplexing.

Frequency-Division Multiplexing

In *frequency-division multiplexing* (FDM), multiple sources that originally occupied the same frequency band are transmitted simultaneously over a single transmission medium. Thus many relatively narrowband channels can be transmitted over a single wideband transmission system.

FDM is an analog multiplexing scheme; the information entering the system is analog and it remains analog throughout transmission. An example of FDM is the AM commercial broadcast band, which occupies a frequency spectrum from 535 to 1605 kHz. Each station carries an audio intelligence signal with a bandwidth of 5 kHz. If the audio from each station were transmitted with the original frequency spectrum, it would be impossible to separate one station's transmissions from another's. Instead, each station amplitude modulates a different carrier frequency and produces a signal with a 10-kHz bandwidth. Because adjacent stations' carrier frequencies are separated by 10 kHz, the total commerical AM band is divided into one hundred and seven 10-kHz frequency slots stacked next to each other in the frequency domain. To receive a particular station, a receiver is simply tuned to the frequency band associated with that station's transmissions. Figure 1-21 shows how commercial AM broadcast station signals are frequency-division multiplexed and transmitted over a single transmission medium (free space). There are many other applications for FDM, such as commercial FM and television broadcasting and high-volume telecommunications systems. Within any of the commercial broadcast bands, each station's transmissions are independent of all the other station's transmissions.

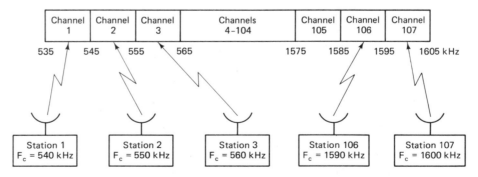

FIGURE 1-21   Frequency-division-multiplexing commercial AM broadcast-band stations.

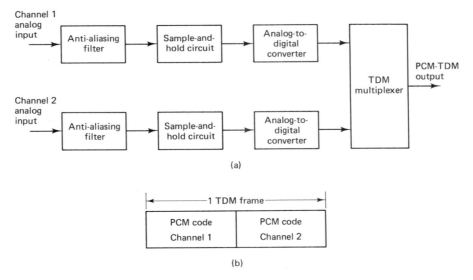

FIGURE 1-22 Two-channel PCM-TDM system: (a) block diagram; (b) TDM frame.

Time-Division Multiplexing

With *time-division multiplexing* (TDM), transmissions from multiple sources occur on the same facility but not at the same time. Transmissions from various sources are *interleaved* in the time domain. The most common type of modulation used with TDM systems is *pulse code modulation* (PCM). PCM is a type of digital transmission where analog signals are periodically *sampled* and converted to a series of binary codes, and the codes are transmitted as binary digital pulses. With a PCM-TDM system, several voice band channels are sampled, converted to PCM codes, then time-division multiplexed onto a single metallic cable pair.

Figure 1-22a shows a simplified block diagram of a two-channel PCM-TDM carrier system. Each channel is alternately sampled and converted to a PCM code. While PCM code for channel 1 is being transmitted, channel 2 is being sampled and converted to a PCM code. While the PCM code from channel 2 is being transmitted, the next sample is taken from channel 1 and converted to a PCM code. This process continues and samples are taken alternately from each channel, converted to PCM codes, and transmitted. The multiplexer is simply a switch with two inputs and one output. Channel 1 and channel 2 are alternately selected and connected to the multiplexer output. The time it takes to transmit one sample from each channel is called the *frame time*.

The PCM code for each channel occupies a fixed time slot (*epoch*) within the total TDM frame. With a two-channel system, the time allocated for each channel is equal to one-half the total frame time. A sample from each channel is taken once during each frame. Therefore, the total frame time is equal to the reciprocal of the sample rate. Figure 1-22b shows the TDM frame allocation for the two-channel system shown in Figure 1-22a.

QUESTIONS

1-1. Define *electronic communications*.

1-2. What three primary components make up a communications system?

1-3. Define *modulation*.

1-4. Define *demodulation*.

1-5. Define *carrier frequency*.

1-6. Explain the relationships among the source information, the carrier, and the modulated wave.

1-7. What are the three properties of an analog carrier that can be varied?

1-8. What organization assigns frequencies for free-space radio propagation in the United States?

1-9. Briefly describe the significance of Hartley's law and give the relationships between information capacity and bandwidth; information capacity and transmission time.

1-10. What are the two primary limitations on the performance of a communications system?

1-11. Describe signal analysis as it pertains to electronic communications.

1-12. Describe a time-domain display of a signal waveform; a frequency-domain display.

1-13. What is meant by the term *even symmetry*? What is another name for even symmetry?

1-14. What is meant by the term *odd symmetry*? What is another name for odd symmetry?

1-15. What is meant by the term *half-wave symmetry*?

1-16. Describe the term *duty cycle*.

1-17. Describe a $(\sin x)/x$ function.

1-18. Define *electrical noise*.

1-19. What is meant by the term *correlated noise*? List and describe two common forms of correlated noise.

1-20. What is meant by the term *uncorrelated noise*? List several types of uncorrelated noise and state their sources.

1-21. Briefly describe thermal noise.

1-22. What are four alternative names for thermal noise?

1-23. Describe the relationship between thermal noise and temperature; thermal noise and bandwidth.

1-24. Define *signal-to-noise ratio*. What does a signal-to-noise ratio of 100 indicate? 100 dB?

1-25. Define *noise figure*. An amplifier has a noise figure of 20 dB; what does this mean?

1-26. What is the noise figure for a totally noiseless device?

1-27. Define *multiplexing*.

1-28. Describe frequency-division multiplexing.

1-29. Describe time-division multiplexing.

PROBLEMS

1-1. For the train of square waves shown below:
  (a) Determine the coefficients for the first five harmonics.
  (b) Draw the frequency spectrum.
  (c) Sketch the time-domain signal for frequency components up to the first five harmonics.

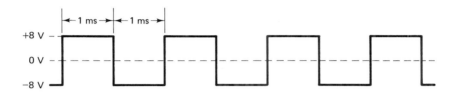

1-2. For the pulse waveform shown below:
  (a) Determine the dc component.
  (b) Determine the peak amplitudes of the first five harmonics.
  (c) Plot the $(\sin x)/x$ function.
  (d) Sketch the frequency spectrum.

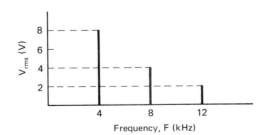

1-3. Determine the percent second-order, third-order, and total harmonic distortion for the output spectrum shown below.

1-4. Determine the percent second-order intermodulation distortion for the A-band, B-band, and second-order intermodulation components shown below.

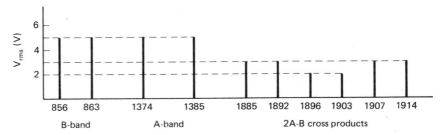

Frequency, F (Hz)

1-5. Determine the second-order cross-product frequencies for the following A- and B-band frequencies: B = 822 and 829 Hz, A = 1356 and 1365 Hz.

1-6. For an amplifier operating at a temperature of 27°C with a bandwidth of 20 kHz, determine:
   (a) The noise power density (N_o) in watts and dBm.
   (b) The total noise power (N) in watts and dBm.
   (c) The rms noise voltage (V_N) for a 50-Ω internal resistance and a 50-Ω load resistor.

1-7. (a) Determine the noise power (N) in watts and dBm for an amplifier operating at a temperature of 400°C with a bandwidth of 1 MHz.
   (b) Determine the decrease in noise power in decibels if the temperature decreased to 100°C.
   (c) Determine the increase in noise power in decibels if the bandwidth doubled.

1-8. Determine the overall noise figure for three cascaded amplifiers, each with individual noise figures of 3 dB and power gains of 20 dB.

1-9. Determine the duty cycle for the pulse waveform shown below.

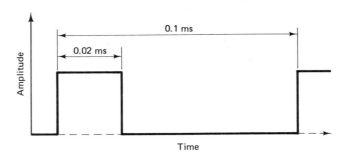

Time

1-10. Determine the % THD for a fundamental frequency voltage $V_{fundamental}$ = 12 V rms and a total voltage for the higher harmonics V_{higher} = 1.2 V rms.

1-11. Determine the percent second-order IMD for a quadratic sum of the A- and B-band components $V_{original}$ = 2.6 V rms and a quadratic sum of the second-order cross products $V_{2nd order}$ = 0.02 V rms.

1-12. If an amplifier with a bandwidth B = 20 kHz and a total noise power $N = 2 \times 10^{-17}$ W, determine:
   (a) Noise density.
   (b) Total noise if the bandwidth is increased to 40 kHz.
   (c) Noise density if the bandwidth is increased to 30 kHz.

Chapter 2

FREQUENCY GENERATION

INTRODUCTION

In electronic communications systems, there are many applications that require repetitive waveforms (both sinusoidal and rectangular). In most of these applications, more than one frequency is required, and very often these frequencies must be synchronized to each other. Therefore, *frequency generation* is an essential part of electronic communications.

OSCILLATORS

By definition, to *oscillate* is to fluctuate between two states or conditions. "To oscillate" is to vibrate, change, or to undergo oscillations. *Oscillating* is the act of fluctuating from one state to another. An *oscillator* is a device that produces *oscillations*. There are many applications for oscillators in electronic communications, such as high-frequency *carrier supplies*, *pilot supplies*, and *clock circuits*.

In electronic applications, an oscillator is a device or a circuit that produces electrical oscillations. An electrical oscillation is a repetitive change in a voltage or current waveform. If an oscillator is *self-sustaining*, the changes in the waveform are *continuous* and *repetitive*; they occur at a periodic rate. A self-sustaining oscillator is also called a *free-running oscillator*. Oscillators that are not self-sustaining require an external input signal or trigger to produce a change in the output waveform. Oscillators that are not self-sustaining are called *triggered* or *one-shot oscillators*. The remainder of this chapter is restricted to self-sustaining oscillators, which require no external input other than a dc supply voltage. Essentially, an oscillator converts a dc input voltage to an ac output voltage. The shape of the output waveform can be a sine wave, a square wave, a sawtooth, or any other shape as long as it repeats at periodic intervals.

Free-running oscillators, once started, generate an ac output signal of which a small portion is fed back to the input, where it is amplified. The amplified input signal appears at the output and the process repeats; a *regenerative* process occurs. The output is dependent on the input and the input is dependent on the output.

According to the *Barkhausen criterion*, for a feedback circuit to sustain oscillations, the net gain around the feedback loop must be unity or greater and the net phase shift around the loop must be a positive integer multiple of 360°.

Essentially, there are four requirements for a feedback oscillator to work: amplification, positive feedback, frequency dependency, and a source of power.

1. *Amplification.* The circuit must be capable of amplifying. In fact, at times it must be capable of infinite gain.
2. *Positive feedback.* There must be a complete path for the output signal to feed back to the input. The feedback signal must be regenerative, which means that it must have the correct phase and amplitude to sustain oscillations. If the phase is incorrect or if the amplitude is insufficient, oscillations will cease. If the amplitude is excessive, the device will saturate. Regenerative feedback is called *positive* feedback, where ''positive'' simply means that its phase aids the oscillation process and is not necessarily indicative of a positive ($+$) or negative ($-$) polarity.
3. *Frequency dependency.* There must be frequency-dependent components to allow the frequency of the oscillator to be set or changed.
4. *Power source.* There must be a source of electrical power, such as a dc power supply.

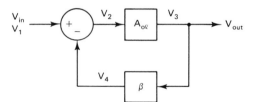

FIGURE 2-1   Model of an amplifier with feedback.

Figure 2-1 shows an electrical model for an amplifier with feedback. It includes an amplifier with an *open-loop gain* (A_{ol}) and a frequency-dependent regenerative feedback path with a *feedback ratio* of β.

From Figure 2-1, the following mathematical relationships are derived:

$$\text{transfer function} = \frac{V_{out}}{V_{in}} = \frac{V_3}{V_1}$$

$$V_2 = V_1 - V_4$$

$$V_3 = A_{ol}V_2$$

$$V_4 = \beta V_3$$

where

V_1 = external input voltage
V_2 = voltage input to amplifier
V_3 = output voltage
V_4 = feedback voltage

Substituting for V_4 gives us

$$V_2 = V_1 - \beta V_3$$

Thus

$$V_3 = (V_1 - \beta V_3)A_{ol}$$

And

$$V_3 = V_1 A_{ol} - V_3 \beta A_{ol}$$

Rearranging and factoring yield

$$V_3 + V_3\beta A_{ol} = V_1 A_{ol} = V_3(1 + \beta A_{ol}) = V_1 A_{ol}$$

Then

$$\frac{V_{out}}{V_{in}} = \frac{V_3}{V_1} = \frac{A_{ol}}{1 + \beta A_{ol}} \tag{2-1}$$

$A_{ol}/(1 + \beta A_{ol})$ is the standard formula used for an amplifier with feedback. If at some frequency βA_{ol} goes to -1, the denominator goes to 0 and V_{out}/V_{in} is infinity. When this happens, the circuit will oscillate and the external input may be removed.

For self-sustained oscillations to occur, a circuit must fulfill the four basic requirements for oscillation outlined previously, meet the criterion of Equation 2-1, and fit the basic feedback circuit model shown in Figure 2-1. Although oscillator action can be accomplished in many different ways, the most common configurations use *RC phase-shift networks*, *LC tank circuits*, *crystals*, or *negative resistance devices*.

Wien-Bridge Oscillator

The *Wien-bridge oscillator* is an *RC* phase-shift oscillator that uses both positive and negative feedback. It is a relatively stable low-frequency oscillator that is easily tuned and is commonly used in signal generators to produce frequencies between 5 Hz and 1 MHz. The Wien-bridge oscillator is the circuit that Mr. Hewlett and Mr. Packard used in their original signal generator design.

Figure 2-2a shows a simple *lead-lag network*. At the frequency of oscillation (F_o), $R = X_c$ and the signal undergoes a $-45°$ phase shift across Z_1 and a $+45°$ phase shift across Z_2. Consequently, at F_o, the total phase shift across the network is exactly

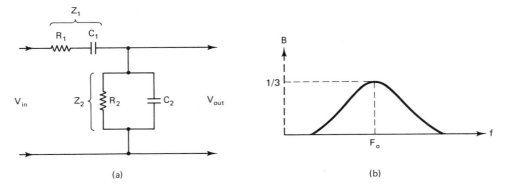

FIGURE 2-2   Lead-lag network: (a) circuit configuration; (b) input-versus-output transfer curve (β).

0°. At frequencies below resonance, the phase shift across the network leads and for frequencies above resonance it lags. At extreme low frequencies, C_1 looks open and there is no output. At extreme high frequencies C_2 looks like a short and there is no output.

A lead-lag network is a *reactive* voltage divider where the input voltage is divided between the series combination of R_1 and C_1 (Z_1) and the parallel combination of R_2 and C_2 (Z_2). Therefore, the lead-lag network is frequency selective and the output voltage is maximum at F_o. The transfer function for the feedback network (β) equals $Z_2/(Z_1 + Z_2)$ and is maximum and equal to $\frac{1}{3}$ at F_o. Figure 1-12b shows a plot of β versus frequency. If $R_1 = R_2$ and $C_1 = C_2$, F_o is determined from the following expression:

$$F_o = \frac{1}{2\pi RC}$$

where

$$R = R_1 = R_2$$
$$C = C_1 = C_2$$

Figure 2-3 shows a Wien-bridge oscillator. The lead-lag network and the resistive voltage divider make up a Wien bridge (hence the name "Wien-bridge oscillator"). When the bridge is balanced, the difference voltage equals zero. The voltage divider provides negative or degenerative feedback that offsets the positive or regenerative feedback from the lead-lag network. The ratio of the resistors in the voltage divider is 2:1, which sets the noninverting gain of amplifier A_1 to 3 ($R_f/R_i + 1$). Thus, at F_o, the signal at the output of A_1 is reduced by a factor of 3 as it passes through the lead-lag network, then amplified by 3 in amplifier A_1. Thus, at F_o, the closed-loop gain is equal to $A\beta$ or 1 ($\frac{1}{3} \times 3$).

To compensate for imbalances in the bridge, *automatic gain control* (AGC) is added to the circuit. A simple way of providing automatic gain is to replace R_I in Figure 2-3 with a variable resistance device such as a FET. The resistance of the FET

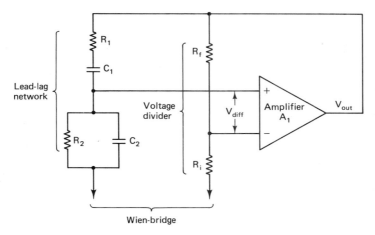

FIGURE 2-3 Wien-bridge oscillator.

is made directly proportional to V_{out}. If V_{out} goes up, the resistance of the FET is decreased and if V_{out} goes down, the resistance of the FET is increased. Therefore, the gain of the amplifier automatically adjusts to changes in the output signal amplitude.

The operation of the circuit shown in Figure 2-3 is as follows. On initial power up, noise (at all frequencies) appears at V_{out} and is fed back through the lead-lag network. Only noise at F_o passes through the lead-lag network with a 0° phase shift and a transfer ratio of $\frac{1}{3}$. Consequently, only a single frequency (F_o) is fed back in phase, undergoes a closed-loop gain of 1, and produces self-sustained oscillations.

LC Oscillators

LC oscillators are oscillators in which the frequency of oscillation is determined by an *LC tank circuit*. Tank circuit operation involves an exchange of energy between *kinetic* and *potential*.

Oscillator action. Figure 2-4 shows the basic oscillator action of an *LC* tank circuit. Initially, S_1 is open (t_0), there is no current flowing, and the charge across C is zero volts. V_{out} equals open-circuit voltage, V_{CC}. When S_1 is closed, current flows, charging C to V_{CC}. V_{out} equals $-V_C + V_{CC} = 0$ V (t_1). S_1 is opened and C discharges through L. The current flowing through L generates a magnetic field around its windings that produces a *counter* electromotive force (CEMF), which opposes the change in current. The inductor controls the rate at which C discharges. When C has completely discharged (t_2) $V_{out} = V_{CC}$ and current flow ceases. The magnetic field around L collapses, reversing the voltage polarity across it, which causes a current to flow that charges C to V_{CC} with the opposite polarity as before. The capacitor controls the rate at which L discharges. When the magnetic field has completely collapsed and C is fully charged, current once again ceases. $V_{out} = +V_C + V_{CC} = 2V_{CC}$ (t_3). C again discharges through

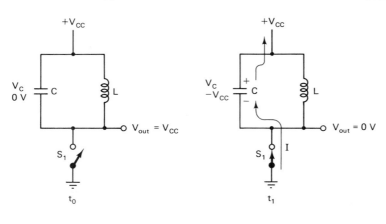

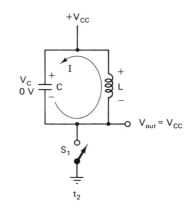

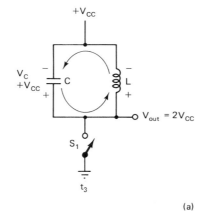

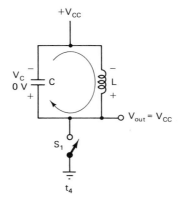

(a)

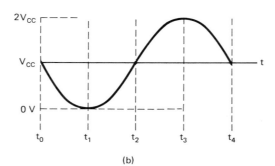

(b)

FIGURE 2-4 *LC* tank circuit: (a) oscillator action and flywheel effect; (b) output waveform.

L except in the opposite direction. When $V_C = 0$ V, current stops and the magnetic field around L again discharges, producing a current that charges C to V_{CC} with the original polarity. $V_{out} = -V_C + V_{CC} = 0$ V (t_4). The process just described repeats itself, producing a continuous sinusoidal waveform at V_{out}. The tank circuit is oscillating. Figure 2-4b shows a plot of V_{out} versus time.

The operation described above is called the *flywheel effect* and involves a continuous exchange of energy between the capacitor and inductor. The frequency of the waveform at V_{out} is determined by the rate at which the capacitor and inductor can take on and give off energy. By definition, resonance occurs when a capacitor and an inductor give off and take on energy at the same rate (i.e., $X_c = X_L$). Therefore, the frequency at V_{out} equals the resonant frequency of the tank circuit and is determined from the following mathematical expression:

$$F_o = \frac{1}{2\pi\sqrt{LC}} \qquad (2\text{-}1)$$

Because the inductor and capacitor in Figure 2-4 are not ideal, they dissipate power whenever current is flowing in the circuit. Consequently, on each successive cycle of V_{out}, the amplitude of the output waveform decreases. This is called a *damped oscillation*. Eventually, all of the energy is dissipated as heat, V_{out} goes to zero volts, and oscillations cease. If oscillations are to continue, it is necessary that S_1 be closed at periodic intervals to supply energy back into the circuit and compensate for power dissipated in the imperfect components.

The two most common oscillator configurations using LC tank circuits are the Hartley and Colpitts oscillators.

Hartley oscillator. Figure 2-5a shows the schematic diagram of a *Hartley oscillator*. The transistor amplifier (Q_1) provides the amplification necessary for unity gain at the resonant frequency. The coupling capacitor (C_c) provides the path for regenerative feedback, L_1 and C_1 are the frequency-determining components, and V_{CC} is the dc supply voltage.

Figure 2-5b shows the dc equivalent circuit for the Hartley oscillator. C_c is a blocking capacitor which isolates the dc base bias voltage and prevents it from being shorted to ground through L_{1b}. C_2 is also a blocking capacitor that prevents the collector supply voltage from being shorted to ground through L_{1a}. The radio frequency choke (RFC) is a dc short.

Figure 2-5c shows the ac equivalent circuit for the Hartley oscillator. C_c is a coupling capacitor for ac and provides a path for regenerative feedback from the tank circuit to the base of Q_1. C_2 couples ac signals from the collector of Q_1 to the tank circuit. The RFC looks open to ac and isolates the dc power supply from ac oscillations.

The Hartley oscillator operates as follows. On initial power-up, a multitude of frequencies appear at the collector of Q_1 and are coupled through C_2 into the tank circuit. The initial noise provides the energy necessary to charge C_1. Once C_1 is partially charged, oscillator action begins. The tank circuit will only oscillate efficiently at its resonant frequency. A portion of the oscillating tank circuit energy is dropped across

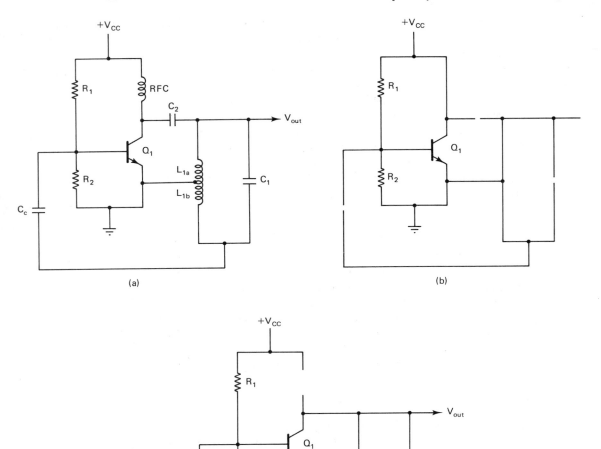

FIGURE 2-5 Hartley oscillator: (a) schematic diagram; (b) dc equivalent circuit; (c) ac equivalent circuit.

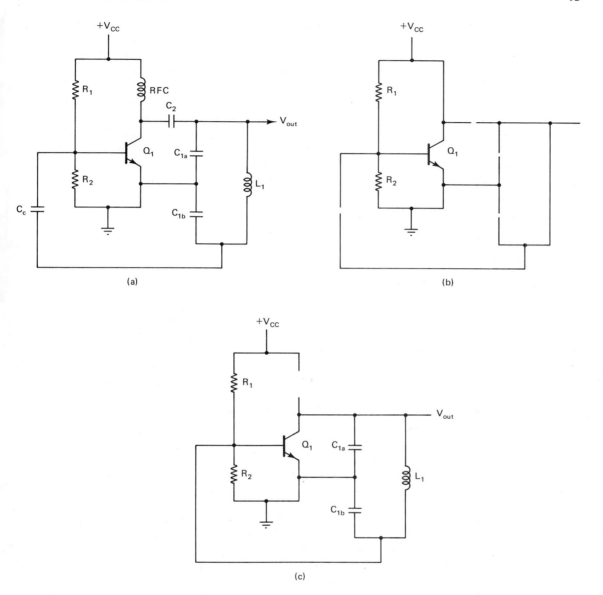

FIGURE 2-6 Colpitts oscillator: (a) schematic diagram; (b) dc equivalent circuit; (c) ac equivalent circuit.

L_{1b} and fed back to the base of Q_1, where it is amplified. The amplified signal appears at the collector $180°$ out of phase with the base signal. An additional $180°$ of phase shift is realized across L_1; consequently, the signal fed back to the base of Q_1 is amplified and phase shifted $360°$. Thus the circuit is regenerative and will sustain oscillations with no external input.

The proportion of oscillating energy that is fed back to the base of Q_1 is determined by the ratio of L_{1b} to the total inductance ($L_{1a} + L_{1b}$). If insufficient energy is fed back, oscillations are damped. If excessive energy is fed back, the transistor saturates. Therefore, the position of the wiper on L_1 is adjusted until the amount of feedback energy is exactly what is required for self-sustained oscillations to continue.

The frequency of oscillation for the Hartley oscillator is closely approximated by the following formula:

$$F_o = \frac{1}{2\pi\sqrt{LC}}$$

where

$$L = L_{1a} + L_{1b}$$
$$C = C_1$$

Colpitts oscillator.   Figure 2-6a shows the schematic diagram of a *Colpitts oscillator*. The operation of the Colpitts oscillator is very similar to that of the Hartley oscillator. Q_1 provides the amplification, C_c provides the regenerative feedback path, L_1, C_{1a}, and C_{1b} are the frequency-determining components, and V_{CC} is the dc supply voltage.

Figure 2-6b shows the dc equivalent circuit for the Colpitts oscillator. C_2 is a blocking capacitor that prevents the collector supply voltage from appearing at the output. The RFC is a dc short.

Figure 2-6c shows the ac equivalent circuit for the Colpitts oscillator. C_c is a coupling capacitor for ac and provides the feedback path for regenerative feedback from the tank circuit to the base of Q_1. The RFC is open to ac and decouples oscillations from the dc power supply.

The operation of the Colpitts oscillator is almost identical to the Hartley oscillator. On initial power-up, noise appears at the collector of Q_1 and supplies energy to the tank circuit, causing it to begin oscillating. C_{1a} and C_{1b} make up an ac voltage divider. The voltage dropped across C_{1b} is fed back to the base of Q_1 through C_c. There is a $180°$ phase shift from the base to the collector of Q_1 and an additional $180°$ phase shift across C_1. Consequently, the total phase shift is $360°$ and the feedback signal is regenerative. The ratio of C_{1a} to C_{1b} determines the amplitude of the feedback signal.

The frequency of oscillation for the Colpitts oscillator is closely approximated by the following formula:

$$F_o = \frac{1}{2\pi\sqrt{LC}}$$

where

$$L = L_1$$
$$C = C_{1a}C_{1b}/(C_{1a} + C_{1b})$$

Crystal Oscillators

Frequency stability. In communications systems, *frequency stability* is of primary importance. Frequency stability is the ability of an oscillator to remain at a fixed frequency. In the *LC* tank circuit and *RC* phase shift oscillators discussed previously, the frequency stability is inadequate for most radio communications applications. For example, commercial FM broadcast stations must maintain their carrier frequency to within ±2 kHz of their assigned frequency. This is approximately a 0.002% tolerance. In commercial AM broadcasting, the maximum allowable carrier shift is only ±20 Hz.

There are several factors that affect the stability of an oscillator. The most obvious are those that directly affect the value of the frequency-determining components. These include changes in the *L*, *C*, and *R* values due to environmental variations such as temperature and humidity and changes in the quiescent operating point of the transistor amplifier. Stability is also affected by ac ripple in the dc power supply. The frequency stability of an *RC* or *LC* oscillator can be greatly improved by regulating the dc power supply and minimizing the environmental variations. Also, special temperature-independent components can be used.

The FCC has established stringent regulations concerning the tolerances of radio-frequency carriers. Whenever the airway (free-space radio propagation) is used as the transmission medium, it is possible that transmissions from one source could interfere with transmissions from other sources if their transmit frequencies or transmission bandwidths overlap. Therefore, it is important that all sources maintain their frequency of operation within a specified tolerance.

Piezoelectric effect. Simply stated, the *piezoelectric effect* occurs when mechanical vibrations in a crystal *lattice* structure generate electrical oscillations, and vice versa. When an alternating voltage is applied across a crystal at or near the natural resonant frequency of the crystal, the crystal will break into mechanical oscillations. This is called exciting a crystal into mechanical vibrations. The magnitude of the mechanical vibration is directly proportional to the magnitude of the applied voltage.

There are a number of natural crystal substances that exhibit piezoelectric properties: quartz, Rochelle salt, and tourmaline. The effect is most pronounced in Rochelle salt, which is why it is the substance commonly used in crystal microphones. Quartz, however, is used more often for frequency control in oscillators because of its *permanence*, low-*temperature coefficient*, and high *mechanical Q*. These properties are explained in subsequent discussions.

Crystal cuts. In nature, complete quartz crystals have a hexagonal cross section with pointed ends, as shown in Figure 2-7a. There are three sets of *axes* associated

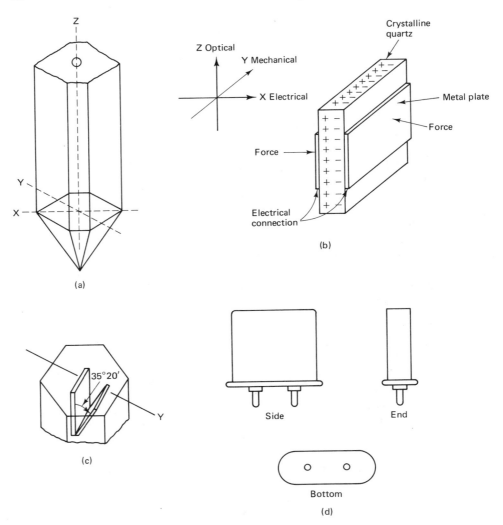

FIGURE 2-7   **Quartz crystal: (a) basic crystal structure; (b) crystal axes; (c) crystal cuts; (d) crystal mountings.**

with a crystal: *optical*, *electrical*, and *mechanical*. The longitudinal axis joining the points at the ends of the crystal is called the optical or Z axis. Electrical stresses applied to the optical axis do not produce the piezoelectric effect. The electrical or X axis passes diagonally through opposite corners of the hexagon. The axis that is perpendicular to the faces of the crystal is the Y or mechanical axis. Figure 2-7b shows the axes and the basic operation of a quartz crystal.

If a thin flat section is cut from a crystal such that the flat sides are perpendicular to an electrical axis, mechanical stresses along the Y axis will produce electrical charges on the flat sides. As the stress changes from compression to tension, and vice versa, the polarity of the charge is reversed. Conversely, if an alternating electrical charge is placed on the flat sides, a mechanical vibration is produced along the Y axis. This, of course, is the piezoelectric effect and is also exhibited when mechanical forces are applied across the faces of a crystal cut with its flat sides perpendicular to the Y axis. When a crystal wafer is cut parallel to the Z axis with its faces perpendicular to the X axis, it is called an X-cut crystal. When the faces are perpendicular to the Y axis, it is called a Y-cut crystal. A variety of cuts can be obtained by rotating the plane of the cut around one or more axes. If the Y cut is made at a 35°20' angle from the vertical axis (Figure 2-7c), an AT cut is obtained. The type, length, and thickness of a cut and the mode of vibration determine the natural resonant frequency of the crystal. Resonant frequencies from a few kilohertz up to approximately 30 MHz are possible.

Crystal *wafers* are generally mounted in crystal *holders*, which include the mounting and housing assemblies. A crystal *unit* refers to the holder and the crystal itself. Figure 2-7d shows a common crystal mounting. Because a crystal's stability is somewhat temperature dependent, a crystal unit may be mounted in an *oven* to maintain a constant operating temperature.

Overtone crystal oscillator.   To increase the frequency of vibration of a quartz crystal, the quartz wafer is sliced thinner. This imposes an obvious physical limitation; the thinner the wafer, the more susceptible it is to damage and the less useful it becomes. The practical limit for fundamental-mode crystal oscillators is approximately 30 MHz. However, it is possible to operate the crystal in an *overtone* mode. That is, the oscillator is tuned to operate at the third, fifth, or even the seventh harmonic of the crystal's fundamental frequency. This increases the limits of crystal oscillators to approximately 200 MHz.

Temperature coefficient.   The natural resonant frequency of a crystal is influenced somewhat by its operating temperature. The magnitude of frequency change (ΔF) is expressed in hertz change per megahertz of crystal frequency per degree Celsius (Hz/MHz/°C). The fractional change in frequency is often given in parts per million (ppm) per °C. For example, a temperature coefficient of $+20$ Hz/MHz/°C is the same as $+20$ ppm/°C. If the direction of the frequency change is the same as the temperature change (i.e., an increase in temperature causes an increase in frequency and a decrease in temperature causes a decrease in frequency), it is called a *positive* temperature coefficient. If the change in frequency is in the opposite direction as the temperature (i.e., an increase in temperature causes a decrease in frequency and a decrease in temperature causes an increase in frequency), it is called a *negative* temperature coefficient. Mathematically, the change in frequency of a crystal is

$$\Delta F = k(F \times \Delta C) \qquad (2\text{-}3)$$

where

ΔF = change in frequency (Hz)
k = temperature coefficient (Hz/MHz/°C)

F = natural crystal frequency (MHz)

ΔC = change in temperature (°C)

The temperature coefficient (k) of a crystal varies depending on the type of crystal cut and its operating temperature. For a range of temperatures from approximately 20 to 50°C, both X- and Y-cut crystals have a temperature coefficient that is nearly constant. X-cut crystals are approximately 10 times more stable than Y-cut crystals. Typically, X-cut crystals have a temperature coefficient that ranges from −10 to −25 Hz/MHz/°C. Y-cut crystals have a temperature coefficient that ranges from approximately −25 to +100 Hz/MHz/°C.

Today there are so called *zero-coefficient* (GT-cut) crystals which have temperature coefficients as low as −1 to +1 Hz/MHz/°C. The GT-cut crystal is almost a perfect zero-coefficient crystal from freezing to boiling but is useful only at frequencies below a few hundred kilohertz.

EXAMPLE 2-1

For a 10-MHz crystal with a positive temperature coefficient of 10 Hz/MHz/°C, determine the frequency of operation if the temperature
(a) Increases 10°C.
(b) Decreases 5°C.

Solution   (a) Substituting into Equation 2-3 yields

$$\Delta F = k(F \times \Delta C)$$

$$= 10(10 \times 10) = 1000 \text{ Hz}$$

$$F = 10 \text{ MHz} + \Delta F = 10 \text{ MHz} + 1 \text{ kHz} = 10.001 \text{ MHz}$$

(b) Again, substituting into Equation 2-3, we have

$$\Delta F = 10(10 \times -5) = -500 \text{ Hz}$$

$$F = 10 \text{ MHz} - 500 \text{ Hz} = 9.9995 \text{ MHz}$$

Crystal equivalent circuit.   Figure 2-8a shows the equivalent electrical circuit for a crystal. Each of the electrical components is equivalent to a mechanical property of the crystal. C_2 is the actual capacitance formed between the electrodes of the crystal, with the crystal itself being the dielectric. C_1 is equivalent to the mechanical *compliance* of the crystal (also called *resilience* or *elasticity*). L_1 is equivalent to the mass of the crystal in vibration, and R is the mechanical *friction* loss. In a crystal, the mechanical *mass-to-friction ratio* (L/R) is quite high. Typical values of L range from 0.1 H to well over 100 H. Consequently, Q factors are quite high for crystals. Q factors in the range 10,000 to 100,000 are not uncommon (as compared to 100 to 1,000 for the discrete inductors used in LC tank circuits). This, of course, attributes to the high stability and accuracy of crystal oscillators as compared to LC tank circuit oscillators. Values for C_1 are typically less than 1 pF, while values for C_2 range between 4 and 40 pF.

Because there is a series and a parallel equivalent circuit for a crystal, there are also two resonant frequencies: a series and a parallel. The series impedance is the

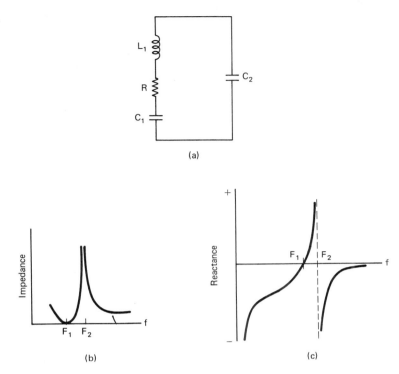

FIGURE 2-8 Crystal equivalent circuit: (a) equivalent circuit; (b) impedance curve; (c) reactance curve.

impedance of R, L, and C_1. The parallel impedance is the impedance of L and C_2. At extreme low frequencies, the series impedance of L, C_1, and R is very high and capacitive $(-)$. This is shown in Figure 2-8c. As the frequency is increased, a point is reached where $X_L = X_{C1}$. At this frequency (F_1), the series impedance is minimum, resistive, and equal to R. As the frequency is increased even further (F_2), the series impedance becomes high and inductive $(+)$. The parallel combination of L and C_2 causes the crystal to act like a parallel resonant circuit (maximum impedance at resonance). The difference between F_1 and F_2 is usually quite small (typically about 1% of the crystal operating frequency). A crystal can operate at either its series or parallel resonant frequency, depending on the circuit application. The relative steepness of the impedance curve (Fig. 2-8b) also attributes to the stability and accuracy of a crystal.

Crystal oscillator circuits.   Although there are many different crystal-based oscillator configurations, the most common are the discrete and IC Pierce and the RLC half-bridge. If you need very good frequency stability and reasonably simple circuitry, the discrete Pierce is a good choice. If low cost and simple digital interfacing capabilities are of primary concern, an IC-based Pierce oscillator will suffice. However, for the best frequency stability, the RLC half-bridge is the best choice.

Discrete Pierce oscillator. The Pierce crystal oscillator has many advantages. Its operating frequency spans the full fundamental crystal range (1 kHz to approximately 30 MHz). It uses relatively simple circuitry requiring few components (most medium-frequency versions require only one transistor). The Pierce oscillator design develops a high output signal power while dissipating very little power in the crystal. Finally, the short-term frequency stability of the Pierce crystal oscillator is excellent (this is because the in-circuit Q is almost as high as the crystal's internal Q). The only drawback to the Pierce oscillator is that it requires a high-gain amplifier (approximately 70). Consequently, you must use a single high-gain transistor of possibly even multiple amplifier stages.

Figure 2-9 shows a discrete 1-MHz Pierce crystal oscillator circuit. Q_1 provides all the gain necessary for self-sustained oscillations to occur. R_1 and C_1 provide a 65° phase lag to the feedback signal. The crystal impedance is basically resistive with a small inductive component. This impedance combined with C_2 provides an additional 115° of *RLC* phase lag. The transistor inverts the signal giving the circuit the necessary 360° phase shift. Because the crystal's load is primarily nonresistive (mostly the series combination of C_1 and C_2), this type of oscillator provides very good short-term frequency stability. Unfortunately, C_1 and C_2 introduce substantial losses, and consequently, the transistor must have a high current gain; this is obviously a drawback.

IC Pierce crystal oscillator. Figure 2-10 shows an IC-based Pierce crystal oscillator. Although it provides less frequency stability, it can be implemented using simple digital IC design and reduce costs substantially over conventional discrete designs.

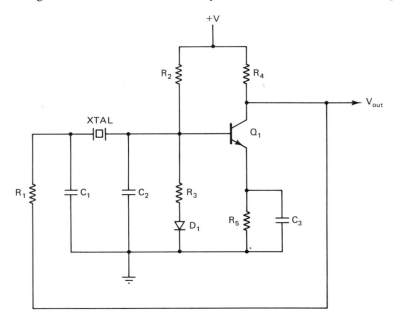

FIGURE 2-9   Discrete Pierce crystal oscillator.

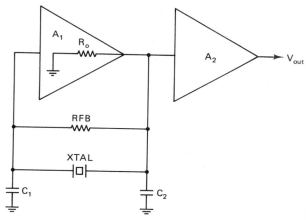

FIGURE 2-10   **IC Pierce crystal oscillator.**

To ensure that oscillations begin, RFB dc biases inverter A_1's input and output for class A operation. A_2 converts the output of A_1 to a full rail-to-rail swing reducing the rise and fall times and buffering A_1's output. The output resistance of A_1 combines with C_1 to provide the RC phase lag needed. CMOS versions operate up to approximately 2 MHz, and ECL versions operate as high as 20 MHz.

RLC half-bridge crystal oscillator. Figure 2-11 shows the Meacham version of the *RLC* half-bridge crystal oscillator. The original Meacham oscillator was developed in the 1940s and used a full four-arm bridge and a negative-temperature-coefficient tungsten lamp. The circuit configuration shown in Figure 2-10 uses only a two-arm bridge and employs a negative-temperature-coefficient thermister. Q_1 serves as a phase splitter and provides two 180° out-of-phase signals. The crystal must operate at its

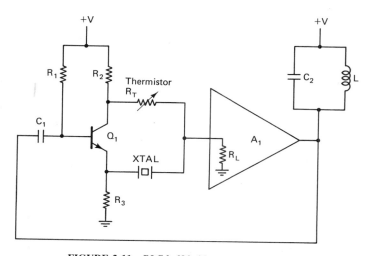

FIGURE 2-11   *RLC* **half-bridge crystal oscillator.**

series resonant frequency, so that its internal impedance is resistive and quite small. When oscillations begin, the signal amplitude increases gradually, decreasing the thermistor resistance until the bridge almost nulls. The amplitude of the oscillations stabilize and determine the final thermister resistance. The *LC* tank circuit at the output is tuned to the crystal's resonant frequency.

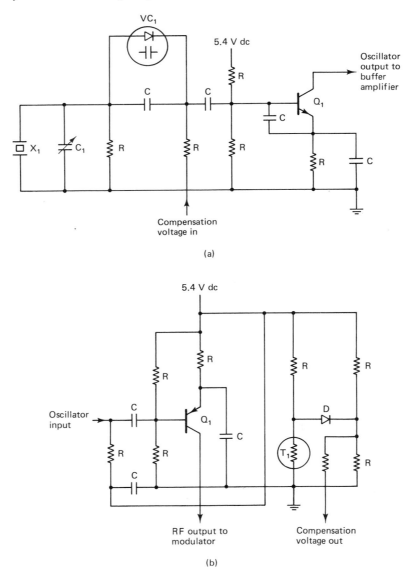

FIGURE 2-12   Crystal oscillator module: (a) schematic diagram; (b) compensation circuit.

Crystal oscillator module. A crystal oscillator *module* consists of a crystal-controlled oscillator and a voltage-variable component such as a varactor diode. The entire oscillator circuit is contained in a single metal *can*. A schematic diagram for a Colpitts crystal oscillator module is shown in Figure 2-12a. X_1 is the crystal itself and Q_1 is the active component for the oscillator. C_1 is a shunt capacitor that allows the crystal oscillator frequency to be varied over a narrow range of operating frequencies. VC_1 is a voltage-variable capacitor (*varicap* or *varactor diode*). A varactor diode is a specially constructed diode that exhibits capacitance when reverse biased, and by varying the reverse-bias voltage, the capacitance of the diode is changed. A varactor diode has a special depletion layer between the *p*- and *n*-type materials that is constructed with various degrees and types of doping material (the term "graded junction" is often used when describing varactor diode fabrication). Figure 2-13 shows the capacitance versus reverse-bias characteristics of a typical varactor diode. The capacitance of a varactor diode is approximated as

$$C_d = \frac{C}{\sqrt{1 + 2/V_r}} \tag{2-4}$$

where

C = diode capacitance with zero reverse bias
V_r = diode reverse bias
C_d = reverse-biased capacitance

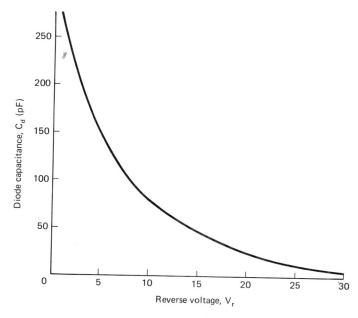

FIGURE 2-13 **Varactor diode characteristics.**

The frequency at which the crystal oscillates can be adjusted slightly by changing the capacitance of VC_1 (i.e., changing the value of the reverse-bias voltage). The varactor diode, in conjunction with a temperature-compensating module, provide instant frequency compensation for variations caused by changes in the temperature. A temperature-compensating module is shown in Figure 2-12b. The compensator module includes a buffer-amplifier (Q_1) and a temperature-compensating network (T_1). T_1 is a negative-temperature-coefficient thermister. When the temperature falls below the threshold value for the thermister, the compensation voltage increases to maintain the proper voltage on the varactor diode in the oscillator module. Compensation modules are available that can compensate for a frequency stability of $+0.0005\%$ from -30 to $+80°C$.

Negative resistance oscillator: tunnel diode.   Although there are several devices currently available that exhibit negative resistance properties, the *tunnel diode* (invented in the 1950s) is one of the most common and thus will be used as an example.

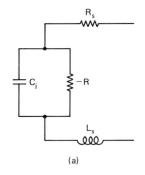

(a)

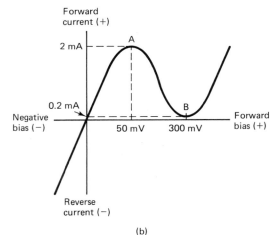

(b)

FIGURE 2-14   Tunnel diode: (a) equivalent circuit; (b) voltage-versus-current characteristics.

The tunnel diode (or *Esaki diode*, as it is sometimes called) is a thin-junction diode which, when slightly forward biased, exhibits negative resistance. Because of its thin junction it has an extremely short transit time and is therefore ideally suited for high-frequency applications.

Figure 2-14a shows the equivalent circuit for a forward-biased tunnel diode. Except for extremely high frequencies (above approximately 2 GHz), the series resistance (R_s) and inductance (L_s) can be ignored. Consequently, for practical applications, the equivalent circuit is reduced to the parallel combination of junction capacitance (C_j) and the negative resistance $(-R)$. The junction capacitance of the tunnel diode is highly dependent on the dc bias voltage and the temperature. As a result, using a tunnel diode with a tuned circuit yields a very low Q and relatively unstable oscillator. However, if a high-Q mechanical cavity is loosely coupled to the diode, a very stable oscillator is obtained which is relatively independent of changes in temperature or dc bias.

A tunnel diode is similar to the conventional *p-n* junction semiconductor diode except that the semiconductor materials are much more heavily doped (perhaps as much as 1000 times more). Because of the heavy doping, the depletion layer is extremely thin (typically about 0.01 μm) and therefore has a very small junction capacitance and has an extremely short transit time. Figure 2-14b shows the voltage-versus-current characteristics of a typical tunnel diode. For forward bias voltages between 0 and 50 mV and greater than 300 mV, the characteristics resemble those of a standard *p-n* junction diode. In the portion of the curve between point *A* (the peak) and point *B* (the valley), an increase in forward bias causes a corresponding reduction in forward current (i.e., negative resistance). If the tunnel diode is operated between points *A* and *B*, self-sustained oscillations can be obtained.

Large-Scale Integration Oscillators

In recent years, the use of large-scale integration integrated circuits for frequency generation has increased at a tremendous rate because they have excellent frequency stability and a wide tuning range. Figure 2-15 shows the specification sheet for the XR-2207 monolithic voltage controlled oscillator. The XR-2207 is a precision oscillator that can provide simultaneous sine and square wave outputs over a frequency range of 0.01 Hz to 1 MHz. It is ideally suited for a wide range of applications in electronic communications including frequency modulation, frequency shift keying, sweep or tone generation, as well as for phase locked loop applications.

Figure 2-16 on page 66 shows the specification sheet for the XR-2209 monolithic precision oscillator. The XR-2209 is a variable frequency oscillator circuit featuring excellent temperature stability and a wide linear sweep range. The circuit provides simultaneous triangle and square-wave outputs over a frequency range of 0.01 Hz to 1 MHz. The XR-2209 is ideally suited for frequency modulation, voltage-to-frequency or current-to-frequency conversion, sweep or tone generation, as well as phase locked loop applications.

Voltage-Controlled Oscillator

GENERAL DESCRIPTION

The XR-2207 is a monolithic voltage-controlled oscillator (VCO) integrated circuit featuring excellent frequency stability and a wide tuning range. The circuit provides simultaneous triangle and squarewave outputs over a frequency range of 0.01 Hz to 1 MHz. It is ideally suited for FM, FSK, and sweep or tone generation, as well as for phase-locked loop applications.

The XR-2207 has a typical drift specification of 20 ppm/°C. The oscillator frequency can be linearly swept over a 1000:1 range with an external control voltage; and the duty cycle of both the triangle and the squarewave outputs can be varied from 0.1% to 99.9% to generate stable pulse and sawtooth waveforms.

FEATURES

Excellent Temperature Stability (20 ppm/°C)
Linear Frequency Sweep
Adjustable Duty Cycle (0.1% to 99.9%)
Two or Four Level FSK Capability
Wide Sweep Range (1000:1 Min)
Logic Compatible Input and Output Levels
Wide Supply Voltage Range (±4V to ±13V)
Low Supply Sensitivity (0.1%/V)
Wide Frequency Range (0.01 Hz to 1 MHz)
Simultaneous Triangle and Squarewave Outputs

APPLICATIONS

FSK Generation
Voltage and Current-to-Frequency Conversion
Stable Phase-Locked Loop
Waveform Generation
  Triangle, Sawtooth, Pulse, Squarewave
FM and Sweep Generation

ABSOLUTE MAXIMUM RATINGS

Power Supply	26 V
Power Dissipation (package limitation)	
Ceramic package	750 mW
Derate above +25°C	6.0 mW/°C
Plastic package	625 mW
Derate above +25°C	5 mW/°C
Storage Temperature Range	-65°C to +150°C

FUNCTIONAL BLOCK DIAGRAM

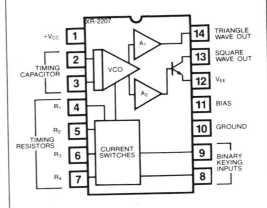

ORDERING INFORMATION

Part Number	Package	Operating Temperature
XR-2207M	Ceramic	-55°C to +125°C
XR-2207N	Ceramic	0°C to +70°C
XR-2207P	Plastic	0°C to +70°C
XR-2207CN	Ceramic	0°C to +70°C
XR-2207CP	Plastic	0°C to +70°C

SYSTEM DESCRIPTION

The XR-2207 utilizes four main functional blocks for frequency generation. These are a voltage controlled oscillator (VCO), four current switches which are activated by binary keying inputs, and two buffer amplifiers for triangle and squarewave outputs. The VCO is actually a current controlled oscillator which gets its input from the current switches. As the output frequency is proportional to the input current, the VCO produces four discrete output frequencies. Two binary input pins determine which timing currents are channelled to the VCO. These currents are set by resistors to ground from each of the four timing terminals.

The triangle ouput buffer provides a low impedance output (typically 10 Ω) while the squarewave is a open-collector type. A programmable reference point allows using the XR-2207 in either single or split supply configurations.

EXAR Corporation, 750 Palomar Avenue, Sunnyvale, CA 94086 • (408) 732-7970 • TWX 910-339-9233

FIGURE 2-15   XR-2207 precision oscillator. (Courtesy of EXAR Corporation.)

XR-2207

ELECTRICAL CHARACTERISTICS

Test Conditions: Test Circuit of Figure 1, $V^+ = V^- = 6V$, $T_A = +25°C$, $C = 5000$ pF, $R_1 = R_2 = R_3 = R_4 = 20$ KΩ, $R_L = 4.7$ KΩ, Binary Inputs grounded, S_1 and S_2 closed unless otherwise specified.

PARAMETERS	XR-2207/XR-2207M			XR-2207C			UNITS	CONDITIONS
	MIN.	TYP.	MAX.	MIN.	TYP.	MAX.		
GENERAL CHARACTERISTICS								
Supply Voltage								
Single Supply	8		26	8		26	V	See Figure 3
Split Supplies	±4		±13	±4		±13	V	
Supply Current								
Single Supply		5	7		5	8	mA	Measured at pin 1, S_1 and S_2 open
Split Supplies								See Figure 2
Positive		5	7		5	8	mA	Measured at pin 1, S_1, S_2 open
Negative		4	6		4	7	mA	Measured at pin 12, S_1, S_2 open
OSCILLATOR SECTION — FREQUENCY CHARACTERISTICS								
Upper Frequency Limit	0.5	1.0		0.5	1.0		MHz	$C = 500$ pF, $R_3 = 2$ KΩ
Lowest Practical Frequency		0.01			0.01		Hz	$C = 50$ μF, $R_3 = 2$ MΩ
Frequency Accuracy		±1	±3		±1	±5	% of f_o	
Frequency Matching		0.5			0.5		% of f_o	
Frequency Stability								
Temperature		20	50		30		ppm/°C	$0° < T_A < 75°C$
Power Supply		0.15			0.15		%/V	
Sweep Range	1000:1	3000:1			1000:1		f_H/f_L	$R_3 = 1.5$ KΩ for f_{H1}
Sweep Linearity								$R_3 = 2$ MΩ for f_L
10:1 Sweep		1	2		1.5		%	$C = 5000$ pF
1000:1 Sweep		5			5			$f_H = 10$ kHz, $f_L = 1$ kHz
FM Distortion		0.1			0.1		%	$f_H = 100$ kHz, $f_L = 100$ Hz
Recommended Range of	1.5		2000	1.5		2000	KΩ	±10% FM Deviation
Timing Resistors								See Characteristic Curves
Impedance at Timing Pins		75			75		Ω	Measured at pins 4, 5, 6, or 7
DC Level at Timing Terminals		10			10		mV	
BINARY KEYING INPUTS								
Switching Threshold	1.4	2.2	2.8	1.4	2.2	2.8	V	Measured at pins 8 and 9, Referenced to pin 10
Input Impedance		5			5		KΩ	
OUTPUT CHARACTERISTICS								
Triangle Output								
Amplitude	4	6		4	6		V_{pp}	Measured at pin 13
Impedance		10			10		Ω	
DC Level		+100			+100		mV	Referenced to pin 10
Linearity		0.1			0.1		%	From 10% to 90% to swing
Squarewave Output								
Amplitude	11	12		11	12		V_{pp}	Measured at pin 13, S_2 closed
Saturation Voltage		0.2	0.4		0.2	0.4	V	Referenced to pin 12
Rise Time		200			200		nsec	$C_L \leq 10$ pF
Fall Time		20			20		nsec	$C_L \leq 10$ pF

PRECAUTIONS

The following precautions should be observed when operating the XR-2207 family of integrated circuits:

1. Pulling excessive current from the timing terminals will adversely effect the temperature stability of the circuit. To minimize this disturbance, it is recommended that the ***total current drawn*** from pins 4, 5, 6, and 7 be limited to ≤6 mA. In addition, permanent damage to the device may occur if the total timing current exceeds 10 mA.
2. Terminals 2, 3, 4, 5, 6, and 7 have very low internal impedance and should, therefore, be protected from accidental shorting to ground or the supply voltages.
3. The keying logic pulse amplitude should not exceed the supply voltage.

2

DEFINITIONS OF TERMS

Frequency Accuracy

The difference between the actual operating frequency and the theoretical frequency determined from the design equations in Table 1, expressed as a percent of the calculated value.

Frequency Matching

The change in operating frequency as different timing terminals are activated for fixed timing resistor and timing capacitor values, expressed as a percent of the original operating frequency.

Binary Input Switching Threshold

The logic level at Pins 8 and 9 activate the binary current switches. The voltage level is referenced to Pin 10.

Frequency Sweep Range

The ratio of the highest and lowest operating frequencies (f_H/F_L) obtainable with a given value of timing capacitor.

Sweep Linearity

The maximum deviation of the sweep characteristics from a best-fit straight line extending over the frequency range.

Triangle Nonlinearity

The maximum deviation from a best-fit straight line extending along the rising and falling edges of the waveform, measured between 10% and 90% of each excursion.

PRINCIPLES OF OPERATION

The XR-2207 oscillator is a modified emitter-couple multivibrator type. As shown in the Functional Block Diagram, the oscillator also contains four current switches which activate the timing terminals, Pins 4, 5, 6, and 7. The oscillator frequency is inversely proportional to the value of timing capacitance, C, between Pins 2 and 3, and directly proportional to the total current, I_T, pulled out of the activated timing terminals.

Figure 3 provides greater detail of the oscillator control mechanism. Timing Pins 4, 5, 6, and 7 correspond to the emitters of switching transistors T_1, T_2, T_3 and T_4 respectively, which are internal to the integrated circuit. The current switches (and corresponding timing terminals) are activated by external logic signals applied to the keying terminals, Pins 8 and 9. The logic table for keying is given in Table 1.

As an example, logic inputs of 0, 0 at Pins 8 and 9 (i.e., both inputs "low") will result in turning on transistor pairs T_3, and only the timing terminal 6 will be activated. Under this condition, the total timing current, I_T, is equal to current I_3 pulled from Pin 6. This current is determined by external resistor R_3, resulting in a frequency $f_O = f_1 = 1/R_3C$ Hz.

It is important to observe that timing Pins 4, 5, 6, and 7 are low impedance points in the circuit. Care must be taken to avoid shorting these pins to the supply voltages or to ground.

EQUIVALENT SCHEMATIC DIAGRAM

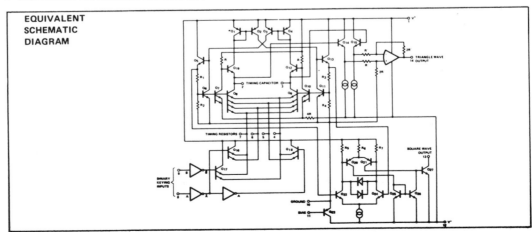

XR-2207

OPERATING INSTRUCTIONS

Split Supply Operation

Figure 4 is the recommended circuit connections for split supply operation. With the generalized connections of Figure 4A, the frequency of operation is determined by timing capacitor, C, and the activated timing resistors (R_1 through R_4). The timing resistors are activated by the logic signals at the binary keying inputs (Pins 8 and 9), as shown in the logic table, Table 1. If a fixed frequency of oscillation is required, the circuit connections can be simplified as shown in Figure 4B. In this connection, the input logic is set at (0,0) and the operating frequency is equal to ($1/R_3C$) Hz.

The squarewave output is obtained at Pin 13 and has a peak-to-peak voltage swing equal to the supply voltages. This output is an "open-collector" type and requires an external pull-up load resistor (nominally 5 KΩ) to the positive supply. The triangle waveform obtained at Pin 14 is centered about ground and has a peak amplitude of $V^+/2$.

The circuit operates with supply voltages ranging from ±4 V to ±13 V. Minimum drift occurs with ±6 volt supplies. For operation with unequal supply voltages, see Figure 7.

The logic levels at the keying inputs (Pins 8 and 9) are referenced to ground. A logic "0" corresponds to a keying voltage $V_k < 1.4$ V, and a logic "1" corresponds to $V_k > 3$ V. An open circuit at the keying inputs also corresponds to a "0" level.

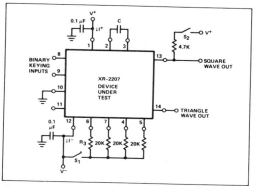

Figure 1. Test Circuit for Split Supply Operation

LOGIC LEVEL		SELECTED TIMING PINS	FREQUENCY	DEFINITIONS
A	B			
0	0	6	f_1	$f_1 = 1/R_3C, \Delta f_1 = 1/R_4C$
0	1	6 and 7	$f_1 + \Delta f_1$	$f_2 = 1/R_2C, \Delta f_2 = 1/R_1C$
1	0	5	f_2	Logic Levels: 0=ground
1	1	4 and 5	$f_2 + \Delta f_2$	1=>3V

Table 1. Logic Table for Binary Keying Controls

Note: For Single Supply Operation, Logic Levels are Referenced to Voltage at Pin 10

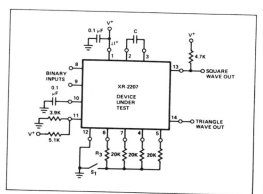

Figure 2. Test Circuit for Single Supply Operation

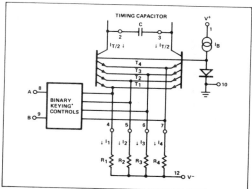

Figure 3. Simplified Schematic of Frequency Control Mechanism

Single Supply Operation

The circuit should be interconnected as shown in Figure 12 for single supply operation. Pin 12 should be grounded, and Pin 11 biased from V^+ through a resistive divider to a value of bias voltage between $V^+/3$ and $V^+/2$. Pin 10 is bypassed to ground through a 1 μF capacitor.

For single supply operation, the dc voltage at Pin 10 and the timing terminals (Pins 4 through 7) are equal and approximately 0.6 V below V_B, the bias voltage at Pin 11. The logic levels at the binary keying terminals are referenced to the voltage at Pin 10.

For a fixed frequency of $f_3 = 1/R_3C$, the external circuit connections can be simplified as shown in Figure 12B.

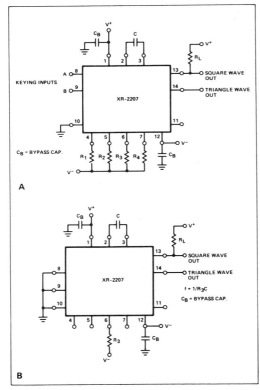

A

B

Figure 4. Split Supply Operation
A: General   B: Fixed Frequency

SELECTION OF EXTERNAL COMPONENTS

Timing Capacitor (Pins 2 and 3)

The oscillator frequency is inversely proportional to the timing capacitor, C, as indicated in Table 1. The minimum capacitance value is limited by stray capacitances and the maximum value by physical size and leakage current considerations. Recommended values range from 100 pF to 100 μF. The capacitor should be non-polar.

Timing Resistors (Pins 4, 5, 6, and 7)

The timing resistors determine the total timing current, I_T, available to charge the timing capacitor. Values for timing resistors can range from 2 KΩ to 2MΩ; however, for optimum temperature and power supply stability, recommended values are 4 KΩ to 200 KΩ (see Figures 8, 10 and 11). To avoid parasitic pick up, timing resistor leads should be kept as short as possible. For noisy environments, unused or deactivated timing terminals should be bypassed to ground through 0.1 μF capacitors.

Supply Voltage (Pins 1 and 12)

The XR-2207 is designed to operate over a power supply range of ±4 V to ±13 V for split supplies, or 8 V to 26 V for single supplies. At high supply voltages, the frequency sweep range is reduced (see Figures 7 and 8). Performance is optimum for ±6 V, or 12 V single supply operation.

Binary Keying Inputs (Pins 8 and 9)

The internal impedance at these pins is approximately 5 KΩ. Keying levels are < 1.4 V for "zero" and >3 V for "one" logic levels referenced to the dc voltage at Pin 10 (see Table 1).

Bias for Single Supply (Pin 11)

For single supply operation, Pin 11 should be externally biased to a potential between $V^+/3$ and $V^+/2$ volts (see Figure 12). The bias current at Pin 11 is nominally 5% of the total oscillation timing current, I_T.

Ground (Pin 10)

For split supply operation, this pin serves as circuit ground. For single supply operation, Pin 10 should be ac grounded through a 1 μF bypass capacitor. During split supply operation, a ground current of $2I_T$ flows out of this terminal, where I_T is the total timing current.

XR-2207

TYPICAL CHARACTERISTICS

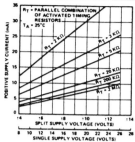

Figure 5. Positive Supply Current,
I⁺ (measured at pin 1) vs.
Supply Voltage*

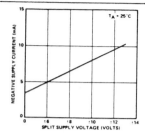

Figure 6. Negative Supply Current,
I⁻ (measured at pin 12) vs.
Supply Voltage

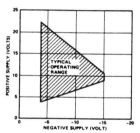

Figure 7. Typical Operating Range
for Split Supply Voltage

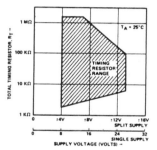

Figure 8. Recommended Timing
Resistor Value vs. Power
Supply Voltage*

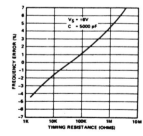

Figure 9. Frequency Accuracy vs.
Timing Resistance

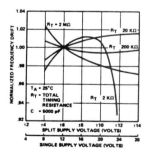

Figure 10. Frequency Drift vs.
Supply Voltage

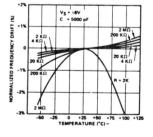

Figure 11. Normalized Frequency
Drift with Temperature

*Note: R_T = Parallel Combination of Activated Timing Resistors

6

XR-2207

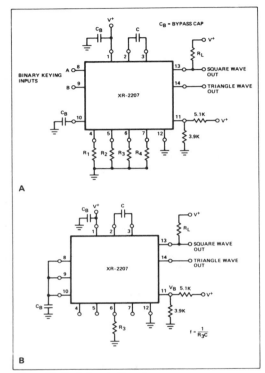

Figure 12. Single Supply Operation

A: General   B: Fixed Frequency

Squarewave Output (Pin 13)

The squarewave output at Pin 13 is an open collector stage capable of sinking up to 20 mA of load current. R_L serves as a pull-up load resistor for this output. Recommended values for R_L range from 1 KΩ to 100 KΩ.

Triangle Output (Pin 14)

The output at Pin 14 is a triangle wave with a peak swing of approximately one-half of the total supply voltage. Pin 14 has a very low output impedance of 10 Ω and is internally protected against short circuits.

Bypass Capacitors

The recommended value for bypass capacitors is 1 μF, although larger values are required for very low frequency operation.

Frequency Control (Sweep and FM)

The frequency of operation is controlled by varying the total timing current, I_T, drawn from the activated timing Pins 4, 5, 6, or 7. The timing current can be modulated by applying a control voltage, V_C, to the activated timing pin through a series resistor R_C, as shown in Figures 13 and 14.

For split supply operation, a *negative* control voltage, V_C, applied to the circuits of Figures 13 and 14 causes the total timing current, I_T, and the frequency, to increase.

As an example, in the circuit of Figure 13, the binary keying inputs are grounded. Therefore, only timing Pin 6 is activated. The frequency of operation, normally $f = \dfrac{1}{R_3C}$ is now proportional to the control voltage, V_C, and determined as:

$$f = \frac{1}{R_3C}\left[1 - \frac{V_CR_3}{R_CV^-}\right] Hz$$

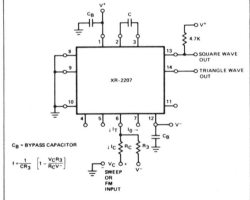

Figure 13. Frequency Sweep Operation

The frequency, f, will increase as the control voltage is made more negative. If $R_3 = 2$ MΩ, $R_C = 2$ KΩ, $C = 5000$ pF, then a 1000:1 frequency sweep would result for a negative sweep voltage $V_C \cong V^-$. The voltage to frequency conversion gain, K, is controlled by the series resistance R_C and can be expressed as:

$$K = \frac{\Delta f}{\Delta V_C} = -\frac{1}{R_CCV^-} Hz/volt$$

7

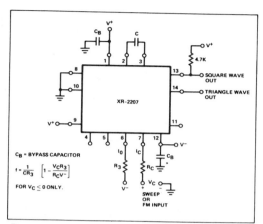

Figure 14. Alternate Frequency Sweep Operation

Figure 15. Sawtooth and Pulse Outputs

The circuit of Figure 13 can operate both with positive and negative values of control voltage. However, for positive values of V_C with small (R_C/R_3) ratio, the direction of the timing current I_T is reversed and the oscillations will stop.

Figure 14 shows an alternate circuit for frequency control where two timing pins, 6 and 7, are activated. The frequency and the conversion gain expressions are the same as before, except that the circuit will operate only with negative values of V_C. For $V_C > 0$, Pin 7 becomes deactivated and the frequency is fixed at

$$f = \frac{1}{R_3 C}$$

CAUTION

For operation of the circuit, total timing current I_T must be less than 6 mA over the frequency control range.

Duty Cycle Control

The duty cycle of the output waveforms can be controlled by frequency shift keying at the end of every half cycle of oscillator output. This is accomplished by connecting one or both of the binary keying inputs (Pins 8 or 9) to the squarewave output at Pin 13. The output waveforms can then be converted to positive or negative pulses and sawtooth waveforms.

Figure 15 is the recommended circuit connection for duty cycle control. Pin 8 is shorted to Pin 13 so that the circuit switches between the "0, 0" and the "1, 0" logic states given in Table 1. Timing Pin 5 is activated when the output is high, and the timing pin is activated when the square-wave output goes to a low state.

The duty cycle of the output waveform is given as:

$$\text{Duty Cycle} = \frac{R_2}{R_2 + R_3}$$

and can be varied from 0.1% to 99.9% by proper choice of timing resistors. The frequency of oscillation, f, is given as:

$$f = \frac{2}{C}\left[\frac{1}{R_2 + R_3}\right]$$

The frequency can be modulated or swept without changing the duty cycle by connecting R_2 and R_3 to a common control voltage V_C, instead of to V^- (see Figure 13). The sawtooth and the pulse output waveforms are shown in Figure 16.

On — Off Keying

The XR-2207 can be keyed on and off by simply activating an open circuited timing pin. Under certain conditions, the circuit may exhibit very low frequency (<1 Hz) residual oscillations in the off state due to internal bias currents. If this effect is undesirable, it can be eliminated by connecting a 10 MΩ resistor from Pin 3 to V^+.

APPLICATIONS

Two-Channel FSK Generator (Modem Transmitter)

The multi-level frequency shift-keying capability of XR-2207 makes it ideally suited for two-channel FSK generation. A recommended circuit connection for this application is shown in Figure 17.

For two-channel FSK generation, the "mark" and "space" frequencies of the respective channels are determined by the timing resistor pairs (R_1, R_2) and (R_3, R_4). Pin 8 is the "channel-select" control in accord with Table 1. For a "high" logic level at Pin 8, the timing resistors R_1 and R_2 are activated. Similarly, for a "low" logic level, timing resistors R_3 and R_4 are enabled.

The "high" and "low" logic levels at Pin 9 determine the respective high and low frequencies within the selected FSK channel.

Recommended component values for various commonly used FSK frequencies are given in Table 2. When only a single FSK channel is used, the remaining channel can be deactivated by connecting Pin 8 to either V^+ or ground. In this case, the unused timing resistors can also be omitted from the circuit.

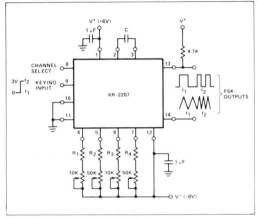

Figure 17. Multi-Channel FSK Generation

The low and high frequencies, f_1 and f_2, for a given FSK channel can be fine tuned using potentiometers connected in series with respective timing resistors. In fine tuning the frequencies, f_1 should be set first with the logic level at Pin 9 in a "low" level.

Typical frequency drift of the circuit for $0°C$ to $+70°C$ operation is $±0.2\%$. Since the frequency stability is directly related to the external timing components, care must be taken to use timing components with low temperature coefficients.

FSK Transceiver (Full-Duplex Modem)

The XR-2207 can be used in conjunction with the XR-210 FSK Demodulator to form a full-duplex FSK transceiver, or modem. A recommended circuit connection for this application is shown in Figure 18. Table 2 shown the recommended component values for 300 Baud (103-type) and 1200 Baud (202-type) modem applications.

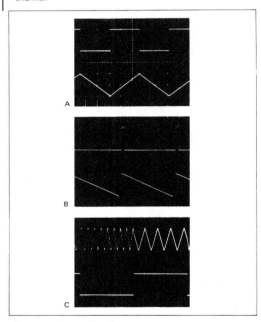

Figure 16. Output Waveforms

A: Squarewave and Triangle Outputs
B: Pulse and Sawtooth Outputs
C: Frequency-Shift Keyed Output
Top: FSK Output with $f_2 = 2f_1$
Bottom: Keying Logic Input

XR-2207

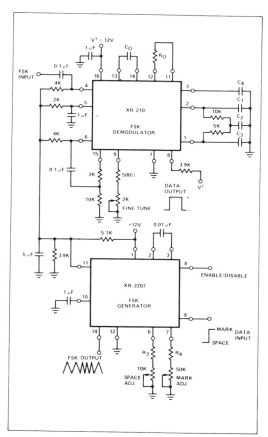

OPERATING CONDITIONS	TYPICAL COMPONENT	VALUE
300 Baud	XR-210	XR-2207
Low Band: $f_1 = 1070$ Hz $f_2 = 1270$ Hz	$R_0 = 5.1$ kΩ, $C_0 = 0.22$ μF $C_1 = C_2 = 0.047$ μF $C_3 = 0.033$ μF	$R_3 = 91$ k $R_4 = 470$ k
High Band: $f_1 = 2025$ Hz $f_2 = 2225$ Hz	$R_0 = 8.2$ kΩ, $C_0 = 0.1$ μF $C_1 = C_2 = C_3 = 0.033$ μF	$R_3 = 47$ k $R_4 = 470$ k
1200 Baud $f_1 = 1200$ Hz $f_2 = 2200$ Hz	$R_0 = 2$ kΩ, $C_0 = 0.14$ μF $C_1 = 0.033$ μF $C_3 = 0.02$ μF $C_2 = 0.01$ μF	$R_3 = 75$ k $R_4 = 91$ k

Table 2. Recommended Component Values for Full Duplex FSK Modem of Figure 18

Figure 18. Full Duplex FSK Modem using XR-210 and XR-2207 (See Table 2 for Component Values)

XR-2209
Precision Oscillator

GENERAL DESCRIPTION

The XR-2209 is a monolithic variable frequency oscillator circuit featuring excellent temperature stability and a wide linear sweep range. The circuit provides simultaneous triangle and squarewave outputs over a frequency range of 0.01 Hz to 1 MHz. The frequency is set by an external RC product. It is ideally suited for frequency modulation, voltage to frequency or current to frequency conversion, sweep or tone generation as well as for phase-locked loop applications when used in conjunction with a phase comparator such as the XR-2208.

The circuit is comprised of three functional blocks: a variable frequency oscillator which generates the basic periodic waveforms and two buffer amplifiers for the triangle and the squarewave outputs.

The oscillator frequency is set by an external capacitor, C, and the timing resistor R. With no sweep signal applied, the frequency of oscillation is equal to 1/RC. The XR-2209 has a typical drift specification of 20 ppm/°C. Its frequency can be linearly swept over a 1000:1 range with an external control signal.

FEATURES

Excellent Temperature Stability (20 ppm/°C)
Linear Frequency Sweep
Wide Sweep Range (1000:1 Min)
Wide Supply Voltage Range (±4V to ±13V)
Low Supply Sensitivity (0.15%/V)
Wide Frequency Range (0.01 Hz to 1 MHz)
Simultaneous Triangle and Squarewave Outputs

APPLICATIONS

Voltage and Current-to-Frequency Conversion
Stable Phase-Locked Loop
Waveform Generation
FM and Sweep Generation

ABSOLUTE MAXIMUM RATINGS

Power Supply	26 volts
Power Dissipation (package limitation)	
Ceramic package	385 mW
Plastic Package	300 mW
Derate above +25°C	2.5 mW/°C
Temperature Range	
Operating	
XR-2209M	−55°C to +125°C
XR-2209C	0°C to +75°C
Storage	−65°C to +150°C

AVAILABLE TYPES

Part Number	Package	Operating Temperature
XR-2209M	Ceramic	−55°C to +125°C
XR-2209CN	Ceramic	0°C to +75°C
XR-2209CP	Plastic	0°C to +75°C

EQUIVALENT SCHEMATIC DIAGRAM

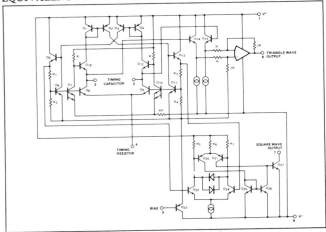

FUNCTIONAL BLOCK DIAGRAM

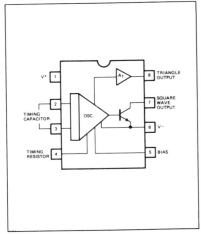

EXAR INTEGRATED SYSTEMS, INC.
750 Palomar Ave., P.O. Box 62229, Sunnyvale, CA 94088
(408) 732-7970   TWX 910-339-9233

FIGURE 2-16   XR-2209 precision oscillator. (Courtesy of EXAR Corporation.)

ELECTRICAL CHARACTERISTICS – PRELIMINARY

Test Conditions: Test Circuit of Figure 1, $V^+ = V^- = 6V$, $T_A = +25°C$, $C = 5000$ pF, $R = 20$ KΩ, $R_L = 4.7$ KΩ. S_1 and S_2 closed unless otherwise specified.

PARAMETERS	XR-2209M MIN.	XR-2209M TYP.	XR-2209M MAX.	XR-2209C MIN.	XR-2209C TYP.	XR-2209C MAX.	UNITS	CONDITIONS
GENERAL CHARACTERISTICS								
Supply Voltage								
Single Supply	8		26	8		26	V	See Figure 2
Split Supplies	±4		±13	±4		±13	V	See Figure 1
Supply Current								
Single Supply		5	7		5	8	mA	Measured at pin 1, S_1, S_2 open See Figure 2
Split Supplies								
Positive		5	7		5	8	mA	Measured at pin 1, S_1, S_2 open
Negative		4	6		4	7	mA	Measured at pin 4, S_1, S_2 open
OSCILLATOR SECTION – FREQUENCY CHARACTERISTICS								
Upper Frequency Limit	0.5	1.0		0.5	1.0		MHz	$C = 500$ pF, $R = 2$ KΩ
Lowest Practical Frequency		0.01			0.01		Hz	$C = 50$ μF, $R = 2$ MΩ
Frequency Accuracy		±1	±3		±1	±5	% of f_o	
Frequency Stability								
Temperature		20	50		30		ppm/°C	$0° < T_A < 75°C$
Power Supply		0.15			0.15		%/V	
Sweep Range	1000:1	3000:1		1000:1			f_H/f_L	$R = 1.5$ KΩ for f_{H1} $R = 2$ MΩ for f_L
Sweep Linearity							%	$C = 5000$ pF
10:1 Sweep		1	2		1.5			$f_H = 10$ kHz, $f_L = 1$ kHz
1000:1 Sweep		5			5			$f_H = 100$ kHz, $f_L = 100$Hz
FM Distortion		0.1			0.1		%	±10% FM Deviation
Recommended Range of Timing Resistors	1.5		2000	1.5		2000	KΩ	See Characteristic Curves
Impedance at Timing Pin		75			75		Ω	Measured at pin 4
OUTPUT CHARACTERISTICS								
Triangle Output								Measured at pin 8
Amplitude	4	6		4	6		Vpp	
Impedance		10			10		Ω	
Linearity		0.1			0.1		%	10% to 90% of swing
Squarewave Output								Measured at pin 7, S_2 closed
Amplitude	11	12		11	12		Vpp	
Saturation Voltage		0.2	0.4		0.2	0.4	V	Referenced to pin 6
Rise Time		200			200		nsec	$C_L \leqslant 10$ pF, $R_L = 4.7$ KΩ
Fall Time		20			20		nsec	$C_L \leqslant 10$ pF

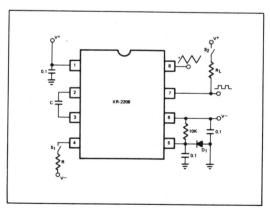

Figure 1. Test Circuit for Split Supply Operation (D_1 = 1N 4148 or Equivalent)

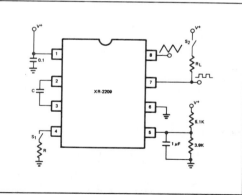

Figure 2. Test Circuit for Single Supply Operation

CHARACTERISTIC CURVES

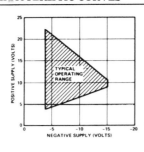

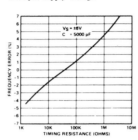

Figure 3. Typical Operating Range For Split Supply Voltage

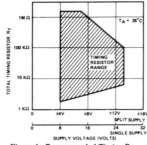

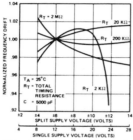

Figure 4. Recommended Timing Resistor Value vs. Power Supply Voltage*

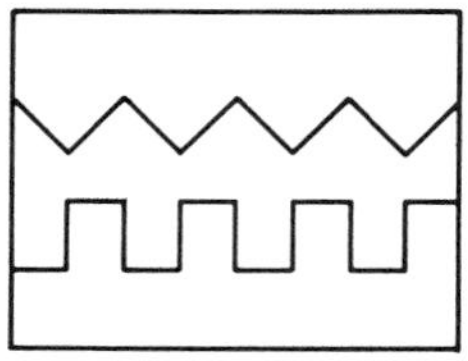

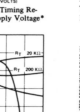

Figure 5: Output Waveforms
Top: Triangle Output (Pin 8)
Bottom: Squarewave Output (Pin 7)

Figure 6. Frequency Accuracy vs. Timing Resistance

Figure 7. Frequency Drift vs. Supply Voltage

*Note: R_T = Timing Resistor at Pin 4

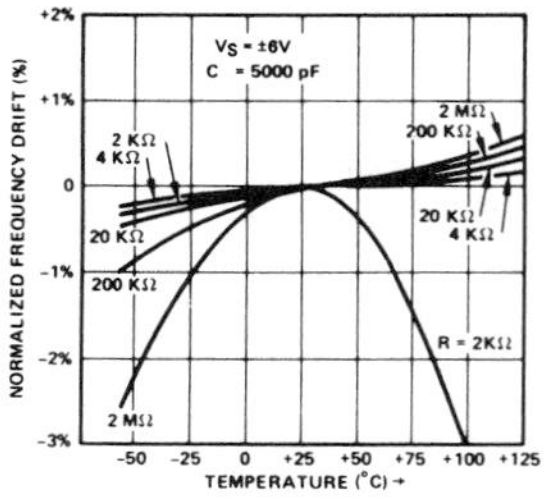

Figure 8. Normalized Frequency Drift With Temperature

RECOMMENDED CIRCUIT CONNECTIONS

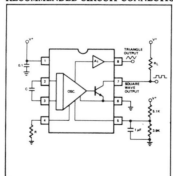

Figure 9. Circuit Connection for Single Supply Operation

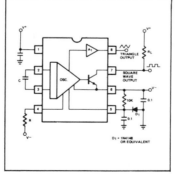

Figure 10. Generalized Circuit Connection for Split Supply Operation

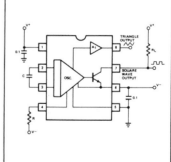

Figure 11. Simplified Circuit Connection for Split Supply Operation with V_{CC} = V_{EE} > ±7V *(Note: Triangle wave output has +0.6V offset with respect to ground.)*

PRECAUTIONS

The following precautions should be observed when operating the XR-2209 family of integrated circuits:

1. Pulling excessive current from the timing terminal will adversely effect the temperature stability of the circuit. To minimize this disturbance, it is recommended that the *total current* drawn from pin 4 be limited to ≤6 mA.

2. Terminals 2, 3, and 4 have very low internal impedance and should, therefore, be protected from accidental shorting to ground or the supply voltages.

3. Triangle waveform linearity is sensitive to parasitic coupling between the square and the triangle-wave outputs (pins 7 and 8). In board layout or circuit wiring care should be taken to minimize stray wiring capacitances between these pins.

DESCRIPTION OF CIRCUIT CONTROLS

TIMING CAPACITOR (PINS 2 and 3)

The oscillator frequency is inversely proportional to the timing capacitor, C. The minimum capacitance value is limited by stray capacitances and the maximum value by physical size and leakage current considerations. Recommended values range from 100 pF to 100 μF. The capacitor should be non-polar.

TIMING RESISTOR (PIN 4)

The timing resistor determines the total timing current, I_T, available to charge the timing capacitor. Values for the timing resistor can range from 1.5 KΩ to 2 MΩ; however, for optimum temperature and power supply stability, recommended values are 4 KΩ to 200 KΩ (see Figures 4, 7, and 8). To avoid parasitic pick up, timing resistor leads should be kept as short as possible.

SUPPLY VOLTAGE (PINS 1 AND 6)

The XR-2209 is designed to operate over a power supply range of ±4V to ±13V for split supplies, or 8V to 26V for single supplies. At high supply voltages, the frequency sweep range is reduced (see Figures 3 and 4). Performance is optimum for ±6V, or 12V single supply operation.

BIAS FOR SINGLE SUPPLY (PIN 5)

For single supply operation, pin 5 should be externally biased to a potential between $V^+/3$ and $V^+/2$ volts (see Figure 9). The bias current at pin 5 is nominally 5% of the total oscillation timing current, I_T, at pin 4. This pin should be bypassed to ground with 0.1 μF capacitor.

SQUAREWAVE OUTPUT (PIN 7)

The squarewave output at pin 7 is a "open-collector" stage capable of sinking up to 20 mA of load current. R_L serves as a pull-up load resistor for this output. Recommended values for R_L range from 1 KΩ to 100 KΩ.

TRIANGLE OUTPUT (PIN 8)

The output at pin 8 is a triangle wave with a peak swing of approximately one-half of the total supply voltage. Pin 8 has a very low output impedance of 10Ω and is internally protected against short circuits.

OPERATING INSTRUCTIONS

SPLIT SUPPLY OPERATION

The recommended circuit for split supply operation is shown in Figure 10. Diode D_1 in the figure assures that the triangle output swing at pin 8 is symmetrical about ground. This circuit operates with supply voltages ranging from ±4V to ±13V. Minimum drift occurs at ±6V supplies. See Figure 3 for operation with unequal supplies.

Simplified Connection

For operation with split supplies in excess of ±7 volts, the simplified circuit connection of Figure 11 can be used. This circuit eliminates the diode D_1 used in Figure 10; however the triangle wave output at pin 8 now has a +0.6 volt DC offset with respect to ground.

SINGLE SUPPLY OPERATION

The recommended circuit connection for single-supply operation is shown in Figure 9. Pin 6 is grounded; and pin 5 is biased from V^+ through a resistive divider, as shown in the figure, and is bypassed to ground with a 1 μF capacitor.

For single supply operation, the DC voltage at the timing terminal, pin 4, is approximately 0.6 volts above V_B, the bias voltage at pin 5.

The frequency of operation is determined by the timing capacitor C and the timing resistor R, and is equal to 1/RC. The squarewave output is obtained at pin 7 and has a peak-to-peak voltage swing equal to the supply voltage. This output is an "open-collector" type and requires an external pull-up load resistor (nominally 5 KΩ) to V^+. The triangle waveform obtained at pin 8 is centered about a voltage level V_O where:

$$V_O = V_B + 0.6V$$

where V_B is the bias voltage at pin 5. The peak-to-peak output swing of triangle wave is approximately equal to $V^+/2$.

FREQUENCY CONTROL (SWEEP AND FM)

The frequency of operation is proportional to the *total* timing current I_T drawn from the timing pin, pin 4. This timing current, and the frequency of operation can be modulated by applying a control voltage, V_C, to the timing pin, through a series resistor, R_S, as shown in Figure 12. If V_C is negative with respect to V_A, the voltage level at pin 4, then an additional current I_O is drawn from the timing pin causing I_T to increase, thus increasing the frequency. Conversely, making V_C higher than V_A causes the frequency to decrease by decreasing I_T.

The frequency of operation, is determined by:

$$f = f_O \left[1 + \frac{R}{R_S} - \frac{V_C}{V_A} \frac{R}{R_S} \right]$$

where $f_O = 1/RC$.

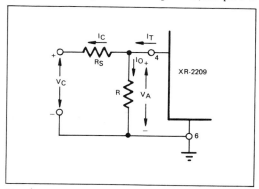

Figure 12. Frequency Sweep Operation

FREQUENCY MULTIPLIERS

In electronic communications systems, it is often necessary or desirable to produce exact multiples of a frequency. A *frequency multiplier* is just such a circuit; it multiplies the frequency of the input signal and generates harmonics of that frequency. A frequency multiplier takes advantage of the nonlinear characteristics of a circuit. In Chapter 1, it was shown that a nonsinusoidal repetitive waveform is made up of a series of harmonically related sine waves. If a pure sine wave is distorted in such a way that it produces a repetitive nonsinusoidal waveform, higher-order harmonics are produced and frequency multiplication is accomplished. The more the signal is distorted, the higher the amplitude of the harmonics. A frequency multiplier consists of an oscillator, a nonlinear amplifier, and a frequency-selective circuit such as an *LC* tank circuit. The schematic diagram of a frequency multiplier is shown in Figure 2-17a. Q_1 is a class C amplifier, and its input voltage and collector current waveforms are shown in Figure 2-17b and c, respectively. It can be seen that the fundamental frequency of the input voltage and output current waveforms are the same. However, the collector current waveform is not a pure sine wave, although it is a repetitive waveform. Consequently, it is made up of a series of harmonically related sine waves. If the resonant frequency of the tank circuit is tuned to a multiple of the input frequency, frequency multiplication is accomplished. The oscillating voltage waveform for the second harmonic is shown in Figure 2-17d. It can be seen that energy is injected into the tank circuit during every other cycle of the output waveform. During each intermediate cycle, the signal is somewhat damped due to the power dissipated in the resistive component of the inductor. However, if the Q factor of the coil is made sufficiently high, the damping factor is negligible and the output is a continuous sine wave at twice the frequency of the input signal.

In Chapter 1 it was shown that the higher the harmonic, the lower its amplitude. Consequently, frequency multipliers, like the one shown in Figure 2-17, are extremely inefficient and are impractical for multiplication factors above 3. A device that is used to generate several different output frequencies is called a *frequency synthesizer*. Frequency synthesizers are explained in Chapter 5.

MIXING

Mixing is the process of combining two or more signals and is an essential process in electronic communications. In essence, there are two ways in which signals are mixed: linearly and nonlinearly.

Linear Summing

Linear summing is when two or more signals combine in a linear device such as a resistive or inductive network or a small-signal amplifier. In the audio recording industry,

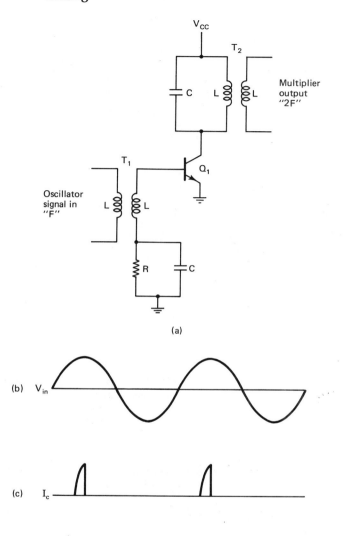

(a)

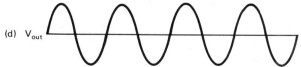

(b) V_{in}

(c) I_c

(d) V_{out}

FIGURE 2-17 Frequency multiplication: (a) schematic drawing; (b) input voltage waveform; (c) collector current waveform; (d) tank circuit oscillating voltage waveform, 2nd harmonic.

linear summing is sometimes called *linear mixing*. However, in communications work mixing almost always implies a nonlinear process.

Single-input frequency.   Figure 2-18a shows the amplification of a single-frequency input signal by a linear amplifier. The output is simply the original input frequency

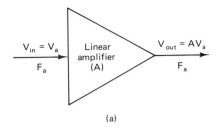

(a)

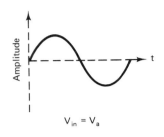

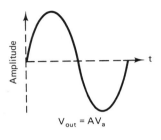

(b)

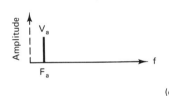

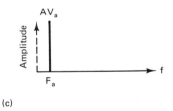

(c)

FIGURE 2-18 Linear amplification of a single input frequency: (a) linear amplification; (b) time domain; (c) frequency domain.

amplified by the gain (A). Figure 2-18b shows the output signal in the time domain, and Figure 2-18c shows the frequency domain. Mathematically, the output is

$$V_{out} = AV_{in} = AV_a \sin 2\pi \, F_a t \qquad (2\text{-}5)$$

where $V_{in} = V_a \sin 2\pi F_a t$.

Multiple-input frequencies.   Figure 2-19a shows two input frequencies combining in a small-signal amplifier. Each input signal is amplified by the gain (A). Therefore, the output is expressed mathematically as

$$V_{out} = AV_{in}$$

where

$$V_{in} = V_a \sin 2\pi F_a t + V_b \sin 2\pi F_b t \qquad (2\text{-}6)$$

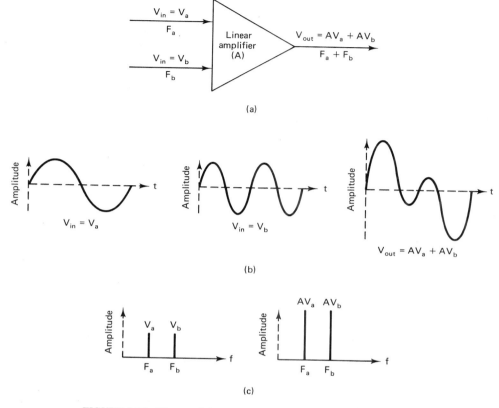

FIGURE 2-19 **Linear mixing: (a) linear amplification; (b) time domain; (c) frequency domain.**

Therefore,

$$V_{\text{out}} = A(V_a \sin 2\pi F_a t + V_b \sin 2\pi F_b t)$$

or

$$V_{\text{out}} = AV_a \sin 2\pi F_a t + AV_b \sin 2\pi F_b t$$

V_{out} is simply an *additive* waveform and is equal to the algebraic sum of V_a and V_b. Figure 2-19b shows the *linear summation* of V_a and V_b in the time domain, and Figure 2-19c shows the linear summation in the frequency domain. If additional input frequencies are applied to the circuit, they are linearly summed with V_a and V_b. In *high-fidelity* audio systems, it is important that the output spectrum contain only the original input frequencies, and therefore linear operation is desired. However, in electronic communications systems where modulation is essential, nonlinear mixing is often required.

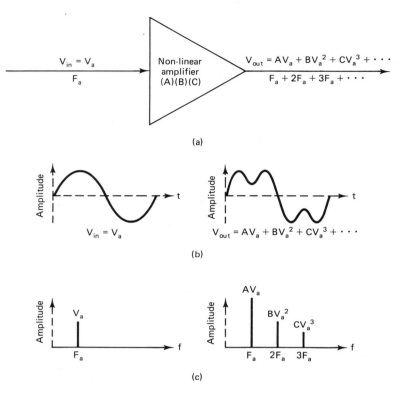

FIGURE 2-20 Nonlinear amplification of a single input frequency: (a) nonlinear amplification; (b) time domain; (c) frequency domain.

Nonlinear Mixing

Nonlinear mixing is when two or more signals are combined in a nonlinear device such as a diode or a large-scale amplifier.

Single-input frequency.   Figure 2-20a shows the amplification of a single-frequency input signal by a nonlinear amplifier. The output from a nonlinear amplifier with a single-frequency input signal is not a single sine wave. Mathematically, the output is in the infinite power series

$$V_{out} = AV_{in} + BV_{in}^2 + CV_{in}^3 + \cdots \qquad (2\text{-}7a)$$

where $V_{in} = V_a \sin 2\pi F_a t$. Therefore

$$V_{out} = A(V_a \sin 2\pi F_a t) + B(V_a \sin 2\pi F_a t)^2 + C(V_a \sin 2\pi F_a t)^3 + \cdots \qquad (2\text{-}7b)$$

where

AV_{in} = linear term or simply the input signal (F_a) amplified by the gain (A)
BV_{in}^2 = quadratic term that generates the second harmonic frequency ($2F_a$)
CV_{in}^3 = cubic term that generates the third harmonic frequency ($3F_a$)

V_{in}^n produces a frequency equal to n times F. BV_{in}^2 generates a frequency equal to $2F_a$, CV_{in}^3 generates a frequency equal to $3F_a$, and so on. Multiples of a *base* frequency are called *harmonics*. As stated in Chapter 1, the original input frequency (F_a) is the first harmonic or the *fundamental* frequency, $2F_a$ is the second harmonic, $3F_a$ the third harmonic, and so on. Figure 2-20b shows the output waveform in the time domain for a nonlinear amplifier with a single input frequency. It can be seen that the output waveform is simply the summation of the input frequency and its higher harmonics (multiples of the fundamental frequency). Figure 2-20c shows the output spectrum in the frequency domain. Note that adjacent harmonics are separated in frequency by a value equal to the fundamental frequency (F_a).

Nonlinear amplification of a single frequency results in the generation of multiples or harmonics of that frequency. If the harmonics are undesired it is called *harmonic distortion*.

A JFET is a special-case nonlinear device that has *square-law* properties. The output from a square-law device is simply

$$V_{out} = AV_{in} + BV_{in}^2 \qquad (2\text{-}8)$$

The output from a square-law device with a single input frequency is the fundamental frequency (i.e., the original input frequency) and its second harmonic. There are no additional harmonics generated. Therefore, there is less harmonic distortion produced with a JFET than with a comparable BJT.

Multiple-input frequencies. Figure 2-21a shows the nonlinear amplification of two input frequencies by a large-signal (nonlinear) amplifier.

Mathematically, the output of a large-signal amplifier with two input frequencies is

$$V_{out} = AV_{in} + BV_{in}^2 + CV_{in}^3 + \cdot\cdot\cdot \qquad (2\text{-}9a)$$

where $V_{in} = V_a \sin 2\pi F_a t + V_b \sin 2\pi F_b t$. Therefore,

$$V_{out} = A(V_a \sin 2\pi F_a t + V_b \sin 2\pi F_b t) + B(V_a \sin 2\pi F_a t + V_b \sin 2\pi F_b t)^2$$
$$+ C(V_a \sin 2\pi F_a t + V_b \sin 2\pi F_b t)^3 + \cdot\cdot\cdot \qquad (2\text{-}9b)$$

The preceding formula is an infinite series and there is no limit to the number of terms it can have. If the binomial theorem is applied to each higher-power term, the formula can be rearranged and written as

$$V_{out} = (AV_a + BV_a^2 + CV_a^3 + \cdot\cdot\cdot) + (AV_b + BV_b^2 + CV_b^3 + \cdot\cdot\cdot)$$
$$+ (2BV_aV_b + 3CV_a^2V_b + 3CV_aV_b^2 + \cdot\cdot\cdot) \qquad (2\text{-}9c)$$

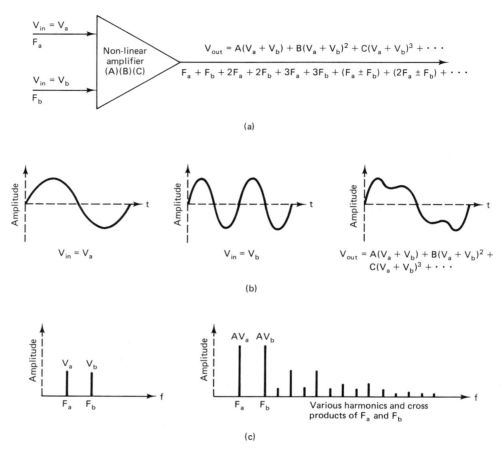

FIGURE 2-21 Nonlinear amplification of two sine waves: (a) nonlinear amplification; (b) time domain; (c) frequency domain.

The terms in the first set of parentheses generate harmonics of F_a ($2F_a$, $3F_a$, etc.). The terms in the second set of parentheses generate harmonics of F_b ($2F_b$, $3F_b$, etc.). The terms in the third set of parentheses generate the cross products ($F_a + F_b$, $F_a - F_b$, $2F_a + F_b$, $2F_a - F_b$, etc.). The cross products are produced from intermodulation among the two original frequencies and their harmonics. The *cross products* are the *sum* and *difference frequencies*; they are the sum and difference of the two original frequencies and the sums and differences of their harmonics. There is an infinite number of harmonic and cross-product frequencies produced when two or more frequencies mix in a nonlinear device. If the cross products are undesired, it is called *intermodulation distortion*. Mathematically, the sum and difference frequencies are

$$\text{cross products} = mF_a \pm nF_b \qquad (2\text{-}10)$$

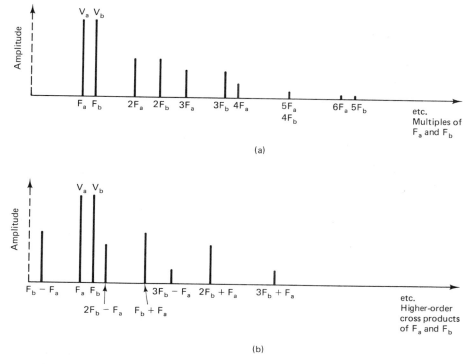

(a)

(b)

FIGURE 2-22 Output spectrum from a nonlinear amplifier with two input frequencies: (a) harmonic distortion; (b) intermodulation distortion.

where *m* and *n* are all positive integers between 1 and infinity. Figure 2-22 shows the output spectrum from a nonlinear amplifier with two input frequencies.

Intermodulation distortion is any unwanted cross product that is generated when two or more frequencies are mixed in a nonlinear device. Consequently, whenever two or more frequencies are amplified by a nonlinear device, harmonic and intermodulation distortion are present in the output.

EXAMPLE 2-2

For a nonlinear amplifier with two input frequencies, 5 kHz and 7 kHz:
(a) Determine the first three harmonics present in the output for each input frequency.
(b) Determine the cross products produced in the output for values of *m* and *n* of 1 and 2.
(c) Draw the output spectrum for the harmonics and cross products determined in parts (a) and (b).

Solution (a) The first three harmonics include the two original frequencies, 5 and 7 kHz; two times each of the original frequencies, 10 and 14 kHz; and three times each of the original frequencies, 15 and 21 kHz.

(b) The cross products for values of m and n of 1 and 2 are determined from Equation 2-10 and are summarized below.

m	n	Cross products
1	1	7 kHz $\pm$ 5 kHz = 2 and 12 kHz
1	2	7 kHz $\pm$ 10 kHz = 3 and 17 kHz
2	1	14 kHz $\pm$ 5 kHz = 9 and 19 kHz
2	2	14 kHz $\pm$ 10 kHz = 4 and 24 kHz

(c) The output spectrum is shown in Figure 2-23.

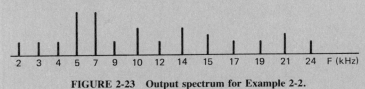

FIGURE 2-23 **Output spectrum for Example 2-2.**

QUESTIONS

2-1. Define *oscillate*; *oscillator*.

2-2. Describe the following terms: *self-sustaining*; *repetitive*; *free-running*; *one-shot*.

2-3. Describe the regenerative process necessary for self-sustained oscillations to occur.

2-4. List and describe the four requirements for a feedback oscillator to work.

2-5. What is meant by the terms *positive feedback* and *negative feedback*?

2-6. Define *open-loop gain*; *closed-loop gain*.

2-7. List the four most common oscillator configurations.

2-8. Describe the operation of a Wien-bridge oscillator.

2-9. Describe oscillator action for an *LC* tank circuit.

2-10. What is the flywheel effect?

2-11. What is meant by *damped oscillations*? What causes them?

2-12. Describe the operation of a Hartley oscillator; a Colpitts oscillator.

2-13. Define *frequency stability*.

2-14. List several factors that affect the frequency stability of an oscillator.

2-15. Describe the piezoelectric effect.

2-16. What is meant by *crystal cut*? List and describe several crystal cuts and contrast their stabilities.

2-17. Describe how an overtone crystal oscillator works.

2-18. What is the advantage of an overtone crystal oscillator over a conventional crystal oscillator?

2-19. What is meant by *positive temperature coefficient*; *negative temperature coefficient*?

2-20. What is meant by *zero coefficient crystal*?

2-21. Sketch the electrical equivalent circuit for a crystal and describe the various components.

2-22. Which crystal oscillator configuration has the best frequency stability? *RLC Half-Bridge*

2-23. Which crystal oscillator configuration is the least expensive and most adaptable to digital interfacing? *Jc Pierce*

2-24. Describe a crystal oscillator module. *1.53*

2-25. What is the predominant advantage of crystal oscillators over *LC* tank circuit oscillators? *freq. Stability*

2-26. Describe the operation of a varactor diode.

2-27. Describe the operation of a negative resistance oscillator.

2-28. Define *frequency multiplier*.

2-29. Define *linear summing*.

2-30. Contrast the input and output frequency spectra for a linear amplifier.

2-31. Define *nonlinear mixing*.

2-32. Contrast the input and output frequency spectra for a nonlinear amplifier.

2-33. When will harmonic distortion and intermodulation distortion occur?

PROBLEMS

2-1. For a 20-MHz crystal with a negative temperature coefficient $k = 8$ Hz/MHz/°C, determine the frequency of operation for the following temperature changes.
   (a) Increase of 10°C.
   (b) Increase of 20°C.
   (c) Decrease of 20°C.

2-2. For the Wien-bridge oscillator shown in Figure 2-3 and the following component values, determine the frequency of oscillation: $R_1 = R_2 = 1$ kΩ; $C_1 = C_2 = 100$ pF.

2-3. For the Hartley oscillator shown in Figure 2-5a and the following component values, determine the frequency of oscillation: $L_{1a} = L_{1b} = 50$ μH; $C_1 = 0.01$ μF.

2-4. For the Colpitts oscillator shown in Figure 2-6 and the following component values, determine the frequency of oscillation: $C_{1a} = C_{1b} = 0.01$ μF; $L_1 = 100$ μH.

2-5. Determine the capacitance for a variactor diode with the following values: $C = 0.005$ μF; $v_r = -2 \ V \ dc$.

2-6. Describe the spectrum shown below. Determine the type of amplifier and the frequency content of the input signal.

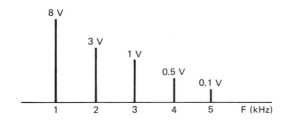

2-6

2-7. Repeat Problem 2-7 for the spectrum shown below.

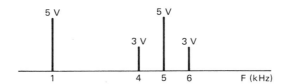

2-8. For a nonlinear amplifier with two input frequencies, 7 kHz and 4 kHz:
  (a) Determine the first three harmonics present in the output for each frequency.
  (b) Determine the cross products produced in the output for values of m and n of 1 and 2.
  (c) Draw the output spectrum for the harmonics and cross products determined in parts (a) and (b).

Chapter 3

AMPLITUDE MODULATION TRANSMISSION

INTRODUCTION

Information signals must be carried between a transmitter and a receiver over some form of transmission medium. However, the information signals are seldom in a form suitable for transmission. *Modulation* is defined as the process of transforming information from its original form to a form that is more suitable for transmission between a transmitter and a receiver. *Demodulation* is the reverse process (i.e., the modulated signal is converted back to its original form). Modulation takes place in a circuit called a *modulator*, and demodulation takes place in a circuit called a *demodulator*.

Amplitude modulation (AM) is the process of changing the amplitude of a high-frequency carrier in accordance with a modulating signal (information). Frequencies that are high enough to be efficiently radiated by an antenna and propagated through free space are commonly called *radio frequencies* or simply RF. In AM, the information is impressed onto the carrier in the form of amplitude changes. Amplitude modulation is a relatively inexpensive, low-quality form of modulation that is used for commercial broadcasting of both audio and video signals. The commercial AM broadcast band extends from 535 to 1605 kHz. Commercial television broadcasting is divided into three bands (two VHF and one UHF). The low-band VHF channels are 2 to 6 (54 to 88 MHz), the high-band VHF channels are 7 to 13 (174 to 216 MHz), and the UHF channels are 14 to 83 (470 to 890 MHz). Amplitude modulation is also used for two-way mobile radio communications such as CB radio (citizens' band, 26.965 to 27.405 MHz).

An AM modulator is a nonlinear device with two inputs: a single-frequency, constant-amplitude carrier and the information. The information acts on or modulates the carrier and may be a single frequency or a complex waveform made up of many frequencies. Because the information acts on the carrier, it is called the *modulating signal*. The carrier is acted upon and is therefore called the *modulated signal*. The resultant is called the *modulated wave*.

AM Envelope

Although mathematically it is not the simplest form of AM, double-sideband full carrier (AM DSBFC) will be discussed first because it is probably the most often used form of AM.

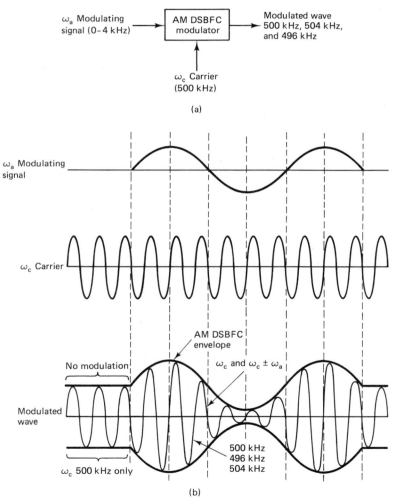

FIGURE 3-1   AM generation: (a) AM DSBFC modulator; (b) producing an AM DSBFC envelope—time domain.

Figure 3-1a shows a AM DSBFC modulator and the relationship among the carrier (ω_c), the modulating signal (ω_a), and the modulated wave (v). Figure 3-1b shows in the time domain how an AM wave is produced from a single-frequency modulating signal. Because the modulated wave contains all of the frequencies that make up the AM signal and is used to carry the information through the system, it is called an AM *envelope*. With no modulating signal, the output wave is simply the amplified carrier. When a modulating signal is applied, the amplitude of the output wave is varied in accordance with the modulating signal. Note the shape of the AM envelope; it is identical to the shape of the modulating signal. Also, the time of one cycle of the envelope is the same as the period of the modulating signal. Consequently, the repetition rate of the envelope is equal to the frequency of the modulating signal.

AM Spectrum and Bandwidth

As stated previously, an AM modulator is a nonlinear device. Therefore, nonlinear mixing occurs in an AM modulator and the output envelope is a complex wave made up of the carrier and the sum ($F_c + F_a$) and difference ($F_c - F_a$) frequencies (i.e., the cross products). The difference between the sum or difference frequency and the carrier frequency is equal to the modulating signal frequency. Therefore, an AM envelope contains both the amplitude and frequency components of the original information signal.

Figure 3-2 shows the frequency spectrum for the AM envelope produced in Figure 3-1. The AM spectrum extends from $F_c - F_a$ to $F_c + F_a$, where F_a is the highest modulating signal frequency. The band of frequencies between $F_c - F_a$ and F_c is called the *lower sideband* (LSB), and any frequency within this band is called a *lower side frequency* (LSF). The band of frequencies between F_c and $F_c + F_a$ is called the *upper sideband* (USB), and any frequency within this band is called an *upper side frequency* (USF). Therefore, the bandwidth (B) of an AM DSBFC wave is equal to the difference between the highest USF and the lowest LSF (i.e., $B = 2F_{a(max)}$). The carrier and all of the frequencies within the upper and lower sidebands must be RF to propagate through the earth's atmosphere.

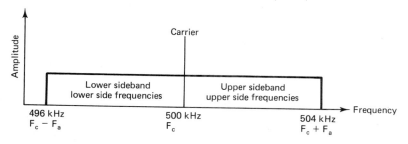

FIGURE 3-2 **Frequency spectrum of the AM DSBFC wave produced in Figure 2-1.**

EXAMPLE 3-1

For an AM modulator with a carrier frequency of 100 kHz and a maximum modulating signal frequency of 5 kHz:
(a) Determine the frequency limits for the upper and lower sidebands.
(b) Determine the upper and lower side frequencies produced when the modulating signal is a single-frequency 3-kHz tone.
(c) Determine the bandwidth.
(d) Draw the output spectrum.

Solution   (a) The lower sideband extends from the lowest possible side frequency to the carrier frequency.

$$\text{LSB} = F_c - F_{a(\max)} \text{ to } F_c$$

$$= (100 - 5) \text{ kHz to } 100 \text{ kHz} = 95 \text{ to } 100 \text{ kHz}$$

The upper sideband extends from the highest possible side frequency to the carrier frequency.

$$\text{USB} = F_c \text{ to } F_c + F_{a(\max)}$$

$$= 100 \text{ kHz to } (100 + 5) \text{ kHz} = 100 \text{ to } 105 \text{ kHz}$$

(b) The USF is the sum of the carrier frequency and the modulating signal frequency, or

$$\text{USF} = F_c + F_a = 100 \text{ kHz} + 3 \text{ kHz} = 103 \text{ kHz}$$

The LSF is the difference between the carrier frequency and the modulating signal frequency, or

$$\text{LSF} = F_c - F_a = 100 \text{ kHz} - 3 \text{ kHz} = 97 \text{ kHz}$$

(c) $B = 2F_a = 2(5 \text{ kHz}) = 10 \text{ kHz}$
(d) The output spectrum is shown in Figure 3-3.

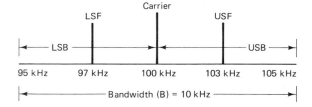

FIGURE 3-3   Output spectrum for Example 3-1.

Phasor Representation of an AM Wave

For a single-frequency modulating signal, an AM envelope is produced from the vector addition of the carrier and the upper and lower side frequencies. Figure 3-4a shows this phasor addition. On a time basis, the carrier and the upper and lower side frequencies all rotate in a counterclockwise direction. However, the USF rotates faster than the carrier ($\omega USF > \omega_c$) and the LSF rotates slower ($\omega LSF < \omega_c$). Consequently, if the

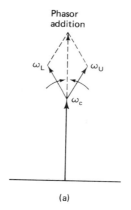

(a)

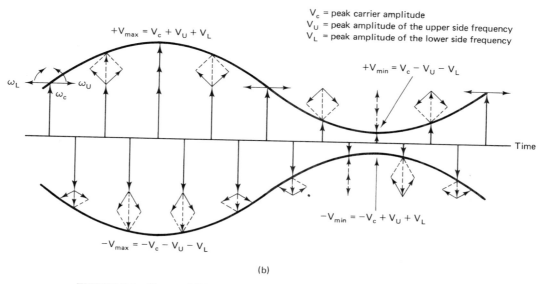

V_c = peak carrier amplitude
V_U = peak amplitude of the upper side frequency
V_L = peak amplitude of the lower side frequency

(b)

FIGURE 3-4 **Phasor addition in an AM DSBFC envelope: (a) phasor addition of the carrier and the upper and lower side frequencies; (b) phasor addition producing an AM envelope.**

carrier vector is held stationary, the vector for the USF will continue in a counterclockwise direction and the vector for the LSF will fall back in a clockwise direction. The carrier, the USF, and the LSF combine, sometimes in phase (adding) and sometimes out of phase (subtracting). For the waveform shown in Figure 3-4b, the maximum positive amplitude of the envelope $(+V_{max})$ occurs when the carrier and the two side frequencies are at their maximum positive values at the same time. The minimum positive amplitude $(+V_{min})$ occurs when the carrier is at its maximum positive value at the same time that the USF and LSF are at their maximum negative values. The maximum negative amplitude

$(-V_{max})$ occurs when the carrier, the USF, and the LSF are at their maximum negative values at the same time. The minimum negative amplitude $(-V_{min})$ occurs when the carrier is at its maximum negative value at the same time that the USF and LSF are at their maximum positive values $(-V_{min})$.

Coefficient of Modulation and Percent Modulation

Coefficient of modulation is a term that describes the amount of amplitude change (modulation) in an AM envelope. *Percent modulation* is simply the coefficient of modulation stated as a percentage. More specifically, percent modulation gives the percentage change in the amplitude of the output wave when the carrier is acted on by a modulating signal. Mathematically, modulation coefficient is

$$m = \frac{E_m}{E_c} \qquad (3\text{-}1a)$$

where

m = modulation coefficient
E_m = peak change in the amplitude of the output wave
E_c = peak amplitude of the unmodulated carrier

Equation 3-1a can be rearranged to solve for E_m or E_c:

$$E_m = mE_c \qquad \text{and} \qquad E_c = \frac{E_m}{m} \qquad (3\text{-}1b)$$

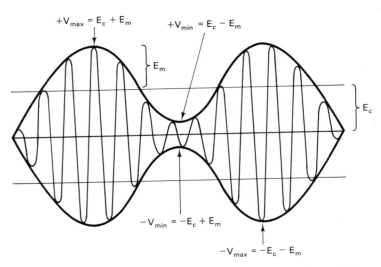

FIGURE 3-5   Modulation coefficient, percent modulation, E_m, and E_c.

and percent modulation (*M*) is

$$M = \frac{E_m}{E_c} \times 100 \qquad \text{or simply} \qquad m \times 100 \qquad (3\text{-}2)$$

The relationship among m, M, E_m, and E_c is shown in Figure 3-5.

If the modulating signal is a pure sine wave and the modulation process is linear (i.e., the positive and negative swings of the envelope's amplitude are equal), percent modulation can be derived as follows:

$$E_m = \tfrac{1}{2}(V_{\text{max}} - V_{\text{min}}) \qquad (3\text{-}3a)$$

and

$$E_c = \tfrac{1}{2}(V_{\text{max}} + V_{\text{min}}) \qquad (3\text{-}3b)$$

Therefore

$$M = \frac{\tfrac{1}{2}(V_{\text{max}} - V_{\text{min}})}{\tfrac{1}{2}(V_{\text{max}} + V_{\text{min}})} \times 100$$

$$= \frac{V_{\text{max}} - V_{\text{min}}}{V_{\text{max}} + V_{\text{min}}} \times 100 \qquad (3\text{-}3c)$$

where

$$V_{\text{max}} = E_c + E_m \qquad (3\text{-}3d)$$
$$V_{\text{min}} = E_c - E_m \qquad (3\text{-}3e)$$

The peak change in the amplitude of the output wave (E_m) is the sum of the voltages from the upper and lower side frequencies. Therefore,

$$E_U = E_L = \frac{E_m}{2} = \frac{\tfrac{1}{2}(V_{\text{max}} - V_{\text{min}})}{2} = \tfrac{1}{4}(V_{\text{max}} - V_{\text{min}}) \qquad (3\text{-}4)$$

where

E_U = peak voltage of the USF
E_L = peak voltage of the LSF

From Equation 3-2, it can be seen that the percent modulation goes to 100% when $E_m = E_c$. This condition is shown in Figure 3-6d. It can also be seen that at 100% modulation, the minimum amplitude of the AM envelope $V_{\text{min}} = 0V$. Figure 3-6c shows a 50% modulated AM envelope, the peak change in the amplitude of the envelope is equal to one-half the amplitude of the unmodulated wave. The maximum percent modulation that can be imposed without causing excessive distortion to the modulated wave is 100%. Sometimes percent modulation is expressed as the peak change in output voltage in respect to the unmodulated carrier voltage (i.e., percent change = $\Delta V_c/V_c \times 100$).

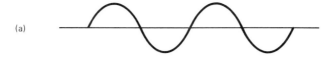

(a)

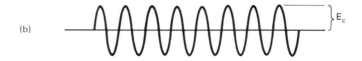

(b)

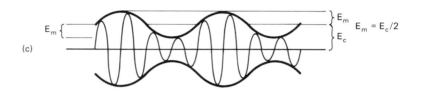

(c)

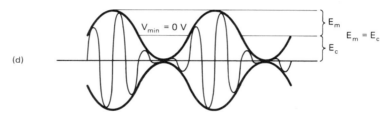

(d)

FIGURE 3-6   Percent modulation of an AM DSBFC envelope: (a) modulating signal; (b) unmodulated carrier; (c) 50% modulated wave; (d) 100% modulated wave.

EXAMPLE 3-2

For the AM envelope shown in Figure 3-7, determine:
(a) The peak amplitude of the upper and lower side frequencies.
(b) The peak amplitude of the unmodulated carrier.
(c) The peak change in the amplitude of the envelope.
(d) The modulation coefficient.
(e) The percent modulation.

Solution   (a) From Equation 3-4,

$$E_U = E_L = \tfrac{1}{4}(V_{max} - V_{min}) = \tfrac{1}{4}(18 - 2) = 4 \text{ V}$$

(b) From Equation 3-3b,

$$E_c = \tfrac{1}{2}(V_{max} + V_{min}) = \tfrac{1}{2}(18 + 2) = 10 \text{ V}$$

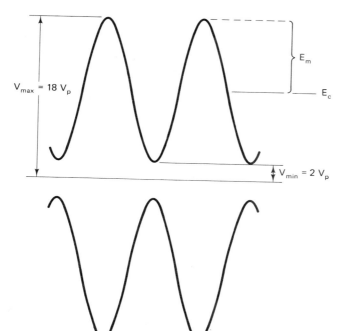

FIGURE 3-7   AM envelope for Example 3-2.

(c) From Equation 3-3a,

$$E_m = \tfrac{1}{2}(V_{max} - V_{min}) = \tfrac{1}{2}(18 - 2) = 8 \text{ V}$$

(d) From Equation 3-1a,

$$m = \frac{E_m}{E_c} = \frac{8}{10} = 0.8$$

(e) From Equation 3-2,

$$M = m \times 100 = 0.8 \times 100 = 80\%$$

and from Equation 3-3c,

$$M = \frac{V_{max} - V_{min}}{V_{max} + V_{min}} \times 100 = \frac{18 - 2}{18 + 2} \times 100 = 80\%$$

MATHEMATICAL ANALYSIS OF AM DOUBLE-SIDEBAND FULL CARRIER

Voltage Distribution in an AM DSBFC Wave

An unmodulated carrier can be described mathematically as

$$v_c = E_c \sin \omega_c t$$

In a previous discussion it was pointed out that the repetition rate of an AM envelope is equal to the frequency of the modulating signal, the amplitude of the AM wave is varied proportional to the amplitude of the modulating signal, and the amplitude of the modulated wave is equal to $E_c + E_m$. Therefore, the instantaneous amplitude of the modulated wave (v_c) can be expressed as

$$v_c = \underbrace{(E_c + E_m \sin \omega_a t)}_{\substack{\text{modulated wave} \\ \text{amplitude}}} \sin \omega_c t \qquad (3\text{-}5a)$$

From Equation 3-1b, mE_c can be substituted for E_m:

$$v_c = \underbrace{(E_c + mE_c \sin \omega_a t)}_{\substack{\text{modulated wave} \\ \text{amplitude}}} \sin \omega_c t \qquad (3\text{-}5b)$$

Multiplying yields

$$v_c = E_c \sin \omega_c t + (mE_c \sin \omega_a t)(\sin \omega_c t)$$

The trigonometric identity for the product of two sines is

$$(\sin A)(\sin B) = -\tfrac{1}{2}\cos (A + B) + \tfrac{1}{2}\cos (A - B)$$

Therefore,

$$v_c = \underbrace{E_c \sin \omega_c t}_{\text{carrier}} - \underbrace{\frac{mE_c}{2} \cos (\omega_c + \omega_a)\, t}_{\text{USF}} + \underbrace{\frac{mE_c}{2} \cos (\omega_c - \omega_a)t}_{\text{LSF}} \qquad (3\text{-}5c)$$

The first component in Equation 3-5c is the carrier (ω_c), the second component is the upper side frequency $(\omega_c + \omega_a)$, and the third component is the lower side frequency $(\omega_c - \omega_a)$. Several interesting properties of an AM DSBFC wave can be pointed out from Equation 3-5c. First, note that the amplitude of the carrier after modulation is equal to the amplitude of the carrier before modulation (E_c). Therefore, the carrier amplitude is unaffected by the modulation process. Second, the amplitude of the USF and LSF are dependent on both the carrier amplitude and the coefficient of modulation. For 100% modulation, $m = 1$ and the amplitude of the USF and the LSF are each equal to one-half the carrier amplitude $(E_c/2)$. Therefore, at 100% modulation,

$$v_{c(\max)} = E_c + \frac{E_c}{2} + \frac{E_c}{2} = 2E_c$$

and

$$v_{c(\min)} = E_c - \frac{E_c}{2} - \frac{E_c}{2} = 0 \text{ V}$$

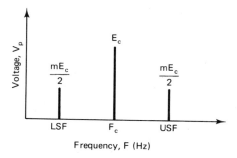

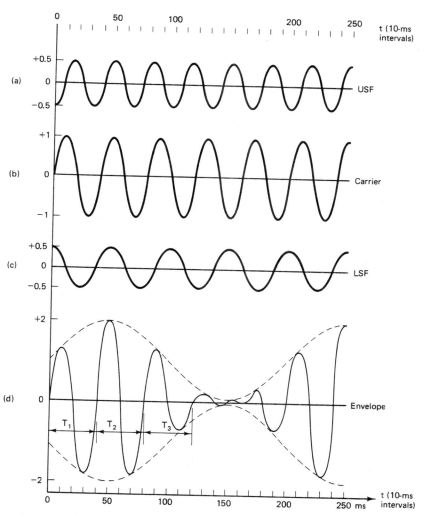

FIGURE 3-9 Generation of an AM DSBFC envelope shown in the time domain: (a) $-\frac{1}{2} \cos 2\pi30t;$ (b) $\sin 2\pi25t;$ (c) $+\frac{1}{2} \cos 2\pi20t;$ (d) summation of (a), (b), and (c).

From the relationships above and using Equation 3-5c, it is evident that as long as we do not exceed 100% modulation, the maximum positive or negative amplitude of an AM envelope $V_{max} = 2E_c$, and the minimum positive or negative amplitude of an AM envelope $V_{min} = 0$. This relationship was shown in Figure 3-6d. Figure 3-8 shows the voltage spectrum for an AM DSBFC wave (note that all of the voltages are given in peak values).

Also, from Equation 3-5c, the relative phase relationship between the carrier and the upper and lower side frequencies is evident. The carrier component of the envelope is a +sine function, the USF a −cosine function, and the LSF a +cosine function. Therefore, the envelope is a repetitive wave. At the beginning of each cycle of the envelope, the carrier is 90° out of phase with both the USF and LSF, and the USF and LSF are 180° out of phase with each other. This relationship can be seen in Figure 3-9.

EXAMPLE 3-3

One input to an AM modulator is a 500-kHz carrier with a peak amplitude of 20 V_p. The second input is a 10-kHz modulating signal that is of sufficient amplitude to cause a peak change in the output wave of ±7.5 V.
(a) Determine the USF and LSF.
(b) Determine the modulation coefficient and the percent modulation.
(c) Determine the peak amplitude of the modulated carrier, the USF, and the LSF.
(d) Determine the maximum and minimum amplitudes of the AM envelope.
(e) Determine the expression for the modulated AM wave.
(f) Draw the output spectrum.
(g) Sketch the modulated AM DSBFC envelope.

Solution   (a) The USF and LSF are the sum and difference frequencies, respectively:

$$USF = F_c + F_a = 500 \text{ kHz} + 10 \text{ kHz} = 510 \text{ kHz}$$

$$LSF = F_c - F_a = 500 \text{ kHz} - 10 \text{ kHz} = 490 \text{ kHz}$$

(b) The modulation coefficient is determined from Equation 3-1a:

$$m = \frac{E_m}{E_c} = \frac{7.5}{20} = 0.375$$

Percent modulation is determined from Equation 3-2:

$$\% M = m \times 100 = 0.375 \times 100 = 37.5\%$$

(c) The peak amplitude of the carrier, USF, and LSF are determined as follows:

$$E_{c(modulated)} = E_{c(unmodulated)} = 20 \text{ V}_p$$

The peak amplitude of the USF and LSF are determined from Equation 3-5:

$$E_L = E_U = \frac{mE_c}{2} = \frac{(0.375) \, 20}{2} = 3.75 \text{ V}_p$$

(d) The maximum and minimum amplitudes of the modulated wave are determined from Equations 3-3d and 3-3e, respectively:

$$V_{max} = E_c + E_m = 20 + 7.5 = 27.5 \text{ V}_p$$

$$V_{min} = E_c - E_m = 20 - 7.5 = 12.5 \text{ V}_p$$

(e) The expression for the modulated wave follows the format of Equation 3-5:

$$v_c = 20 \sin(2\pi 500kt) - 3.75 \cos(2\pi 510kt) + 3.75 \cos(2\pi 490kt)$$

(f) The output spectrum is shown in Figure 3-10.

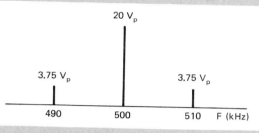

FIGURE 3-10 Output spectrum for Example 3-3.

(g) The modulated envelope is shown in Figure 3-11.

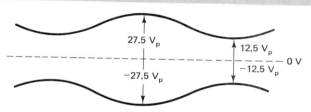

FIGURE 3-11 AM envelope for Example 3.3.

Generating an AM DSBFC Envelope in the Time Domain

Figure 3-9 showed how an AM DSBFC envelope is produced from the algebraic addition of the carrier and the upper and lower side frequency waveforms. For simplicity, the following waveforms are used for the modulating and carrier input signals:

$$\text{carrier} = v_c = E_c \sin(2\pi 25t) \qquad \text{(3-6a)}$$

$$\text{modulating signal} = v_a = E_a \sin(2\pi 5t) \qquad \text{(3-6b)}$$

substituting Equations 3-6a and 3-6b into Equation 3-5c, the expression for the modulated wave $M(t)$ is

$$M(t) = \underbrace{E_c \sin(2\pi 25t)}_{\text{carrier}} - \underbrace{\frac{mE_c}{2} \cos(2\pi 30t)}_{\text{USF}} + \underbrace{\frac{mE_c}{2} \cos(2\pi 20t)}_{\text{LSF}}$$

The waveforms for 100% modulation and an unmodulated carrier voltage $E_c = 1V_p$ are shown in Figure 3-9. Table 3-1 lists the instantaneous voltages for the carrier, the upper and lower side frequencies, and the total envelope at 10-ms intervals.

TABLE 3-1   INSTANTANEOUS VOLTAGES

USF, $-\frac{1}{2}\cos 2\pi 30t$	Carrier, $\sin 2\pi 25t$	LSF, $+\frac{1}{2}\cos 2\pi 25t$	Envelope, $M(t)$	Time, t (ms)
−0.5	0	+0.5	0	0
+0.155	+1	+0.155	+1.31	10
+0.405	0	−0.405	0	20
−0.405	−1	−0.405	−1.81	30
−0.155	0	+0.155	0	40
+0.5	+1	+0.5	2	50
−0.155	0	+0.155	0	60
−0.405	−1	−0.405	−1.81	70
+0.405	0	−0.405	0	80
+0.155	+1	+0.155	+1.31	90
−0.5	0	+0.5	0	100
+0.155	−1	+0.155	−0.69	110
+0.405	0	−0.405	0	120
−0.405	+1	−0.405	+0.19	130
−0.155	0	+0.155	0	140
+0.5	−1	+0.5	0	150
−0.155	0	+0.155	0	160
−0.405	+1	−0.405	+0.19	170
+0.405	0	−0.405	0	180
+0.155	−1	+0.155	−0.69	190
−0.5	0	+0.5	0	200
+0.155	+1	+0.155	+1.31	210
+0.405	0	−0.405	0	220
−0.405	−1	−0.405	−1.81	230
+0.405	0	−0.405	0	240
+0.155	+1	+0.155	+1.31	250

In Figure 3-9, note that the time between zero crossings is constant for the envelope (i.e., $T_1 = T_2 = T_3$, etc). Also note that the peak amplitudes of successive peaks are not equal. This indicates that a cycle within the envelope is not a pure sine wave; as mentioned previously, the modulated wave is the summation of the carrier and the upper and lower side frequencies. Thus, with AM DSBFC, the amplitude of the carrier is not varied, but rather, the amplitude of the envelope is varied in accordance with the modulating signal.

Power Distribution In An AM DSBFC Wave

In any electrical circuit, the power dissipated is equal to the rms voltage squared, divided by the resistance (i.e., $P = E^2/R$). Thus, the power developed across a load by an unmodulated carrier is equal to the carrier voltage squared, divided by the load resistance. Therefore, in an AM DSBFC wave, the unmodulated carrier power is

$$P_c = \frac{(E_c)^2}{R}$$

where

P_c = unmodulated carrier power
E_c = unmodulated carrier voltage
R = load resistance

The total power in an AM DSBFC wave is equal to the sum of the powers of the carrier, the upper side frequency, and the lower side frequency. (For a complex modulating signal, the total power equals the sum of the carrier and the upper and lower sideband powers.) Mathematically, the total power in an AM envelope is

$$P_t = \underbrace{\frac{(E_c)^2}{R}}_{\substack{\text{modulated}\\\text{carrier}\\\text{power}}} + \underbrace{\frac{\left(\dfrac{mE_c}{2}\right)^2}{R}}_{\substack{\text{USF}\\\text{power}}} + \underbrace{\frac{\left(\dfrac{mE_c}{2}\right)^2}{R}}_{\substack{\text{LSF}\\\text{power}}} \qquad (3\text{-}7)$$

where

P_t = total power of an AM envelope

$\dfrac{(E_c)^2}{R}$ = modulated carrier power

$\dfrac{\left(\dfrac{mE_c}{2}\right)^2}{R} = \dfrac{m^2 E_c^{\,2}}{4R}$ = USF and LSF power

From Equation 3-7, it can be seen that the term for the modulated carrier power is the same as the expression for the unmodulated carrier power. Thus it is evident that the carrier power is unaffected by the modulation process. Also, because the total power in the AM wave is the sum of the carrier power and the power in the side frequencies (or sidebands), the total power in an AM envelope increases with modulation. Equation 3-7 can be expanded and rewritten as

$$P_t = P_c + \frac{m^2}{4}\frac{E_c^2}{R} + \frac{m^2}{4}\frac{E_c^2}{R} \qquad (3\text{-}8)$$

and because $E_c^2/R = P_c$,

$$P_t = \underbrace{P_c}_{\substack{\text{carrier}\\\text{power}}} + \underbrace{\frac{m^2 P_c}{4}}_{\substack{\text{USF}\\\text{power}}} + \underbrace{\frac{m^2 P_c}{4}}_{\substack{\text{LSF}\\\text{power}}} \tag{3-9}$$

Combining the powers from the upper and lower side frequencies gives us

$$P_t = P_c + \frac{m^2 P_c}{2} \tag{3-10}$$

Factoring P_c yields

$$P_t = P_c \left(1 + \frac{m^2}{2} \right) \tag{3-11}$$

Equation 3-11 is the general expression for the total power in an AM envelope. From Equation 3-9 it can be seen that the power in the USF is equal to the power in the LSF. Therefore,

$$P(\text{LSF}) = P(\text{USF}) = \frac{m^2 P_c}{4} \tag{3-12}$$

Thus the total power of both sidebands is

$$P_{\text{sb}} = \frac{2\, m^2 P_c}{4} = \frac{m^2 P_c}{2} \tag{3-13}$$

Equations 3-8 through 3-13 use the peak voltages of the carrier and the side frequencies and, consequently, solve for peak powers. To convert to rms, simply multiply each voltage by 0.707 or divide the total peak power by 2.

Figure 3-12 shows the power spectrum for an AM DSBFC wave. Note that with 100% modulation, the maximum power in either the USF or the LSF is equal to only one-fourth the power in the carrier. Thus the maximum total sideband power P(USB) + P(LSB) is equal to one-half of the carrier power. This is one significant disadvantage

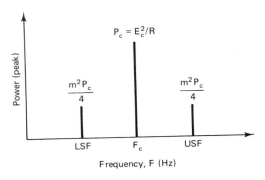

FIGURE 3-12 Power spectrum for an AM DSBFC wave.

of AM DSBFC transmission; the information is contained in the side frequencies, although most of the envelope power is wasted in the carrier. Actually, the power in the carrier is not totally wasted; it allows for the use of simple and inexpensive demodulators in the receiver using only a single diode, which is the predominant advantage of AM DSBFC.

EXAMPLE 3-4

For an AM DSBFC wave with a peak unmodulated carrier voltage of 10 V and a load resistance of 10 Ω:
(a) Determine the peak and rms powers of the carrier, the USF, and the LSF.
(b) Determine the total power of the modulated wave for a modulation coefficient of 0.5.
(c) Draw the power spectrum.

Solution   (a) The peak modulated carrier power is:

$$P_c = \frac{E_c^2}{R} = \frac{10^2}{10} = \frac{100}{10} = 10 \text{ W}_p$$

The rms carrier power is

$$P_c = \frac{(0.707E_c)^2}{R} = \frac{7.07^2}{10} = \frac{50}{10} = 5 \text{ W rms}$$

The peak power of the USF and LSF are equal and determined from Equation 3-12:

$$P(\text{USF}) = P(\text{LSF}) = \frac{m^2P_c}{4} = \frac{(0.5)^2(10)}{4} = 0.625 \text{ W}_p$$

The rms power is

$$P(\text{USF}) = P(\text{LSF}) = \frac{(0.5)^2(5)}{4} = 0.3125 \text{ W rms}$$

The total sideband power is then

$$P_{sb} = P(\text{USF}) + P(\text{LSF}) = 2(0.625) = 1.25 \text{ W}_p$$

$$P_{sb} = P(\text{USF}) + P(\text{LSF}) = 2(0.3125) = 0.625 \text{ W rms}$$

or from Equation 3-13:

$$P_{sb} = \frac{m^2P_c}{2} = \frac{(0.5)^2(5)}{2} = 0.625 \text{ W rms}$$

(b) The total peak power of the modulated wave is the sum of the carrier and side frequency powers:

$$P_t = P_c + P(\text{USF}) + P(\text{LSF})$$

$$= 10 + 0.625 + 0.625 = 11.25 \text{ W}_p$$

and the total rms power is

$$P_t = 5 + 0.3125 + 0.3125 = 5.625 \text{ W rms}$$

or from Equation 3-11,

$$P_t = P_c \left(1 + \frac{m^2}{2}\right) = 10 \left(1 + \frac{0.5^2}{2}\right) = 11.25 \text{ W}_p$$

(c) The peak power spectrum is shown in Figure 3-13.

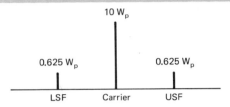

FIGURE 3-13   **Power spectrum for AM DSBFC.**

CIRCUITS FOR GENERATING AM DSBFC

Low-Power AM DSBFC Transistor Modulator

Figure 3-14 shows a simple low-power transistor AM modulator circuit which has only a single active component (Q_1). The carrier (ω_c) is applied to the base, and the modulating signal (ω_a) is applied to the emitter. Therefore, this is called *emitter modulation*. With

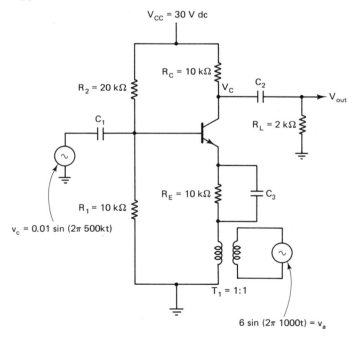

FIGURE 3-14   **Single transistor, emitter modulator.**

emitter modulation, it is important that the transistor be biased *class A* with a centered Q point.

Circuit operation. Figure 3-14 shows that the amplitude of the carrier (10 mV$_p$) is appreciably smaller than the amplitude of the modulating signal (6 V$_p$). If the modulating signal is removed or held constant at 0 V, Q_1 is a linear class A amplifier. The base input signal is simply amplified and inverted 180° at the collector. The amplification by Q_1 is determined by the ratio of the ac collector resistance (r_c) to the ac emitter resistance (r_e') (i.e., $A_v = rc/r_e'$). For the circuit shown in Figure 3-14, r_c and r_e' are determined as follows:

$$rc = \text{parallel combination of } R_C \text{ and } R_L$$

$$= R_L \| R_C = 10 \text{ k}\Omega \| 2 \text{ k}\Omega$$

$$= \frac{(10,000)(2000)}{10,000 + 2000} = 1667$$

$$r_e' = \frac{25 \text{ mV}}{I_e}$$

$$I_E = \frac{V_{th} - V_{be}}{(R_{th}/B) + R_e}$$

$$V_{th} = \frac{V_{cc}(R_1)}{R_1 + R_2} = \frac{30(10 \text{ k}\Omega)}{30 \text{ k}\Omega} = 10 \text{ V}$$

$$R_{th} = R_1 \| R_2 = \frac{(20,000)(10,000)}{20,000 + 10,000} = 6667$$

Thus for $\beta = 100$,

$$I_E = \frac{10 - 0.7}{(6667/\beta) + 10,000} = 0.924 \text{ mA}$$

Therefore,

$$r_e' = \frac{25 \text{ mV}}{I_E} = \frac{0.025}{0.000924} = 27$$

and

$$A_q = A_v = \frac{rc}{r_e'} = \frac{1667}{27} = 61.7$$

where Aq = quiescent gain.

With no input modulating signal, Q_1 is a linear small-signal amplifier with a *quiescent* voltage gain (A_q) of 61.7. With the carrier input voltage shown,

$$V_{out} = A_q V_{in} = 61.7(0.01 \text{ V}) = 0.617 \text{ V}_p$$

When the modulating signal (V_a) is applied to the circuit, its voltage combines with the dc Thévenin voltage. The result is a bias voltage that has a constant term and a term that varies at a low-frequency sinusoidal rate. Thus

$$V_{bias} = \underbrace{V_{th}}_{\substack{\text{constant} \\ \text{term}}} + \underbrace{V_a \sin \omega_a t}_{\substack{\text{sinusoidal} \\ \text{term}}}$$

and

$$V_{bias} = 10 + 6 \sin (2\pi 1000t)$$

To analyze the operation of this circuit, it is not necessary to consider every possible value of V_{bias}. Instead, several key values are calculated and the others interpolated from them. The three most significant values for V_{bias} are when the sinusoidal voltage term is 0, maximum positive, and maximum negative. When $V_a = 0$ V, the circuit bias is equal to its quiescent value (V_{th}). For this example, the voltage gain at quiescence (A_q) is equal to 61.7.

When the sinusoidal term is maximum and negative:

$$\sin (2\pi 1000t) = -1$$

$$V_{bias} = V_{th} - 6(-1) = 10 + 6 = 16 \text{ V}$$

$$I_E = \frac{16 - 0.7}{(6667/100) + 10,000} = 1.52 \text{ mA}$$

$$r_e' = \frac{0.025}{0.00152} = 16.45$$

$$A_v = A_{max} = \frac{1667}{16.45} = 101.3$$

$$V_{out} = V_{in}A_{max} = (0.01)(101.3) = 1.013 \text{ V}_p$$

When the sinusoidal term is maximum and positive:

$$\sin 2\pi 1000t = +1$$

$$V_{bias} = V_{th} - 6(1) = 10 - 6 = 4 \text{ V}$$

$$I_E = \frac{4 - 0.7}{(6667/100) + 10,000} = 0.328 \text{ mA}$$

$$r_e' = \frac{0.025}{0.000328} = 76.3$$

$$A_v = A_{min} = \frac{1667}{76.3} = 21.9$$

$$V_{out} = V_{in}A_{min} = 0.01(21.9) = 0.219 \text{ V}_p$$

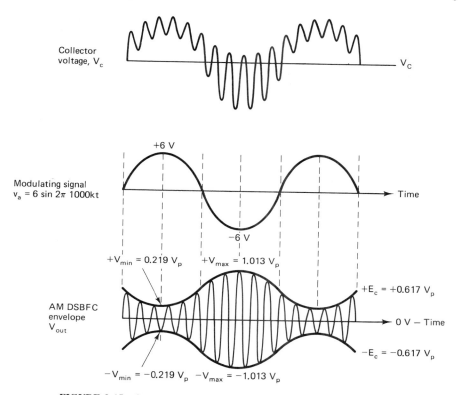

FIGURE 3-15 Output waveform for the circuit shown in Figure 3-14.

In this example, A_v varies at a sinusoidal rate equal to ω_a from a quiescent value $A_q = 61.7$ to a maximum value $A_{max} = 101.3$ and a minimum value $A_{min} = 21.9$. The collector voltage (V_c) and output voltage (V_{out}) are shown in Figure 3-15. The collector waveform is an AM DSBFC envelope riding on top of the low-frequency modulating signal. The average voltage is equal to the quiescent collector voltage V_C. The dc collector voltage and the low-frequency modulating signal component are removed from the waveform by coupling capacitor C_2. Consequently, the output waveform is an AM DSBFC envelope with an average voltage equal to 0 V, an unmodulated carrier amplitude of 0.617 V_p, a maximum positive or negative amplitude of 1.013 V_p, and a minimum positive or negative amplitude of 0.219 V_p. It is interesting to note that with emitter modulation the maximum amplitude of the envelope occurs when the modulating signal is maximum negative, and the minimum amplitude occurs when the modulating signal is maximum positive.

From the preceding example it can be seen that A_v varies at a sinusoidal rate equal to ω_a. Therefore, the voltage gain can be expressed mathematically as

$$A_v = A_q(1 + m \sin \omega_a t) \qquad (3\text{-}14a)$$

$\sin \omega_a t$ goes from a maximum value of $+1$ to a minimum value of -1. Thus

$$A_v = A_q(1 \pm m) \qquad (3\text{-}14b)$$

where

m = modulation coefficient

A_q = quiescent gain

Therefore, at 100% modulation ($m = 1$)

$$A_{max} = A_q(1 + 1) = 2A_q$$

$$A_{min} = A_q(1 - 1) = 0$$

and the maximum and minimum amplitudes of V_{out} are

$$V_{out(max)} = A_{max}(V_{in}) = 2A_q(V_{in})$$

$$V_{out(min)} = A_{min}(V_{in}) = 0(V_{in}) = 0$$

For the preceding example, the modulation coefficient is

$$m = \frac{V_{max} - V_{min}}{V_{max} + V_{min}} = \frac{1.013 - 0.2195}{1.013 + 0.2195} = 0.645$$

and

$$\% \text{ modulation} = 0.645(100) = 64.5\%$$

Substituting into Equation 3-14b yields

$$A_{max} = A_q(1 + m) = 61.7(1 + 0.645) = 101.5$$

$$A_{min} = A_q(1 - m) = 61.7(1 - 0.645) = 21.9$$

Also, in the preceding example, the voltage gain changed symmetrically with modulation. In other words, the amount of increase in gain is equal to the amount of decrease in gain. The gain is approximately $A_q \pm 40$. This is called linear modulation, and is desired. Essentially, a conventional AM DSBFC receiver reproduces the original modulating signal from the shape of the envelope. If modulation is not linear, the envelope does not accurately represent the shape of the modulating signal and the demodulator will produce a distorted output waveform.

EXAMPLE 3-6

For a low-power transistor AM modulator similar to the one shown in Figure 3-10 with a modulation coefficient of 0.8, a quiescent gain of 100, and an input carrier amplitude of 0.005 V_p:
(a) Determine the maximum and minimum gains for the transistor.
(b) Determine the maximum and minimum amplitude of V_{out}.
(c) Sketch the output AM envelope.

Solution   (a) substituting into Equation 3-14b

$$A_{max} = A_q(1 + m) = 100(1 + 0.8) = 180$$

$$A_{min} = A_q(1 - m) = 100(1 - 0.8) = 20$$

(b) $V_{out(max)} = A_{max}(V_{in})$,
$V_{out(max)} = 180(0.005) = 0.9 \text{ V}_p$
$V_{out(min)} = A_{min}(V_{in})$
$V_{out(min)} = 20(0.005) = 0.1 \text{ V}_p$

(c) The AM DSBFC output waveform is shown in Figure 3-16.

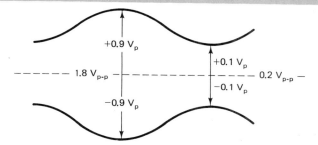

FIGURE 3-16   AM envelope for Example 3-6.

With no modulating signal, the transistor modulator shown in Figure 3-14 is a linear small-signal amplifier. When a modulating signal is applied, the *Q-point* of the amplifier is driven first toward *saturation,* then toward *cutoff* (i.e., the transistor is forced to operate over a nonlinear portion of its operating curve). Therefore, a linear amplifier operates nonlinearly when a modulating signal is applied.

The transistor modulator shown in Figure 3-14 is adequate for low-power applications but is not a practical circuit for high-power applications. This is because the transistor is operating class A, which is extremely inefficient. With AM, any amplifiers that follow the modulator circuit must be linear. If they are not linear, intermodulation between the upper and lower side frequencies and the carrier will generate additional cross-product frequencies that could interfere with signals from other transmitters.

Low-Level and High-Level Modulators

The terms *low-level* and *high-level* modulation simply refer to the point in a transmitter where the modulation is performed. With low-level modulation, the modulation takes place prior to the output element of the *final* stage. In other words, prior to the collector of the output transistor in a transistorized transmitter, prior to the drain of the output FET in a FET transmitter, or prior to the plate of the final stage in a tube transmitter.

The modulator shown in Figure 3-14 is a low-level modulator regardless of whether it is the final stage or not because modulation takes place in the emitter, which is not the output element. If the modulation took place in the collecter and the modulator were the final stage, it would be a high-level modulator.

An advantage of low-level modulation is that less modulating signal power is required. In high-level modulators, the modulation takes place in the final element of the final stage, where the carrier is at its maximum amplitude and requires a much higher modulating signal amplitude to achieve a reasonable percent modulation. With high-level modulation, the final audio amplifier must supply all of the sideband power, which could be as much as 50% of the power in the carrier. An obvious disadvantage of low-level modulation is in high-power applications when all of the amplifiers that follow the modulator stage must operate class A, which is extremely inefficient.

Medium-Power AM DSBFC Transistor Modulator

Early medium- and high-power AM transmitters were limited to those that used tubes for the active devices. However, since the mid-1970s, solid-state AM transmitters have been available with output powers as high as several thousand watts. This is usually not accomplished with a single output stage, but rather, by placing several output stages in parallel such that their output signals combine in phase and are, thus, additive.

Figure 3-17a shows a simplified single-transistor medium-power AM DSBFC modulator. The modulation takes place in the collector circuit, which is the output element of the transistor. Therefore, if this is the final stage of the transmitter (i.e., there are no amplifier stages following it), it is a high-level modulator. If the output of the modulator goes through additional amplifiers prior to transmission, it is a low-level modulator. High- and low-level transmitter configurations are discussed in a later section of this chapter.

To achieve high efficiency, medium- and high-power AM modulators generally operate *class C*. Therefore, a theoretical efficiency of as high as 80% is possible. The circuit shown in Figure 3-17a is a class C transistor amplifier with two inputs: an unmodulated carrier (v_c) and a single-frequency modulating signal (v_a). Because the transistor is biased class C, it is operating nonlinear and therefore is capable of nonlinear mixing (modulation). This circuit is a *collector modulator*; the modulating signal is applied to the collector. The RFC is a radio-frequency choke. At low frequencies the RFC looks like a short and at high frequencies it looks like an open. Consequently, it allows the low-frequency intelligence signal to modulate the collector of Q_1 and, at the same time, prevents the high-frequency carrier from entering the dc power supply.

Circuit operation.   For the following explanation, refer to the circuit shown in Figure 3-17a and the waveforms shown in Figure 3-17b. When the amplitude of the carrier exceeds the barrier potential of the base–emitter junction (approximately 0.7 V for a silicon transistor), Q_1 turns on and collector current saturates. When the carrier voltage drops below 0.7 V, the transistor turns off and collector current ceases. Thus Q_1 is switched between saturation and cutoff and collector current flows for less than 180° of each input cycle. Thus Q_1 operates class C. Each successive positive half-cycle of the input signal turns Q_1 on and allows collector current to flow, which produces a negative going voltage waveform at the collector of Q_1. The collector current and

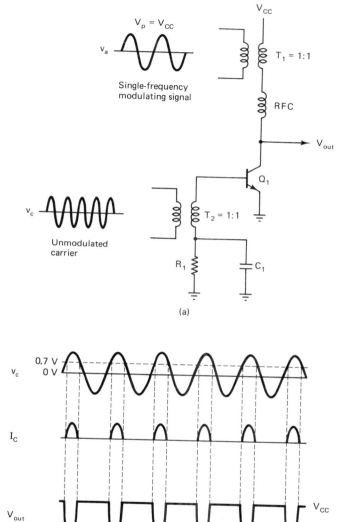

FIGURE 3-17   Simplified medium-power transistor AM DSBFC modulator: (a) schematic diagram; (b) collector waveforms with no modulating signal; (c) collector waveforms with modulating signal.

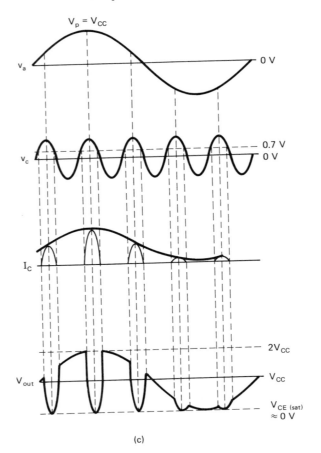

(c)

FIGURE 3-17 (*continued*)

voltage waveforms are shown in Figure 3-17b. The output voltage waveform resembles a repetitive half-wave rectified signal with a fundamental frequency equal to ω_c.

When a modulating signal is applied to the collector in series with V_{CC}, it adds to and subtracts from V_{CC}. The waveforms shown in Figure 3-17c are produced when the maximum peak modulating signal amplitude equals V_{CC}. It can be seen that the output waveform swings from a maximum value of $2V_{CC}$ to approximately 0 V ($V_{CE(sat)}$). The change in collector voltage is equal to V_{CC}. Again, the waveform resembles a half-wave rectified signal superimposed onto a low-frequency ac component.

Because Q_1 is operating nonlinear, the collector waveform contains the two original input frequencies (ω_a and ω_c) and their sum and difference frequencies ($\omega_c \pm \omega_a$). The output waveform also contains all of the higher-order harmonics and intermodulation products (i.e., $2\omega_c$, $2\omega_a$, $3\omega_c$, $3\omega_a$, $2\omega_c + \omega_a$, $3\omega_c + \omega_a$, etc). Consequently, before the output signal is transmitted, it must be bandlimited to $\omega_c \pm \omega_a$.

A more practical circuit for medium-power AM DSBFC generation is shown in Figure 3-18a with the corresponding waveforms shown in Figure 3-18b. This circuit is also a collector modulator with a maximum peak modulating signal amplitude equal to V_{CC}. Operation of this circuit is almost identical to the circuit shown in Figure 3-17a except for the addition of the *tuned* circuit (C_1 and L_1) in the collector of Q_1. Because the transistor is operating between saturation and cutoff, collector current is not dependent on base drive voltage. The voltage developed across the *tank* circuit is determined by the ac component of the collector current and the impedance of the tank circuit at *resonance* (which depends on the Q of the coil). The waveforms for the modulating signal, the carrier, and the collector current are identical to those of the previous example. The output voltage across R_L is a symmetrical, AM DSBFC envelope with an average voltage of 0 V, a maximum positive peak amplitude equal to $2V_{CC}$, and a maximum negative peak amplitude equal to $-2V_{CC}$. The positive half-cycle of the output waveform

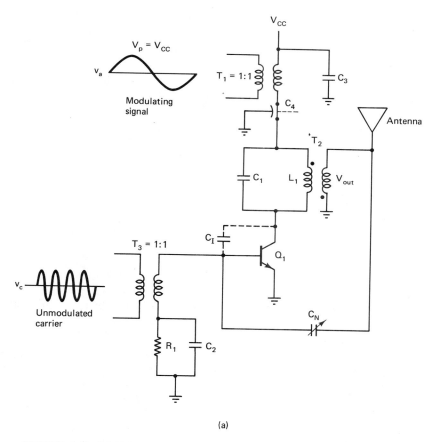

(a)

FIGURE 3-18 Medium-power transistor AM DSBFC modulator: (a) schematic diagram; (b) collector and output waveforms.

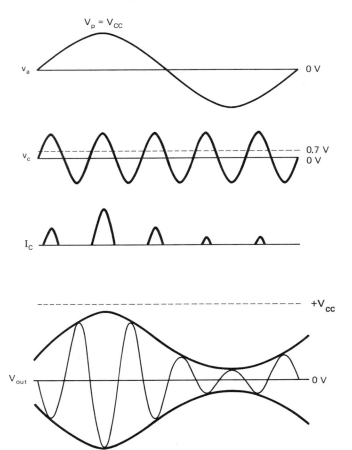

(b)

FIGURE 3-18 (*continued*)

is produced by the *flywheel effect* in the tank circuit. When Q_1 is conducting, C_1 charges to $V_{CC} + v_a$ (a maximum value of $2V_{CC}$) and, when Q_1 is off, C_1 discharges through L_1. When L_1 discharges, C_1 charges to $-2V_{CC}$. This produces the positive half cycle of the AM envelope. The resonant frequency of the tank circuit is equal to the carrier frequency and the bandwidth extends from $F_c - F_a$ to $F_c + F_a$. Consequently, the modulating signal, the harmonics, and all of the higher-order cross products are removed from the waveform, leaving a symetrical AM DSBFC envelope. One hundred percent modulation occurs when the peak amplitude of the modulated wave equals $2V_{CC}$.

There are several components shown in Figure 3-18a that have not been explained. R_1 is the bias resistor for Q_1. R_1, in conjunction with C_2 and the barrier potential of the transistor, determine the turn-on voltage for Q_1. Consequently, Q_1 can be biased to turn on only during the most positive peaks of the input carrier signal. This produces a narrower collector current waveform and enhances class C efficiency.

C_3 is a bypass capacitor. It looks like a short circuit to audio frequencies and thus prevents the information from entering the dc power supply. C_I is the base-to-collector *junction* capacitance of Q_1. At radio frequencies, the relatively small junction capacitances within a transistor are significant. If the capacitive reactance of C_I is sufficiently small, the collector signal may be returned to the base with sufficient amplitude to cause Q_1 to oscillate. Therefore, a signal of equal amplitude and frequency and 180° out of phase must be fed back to the base to cancel or *neutralize* the *intercapacitive feedback*. C_N is a *neutralizing capacitor*. Its purpose is to provide a feedback path for a signal that is equal in amplitude but 180° out of phase with the signal fed back through C_I. C_4 is an RF bypass capacitor. Its purpose is to isolate the dc power supply from RF frequencies. Its operation is quite simple; at the carrier frequency, C_4 looks like a short circuit preventing the carrier from leaking into the power supply or the audio circuitry and distributing throughout the transmitter.

High-Power AM DSBFC Transistor Modulators

Medium-power collector modulators produce a more linear (symmetrical) envelope than low-power base modulators. Also, collector modulators are more power efficient. However, collector modulators require a higher modulating signal drive power, and collector modulators cannot achieve 100% modulation because the collector current saturates and does not allow a full output voltage swing of $\pm V_{CC}$. Therefore, to achieve linear modulation, operate at maximum efficiency, develop a high output power, and require as little modulating signal drive power as possible, base and collector modulation are sometimes used simultaneously.

Circuit operations. Figure 3-19 shows a modulator that uses a combination of base and collector modulation. The modulating signal is simultaneously fed into the collectors of the push-pull modulators (Q_2 and Q_3) and the collector of the driver amplifier (Q_1). Therefore, collector modulation occurs in Q_1, Q_2, and Q_3. The carrier signal supplied to the base circuits of Q_2 and Q_3 has already been partially modulated. Thus the modulating signal power can be reduced, and the modulators are not required to operate over their entire operating curve to achieve 100% modulation.

Vacuum-Tube AM DSBFC Modulators

With the advent of medium- and high-power solid-state AM transmitters, *vacuum-tube* modulators have somewhat gone by the wayside. Vacuum-tube circuits have many disadvantages compared to solid-state circuits. Vacuum tubes require higher dc supply voltages (often, both positive and negative polarities for a single circuit), a separate *filament* (heater) supply voltage, very high modulating signal amplitudes, and vacuum tubes take up considerably more space and weigh much more than their solid-state counterparts.

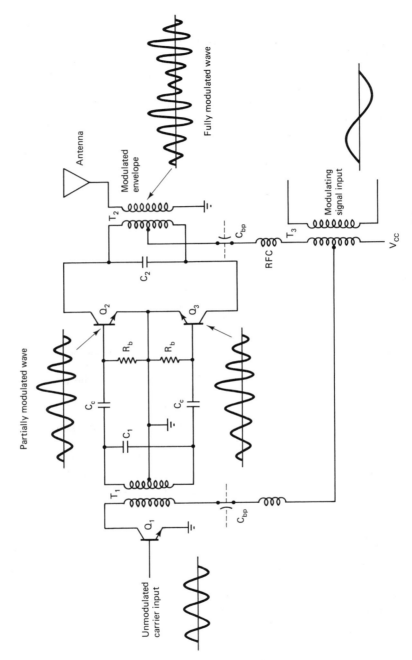

FIGURE 3-19   High-power AM DSBFC transistor modulator.

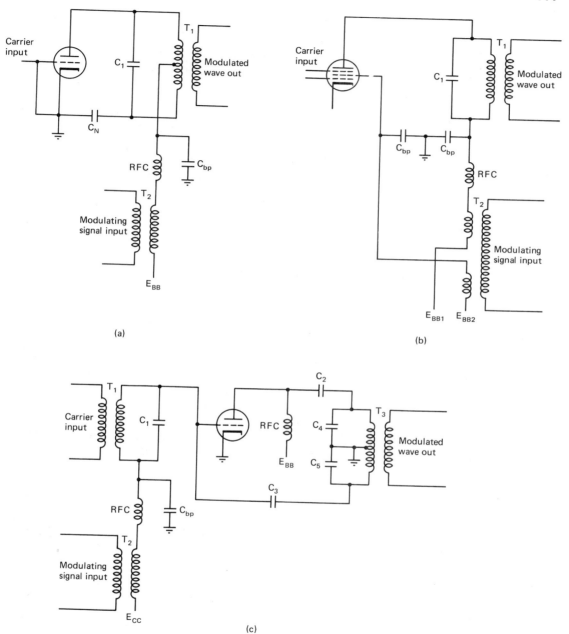

FIGURE 3-20 Vacuum-tube AM DSBFC modulator circuits: (a) triode plate modulator; (b) multi-grid plate modulator; (c) grid-bias modulator; (d) suppressor-grid modulator; (e) screen-grid modulator.

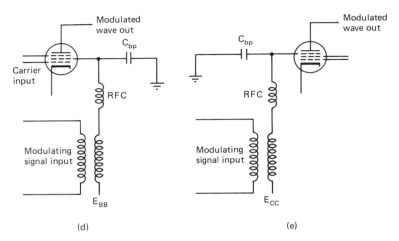

(d)   (e)

FIGURE 3-20 (*continued*)

Consequently, a vacuum-tube modulator (or, for that matter, any other vacuum-tube circuit) is seldom used today. Essentially, vacuum-tube modulators are used in applications where extreme high powers are required, such as in commercial AM and TV broadcast-band transmitters. For the more common low- and medium-power applications, such as two-way radio communications, solid-state circuitry dominates.

As a historical perspective (i.e., a look back into the "cavetronic" era), several vacuum-tube AM DSBFC modulators are shown in Figure 3-20. Assuming that the reader is familiar with vacuum-tube terminology (i.e., *plate*, *grid*, *cathode*, etc.), the operation of a vacuum-tube modulator is quite similar to a transistor modulator. A plate modulator is analogous to a collector modulator and a grid-bias modulator to a base modulator. Because there are more elements in a vacuum tube, there are more modulator configurations possible with vacuum tubes than with transistors (i.e., *suppressor* and *screen-grid modulators*).

High-level vacuum-tube triode modulators are class C biased and use plate modulation (Figure 3-20a). A high-power modulating signal is transformer coupled into the plate circuit in series with the plate supply voltage. Consequently, the modulating signal adds to and subtracts from E_{BB}. A relatively low-power carrier signal is coupled into the grid circuit, where it biases the tube into conduction only during its most positive peaks. Therefore, as with a transistor collector modulator, narrow pulses of plate current flow and supply the energy necessary to sustain oscillations in the plate tank circuit.

Figure 3-20b shows a *multigrid* vacuum-tube plate modulator. The vacuum tube is a pentode. It has five elements: plate, cathode, control grid, suppressor grid, and screen grid. *Pentode* vacuum-tube modulators are capable of high output powers and high efficiencies. However, to achieve 100% modulation, it is necessary to modulate the screen grid as well as the plate.

Figure 3-20c shows a vacuum-tube grid-bias modulator. The triode can be replaced with a *tetrode* or a pentode. The grid-bias modulator requires much less modulating power than the triode or pentode plate modulators, but it has poorer modulation linearity, more distortion, lower plate efficiency, and produces a lower output power.

Figures 3-20d and e show suppressor-grid and screen-grid modulators, respectively. These two modulator configurations fall somewhere between the grid-bias and plate modulators discussed previously.

AM DOUBLE-SIDEBAND FULL-CARRIER TRANSMITTERS

Low-Level Transmitters

Figure 3-21 shows a block diagram for a low-level AM DSBFC transmitter. For voice or music transmission, the modulating signal source can be a microphone, a magnetic tape or disk, or a phonograph. The *preamplifier* is typically a sensitive, class A linear voltage amplifier with a high input impedance. The preamplifier's function is to raise the signal voltage from the source to a usable level and, at the same time, produce a minimum amount of nonlinear distortion. The modulating driver is also a linear class A amplifier. The modulating driver simply amplifies the signal from the preamplifier to an adequate level to modulate the carrier sufficiently. More than one driver amplifier may be required.

The RF oscillator can be any of the oscillator configurations discussed. Cost and stability generally dictate which oscillator configuration is chosen. However, the FCC has stringent requirements on transmitter stability. The *buffer amplifier* is a low-gain, high-input impedance linear amplifier. Its function is to isolate the oscillator from the high-power amplifiers. The buffer provides a relatively constant load to the oscillator, which helps to reduce short-term frequency variations. An *emitter follower* or an *integrated-circuit op-amp* is commonly used for the buffer amplifier. The modulator can use either emitter or collector modulation. The intermediate and final power amplifiers are either linear class A or class B push-pull amplifiers. This is required with low-level transmitters to maintain the amplitude linearity of the AM DSBFC envelope. The

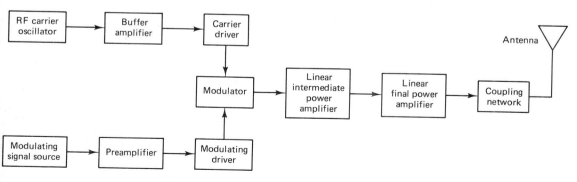

FIGURE 3-21 Block diagram of a low-level AM DSBFC transmitter.

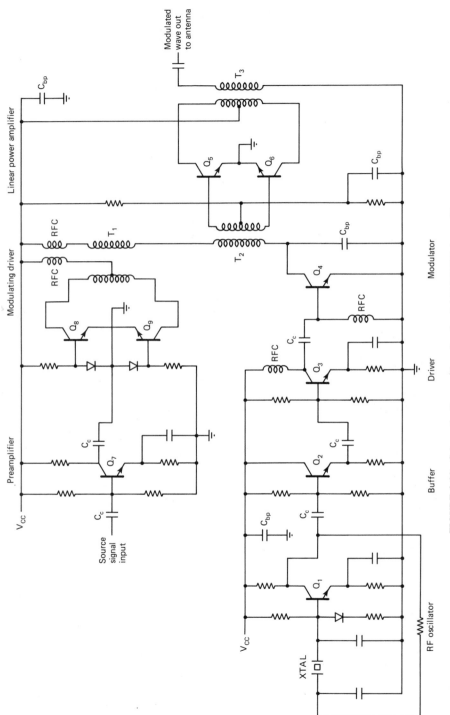

FIGURE 3-22 Schematic diagram for a low-level AM DSBFC transmitter.

antenna coupling network matches the output impedance of the final amplifier to the transmission line and antenna.

Low-level transmitters like the one shown in Figure 3-21 are used predominantly for low-power, low-capacity AM systems such as wireless intercoms, remote control units, and short-range walkie-talkies.

Figure 3-22 shows a schematic diagram for a simple low-level AM transmitter. The RF oscillator is the Pierce crystal oscillator that was discussed in Chapter 2. The buffer amplifier is an emitter follower which provides a constant high-impedance load to the oscillator and supplies the unmodulated carrier to the class A driver amplifier. The low-level source signal is amplified first by the small-signal class A preamplifier and then by the push-pull class B modulating driver. Class C collector modulation is used with this transmitter. Finally, the modulated wave is amplified to the desired transmit signal power by the class B power amplifier. A *class B* final power amplifier is chosen because it can develop more output power than its class A counterpart, it has a higher power efficiency, and it provides linear amplification of the modulated envelope. Note the presence of several radio-frequency (RF) bypass capacitors and radio-frequency chokes (RFCs). Also, note the biasing components used in the two class B amplifier configurations. The diodes in the base circuits of the modulating driver compensate for variations in the barrier potential of the transistors due to changes in the environmental temperature.

High-Level Transmitters

Figure 3-23 shows a block diagram for a high-level AM DSBFC transmitter. The modulating signal is processed the same as it is in the low-level transmitter except with the addition of a modulating signal power amplifier. With high-level modulation, the modulating signal power must be considerably higher than with low-level modulation. This is because the carrier is at full power when modulation occurs and, consequently, requires a high-amplitude modulating signal to produce 100% AM modulation.

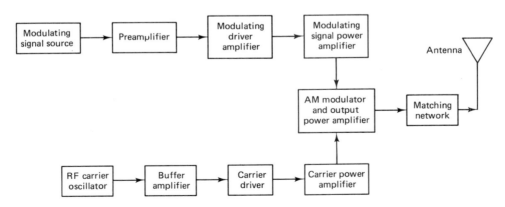

FIGURE 3-23 Block diagram of a high-level AM DSBFC transmitter.

The RF oscillator, its associated buffer, and the carrier driver are also the same as in a low-level transmitter. However, with high-level transmitters, the RF carrier undergoes additional power amplification prior to the modulator stage, and the final power amplifier is the modulator. Consequently, the modulator is generally a drain-, plate-, or collector-modulated class C power amplifier.

Figure 3-24 shows a partial schematic diagram for a high-level AM DSBFC transmitter. Q_1 is the active component in a standard crystal oscillator configuration. L_1 and the input impedance to Q_2 present a constant-impedance load to the oscillator. Q_2 is also the driver for the carrier. Class C amplifier, Q_3, is the modulator and also the final RF power amplifier. The modulating signal is coupled into the collector of Q_3 superimposed onto 13.5 V dc. The 13.5 V dc is the collector supply voltage for Q_3 and is present only when there is an input modulating signal. The 13.5 V dc is *keyed* on and off with the modulating signal. Consequently, the modulator has an output only when modulation is present. Although high-level modulation is accomplished in Q_3, in order to obtain 100% modulation, the collector of Q_2 is also collector modulated. The modulating signal is applied directly to the collector of Q_3 on top of the keyed 13.5 V dc. At the same time, dual diodes D_1 and D_2 allow Q_1 to be modulated. When the modulating signal voltage drops below 13.5 V, D_2 is reverse biased and shuts off, thus

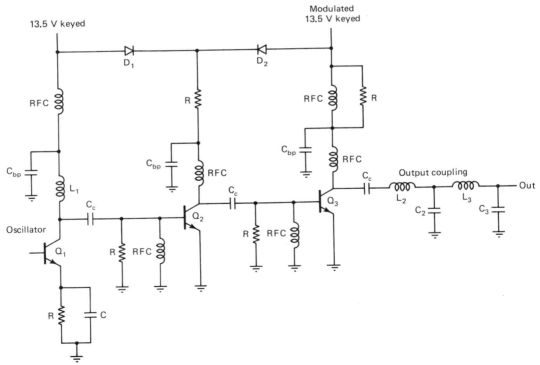

FIGURE 3-24   Schematic diagram for a high-level AM DSBFC transmitter.

removing modulation from Q_2. D_1 turns on and supplies the collector voltage for Q_2. When the modulating signal voltage rises above 13.5 V, D_2 turns on and applies modulation to the collector of Q_2 and shuts off D_1. This configuration allows the collector supply voltage for Q_3 to fluctuate above and below 13.5 V with the modulating signal and, at the same time, Q_2 can supply a partially modulated signal to the base of Q_3.

The output coupling circuit (L_2, L_3, C_2, and C_3) is an RF *trap*. An RF trap is simply a high-Q bandstop filter with a center frequency equal to twice the RF carrier frequency. Remember, the modulator is a nonlinear amplifier and, consequently, produces harmonic distortion. The second harmonic is the most significant harmonic produced and must be suppressed so that it does not interfere with transmissions from other transmitters an *octave* above in the radio spectrum.

In a high-level transmitter, the modulator circuit has three primary functions. It provides the circuitry necessary for AM modulation (nonlinearity), it is the final RF power amplifier (class C), and it is a frequency up-converter. An up-converter simply translates the low-frequency modulating signals to RF frequencies that can be more efficiently transmitted.

TRAPEZOIDAL PATTERNS

Trapezoidal patterns are used for testing AM transmitters and observing their modulation characteristics (i.e., percent modulation and modulation linearity). Although the modulation characteristics can be seen with an oscilloscope by observing the AM envelope, a

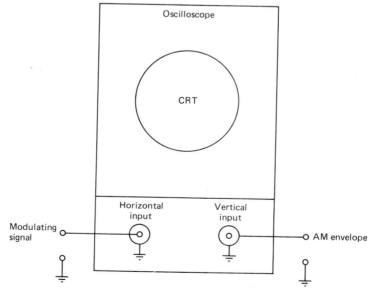

FIGURE 3-25   Test setup for displaying a trapezoidal pattern on an oscilloscope.

trapezoidal pattern is more easily interpreted, and is interpreted faster and more accurately. Figure 3-25 shows the basic test setup for producing a trapezoidal pattern on the CRT of a standard oscilloscope. The AM envelope is applied to the vertical input, and the modulating signal is applied to the external horizontal input (the internal horizontal sweep is disabled). Therefore, the horizontal sweep rate is determined by the modulating signal frequency, and the magnitude of the horizontal deflection is proportional to the amplitude of the modulating signal. The vertical deflection is totally dependent on the amplitude and rate of change of the AM envelope. In essence, the electron beam that is emitted from the cathode of the CRT is acted on in both the horizontal and vertical planes simultaneously.

Figure 3-26 shows how the AM envelope and the modulating signal produce a

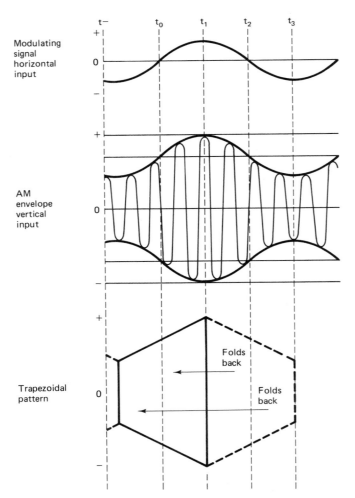

FIGURE 3-26 Producing a trapezoidal pattern.

trapezoidal pattern. With an oscilloscope, 0 V applied to the vertical input will center the electron beam vertically on the CRT, and 0 V applied to the external horizontal input will center the electron beam horizontally. Positive and negative voltages will deflect the beam up and down or right and left, respectively, for the vertical and horizontal inputs. If we begin when both the AM envelope and the modulating signal are 0 V (t_0), the electron beam is located in the exact center of the CRT. As the modulating signal goes positive, the beam is deflected to the right. At the same time the AM envelope is going positive, deflecting the beam upward. The beam deflects to the right until the modulating signal reaches its maximum positive value (t_1). While the beam is moving toward the right, the beam is deflected up and down as the AM envelope

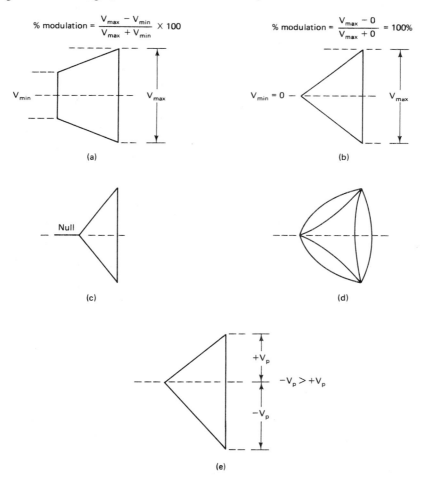

FIGURE 3-27 Trapezoidal patterns: (a) linear 50% AM modulation; (b) 100% AM modulation; (c) more than 100% AM modulation; (d) improper phase relationship; (e) nonlinear AM envelope.

alternately swings positive and negative. Notice that on each successive alternation, the AM envelope reaches a higher magnitude than the previous alternation. The AM envelope reaches its maximum value at the same time as the modulating signal. Therefore, as the CRT beam is deflected to the right, its peak-to-peak vertical deflection increases. As the modulating signal becomes less positive, the beam is deflected to the left (toward the center of the CRT). At the same time, the AM envelope alternately swings positive and negative, deflecting the beam up and down except now each successive alternation is lower in amplitude than the previous alternation. Consequently, as the beam moves horizontally back toward the center of the CRT, the vertical deflection decreases. The modulating signal and the AM envelope pass through 0 V at the same time and the beam is again in the center of the CRT (t_2). As the modulating signal goes negative, the beam is deflected to the left side of the CRT. At the same time, the AM envelope is decreasing in magnitude on each successive alternation. The modulating signal reaches its maximum negative value at the same time as the AM envelope (t_3). The trapezoidal pattern shown between times t_1 and t_3 folds back on top of the pattern displayed during times $t-$ and t_0. Thus a complete trapezoidal pattern is displayed on the screen during both the left-to-right and the right-to-left horizontal sweeps.

From Figure 3-26 it can be seen that the time of one cycle of the modulating signal equals the time of one cycle of the AM envelope and also the time for one complete horizontal alternation of the trapezoidal pattern. If the AM envelope is linear (the top half of the envelope is a mirror image of the bottom half), a linear trapezoidal pattern like the ones shown in Figure 3-27 are produced. Also, at 100% modulation, the peak-to-peak amplitude of the envelope goes to zero and the trapezoidal pattern comes to a point at one end (Figure 3-27b). If the percent modulation exceeds 100%, the pattern shown in Figure 3-27c is produced. The pattern shown in Figure 3-27a is for a 50% modulated wave. If the modulating signal and the AM envelope are out of phase, the pattern shown in Figure 3-27d is produced. If the magnitude of the positive and negative alternations of the AM envelope are not equal, the pattern shown in Figure 3-27e is produced. If the phase of the modulating signal is shifted 180° (inverted), the trapezoidal patterns would simply point in the opposite direction. As you can see, percent modulation and modulation linearity are more easily observed with a trapezoidal pattern than with a display of the AM envelope.

CARRIER SHIFT

Carrier shift is a term that is often misunderstood or misinterpreted. Carrier shift (sometimes called *upward* or *downward modulation*) has absolutely nothing to do with the frequency of the carrier. Carrier shift is a form of amplitude distortion introduced when the positive and negative alternations in the AM envelope are not equal (i.e., nonlinear modulation). Carrier shift may be either positive or negative. If the positive alternation of the envelope has a larger amplitude than the negative alternation, positive carrier shift results. If the negative alternation has a larger amplitude than the positive alternation, negative carrier shift results. Carrier shift is an indication of the average value of an

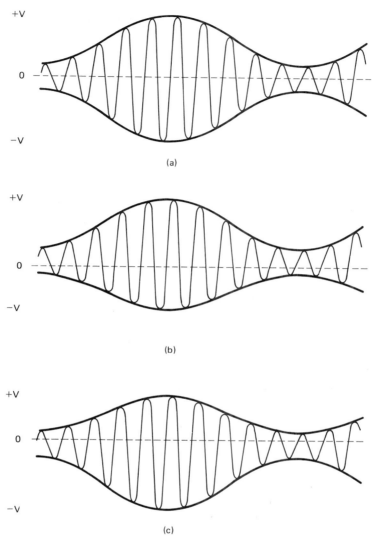

FIGURE 3-28 Carrier shift: **(a) linear modulation; (b) positive carrier shift; (c) negative carrier shift.**

AM envelope. If the positive and negative halves of the envelope are equal, the average voltage is 0. If the positive half is larger than the negative half, the average voltage is positive, and if the negative half is larger than the positive half, the average voltage is negative. Figure 3-28a shows a symmetrical AM envelope (no carrier shift); the average value for the envelope is 0 V. Figure 3-28b and c show positive and negative carrier shift, respectively.

QUESTIONS

3-1. Define *amplitude modulation*.

3-2. Define *RF*.

3-3. How many inputs are there to an amplitude modulator? What are they?

3-4. In an AM system, what is meant by the *modulating signal*; the *modulated signal*; the *modulated wave*?

3-5. Describe an AM envelope. Why is it called an envelope?

3-6. Describe upper and lower sidebands and upper and lower side frequencies.

3-7. Define *modulation coefficient*.

3-8. Define *percent modulation*.

3-9. What is the highest coefficient of modulation and percent modulation that can occur without causing excessive distortion?

3-10. Describe the meaning of each term in the following equation:

$$e_c = E_c \sin \omega_c t - \frac{mE_c}{2} \cos (\omega_c + \omega_a)t + \frac{mE_c}{2} \cos (\omega_c - \omega_a)t$$

3-11. What effect does modulation have on the amplitude of the carrier?

3-12. Describe the significance of the following formula:

$$P_t = P_c + \frac{m^2}{2} P_c$$

3-13. What does AM DSBFC stand for?

3-14. Describe the relationship between the carrier and sideband powers in an AM DSBFC modulator.

3-15. What is the predominant disadvantage of AM double-sideband transmission?

3-16. What is the predominant advantage of AM double-sideband transmission?

3-17. What is the maximum efficiency that can be achieved with AM DSBFC?

3-18. Why do any amplifiers that follow the modulator in an AM DSBFC system have to be linear?

3-19. What is the primary disadvantage of a low-power emitter-biased transistor modulator such as the one shown in Figure 3-14?

3-20. Describe the difference between a low- and a high-level modulator.

3-21. List the advantages of low-level modulation; high-level modulation.

3-22. What is the advantage of using a trapezoidal pattern to evaluate an AM envelope?

PROBLEMS

3-1. If a 20-V_p carrier changes in amplitude ± 5 V_p, determine the modulation coefficient and percent modulation.

3-2. For a maximum positive envelope voltage of 12 V_p and a minimum positive amplitude of 4 V_p, determine the modulation coefficient and percent modulation.

3-3. For an envelope with $+V_{max} = 40$ V_p and $+V_{min} = 10$ V_p, determine:
  (a) The unmodulated carrier amplitude.
  (b) The peak change in amplitude of the modulated wave.
  (c) The modulation coefficient.
  (d) The percent modulation.

3-4. Describe the following expression for an amplitude-modulated wave in terms of frequency content and voltage amplitude:

$$e_c = 10 \sin 2\pi 500kt - 5 \cos 2\pi 515kt + 5 \cos 2\pi 485kt$$

3-5. For an unmodulated carrier amplitude $V_c = 16$ V_p and a modulation coefficient $m = 0.4$, determine the amplitudes of the carrier and side frequencies.

3-6. Sketch the envelope for Problem 3-5 (label all pertinent voltages).

3-7. For the AM envelope shown below, determine:
  (a) The peak amplitude of the upper and lower side frequencies.
  (b) The peak amplitude of the carrier.
  (c) The peak change in the amplitude of the AM envelope.
  (d) The modulation coefficient.
  (e) The percent modulation.

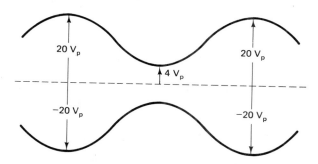

3-8. One input to an AM DSBFC modulator is a 800-kHz carrier with a peak amplitude $V_c = 40$ V. The second input is a 25-kHz modulating signal whose amplitude is sufficient to produce a ± 10 V_p change in the envelope.
  (a) Determine the USF and LSF frequencies.
  (b) Determine the modulation coefficient.
  (c) Determine the percent modulation.
  (d) Determine the maximum and minimum positive peak amplitudes of the envelope.
  (e) Draw the output spectrum.
  (f) Sketch the envelope (label all pertinent voltages).

3-9. For a modulation coefficient $m = 0.2$ and a peak carrier power $P_c = 1000$ W, determine:
  (a) The sideband power.
  (b) The total transmit power.

3-10. For an AM DSBFC wave with a peak unmodulated carrier voltage $V_c = 25$ V and a load resistance $R_1 = 50$ Ω:
  (a) Determine the peak and rms powers of the unmodulated carrier.
  (b) Determine the peak and rms powers of the modulated carrier, the USF, and the LSF for a modulation coefficient $m = 0.6$.

(c) Determine the total power in the modulated wave.

(d) Draw the output power spectrum.

3-11. Determine the quiescent, maximum, and minimum voltage gains for the emitter modulator shown below with the indicated carrier and modulating signal amplitudes.

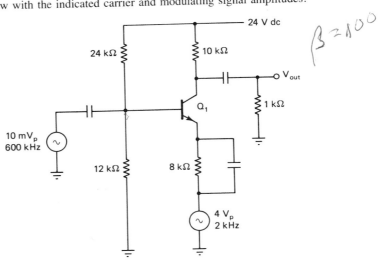

3-12. Sketch the output envelope and draw the output frequency spectrum for the circuit of Problem 3-11.

3-13. For a low-power transistor AM modulator with a modulation coefficient $m = 0.4$, a quiescent gain $A_q = 80$, and an input carrier amplitude of 0.002 V_p:

(a) Determine the maximum and minimum gains for the amplifier.

(b) Determine the maximum and minimum amplitudes for V_{out}.

(c) Sketch the AM envelope.

3-14. For the trapezoidal pattern shown below, determine:

(a) The modulation coefficient.

(b) The percent modulation.

(c) The carrier amplitude.

(d) The upper and lower side frequency amplitudes.

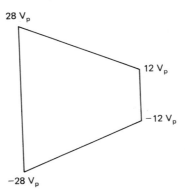

3-15. For an AM modulator with a carrier frequency of 200 kHz and a maximum modulating signal frequency of 10 kHz:

(a) Determine the frequency limits for the upper and lower sidebands.

(b) Determine the upper and lower side frequencies produced when the modulating signal is a 7-kHz tone.

(c) Determine the bandwidth.

(d) Draw the output spectrum.

Chapter 4

AMPLITUDE MODULATION RECEPTION

INTRODUCTION

In essence, AM reception is the reverse process of AM transmission. An AM receiver simply converts an amplitude-modulated wave back to the original source information (i.e., it demodulates the AM wave). However, the demodulation process can be quite different from the modulation process. To understand the demodulation process, it is necessary to have a basic understanding of the terminology associated with receivers and receiver circuits.

Figure 4-1 shows a simplified block diagram of an AM receiver. The RF section is the first stage of the receiver and is therefore often called the receiver *front end*. The primary functions of the RF section are bandlimiting and amplification of the received RF signals. The RF section comprises one or more of the following circuits: *antenna*, *antenna coupling network*, *receiver input filter* (*preselector*), and one or more *RF amplifiers*. The *mixer/converter* section down-converts the received RF frequencies to *intermediate frequencies* (IF). The *IF section* includes several *cascaded* amplifiers and bandpass filters. The primary functions of the IF section are amplification and selectivity. The *AM detector* demodulates the AM wave and recovers the original source information from the envelope. The audio section simply amplifies the recovered information to a usable level.

Receiver Parameters

Selectivity. The *selectivity* of a receiver is a measure of the ability of the receiver to accept a given band of frequencies and reject all others. For example, with the AM

126

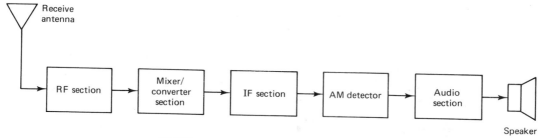

FIGURE 4-1 Simplified block diagram of an AM receiver.

commercial broadcast band, each station's transmitter is allocated a 10-kHz bandwidth (the carrier ± 5 kHz). Therefore, for a receiver to select a single channel, the input to the receiver must be bandlimited to the desired 10-kHz passband. If the passband of the receiver is greater than 10 kHz, more than one channel can be received and demodulated simultaneously. If the passband of the receiver is less than 10 kHz, a portion of the source information is rejected and information is lost.

Selectivity is often defined as the bandwidth of a receiver at some predetermined *attenuation factor* (commonly −60 dB) to the bandwidth at the −3 dB (half-power) points. This ratio is often called the shape factor (SF) and is determined by the number of *poles* and the *Q-factors* of the receiver's input filter. The *shape factor* defines the shape of the gain-versus-frequency plot for a filter and is expressed mathematically as

$$SF = \frac{B(-60 \text{ dB})}{B(-3 \text{ dB})} \qquad (4\text{-}1a)$$

For a perfect filter, the attenuation factor is infinite and the bandwidth at the −3-dB frequencies is equal to the bandwidth at the −60-dB frequencies. Therefore, the ideal shape factor is 1. Selectivity is often given as a percentage and expressed mathematically as

$$\% \text{ selectivity} = SF \times 100$$
$$= \frac{B(-60 \text{ dB})}{B(-3 \text{ dB})} \times 100 \qquad (4\text{-}1b)$$

EXAMPLE 4-1

Determine the shape factor and the percent selectivity for the gain-versus-frequency plot shown in Figure 4-2.

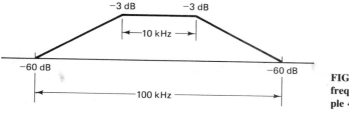

FIGURE 4-2 Gain-vs.-frequency plot for Example 4-1.

Solution   The shape factor is determined from Equation 4-1a:

$$SF = \frac{100 \text{ kHz}}{10 \text{ kHz}} = 10$$

The percent selectivity is determined from Equation 4-1b:

$$\% \text{ selectivity} = 10 \times 100 = 1000\%$$

Bandwidth improvement.   As stated in Chapter 1 and given in Equation 1-17, noise is directly proportional to bandwidth. Therefore, if the bandwidth is reduced, the noise is also reduced by the same proportion. Essentially, *bandwidth improvement* is the noise reduction ratio achieved by reducing the bandwidth. As a signal propagates from the antenna through the RF section, the mixer/converter section, and the IF section; the bandwidth is reduced. Therefore, the noise is also reduced. Effectively, this is the same as reducing (improving) the noise figure of the receiver. The bandwidth improvement factor is the ratio of the RF bandwidth to the IF bandwidth. Mathematically, bandwidth improvement is

$$BI = \frac{B(\text{RF})}{B(\text{IF})} \tag{4-2}$$

where

$$BI = \text{bandwidth improvement factor}$$
$$B(\text{RF}) = \text{RF bandwidth}$$
$$B(\text{IF}) = \text{IF bandwidth}$$

and the corresponding reduction in the noise figure is

$$NF \text{ (improvement)} = 10 \log BI$$

EXAMPLE 4-2

Determine the improvement in the noise figure for a receiver with an RF bandwidth equal to 200 kHz and an IF bandwidth equal to 10 kHz.

Solution

$$BI = \frac{B(\text{RF})}{B(\text{IF})} = \frac{200 \text{ kHz}}{10 \text{ kHz}} = 20$$

The noise-figure improvement is

$$10 \log BI = 10 \log 20 = 13 \text{ dB}$$

Consequently, a reduction in the bandwidth by a factor of 20 also reduces the noise by a factor of 20.

Sensitivity.   The *sensitivity* of a receiver is the minimum RF signal level that can be received and still produce a usable demodulated signal. Generally, the signal-to-noise ratio and the power of the signal at the output of the audio section determine whether the demodulated signal is usable or not. For commercial AM broadcast band receivers, 10 db *S/N* with $\frac{1}{2}$ W of audio power is generally considered usable. However, for broadband microwave receivers, 40 dB *S/N* and approximately 5 mW of audio power is necessary. The sensitivity of a receiver is usually stated in microvolts of received signal. For example, a common sensitivity level for a commercial AM receiver is approximately 50 μV, and a two-way mobile radio receiver has a sensitivity between 0.1 and 10 μV. Receiver sensitivity is also called *receiver threshold*. The sensitivity of an AM receiver depends on the noise power present at the input to the receiver, the receiver's noise figure, the sensitivity of the AM detector, and the bandwidth improvement factor. The best ways to improve the sensitivity of a receiver is to reduce the noise level (i.e., artificially cool the receiver's front end, reduce the receiver's bandwidth, or improve the receiver's noise figure).

Fidelity. *Fidelity* is a measure of the ability of a communications system to produce, at the output of the receiver, an exact replica of the original source information. Any frequency, phase, or amplitude variation that is present in the demodulated waveform that was not present in the original source information is distortion.

Essentially, there are three forms of distortion that reduce the fidelity of a system: *amplitude*, *frequency*, and *phase*. Phase distortion is not particularly important for voice transmission because the human ear is relatively insensitive to phase variations. However, phase distortion can be devastating to data transmissions. The predominant cause of phase distortion is filtering. Frequencies at or near the break frequency of a filter undergo severe phase shift. Actually, absolute phase shift can be tolerated as long as all frequencies undergo the same amount of phase shift. Differential phase shift occurs when different frequencies undergo different amounts of phase shift. Phase shift is analogous to propagation delay. If all frequencies are not delayed by the same amount of time, the frequency/phase relationship of the received waveform is not consistent with the original source information.

Amplitude distortion occurs when the frequency-versus-amplitude characteristics of a signal at the output of a receiver differ from those of the original source information. Amplitude distortion is the result of *nonuniform* gain in amplifiers and filters.

Frequency distortion occurs when there are frequencies present in the received signal that were not present in the original source information. Frequency distortion is a result of harmonic and intermodulation distortion and is caused by nonlinear amplification. *Third-order intercept distortion* is the predominant form of frequency distortion. Third-order intercept distortion is a special case of intermodulation distortion. Third-order intermodulation components are the cross-product frequencies produced when the second harmonic of one signal is multiplied by the fundamental frequency of another signal (i.e., $2F_1 \pm F_2$, $2F_2 \pm F_1$, etc.). Frequency distortion can be reduced by using *square-law devices*, such as MOSFETs, in the front end of a receiver. Square-law devices have the unique advantage over BJTs in that they produce only second-order harmonic and intermodulation components.

Insertion loss.   *Insertion loss* is a parameter that is associated with the frequencies that fall within the passband of a filter and is generally defined as the ratio of the power transferred to a load with a filter in the circuit to the power transferred to the load without a filter. Because filters are generally constructed from lossy components such as resistors and imperfect capacitors, even signals within the passband of the filter are attenuated (reduced in value). Typical filter insertion losses are between a few tenths of a decibel to several decibels.

Noise temperature and equivalent noise temperatures.   Because thermal noise is directly proportional to temperature, it stands to reason that noise can be expressed in degrees as well as watts or volts. Rearranging Equation 1-17 yields

$$T = \frac{N}{KB} \tag{4-3}$$

where

T = Environmental temperature (K)
N = noise power (W)
K = Boltzmann constant (1.38×10^{-23} J/K)
B = bandwidth (Hz)

Equivalent noise temperature (T_e) is a hypothetical value that cannot be directly measured. T_e is a parameter that is often used in low-noise, sophisticated receivers rather than noise figure. T_e is an indication of the reduction in the *S/N* ratio as a signal propagates through a receiver. The lower the equivalent noise temperature, the better the quality of the receiver. Typical values for T_e range from 20° for "cool" receivers to 1000° for noisy receivers. The overall equivalent noise temperature for a receiver is simply the sum of the T_e values for each receiver stage. Mathematically, T_e is

$$T_e = T(F - 1) \tag{4-4}$$

where

T_e = equivalent noise temperature (K)
T = environmental temperature (K)
F = noise figure (unitless)

AM RECEIVERS

Essentially, there are two types of radio receivers: coherent and noncoherent. With a coherent or *synchronous receiver*, the carrier frequency used for demodulation in the receiver is synchronized to the carrier frequency used in the transmitter (the receiver has some means of recovering and synchronizing to the transmitter's carrier). With *noncoherent* or *asynchronous receivers*, either there is no carrier frequency generated in the receiver or the carrier frequency used for demodulation is totally independent

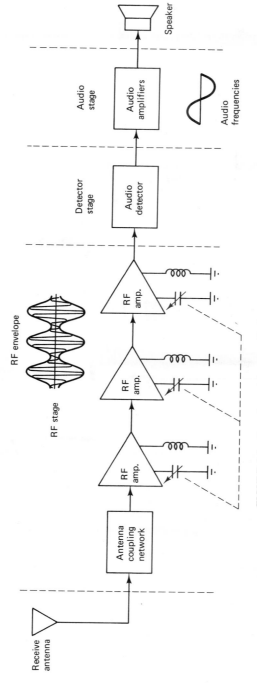

FIGURE 4-3 Noncoherent tuned radio-frequency receiver block diagram.

from the transmitter's carrier frequency. *Noncoherent detection* is often called *envelope detection* because the information is recovered by detecting the shape of the modulated envelope.

Tuned Radio-Frequency AM Receiver

The *tuned radio-frequency AM receiver* (TRF) was one of the earliest types of AM receivers and it is still probably the simplest design available. A block diagram of a noncoherent TRF is shown in Figure 4-3 (it is noncoherent because there is no synchronous carrier regenerated in the receiver). A TRF is essentially a three-stage receiver: an RF stage, a detector stage, and an audio stage. Generally, two or three RF amplifiers are required to develop sufficient signal amplitude to drive the detector stage. The detector converts the RF signals directly to audio signals, and the audio stage amplifies the audio signals to a usable level.

Tuning a TRF introduces three disadvantages that limit its usefulness to single-station applications. The primary disadvantage of a TRF is that its selectivity (bandwidth) varies when it is tuned over a wide range of input frequencies. The bandwidth of the RF input filter varies with the center frequency of the tuned circuit. This is caused by a phenomenon called the *skin effect*. At RF, current flow is limited to the outermost area of a conductor, and therefore the resistance of the conductor increases with frequency. (The skin effect is explained in detail in Chapter 9.) Consequently, the Q of the tank circuit (X_L/R) remains relatively constant over a wide range of frequencies and the bandwidth (F/Q) increases with frequency. As a result, the selectivity of the filter changes over any appreciable range of RF. If the bandwidth of the input filter is set to the desired value for low-band RF signals, it will be excessive for the high-band RF signals, and if the bandwidth is set to the desired value for high-band RF signals, it will be insufficient for the low-band RF signals. The second disadvantage of the TRF is that its gain is not uniform over a very wide frequency range. This is due to the nonuniform L/C ratio of the transformer-coupled tank circuits in the RF amplifiers (i.e., the ratio of the inductance to capacitance in one tuned amplifier is not the same as that of the other tuned amplifiers). The third disadvantage of the TRF is that it requires *multistage tuning*. To change stations, each RF filter must be tuned simultaneously to the new frequency band with a single adjustment, which requires exactly the same characteristics for each tuned circuit. With the development of the *superheterodyne receiver*, the TRF concept has essentially gone by the wayside except for single-station receivers and therefore does not warrant any further discussion.

EXAMPLE 4-3

For an AM commercial broadcast-band receiver (535 to 1605 kHz) with an input filter Q-factor of 54, determine the bandwidth at the low and high ends of the RF spectrum.

Solution   The bandwidth at the low end is centered around a carrier frequency of 540 kHz and is

$$B = \frac{F_c}{Q} = \frac{540 \text{ kHz}}{54} = 10 \text{ kHz}$$

The bandwidth at the high end is centered around a carrier frequency of 1600 kHz and is

$$B = \frac{1600 \text{ kHz}}{54} = 29{,}630 \text{ Hz}$$

The -3-dB bandwidth at the low end of the frequency spectrum is exactly 10 kHz, which is the desired value. However, the bandwidth at the high end is almost 30 kHz, which is three times the desired value. Consequently, when tuning for stations at the high end of the frequency spectrum, three stations would be received simultaneously.

To achieve a bandwidth of 10 kHz at the high end of the spectrum, a Q of 160 is required (1600 kHz/10 kHz). With a Q of 160 the bandwidth at the low end is

$$B = \frac{540 \text{ kHz}}{160} = 3375 \text{ Hz}$$

3375 Hz is too selective because approximately 66.25% of the information bandwidth is rejected.

Superheterodyne AM Receiver

The nonuniform selectivity of the TRF led to the development of the superheterodyne receiver at the end of World War I. The quality of the superheterodyne receiver has improved greatly since its original design and the basic receiver configuration is still used extensively in radio communications receivers today. The superheterodyne receiver has remained in use because its gain, selectivity, and sensitivity characteristics are superior to those of other receiver configurations.

Heterodyne means to mix two frequencies together in a nonlinear device or to translate one frequency to another using nonlinear mixing. A block diagram of a noncoherent superheterodyne receiver is shown in Figure 4-4 (it is noncoherent because the *local oscillator* is totally independent of the transmitter's carrier oscillator). Essentially, there are five primary sections that make up a superheterodyne receiver: the RF section, the mixer/converter section, the IF section, the detector section, and the audio amplifier section.

RF section. The RF section usually has a *preselector* stage and an RF amplifier stage. They can be two separate circuits or a single combined circuit. The preselector is a broad-tuned bandpass filter with an adjustable center frequency that is tuned to the desired RF carrier. The primary purpose of the preselector is to provide enough initial bandlimiting to prevent a specific unwanted radio frequency called the *image frequency* from entering the receiver. (Image frequency is explained later in this chapter.) The preselector also reduces the noise bandwidth of the receiver and provides the initial step toward reducing the overall receiver bandwidth to the minimum bandwidth required to pass the information signal. The RF amplifier determines the sensitivity of the receiver

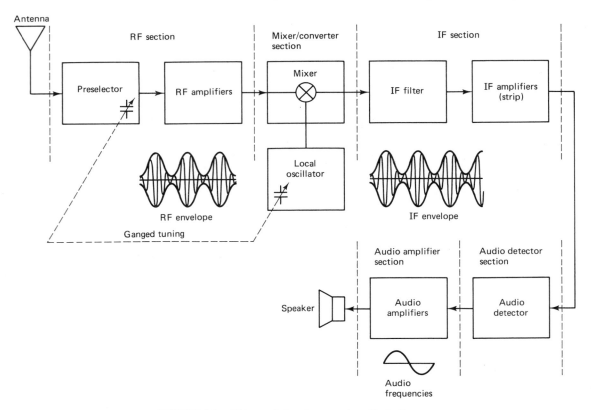

FIGURE 4-4   AM superheterodyne receiver block diagram.

(i.e., sets the signal threshold value). Also, because the RF amplifier is the first active device encountered by a received signal, it is the primary contributor of noise and therefore a predominant factor in determining the noise figure for the receiver. A receiver can have zero or only one RF amplifier, or there could be several, depending on the sensitivity desired.

Mixer/converter section. The mixer/converter section includes an RF oscillator stage (commonly called a local oscillator) and a mixer/converter stage (commonly called the *first detector*). The local oscillator can be any of the oscillator circuits discussed in Chapter 2, depending on the stability and accuracy desired. The mixer stage is a nonlinear device and its purpose is to convert RF to IF (i.e., frequency translation). Heterodyning takes place in the mixer stage and RF is down-converted to IF. Although the carrier and sideband frequencies are translated from RF to IF values, the shape of the AM envelope remains the same, and therefore the original information contained in the envelope remains unchanged. It is important to note that although the carrier and the upper and lower side frequencies are changed, the bandwidth of the spectrum is unchanged by the heterodyning process. The most common intermediate frequency used in AM broadcast-band receivers is 455 kHz.

IF section. The IF section comprises a series of IF amplifiers (commonly called an *IF strip*). Most of the receiver gain and selectivity is achieved in the IF section. The IF is constant for all stations and is chosen so that its frequency is less than any of the RF signals to be received. The IF is always lower because it is easier and less expensive to construct high-gain amplifiers for IF than for RF. Also, low-frequency IF amplifiers are less likely to oscillate than their RF counterparts. Therefore, it is not uncommon to see a receiver with five or six IF amplifiers and a single RF amplifier or possibly no RF amplification.

Detector section. The purpose of the detector section is to convert the IF envelope back to the original source information. The audio detector can be as simple as a diode or as complex as a phase-locked loop or balanced demodulator.

Audio section. The *audio section* comprises several cascaded audio amplifiers. The number of amplifiers depends on the audio signal power desired.

Operation of the superheterodyne receiver. During the demodulation process in a superheterodyne receiver, the received signals undergo two frequency *translations*: first the RF is converted to IF, then the IF is converted to audio.

Frequency conversion. *Frequency conversion* in the mixer/converter stage is identical to frequency conversion in the modulator stage of a transmitter except that in the receiver, the frequencies are *down-converted* rather than *up-converted*. In the mixer/converter, RF signals are mixed with the local oscillator frequency in a nonlinear device. The output of the mixer contains an infinite number of harmonic and cross-product frequencies which include the sum and difference frequencies between the desired RF and the local oscillator frequency. The IF filters are tuned to the difference frequencies. The local oscillator is designed such that its frequency of oscillation is always the IF above or below the desired RF. Therefore, the difference between the RF and the local oscillator frequency is the IF. The adjustment for the center frequency of the preselector and adjustment for the local oscillator frequency are *gang tuned*. Gang tuning means that the two adjustments are mechanically tied together so that a single adjustment will change the center frequency of the preselector and, at the same time, change the local oscillator frequency. When the local oscillator frequency is tuned above the RF, it is called *high-side* or *high-beat* injection. When the local oscillator frequency is tuned below the RF, it is called *low-side* or *low-beat* injection. In AM broadcast-band receivers, high-side injection is always used (the reason for this is explained later). Mathematically, the local oscillator frequency is:

For high-side injection:

$$F_{lo} = F_{rf} + F_{if}$$

(4-5a)

For low-side injection:

$$F_{lo} = F_{rf} - F_{if} \qquad\qquad (4\text{-}5b)$$

where

F_{lo} = local oscillator frequency
F_{rf} = radio frequency
F_{if} = intermediate frequency

EXAMPLE 4-4

For an AM superheterodyne receiver that uses high-side injection and has a local oscillator frequency of 1355 kHz, determine the IF carrier, upper side frequency, and lower side frequency for an RF envelope which is made up of a carrier and upper and lower side frequencies of 900, 905, and 895 kHz, respectively.

Solution (Refer to Figure 4-5)   Because high-side injection is used, the IF is the difference between the RF and the local oscillator frequency. Rearranging Equation 4-5a yields

$$F_{if} = F_{lo} - F_{rf}$$

$$= 1355 \text{ kHz} - 900 \text{ kHz} = 455 \text{ kHz}$$

The IF upper and lower side frequencies are

$$\text{IF(usf)} = F_{lo} - F_{rf}(\text{lsf})$$

$$= 1355 \text{ kHz} - 895 \text{ kHz} = 460 \text{ kHz}$$

$$\text{IF(lsf)} = F_{lo} - F_{rf}(\text{usf})$$

$$= 1355 \text{ kHz} - 905 \text{ kHz} = 550 \text{ kHz}$$

Note that the side frequencies undergo a reversal during the heterodyning process (i.e., the RF upper side frequency is translated to the IF lower side frequency, and the RF lower side frequency is translated to the IF upper side frequency—this is commonly called *sideband inversion*).

Local oscillator tracking. *Tracking* is the ability of the local oscillator in a receiver to remain the IF above or below the preselector center frequency. With high-side injection, the local oscillator should track above the incoming RF by a fixed frequency which is equal to the IF.

Figure 4-6a shows the schematic diagram for the tuned circuits in a preselector and local oscillator. The tuned circuit in the preselector is tunable from 540 to 1600 kHz (a ratio of 2.96:1), and the tuned circuit in the local oscillator is tunable from 995 to 2055 kHz (a ratio of 2.15:1). Because the resonant frequency of a tuned circuit is inversely proportional to the square root of the capacitance, the capacitor in the preselector must change by a factor of 8.8, while at the same time, the capacitor in the local oscillator must change by a factor of only 4.6. The local oscillator should track 455 kHz above the preselector over the entire RF spectrum and there should be a single tuning control. Fabricating such a unit is difficult, if not impossible. Therefore,

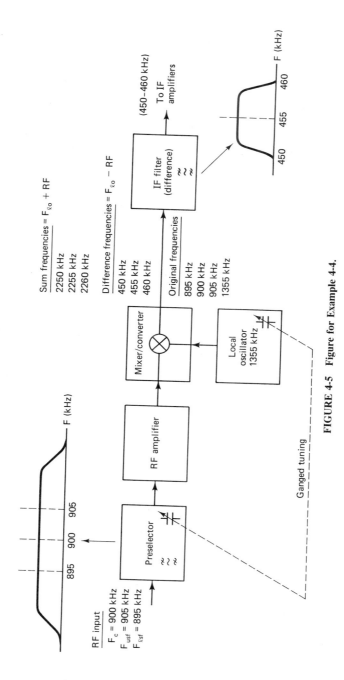

FIGURE 4-5 Figure for Example 4-4.

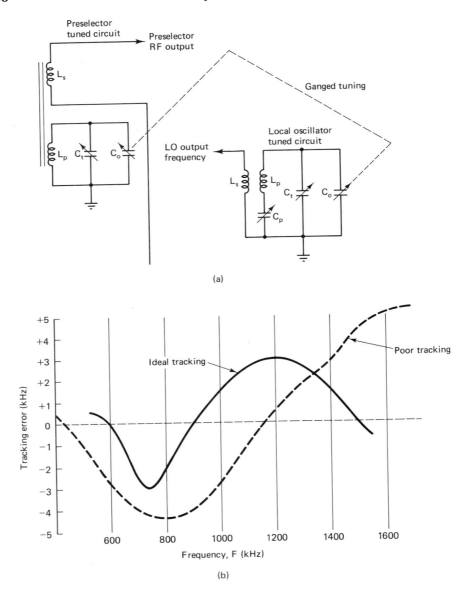

(a)

(b)

FIGURE 4-6 Receiver tracking: (a) preselector and local oscillator schematic; (b) tracking curve.

perfect tracking over the entire RF spectrum is unlikely. The difference between the actual tracking frequency and the IF is the tracking error. The *tracking error* is not uniform over the entire RF spectrum. A maximum tracking error of ±3 kHz is about the best that can be expected from a domestic AM broadcast-band receiver with a 455-kHz IF. Figure 4-6b shows a typical tracking curve. A tracking error of +3 kHz corresponds to a tracking frequency of 458 kHz, and a tracking error of −3 kHz corresponds to a tracking frequency of 452 kHz.

The tracking error is reduced by a technique called *three-point tracking*. The preselector and the local oscillator each have a *trimmer* capacitor (C_t) in parallel with the primary tuning capacitor (C_o) that compensates for minor tracking errors at the high end of the spectrum, and the local oscillator has an additional *padder* capacitor (C_p) placed in series with the tuning coil that compensates for minor tracking errors at the low end of the spectrum. With three-point tracking, the tracking error is adjusted to 0 Hz at approximately 600, 950, and 1500 kHz.

With low-side injection, the local oscillator is tunable from 85 to 1145 kHz (a ratio of 13.5 : 1). Consequently, the capacitance must change by a factor of 182. Standard variable capacitors seldom tune over more than a 10 : 1 ratio. This is why low-side injection is impractical for standard AM broadcast-band receivers.

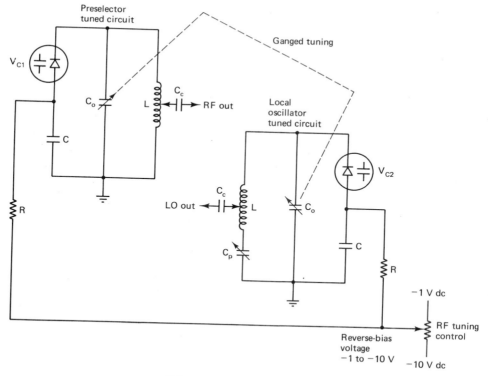

FIGURE 4-7 Electronic tuning.

Ganged capacitors are relatively large and expensive and they are somewhat difficult to compensate. Consequently, they are gradually being replaced with solid-state electronically tuned circuits. Electronically tuned circuits are smaller, less expensive, more accurate, more immune to environmental changes, more easily compensated, and more easily adapted to digital remote control and pushbutton tuning than are their mechanical counterparts. As with the crystal oscillator modules explained in Chapter 2, electronic tuned circuits use solid-state variable capacitance diodes (varactor diodes). Figure 4-7 shows the schematic diagrams for an electronically tuned preselector and local oscillator. The -1- to -10-V reverse bias comes from a single tuning control. By changing the position of the wiper on a precision variable resistor, the dc reverse bias for the two tuning diodes (VC_1 and VC_2) is changed. The diode capacitance and, consequently, the resonant frequency of the tuned circuit vary with the reverse bias. Three-point compensation with electronic tuning is accomplished the same as with mechanical tuning.

In a superheterodyne receiver, most of the receiver's selectivity is accomplished in the IF stage. For maximum noise reduction, the bandwidth of the IF filters is equal to the minimum bandwidth required to pass the information signal, which with double-sideband transmission is equal to two times the highest modulating frequency. For a maximum modulating signal frequency of 5 kHz, the minimum IF bandwidth is 10 kHz. For a 455-kHz IF and perfect tracking, a 450- to 460-kHz bandwidth is required. In reality, some RF stations are tracked 3 kHz above and some 3 kHz below 455 kHz. Therefore, the IF bandwidth must be expanded to allow the IF from the off-track stations to pass through the IF filters.

EXAMPLE 4-5

For the tracking curve shown in Figure 4-8a, a 455-kHz IF, and a maximum modulating signal frequency of 5 kHz; determine the minimum IF bandwidth.

Solution   The maximum IF occurs for the RF carrier with the most positive tracking error (1400 kHz) and a 5-kHz modulating signal:

$$IF(maximum) = IF + tracking\ error + modulating\ signal$$

$$= 455\ kHz + 3\ kHz + 5\ kHz = 463\ kHz$$

The minimum IF occurs for the RF carrier with the most negative tracking error (800 kHz) and a 5-kHz modulating signal:

$$IF(minimum = IF + tracking\ error - modulating\ signal$$

$$= 455\ kHz + (-3\ kHz) - 5\ kHz = 447\ kHz$$

The minimum IF bandwidth necessary to pass the two sidebands is the difference between the maximum and minimum IFs:

$$minimum\ bandwidth = 463\ kHz - 447\ kHz = 16\ kHz$$

Figure 4-8b shows the IF bandpass characteristics for Example 4-5.

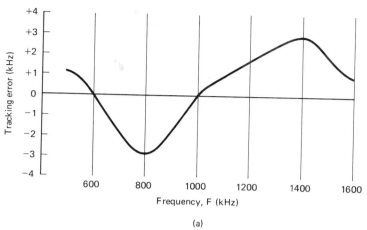

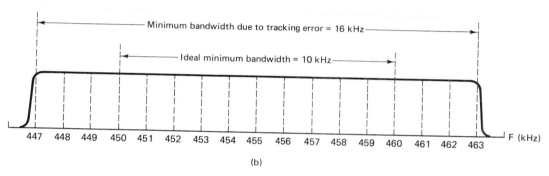

FIGURE 4-8 Tracking error for Example 4-5: (a) tracking curve; (b) bandpass characteristics.

Image frequency. An *image frequency* is any frequency other than the selected RF which, if allowed to enter a receiver and mix with the local oscillator will produce a cross-product frequency that is equal to the IF. Each RF carrier has an image frequency. Once an image frequency has been mixed down to IF, it cannot be filtered out or suppressed. If the selected RF carrier and its image frequency enter a receiver at the same time, they both mix with the local oscillator frequency in the mixer/converter and produce a difference frequency equal to the IF. Consequently, two different stations are received and demodulated simultaneously producing two audio signals. For a radio frequency to produce a cross product equal to the IF, it must be displaced from the local oscillator frequency by a value equal to the IF. With high-side injection, the selected RF is the IF below the local oscillator frequency. Therefore, the image frequency is the radio frequency that is the IF above the local oscillator frequency. Mathematically, for high-side injection, the image frequency is

$$F_{\text{image}} = F_{\text{lo}} + F_{\text{if}} \qquad (4\text{-}6a)$$

and since the desired RF equals the local oscillator frequency minus the IF:

$$F_{\text{image}} = F_{\text{rf}} + 2F_{\text{if}} \qquad (4\text{-}6b)$$

Figure 4-9 shows the relative frequency spectrum for the RF, the IF, the local oscillator frequency, and the image frequency in a superheterodyne receiver using high-side injection.

From Figure 4-9 it can be seen that the higher the IF, the farther away in the frequency spectrum the image frequency is from the selected RF. Therefore, for better *image frequency rejection*, a high IF is preferred. However, the higher the IF, the more difficult it is to build stable amplifiers with high gain. Therefore, there is a trade-off when selecting the IF for a radio receiver between image frequency rejection and IF gain.

Image frequency rejection ratio. The *image frequency rejection ratio* (IFRR) is a numerical measure of the ability of a preselector to reject the image frequency. For a single-tuned circuit, the ratio of its gain at the selected RF to the gain at the image frequency is the IFRR. Mathematically, IFRR is

$$\text{IFRR} = \sqrt{1 + Q^2\rho^2} \qquad (4\text{-}7a)$$

where

$$\rho = \frac{F(\text{image})}{F(\text{RF})} - \frac{F(\text{RF})}{F(\text{image})}$$

$$Q = \text{quality factor of the tuned circuit} \qquad (4\text{-}7b)$$
$$\text{IFRR (dB)} = 20 \log \text{IFRR}$$

If there is more than one tuned circuit in the front end of the receiver (perhaps a preselector filter and a separately tuned RF amplifier), the IFRR is simply the product of the two ratios.

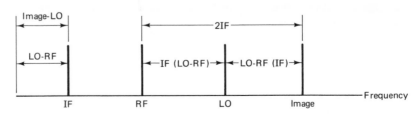

FIGURE 4-9   Image frequency.

EXAMPLE 4-6

For an AM broadcast-band superheterodyne receiver with an IF, RF, and local oscillator frequency of 455, 600, and 1055 kHz, respectively:
(a) Determine the image frequency.
(b) Calculate the IFRR for a preselector Q of 100.

Solution   (a) From Equation 4-6a,

$$F_{image} = F_{lo} + F_{if}$$

$$= 1055 \text{ kHz} + 455 \text{ kHz} = 1510 \text{ kHz}$$

or from Equation 4-6b,

$$F_{image} = F_{rf} + 2F_{if}$$

$$= 600 \text{ kHz} + 2(455 \text{ kHz}) = 1510 \text{ kHz}$$

(b) From Equations 4-7a and 4-7b,

$$\rho = \frac{1510 \text{ kHz}}{600 \text{ kHz}} - \frac{600 \text{ kHz}}{1510 \text{ kHz}}$$

$$= 2.51 - 0.397 = 2.113$$

$$\text{IFRR} = \sqrt{1 + (100^2)(2.113^2)}$$

$$= 211.3 \text{ or } 46.5 \text{ dB}$$

See Figure 4-10.

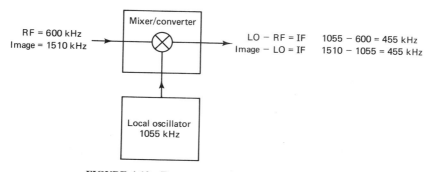

FIGURE 4-10   Frequency conversion for Example 4-6.

   Once an image frequency has been down-converted to IF, it cannot be removed. Therefore, to reject the image frequency, it has to be removed prior to the mixer/converter stage. Image frequency rejection is the primary purpose of the RF preselector. If the bandwidth of the preselector is sufficiently narrow, the image frequency is prevented from entering the receiver. Figure 4-11 illustrates how proper RF and IF filtering can prevent an image frequency from interfering with the selected RF carrier.

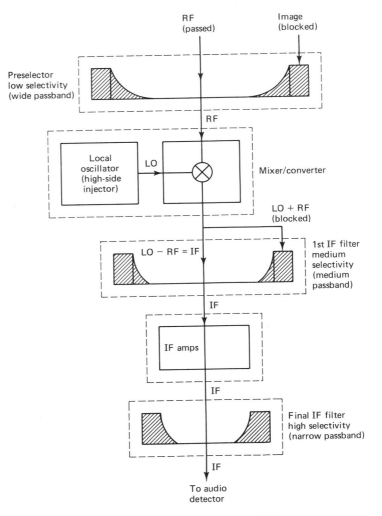

FIGURE 4-11   Image frequency rejection.

The ratio of the RF to the IF is also an important consideration for image frequency rejection. The closer the RF is to the IF, the closer the RF is to the image frequency.

EXAMPLE 4-7

For a citizens' band receiver using high-side injection with an RF of 27 MHz and an IF of 455 kHz determine:

(a) The local oscillator frequency.

(b) The image frequency.

(c) The IFRR for a preselector Q of 100.

(d) The preselector Q required to achieve the same IFRR as that achieved for an RF of 600 kHz.

Solution   (a) From Equation 4-5a,

$$F_{lo} = 27 \text{ MHz} + 455 \text{ kHz} - 27.455 \text{ MHz}$$

(b) From Equation 4-6a,

$$F_{image} = 27.455 \text{ MHz} + 455 \text{ kHz} = 27.91 \text{ MHz}$$

(c) From Equations 4-7a and 4-7b,

$$IFRR = 6.7 \text{ or } 16.5 \text{ dB}$$

(d) Rearranging Equation 4-7a yields

$$Q = \sqrt{\frac{IFRR^2 - 1}{\rho^2}} = \sqrt{\frac{211.3^2 - 1}{0.0663^2}} = 3187$$

From Examples 4-6 and 4-7 it can be seen that the higher the RF carrier, the more difficult it is to prevent the image frequency from entering the receiver. For the same IFRR, the higher RF carrier requires a much higher-quality filter in the preselector. This is illustrated in Figure 4-12.

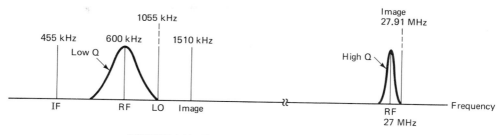

FIGURE 4-12   Frequency spectrum for Example 4-7.

AM RECEIVER CIRCUITS

RF Amplifier Circuits

An *RF amplifier* is a high-gain, low-noise, tuned amplifier that when used, is the first active stage encountered by a received signal. The primary purposes of an RF stage are selectivity, amplification, and sensitivity. Therefore, characteristics that are desirable in an RF amplifier are:

1. Low thermal noise
2. Low noise figure

3. Moderate to high gain
4. Low intermodulation and harmonic distortion
5. Moderate selectivity
6. High image frequency rejection ratio

Two of the most important parameters for a receiver are amplification and noise figure, of which both are dependent on the performance of the RF stage. An AM demodulator detects amplitude variations in its input signal and converts them to changes in its output signal. Consequently, amplitude variations caused by noise are converted to erroneous fluctuations in the detector output and the quality of the receive signal is degraded. The more gain that a signal experiences as it passes through a receiver, the more pronounced are its amplitude variations at the demodulator input, and the less noticeable are the variations caused by noise. The narrower the bandwidth, the less noise propagated through the receiver and, consequently, the less noise demodulated by the detector. From Equation 1-19 ($V_N = \sqrt{4RKTB}$), noise voltage is directly proportional to the square root of the temperature, the bandwidth, and the resistance. Therefore, if these three parameters are minimized, the thermal noise is reduced. The temperature of an RF stage can be reduced by artificially cooling the front end of a receiver. The bandwidth of an RF amplifier is reduced by using tuned amplifiers, and the resistance is reduced by using specially constructed solid-state components for the active device.

Noise figure is essentially a measure of the gain of the amplifier to the noise added by the amplifier. Therefore, the noise figure is improved (reduced) either by reducing the internal noise or by increasing the amplifier's gain.

Intermodulation and harmonic distortion are both forms of nonlinear distortion that reduce the noise figure by adding correlated noise to the total noise spectrum. The more linear an amplifier's operation, the less nonlinear distortion produced, and the better the receiver's noise figure. The IFRR ratio of an RF amplifier combines with the IFRR of the preselector to reduce the receiver input bandwidth sufficiently and prevent the image frequency from entering the mixer/converter stage. Consequently, moderate selectivity is all that is required from the RF stage.

Figure 4-13 shows several commonly used RF amplifier circuits. Keep in mind that RF is a relative term. RF simply means that the frequency is high enough to be efficiently radiated by an antenna and propagated through free space as an electromagnetic wave. RF for the AM broadcast band is between 535 and 1605 kHz, whereas RF for microwave radio is in excess of 1 GHz (1000 MHz). A common IF frequency used for FM broadcast band receivers is 10.7 MHz, which is considerably higher than the RF frequencies associated with the AM broadcast band. RF is simply the radiated or received frequency, and IF is an intermediate frequency within a transmitter or a receiver. Therefore, many of the considerations for RF amplifiers also apply to IF amplifiers such as neutralization, filtering, and coupling.

Figure 4-13a shows a schematic diagram for a bipolar RF amplifier. C_a, C_b, C_c, and L_1 form the coupling circuit from the antenna. Q_1 is class A biased to reduce nonlinear distortion. The collector circuit is transformer coupled to the mixer/converter through T_1, which is double tuned for more selectivity. C_x and C_y are RF bypass capaci-

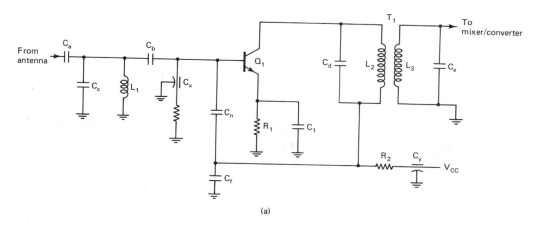

(a)

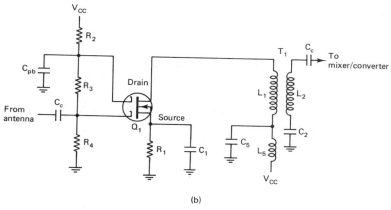

(b)

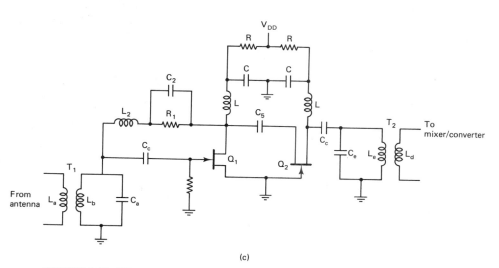

(c)

FIGURE 4-13 RF amplifier configurations: (a) bipolar transistor RF amplifier; (b) DEMOS-FET RF amplifier; (c) cascaded RF amplifier.

147

tors. Their symbols indicate that they are specially constructed *feedthrough* capacitors. Feedthrough capacitors offer less inductance, which prevents radiation from their leads. C_n is a neutralization capacitor. A portion of the collector signal is fed back to the base circuit to offset (or neutralize) the signal fed back through the collector-to-base lead capacitance of the transistor to prevent oscillations. C_f, in conjunction with C_n, form an ac voltage divider for the feedback signal. This neutralization configuration is called *off-ground* neutralization.

Figure 4-13b shows an RF amplifier using dual-gate field-effect transistors. This configuration uses DEMOS (depletion–enhancement metal-oxide semiconductor) FETs. The FETs feature high input impedance and low noise. A FET is a square-law device that generates only second-order harmonic and intermodulation distortion components, therefore producing less nonlinear distortion than a bipolar transistor. Q_1 is again biased class A for linear operation. T_1 is single tuned to the desired RF carrier to enhance the receiver's selectivity and improve the IFRR. L_5 is a RFC choke, and in conjunction with C_5, decouples the RF signals from the dc power supply.

Figure 4-13c shows the schematic diagram for a special RF amplifier configuration called a *cascoded* amplifier. A cascoded amplifier offers higher gain and less noise than conventional amplifiers. The active devices can be either bipolar transistors or FETs. Q_1 is a common-gate amplifier whose output is impedance-coupled to the source of Q_2. Because of the low input impedance of Q_2, Q_1 does not need to be neutralized; however, neutralization reduces the noise even further. Therefore, L_2, R_1, and C_2 provide the feedback path for neutralization. Q_2 is also a common-gate amplifier and because of its low input impedance requires no neutralization.

Mixer/Converter Circuits

As stated previously, the purpose of the mixer/converter stage is to down-convert the incoming RF to IF. This is done by mixing the RF with a local oscillator frequency in

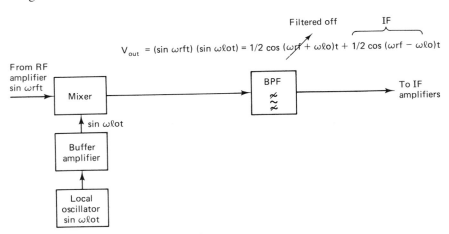

FIGURE 4-14   **Mixer/converter block diagram.**

a nonlinear device. In essence, this is heterodyning. A mixer is a nonlinear amplifier similar to a modulator except that the output is tuned to the difference between the RF and the local oscillator frequency. Figure 4-14 shows a block diagram for a mixer/converter stage. The output of a mixer is the product of the RF and the LO frequency and is expressed mathematically as

$$v_{out} = (\sin \omega_{rf}t)(\sin \omega_{lo}t)$$

where

ω_{rf} = RF input signal
ω_{lo} = LO input signal

Therefore, using the trigonometric identity for the product of two sines, the output of the mixer is

$$v_{out} = - \underbrace{\tfrac{1}{2}\cos(\omega_{rf} + \omega_{lo})t}_{\text{sum frequency}} + \underbrace{\tfrac{1}{2}\cos(\omega_{rf} - \omega_{lo})t}_{\substack{\text{difference}\\\text{frequency}}}$$

The difference frequency ($\omega_{rf} - \omega_{lo}$) is the IF.

Although any nonlinear device can be used for a mixer, a transistor or a FET is preferred over a simple diode because they are also capable of amplification. However, because the actual output signal from a mixer is a cross-product frequency, a mixer has a net loss. This loss is called *conversion* loss (or sometimes, conversion gain) because a frequency conversion has occurred and, at the same time, the output signal (IF) is lower in amplitude than the input signal (RF). The conversion loss is generally about 6 dB (which corresponds to a conversion gain of −6 dB). Essentially, the conversion gain is the difference between the IF output level with an RF input to the IF output level with an IF input.

Figure 4-15 shows the schematic diagrams for several common mixer/converter circuits. Figure 4-15a shows what is probably the simplest mixer circuit available (other than diode mixers). The mixer shown in Figure 4-15a is used exclusively for inexpensive AM broadcast band receivers. RF from the antenna is filtered by the preselector tuned circuit (L_1 and C_1), then transformer coupled to the base of Q_1. The active device for the mixer (Q_1) is also the gain device for the local oscillator. This configuration is commonly called a *self-excited* mixer because the mixer excites itself by feeding energy back to the local oscillator tank circuit (C_2 and L_2) to sustain oscillations. When power is initially applied, Q_1 amplifies both the incoming RF and any noise present and supplies the oscillator tank circuit with enough energy to begin oscillator action. The LO frequency is the resonant frequency of the tank circuit. A portion of the resonant tank circuit energy is coupled through L3 and L5 to the emitter of Q_1. This signal drives Q_1 into its nonlinear region and, consequently, produces sum and difference frequencies at the collector. The difference frequency is the IF. The output tank circuit (L_3 and C_3) is tuned to the IF. Therefore, the IF signal is transformer coupled to the first IF amplifier. The process is regenerative as long as there is an incoming RF signal. The tuning

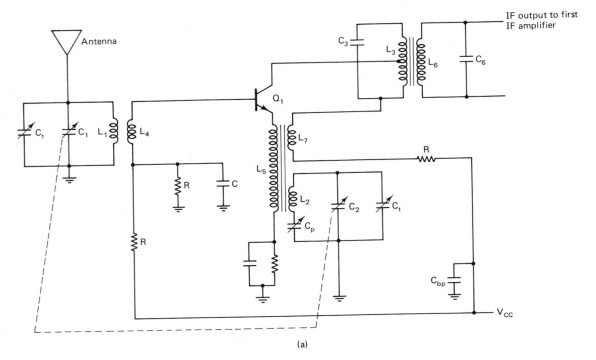

(a)

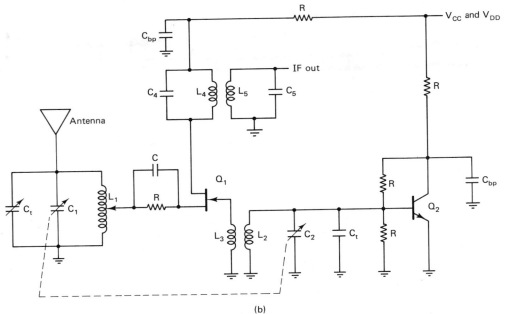

(b)

FIGURE 4-15

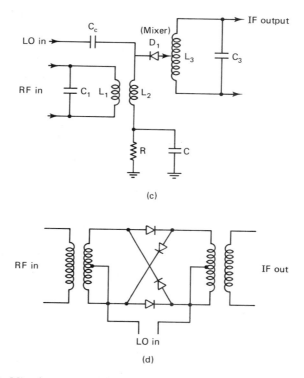

(c)

(d)

FIGURE 4-15 Mixer/converter circuits: (a) self-excited mixer; (b) separately excited mixer; (c) diode mixer; (d) balanced diode mixer.

capacitors in the RF and LO tank circuits are ganged together into a single tuning control. C_p and C_t are used for three-point tracking. This configuration has poor selectivity and poor image frequency rejection because there is no amplifier tuned to the RF signal and, consequently, the only RF selectivity is in the preselector. In addition there is essentially no RF gain and the transistor nonlinearities produce harmonic and intermodulation components that fall within the IF passband.

The mixer/converter circuit shown in Figure 4-15b is a separately excited mixer. Its operation is essentially the same as the self-excited mixer except that the local oscillator and the mixer each have their own gain device. The mixer itself is a FET which has nonlinear characteristics that are better suited for IF conversion than those of a bipolar transistor. This circuit is commonly used for high-frequency (HF) and very high-frequency (VHF) receivers.

The mixer/converter circuit shown in Figure 4-15c is a single-diode mixer. The concept is quite simple: the RF and LO signals are coupled into the diode, which is a nonlinear device. Therefore, nonlinear mixing occurs and the sum and difference frequencies are produced. The output tank circuit (C_3 and L_3) is tuned to the difference (IF) frequency. A single-diode mixer is inefficient because it has no gain. However, a diode mixer is commonly used for the audio detector in an AM receiver and to produce the audio subcarrier in a television receiver.

Figure 4-15d shows the schematic diagram for a *balanced diode mixer*. Balanced mixers are one of the most important circuits used in communications systems today. Balanced mixers are also called *balanced modulators*, *product modulators*, and *product detectors*. Balanced mixers are used extensively in both transmitters and receivers for AM, FM, and many of the digital modulation schemes, such as PSK and QAM. There are two inherent advantages that balanced mixers have over other types of mixers: noise reduction and carrier suppression. A detailed explanation of both descrete and integrated circuit balanced mixers is given in Chapter 5.

IF Amplifier Circuits

IF (intermediate-frequency) amplifiers are relatively high-gain tuned amplifiers that are very similar to RF amplifiers except IF amplifiers operate at a fixed frequency with a fixed passband. Consequently, it is easy to design and build IF amplifiers that are stable, do not radiate, and are easily neutralized. Because IF amplifiers operate at a fixed frequency, successive amplifiers can be inductively coupled with *double-tuned* circuits (with double-tuned circuits, both the primary and secondary sides of the transformer are tuned tank circuits). Therefore, it is easier to achieve an optimum (low) shape factor and good selectivity. Most of a receiver's gain and selectivity is achieved in the IF amplifier section. An IF stage generally has between two and five IF amplifiers. Figure 4-16 shows a schematic diagram for a three-stage IF section. T_1 and T_2 are double-tuned transformers; and L_1, L_2, and L_3 are tapped to reduce the effects of loading. The base of Q_3 is fed from the tapped capacitor pair, C_9 and C_{10}, for the same reason. C_1 and C_6 are neutralization capacitors.

Inductive coupling.   *Inductive* or *transformer coupling* is the most common technique used for coupling RF and IF amplifiers. With inductive coupling, voltage that is induced in the primary windings of a transformer is transferred to the secondary windings. The proportion of the primary voltage that is coupled across to the secondary depends on several factors, including the number of turns in the primary and secondary windings (i.e., the turns ratio), the amount of *magnetic flux* in the primary winding, the *coefficient of coupling*, and the speed at which the flux is changing. Mathematically, the voltage induced in the secondary is

$$E_s = \omega M I_p \qquad (4\text{-}8)$$

where

E_s = voltage induced in the secondary
ω = angular velocity ($2\pi F$)
M = mutual inductance
I_p = primary current

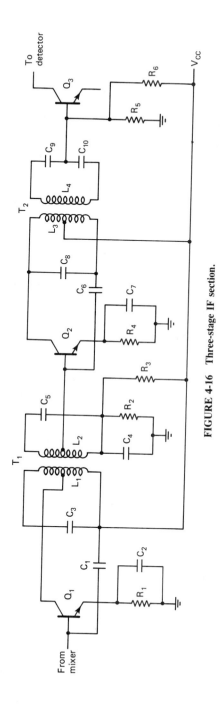

FIGURE 4-16 Three-stage IF section.

The ability of a coil to induce a voltage within its own windings is called *self-inductance* or simply *inductance* (*L*). When one coil induces a voltage in another coil, the two coils are said to be *coupled together*. The ability of one coil to induce a voltage in another coil is called *mutual inductance* (*M*). Mutual inductance in a transformer is caused by the magnetic lines of force (flux) that are produced in the primary windings cutting through the secondary windings and is directly proportional to the coefficient of coupling. Coefficient of coupling is the ratio of the secondary flux to the primary flux and is expressed mathematically as

$$k = \frac{\phi_s}{\phi_p} \qquad \qquad (4\text{-}9)$$

where

k = coefficient of coupling
ϕ_s = secondary flux
ϕ_p = primary flux

If all of the flux produced in the primary windings cuts through the secondary windings, the coefficient of coupling is 1. If none of the primary flux cuts through the secondary windings, the coefficient of coupling is 0. A coefficient of coupling of 1 is nearly impossible to achieve unless the two coils are wound around a common high-permeable iron core. Typically, the coefficient of coupling for standard RF and IF transformers is much less than 1. The transfer of flux from the primary windings to the secondary windings is called *flux linkage* and is directly proportional to the coefficient of coupling. The mutual inductance of a transformer is directly proportional to the coefficient of coupling and the square root of the product of the primary and secondary inductances. Mathematically, mutual inductance is

$$M = k\sqrt{L_s L_p} \qquad \qquad (4\text{-}10)$$

where

M = mutual inductance
L_s = inductance of the secondary winding
L_p = inductance of the primary winding
k = coefficient of coupling

Transformer-coupled amplifiers are divided into two general categories: single and double tuned.

Single-tuned transformers. Figure 4-17a shows a schematic diagram for a *single-tuned inductively* coupled amplifier. This configuration is called *untuned primary-tuned secondary*. The primary side of T_1 is simply the inductance of the primary windings, whereas a capacitor is in parallel with the secondary windings creating a tuned secondary. The transformer windings are not tapped because the loading effect of the FET is insignificant. Figure 4-17b shows the response curve for an untuned primary–tuned secondary transformer. E_s increases until the resonant frequency (F_o) of the secondary is reached,

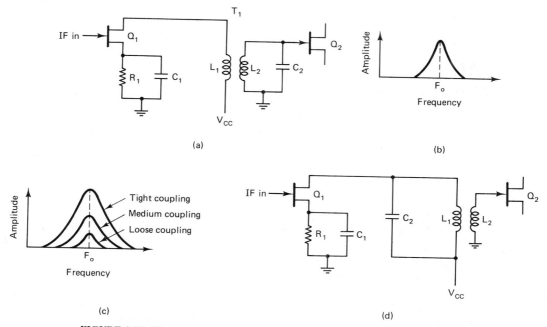

FIGURE 4-17   **Single-tuned transformer: (a) schematic diagram; (b) response curve; (c) effects of coupling; (d) tuned primary, untuned secondary.**

then E_s begins to decrease. The peaking of the response curve at F_o is caused by the reflected impedance. The impedance of the secondary is reflected back into the primary due to the mutual inductance between the two windings. For frequencies below resonance, the increase in ωM is greater than the decrease in I_p; therefore, E_s increases. For frequencies above resonance, the increase in ωM is less than the decrease in I_p; therefore, E_s decreases.

Figure 4-17c shows the effect of coupling on the response curve of an untuned primary–tuned secondary transformer. With *loose* coupling (low coefficient of coupling), the secondary voltage is relatively low and the bandwidth is narrow. As the degree of coupling is increased (coefficient of coupling increases), the secondary induced voltage increases and the bandwidth widens. Therefore, for a high degree of selectivity, loose coupling is desired, however, signal amplitude must be sacrificed. For high gain and a broad bandwidth, *tight* coupling is desired. Another single-tuned amplifier configuration is the *tuned primary–untuned secondary*, which is shown in Figure 4-17d.

Double-tuned transformers. Figure 4-18a shows a schematic diagram for a *double-tuned* inductively coupled amplifier. This configuration is called a *tuned primary–tuned secondary*; there is a capacitor in parallel with both the primary and secondary windings of T_1. Figure 4-18b shows the effect of coupling on the response curve of a double-tuned inductively coupled transformer. The response curve closely resembles that of a single-tuned circuit up to a coefficient of coupling called *critical coupling*. Critical

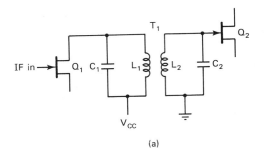

(a)

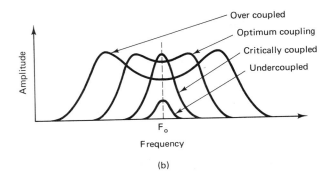

(b)

FIGURE 4-18 Double-tuned transformer: (a) schematic diagram; (b) response curve.

coupling is the point where the reflected resistance is equal to the primary resistance. At critical coupling, the primary's Q is halved, and consequently, the bandwidth is doubled. If the coefficient of coupling is increased beyond critical coupling, the response at the resonant frequency decreases and two new peaks occur on either side of the resonant frequency. This double peaking is caused from the reactive element of the reflected impedance being significant enough to change the resonant frequency of the primary tuned circuit. If the coefficient of coupling is increased further, the dip at resonance becomes more pronounced and the two peaks are spread even farther away from the resonant frequency. Increasing coupling beyond the critical value broadens the bandwidth but at the same time, produces a ripple in the response curve. An ideal response curve has a rectangular shape (i.e., a flat top with steep skirts). From Figure 4-18b it can be seen that a coefficient of coupling approximately 50% greater than the critical value yields a good compromise between flat response and steep *skirts*. This value of coupling is called *optimum coupling* and is expressed mathematically as

$$K_{\text{opt}} = 1.5k_c \qquad (4\text{-}11a)$$

where

K_{opt} = optimum coupling

k_c = critical coupling = $\dfrac{1}{\sqrt{Q_p Q_s}}$

The bandwidth of a single double-tuned amplifier is

$$BW_{dt} = kF_o \qquad (4\text{-}11b)$$

Bandwidth reduction. When several tuned amplifiers are cascaded, the total response is the product of all of the amplifier's individual responses. Figure 4-19a shows a response curve for a tuned amplifier. The gain at F_1 and F_2 is 0.707 of the

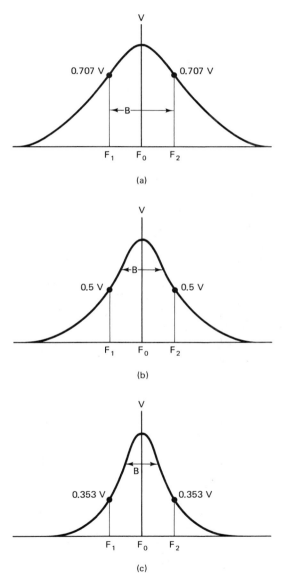

(a)

(b)

(c)

FIGURE 4-19 Bandwidth reduction: (a) single-tuned stage; (b) two cascaded stages; (c) three cascaded stages.

gain at F_o. If two identical tuned amplifiers are cascaded, the gain at F_1 and F_2 will be 0.5 of the gain at F_o ($0.707 \times 0.707 = 0.5$), and if three identical tuned amplifiers are cascaded, the gain at F_1 and F_2 is reduced to 0.353 of the gain at F_o. Consequently, as additional tuned amplifiers are added, the overall bandwidth of the stage is reduced. This bandwidth reduction is shown in Figure 4-19b and c. Mathematically, the overall bandwidth of n single-tuned stages is given as

$$BW_n = BW_1 \sqrt{2^{1/n} - 1} \qquad (4\text{-}12\text{a})$$

where

$\qquad BW_n$ = bandwidth of n stages
$\qquad BW_1$ = bandwidth of one single-tuned stage
$\qquad n$ = number of stages

The bandwidth for n double-tuned stages is

$$BW_{ndt} = BW_{1dt} \sqrt[4]{2^{1/n} - 1} \qquad (4\text{-}12\text{b})$$

where

$\qquad BW_{ndt}$ = overall bandwidth of n double-tuned amplifiers
$\qquad BW_{1dt}$ = bandwidth of a single double-tuned amplifier
$\qquad n$ = number of stages

EXAMPLE 4-8

Determine the overall bandwidth for (a) two single-tuned amplifiers each with a bandwidth of 10 kHz, (b) three single-tuned amplifiers each with a bandwidth of 10 kHz, and (c) four single-tuned amplifiers each with a bandwidth of 10 kHz.

(d) Determine the bandwidth for a double-tuned amplifier with optimum coupling, a k_c of 0.02 and a resonant frequency of 1 MHz.

(e) Repeat parts (a), (b), and (c) for the double-tuned amplifier of part (d).

Solution   (a) From Equation 4-12a,

$$BW_2 = 10 \text{ kHz} \sqrt{2^{1/2} - 1}$$

$$= 6436 \text{ Hz}$$

(b) Again, from Equation 4-12a,

$$BW_3 = 10 \text{ kHz} \sqrt{2^{1/3} - 1}$$

$$= 5098 \text{ Hz}$$

(c) Again, from Equation 4-12a,

$$BW_4 = 10 \text{ kHz} \sqrt{2^{1/4} - 1}$$

$$= 4350 \text{ Hz}$$

(d) From Equation 4-11a,

$$K_{opt} = 1.5(0.02) = 0.03$$

From Equation 4-12b,

$$BW_{dt} = 0.03(1 \text{ MHz}) = 30 \text{ kHz}$$

(e) From Equation 4-12b,

n	BW (Hz)
2	24,067
3	21,420
4	19,786

AM Detector Circuits

The function of an AM detector is to demodulate the IF AM envelope and recover or reproduce the original source information. The recovered signal should contain the same frequencies as the original modulating signal and have the same relative amplitude characteristics.

Peak detector. Figure 4-20a shows a schematic diagram for a simple noncoherent AM demodulator which is commonly called a *peak detector*. Because a diode is a nonlinear device, nonlinear mixing occurs in D_1 when two or more frequencies are applied to its input. Therefore, the output contains the original input frequencies, their harmonics, and their cross products (the sum and difference frequencies). If a 300-kHz carrier is amplitude modulated by a 2-kHz sine wave, the modulated wave is made up of a LSF, carrier, and USF of 298, 300, and 302 kHz, respectively. If the resultant envelope is the input to the AM detector shown in Figure 4-20a, the output will comprise the three input frequencies, the harmonics of all three frequencies, and the cross products of all possible combinations of the three frequencies and their harmonics. Mathematically, the output is

$$V_{out} = \text{input frequencies} + \text{harmonics} + \text{sums and differences}$$

Because the *RC* network is a lowpass filter, only the difference frequencies are passed on to the audio section. Therefore, the output is simply

$$V_{out} = 300 - 298 = 2 \text{ kHz}$$
$$302 - 300 = 2 \text{ kHz}$$
$$302 - 298 = 4 \text{ kHz}$$

Therefore,

$$V_{out} = 2 \text{ kHz and } 4 \text{ kHz}$$

Because of the relative amplitude characteristics of the LSF, carrier, and USF, the difference between the carrier and either the upper or lower side frequency is the

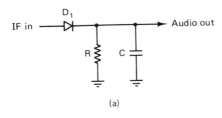

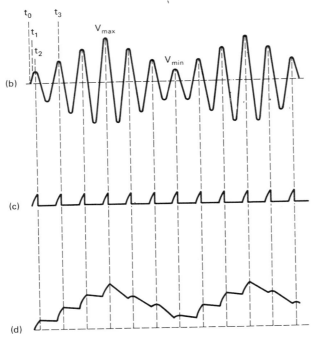

FIGURE 4-20 Peak detector: (a) schematic diagram; (b) AM input waveform; (c) diode current waveform; (d) output voltage waveform.

predominant output signal. Consequently, the original modulating signal (a 2-kHz sine wave) is recovered by the peak detector.

In the preceding analysis, the diode detector was analyzed as a simple mixer, which it is. Essentially, the difference between an AM modulator and an AM demodulator is that the output of a modulator is tuned to the sum frequency (up-converter), whereas the output of a demodulator is tuned to the difference frequency (down-converter). The demodulator shown in Figure 4-20a is commonly called a diode detector because the nonlinear device is a single diode, or a peak detector because it detects the peaks of the input envelope, or a *shape* or *envelope detector* because it detects the shape of the input envelope. Essentially, the carrier "*captures*" the diode and forces it to turn on and off (rectify) synchronously (both frequency and phase). Thus the sidebands can mix with the carrier and the original baseband signals are recovered.

Figure 4-20b, c, and d show a detector input voltage waveform, the corresponding diode current waveform, and the detector output voltage waveform. At time $t = 0$, the diode is reverse biased and off ($i_d = 0$ A), the capacitor is completely discharged ($V_C = 0$ V), and thus the output is 0 V. The diode remains off until the input voltage exceeds the barrier potential of D_1 (approximately 0.3 V). When V_{in} reaches 0.3 V ($t = 1$), the diode turns on and diode current begins to flow charging the capacitor. The capacitor voltage remains 0.3 V below the input voltage until V_{in} reaches its peak. When the input voltage begins to decrease, the diode turns off and i_d goes to 0 A ($t = 2$). The capacitor begins to discharge through the resistor but the RC time constant is made sufficiently long so that the capacitor cannot discharge as rapidly as V_{in} is decreasing. The diode remains off until the next input cycle when V_{in} goes 0.3 V more positive than V_C ($t = 3$). At this time the diode turns on, current flows, and the capacitor begins to charge again. It is relatively easy for the capacitor to charge to the new value because the R_C charging time constant is $R_d C$, where R_d is the "on" resistance of the diode, which is quite small. This sequence repeats itself on each successive positive peak of V_{in} and the capacitor voltage follows the positive peaks of V_{in} (hence the name "peak detector"). The output waveform resembles the shape of the input envelope (hence the name "shape detector"). The output waveform has a high-frequency ripple that is equal to the carrier frequency. This is due to the diode turning on during the positive peaks of the envelope. The ripple is easily removed by the audio amplifiers because the carrier frequency is much higher than the highest audio frequency. The circuit shown in Figure 4-20 responds only to the positive peaks of V_{in} and is, therefore, called a positive peak detector. By simply turning the diode around, the circuit becomes a negative peak detector. The output voltage reaches its peak positive amplitude at the same time that the input envelope reaches its maximum positive value (V_{max}), and the output voltage goes to its minimum peak amplitude at the same time that the input voltage goes to its minimum value (V_{min}). For 100% modulation, V_{out} swings from 0 V to $V_{max} - 0.3$ V.

Figure 4-21 shows the input and output waveforms for a peak detector with various

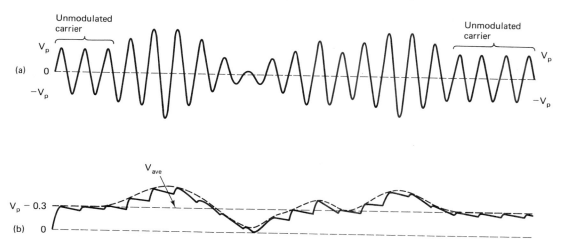

FIGURE 4-21   Positive peak detector: (a) input waveform; (b) output waveform.

percentages of modulation. With no modulation, a peak detector is simply a half-wave rectifier and the output voltage is approximately equal to the peak input voltage. As the percent modulation changes, the variations in the output voltage increase and decrease proportionally; the output waveform follows the shape of the AM envelope. However, regardless of whether there is modulation present or not, the average value of the output voltage is approximately equal to the peak value of the unmodulated carrier. Therefore, the output voltage variations reflect only the changes in the envelope (i.e., the detector is removing the audio information from the envelope). With no modulating signal, the output of the detector is a constant dc voltage approximately equal to the peak amplitude of the unmodulated carrier.

Detector distortion.   When the positive peaks of the input waveform are increasing, it is important that the capacitor hold its charge between successive peaks (i.e., a relatively long *RC* time constant is necessary). However, when the positive peaks are decreasing in amplitude, it is important that the capacitor discharge between peaks to a value less than the next peak (i.e., a short *RC* time constant is necessary). Obviously, a trade-off between a long and a short time constant is in order. If the *RC* time constant

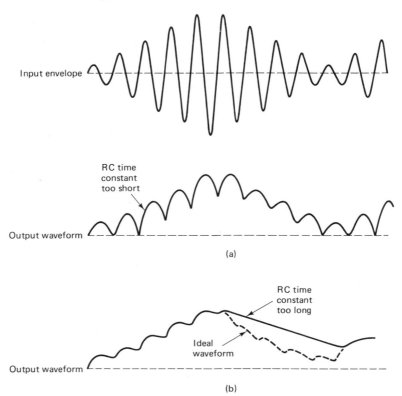

FIGURE 4-22 **Detector distortion: (a) rectifier distortion; (b) diagonal clipping.**

is too short, the output waveform resembles a half-wave rectified signal. This is sometimes called *rectifier distortion* and is shown in Figure 4-22a. If the *RC* time constant is too long, the slope of the output waveform cannot follow the trailing slope of the envelope. This type of distortion is called *diagonal clipping* and is shown in Figure 4-22b.

The *RC* network following the diode in a peak detector is a low-pass filter. The slope of the envelope is dependent on both the modulating signal frequency and the modulation coefficient (*m*). Therefore, the maximum slope (fastest rate of change) occurs when the envelope is crossing its zero axis in the negative direction (point *X* in Figure 4-22). The highest modulating signal frequency that can be demodulated without attenuation is given as

$$F_{a(\max)} = \frac{\sqrt{(1/m^2) - 1}}{2\pi RC}$$ (4-13a)

where

$F_{a(\max)}$ = maximum modulating frequency
m = modulation coefficient
RC = RC time constant

For 100% modulation, the numerator in Equation 4-13a goes to 0, which essentially means that all modulating signal frequencies are attenuated as they are demodulated. Typically, the modulating signal amplitude in a transmitter is limited or compressed such that approximately 90% modulation is the maximum that can be achieved. For 70.7% modulation, Equation 4-13a reduces to

$$F_{a(\max)} = \frac{1}{2\pi RC}$$ (4-13b)

Equation 4-13b is commonly used when designing peak detectors to determine an approximate maximum modulating signal frequency.

AUTOMATIC GAIN CONTROL AND SQUELCH

Automatic Gain Control Circuits

An *automatic gain control circuit* (AGC) compensates for minor variations in the received signal level. The AGC circuit automatically increases the receiver gain for weak RF input signals, and automatically decreases the receiver gain for strong RF signals. Weak signals can be buried in the receiver noise and, consequently, masked from the audio detector. Excessively strong signals can overdrive the RF and/or IF amplifiers and produce excessive nonlinear distortion. There are several types of AGC, including direct or simple AGC, delayed AGC, and forward AGC.

Simple AGC. Figure 4-23 shows a block diagram for an AM superheterodyne receiver with *simple AGC*. The automatic gain control circuit monitors the received

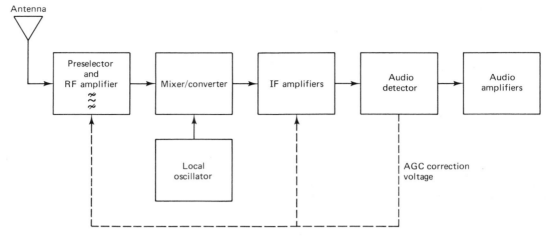

FIGURE 4-23   AM receiver with simple AGC.

signal level and sends a signal back to the RF and/or IF amplifiers to adjust their gain automatically. AGC is a form of *degenerative* or negative feedback. The purpose of AGC is to allow a receiver to detect and demodulate, equally well, signals that are transmitted from different stations whose output power and distance from the receiver varies. For example, an AM radio in a vehicle does not receive the same signal level from all of the transmitting stations or, for that matter, from a single station when the automobile is moving. The AGC circuit sends a voltage back to the RF and/or IF amplifiers to adjust the receiver gain and keep the IF carrier power at the input to the AM detector at a constant level. The AGC circuit is not a form of *automatic volume*

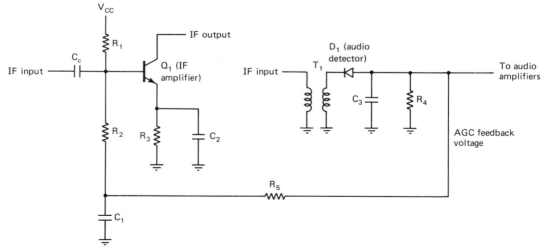

FIGURE 4-24   Simple AGC circuit.

control circuit; AGC is independent of modulation and totally unaffected by normal changes in the audio modulating signal amplitude.

Figure 4-24 shows a schematic diagram for a simple AGC circuit. As you can see, an AGC circuit is essentially a peak detector. In fact, very often the AGC correction voltage is taken from the output of the audio detector. In Figure 4-21 it was shown that the average dc voltage at the output of a peak detector is approximately equal to the peak unmodulated carrier amplitude and is totally independent of modulation. If the carrier amplitude increases, the AGC voltage increases and if the carrier amplitude decreases, the AGC voltage decreases. The circuit shown in Figure 4-24 is a negative peak detector, and therefore the output is a negative voltage. The higher the amplitude of the input carrier, the more negative the output voltage. The negative voltage from the AGC detector is fed back to the IF stage, where it controls the bias voltage on the base of Q_1. When the carrier amplitude increases, the voltage on the base of Q_1 goes more negative, causing the emitter current to decrease. As a result, r'_e increases and the amplifier gain (r_c/r'_e) decreases, causing the carrier amplitude to decrease. If the carrier amplitude decreases, the AGC voltage goes less negative, the emitter current increases, r'_e decreases, and the amplifier gain increases. Capacitor C_1 is an audio bypass capacitor that prevents changes in the AGC voltage due to modulation from affecting the bias or gain of Q_1.

Delayed AGC. Simple AGC is used in most inexpensive broadcast-band receivers. However, with simple AGC, the AGC bias begins to increase as soon as the received signal level exceeds the thermal noise of the receiver. Consequently, the receiver becomes less sensitive (this is sometimes called *automatic desensing*). *Delayed AGC* prevents the AGC voltage from reaching the RF and/or IF amplifiers until the RF level exceeds a predetermined level. Once the carrier signal has exceeded the threshold level, the delayed AGC voltage is proportional to the signal strength. Figure 4-25a shows the response characteristics for both simple and delayed AGC. It can be seen that with delayed AGC, the RF signal is unaffected until the AGC threshold level is exceeded, whereas with simple AGC, the RF signal is immediately affected. Delayed AGC is used with more sophisticated communications receivers. Figure 4-25b shows IF gain versus RF input signal level for both simple and delayed AGC.

Forward AGC. An inherent problem with both simple and delayed AGC is the fact that they are both forms of *post-AGC* (after-the-fact compensation). With post-AGC, the circuit that monitors the carrier level and provides the AGC correction voltage is located after the IF amplifiers, and therefore the simple fact that the AGC voltage changed indicates that it may be too late (the carrier level has already changed). Therefore, simple and delayed AGC cannot compensate for rapid changes in the carrier amplitude. *Forward AGC* is similar to conventional AGC except that the carrier is monitored closer to the front end of the receiver and the correction voltage is fed to IF and/or RF amplifiers further back in the receiver. Consequently, when a signal change is detected, the change can be compensated for in succeeding stages. Figure 4-26 shows an AM superheterodyne

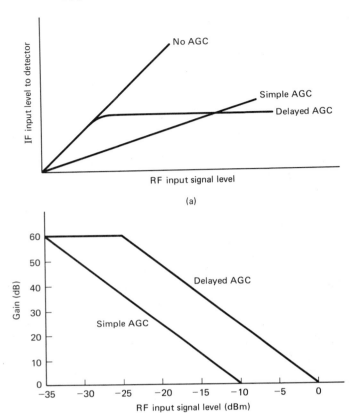

(a)

(b)

FIGURE 4-25 Automatic gain control (AGC): (a) response characteristics; (b) IF gain versus RF input signal level.

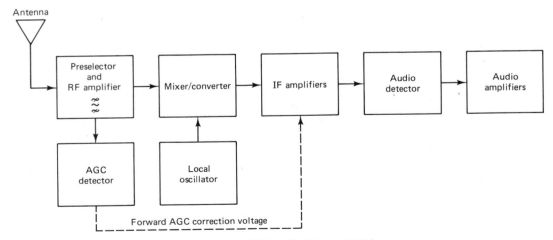

FIGURE 4-26 Forward AGC.

receiver with forward AGC. For a more sophisticated method of accomplishing AGC, see Chapter 8 under the heading ''Two-Way FM Receivers.''

Squelch Circuits

The purpose of a *squelch circuit* is to ''quiet'' a receiver in the absence of a received RF signal. If an AM receiver is off-tuned to a location in the RF spectrum where there is no RF signal, the AGC circuit adjusts the receiver for maximum gain. Consequently, the receiver amplifies and demodulates its own internal noise. This is the familiar crackling and sputtering heard on the speaker. In domestic AM systems, each station is continuously transmitting a carrier regardless of whether there is any modulation or not. Therefore, the only time the idle receiver noise is heard is when tuning between stations. However, in two-way radio communications systems, the carrier in the transmitter is generally turned off unless a modulating signal is present. Therefore, during idle times, a receiver is simply amplifying and demodulating noise. A squelch circuit keeps the audio section turned off or muted in the absence of a received signal; the receiver is squelched. A disadvantage of a squelch circuit is that the receiver is *desensitized* and will not pick up weak RF signals.

Figure 4-27 shows a schematic diagram for a squelch circuit. This squelch circuit

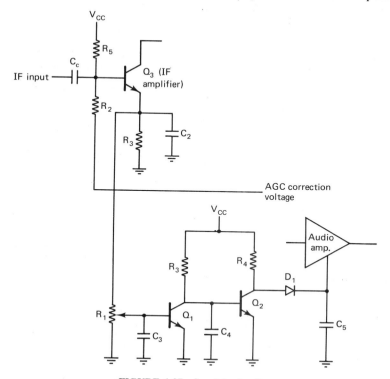

FIGURE 4-27 Squelch circuit.

uses the AGC voltage to determine how much RF signal is being received. The more AGC, the stronger the RF signal. When the AGC voltage drops below a preset level, the squelch circuit disables the audio section and mutes the receiver. In Figure 4-27 it can be seen that the squelch detector simply uses a resistive voltage divider to monitor the AGC voltage. When the RF signal drops below the squelch level, Q_1 turns on, Q_2 turns off, and D_1 turns on and shuts off the audio amplifier. When the RF level increases above squelch level, the AGC voltage goes more negative, turning off Q_1 and turning on Q_2. D_1 turns off, which, in turn, enables the audio amplifiers. The squelch threshold voltage is adjusted with R_1.

For a more sophisticated method of squelching a receiver, see Chapter 8 under the heading "Two-Way FM Receivers."

DOUBLE-CONVERSION AM RECEIVERS

In a previous section it was pointed out that for good image frequency rejection, a relatively high IF is selected. However, for high-gain, selective IF amplifiers that are easily neutralized, a low IF is more desirable. The solution is to use two intermediate frequencies. The first IF is relatively high for good image frequency rejection, and the second IF is relatively low for easy amplification. Figure 4-28 shows a block diagram for a *double-conversion* AM receiver. The first IF is 10.625 MHz, which pushes the image frequency 21.25 MHz away from the desired RF. The first IF is immediately down-converted to 455 kHz and fed to a series of high-gain IF amplifiers. Figure 4-29 illustrates the filtering requirements for a double-conversion AM receiver.

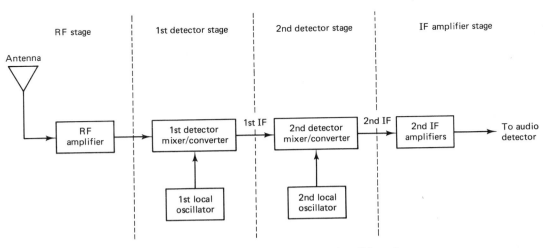

FIGURE 4-28   Double conversion AM receiver.

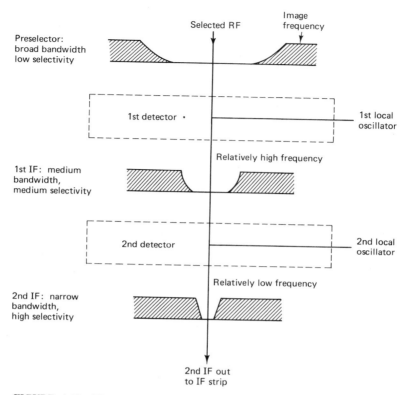

FIGURE 4-29   Filtering requirements for the double-conversion AM receiver shown in Figure 4–28.

QUESTIONS

4-1. What is meant by the *front end* of a receiver?

4-2. What are the primary functions of the front end?

4-3. Define *selectivity*; *shape factor*. What is the relationship between receiver noise and selectivity?

4-4. Describe *bandwidth improvement*. What is the relationship between bandwidth improvement and receiver noise?

4-5. Define *sensitivity*.

4-6. What is the relationship among receiver noise, bandwidth, and temperature?

4-7. Define *fidelity*.

4-8. List and describe the three types of distortion that reduce the fidelity of a receiver.

4-9. Define *insertion loss*.

4-10. Define *noise temperature*; *equivalent noise temperature*.

4-11. Describe the difference between a coherent and a noncoherent receiver.

4-12. Draw the block diagram for a TRF receiver and briefly describe its operation.

4-13. What are the three predominant disadvantages of a TRF receiver?

4-14. Draw the block diagram for an AM superheterodyne receiver and describe its operation and the primary functions of each stage.

4-15. Define *heterodyning*.

4-16. What are meant by the terms *high-* and *low-side injection*?

4-17. Define *local oscillator tracking*; *tracking error*.

4-18. Describe three-point tracking.

4-19. What is meant by *gang tuning*?

4-20. Define *image frequency*.

4-21. Describe image frequency rejection ratio.

4-22. List six characteristics that are desirable in an RF amplifier.

4-23. What advantage do FET RF amplifiers have over BJT amplifiers?

4-24. Define *neutralization*. Describe the neutralization process.

4-25. What is a cascoded amplifier?

4-26. Define *conversion gain*. What is another name for conversion gain?

4-27. What is the advantage of having a high IF; a low IF?

4-28. Define the following terms: *inductive coupling*; *self-inductance*; *mutual inductance*; *coefficient of coupling*; *critical coupling*; *optimum coupling*.

4-29. Describe loose coupling; tight coupling.

4-30. Describe the operation of a peak detector.

4-31. Describe rectifier distortion and what causes it. Describe diagonal clipping.

4-32. Describe automatic gain control.

4-33. Describe the following: simple AGC; delayed AGC; forward AGC.

4-34. Describe the purpose of a squelch circuit.

4-35. Explain the operation of a double-conversion superheterodyne receiver.

PROBLEMS

4-1. Determine the shape factor and percent selectivity for the gain-versus-frequency plot shown below.

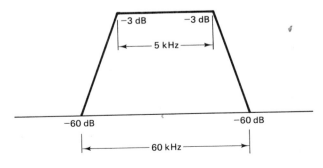

4-2. Determine the improvement in the noise figure for a receiver with an RF bandwidth equal to 40 kHz and an IF bandwidth equal to 16 kHz.

4-3. Determine the equivalent noise temperature for an amplifier with a noise figure F = 6 dB and an environmental temperature $T = 27°C$.

4-4. For an AM commercial broadcast band receiver with an input filter Q-factor of 85, determine the bandwidth at the low and high ends of the RF spectrum.

4-5. For an AM superheterodyne receiver using high-side injection with a local oscillator frequency of 1200 kHz, determine the IF carrier and upper and lower side frequencies for an RF envelope which is made up of a carrier and upper and lower side frequencies of 600, 604, and 596 kHz, respectively.

4-6. For a receiver with a ± 2.5-kHz tracking error and a maximum modulating signal frequency $F_a = 6$ kHz, determine the minimum IF bandwidth.

4-7. For a receiver with IF, RF, and local oscillator frequencies of 455, 900, and 1355 kHz, respectively, determine:
   (a) The image frequency.
   (b) The IFRR for a preselector $Q = 80$.

4-8. For a citizens' band receiver using high-side injection with an RF carrier of 27.04 MHz and a 10.645-MHz first IF, determine:
   (a) The local oscillator frequency.
   (b) The image frequency.

4-9. For a three-stage double-tuned RF amplifier with an RF equal to 800 kHz and a coefficient of coupling $k_{opt} = 0.025$, determine:
   (a) The bandwidth for each individual stage.
   (b) The overall bandwidth of the three stages.

4-10. Determine the maximum modulating signal frequency for a peak detector with the following parameters: $C = 1000$ pF, $R = 10$ kΩ, and $m = 0.5$. Repeat the problem for $m = 0.707$.

Chapter 5

PHASE-LOCKED LOOPS AND FREQUENCY SYNTHESIZERS

INTRODUCTION

With sophisticated electronic communications systems, such as single-sideband and frequency modulation systems, there are several circuits that are universally used. They include the *phase-locked loop* and the *frequency synthesizer*. Therefore, it is necessary that these circuits be discussed and understood before examining the more complicated communications systems.

PHASE-LOCKED LOOP

The phase-locked loop (PLL) is used extensively in electronic communications for modulation, demodulation, and frequency generation. PLLs are used in both transmitters and receivers with both analog and digital modulation and with the transmission of digital pulses. Phase-locked loops were first used in 1932 in receivers for synchronous detection of radio signals. However, for many years, PLLs were avoided because of their complexity and expense. With the advent of *large-scale integration* (LSI), phase-locked loops take up little space, are easy to use, and are more reliable. Therefore, they have become a universal building block with numerous applications.

Essentially, a PLL is a *closed-loop* feedback control system where the feedback signal is a frequency rather than simply a voltage. The basic phase-locked loop consists of a *phase comparator* (*frequency multiplier*), a *voltage-controlled oscillator*, and a *low-gain amplifier* (op-amp). The block diagram for a PLL is shown in Figure 5-1.

172

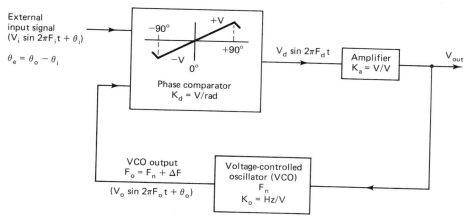

External
input signal
$(V_i \sin 2\pi F_i t + \theta_i)$

$\theta_e = \theta_o - \theta_i$

$V_d \sin 2\pi F_d t$

Amplifier
$K_a = V/V$

V_{out}

Phase comparator
$K_d = V/rad$

VCO output
$F_o = F_n + \Delta F$

$(V_o \sin 2\pi F_o t + \theta_o)$

Voltage-controlled
oscillator (VCO)
F_n
$K_o = Hz/V$

FIGURE 5-1 **Block diagram: phase-locked loop.**

Voltage-Controlled Oscillator

A *voltage-controlled oscillator* (VCO) is an oscillator (more specifically, a *free-running multivibrator*) with a stable frequency of oscillation that is dependent on an external bias voltage. The output from a VCO is a frequency and its input is a dc bias or *control* voltage. When a dc or a slowly changing ac voltage is applied to the VCO, its output frequency changes or deviates accordingly. Figure 5-2 shows the transfer curve (output frequency-versus-input bias voltage characteristics) for a typical VCO. The output frequency (F_o) with 0 V input bias is the VCO's *natural* frequency (F_n), and a change in the output frequency caused by a change in the input voltage is called frequency deviation (ΔF). Consequently, $F_o = F_n + \Delta F$. For a symmetrical ΔF, the natural fre-

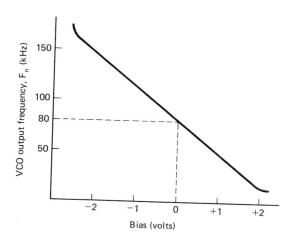

FIGURE 5-2 **VCO input versus output characteristics.**

quency of the VCO should be centered within the linear portion of the input/output curve. The transfer function for a VCO is

$$K_o = \frac{\Delta F}{\Delta V} \qquad (5\text{-}1)$$

where

K_o = input/output transfer function (Hz/V)
ΔV = input control voltage (V)
ΔF = frequency deviation (Hz)

Phase Comparator

A *phase comparator* is a nonlinear mixer with two inputs: an externally generated frequency (F_i) and the VCO output signal (F_o). The phase comparator output is the product of F_o and F_i and therefore contains their sum and difference frequencies. Figure 5-3a shows the schematic diagram for a simple phase comparator. V_o is applied simultaneously to the two halves of input transformer T_1. D_1, R_1, and C_1 make up a half-wave rectifier, as do D_2, R_2, and C_2 (note that $C_1 = C_2$ and $R_1 = R_2$). During the positive alternation of V_o, D_1 and D_2 are forward biased and "on," charging C_1 and C_2 to equal values but with opposite polarities. Therefore, the average output voltage is $V_d = V_{C1} + (-V_{C2}) = 0$ V. This is shown in Figure 5-3b. During the negative half-cycle of V_o, D_1 and D_2 are reverse biased and "off." Therefore, C_1 and C_2 discharge equally through R_1 and R_2, respectively, keeping the output voltage at 0 V. This is shown in Figure 5-3c. The two half-wave rectifiers produce equal-magnitude output voltages with opposite polarities. Therefore, the output voltage due to V_o is constant and equal to 0 V. The corresponding input and output waveforms for a square-wave VCO signal are shown in Figure 5-3d.

Circuit operation.   When an external signal ($V_i \sin 2\pi F_i t$) is applied to the phase comparator, its voltage adds to V_o, causing C_1 and C_2 to charge and discharge, producing a proportional change in the output voltage. Figure 5-4a shows the unfiltered output waveform when $F_o = F_i$ and V_o leads V_i by 90°. For the phase comparator to operate properly, V_o is made much larger than V_i. Therefore, D_1 and D_2 are switched "on" only during the positive alternation of V_o and "off" during the negative alternation. During the first half of the "on" time, the voltage applied to $D_1 = V_o - V_i$ and $D_2 = V_o + V_i$. Therefore, C_1 is discharging and C_2 is charging. During the second half of the "on" time, the voltage applied to $D_1 = V_o + V_i$ and $D_2 = V_o - V_i$ and C_1 is charging while C_2 is discharging. During the "off" time, C_1 and C_2 are neither charging nor discharging. For each cycle of V_o, C_1 and C_2 charge and discharge equally and the average output voltage V_d remains at 0 V. Thus the average value of V_d is unaffected by V_i.

Figure 5-4b shows the unfiltered output voltage waveform when V_o leads V_i by 45°. V_i is positive for 75% of the "on" time and negative for the remaining 25%. As

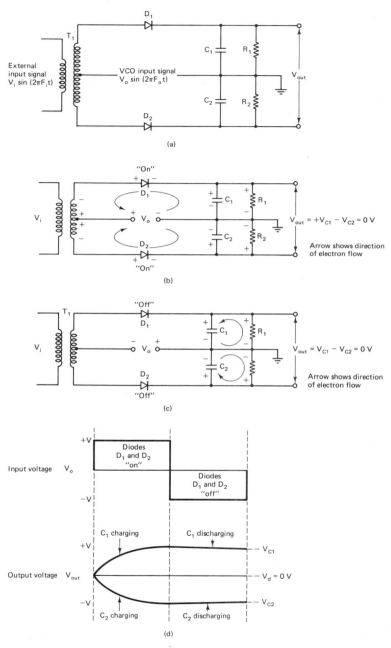

FIGURE 5-3 Phase comparator: (a) schematic diagram; (b) output voltage due to positive half-cycle of V_o; (c) output voltage due to negative half-cycle of V_o; (d) input and output voltage waveforms.

a result, the average output voltage for one cycle of V_o is positive and approximately equal to 0.3 V. Figure 5-4c shows the unfiltered output waveform when V_o and V_i are in phase. During the entire "on" time, V_i is positive. Consequently, the output voltage is positive and approximately equal to 0.636 V. Figures 5-4d and e show the unfiltered output waveform when V_o leads V_i by 135° and 180°, respectively. It can be seen that the output voltage goes negative when V_o leads V_i by more than 90° and reaches its maximum value when V_o leads V_i by 180°. In essence, a phase comparator rectifies the difference voltage between V_o and V_i and integrates it to produce an output voltage that is proportional to the difference between their phases. Simply stated, the magnitude

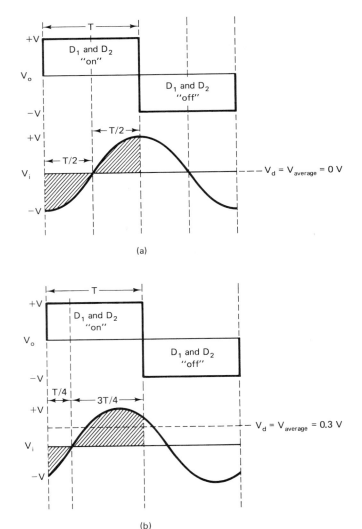

(a)

(b)

FIGURE 5-4 Phase comparator output voltage waveforms: (a) V_o leads V_i by 90°; (b) V_o leads V_i by 45°; (c) V_o and V_i in phase; (d) V_o leads V_i by 135°; (e) V_o leads V_i by 180°.

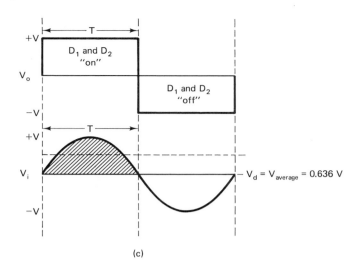

(c)

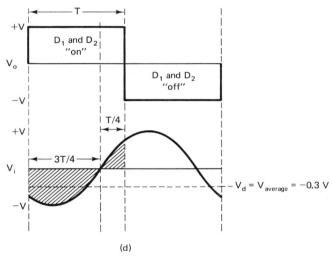

(d)

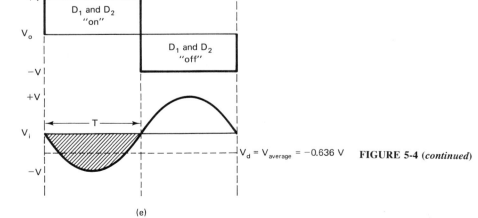

FIGURE 5-4 (continued)

(e)

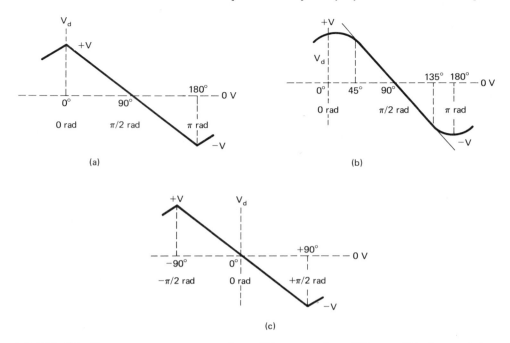

FIGURE 5-5 Phase comparator output voltage (V_d) versus phase difference (θ_e) characteristics: (a) square-wave inputs; (b) sinusoidal inputs; (c) square-wave inputs, phase bias reference.

and polarity of the phase comparator output voltage is proportional to the difference in phase between V_o and V_i.

Figure 5-5 shows the output voltage (V_d)-versus-input phase difference (θ_e) characteristics for the phase comparator shown in Figure 5-3. Figure 5-5a shows the curve for a square-wave phase comparator. The curve has a triangular shape with a negative slope from 0 to 180°. V_d is maximum positive when V_o and V_i are in phase, 0 V when V_o leads V_i by 90°, and maximum negative when V_o leads V_i by 180°. If V_o advances more than 180°, the output voltage becomes less negative and, if V_o lags behind V_i, the output voltage becomes less positive. Therefore, the maximum phase difference that the phase comparator can track is 90° ± 90° or from 0 to 180°. The input to the phase comparator is the difference between the phase of F_o (θ_o) and the phase of the external input frequency (θ_i). The phase difference is called the *phase error* (θ_e), where $\theta_e = \theta_o - \theta_i$. The output of the phase comparator is a dc voltage which is linear between 0 and 180° (0 to πr). Therefore, the gain for a square-wave comparator for $\theta_e = 0$ to πr is given as

$$K_d = \frac{V_d}{\theta_e} = 2 \text{ V}/\pi \text{ (volts/radian)} \qquad (5\text{-}2)$$

where

K_d = gain (V/rad)
V_d = dc output voltage (V)
θ_e = phase error, $\theta_o - \theta_i$ (rad)
$\quad$ = π radians
V = peak input voltage

Figure 5-5b shows the curve for an analog phase comparator with sinusoidal characteristics. The phase error-versus-output voltage is linear only from 45° to 135°. Therefore, the gain is given as

$$K_d = \frac{V_d}{\theta_e} = V \text{ (volts/radian)} \qquad (5\text{-}3)$$

From Figures 5-5a and b, it can be seen that the phase comparator output voltage $V_d = 0$ V when $F_o = F_i$ and V_o and V_i are 90° out of phase. Therefore, if F_i is initially equal to F_n, a 90° phase difference is required to hold $V_d = 0$ V. This is understandable because the VCO does not require any correction bias. The 90° phase difference is always there and is equivalent to a bias or offset phase. Generally, the phase bias is considered the reference phase, which can be deviated $\pm\pi/2$ rad. Therefore, V_d goes from its maximum positive value at $-90°$ ($-\pi/2$ rad) and to its maximum negative value at $+90°$ ($+\pi/2$ rad). Figure 5-5c shows the V_d-versus-0_e characteristics for square-wave inputs with the phase bias as the reference.

Figure 5-6a shows the unfiltered output waveform when V_i leads V_o by 90°. Note that the average value of $V_d = 0$ V (the same as when V_o lead V_1 by 90°). When frequency *lock* occurs, it is uncertain whether V_o will lock onto V_i with a + or − 90° phase bias. Therefore, there is a 180° phase ambiguity in the phase of V_o. Figure 5-6b shows the output voltage-versus-phase difference characteristics for square-wave inputs when $F_o = F_n$ and V_o has locked onto V_i, lagging by 90°. Note that the opposite conditions to those shown in Figure 5-5 prevail; the maximum positive and negative voltages occur for the opposite direction phase error and the slope is positive rather than negative from $-\pi/2$ to $+\pi/2$ rad. When frequency lock occurs, the PLL produces a coherent frequency ($F_o = F_i$), but the phase of the recovered signal is uncertain (either F_o leads F_i by 90° $\pm\theta_e$, or vice versa).

Loop Operation

For the following explanations, refer to Figure 5-7a.

Loop acquisition. An external input signal ($V_i \sin 2\pi F_i t$) enters the phase comparator and mixes with the VCO output signal (a square wave with peak amplitude, V_o, and a fundamental frequency, F_o). Initially, $F_o \neq F_i$ and the loop is *unlocked*. Because the phase comparator is a nonlinear device, F_i and F_o mix and generate cross-product frequencies ($F_o + F_i$ and $F_o - F_i$). Therefore, the primary output frequencies from the phase comparator are F_i, F_o, $F_i + F_o$, and $F_i - F_o$. The LPF blocks F_o, F_i,

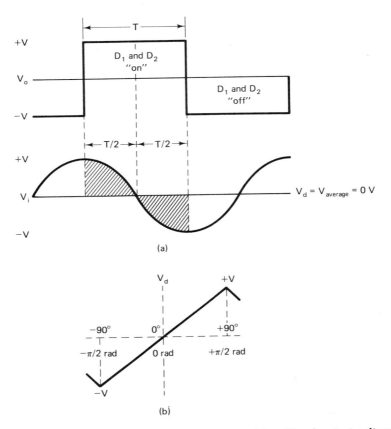

FIGURE 5-6 Phase comparator output voltage: (a) unfiltered output voltage waveform when V_i leads V_o by 90°; (b) output voltage versus phase difference characteristics.

and $F_o + F_i$; thus the input signal to the op-amp is the difference frequency $F_d = F_o - F_i$ (sometimes called the *beat* frequency). F_d is amplified by the op-amp, then applied to the input of the VCO, where it deviates F_o by an amount proportional to the polarity and frequency of F_d. The change in frequency $= \Delta F$ and the VCO output frequency $F_o = F_n + \Delta F$. As F_o changes frequency, the amplitude and frequency of F_d change proportionately until $F_o = F_i$. At this time the output from the phase comparator $V_d = F_o - F_i = 0$ Hz (dc) and the loop is said to be *locked* ($\Delta F = F_n - F_i$). Figure 5-7b shows the beat frequency produced when F_o is swept by F_d. It can be seen that once lock has occurred, F_d is a dc voltage which is necessary to bias the VCO and keep $F_o = F_i$. In essence, the phase comparator is a frequency comparator until frequency *acquisition* (*zero beat*) is achieved; then it becomes a phase comparator. Once the loop is locked, the difference in phase between F_o and F_i is converted to a dc bias voltage and fed back to the VCO to hold lock. Therefore, it is necessary that a phase error ($\theta_o - \theta_i$) is maintained. The change in F_o required to achieve lock $\Delta F = F_n - F_i$, and the

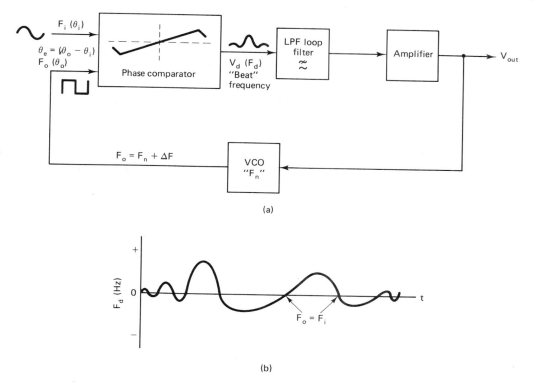

FIGURE 5-7   Loop operation of a PLL: (a) block diagram; (b) beat frequency.

time required to achieve lock (*acquisition* or *pull-in* time) for a PLL with no *loop* filter (loop filters are explained later in this chapter) is approximately equal to $1/K_v$ seconds, where K_v is the *open-loop gain* of the PLL. Once the loop is locked, any change in the frequency of F_i is seen as a phase error and the comparator produces a corresponding change in V_d. The change in V_d is amplified and fed back to the VCO to reestablish lock. Thus the loop dynamically adjusts itself to follow input frequency changes.

The input frequency range within which the PLL will lock onto the input signal is called the *capture range*. For a typical PLL, the capture range is between 1.1 and 1.7 times the natural frequency of the VCO. Once the loop is locked, V_d is proportional to the difference in phase between V_o and V_i ($\theta_o - \theta_i$). Mathematically, the output from the phase comparator is (considering only the fundamental frequency for V_o and excluding the 90° phase bias)

$$V_d = (\sin 2\pi F_o t + \theta_o) \times (V \sin 2\pi F_i t + \theta_i)$$

$$= \frac{V}{2} \sin (2\pi F_o t + \theta_o - 2\pi F_i t - \theta_i)$$

$$- \frac{V}{2} \sin (2\pi F_o t + \theta_o + 2\pi F_i t + \theta_i)$$

When $F_o = F_i$,

$$V_d = \frac{V}{2} \sin (\theta_o - \theta_i)$$

where $\theta_o - \theta_i = \theta_e$ (phase error). θ_e is the phase error required to change the VCO output frequency from F_n to F_o (a change $= \Delta F$) and is often called the static phase error.

Capture range is defined as the range of input frequencies (centered on the VCO center frequency, F_o) over which the PLL will lock onto the incoming signal. *Pull-in range* is the peak capture range (i.e., capture range $= 2 \times$ pull-in range). Capture and pull-in range are shown in frequency diagram form in Figure 5-8.

Loop gain. The *loop gain* for a PLL is simply the product of the individual gains around the closed loop. In Figure 5-6a the open-loop gain is the product of the phase comparator gain, the low-pass filter gain, the op-amp gain, and the VCO gain. Mathematically, the open-loop gain is

$$K_v = (K_d)(K_f)(K_a)(K_o) \tag{5-4a}$$

where

$$K_d = \text{phase comparator gain (V/rad)}$$
$$K_f = \text{filter gain (V/V)}$$
$$K_a = \text{op-amp gain (V/V)}$$
$$K_o = \text{VCO gain (Hz/V)}$$

and

$$K_v = \text{open-loop gain}$$

$$= \left(\frac{V}{\text{rad}}\right) \left(\frac{V}{V}\right) \left(\frac{V}{V}\right) \left(\frac{\text{Hz}}{V}\right) = \frac{\text{Hz}}{\text{rad}}$$

or

$$K_v = \frac{F \text{ cycles/s}}{\text{rad}} = \frac{F \text{ cycles}}{\text{rad-s}} \times \frac{2\pi \text{rad}}{\text{cycle}} = 2\pi F s^{-1}$$

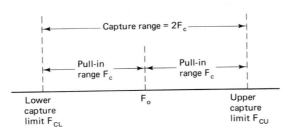

FIGURE 5-8   Capture range.

Expressed in decibels gives us

$$K_v(\text{dB}) = 20 \log k_v \tag{5-4b}$$

From Equation 5-4 and Figure 5-6a, the following relationships are derived:

$$V_d = (\theta_e)(K_d) \tag{5-5}$$
$$V_{\text{out}} = (V_d)(K_f)(K_a) \tag{5-6}$$
$$\Delta F = (V_{\text{out}})(K_o) \tag{5-7}$$

The *closed-loop gain* of a locked PLL is simply unity or 1 (0 dB). This is because there is always 100% feedback in a closed-loop PLL.

Hold-in range. The *hold-in range* for a PLL is the range of input frequencies over which the PLL will remain locked. This presumes that the PLL was initially locked. The hold-in range is often called the *tracking range*; it is the range of frequencies in which the VCO will accurately track or follow the input frequency. The hold-in range is limited by the peak-to-peak swing in V_d and is dependent on the phase comparator, op-amp, and VCO gains. From Figure 5-5c it can be seen that the phase comparator output voltage (V_d) is corrective for $\pm 90°$ ($\pm \pi/2$ rad). Beyond these limits, the polarity of V_d reverses and actually chases the VCO frequency away from the input frequency. Therefore, the maximum phase error (θ_e) that is allowed is $\pm \pi/2$ radians. Consequently, the maximum change in the VCO natural frequency is

$$\Delta F_{\text{max}} = \pm(\pi/2 \text{ rad})(K_d)(K_f)(K_a)(K_o)$$

or

$$\Delta F_{max} = \pm(\pi/2 \text{ rad})(K_v) \tag{5-8}$$

where ΔF_{max} is the hold-in range.

EXAMPLE 5-1

For the PLL shown in Figure 5-7a, a VCO natural frequency $F_n = 200$ kHz, an input frequency $F_i = 210$ kHz, and the following circuit gains: $K_d = 0.2$ V/rad, $K_f = 1$, $K_a = 5$, and $K_o = 20$ kHz/V; determine:

  (a) The open-loop gain, K_v.
  (b) The change in frequency required to achieve lock, ΔF.
  (c) V_{out}.
  (d) V_d.
  (e) The static phase error, θ_e.
  (f) The hold-in range, ΔF_{max}.

Solution   (a) From Equation 5-4a

$$K_v = \frac{0.2 \text{ V}}{\text{rad}} \frac{1 \text{ V}}{\text{V}} \frac{5 \text{ V}}{\text{V}} \frac{20 \text{ kHz}}{\text{V}} = 20 \frac{\text{kHz}}{\text{rad}}$$

$$20 \frac{\text{kHz}}{\text{rad}} = \frac{20 \text{ kilocycles}}{\text{rad-s}} \times \frac{2\pi \text{ rad}}{\text{cycle}} = 125.6 \text{ ks}^{-1}$$

Expressed as a decibel, we have

$$K_v \text{ (dB)} = 20 \log 125.6 \text{ ks}^{-1} = 102 \text{ dB}$$

(b) $\Delta F = F_i - F_o = 210 - 200 \text{ kHz} = 10 \text{ kHz}$

(c) Rearranging Equation 5-1 gives us

$$V_{out} = \frac{\Delta F}{K_o} = \frac{10 \text{ kHz}}{20 \text{ kHz/V}} = 0.5\text{V}$$

(d) $V_d = \dfrac{V_{out}}{(K_f)(K_a)} = \dfrac{0.5 \text{ V}}{(1)(5)} = 0.1 \text{ V}$

(e) Rearranging Equation 5-3 gives us

$$\theta_e = \frac{V_d}{K_d} = \frac{0.1 \text{ V}}{0.2 \text{ V/rad}} = 0.5 \text{ rad or } 28.65°$$

(f) $\Delta F_{max} = \dfrac{(\pi/2 \text{ rad})(20 \text{ kHz})}{\text{rad}} = \pm 31.4 \text{ kHz}$

Lock range is the range of frequencies over which the loop will stay locked onto the input signal once lock has been established. Hold-in range is half the lock range. The relationship between lock and hold-in range is shown in Figure 5-9.

The lock range is expressed in rad/s ($2\omega_L$) and is related to the open-loop voltage gain K_v ($K_v = K_f K_o K_d$ for a simple loop with LPF, phase comparator, and VCO, or $K_f K_o K_d K_a$ for a loop with an amplifier, as in Figure 5-7) as follows:

$$2\omega_L = 2K_v = 2K_f K_o K_d \text{ for a simple loop or}$$
$$2K_f K_o K_d K_a \text{ for the loop of Figure 5-7}$$

The lock range in rad/s is twice the dc voltage gain (open loop) and is independent of the LPF response. The capture range $2\omega_c$ depends on the lock range and on the LPF response, so that it changes with the type of LP filter used, and with the filter cutoff frequency. For a simple RC LPF, it is given by

$$2\omega_C = 2 \left(\frac{\omega_L}{RC}\right)^{1/2}$$

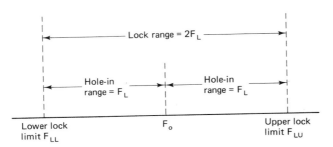

Lock range = 2F$_L$

Hole-in range = F$_L$ Hole-in range = F$_L$

Lower lock limit F$_{LL}$ F$_o$ Upper lock limit F$_{LU}$

FIGURE 5-9   Lock range.

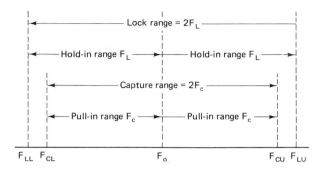

FIGURE 5-10   Capture and lock range.

The capture range is never greater than and is almost always less than the lock range. The relationships among capture, lock, hold-in, and pull-in range are shown in Figure 5-10. Note that lock range ≥ capture range and hold-in range ≥ pull-in range.

Closed-loop frequency response.   The closed-loop frequency response for an *uncompensated* (unfiltered) PLL is shown in Figure 5-11. As shown previously, the open-loop gain of a PLL for a frequency of 1 rad/s = K_v. The frequency response shown in Figure 5-11 is for the circuit and PLL parameters given in Example 5-1. It can be seen that the open-loop gain (K_v) at 1 rad/s = 102 dB, and the open-loop gain equals 1 or 0 dB at the loop cutoff frequency (ω_v). Also, the closed-loop gain is unity up to ω_v, where it drops to −3 dB and continues to roll off at 6 dB/octave (20 dB/decade). Also, $\omega_v = K_v = 125.6$ krad/s, which is the single-sided bandwidth of the uncompensated closed loop.

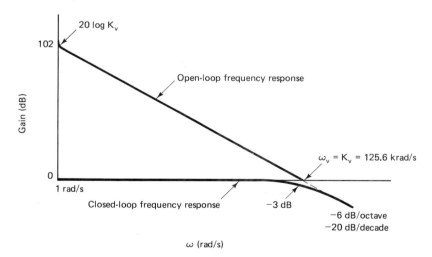

FIGURE 5-11   Frequency response for an uncompensated PLL.

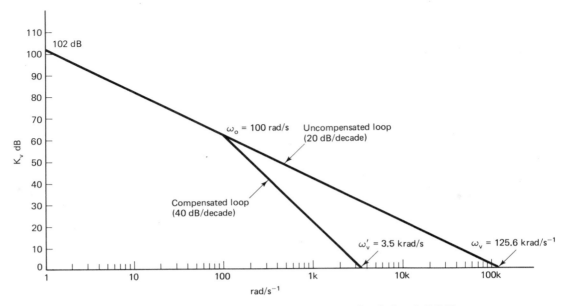

FIGURE 5-12 Loop frequency response for single-pole RC filter.

From Figure 5-11 it can be seen that the frequency response for an uncompensated PLL is identical to that of a single-pole (first-order) low-pass filter with a break frequency $\omega_c = 1$ rad/s. In essence, a PLL is a low-pass *tracking* filter that follows input frequency changes that fall within a bandwidth equal to $\pm K_v$.

If additional bandlimiting is required, a low-pass filter can be added between the phase comparator and the op-amp as shown in Figure 5-7. This filter can be either a single- or multiple-pole filter. Figure 5-12 shows the loop frequency response for a simple single-pole *RC* filter with a cutoff frequency $\omega_c = 1$ krad/s. The frequency response follows that of Figure 5-11 up to the loop filter break frequency, then the response rolls off at 12 dB/octave (40 db/decade). As a result, the compensated unity-gain frequency (ω_c') is reduced (the PLL bandwidth is reduced to approximately ± 3.5 krad/s).

EXAMPLE 5-2

Plot the frequency response for a PLL with an open-loop gain $K_v = 15$ kHz/rad ($\omega_v = 94.2$ krad/s). On the same log paper, plot the response with the addition of a single-pole loop filter with a cutoff frequency $\omega_c = 1.59$ Hz/rad (10 rad/s) and a two/pole loop filter with the same cutoff frequency.

Solution The specified frequency response curves are shown in Figure 5-13. It can be seen that with the single-pole filter the compensated loop response $= \omega_v' = 1$ krad/s and with the two-pole filter $\omega_v'' = 200$ rad/s.

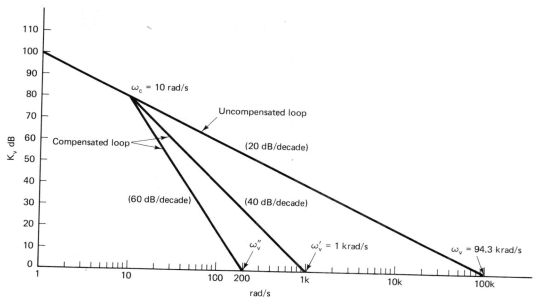

FIGURE 5-13 Loop frequency response for Example 5-2.

The bandwidth of the loop filter (or for that matter, whether a loop filter is needed) depends on the specific application. In Chapters 6, 7, and 8, several applications for PLLs are given with varying requirements. Figure 5-14 shows the specification sheets for the XR2211 integrated-circuit phase-locked loop.

FREQUENCY SYNTHESIZERS

Synthesize means to form an entity by combining parts or elements. A *frequency synthesizer* is used to generate many output frequencies through the addition, subtraction, multiplication, and division of a smaller number of fixed frequency sources. Simply stated, a frequency synthesizer is a crystal-controlled variable-frequency generator. The objective of a synthesizer is twofold. It should produce as many frequencies as possible from a minimum number of sources and each frequency should be as accurate and stable as every other frequency. The ideal frequency synthesizer can generate hundreds or even thousands of different frequencies from a single-crystal oscillator. A frequency synthesizer may be capable of simultaneously generating more than one output frequency, where each frequency is synchronous to a single reference or master oscillator frequency. Frequency synthesizers are used extensively in test and measurement equipment (audio and RF signal generators), tone-generating equipment (touch tone), remote control units (electronic tuners), and multichannel communications systems (telephony).

FSK Demodulator / Tone Decoder

GENERAL DESCRIPTION

The XR-2211 is a monolithic phase-locked loop (PLL) system especially designed for data communications. It is particularly well suited for FSK modem applications. It operates over a wide supply voltage range of 4.5 to 20 V and a wide frequency range of 0.01 Hz to 300 kHz. It can accommodate analog signals between 2 mV and 3 V, and can interface with conventional DTL, TTL, and ECL logic families. The circuit consists of a basic PLL for tracking an input signal within the pass band, a quadrature phase detector which provides carrier detection, and an FSK voltage comparator which provides FSK demodulation. External components are used to independently set center frequency, bandwidth, and output delay. An internal voltage reference proportional to the power supply provides ratio metric operation for low system performance variations with power supply changes.

The XR-2211 is available in 14 pin DTL ceramic or plastic packages specified for commercial or military temperature ranges.

FEATURES

Wide Frequency Range	0.01 Hz to 300 kHz
Wide Supply Voltage Range	4.5 V to 20 V
DTL/TTL/ECL Logic Compatibility	
FSK Demodulation, with Carrier Detection	
Wide Dynamic Range	2 mV to 3 V rms
Adjustable Tracking Range (±1% to ±80%)	
Excellent Temp. Stability	20 ppm/°C, typ.

APPLICATIONS

FSK Demodulation
Data Synchronization
Tone Decoding
FM Detection
Carrier Detection

ABSOLUTE MAXIMUM RATINGS

Power Supply	20 V
Input Signal Level	3 V rms
Power Dissipation	
Ceramic Package	750 mW
Derate above T_A = +25°C	6 mV/°C
Plastic Package	625 mW
Derate above T_A = +25°C	5.0 mW/°C

FUNCTIONAL BLOCK DIAGRAM

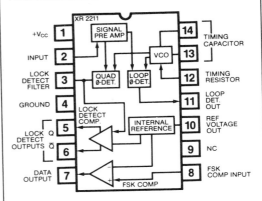

ORDERING INFORMATION

Part Number	Package	Operating Temperature
XR-2211M	Ceramic	–55°C to +125°C
XR-2211CN	Ceramic	0°C to + 75°C
XR-2211CP	Plastic	0°C to + 75°C
XR-2211N	Ceramic	–40°C to + 85°C
XR-2211P	Plastic	–40°C to + 85°C

SYSTEM DESCRIPTION

The main PLL within the XR-2211 is constructed from an input preamplifier, analog multiplier used as a phase detector, and a precision voltage controlled oscillator (VCO). The preamplifier is used as a limiter such that input signals above typically 2MV RMS are amplified to a constant high level signal. The multipling-type phase detector acts as a digital exclusive or gate. Its output (unfiltered) produces sum and difference frequencies of the input and the VCO output, f input + f input (2 f input) and f input - f input (0 Hz) when the phase detector output to remove the "sum" frequency component while passing the difference (DC) component to drive the VCO. The VCO is actually a current controlled oscillator with its nominal input current (f_0) set by a resistor (R_0) to ground and its driving current with a resistor (R_1) from the phase detector.

The other sections of the XR-2211 act to: determine if the VCO is driven above or below the center frequency (FSK comparator); produced both active high and active low outputs to indicate when the main PLL is in lock (quadrature phase detector and lock detector comparator).

 EXAR Integrated Systems, Inc., 750 Palomar Avenue, Sunnyvale, CA 94086 • (408) 732-7970 • TWX 910-339-9233

FIGURE 5-14   XR-2211 Phase-locked loop. (Courtesy of EXAR Corporation.)

XR-2211

ELECTRICAL CHARACTERISTICS

Test Conditions: Test Circuit of Figure 1, $V^+ = V^- = 6V$, $T_A = +25°C$, $C = 5000$ pF, $R_1 = R_2 = R_3 = R_4 = 20$ KΩ, $R_L = 4.7$ KΩ. Binary Inputs grounded, S_1 and S_2 closed unless otherwise specified.

PARAMETERS	XR-2211/2211M			XR-2211C			UNITS	CONDITIONS
	MIN.	TYP.	MAX.	MIN.	TYP.	MAX.		
GENERAL								
Supply Voltage	4.5		20	4.5		20	V	
Supply Current		4	7		5	9	mA	$R_0 \geqslant 10$ KΩ See Fig. 4
OSCILLATOR SECTION								
Frequency Accuracy		±1	±3		±1		%	Deviation from $f_0 = 1/R_0C_0$
Frequency Stability								$R_1 = \frac{1}{2}$
Temperature		±20	±50		±20		ppm/°C	See Fig. 8.
Power Supply		0.05	0.5		0.05		%/V	$V^+ = 12 \pm 1$ V. See Fig. 7.
		0.2			0.2		%/V	$V^+ = 5 \pm 0.5$ V. See Fig. 7.
Upper Frequency Limit	100	300			300		kHz	$R_0 = 8.2$ KΩ, $C_0 = 400$ pF
Lowest Practical								
Operating Frequency		0.01			0.01		Hz	$R_0 = 2$ MΩ, $C_0 = 50$ μF
Timing Resistor, R_0								See Fig. 5.
Operating Range	5		2000	5		2000	KΩ	
Recommended Range	15		100	15		100	KΩ	See Fig. 7 and 8.
LOOP PHASE DETECTOR SECTION								
Peak Output Current	±150	±200	±300	±100	±200	±300	μA	Measured at Pin 11.
Output Offset Current		±1			±2		μA	
Output Impedance		1			1		MΩ	
Maximum Swing	±4	±5		±4	±5		V	Referenced to Pin 10.
QUADRATURE PHASE DETECTOR								
								Measured at Pin 3.
Peak Output Current	100	150			150		μA	
Output Impedance		1			1		MΩ	
Maximum Swing		11			11		V pp	
INPUT PREAMP SECTION								
								Measured at Pin 2.
Input Impedance		20			20		KΩ	
Input Signal								
Voltage Required to								
Cause Limiting		2	10		2		mV rms	
VOLTAGE COMPARATOR SECTIONS								
Input Impedance		2			2		MΩ	Measured at Pins 3 and 8.
Input Bias Current		100			100		nA	
Voltage Gain	55	70		55	70		dB	$R_L = 5.1$ KΩ
Output Voltage Low		300			300		mV	$I_C = 3$ mA
Output Leakage Current		0.01			0.01		μA	$V_O = 12$ V
INTERNAL REFERENCE								
Voltage Level	4.9	5.3	5.7	4.75	5.3	5.85	V	Measured at Pin 10.
Output Impedance		100			100		Ω	

FIGURE 5-14 (*continued*)

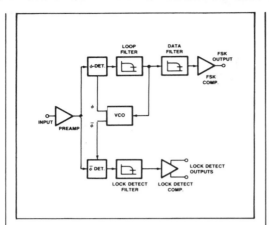

Figure 1: Functional Block Diagram of a Tone and FSK Decoding System Using XR-2211

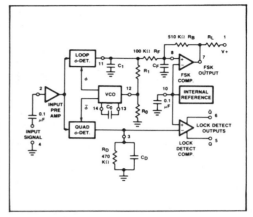

Figure 2: Generalized Circuit Connection for FSK and Tone Detection

Reference Voltage, V_R (Pin 10): This pin is internally biased at the reference voltage level, V_R: V_R = V+/2 – 650 mV. The dc voltage level at this pin forms an internal reference for the voltage levels at Pins 5, 8, 11 and 12. Pin 10 *must* be bypassed to ground with a 0.1 μF capacitor for proper operation of the circuit.

Loop Phase Detector Output (Pin 11): This terminal provides a high impedance output for the loop phase detector. The PLL loop filter is formed by R_1 and C_1 connected to Pin 11 (see Figure 2). With no input signal, or with no phase error within the PLL, the dc level at Pin 11 is very nearly equal to V_R. The peak voltage swing available at the phase detector output is equal to $\pm V_R$.

VCO Control Input (Pin 12): VCO free-running frequency is determined by external timing resistor, R_0, connected from this terminal to ground. The VCO free-running frequency, f_0, is:

$$f_0 = \frac{1}{R_0 C_0} \text{ Hz}$$

where C_0 is the timing capacitor across Pins 13 and 14. For optimum temperature stability, R_0 must be in the range of 10 KΩ to 100 KΩ see Figure 8).

This terminal is a low impedance point, and is internally biased at a dc level equal to V_R. The maximum timing current drawn from Pin 12 must be limited to $\leqslant$3 mA for proper operation of the circuit.

VCO Timing Capacitor (Pins 13 and 14): VCO frequency is inversely proportional to the external timing capacitor, C_0, connected across these terminals (see Figure 5). C_0 must be nonpolar, and in the range of 200 pF to 10 μF.

VCO Frequency Adjustment: VCO can be fine-tuned by connecting a potentiometer, R_X, in series with R_0 at Pin 12 (see Figure 9).

VCO Free-Running Frequency, f_0: XR-2211 does not have a separate VCO output terminal. Instead, the VCO outputs are internally connected to the phase detector sections of the circuit. However, for set-up or adjustment purposes, VCO free-running frequency can be measured at Pin 3 (with C_D disconnected), with no input and with Pin 2 shorted to Pin 10.

DESIGN EQUATIONS
(See Figure 2 for definition of components.)

1. VCO Center Frequency, f_0:

   $f_0 = 1/R_0 C_0$ Hz

2. Internal Reference Voltage, V_R (measured at Pin 10):

   V_R = V+/2 – 650 mV

3. Loop Low-Pass Filter Time Constant, τ:

   $\tau = R_1 C_1$

FIGURE 5-14 (*continued*)

4. Loop Damping, ζ:

$$\zeta = 1/4 \sqrt{\frac{C_0}{C_1}}$$

5. Loop Tracking Bandwidth, $\pm\Delta f/f_0$:

$\Delta f/f_0 = R_0/R_1$

6. FSK Data Filter Time Constant, τF:

$\tau F = R_F C_F$

7. Loop Phase Detector Conversion Gain, $K\phi$: ($K\phi$ is the differential dc voltage across Pins 10 and 11, per unit of phase error at phase detector input):

$K\phi = -2V_R/\pi$ volts/radian

8. VCO Conversion Gain, K_0: (K_0 is the amount of change in VCO frequency, per unit of dc voltage change at Pin 11):

$K_0 = -1/V_R C_0 R_1$ Hz/volt

9. Total Loop Gain, K_T:

$K_T = 2\pi K\phi K_0 = 4/C_0 R_1$ rad/sec/volt

10. Peak Phase Detector Current I_A:

$I_A = V_R$ (volts)/25 mA

APPLICATIONS INFORMATION

FSK DECODING:

Figure 9 shows the basic circuit connection for FSK decoding. With reference to Figures 2 and 9, the functions of external components are defined as follows: R_0 and C_0 set the PLL center frequency, R_1 sets the system bandwidth, and C_1 sets the loop filter time constant and the loop damping factor. C_F and R_F form a one-pole post-detection filter for the FSK data output. The resistor R_B (= 510 KΩ) from Pin 7 to Pin 8 introduces positive feedback across the FSK comparator to facilitate rapid transition between output logic states.

Recommended component values for some of the most commonly used FSK bands are given in Table 1.

Design Instructions:

The circuit of Figure 9 can be tailored for any FSK decoding application by the choice of five key circuit components: R_0, R_1, C_0, C_1 and C_F. For a given set of FSK mark and space frequencies, f_1 and f_2, these parameters can be calculated as follows:

a) Calculate PLL center frequency, f_0:

$$f_0 = \frac{f_1 + f_2}{2}$$

b) Choose value of timing resistor R_0, to be in the range of 10 KΩ to 100 KΩ. This choice is arbitrary. The recommended value is $R_0 \equiv 20$ KΩ. The final value of R_0 is normally fine-tuned with the series potentiometer, R_X.

c) Calculate value of C_0 from design equation (1) or from Figure 6:

$$C_0 = 1/R_0 f_0$$

d) Calculate R_1 to give a Δf equal to the mark space deviation:

$$R_1 = R_0 [f_0/(f_1 - f_2)]$$

e) Calculate C_1 to set loop damping. (See design equation no. 4.):

Normally, $\zeta \approx 1/2$ is recommended.

Then: $C_1 = C_0/4$ for $\zeta = 1/2$

f) Calculate Data Filter Capacitance, C_F:

For $R_F = 100$ KΩ, $R_B = 510$ KΩ, the recommended value of C_F is:

$$C_F \approx 3/(\text{Baud Rate}) \mu F$$

Note: All calculated component values except R_0 can be rounded to the nearest standard value, and R_0 can be varied to fine-tune center frequency, through a series potentiometer, R_X. (See Figure 9.)

FIGURE 5-14 (*continued*)

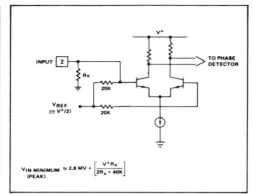

Figure 3: Desensitizing Input Stage

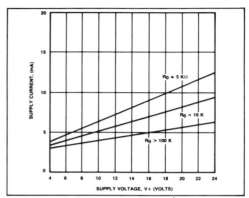

Figure 4: Typical Supply Current vs V$^+$ (Logic Outputs Open Circuited).

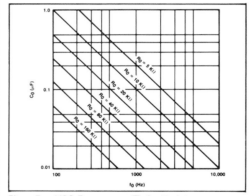

Figure 5: VCO Frequency vs Timing Resistor

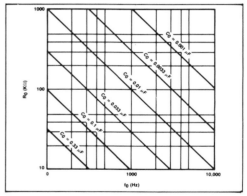

Figure 6: VCO Frequency vs Timing Capacitor

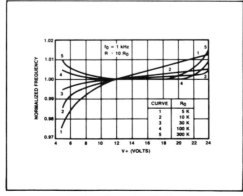

Figure 7: Typical f_0 vs Power Supply Characteristics

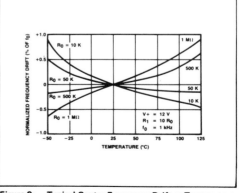

Figure 8: Typical Center Frequency Drift vs Temperature

FIGURE 5-14 (*continued*)

XR-2211

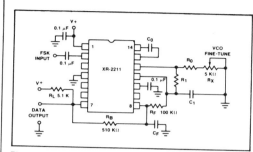

Figure 9: Circuit Connection for FSK Decoding

Design Example:

75 Baud FSK demodulator with mark space frequencies of 1110/1170 Hz:

Step 1: Calculate f_0: $f_0 = (1110 + 1170) (1/2) = 1140$ Hz
Step 2: Choose $R_0 = 20$ KΩ (18 KΩ fixed resistor in series with 5 KΩ potentiometer)
Step 3: Calculate C_0 from Figure 6: $C_0 = 0.044$ μF
Step 4: Calculate R_1: $R_1 = R_0$ (2240/60) = 380 KΩ
Step 5: Calculate C_1: $C_1 = C_0/4 = 0.011$ μF

Note: All values except R_0 can be rounded to *nearest* standard value.

Table 1. Recommended Component Values for Commonly Used FSK Bands. (See Circuit of Figure 9.)

FSK BAND	COMPONENT VALUES	
300 Baud $f_1 = 1070$ Hz $f_2 = 1270$ Hz	$C_0 = 0.039$ μF $C_1 = 0.01$ μF $R_1 = 100$ KΩ	$C_F = 0.005$ μF $R_0 = 18$ KΩ
300 Baud $f_1 = 2025$ Hz $f_2 = 2225$ Hz	$C_0 = 0.022$ μF $C_1 = 0.0047$ μF $R_1 = 200$ KΩ	$C_F = 0.005$ μF $R_0 = 18$ KΩ
1200 Baud $f_1 = 1200$ Hz $f_2 = 2200$ Hz	$C_0 = 0.027$ μF $C_1 = 0.01$ μF $R_1 = 30$ KΩ	$C_F = 0.0022$ μF $R_0 = 18$ KΩ

FSK DECODING WITH CARRIER DETECT:

The lock detect section of XR-2211 can be used as a carrier detect option, for FSK decoding. The recommended circuit connection for this application is shown in Figure 10. The open collector lock detect output, Pin 6, is shorted to data output (Pin 7). Thus, data output will be disabled at "low" state, until there is a carrier within the detection band of the PPL, and the Pin 6 output goes "high," to enable the data output.

The minimum value of the lock detect filter capacitance C_D is inversely proportional to the capture range, $\pm\Delta f_C$. This is the range of incoming frequencies over which the loop can acquire lock and is always less than the tracking range. It is further limited by C_1. For most applications, $\Delta f_C > \Delta f/2$. For $R_D = 470$ KΩ, the approximate minimum value of C_D can be determined by:

$$C_D \, (\mu F) \geqslant 16/\text{capture range in Hz}.$$

With values of C_D that are too small, chatter can be observed on the lock detect output as an incoming signal frequency approaches the capture bandwidth. Excessively large values of C_D will slow the response time of the lock detect output.

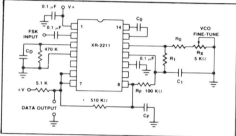

Figure 10: External Connectors for FSK Demodulation with Carrier Detect Capability

Note: Data Output is "Low" When No Carrier is Present.

TONE DETECTION:

Figure 11 shows the generalized circuit connection for tone detection. The logic outputs, Q and $\overline{Q}$ at Pins 5 and 6 are normally at "high" and "low" logic states, respectively. When a tone is present within the detection band of the PLL, the logic state at these outputs become reversed for the duration of the input tone. Each logic output can sink 5 mA of load current.

Both logic outputs at Pins 5 and 6 are open collector type stages, and require external pull-up resistors R_{L1} and R_{L2}, as shown in Figure 11.

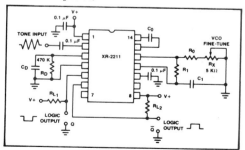

Figure 11: Circuit Connection for Tone Detection.

FIGURE 5-14 (*continued*)

With reference to Figures 2 and 11, the functions of the external circuit components can be explained as follows: R_0 and C_0 set VCO center frequency; R_1 sets the detection bandwidth; C_1 sets the low pass-loop filter time constant and the loop damping factor. R_{L1} and R_{L2} are the respective pull-up resistors for the Q and $\overline{Q}$ logic outputs.

Design Instructions:

The circuit of Figure 11 can be optimized for any tone detection application by the choice of the 5 key circuit components: R_0, R_1, C_0, C_1 and C_D. For a given input, the tone frequency, f_S, these parameters are calculated as follows:

a) Choose R_0 to be in the range of 15 KΩ to 100 KΩ. This choice is arbitrary.

b) Calculate C_0 to set center frequency, f_0 equal to f_s (see Figure 6): $C_0 = 1/R_0 f_S$

c) Calculate R_1 to set bandwidth $\pm\Delta f$ (see design equation no. 5):

$$R_1 = R_0(f_0/\Delta f)$$

Note: The total detection bandwidth covers the frequency range of $f_0 \pm \Delta f$.

d) Calculate value of C_1 for a given loop damping factor:

$$C_1 = C_0/16\zeta 2$$

Normally $\zeta \approx 1/2$ is optimum for most tone detector applications, giving $C_1 = 0.25\ C_0$.

Increasing C_1 improves the out-of-band signal rejection, but increases the PLL capture time.

e) Calculate value of filter capacitor C_D. To avoid chatter at the logic output, with $R_D = 470$ KΩ, C_D must be:

$$C_D(\mu F) \geqslant (16/\text{capture range in Hz})$$

Increasing C_D slows down the logic output response time.

Design Examples:

Tone detector with a detection band of 1 kHz ± 20 Hz:

a) Choose R_0 = 20 KΩ (18 KΩ in series with 5 KΩ potentiometer).

b) Choose C_0 for f_0 = 1 kHz (from Figure 6): C_0 = 0.05 μF.

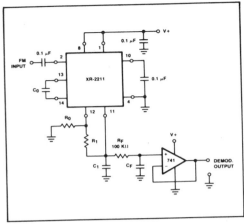

Figure 12: Linear FM Detector Using XR-2211 and an External Op Amp. (See section on Design Equation for Component Values.)

c) Calculate R_1: R_1 = (R_0) (1000/20) = 1 MΩ.
d) Calculate C_1: for ζ = 1/2, C_1 = 0.25, C_0 = 0.013 μF.
e) Calculate C_D: C_D = 16/38 = 0.42 μF.
f) Fine-tune center frequency with 5 KΩ potentiometer, R_X.

LINEAR FM DETECTION:

XR-2211 can be used as a linear FM detector for a wide range of analog communications and telemetry applications. The recommended circuit connection for this application is shown in Figure 12. The demodulated output is taken from the loop phase detector output (Pin 11), through a post-detection filter made up of R_F and C_F, and an external buffer amplifier. This buffer amplifier is necessary because of the high impedance output at Pin 11. Normally, a non-inverting unity gain op amp can be used as a buffer amplifier, as shown in Figure 12.

The FM detector gain, i.e., the output voltage change per unit of FM deviation can be given as:

$$V_{out} = R_1\ V_R/100\ R_0\ \text{Volts/\%deviation}$$

where V_R is the internal reference voltage (V_R = V+/2 − 650 mV). For the choice of external components R_1, R_0, C_D, C_1 and C_F, see section on design equations.

FIGURE 5-14 (*continued*)

XR-2211

PRINCIPLES OF OPERATION

Signal Input (Pin 2): Signal is ac coupled to this terminal. The internal impedance at Pin 2 is 20 KΩ. Recommended input signal level is in the range of 10 mV rms to 3 V rms.

Quadrature Phase Detector Output (Pin 3): This is the high impedance output of quadrature phase detector and is internally connected to the input of lock detect voltage comparator. In tone detection applications, Pin 3 is connected to ground through a parallel combination of R_D and C_D (see Figure 2) to eliminate the chatter at lock detect outputs. If the tone detect section is not used, Pin 3 can be left open circuited.

Lock Detect Output, Q (Pin 5): The output at Pin 5 is at "high" state when the PLL is out of lock and goes to "low" or conducting state when the PLL is locked. It is an open collector type output and requires a pull-up resistor, R_L, to V+ for proper operation. At "low" state, it can sink up to 5 mA of load current.

Lock Detect Complement, $\overline{Q}$ (Pin 6): The output at Pin 6 is the logic complement of the lock detect output at Pin 5. This output is also an open collector type stage which can sink 5 mA of load current at low or "on" state.

FSK Data Output (Pin 7): This output is an open collector logic stage which requires a pull-up resistor, R_L, to V+ for proper operation. It can sink 5 mA of load current. When decoding FSK signals, FSK data output is at "high" or "off" state for low input frequency, and at "low" or "on" state for high input frequency. If no input signal is present, the logic state at Pin 7 is indeterminate.

FSK Comparitor Input (Pin 8): This is the high impedance input to the FSK voltage comparator. Normally, an FSK post-detection or data filter is connected between this terminal and the PLL phase detector output (Pin 11). This data filter is formed by R_F and C_F of Figure 2. The threshold voltage of the comparator is set by the internal reference voltage, V_R, available at Pin 10.

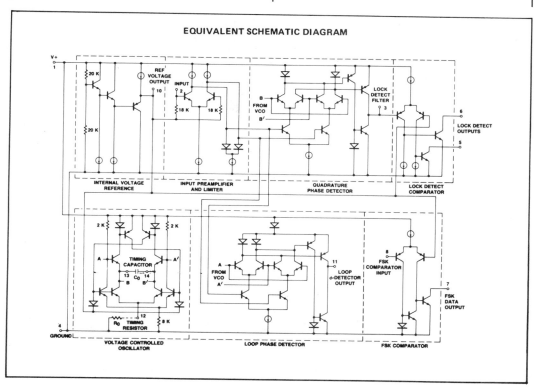

EQUIVALENT SCHEMATIC DIAGRAM

Rev. 8/83

FIGURE 5-14 *(continued)*

195

Essentially, there are two methods of frequency synthesis: direct and indirect. With *direct frequency synthesis*, multiple output frequencies are generated by mixing the outputs from two or more crystal-controlled oscillators or by dividing/multiplying the output frequency from a single-crystal oscillator. With *indirect frequency synthesis*, a feedback-controlled divider/multiplier (such as a PLL) is used to generate multiple output frequencies. Indirect frequency synthesis is slower and more susceptible to noise; however, it is less expensive and requires fewer and less complicated filters than direct synthesis.

Direct Frequency Synthesizers

Multiple-crystal frequency synthesizer. Figure 5-15 shows a block diagram for a *multiple-crystal frequency synthesizer* that uses nonlinear mixing (heterodyning) and filtering to produce 128 different frequencies from 20 crystals and two oscillator modules. For the crystal values shown, a range of frequencies from 510 to 1890 kHz in 10-kHz steps are synthesized. A synthesizer such as this can be used to generate the carrier frequencies for the 106 AM broadcast-band stations (540 to 1600 kHz). For the switch positions shown, the 160- and 700-kHz oscillators are selected, and the outputs from the balanced mixer are their sum and difference frequencies (700 kHz ± 160

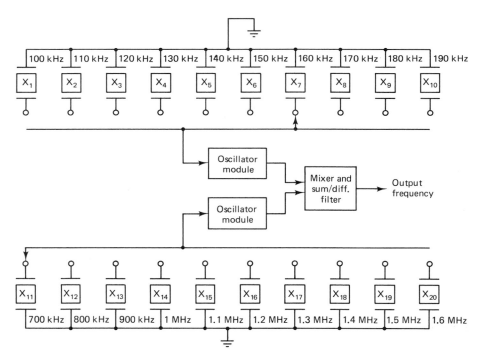

FIGURE 5-15 Multiple-crystal frequency synthesizer.

kHz = 540 and 860 kHz). The output filter is tuned to 540 kHz, which is the carrier frequency for channel 1. To generate the carrier frequency for channel 106, the 100-kHz crystal is selected with either the 1700-kHz (difference) or 1500-kHz (sum) crystal. The minimum frequency separation between output frequencies for a synthesizer is called *resolution*. The resolution for the synthesizer shown in Figure 5-15 is 10 kHz.

Single-crystal frequency synthesizer. Figure 5-16 shows a block diagram for a *single-crystal frequency synthesizer* that again uses frequency addition, multiplication, and division to generate frequencies (in 1-Hz steps) from 1 to 999,999 Hz. A 100-kHz crystal is the source for the master oscillator from which all frequencies are derived.

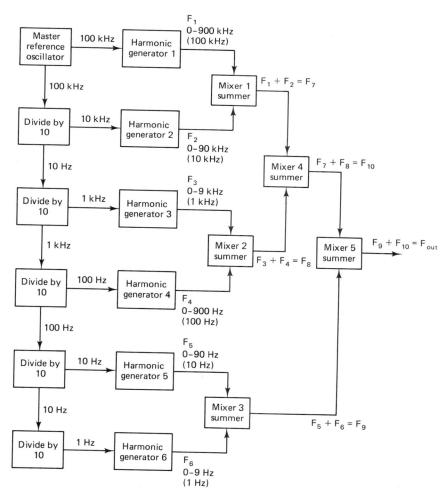

FIGURE 5-16 Single-crystal frequency synthesizer.

The master oscillator frequency is a base frequency which is repeatedly divided by 10 to generate 5 additional base frequencies (10 kHz, 1 kHz, 100 Hz, 10 Hz, and 1 Hz). Each of the base frequencies is fed to a separate harmonic generator (frequency multiplier) which consists of a nonlinear amplifier and a filter. The filter is tunable to each of the first nine harmonics of its base frequency. Therefore, the possible output frequencies for harmonic generator 1 are 100 to 900 kHz in 100-kHz steps, for harmonic generator 2 are 10 to 90 kHz in 10-kHz steps, and so on. The resolution for the synthesizer is determined by how many times the master crystal oscillator frequency is divided. For the synthesizer shown in Figure 5-16, the resolution is 1 Hz. The mixers used are balanced modulators with output filters that are tuned to the sum of the two input frequencies. For example, the harmonics selected in Figure 5-16 produce a 246,313-Hz output frequency. Table 5-1 lists the selector switch positions for each harmonic generator and the input and output frequencies from each mixer. It can be seen that the five mixers simply sum the output frequencies from the six harmonic generators with three levels of mixing (addition).

TABLE 5-1   SWITCH POSITIONS AND HARMONICS

Harmonic generator	Output frequency	Mixer	Output frequency
1	200 kHz	1	240 kHz
2	40 kHz		
3	6 kHz	2	6300 Hz
4	300 Hz		
5	10 Hz	3	13 Hz
6	3 Hz		
		4	246.3 kHz
		5	246.313 kHz

Switched crystal frequency synthesizer.   In the early model citizens' band (CB) radios, each channel that a transceiver was equipped for required a separate crystal. Consequently, a transceiver equipped for 23 class D channels, required a minimum of 23 crystals. Figure 5-17 shows a block diagram for a switched crystal frequency synthesizer that is used to generate 23 RF and 5 IF frequencies (four first IFs and one second IF). As you can see, a *switched crystal frequency synthesizer* is a form of *multiple-crystal synthesizer*. Although this technique of frequency synthesis is gradually being replaced with more modern PLL frequency synthesizers, it does illustrate the concept of frequency synthesis quite well. The 28 frequencies are generated with 14 crystals; 6 crystals for the synthesizer oscillator, and 4 each in the transmit and receive oscillators. The three selector switches (S_1, S_2, and S_3) are mechanically ganged to a common tuning switch (S_1). S_1 selects the same crystal for four successive switch positions. Therefore, each crystal in the synthesizer oscillator is used to generate four successive channel frequencies except XTAL 6, which is used to generate only three (i.e., XTAL 1 is for

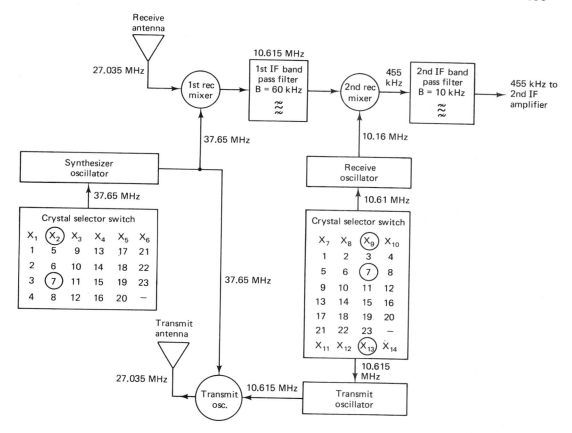

FIGURE 5-17 Switched crystal frequency synthesizer.

channels 1 to 4, XTAL 2 is for channels 5 to 8, etc.). S_2 and S_3 sequentially select the four second receive and transmit oscillator crystals, respectively, as the channel selector is switched through the 23 channel positions. Therefore, each crystal in the transmit and second receive oscillators is used for six channels except XTALs 9 and 13, which are used for only five. However, the channels are not successive (i.e., XTALs 7 and 11 are selected for channels 1, 5, 9, 13, 17, and 21; and XTALs 8 and 12 are selected for channels 2, 6, 10, 14, 18, and 22; etc.). In the receiver, the first four RF carrier frequencies combine in the first receive mixer with XTAL 1 frequency and produce four different first IF frequencies (10.635, 10.625, 10.615, and 10.595 MHz, respectively). However, for all 23 channels, the first IF mixes with the receive oscillator frequency in the second receive mixer to produce a difference frequency of 455 kHz, which is the second IF for all 23 channels. Generating the 23 transmit RF carriers is accomplished in a similar manner. For each channel, one of the six synthesizer XTAL frequencies combines in the transmit mixer with one of the four transmit XTAL frequencies to produce a different RF carrier frequency. The transmit frequency is the difference

TABLE 5-2 CRYSTAL AND SWITCH ASSIGNMENTS FOR THE TRANSCEIVER SHOWN IN FIGURE 5-17

Synthesizer oscillator			Receive oscillator			Transmit oscillator		
Channels	Crystal	Frequency	Channels	Crystal	Frequency	Channels	Crystal	Frequency
1–4	X1	37.6 MHz	1, 5, 9, 13, 17, 21	X7	10.18 MHz	1, 5, 9, 13, 17, 21	X11	10.63 MHz
5–8	X2	37.65 MHz	2, 6, 10, 14, 18, 22	X8	10.17 MHz	2, 6, 10, 14, 18, 22	X12	10.625 MHz
9–12	X3	37.7 MHz	3, 7, 11, 15, 19, 23	X9	10.16 MHz	3, 7, 11, 15, 19, 23	X13	10.615 MHz
13–16	X4	37.75 MHz	4, 8, 12, 16, 20	X10	10.14 MHz	4, 8, 12, 16, 20	X14	10.59 MHz
17–20	X5	37.8 MHz						
21–23	X6	37.85 MHz						

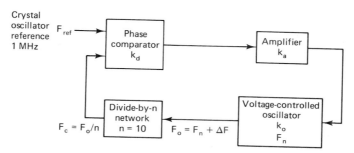

FIGURE 5-18 Single-loop PLL frequency synthesizer.

between the synthesizer and transmit oscillator frequencies. Table 5-2 lists the crystals, selector switch positions, and first and second IFs used for each of the 23 channels. The switch positions and frequencies shown in Figure 5-17 are for transmitting and receiving channel 7.

Indirect Frequency Synthesizers

Phased-locked-loop frequency synthesizers. In recent years, PLL frequency synthesizers have rapidly become the most popular method for frequency synthesis. Figure 5-18 shows a block diagram for a simple *single-loop* PLL frequency synthesizer. The stable frequency reference is a crystal-controlled oscillator. The range of frequencies generated and the resolution are dependent on the divider network and the open-loop gain. The frequency *divider* is a divide-by-n circuit, where n is a whole integer number. The simplest form of divider circuit is a *programmable digital up-down counter* with an output frequency $F_c = F_o/n$. With this arrangement, once lock has occurred, $F_c = F_{ref}$ (where $F_c = F_n + \Delta F$) and the VCO and synthesizer output frequency $F_o = nF_{ref}$. Thus the synthesizer is essentially a times-n frequency multiplier. The frequency divider reduces the open-loop gain by a factor of n. Consequently, the other circuits around the loop must have relatively high gains. The open-loop gain for the frequency synthesizer shown in Figure 5-18 is

$$K_v = \frac{(K_d)(K_a)(K_o)}{n} \tag{5-9a}$$

From Equation 5-9a it can be seen that as n changes, the open-loop gain changes in inverse proportion to n. A way to remedy this problem is to program the amplifier gain as well as the divider ratio. Thus the open-loop gain is

$$K_v = \frac{K_d n K_a K_o}{n} = K_d K_a K_o \tag{5-9b}$$

For the reference frequency and divider circuit shown in Figure 5-18, the range of output frequencies is

$$F_o = nF_{\text{ref}}$$
$$= F_{\text{ref}} \text{ to } 10F_{\text{ref}}$$
$$= 1 \text{ to } 10 \text{ MHz}$$

Figure 5-19 shows a block diagram for a CB *transceiver* that uses three crystal oscillators and a PLL frequency synthesizer to generate the 23 RF carriers and two IFs. The divide-by-1024 network is used to reduce the resolution of the PLL to 10 kHz and, at the same time, provide the 10.24-MHz beat frequency required to generate the 455-kHz second IF. Unlike the CB transceiver shown in Figure 5-17, the first IF is the same for all 23 channels (10.695 MHz). The programmable divider is controlled by a binary code generated in the channel selector switch. The VCO output frequency (F_o) varies from 37.66 MHz (channel 1) to 37.94 MHz (channel 23) and mixes with the receive crystal oscillator frequency (36.38 MHz) to generate a difference frequency $F_1 = 1.28$ to 1.56 MHz, which is the input frequency to the programmable divider. The programmable divider is programmed for values of n from 128 for channel 1 to

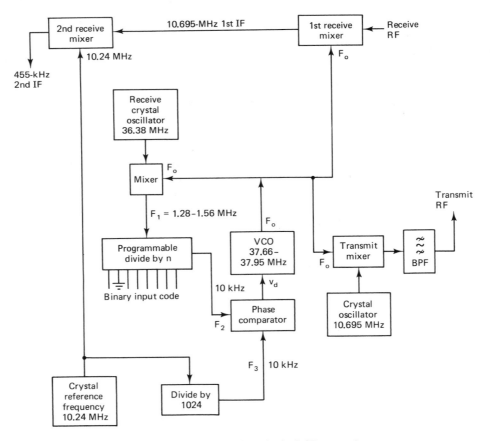

FIGURE 5-19   23-Channel synthesized CB transceiver.

156 for channel 23. Consequently, $F_2 = 10$ kHz for all 23 channels. F_2 and the crystal reference frequency (F_3) are compared in the phase comparator. The phase comparator output voltage is used to tune the VCO and lock it onto the crystal reference frequency. The VCO output frequency is also mixed with the received RF carrier to generate a 10.695-MHz difference frequency, which is the first IF. To generate the 23 transmit RF carrier frequencies, the VCO output frequency is mixed with the output frequency from crystal oscillator 3 (10.695 MHz). The output of the transmit mixer is tuned to their difference frequency. Table 5-3 lists the 23 channel carrier frequencies and their corresponding values for n and F_1.

Prescaled frequency synthesizer.   Figure 5-20 shows the block diagram for a frequency synthesizer that uses *prescaling* to achieve *fractional division*. Prescaling is also necessary for generating frequencies greater than 100 MHz because programmable counters are not available that operate at such high frequencies. The synthesizer shown in Figure 5-20 uses a *two-modulus* prescaler. The prescaler has two modes of operation. One mode provides an output for every input pulse (P) and the other mode provides an output for every $P + 1$ input pulses. Whenever the m-register contains a nonzero number, the prescaler counts in the $P + 1$ mode. Consequently, once the m and n registers have been initially loaded, the prescaler will count down $(P + 1)m$ times

TABLE 5-3

Channel	Frequency (MHz)	VCO frequency (MHz)	n	F_1 (MHz)
1	26.965	37.66	128	1.28
2	26.975	37.67	129	1.29
3	26.985	37.68	130	1.30
4	27.005	37.70	132	1.32
5	27.015	37.71	133	1.33
6	27.025	37.72	134	1.34
7	27.035	37.73	135	1.35
8	27.055	37.75	137	1.37
9	27.065	37.76	138	1.38
10	27.075	37.77	139	1.39
11	27.085	37.78	140	1.40
12	27.105	37.80	142	1.42
13	27.115	37.81	143	1.43
14	27.125	37.82	144	1.44
15	27.135	37.83	145	1.45
16	27.155	37.85	147	1.47
17	27.165	37.86	148	1.48
18	27.175	37.87	149	1.49
19	27.185	37.88	150	1.50
20	27.205	37.90	152	1.52
21	27.215	37.91	153	1.53
22	27.225	37.92	154	1.54
23	27.245	37.94	156	1.56

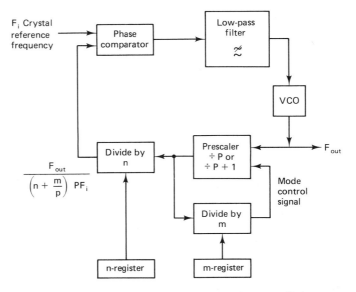

FIGURE 5-20   Frequency synthesizer using prescaling.

until the m counter goes to zero, the prescaler operates in the P mode and the n counter counts down $(n - m)$ times. At this time, both the m and n counters are reset to their initial values which have been stored in the m and n registers, respectively, and the process repeats. Mathematically, the synthesizer output frequency (F_o) is

$$F_o = \left(n + \frac{m}{P} \right) PF_i$$

Multifrequency synthesizer.   Figure 5-21a shows a block diagram for a frequency synthesizer that is used to generate 12 carrier frequencies and a pilot for a multiple-channel telecommunications system. A 64-kHz pilot is received, divided by 16, then fed to the input of a PLL. The PLL generates a coherent 4-kHz base frequency from which all local carrier frequencies are derived. The 4-kHz base frequency is multiplied and heterodyned in a harmonic generator to produce the necessary output frequencies. A filter network is used to select 12 channel carrier frequencies which are all multiples of 4 kHz. In addition, the 4-kHz base frequency is used to generate a 104.08-kHz *pilot* that is synchronous with the 12 channel carrier frequencies. Each channel carrier is separated from adjacent carrier frequencies by 4 kHz. Therefore, twelve 4-kHz voice channels are stacked on top of each other in the frequency domain. The composite signal modulates a single high-frequency carrier which is transmitted as one radio channel. This technique is called *frequency-division multiplexing*. Figure 5-21b shows the composite spectrum for the 12 channel carrier frequencies and one pilot frequency generated by the frequency synthesizer shown in Figure 5-21a.

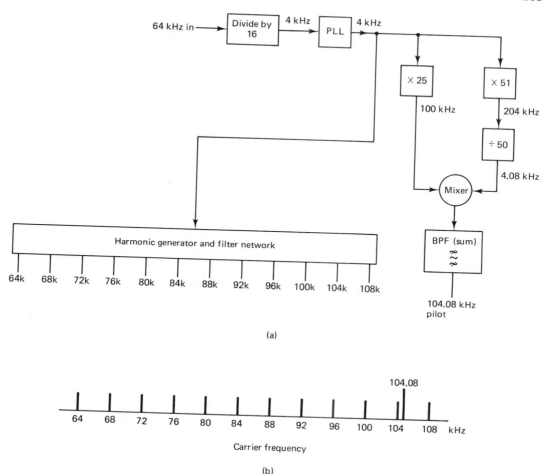

FIGURE 5-21   Multichannel communications system—frequency generation:
(a) block diagram; (b) carrier and pilot spectrum.

LARGE-SCALE INTEGRATION PROGRAMMABLE TIMERS

In recent years, the use of large-scale integration integrated circuits for frequency synthesis has increased at a tremendous rate primarily because of their simplicity and high degree of accuracy. Figure 5-22 shows the specification sheet for the XR-2240 programmable timer/counter. The XR-2240 is a monolithic controller capable of producing precision timing intervals ranging from microseconds through several days. The XR-2240 can generate 256 frequencies or pulse patterns from a single RC setting and can be synchronized with external clock signals. The XR-2240 is ideally suited for precision timing, sequential timing, pattern generation, and frequency synthesis.

Programmable Timer/Counter

GENERAL DESCRIPTION

The XR-2240 Programmable Timer/Counter is a mono-lithic controller capable of producing ultra-long time delays without sacrificing accuracy. In most applications, it provides a direct replacement for mechanical or electromechanical timing devices and generates programmable time delays from micro-seconds up to five days. Two timing circuits can be cascaded to generate time delays up to three years.

As shown in Figure 1, the circuit is comprised of an internal time-base oscillator, a programmable 8-bit counter and a control flip-flop. The time delay is set by an external R-C network and can be programmed to any value from 1 RC to 255 RC.

In astable operation, the circuit can generate 256 separate frequencies or pulse-patterns from a single RC setting and can be synchronized with external clock signals. Both the control inputs and the outputs are compatible with TTL and DTL logic levels.

FEATURES

Timing from micro-seconds to days
Programmable delays: 1RC to 255 RC
Wide supply range; 4V to 15V
TTL and DTL compatible outputs
High accuracy: 0.5%
External Sync and Modulation Capability
Excellent Supply Rejection: 0.2%/V

APPLICATIONS

Precision Timing   Frequency Synthesis
Long Delay Generation   Pulse Counting/Summing
Sequential Timing   A/D Conversion
Binary Pattern Generation   Digital Sample and Hold

ABSOLUTE MAXIMUM RATINGS

Supply Voltage   18V
Power Dissipation
  Ceramic Package   750 mW
    Derate above +25°C   6 mw/°C
  Plastic Package   625 mW
    Derate above +25°C   5 mW/°C
Operating Temperature
  XR-2240M   −55°C to +125°C
  XR-2240C   0°C to +70°C
Storage Temperature   −65°C to +150°C

FUNCTIONAL BLOCK DIAGRAM

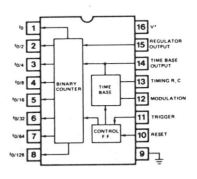

ORDERING INFORMATION

Part Number	Package	Operating Temperature
XR-2240M	Ceramic	−55°C to +125°C
XR-2240N	Ceramic	0°C to +70°C
XR-2240CN	Ceramic	0°C to +70°C
XR-2240P	Plastic	0°C to +70°C
XR-2240CP	Plastic	0°C to +70°C

SYSTEM DESCRIPTION

The XR-2240 is a combination timer/counter capable of generating accurate timing intervals ranging from microseconds through several days. The time base works as an astable multivibrator with a period equal to RC. The eight bit counter can divide the time base output by any integer value from 1 to 255. The wide supply voltage range of 4.5 to 15 V, TTL and DTL logic compatibility, and 0.5% accuracy allow wide applicability. The counter may operate independently of the time base. Counter outputs are open collector and may be wire-OR connected.

The circuit is triggered or reset with positive going pulses. By connecting the reset pin (Pin 10) to one of the counter outputs, the time base will halt at timeout. If none of the outputs are connected to the reset, the circuit will continue to operate in the astable mode. Activating the trigger terminal (Pin 11) while the timebase is stopped will set all counter outputs to the low state and start the timebase.

EXAR Corporation, 750 Palomar Avenue, Sunnyvale, CA 94086 • (408) 732-7970 • TWX 910-339-9233

FIGURE 5-22   XR-2240 programmable timer/counter. (Courtesy of EXAR Corporation.)

XR-2240

ELECTRICAL CHARACTERISTICS

Test Conditions: See Figure 2, V^+ = 5V, T_A = 25°C, R = 10 kΩ, C = 0.1 μF, unless otherwise noted.

PARAMETERS	XR-2240			XR-2240C			UNIT	CONDITIONS
	MIN	TYP	MAX	MIN	TYP	MAX		
GENERAL CHARACTERISTICS								
Supply Voltage	4		15	4		15	V	For V^+ < 4.5V, Short Pin 15 to Pin 16
Supply Current								
Total Circuit		3.5	6		4	7	mA	V^+ = 5V, V_{TR} = 0, V_{RS} = 5V
		12	16		13	18	mA	V^+ = 15V, V_{TR} = 0, V_{RS} = 5V
Counter Only		1			1.5		mA	See Figure 3
Regulator Output, V_R	4.1	4.4		3.9	4.4		V	Measured at Pin 15, V^+ = 5V
	6.0	6.3	6.6	5.8	6.3	6.8	V	V^+ = 15V, See Figure 4
TIME BASE SECTION								
Timing Accuracy*		0.5	2.0		0.5	5	%	See Figure 2
Temperature Drift		150	300		200		ppm/°C	V_{RS} = 0, V_{TR} = 5V
		80			80		ppm/°C	V^+ = 5V 0°C ≤ T ≤ 75°C
Supply Drift		0.05	0.2		0.08	0.3	%/V	V^+ = 15V
Max. Frequency	100	130			130		kHz	V^+ ≥ 8 Volts, See Figure 11
								R = 1 kΩ, C = 0.007 μF
Modulation Voltage Level								Measured at Pin 12
	3.00	3.50	4.0	2.80	3.50	4.20	V	V^+ 5V
		10.5			10.5		V	V^+ = 15V
Recommended Range of Timing Components								See Figure 8
Timing Resistor, R	0.001		10	0.001		10	MΩ	
Timing Capacitor, C	0.007		1000	0.01		1000	μF	
TRIGGER/RESET CONTROLS								
Trigger								
Trigger Threshold		1.4	2.0		1.4	2.0	V	Measures at Pin 11, V_{RS} = 0
Trigger Current		8			10		μA	
Impedance		25			25		kΩ	V_{RS} = 0, V_{TR} = 2V
Response Time**		1			1		μsec.	
Reset								
Reset Threshold		1.4	2.0		1.4	2.0	V	
Reset Current		8			10		μA	V_{TR} = 0, V_{RS} = 2V
Impedance		25			25		kΩ	
Response Time**		0.8			0.8		μsec.	
COUNTER SECTION								
Max. Toggle Rate	0.8	1.5			1.5		MHz	See Figure 4, V^+ = 5V
								V_{RS} = 0, V_{TR} = 5V
Input:								Measured at Pin 14
Impedance		20			20		kΩ	
Threshold	1.0	1.4		1.0	1.4		V	
Output:								Measured at Pins 1 thru 8
Rise Time		180			180		nsec.	R_L = 3k, C_L = 10 pF
Fall Time		180			180		nsec.	
Sink Current	3	5		2	4		mA	V_{OL} ≤ 0.4V
Leakage Current		0.01	8		0.01	15	μA	V_{OH} = 15V

*Timing error solely introduced by XR-2240, measured as % of ideal time-base period of T = 1.00 RC.
**Propagation delay from application of trigger (or reset) input to corresponding state change in counter output at pin 1.

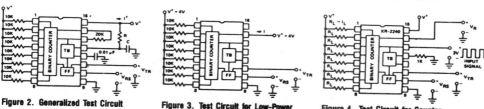

Figure 2. Generalized Test Circuit

Figure 3. Test Circuit for Low-Power Operation (Time-Base Powered Down)

Figure 4. Test Circuit for Counter Section

FIGURE 5-22 (*continued*)

PRINCIPLES OF OPERATION

The timing cycle for the XR-2240 is initiated by applying a positive-going trigger pulse to pin 11. The trigger input actuates the time-base oscillator, enables the counter section, and sets all the counter outputs to "low" state. The time-base oscillator generates timing pulses with its period, T, equal to 1 RC. These clock pulses are counted by the binary counter section. The timing cycle is completed when a positive-going reset pulse is applied to pin 10.

Figure 5 gives the timing sequence of output waveforms at various circuit terminals, subsequent to a trigger input. When the circuit is at reset state, both the time-base and the counter sections are disabled and all the counter outputs are at "high" state.

In most timing applications, one or more of the counter outputs are connected back to the reset terminal, as shown in Figure 6, with S_1 closed. In this manner, the circuit will start timing when a trigger is applied and will automatically reset itself to complete the timing cycle when a programmed count is completed. If none of the counter outputs are connected back to the reset terminal (switch S_1 open), the circuit would operate in its astable or free-running mode, subsequent to a trigger input.

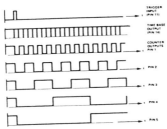

Figure 5. Timing Diagram of Output Waveforms

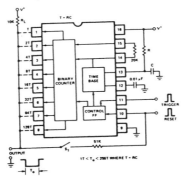

Figure 6. Generalized Circuit Connection for Timing Applications (Switch S_1 Open for Astable Operations, Closed for Monostable Operations)

PROGRAMMING CAPABILITY

The binary counter outputs (pins 1 through 8) are open-collector type stages and can be shorted together to a common pull-up resistor to form a "wired-or" connection. The combined output will be "low" as long as any one of the outputs is low. In this manner, the time delays associated with each counter output can be *summed* by simply shorting them together to a common output bus as shown in Figure 6. For example, if only pin 6 is connected to the output and the rest left open, the total duration of the timing cycle, T_O, would be 32T. Similarly, if pins 1, 5, and 6 were shorted to the output bus, the total time delay would be $T_O = (1 + 16 + 32) T = 49T$. In this manner, by proper choice of counter terminals connected to the output bus, one can program the timing cycle to be: $1T \leq T_O \leq 255T$, where $T = RC$.

TRIGGER AND RESET CONDITIONS

When power is applied to the XR-2240 with no trigger or reset inputs, the circuit reverts to "reset" state. Once triggered, the circuit is immune to additional trigger inputs, until the timing cycle is completed or a reset input is applied. If both the reset and the trigger controls are activated simultaneously trigger overrides reset.

DESCRIPTION OF CIRCUIT CONTROLS

COUNTER OUTPUTS (PINS 1 THROUGH 8)

The binary counter outputs are buffered "open-collector" type stages, as shown in Figure 15. Each output is capable of sinking $\approx$ 5 mA of load current. At reset condition, all the counter outputs are at high or non-conducting state. Subsequent to a trigger input, the outputs change state in accordance with the timing diagram of Figure 5.

The counter outputs can be used individually, or can be connected together in a "wired-or" configuration, as described in the Programming section.

RESET AND TRIGGER INPUTS (PINS 10 AND 11)

The circuit is reset or triggered with positive-going control pulses applied to pins 10 and 11. The threshold level for these controls is approximately two diode drops ($\approx$ 1.4V) above ground.

Minimum pulse widths for reset and trigger inputs are shown in Figure 10. Once triggered, the circuit is immune to additional trigger inputs until the end of the timing cycle.

MODULATION AND SYNC INPUT (PIN 12)

The period T of the time-base oscillator can be modulated by applying a dc voltage to this terminal (see Figure 13). The time-base oscillator can be synchronized to an external clock by applying a sync pulse to pin 12, as shown in Figure 16. Recommended sync pulse widths and amplitudes are also given in the figure.

FIGURE 5-22 (*continued*)

XR-2240

TYPICAL CHARACTERISTICS

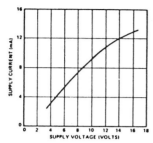

Figure 7. Supply Current vs. Supply Voltage in Reset Condition (Supply Current Under Trigger Condition is ≈ 0.7 mA less)

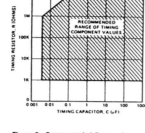

Figure 8. Recommended Range of Timing Component Values.

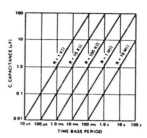

Figure 9. Time-Base Period, T, as a Function of External RC

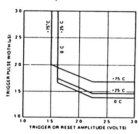

Figure 10. Minimum Trigger and Reset Pulse Widths at Pins 10 and 11

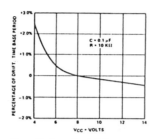

Figure 11. Power Supply Drift

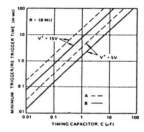

Figure 12.
A) Minimum Trigger Delay Time Subsequent to Application of Power
B) Minimum Re-trigger Time, Subsequent to a Reset Input

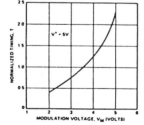

Figure 13. Normalized Change in Time-Base Period As a Function of Modulation Voltage at Pin 12

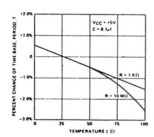

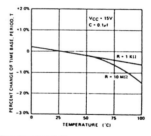

Figure 14. Temperature Drift of Time-Base Period, T

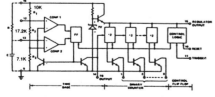

Figure 15. Simplified Circuit Diagram of XR-2240

Figure 16. Operation with External Sync Signal.
(a) Circuit for Sync Input
(b) Recommended Sync Waveform

FIGURE 5-22 (*continued*)

209

HARMONIC SYNCHRONIZATION

Time-base can be synchronized with *integer multiples or harmonics* of input sync frequency, by setting the time-base period, T, to be an integer multiple of the sync pulse period, T_S. This can be done by choosing the timing components R and C at pin 13 such that:

$$T = RC = (T_S/m) \text{ where}$$

m is an integer, $1 \leq m \leq 10$.

Figure 17 gives the typical pull-in range for harmonic synchronization, for various values of harmonic modulus, m. For m < 10, typical pull-in range is greater than ±4% of time-base frequency.

TIMING TERMINAL (PIN 13)

The time-base period T is determined by the external R-C network connected to this pin. When the time-base is triggered, the waveform at pin 13 is an exponential ramp with a period $T = 1.0 \, RC$.

TIME-BASE OUTPUT (PIN 14)

Time-Base output is an open-collector type stage, as shown in Figure 15 and requires a 20 KΩ pull-up resistor to Pin 15 for proper operation of the circuit. At reset state, the time-base output is at "high" state. Subsequent to triggering, it produces a negative-going pulse train with a period $T = RC$, as shown in the diagram of Figure 5.

Time-base output is internally connected to the binary counter section and also serves as the input for the external clock signal when the circuit is operated with an external time-base.

The counter input triggers on the negative-going edge of the timing or clock pulses applied to pin 14. The trigger threshold for the counter section is ≈ +1.5 volts. The counter section can be disabled by clamping the voltage level at pin 14 to ground.

Note:
Under certain operating conditions such as high supply voltages ($V^+ > 7V$) and small values of timing capacitor ($C < 0.1 \, \mu F$) the pulse-width of the time-base output at pin 14 may be too narrow to trigger the counter section. This can be corrected by connecting a 300 pF capacitor from pin 14 to ground.

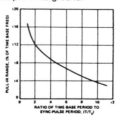

Figure 17. Typical Pull-in Range for Harmonic Synchronization

REGULATOR OUTPUT (PIN 15)

This terminal can serve as a V^+ supply to additional XR-2240 circuits when several timer circuits are cascaded (See Figure 20), to minimize power dissipation. For circuit operation with external clock, pin 15 can be used as the V^+ terminal to power-down the internal time-base and reduce power dissipation. The output current shall not exceed 10 mA.

When the internal time-base is used with $V^+ \leq 4.5V$, pin 15 should be shorted to pin 16.

APPLICATIONS INFORMATION

PRECISION TIMING (Monostable Operation)

In precision timing applications, the XR-2240 is used in its monostable or "self-resetting" mode. The generalized circuit connection for this application is shown in Figure 18.

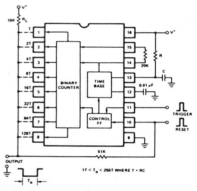

Figure 18. Circuit for Monostable Operation ($T_O = NRC$ where $1 \leq N \leq 255$)

The output is normally "high" and goes to "low" subsequent to a trigger input. It stays low for the time duration T_O and then returns to the high state. The duration of the timing cycle T_O is given as:

$$T_O = NT = NRC$$

where $T = RC$ is the time-base period as set by the choice of timing components at pin 13 (See Figure 9). N is an integer in the range of:

$$1 \leq N \leq 255$$

as determined by the combination of counter outputs (pins 1 through 8) connected to the output bus, as described below.

PROGRAMMING OF COUNTER OUTPUTS: The binary counter outputs (pins 1 through 8) are open-collector type stages and can be shorted together to a common pull-up resistor to form a "wired-or" connection where

FIGURE 5-22 (*continued*)

the combined output will be "low" as long as any one of the outputs is low. In this manner, the time delays associated with each counter output can be summed by simply shorting them together to a common output bus as shown in Figure 18. For example if only pin 6 is connected to the output and the rest left open, the total duration of the timing cycle, T_O, would be 32T. Similarly, if pins 1, 5, and 6 were shorted to the output bus, the total time delay would be $T_O = (1 + 16 + 32) T = 49T$. In this manner, by proper choice of counter terminals connected to the output bus, one can program the timing cycle to be: $1T \leq T_O \leq 255T$.

ULTRA-LONG DELAY GENERATION

Two XR-2240 units can be cascaded as shown in Figure 19 to generate extremely long time delays. In this application, the reset and the trigger terminals of both units are tied together and the time base of Unit 2 disabled. In this manner, the output would normally be high when the system is at reset. Upon application of a trigger input, the output would go to a low stage and stay that way for a total of $(265)^2$ or 65,536 cycles of the time-base oscillator.

PROGRAMMING: Total timing cycle of two cascaded units can be programmed from $T_O = 256RC$ to $T_O = 65,536RC$ in 256 discrete steps by selectively shorting any one or the combination of the counter outputs from Unit 2 to the output bus.

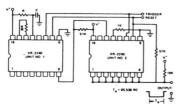

Figure 19. Cascaded Operation for Long Delay Generation

LOW-POWER OPERATION

In cascaded operation, the time-base section of Unit 2 can be powered down to reduce power consumption, by using the circuit connection of Figure 20. In this case, the V^+ terminal (pin 16) of Unit 2 is left open-circuited, and the second unit is powered from the regulator output of Unit 1, by connecting pin 15 of both units.

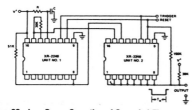

Figure 20. Low-Power Operation of Cascaded Timers

ASTABLE OPERATION

The XR-2240 can be operated in its astable or free-running mode by disconnecting the reset terminal (pin 10) from the counter outputs. Two typical circuit connections for this mode of operation are shown in Figure 21. In the circuit connection of Figure 21(a), the circuit operates in its free-running mode, with external trigger and reset signals. It will start counting and timing subsequent to a trigger input until an external reset pulse is applied. Upon application of a positive-going reset signal to pin 10, the circuit reverts back to its rest state. The circuit of Figure 21(a) is essentially the same as that of Figure 6, with the feedback switch S_1 open.

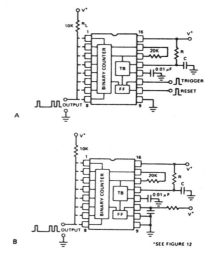

Figure 21. Circuit Connections for Astable Operation
(a) Operation with External Trigger and Reset Controls
(b) Free-running or Continuous Operation

The circuit of Figure 21(b) is designed for continuous operation. The circuit self-triggers automatically when the power supply is turned on, and continues to operate in its free-running mode indefinitely.

In astable or free-running operation, each of the counter outputs can be used individually as synchronized oscillators; or they can be interconnected to generate complex pulse patterns.

BINARY PATTERN GENERATION

In astable operation, as shown in Figure 21, the output of the XR-2240 appears as a complex pulse pattern. The waveform of the output pulse train can be determined directly from the timing diagram of Figure 5 which shows the phase relations between the counter outputs. Figure 22 shows some of these complex pulse patterns. The pulse pattern repeats itself at a rate equal to the period of the *highest* counter bit connected

FIGURE 5-22 (*continued*)

to the common output bus. The minimum pulse width contained in the pulse train is determined by the *lowest* counter bit connected to the output.

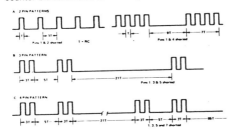

Figure 22. Binary Pulse Patterns Obtained by Shorting Various Counter Outputs

OPERATION WITH EXTERNAL CLOCK

The XR-2240 can be operated with an external clock or time-base, by disabling the internal time-base oscillator and applying the external clock input to pin 14. The recommended circuit connection for this application is shown in Figure 23. The internal time-base can be deactivated by connecting a 1 KΩ resistor from pin 13 to ground. The counters are triggered on the negative-going edges of the external clock pulse. For proper operation, a minimum clock pulse amplitude of 3 volts is required. Minimum external clock pulse width must be ≥ 1 μS.

For operation with supply voltages of 6V or less, the internal time-base section can be powered down by open-circuiting pin 16 and connecting pin 15 to V⁺. In this configuration, the internal time-base does not draw any current, and the over-all current drain is reduced by ≈ 3 mA.

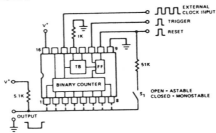

Figure 23. Operation with External Clock

FREQUENCY SYNTHESIZER

The programmable counter section of XR-2240 can be used to generate 255 discrete frequencies from a given time base setting using the circuit connection of Figure 24. The output of the circuit is a positive pulse train with a pulse width equal to T, and a period equal to (N + 1) T where N is the programmed count in the counter.

The modulus N is the *total count* corresponding to the counter outputs connected to the output bus. Thus, for example, if pins 1, 3 and 4 are connected together to the output bus, the total count is: N = 1 + 4 + 8 = 13; and the period of the output waveform is equal to (N + 1) T or 14T. In this manner, 256 different frequencies can be synthesized from a given time-base setting.

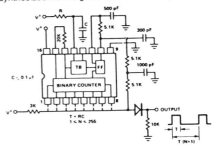

Figure 24. Frequency Synthesis from Internal Time-Base

SYNTHESIS WITH HARMONIC LOCKING: The harmonic synchronization property of the XR-2240 time-base can be used to generate a wide number of discrete frequencies from a given input reference frequency. The circuit connection for this application is shown in Figure 25. (See Figures 16 and 17 for external sync waveform and harmonic capture range.) If the time base is synchronized to (m)th harmonic of input frequency where 1 ≤ m ≤ 10, as described in the section on "Harmonic Synchronization", the frequency f_O of the output waveform in Figure 25 is related to the input reference frequency f_R as:

$$f_O = f_R \frac{m}{(N + 1)}$$

where m is the harmonic number, and N is the programmed counter modulus. For a range of 1 ≤ N ≤ 255, the circuit of Figure 25 can produce 1500 separate frequencies from a single fixed reference.

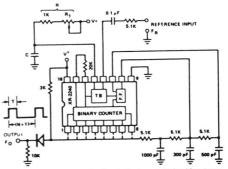

Figure 25. Frequency Synthesis by harmonic Locking to an External Reference

FIGURE 5-22 (*continued*)

One particular application of the circuit of Figure 25 is generating frequencies which are not harmonically related to a reference input. For example, by choosing the external R-C to set m = 10 and setting N = 5, one can obtain a 100 Hz output frequency synchronized to 60 Hz power line frequency.

STAIRCASE GENERATOR

The XR-2240 Timer/Counter can be interconnected with an external operational amplifier and a precision resistor ladder to form a staircase generator, as shown in Figure 26. Under reset condition, the output is low. When a trigger is applied, the op. amp. output goes to a high state and generates a negative going staircase of 256 equal steps. The time duration of each step is equal to the time-base period T. The staircase can be stopped at any desired level by applying a "disable" signal to pin 14, through a steering diode, as shown in Figure 26. The count is stopped when pin 14 is clamped at a voltage level less than 1.4V.

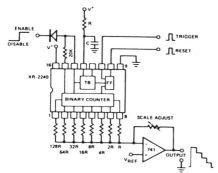

Figure 26. Staircase Generator

DIGITAL SAMPLE/HOLD

Figure 27 shows a digital sample and hold circuit using the XR-2240. The principle of operation of the circuit is similar to the staircase generator described in the previous section. When a "strobe" input is applied, the RC low-pass network between the reset and the trigger inputs of XR-2240 causes the timer to be first reset and then triggered by the same strobe input. This strobe input also sets the output of the bistable latch to a high state and activates the counter.

The circuit generates a staircase voltage at the output of the op. amp. When the level of the staircase reaches that of the analog input to be sampled, comparator changes state, activates the bistable latch and stops the count. At this point, the voltage level at the op. amp. output corresponds to the sampled analog input. Once the input is sampled, it will be held until the next strobe signal. Minimum re-cycle time of the system is ≈6 msec.

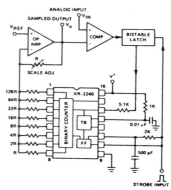

Figure 27. Digital Sample and Hold Circuit

ANALOG-TO-DIGITAL CONVERTER

Figure 28 shows a simple 8-bit A/D converter system using the XR-2240. The operation of the circuit is very similar to that described in connection with the digital sample/hold system of Figure 15. In the case of A/D conversion, the digital output is obtained in parallel format from the binary counter outputs, with the output at pin 8 corresponding to the most significant bit (MSB). The re-cycle time of the A/D converter is ≈6 msec.

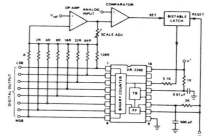

Figure 28. Analog-to-Digital Converter

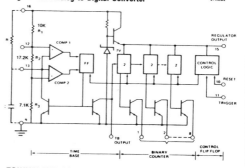

EQUIVALENT SCHEMATIC DIAGRAM

FIGURE 5-22 (*continued*)

QUESTIONS

5-1. Describe a phase-locked loop.

5-2. List the advantages of an integrated-circuit PLL over a discrete PLL.

5-3. Draw the block diagram for a PLL, list the functions of each block, and describe its basic operation.

5-4. Describe the operation of a voltage-controlled oscillator.

5-5. Describe the operation of a phase comparator.

5-6. Describe how loop acquisition is accomplished with a PLL from an initial unlocked condition until lock is achieved.

5-7. Define the following terms: *beat frequency*; *zero beat*; *acquisition time*; *open-loop gain*.

5-8. Contrast the following terms: *capture range*; *pull-in range*; *closed-loop gain*; *hold-in range*; *tracking range*; *lock range*.

5-9. Define the following terms: *uncompensated PLL*; *loop cutoff frequency*; *tracking filter*.

5-10. Define *synthesize*. What is a frequency synthesizer?

5-11. Describe direct and indirect frequency synthesis.

5-12. What is meant by the *resolution* of a frequency synthesizer?

PROBLEMS

5-1. For the VCO input-versus-output characteristics curve shown below, determine:
  (a) The frequency of operation for a −2-V input signal.
  (b) The frequency deviation for a ±2-V_p input signal.
  (c) The transfer function, K_o, for the linear portion of the curve (−3 to +3 V).

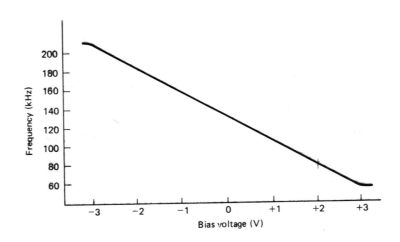

5-2. For the output voltage-versus-phase difference (θ_e) characteristic curve shown below, determine:

  (a) The output voltage for a $-45°$ phase difference.
  (b) The output voltage for a $+60°$ phase difference.
  (c) The maximum peak output voltage.
  (d) The transfer function, K_d.

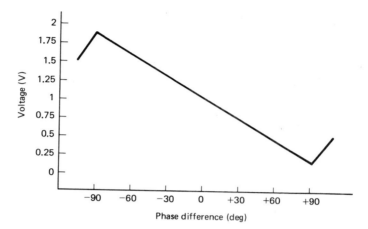

5-3. For the PLL shown in Figure 5-7a a VCO natural frequency $F_n = 150$ kHz, an input frequency $F_i = 160$ kHz, and the following circuit gains $K_d = 0.2$ V/rad, $K_f = 1$, $K_a = 4$, and $K_o = 15$ kHz/V, determine:

  (a) The open-loop gain K_v.
  (b) ΔF.
  (c) V_{out}.
  (d) V_d.
  (e) θ_e.
  (f) The hold-in range, ΔF_{max}.

5-4. Plot the frequency response for a PLL with an open-loop gain $K_v = 20$ kHz/rad. On the same log paper, plot the response with a single-pole loop filter with a cutoff frequency $\omega_c = 100$ rad/s, and a two-pole loop filter with the same cutoff frequency.

5-5. Determine the change in frequency (ΔF) for a VCO with a transfer function $K_o = 2.5$ kHz/V and a dc input voltage change $\Delta V = 0.8$ V.

5-6. Determine the voltage at the output of a phase comparator with a transfer function $K_d = 0.5$ V/rad and a phase error $\theta_e = 0.75$ rad.

5-7. Determine the hold-in range (ΔF_{max}) for a PLL with a loop gain $K_v = 110$ kHz/rad.

5-8. Determine the phase error necessary to produce a VCO frequency shift $\Delta F_o = 10$ kHz for a loop gain $K_v = 40$ kHz/rad.

5-9. Determine two output frequencies for the multiple-crystal frequency synthesizer shown in Figure 5-15 if crystals X8 and X18 are selected.

5-10. Determine the output frequency from the single-crystal frequency synthesizer shown in Figure 5-16 for the following harmonics.

Harmonic Generator	Harmonic
1	6
2	4
3	7
4	1
5	2
6	6

5-11. Determine which channel the switched-crystal frequency synthesizer shown in Figure 5-17 is tuned for the following crystal selections: X4 and X8.

5-12. Determine the first and second intermediate frequencies for Problem 5-11.

5-13. Determine F_c for the PLL frequency synthesizer shown in Figure 5-18 for a natural frequency $F_n = 200$ kHz, $\Delta F = 0$ Hz, and $n = 20$.

Chapter 6

SINGLE-SIDEBAND COMMUNICATIONS SYSTEMS

INTRODUCTION

Conventional AM DSBFC communications systems, such as those discussed in Chapters 3 and 4, have several inherent and outstanding disadvantages. First, in conventional AM systems, at least two-thirds of the transmitted power is in the carrier. However, there is no information in the carrier; the information is contained in the sidebands. Also, the information contained in the upper sideband is identical to the information contained in the lower sideband. Therefore, transmitting both sidebands is redundant. Consequently, conventional AM is both power and bandwidth inefficient, which are two of the most important considerations when designing electronic communications systems.

SINGLE-SIDEBAND SYSTEMS

There are many different types of *sideband* communications systems. Some of these systems conserve bandwidth, some conserve power, and some conserve both bandwidth and power. Figure 6-1 compares the frequency spectra and relative power distributions for conventional AM and several of the more common single-sideband (SSB) systems.

AM Single-Sideband Full Carrier

AM single-sideband full carrier (SSBFC) is a form of amplitude modulation in which the carrier is transmitted at *full* power, but one of the sidebands is removed. Therefore,

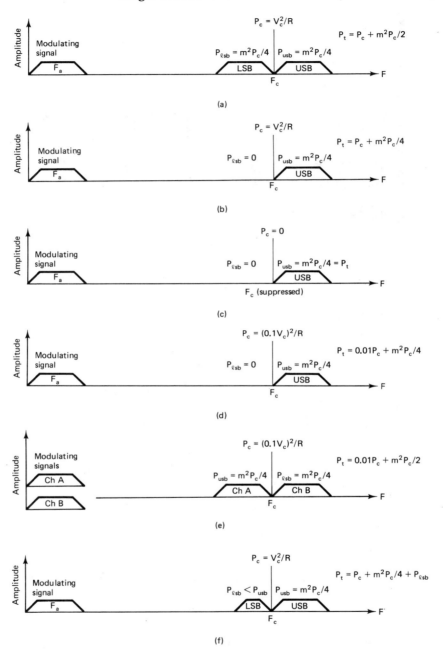

FIGURE 6-1 Single-sideband systems: (a) conventional DSBFC AM; (b) full-carrier single sideband; (c) suppressed-carrier single sideband; (d) reduced-carrier single sideband; (e) independent sideband; (f) vestigal sideband.

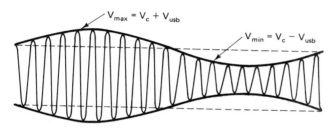

FIGURE 6-2 SSBFC waveform, 100% modulation.

SSBFC transmission requires only half as much bandwidth as conventional AM. The frequency spectrum and relative power distribution for SSBFC are shown in Figure 6-1b. Note that with 100% modulation, the carrier power (P_c) constitutes at least four-fifths (80%) of the total transmitted power (P_t), and only one-fifth (20%) of the total power is in the sideband. With conventional AM, two-thirds (67%) of the total power is in the carrier and one-third (33%) is in the sidebands. Therefore, although SSBFC requires less total power (P_t), it actually utilizes a smaller percentage of the total power for the information carrying portion of the signal (i.e., the sidebands). Figure 6-2 shows the envelope for a 100% modulated SSBFC wave for a single frequency-modulating signal. Comparing to Figure 3-6c shows that the SSBSC waveform is identical to that of a 50% modulated double-sideband wave. Recall from Chapter 3 that the maximum positive and negative peaks in a conventional AM envelope occur when the carrier and both of the side frequencies reach their respective peaks at the same time, and the peak change in the envelope is equal to the sum of the amplitudes of the upper and lower side frequencies. With single-sideband transmission, there is only a single side frequency (either the upper or lower) to add to the carrier. Therefore, the positive and negative excursions in the envelope are only half what they are with double-sideband transmission. Consequently, with single-sideband full-carrier transmission, the demodulated signals have only half the amplitude as a double-sideband demodulated wave. Thus a trade-off is made with SSBFC. SSBFC requires less bandwidth than DSBFC but also produces a lower amplitude-demodulated signal. If the bandwidth is halved, the total noise is reduced by 3 dB. However, if one sideband is removed, the power in the information portion of the wave is also halved. Consequently, the S/N is the same. With SSBFC, the repetition rate of the envelope is the frequency of the modulating signal. Therefore, as with double-sideband transmission, the information is contained in the shape of the modulated envelope.

AM Single-Sided Suppressed Carrier

AM single-sideband suppressed carrier (SSBSC) is a form of amplitude modulation in which the carrier is totally suppressed and one of the sidebands is removed. Therefore, SSBSC requires half as much bandwidth as conventional AM and considerably less transmitted power. The frequency spectrum and relative power distribution for SSBSC are shown in Figure 6-1c. It can be seen that the sideband power (P_{sb}) makes up 100% of the total transmitted power. Figure 6-3 shows a SSBSC waveform for a single-fre-

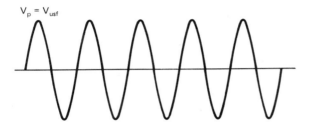

$V_p = V_{usf}$

FIGURE 6-3   SSBSC waveform.

quency modulating signal. As you can see, the waveform is not an envelope; it is simply a single frequency equal to the carrier frequency plus or minus the modulating signal frequency, depending on whether the USF or LSF is suppressed.

AM Single-Sideband Reduced Carrier

AM single-sideband reduced carrier (SSBRC) is a form of amplitude modulation where one sideband is totally removed and the carrier voltage is reduced to approximately 10% of its unmodulated amplitude. Consequently, the transmitted sideband can have as much as 86% of the total transmitted power. Generally, the carrier is totally suppressed during modulation, then reinserted later at a reduced amplitude. Consequently, SSBRC is sometimes called single-sideband *reinserted* carrier. The reinserted carrier is called a *pilot* carrier and is reinserted for demodulation purposes, which is explained later in this chapter. The frequency spectrum and relative power distribution for SSBRC are shown in Figure 6-1d. The figure shows that the sideband power constitutes approximately 100% of the total transmitted power. Figure 6-4a shows the transmitted waveform for a single-frequency modulating signal when the carrier level equals the sideband level, and Figure 6-4b shows the waveform when the carrier level is less than the sideband level. Note that the repetition rate of the envelope is equal to the modulating signal frequency (F_a). To demodulate this signal with conventional peak or envelope detectors, the carrier must be separated, amplified, then reinserted in the receiver. Therefore, suppressed-carrier transmission is sometimes called *exalted* carrier because the carrier is exalted or elevated in the receiver prior to demodulation. With an exalted carrier system, the separate carrier amplification must be sufficient to raise the amplitude of the carrier to a value greater than that of the sideband signal. SSBRC requires half as much bandwidth as conventional AM and, because the carrier is transmitted at a reduced level, also conserves power.

AM Independent Sideband

AM independent sideband (ISB) is a form of amplitude modulation where a single carrier frequency is separately modulated by two independent modulating signals. In essence, ISB is a form of double-sideband transmission in which the transmitter has two single-sideband suppressed-carrier modulators. One modulator removes the upper

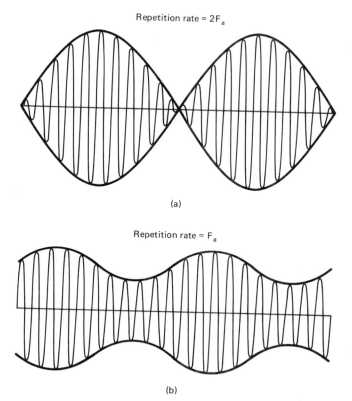

Repetition rate = $2F_a$

(a)

Repetition rate = F_a

(b)

FIGURE 6-4 SSBRC waveform: (a) carrier level equal to the sideband level; (b) carrier level less than the sideband level.

sideband and the other removes the lower sideband. The single-sideband output signals from the two modulators are combined to form a double-sideband signal in which the two sidebands are totally independent from each other except that they are symmetrical about a common carrier frequency. ISB transmission consists of two independent sidebands with one positioned above and one positioned below the suppressed carrier. For demodulation purposes, the carrier is generally reinserted at a reduced level. Figure 6-1e shows the frequency spectrum and power distribution for ISB, and Figure 6-5 shows the transmitted waveform for two independent single-frequency sources. The two sources are of equal frequency; therefore, the waveform is identical to a double-sideband suppressed-carrier waveform with a repetition rate equal to twice the modulating signal frequency ($2F_a$). ISB conserves both transmitted power and bandwidth as two information sources are transmitted within the frequency spectrum required for a single source with conventional AM DSBFC. ISB is one technique that is used in the United States for stereo AM transmission. One channel (the left) is transmitted in the lower sideband, and the other channel (the right) is transmitted in the upper sideband.

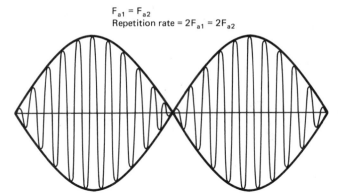

$F_{a1} = F_{a2}$
Repetition rate = $2F_{a1} = 2F_{a2}$

FIGURE 6-5   ISB waveform.

AM Vestigial Sideband

AM vestigial sideband (VSB) is a form of amplitude modulation where the carrier is transmitted at full power along with one complete sideband and part of the other sideband. VSB is a form of AM double sideband where the lower frequency-modulating signals are transmitted double sideband and thus appreciate the benefit of 100% AM modulation. The higher-frequency modulating signals are transmitted single sideband and thus can achieve only 50% AM modulation. Consequently, the lower frequencies are emphasized in the demodulator and produce larger amplitude signals than the higher frequencies. The frequency spectrum and relative power distribution for VSB are shown in Figure 6-1f. Probably the most widely known VSB system is the picture portion of a commercial television broadcasting signal.

Comparison of Single-Sideband to Conventional AM

Bandwidth conservation is an obvious advantage of single-sideband transmission over conventional double-sideband AM. The preceding discussions and the frequency spectra of Figure 6-1 show that single sideband requires only half as much bandwidth as conventional AM, and also that single-sideband transmission conserves transmitted power. However, the transmitted power necessary to produce a given receive signal-to-noise ratio (*S/N*) is a convenient means of comparing the power requirements and relative performance of sideband and conventional AM systems. The receive *S/N* determines the degree of intelligibility of a received signal.

Peak envelope power (PEP) is the rms power developed at the crest of the modulation envelope. With conventional AM, the envelope contains 1 unit of carrier power and 0.25 units of power in each sideband for a total transmitted PEP = 1.5 units. A single-sideband transmitter rated at 0.5 unit of PEP will produce the same *S/N* at the output of the receiver as 1.5 units of carrier plus sideband power from a conventional AM system. In other words, the same performance is achieved with SSB using only one-third the transmitted power. Table 6-1 compares conventional AM with single side-

TABLE 6-1   CONVENTIONAL AM VERSUS SINGLE-SIDEBAND

	Conventional AM	SSB
Rated power (in units)	$P_t = 1 + 0.5 = 1.5$	$P_t = 0.5 = \text{PEP}$
Voltage vector (100% mod.)		$\text{PEV} = 0.7$
RF envelope	PEV = 2 PEP = 4	PEV = 0.7 PEP = 0.5
Demodulated information signal	USB + LSB = 1	0.7
Arbitrary noise voltage per kHz of bandwidth	0.1 V/kHz	0.07 V/kHz
S/N ratio = 20 log S/N	$20 \log \frac{1}{0.1} = 20 \text{ dB}$	$20 \log \frac{0.7}{0.07} = 20 \text{ dB}$

band for a single frequency-modulating signal. The voltage vectors for the power requirements stated are shown. It can be seen that it requires 0.5 unit of voltage per sideband
and 1 unit for the carrier with conventional AM for a total of 2 PEV units (peak
envelope volts) and only 0.7 PEV for single sideband. The RF envelopes for AM and

sideband transmission are also shown, which correspond to the voltage and power relationships outlined previously. The demodulated signal at the output of a conventional AM receiver is proportional to the quadratic sum of the upper and lower sideband signals, which equals 1 PEV unit. For sideband reception, the demodulated signal is 0.707(1) = 0.7 PEV. If the noise voltage for conventional AM is arbitrarily chosen as 0.1 V/kHz, the noise voltage for a single-sideband waveform with half the bandwidth is 0.7 V/kHz. Consequently, the *S/N* performance for single sideband is equal to that of conventional AM.

CIRCUITS FOR GENERATING SSB

In the preceding section it was shown that with most single-sideband systems, the carrier is either totally suppressed or reduced to only a fraction of its original value. To remove the carrier from a modulated wave or to reduce its amplitude is difficult, if not impossible, without also removing a portion of each sideband (conventional notch filters simply do not have sufficient Q-factors). Therefore, modulator circuits that inherently remove the carrier during the modulation process have been developed. A circuit that produces a double-sideband signal with a suppressed carrier is the *balanced* modulator. The balanced modulator has rapidly become one of the most widely used circuits in electronic communications. In addition to suppressed-carrier AM systems, the balanced modulator is widely used in FM and PM systems. The balanced modulator is also used extensively in data modems and digital modulation systems such as phase shift keying (PSK) and quadrature amplitude modulation (QAM).

Balanced Ring Modulator

Figure 6-6a shows the schematic diagram for a *balanced ring modulator* constructed with diodes. Semiconductor diodes are ideally suited for use in balanced modulator circuits because they are stable, require no external power source, have a long life, and require very little maintenance. The balanced ring modulator is sometimes called a *balanced lattice modulator* or simply a *balanced modulator*. The balanced modulator has two inputs: a single frequency carrier and the modulating signal, which may be a single frequency or a complex waveform. For the balanced modulator to operate properly, the amplitude of the carrier must be sufficiently greater than the amplitude of the modulating signal (approximately six to seven times greater). This is to ensure that the carrier controls the ''on'' or ''off'' condition of the four diode switches (D_1 to D_4).

Circuit operation.   Essentially, diodes D_1 to D_4 are electronic switches that control whether the modulating signal is passed from transformer T_1 to transformer T_2 as is or with a 180° phase reversal. With the carrier (V_c) polarity as shown in Figure 6-6b, diode switches D_1 and D_2 are forward biased and ''on,'' while diode switches D_3 and D_4 are reverse biased and ''off.'' Consequently, the modulating signal (V_a) is transferred across the closed switches to T_2 as is. When the polarity of the carrier

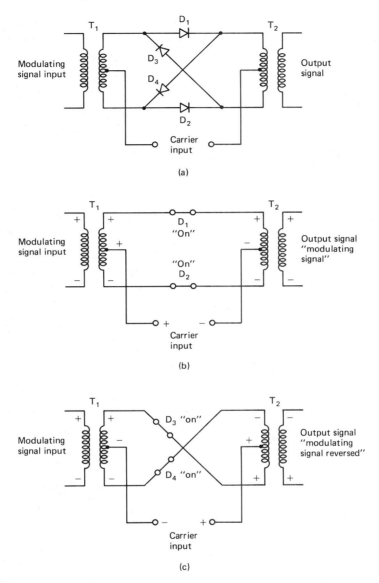

FIGURE 6-6 **Balanced ring modulator: (a) schematic diagram; (b) D_1 and D_2 biased "on";**
(c) D_3 and D_4 biased "on."

reverses, as shown in Figure 6-6c, diode switches D_1 and D_2 are reverse biased and "off," while diode switches D_3 and D_4 are forward biased and "on." Consequently, the modulating signal undergoes a 180° phase reversal before reaching T_2. Carrier current flows from its source to the center taps of T_1 and T_2, where it splits and goes in opposite directions through the upper and lower halves of the transformers. Thus their magnetic fields cancel in the secondary and the carrier component is removed. If the diodes are not perfectly matched or if the transformers are not tapped in their exact centers, the circuit is out of balance and the carrier is not totally suppressed. It is virtually impossible to achieve perfect balance; therefore, there is always a small carrier component in the output signal. This is commonly called *carrier leak*. The amount of carrier suppression is typically between 40 and 60 dB.

Figure 6-7 shows the balanced modulator waveforms for a single frequency-modulating signal, the carrier, the output waveform before filtering, and the output waveform after filtering. It can be seen that D_1 and D_2 conduct during the positive half-cycles of V_c, and D_3 and D_4 conduct during the negative half-cycles. The output of the modulator

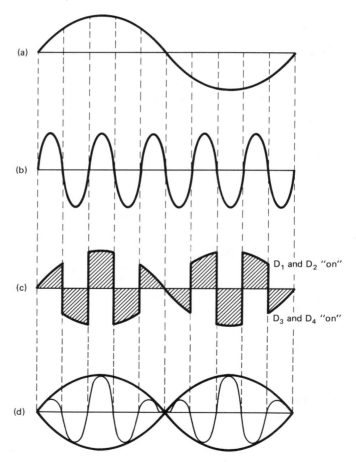

FIGURE 6-7   Balanced modulator waveforms: (a) modulating signal; (b) carrier signal; (c) output waveform before filtering; (d) output waveform after filtering.

consists of a series of RF pulses whose repetition rate is determined by the switching or RF carrier frequency and whose amplitude is controlled by the level of the modulating signal. Consequently, the output waveform takes the shape of the modulating signal except with alternating positive and negative polarities that correspond to the polarity of V_c.

Mathematical analysis. A balanced modulator is a *product* modulator; the output signal is the product of the two input signals. The carrier frequency is multiplied by the modulating signal. Mathematically, the output of a balanced modulator for a single frequency-modulating signal is

$$\text{output} = \underbrace{(\sin \omega_a t)}_{\substack{\text{modulating} \\ \text{signal}}} \times \underbrace{(\sin \omega_c t)}_{\substack{\text{carrier} \\ \text{signal}}}$$

Using the trigonometric identity for the product of two sine functions, the output is

$$\text{output} = \overbrace{\underbrace{\frac{1}{2} \cos (\omega_c - \omega_a)t}_{\substack{\text{difference} \\ \text{frequency}}} - \underbrace{\frac{1}{2} \cos (\omega_c + \omega_a)t}_{\substack{\text{sum} \\ \text{frequency}}}}^{\text{cross products}}$$

From the preceding mathematical operation, it can be seen that the output signal is simply two cosine waves; one is the sum frequency ($\omega_c + \omega_a$) and the other is the difference frequency ($\omega_c - \omega_a$). Both the carrier and the modulating signal are suppressed.

FET Push-Pull Balanced Modulator

Figure 6-8 shows a schematic diagram for a balanced modulator that uses FETs (field-effect transistors) rather than switching diodes for the nonlinear devices. A FET is a nonlinear device that exhibits square-law properties. That is, its output contains a term that is proportional to the input signal squared, which includes, of course, the second-order cross-product frequencies (i.e., the primary sum and difference frequencies). Like the diode balanced modulator, the FET modulator is a product modulator and produces only the cross-product frequencies at its output. The FET balanced modulator is similar to a standard push-pull amplifier except that the modulator circuit has two inputs (the carrier and the modulating signal).

Circuit operation. The carrier is fed into the circuit in such a way that it is applied simultaneously and in phase to the gates of both FET amplifiers (Q_1 and Q_2). The carrier produces currents in the top and bottom halves of the output transformer that are equal in magnitude but 180° out of phase. Therefore, they cancel and no carrier component appears in the output. The modulating signal is applied to the circuit in

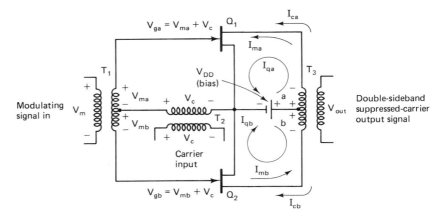

FIGURE 6-8   FET balanced modulator. For the polarities shown:

$$I_{ta} = I_{qa} + I_{ca} + I_{ma}$$
$$I_{tb} = -I_{qb} - I_{cb} + I_{mb}$$
$$I_t = I_{ma} + I_{mb} = 2I_m$$

V_{out} **is proportional to the modulating current (I_{ma} and I_{mb}).**

such a way that it is applied simultaneously to the two gates 180° out of phase. The modulating signal causes an increase in the drain current in one FET and, at the same time, causes a decrease in the drain current in the other FET.

Figure 6-9 shows the phasor diagrams for the currents produced in the output transformer of a FET balanced modulator. Figure 6-9a shows that the quiescent dc drain currents from Q_1 and Q_2 (I_{qa} and I_{qb}) pass through their respective halves of the primary winding of T_3 180° out of phase with each other. Figure 6-9a also shows that an increase in drain current due to the carrier signal (I_{ca} and I_{cb}) adds to the quiescent current in both halves of the transformer windings, producing currents (I_{ta} and I_{tb}) that are equal and simply the sum of the quiescent and carrier currents. I_{ta} and I_{tb} are equal but travel in opposite directions; consequently, they cancel each other. Figure 6-9b shows the phasor sum of the quiescent and carrier currents when the carrier currents travel in the opposite direction to the quiescent currents. The total currents in both halves of the winding are still equal in magnitude, but now they are equal to the difference between the quiescent and carrier currents. Figure 6-9c shows the phasor diagram when a current component is added due to a modulating signal. The modulating signal currents (I_{ma} and I_{mb}) produce in their respective halves of the output transformer currents that are in phase with each other. However, it can be seen that in one half of the windings the total current is equal to the sum of the dc current, the carrier current, and the modulating signal current; and in the other half of the windings the total current is equal to the difference between the dc and carrier currents and the modulating signal current. Thus the dc and carrier currents cancel in the secondary windings, while the sum and difference components remain. The continuously changing carrier and modulating signal currents produce the cross-product frequencies.

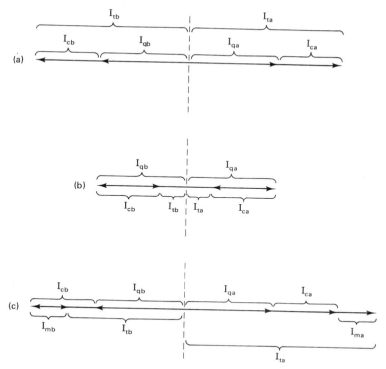

FIGURE 6-9 **FET balanced modulator phasor diagrams: (a) in-phase sum of dc and carrier currents; (b) out-of-phase sum of dc and carrier currents; (c) sum of dc, carrier, and modulating signal currents.**

The carrier and modulating signal polarities shown in Figure 6-8 produce an output current that is proportional to the carrier and modulating signal voltages. The carrier signal (V_c) produces a current in both FETs (I_{ca} and I_{cb}) that is in the same direction as the quiescent currents (I_{qa} and I_{qb}). The modulating signal (V_{ma} and V_{mb}) produces a current in Q_1 (I_{ma}) that is in the same direction as I_{ca} and I_{qa} and a current in Q_2 (I_{mb}) that is in the opposite direction as I_{cb} and I_{qb}. Therefore, the total current through the a-side of T_3 is $I_{ta} = I_{ca} + I_{qa} + I_{ma}$ and the total current through the b-side of T_3 is $I_{tb} = -I_{cb} - I_{qb} + I_{mb}$. Thus the net current through the primary winding of T_3 is $I_{ta} + I_{tb} = I_{ma} + I_{mb}$. For a modulating signal with the opposite polarity, the drain current in Q_2 will increase and in Q_1 it will decrease. Ignoring the quiescent dc current (I_{qa} and I_{qb}), the drain current in one FET is the sum of the carrier and modulating signal currents ($I_c + I_m$), and the drain current in the other FET is the difference between the carrier and modulating signal currents ($I_c - I_m$).

T_1 is an audio transformer while T_2 and T_3 are radio-frequency transformers. Therefore, any audio component that appears at the drain circuits of Q_1 and Q_2 is not passed on to the output. To achieve total carrier suppression, Q_1 and Q_2 must be perfectly matched and T_1 and T_3 must be tapped in their exact centers. With both the diode and FET balanced modulators, between 40 and 60 dB of carrier suppression is typical.

Balanced Bridge Modulator

Figure 6-10a shows the schematic diagram for a *balanced bridge modulator*. The operation of the bridge modulator, like that of the balanced ring modulator, is completely dependent on the switching action of diodes D_1 through D_4 under the influence of the carrier and modulating signal voltages. Again, the carrier voltage controls the "on" or "off" condition of the diodes and therefore must be appreciably larger than the modulating signal voltage.

Circuit operation.   For the carrier polarities shown in Figure 6-10b, all four diodes are reverse biased and "off." Consequently, the audio signal voltage is transferred

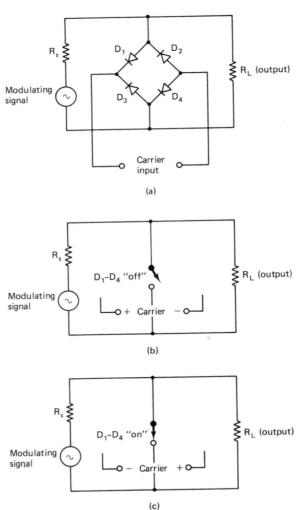

FIGURE 6-10   Balanced bridge modulator: (a) schematic diagram; (b) diodes biased "off"; (c) diodes biased "on."

directly to the load resistor (R_L). Figure 6-10c shows the equivalent circuit for a carrier with the opposite polarity. All four diodes are forward biased and "on," and the load resistor is essentially shorted out. As the carrier voltage changes from positive to negative, and vice versa, the output waveform contains a series of pulses that are comprised mainly of the upper and lower sideband frequencies.

Linear Integrated Circuit (LIC) Balanced Modulator

LIC balanced modulators are available, such as the LM1496, that can provide carrier suppression of 50 dB at 10 MHz and up to 65 dB at 500 kHz. The LM1496 balanced modulator integrated circuit is a *double-balanced modulator-demodulator* which produces an output signal that is proportional to the product of its input signals (the carrier and modulating signal). Integrated circuits are ideally suited for applications that require a balanced operation.

Circuit operation. Figure 6-11 shows a simplified schematic diagram for a *differential* amplifier, which is the fundamental circuit of an LIC balanced modulator because of its excellent *common-mode rejection ratio* (typically 85 dB). When a carrier signal is applied to the base of Q_1, the emitter current in Q_1 and Q_2 will vary by the

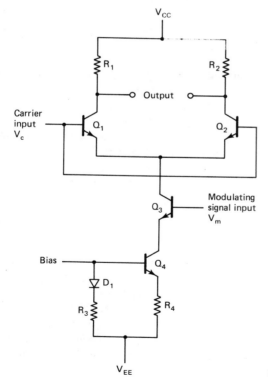

FIGURE 6-11   Differential amplifier schematic.

same amount. Because the emitter current for both Q_1 and Q_2 comes from a common current source (Q_4), any increase in Q_1's emitter current results in a corresponding decrease in Q_2's emitter current, and vice versa. Similarly, when a carrier signal is applied to the base of Q_2, the emitter currents in Q_1 and Q_2 will vary by the same amount except in opposite directions. Consequently, if the same carrier signal is fed simultaneously to the base of Q_1 and Q_2, the respective increases and decreases in the emitter currents of Q_1 and Q_2 are equal and cancel each other. Therefore, the collector currents and output voltage remain unchanged. If a modulating signal is applied to the base of Q_3, it causes a corresponding increase or decrease (depending on its polarity) in the collector currents of Q_1 and Q_2. However, the carrier and modulating signal frequencies mix in Q_1 and Q_2 and produce cross-product frequencies in the output. Therefore, the carrier and modulating signal frequencies are canceled in the balanced transistors, while the sum and difference frequencies appear at the output.

Figure 6-12a shows the schematic diagram for a typical AM DSBSC modulator using the LM1496 integrated circuit. The carrier signal is applied to pin 10, which, in conjunction with pin 8, provides an input to a quad cross-coupled differential output amplifier. This configuration is used to ensure that full-wave multiplication of the carrier and modulating signal occurs. The modulating signal is applied to pin 1, which, in conjunction with pin 4, provides a differential input to the current driving transistors for the output difference amplifiers. The 1-kΩ resistor connected between pins 2 and 3 sets the gain of the difference amplifier. The 50-kΩ potentiometer, in conjunction with V_{EE} (-8 V dc), is used to balance the bias currents for the difference amplifiers and null the carrier. Pins 6 and 12 are single-ended outputs which contain carrier and sideband components. When one of the outputs is inverted and added to the other output, the carrier is suppressed and a double-sideband suppressed carrier wave is produced. Such a process is accomplished in the op-amp subtractor circuit. The subtractor inverts the signal at the inverting ($-$) input and adds it to the signal at the noninverting ($+$) input. Thus a double-sideband suppressed carrier wave appears at the output of the op-amp. The 6.8-kΩ resistor connected to pin 5 is a bias resistor for the internal constant current supply. The specification sheet for Motorola's 1496 integrated-circuit balanced modulator chip is shown in Figure 6-12b.

SINGLE-SIDEBAND TRANSMITTERS

The transmitters used for SSBSC and SSBRC transmission are identical except that the reinserted carrier transmitters have an additional circuit which adds a low-amplitude carrier to the single-sideband waveform after suppressed carrier modulation has been performed. The reinserted carrier is called a pilot carrier. The circuit where the carrier is reinserted is called a linear summer if it is a resistive network and a *hybrid coil* if the SSB waveform and the pilot are inductively combined in a transformer bridge. There are three techniques that are commonly used for single-sideband generation: the *filter method*, the *phase shift method*, and the so-called "*third method*."

SSB Transmitter: Filter Method

Figure 6-13 shows a block diagram for a SSB transmitter that uses balanced modulators to suppress the carrier and bandpass filters to suppress the unwanted sideband. The modulating signal is an audio spectrum that extends from 0 to 5 kHz. In the first balanced modulator (balanced modulator 1), the modulating signal mixes with a low-frequency (LF) 100-kHz carrier to produce a single-sideband suppressed-carrier spectrum that is centered around the suppressed 100-kHz IF carrier. Bandpass filter 1 (BPF 1) is tuned to a 5-kHz bandwidth centered around 102.5 kHz, which is the center frequency of the upper sideband spectrum. The pilot or reduced amplitude carrier is added to the single-sideband waveform in the carrier reinsertion stage, which is simply a linear summer. The summer is a simple adder circuit that combines the 100-kHz pilot carrier to the 100- to 105-kHz upper sideband spectrum. Thus the output of the summer is a SSBRC waveform. (If suppressed-carrier transmission is used, the carrier pilot and summer circuit are omitted.) The low-frequency IF is converted to the final operating frequency through a series of frequency translations. First, the SSBRC waveform is mixed in balanced modulator 2 with a 2-MHz medium-frequency carrier (MF). The output is a double-sideband suppressed-carrier signal where the upper and lower sidebands each contain the original SSBRC signal spectrum. The upper and lower sidebands are

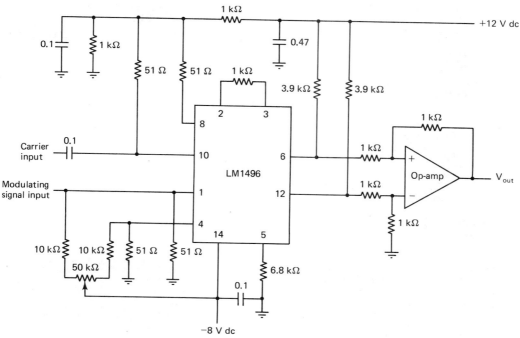

FIGURE 6-12 AM DSBSC modulator using the LM1496 linear integrated circuit: (a) schematic diagram; (b) 1496 specification sheets. (Copyright Motorola, Inc., 1982. Used by permission.)

MOTOROLA
SEMICONDUCTORS
P.O. BOX 20912 • PHOENIX, ARIZONA 85036

MC1496
MC1596

Specifications and Applications Information

BALANCED MODULATOR -- DEMODULATOR

SILICON MONOLITHIC INTEGRATED CIRCUIT

BALANCED MODULATOR — DEMODULATOR

. . . designed for use where the output voltage is a product of an input voltage (signal) and a switching function (carrier). Typical applications include suppressed carrier and amplitude modulation, synchronous detection, FM detection, phase detection, and chopper applications. See Motorola Application Note AN-531 for additional design information.

- Excellent Carrier Suppression — 65 dB typ @ 0.5 MHz
  — 50 dB typ @ 10 MHz
- Adjustable Gain and Signal Handling
- Balanced Inputs and Outputs
- High Common-Mode Rejection — 85 dB typ

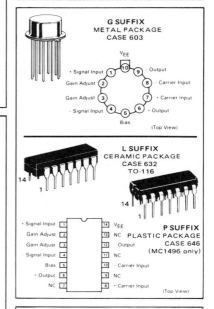

G SUFFIX
METAL PACKAGE
CASE 603

V_{EE}

- Signal Input (1) — Output
Gain Adjust (2) — Carrier Input
Gain Adjust (3) + Carrier Input
- Signal Input (4) + Output
Bias
(Top View)

L SUFFIX
CERAMIC PACKAGE
CASE 632
TO-116

P SUFFIX
PLASTIC PACKAGE
CASE 646
(MC1496 only)

Pin		Pin	
+ Signal Input	1	14	V_{EE}
Gain Adjust	2	13	NC
Gain Adjust	3	12	- Output
- Signal Input	4	11	NC
Bias	5	10	- Carrier Input
+ Output	6	9	NC
NC	7	8	+ Carrier Input

(Top View)

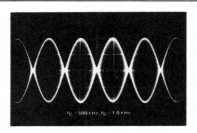

FIGURE 1 —
SUPPRESSED-CARRIER
OUTPUT WAVEFORM

f_C = 500 kHz, f_S = 1.0 kHz

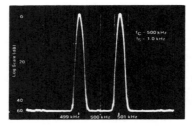

FIGURE 2 —
SUPPRESSED-CARRIER
SPECTRUM

f_C = 500 kHz
f_S = 1.0 kHz

ORDERING INFORMATION

Device	Temperature Range	Package
MC1496G	0 to +70°C	Metal Can
MC1496L	0 to +70°C	Ceramic DIP
MC1496P	0 to +70°C	Plastic DIP
MC1596G	-55 to +125°C	Metal Can
MC1596L	-55 to +125°C	Ceramic DIP

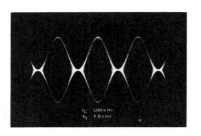

FIGURE 3 —
AMPLITUDE-MODULATION
OUTPUT WAVEFORM

f_C = 500 kHz
f_S = 1.0 kHz

FIGURE 4 — AMPLITUDE-MODULATION SPECTRUM

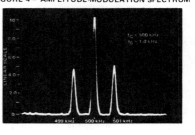

f_C = 500 kHz
f_S = 1.0 kHz

DS 9132 R2

FIGURE 6-12(b)

MAXIMUM RATINGS* (T_A = +25°C unless otherwise noted)

Rating	Symbol	Value	Unit
Applied Voltage ($V_6 - V_7$, $V_8 - V_1$, $V_9 - V_7$, $V_9 - V_8$, $V_7 - V_4$, $V_7 - V_1$, $V_8 - V_4$, $V_6 - V_8$, $V_2 - V_5$, $V_3 - V_5$)	ΔV	30	Vdc
Differential Input Signal	$V_7 - V_8$ $V_4 - V_1$	+5.0 ±(5+I_5R_e)	Vdc
Maximum Bias Current	I_5	10	mA
Thermal Resistance, Junction to Air Ceramic Dual In-Line Package Plastic Dual In-Line Package Metal Package	$R_{\theta JA}$	 180 100 200	°C/W

Operating Temperature Range
           MC1496
           MC1596
	T_A	
		 0 to +70
		 -55 to +125
			°C

Storage Temperature Range	T_{stg}	-65 to +150	°C

ELECTRICAL CHARACTERISTICS* (V_{CC} = +12 Vdc, V_{EE} = -8.0 Vdc, I_5 = 1.0 mAdc, R_L = 3.9 kΩ, R_e = 1.0 kΩ,
T_A = +25°C unless otherwise noted) (All input and output characteristics are single-ended unless otherwise noted.)

Characteristic	Fig	Note	Symbol	MC1596 Min	MC1596 Typ	MC1596 Max	MC1496 Min	MC1496 Typ	MC1496 Max	Unit

Carrier Feedthrough
   V_C = 60 mV(rms) sine wave and f_C = 1.0 kHz
   offset adjusted to zero f_C = 10 MHz
	5	1	V_{CFT}	
				 —
				 —
					
					 40
					 140
						
						 —
						 —
							
							 —
							 —
								
								 40
								 140
									
									 —
									 —
										µV(rms)

V_C = 300 mVp-p square wave:
   offset adjusted to zero f_C = 1.0 kHz
   offset not adjusted f_C = 1.0 kHz
				
				 —
				 —
					
					 0.04
					 20
						
						 0.2
						 100
							
							 —
							 —
								
								 0.04
								 20
									
									 0.4
									 200
										mV(rms)

Carrier Suppression
   f_S = 10 kHz, 300 mV(rms)
   f_C = 500 kHz, 60 mV(rms) sine wave
   f_C = 10 MHz, 60 mV(rms) sine wave
	5	2	V_{CS}	
				
				 50
				 —
					
					 65
					 50
						
						 —
						 —
							
							 40
							 —
								
								 65
								 50
									
									 —
									 —
										dB
										
										 k

Transadmittance Bandwidth (Magnitude) (R_L = 50 ohms)
   Carrier Input Port, V_C = 60 mV(rms) sine wave
   f_S = 1.0 kHz, 300 mV(rms) sine wave
   Signal Input Port, V_S = 300 mV(rms) sine wave
   $|V_C|$ = 0.5 Vdc
	8	8	BW_{3dB}	
				
				 —
				
				 —
					
					 300
					
					 80
						
						 —
						
						 —
							
							 —
							
							 —
								
								 300
								
								 80
									
									 —
									
									 —
										MHz

Signal Gain V_S = 100 mV(rms), f = 1.0 kHz; $|V_C|$ = 0.5 Vdc	10	3	A_{VS}	2.5	3.5	—	2.5	3.5	—	V/V

Single-Ended Input Impedance, Signal Port, f = 5.0 MHz
   Parallel Input Resistance
   Parallel Input Capacitance
	6	—	
			 r_{ip}
			 c_{ip}
				
				 —
				 —
					
					 200
					 2.0
						
						 —
						 —
							
							 —
							 —
								
								 200
								 2.0
									
									 —
									 —
										kΩ
										 pF

Single-Ended Output Impedance, f = 10 MHz
   Parallel Output Resistance
   Parallel Output Capacitance
	6	—	
			 r_{op}
			 c_{op}
				
				 —
				 —
					
					 40
					 5.0
						
						 —
						 —
							
							 —
							 —
								
								 40
								 5.0
									
									 —
									 —
										kΩ
										 pF

Input Bias Current $I_{bS} = \frac{I_1 + I_4}{2}$; $I_{bC} = \frac{I_7 + I_8}{2}$	7	—	 I_{bS} I_{bC}	 — —	 12 12	 25 25	 — —	 12 12	 30 30	µA
Input Offset Current $I_{ioS} = I_1 - I_4$; $I_{ioC} = I_7 - I_8$	7	—	 $|I_{ioS}|$ $|I_{ioC}|$	 — —	 0.7 0.7	 5.0 5.0	 — —	 0.7 0.7	 7.0 7.0	µA
Average Temperature Coefficient of Input Offset Current (T_A = -55°C to +125°C)	7	—	$|TC_{Iio}|$	—	2.0	—	—	2.0	—	nA/°C
Output Offset Current ($I_6 - I_9$)	7	—	$|I_{oo}|$	—	14	50	—	14	80	µA
Average Temperature Coefficient of Output Offset Current (T_A = -55°C to +125°C)	7	—	$|TC_{Ioo}|$	—	90	—	—	90	—	nA/°C
Common-Mode Input Swing, Signal Port, f_S = 1.0 kHz	9	4	CMV	—	5.0	—	—	5.0	—	Vp-p
Common-Mode Gain, Signal Port, f_S = 1.0 kHz, $|V_C|$ = 0.5 Vdc	9	—	ACM	—	-85	—	—	-85	—	dB
Common-Mode Quiescent Output Voltage (Pin 6 or Pin 9)	10	—	V_o	—	8.0	—	—	8.0	—	Vdc
Differential Output Voltage Swing Capability	10	—	V_{out}	—	8.0	—	—	8.0	—	Vp-p

Power Supply Current
   $I_6 + I_9$
   I_{10}
	7	6	
			 I_{CC}
			 I_{EE}
				
				 —
				 —
					
					 2.0
					 3.0
						
						 3.0
						 4.0
							
							 —
							 —
								
								 2.0
								 3.0
									
									 4.0
									 5.0
										mAdc

DC Power Dissipation	7	5	P_D	—	33	—	—	33	—	mW

* Pin number references pertain to this device when packaged in a metal can. To ascertain the corresponding pin numbers for plastic or ceramic packaged devices refer to the first page of this specification sheet.

 MOTOROLA *Semiconductor Products Inc.*

FIGURE 6-12(b) *(continued)*

GENERAL OPERATING INFORMATION *

Note 1 — Carrier Feedthrough

Carrier feedthrough is defined as the output voltage at carrier frequency with only the carrier applied (signal voltage = 0).

Carrier null is achieved by balancing the currents in the differential amplifier by means of a bias trim potentiometer (R_1 of Figure 5).

Note 2 — Carrier Suppression

Carrier suppression is defined as the ratio of each sideband output to carrier output for the carrier and signal voltage levels specified.

Carrier suppression is very dependent on carrier input level, as shown in Figure 22. A low value of the carrier does not fully switch the upper switching devices, and results in lower signal gain, hence lower carrier suppression. A higher than optimum carrier level results in unnecessary device and circuit carrier feedthrough, which again degenerates the suppression figure. The MC1596 has been characterized with a 60 mV(rms) sinewave carrier input signal. This level provides optimum carrier suppression at carrier frequencies in the vicinity of 500 kHz, and is generally recommended for balanced modulator applications.

Carrier feedthrough is independent of signal level, V_S. Thus carrier suppression can be maximized by operating with large signal levels. However, a linear operating mode must be maintained in the signal-input transistor pair — or harmonics of the modulating signal will be generated and appear in the device output as spurious sidebands of the suppressed carrier. This requirement places an upper limit on input-signal amplitude (see Note 3 and Figure 20). Note also that an optimum carrier level is recommended in Figure 22 for good carrier suppression and minimum spurious sideband generation.

At higher frequencies circuit layout is very important in order to minimize carrier feedthrough. Shielding may be necessary in order to prevent capacitive coupling between the carrier input leads and the output leads.

Note 3 — Signal Gain and Maximum Input Level

Signal gain (single-ended) at low frequencies is defined as the voltage gain,

$$A_{VS} = \frac{V_o}{V_S} = \frac{R_L}{R_e + 2r_e} \quad \text{where } r_e = \frac{26 \text{ mV}}{I_5 \text{ (mA)}}$$

A constant dc potential is applied to the carrier input terminals to fully switch two of the upper transistors "on" and two transistors "off" ($V_C = 0.5$ Vdc). This in effect forms a cascode differential amplifier.

Linear operation requires that the signal input be below a critical value determined by R_E and the bias current I_5.

$$V_S \leq I_5 \, R_E \text{ (Volts peak)}$$

Note that in the test circuit of Figure 10, V_S corresponds to a maximum value of 1 volt peak.

Note 4 — Common-Mode Swing

The common-mode swing is the voltage which may be applied to both bases of the signal differential amplifier, without saturating the current sources or without saturating the differential amplifier itself by swinging it into the upper switching devices. This swing is variable depending on the particular circuit and biasing conditions chosen (see Note 6).

Note 5 — Power Dissipation

Power dissipation, P_D, within the integrated circuit package should be calculated as the summation of the voltage-current products at each port, i.e. assuming $V_9 = V_6$, $I_5 = I_6 = I_9$ and ignoring base current, $P_D = 2 I_5 (V_6 - V_{10}) + I_5 (V_5 - V_{10})$ where subscripts refer to pin numbers.

Note 6 — Design Equations

The following is a partial list of design equations needed to operate the circuit with other supply voltages and input conditions. See Note 3 for R_e equation.

A. Operating Current

The internal bias currents are set by the conditions at pin 5. Assume:

$$I_5 = I_6 = I_9$$

$$I_B \ll I_C \text{ for all transistors}$$

then:

$$R_5 = \frac{V^- - \phi}{I_5} - 500 \ \Omega \quad \text{where: } R_5 \text{ is the resistor between pin}$$
5 and ground
$$\phi = 0.75 \text{ V at } T_A = +25^{\circ}C$$

The MC1596 has been characterized for the condition $I_5 = 1.0$ mA and is the generally recommended value.

B. Common-Mode Quiescent Output Voltage

$$V_6 = V_9 = V^+ - I_5 \, R_L$$

Note 7 — Biasing

The MC1596 requires three dc bias voltage levels which must be set externally. Guidelines for setting up these three levels include maintaining at least 2 volts collector-base bias on all transistors while not exceeding the voltages given in the absolute maximum rating table;

$$30 \text{ Vdc} \geq [(V_6, V_9) - (V_7, V_8)] \geq 2 \text{ Vdc}$$

$$30 \text{ Vdc} \geq [(V_7, V_8) - (V_1, V_4)] \geq 2.7 \text{ Vdc}$$

$$30 \text{ Vdc} \geq [(V_1, V_4) - (V_5)] \geq 2.7 \text{ Vdc}$$

The foregoing conditions are based on the following approximations:

$$V_6 = V_9, \quad V_7 = V_8, \quad V_1 = V_4$$

Bias currents flowing into pins 1, 4, 7, and 8 are transistor base currents and can normally be neglected if external bias dividers are designed to carry 1.0 mA or more.

Note 8 — Transadmittance Bandwidth

Carrier transadmittance bandwidth is the 3-dB bandwidth of the device forward transadmittance as defined by:

$$Y_{21C} = \frac{i_o \text{ (each sideband)}}{v_s \text{ (signal)}} \bigg|_{V_o = 0}$$

Signal transadmittance bandwidth is the 3-dB bandwidth of the device forward transadmittance as defined by:

$$Y_{21S} = \frac{i_o \text{ (signal)}}{v_s \text{ (signal)}} \bigg|_{V_C = 0.5 \text{ Vdc}, V_o = 0}$$

*Pin number references pertain to this device when packaged in a metal can. To ascertain the corresponding pin numbers for plastic or ceramic packaged devices refer to the first page of this specification sheet.

 MOTOROLA *Semiconductor Products Inc.*

FIGURE 6-12(b) *(continued)*

Note 9 — Coupling and Bypass Capacitors C_1 and C_2

Capacitors C_1 and C_2 (Figure 5) should be selected for a reactance of less than 5.0 ohms at the carrier frequency.

Note 10 — Output Signal, V_o

The output signal is taken from pins 6 and 9, either balanced or single-ended. Figure 12 shows the output levels of each of the two output sidebands resulting from variations in both the carrier and modulating signal inputs with a single-ended output connection.

Note 11 — Negative Supply, V_{EE}

V_{EE} should be dc only. The insertion of an RF choke in series with V_{EE} can enhance the stability of the internal current sources.

Note 12 — Signal Port Stability

Under certain values of driving source impedance, oscillation may occur. In this event, an RC suppression network should be connected directly to each input using short leads. This will reduce the Q of the source-tuned circuits that cause the oscillation.

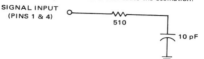

An alternate method for low-frequency applications is to insert a 1 k-ohm resistor in series with the inputs, pins 1 and 4. In this case input current drift may cause serious degradation of carrier suppression.

TEST CIRCUITS

FIGURE 5 — CARRIER REJECTION AND SUPPRESSION

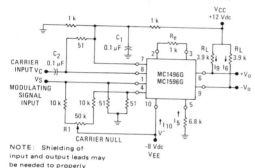

NOTE: Shielding of input and output leads may be needed to properly perform these tests.

FIGURE 6 — INPUT-OUTPUT IMPEDANCE

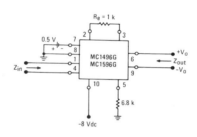

FIGURE 7 — BIAS AND OFFSET CURRENTS

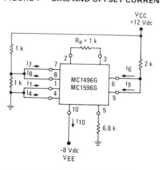

FIGURE 8 — TRANSCONDUCTANCE BANDWIDTH

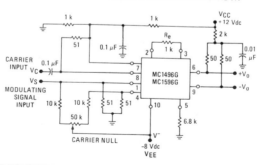

NOTE: Pin number references pertain to this device when packaged in a metal can. To ascertain the corresponding pin numbers for plastic or ceramic packaged devices refer to the first page of this specification sheet.

 MOTOROLA *Semiconductor Products Inc.*

FIGURE 6-12(b) *(continued)*

237

TEST CIRCUITS (continued)

FIGURE 9 – COMMON-MODE GAIN

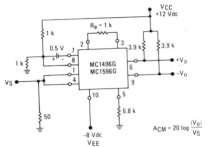

$$A_{CM} = 20 \log \frac{|V_O|}{V_S}$$

FIGURE 10 – SIGNAL GAIN AND OUTPUT SWING

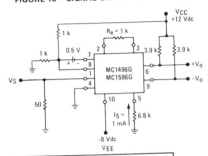

NOTE: Pin number references pertain to this device when packaged in a metal can. To ascertain the corresponding pin numbers for plastic or ceramic packaged devices refer to the first page of this specification sheet.

TYPICAL CHARACTERISTICS (continued)

Typical characteristics were obtained with circuit shown in Figure 5, f_C = 500 kHz (sine wave),
V_C = 60 mV(rms), f_S = 1 kHz, V_S = 300 mV(rms), T_A = +25°C unless otherwise noted.

FIGURE 11 – SIDEBAND OUTPUT versus CARRIER LEVELS

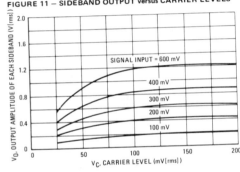

FIGURE 12 – SIGNAL-PORT PARALLEL-EQUIVALENT INPUT RESISTANCE versus FREQUENCY

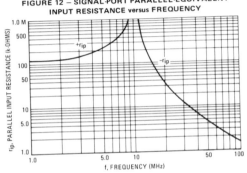

FIGURE 13 – SIGNAL-PORT PARALLEL-EQUIVALENT INPUT CAPACITANCE versus FREQUENCY

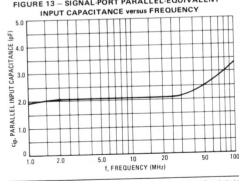

FIGURE 14 – SINGLE-ENDED OUTPUT IMPEDANCE versus FREQUENCY

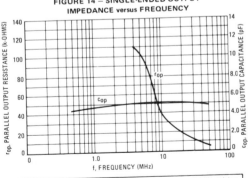

 MOTOROLA *Semiconductor Products Inc.*

FIGURE 6-12(b) *(continued)*

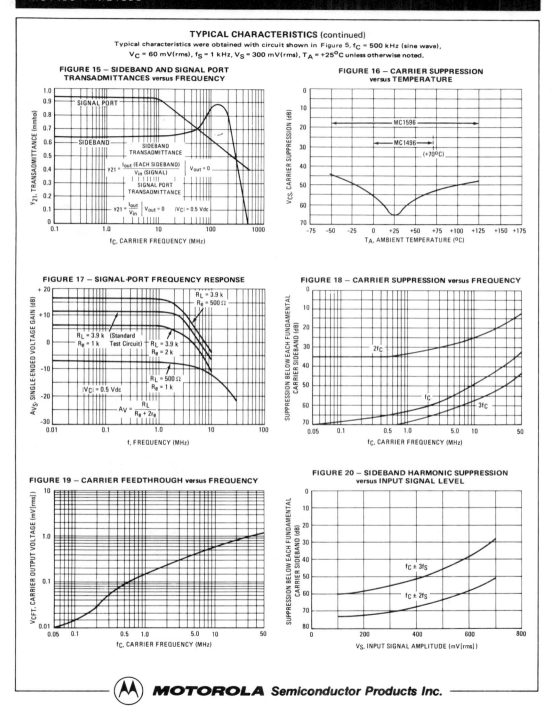

TYPICAL CHARACTERISTICS (continued)

Typical characteristics were obtained with circuit shown in Figure 5, f_C = 500 kHz (sine wave), V_C = 60 mV(rms), f_S = 1 kHz, V_S = 300 mV(rms), T_A = +25°C unless otherwise noted.

FIGURE 15 – SIDEBAND AND SIGNAL PORT TRANSADMITTANCES versus FREQUENCY

FIGURE 16 – CARRIER SUPPRESSION versus TEMPERATURE

FIGURE 17 – SIGNAL-PORT FREQUENCY RESPONSE

FIGURE 18 – CARRIER SUPPRESSION versus FREQUENCY

FIGURE 19 – CARRIER FEEDTHROUGH versus FREQUENCY

FIGURE 20 – SIDEBAND HARMONIC SUPPRESSION versus INPUT SIGNAL LEVEL

(M) MOTOROLA *Semiconductor Products Inc.*

FIGURE 6-12(b) *(continued)*

239

TYPICAL CHARACTERISTICS (continued)

FIGURE 21 – SUPPRESSION OF CARRIER HARMONIC SIDEBANDS versus CARRIER FREQUENCY

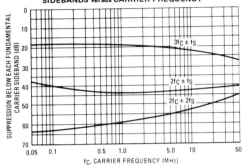

FIGURE 22 – CARRIER SUPPRESSION versus CARRIER INPUT LEVEL

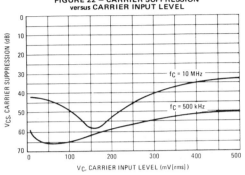

OPERATIONS INFORMATION

The MC1596/MC1496, a monolithic balanced modulator circuit, is shown in Figure 23.

This circuit consists of an upper quad differential amplifier driven by a standard differential amplifier with dual current sources. The output collectors are cross-coupled so that full-wave balanced multiplication of the two input voltages occurs. That is, the output signal is a constant times the product of the two input signals.

Mathematical analysis of linear ac signal multiplication indicates that the output spectrum will consist of only the sum and difference of the two input frequencies. Thus, the device may be used as a balanced modulator, doubly balanced mixer, product detector, frequency doubler, and other applications requiring these particular output signal characteristics.

The lower differential amplifier has its emitters connected to the package pins so that an external emitter resistance may be used. Also, external load resistors are employed at the device output.

Signal Levels

The upper quad differential amplifier may be operated either in a linear or a saturated mode. The lower differential amplifier is operated in a linear mode for most applications.

For low-level operation at both input ports, the output signal will contain sum and difference frequency components and have an amplitude which is a function of the product of the input signal amplitudes.

For high-level operation at the carrier input port and linear operation at the modulating signal port, the output signal will contain sum and difference frequency components of the modulating signal frequency and the fundamental and odd harmonics of the carrier frequency. The output amplitude will be a constant times the modulating signal amplitude. Any amplitude variations in the carrier signal will not appear in the output.

FIGURE 23 – CIRCUIT SCHEMATIC

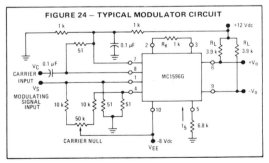

FIGURE 24 – TYPICAL MODULATOR CIRCUIT

NOTE: Pin number references pertain to this device when packaged in a metal can. To ascertain the corresponding pin numbers for plastic or ceramic packaged devices refer to the first page of this specification sheet.

 MOTOROLA *Semiconductor Products Inc.*

FIGURE 6-12(b) *(continued)*

OPERATIONS INFORMATION (continued)

The linear signal handling capabilities of a differential amplifier are well defined. With no emitter degeneration, the maximum input voltage for linear operation is approximately 25 mV peak. Since the upper differential amplifier has its emitters internally connected, this voltage applies to the carrier input port for all conditions.

Since the lower differential amplifier has provisions for an external emitter resistance, its linear signal handling range may be adjusted by the user. The maximum input voltage for linear operation may be approximated from the following expression:

$$V = \left(I_5\right)\left(R_E\right) \text{volts peak.}$$

This expression may be used to compute the minimum value of R_E for a given input voltage amplitude.

FIGURE 25 – TABLE 1
VOLTAGE GAIN AND OUTPUT FREQUENCIES

Carrier Input Signal (V_C)	Approximate Voltage Gain	Output Signal Frequency(s)
Low-level dc	$\dfrac{R_L \, V_C}{2(R_E + 2r_e)\left(\dfrac{KT}{q}\right)}$	f_M
High-level dc	$\dfrac{R_L}{R_E + 2r_e}$	f_M
Low-level ac	$\dfrac{R_L \, V_C(\text{rms})}{2\sqrt{2}\left(\dfrac{KT}{q}\right)(R_E + 2r_e)}$	$f_C \pm f_M$
High-level ac	$\dfrac{0.637 \, R_L}{R_E + 2r_e}$	$f_C \pm f_M, \, 3f_C \pm f_M,$ $5f_C \pm f_M, \, \ldots$

The gain from the modulating signal input port to the output is the MC1596/MC1496 gain parameter which is most often of interest to the designer. This gain has significance only when the lower differential amplifier is operated in a linear mode, but this includes most applications of the device.

As previously mentioned, the upper quad differential amplifier may be operated either in a linear or a saturated mode. Approximate gain expressions have been developed for the MC1596/MC1496 for a low-level modulating signal input and the following carrier input conditions:

1) Low-level dc
2) High-level dc
3) Low-level ac
4) High-level ac

These gains are summarized in Table 1, along with the frequency components contained in the output signal.

NOTES:
1. Low-level Modulating Signal, V_M, assumed in all cases. V_C is Carrier Input Voltage.
2. When the output signal contains multiple frequencies, the gain expression given is for the output amplitude of each of the two desired outputs, $f_C + f_M$ and $f_C - f_M$.
3. All gain expressions are for a single-ended output. For a differential output connection, multiply each expression by two.
4. R_L = Load resistance.
5. R_E = Emitter resistance between pins 2 and 3.
6. r_e = Transistor dynamic emitter resistance, at +25°C;

$$r_e \approx \frac{26 \text{ mV}}{I_5 \text{ (mA)}}$$

7. K = Boltzmann's Constant, T = temperature in degrees Kelvin, q = the charge on an electron.

$$\frac{KT}{q} \approx 26 \text{ mV at room temperature}$$

APPLICATIONS INFORMATION

Double sideband suppressed carrier modulation is the basic application of the MC1596/MC1496. The suggested circuit for this application is shown on the front page of this data sheet.

In some applications, it may be necessary to operate the MC1596/MC1496 with a single dc supply voltage instead of dual supplies. Figure 26 shows a balanced modulator designed for operation with a single +12 Vdc supply. Performance of this circuit is similar to that of the dual supply modulator.

AM Modulator

The circuit shown in Figure 27 may be used as an amplitude modulator with a minor modification.

All that is required to shift from suppressed carrier to AM operation is to adjust the carrier null potentiometer for the proper amount of carrier insertion in the output signal.

However, the suppressed carrier null circuitry as shown in Figure 27 does not have sufficient adjustment range. Therefore, the modulator may be modified for AM operation by changing two resistor values in the null circuit as shown in Figure 28.

Product Detector

The MC1596/MC1496 makes an excellent SSB product detector (see Figure 29).

This product detector has a sensitivity of 3.0 microvolts and a dynamic range of 90 dB when operating at an intermediate frequency of 9 MHz.

The detector is broadband for the entire high frequency range. For operation at very low intermediate frequencies down to 50 kHz the 0.1 µF capacitors on pins 7 and 8 should be increased to 1.0 µF. Also, the output filter at pin 9 can be tailored to a specific intermediate frequency and audio amplifier input impedance.

As in all applications of the MC1596/MC1496, the emitter resistance between pins 2 and 3 may be increased or decreased to adjust circuit gain, sensitivity, and dynamic range.

This circuit may also be used as an AM detector by introducing carrier signal at the carrier input and an AM signal at the SSB input.

The carrier signal may be derived from the intermediate frequency signal or generated locally. The carrier signal may be introduced with or without modulation, provided its level is sufficiently high to saturate the upper quad differential amplifier. If the carrier signal is modulated, a 300 mV(rms) input level is recommended.

 MOTOROLA *Semiconductor Products Inc.*

FIGURE 6-12(b) *(continued)*

APPLICATIONS INFORMATION (continued)

Doubly Balanced Mixer

The MC1596/MC1496 may be used as a doubly balanced mixer with either broadband or tuned narrow band input and output networks.

The local oscillator signal is introduced at the carrier input port with a recommended amplitude of 100 mV(rms).

Figure 30 shows a mixer with a broadband input and a tuned output.

Frequency Doubler

The MC1596/MC1496 will operate as a frequency doubler by introducing the same frequency at both input ports.

Figures 31 and 32 show a broadband frequency doubler and a tuned output very high frequency (VHF) doubler, respectively.

Phase Detection and FM Detection

The MC1596/MC1496 will function as a phase detector. High-level input signals are introduced at both inputs. When both inputs are at the same frequency the MC1596/MC1496 will deliver an output which is a function of the phase difference between the two input signals.

An FM detector may be constructed by using the phase detector principle. A tuned circuit is added at one of the inputs to cause the two input signals to vary in phase as a function of frequency. The MC1596/MC1496 will then provide an output which is a function of the input signal frequency.

NOTE: Pin number references pertain to this device when packaged in a metal can. To ascertain the corresponding pin numbers for plastic or ceramic packaged devices refer to the first page of this specification sheet.

TYPICAL APPLICATIONS

FIGURE 26 – BALANCED MODULATOR
(+12 Vdc SINGLE SUPPLY)

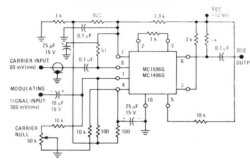

FIGURE 27 – BALANCED MODULATOR-DEMODULATOR

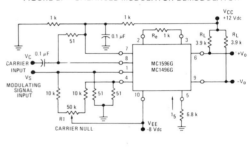

FIGURE 28 – AM MODULATOR CIRCUIT

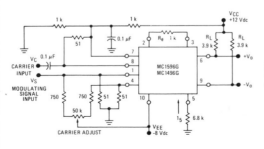

FIGURE 29 – PRODUCT DETECTOR
(+12 Vdc SINGLE SUPPLY)

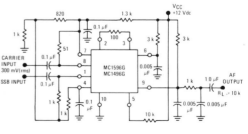

 MOTOROLA *Semiconductor Products Inc.*

FIGURE 6-12(b) *(continued)*

TYPICAL APPLICATIONS (continued)

FIGURE 30 – DOUBLY BALANCED MIXER
(BROADBAND INPUTS, 9.0 MHz TUNED OUTPUT)

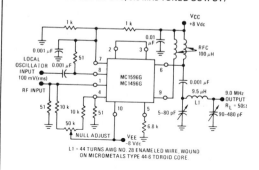

L1 = 44 TURNS AWG NO. 28 ENAMELED WIRE, WOUND
ON MICROMETALS TYPE 44-6 TOROID CORE

FIGURE 31 – LOW-FREQUENCY DOUBLER

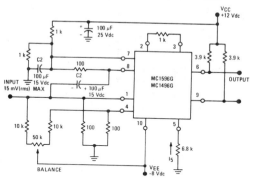

FIGURE 32 – 150 to 300 MHz DOUBLER

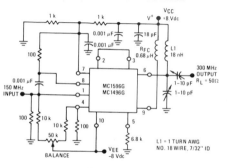

L1 = 1 TURN AWG
NO. 18 WIRE, 7/32" ID

DEFINITIONS

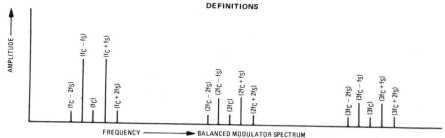

f_C — CARRIER FUNDAMENTAL
f_S — MODULATING SIGNAL
$f_C \pm f_S$ — FUNDAMENTAL CARRIER SIDEBANDS

$f_C \pm n f_S$ — FUNDAMENTAL CARRIER SIDEBAND HARMONICS
$n f_C$ — CARRIER HARMONICS
$n f_C \pm n f_S$ — CARRIER HARMONIC SIDEBANDS

NOTE: Pin number references pertain to this device when packaged in a metal can. To ascertain the corresponding pin numbers for plastic or ceramic packaged devices refer to the first page of this specification sheet.

 MOTOROLA *Semiconductor Products Inc.*

FIGURE 6-12(b) *(continued)*

OUTLINE DIMENSIONS

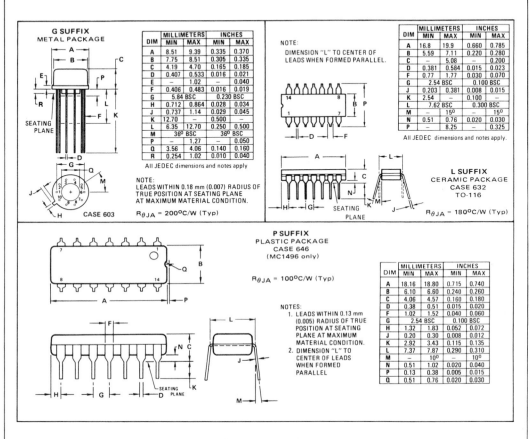

G SUFFIX
METAL PACKAGE

DIM	MILLIMETERS		INCHES	
	MIN	MAX	MIN	MAX
A	8.51	9.39	0.335	0.370
B	7.75	8.51	0.305	0.335
C	4.19	4.70	0.165	0.185
D	0.407	0.533	0.016	0.021
E	–	1.02	–	0.040
F	0.406	0.483	0.016	0.019
G	5.84 BSC		0.230 BSC	
H	0.712	0.864	0.028	0.034
J	0.737	1.14	0.029	0.045
K	12.70	–	0.500	–
L	6.35	12.70	0.250	0.500
M	36° BSC		36° BSC	
P	–	1.27	–	0.050
Q	3.56	4.06	0.140	0.160
R	0.254	1.02	0.010	0.040

All JEDEC dimensions and notes apply

NOTE:
LEADS WITHIN 0.18 mm (0.007) RADIUS OF TRUE POSITION AT SEATING PLANE AT MAXIMUM MATERIAL CONDITION.

$R_{\theta JA}$ = 200°C/W (Typ)

SEATING PLANE

CASE 603

NOTE:
DIMENSION "L" TO CENTER OF LEADS WHEN FORMED PARALLEL.

DIM	MILLIMETERS		INCHES	
	MIN	MAX	MIN	MAX
A	16.8	19.9	0.660	0.785
B	5.59	7.11	0.220	0.280
C	–	5.08	–	0.200
D	0.381	0.584	0.015	0.023
F	0.77	1.77	0.030	0.070
G	2.54 BSC		0.100 BSC	
J	0.203	0.381	0.008	0.015
K	2.54	–	0.100	–
L	7.62 BSC		0.300 BSC	
M	–	15°	–	15°
N	0.51	0.76	0.020	0.030
P	–	8.25	–	0.325

All JEDEC dimensions and notes apply.

L SUFFIX
CERAMIC PACKAGE
CASE 632
TO-116

SEATING PLANE

$R_{\theta JA}$ = 180°C/W (Typ)

P SUFFIX
PLASTIC PACKAGE
CASE 646
(MC1496 only)

$R_{\theta JA}$ = 100°C/W (Typ)

NOTES:
1. LEADS WITHIN 0.13 mm (0.005) RADIUS OF TRUE POSITION AT SEATING PLANE AT MAXIMUM MATERIAL CONDITION.
2. DIMENSION "L" TO CENTER OF LEADS WHEN FORMED PARALLEL

DIM	MILLIMETERS		INCHES	
	MIN	MAX	MIN	MAX
A	18.16	18.80	0.715	0.740
B	6.10	6.60	0.240	0.260
C	4.06	4.57	0.160	0.180
D	0.38	0.51	0.015	0.020
F	1.02	1.52	0.040	0.060
G	2.54 BSC		0.100 BSC	
H	1.32	1.83	0.052	0.072
J	0.20	0.30	0.008	0.012
K	2.92	3.43	0.115	0.135
L	7.37	7.87	0.290	0.310
M	–	10°	–	10°
N	0.51	1.02	0.020	0.040
P	0.13	0.38	0.005	0.015
Q	0.51	0.76	0.020	0.030

SEATING PLANE

THERMAL INFORMATION

The maximum power consumption an integrated circuit can tolerate at a given operating ambient temperature, can be found from the equation:

$$P_{D(T_A)} = \frac{T_{J(max)} - T_A}{R_{\theta JA}(Typ)} \geq V_I I_S \cdot V_O I_O$$

Where: $P_{D(T_A)}$ = Power Dissipation allowable at a given operating ambient temperature.

$T_{J(max)}$ = Maximum Operating Junction Temperature as listed in the Maximum Ratings Section

T_A = Maximum Desired Operating Ambient Temperature

$R_{\theta JA}(Typ)$ = Typical Thermal Resistance Junction to Ambient

I_S = Total Supply Current

 MOTOROLA Semiconductor Products Inc.

FIGURE 6-12(b) (*continued*)

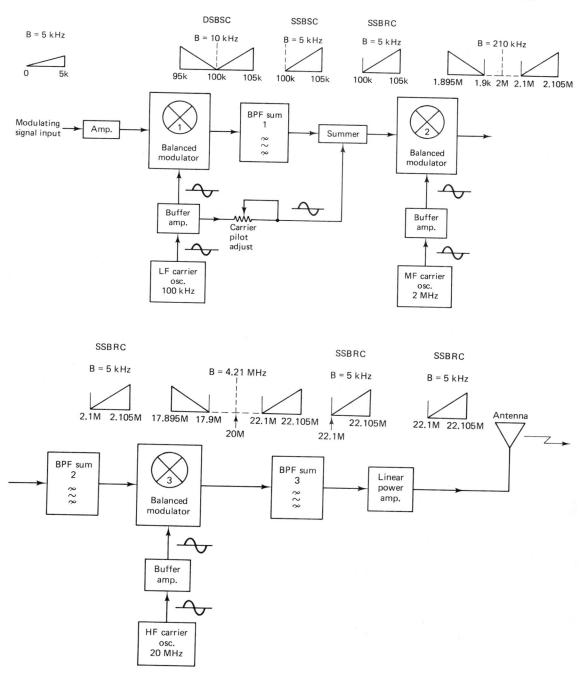

FIGURE 6-13 Single-sideband transmitter: filter method.

separated by a 200-kHz frequency band that is void of information. BPF 2 is centered on 2.1025 MHz with a 5 kHz bandwidth. Therefore, the output of BPF 2 is once again a single-sideband reduced-carrier waveform. Its spectrum comprises a reduced 2.1-MHz second IF carrier and a 5-kHz upper sideband. The output of BPF 2 is mixed with a 20-MHz high-frequency (HF) carrier in balanced modulator 3. The output is a double-sideband suppressed-carrier signal where the upper and lower sidebands again each contain the original SSBRC signal spectrum. The sidebands are separated by a 4.2-MHz frequency band that is void of information. BPF 3 is centered on 22.1025 MHz with a 5 kHz bandwidth. Therefore, the output of BPF 3 is once again a single-sideband reduced-carrier waveform with a reduced 22.1-MHz RF carrier and a 5-kHz upper sideband. The output waveform is amplified in the linear power amplifier and transmitted.

In the transmitter just described, the original modulating signal spectrum was up-converted in three modulation steps to a final carrier frequency of 22.1 MHz and a single upper sideband that extended to 22.105 MHz. After each up-conversion (frequency translation), the desired sideband is separated from the double-sideband spectrum with a BPF. The same final output spectrum can be produced with a single heterodyning process: one balanced modulator, one bandpass filter, and a single HF carrier supply. Figure 6-14 shows the block diagram and output spectrum for a single-step up-conversion transmitter. The output of the balanced modulator is a double-sideband spectrum centered around a suppressed carrier frequency of 22.1 MHz. To separate the 5-kHz upper sideband from the composite spectrum, a multiple-pole BPF with an extremely high Q is required. A BPF that meets this criterion is in itself difficult to construct but suppose that this were a multiple-channel transmitter and the carrier frequency were tunable; then the BPF must also be tunable. Constructing a tunable BPF in the MHz range with a passband of only 5 kHz is beyond economic and engineering feasibility. The only BPF in the transmitter shown in Figure 6-13 that had to separate sidebands that were immediately adjacent was BPF 1. To construct a multiple-pole, steep-skirted BPF at 100 kHz is a relatively simple task, as only a moderate Q is required. The sidebands separated by BPF 2 are 200 kHz apart; thus a low-Q filter with gradual roll-off characteristics can be used with no danger of passing any portion of the undesired sideband. BPF 3 separates sidebands that are 4.2 MHz apart. If multiple channels are used and the HF carrier is tunable, a single broadband filter can be used for BPF 3 with no danger of the undesired sideband leaking through the filter. For single-channel operation, the single conversion transmitter is the simplest design, but for multiple-channel operation, the three-conversion system is more practical. Figure 6-14b and c show the output spectrum and filtering requirements for both methods.

Single-sideband filters.   It is evident that filters are an essential part of a single-sideband system. Transmitters as well as receivers have requirements for highly selective networks for limiting both the signal and noise spectrums. Conventional *LC* filters do not have a high enough Q for most sideband transmitters. Therefore, most filters used for sideband generation are constructed from either *crystal* or *ceramic* material, or they use *mechanical* filters.

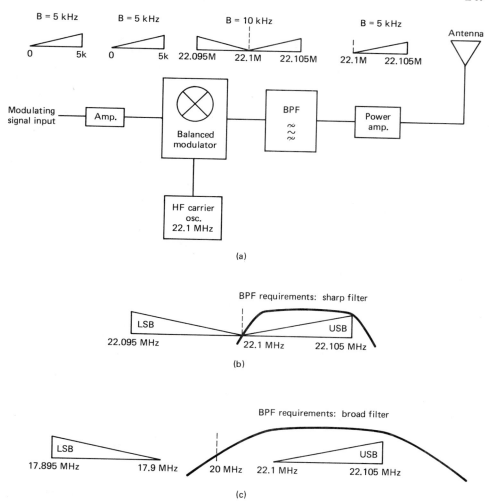

FIGURE 6-14 Single conversion SSBSC transmitter—filter method: (a) block diagram; (b) output spectrum and filtering requirements for a single-conversion transmitter; (c) output spectrum and filtering requirements for a three-conversion transmitter.

Crystal filters. The *crystal lattice filter* is commonly used in single-sideband systems. The schematic diagram for a typical crystal lattice bandpass filter is shown in Figure 6-15a. The lattice comprises two sets of matched crystal pairs (X_1 and X_2, X_3 and X_4) connected between tuned input and output transformers. Crystals X_1 and X_2 are series connected, while X_3 and X_4 are connected in parallel. X_1 and X_2 are cut to operate at the filter lower cutoff frequency, while X_3 and X_4 are cut to operate at the upper cutoff frequency. The input and output transformers are tuned to the center of the desired passband; this tends to spread the difference between the series and parallel

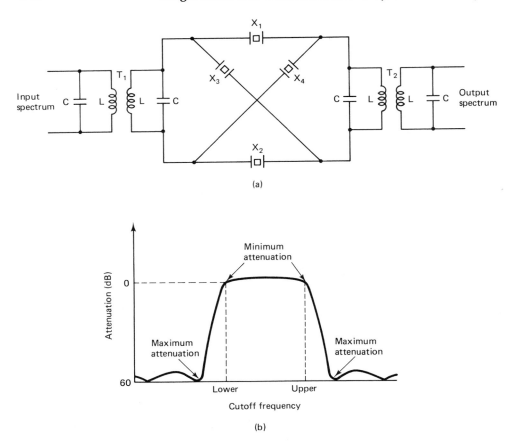

FIGURE 6-15 **Crystal lattice filter: (a) schematic diagram; (b) characteristic curve.**

resonant crystal frequencies. When the reactances of the bridge arms are equal and have the same sign (inductive or capacitive), the input signals propagating through the two possible paths cancel each other. When the reactances are equal with opposite signs (one inductive and one capacitive), the maximum signal is transmitted through the network. Figure 6-15b shows a typical characteristic curve for a crystal lattice filter. Crystal filters are available with a Q as high as 100,000. The filter shown in Figure 6-15a is a single-element filter. However, in order for a crystal filter to adequately pass a specific band and reject all other frequencies, at least two elements are required. Typical insertion losses for crystal filters are between 1.5 and 3 dB.

  Ceramic filter. *Ceramic filters* are made from lead zinconate–titanate, which exhibits the piezoelectric effect. Therefore, they operate quite similar to crystal filters except that ceramic filters do not have as high a Q-factor. Typically, Q values for ceramic filters go up to about 2000. Ceramic filters are less expensive, smaller, and more rugged than their crystal counterparts. However, ceramic filters have more loss than crystal filters. The insertion loss for ceramic filters is typically between 2 and 4 dB.

Mechanical filters. A *mechanical filter* is a *mechanically resonant* device. It receives electrical energy, converts it to mechanical vibrations, then converts the vibrations back to electrical energy at its output. Essentially, there are four elements that comprise a mechanical filter: an input transducer that converts the input electrical energy to mechanical vibrations, a series of mechanical resonant metal disks that vibrate at the desired resonant frequency, a coupling rod that couples the metal disks together, and an output transducer that converts the mechanical vibrations back to electrical energy. Figure 6-16 shows the electrical equivalent circuit for a mechanical filter. The series resonant circuits (L and C) represent the metal disks, coupling capacitor C_1 represents the coupling rod, and R represents the matching mechanical loads. The resonant frequency of the filter is determined by the series LC disks and C_1 determines the bandwidth. Mechanical filters are more rugged than either ceramic or crystal filters and have comparable frequency response characteristics. However, mechanical filters are larger and heavier and therefore are impractical for mobile communications equipment.

SSB Transmitter: Phase-Shift Method

With the phase-shift method of single-sideband generation, the undesired sideband is canceled in the output of the modulator; therefore, sharp filtering is unnecessary. Figure 6-17 shows a block diagram for a SSB transmitter that uses the phase-shift method to eliminate the upper sideband. Essentially, there are two separate double-sideband modulators (balanced modulators 1 and 2). The modulating signal and carrier are applied to the two modulators 90° out of phase. The outputs from the two balanced modulators are double-sideband suppressed carrier signals with the proper phase such that when they are combined in the linear summer the upper sidebands cancel.

Phasor representation.   The phasors shown in Figure 6-17 illustrate how the upper sideband is eliminated by rotating both the carrier and the modulating signal 90° prior to modulation. The output phasor from balanced modulator 1 shows the relative position and direction of rotation of the upper (ω_u) and lower (ω_l) side frequencies to the suppressed carrier (ω_c). The output of balanced modulator 2 is essentially the same phasor except that the phase of the carrier and the modulating signal are each rotated 90° from the reference. The output of the linear summer shows the sum of the output

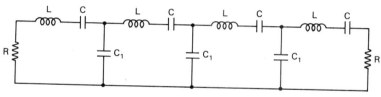

FIGURE 6-16   **Mechanical filter equivalent circuit.**

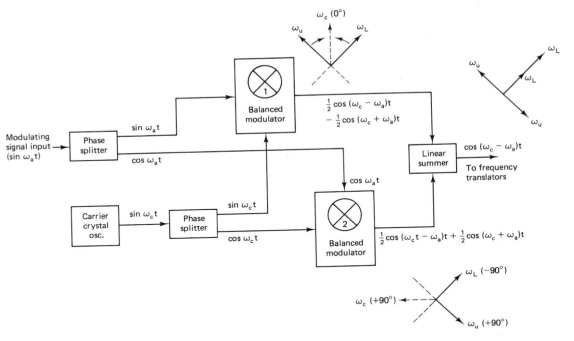

FIGURE 6-17   SSB transmitter: phase-shift method.

phasors from the two balanced modulators. The two phasors for the lower sideband are in phase and reinforce, whereas the phasors for the upper sideband are 180° out of phase and cancel. Consequently, the upper sideband is removed at the output of the linear summer.

Mathematical analysis. In Figure 6-17 the input modulating signal ($\sin \omega_a t$) is fed directly to balanced modulator 1 and shifted 90° (to $\cos \omega_a t$) and fed to balanced modulator 2. The low-frequency carrier ($\sin \omega_c t$) is also fed directly to balanced modulator 1 and shifted 90° and fed to balanced modulator 2. The balanced modulators are product modulators and their outputs are expressed mathematically as

$$\text{output from balanced modulator 1} = (\sin \omega_a t) \times (\sin \omega_c t)$$
$$= \tfrac{1}{2} \cos (\omega_c - \omega_a)t - \tfrac{1}{2} \cos (\omega_c + \omega_a)t$$

$$\text{output from balanced modulator 2} = (\cos \omega_a t) \times (\cos \omega_c t)$$
$$= \tfrac{1}{2} \cos (\omega_c - \omega_a)t + \tfrac{1}{2} \cos (\omega_c + \omega_a)t$$

and the output from the linear summer is

$$\frac{\frac{1}{2}\cos(\omega_c - \omega_a)t - \frac{1}{2}\cos(\omega_c + \omega_a)t}{+ \frac{1}{2}\cos(\omega_c - \omega_a)t + \frac{1}{2}\cos(\omega_c + \omega_a)t}$$
$$\underbrace{\cos(\omega_c - \omega_a)t}_{\text{lower sideband}} \qquad \text{canceled}$$
(difference frequencies)

SSB Transmitter: The Third Method

The so-called *third method* of single-sideband generation, developed by D. K. Weaver in the 1950s, is similar to the phase shift method described previously in that it uses phase shifting and summing to cancel the undesired sideband. However, it has an advantage in that the information signal is initially modulated onto an audio subcarrier, thus eliminating the need for a *wideband* phase shifter (which is difficult to build in practice). The block diagram for a third-method SSB modulator is shown in Figure 6-18. Notice that all of the inputs to the phase shifters are single-frequencies (ω_o, $\omega_o + 90°$, ω_c, and $\omega_c + 90°$). The input audio mixes with the audio subcarrier in balanced modulators

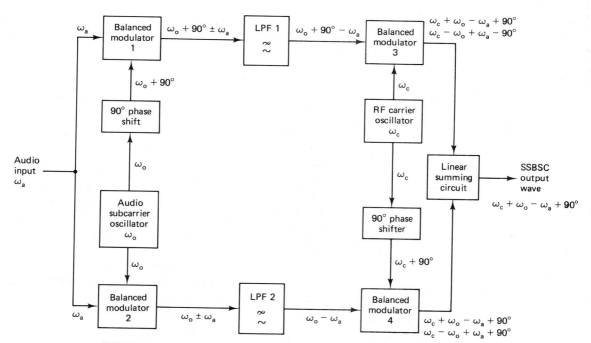

FIGURE 6-18 Single-sideband suppressed-carrier modulator: the "third method."

1 and 2, which are supplied with quadrature (90° out of phase) subcarrier signals. The output from balanced modulator 2 contains the upper and lower sidebands ($\omega_o \pm \omega_a$), while the output from balanced modulator 1 contains the upper and lower sidebands, each shifted in phase 90° ($\omega_o + 90° \pm \omega_a$). The upper sidebands are removed by their respective low-pass filters, which have an upper cutoff frequency equal to that of the suppressed audio subcarrier. The output from LPF 1 ($\omega_o + 90° - \omega_a$) is mixed with the RF carrier (ω_c) in balanced modulator 3, and the output from LPF 1 ($\omega_o - \omega_a$) is mixed with a 90° phase-shifted RF carrier ($\omega_c + 90°$) in balanced modulator 4. The RF carriers are, of course, suppressed in balanced modulators 3 and 4. Therefore, the output from balanced modulator 3 ($\omega_c + \omega_o - \omega_a + 90°$) + ($\omega_c - \omega_o + \omega_a - 90°$) is combined in the linear summer with the output from balanced modulator 4 ($\omega_c + \omega_o - \omega_a + 90°$) + ($\omega_c - \omega_o + \omega_a + 90°$). The output from the summer is

$$
\begin{array}{c}
(\omega_c + \omega_o - \omega_a + 90°) + (\omega_c - \omega_o + \omega_a - 90°) \\
+ \; (\omega_c + \omega_o - \omega_a + 90°) + (\omega_c - \omega_o + \omega_a + 90°) \\
\hline
(\omega_c + \omega_o - \omega_a + 90°) \qquad \text{canceled}
\end{array}
$$

The final RF output frequency is $F_c + F_o - F_a$, which is essentially the lower sideband of RF carrier $F_c + F_o$. The +90° offset phase is an absolute phase shift and is, consequently, insignificant. If the RF upper sideband is desired, simply interchange the carrier inputs to balanced modulators 3 and 4, in which case the final RF carrier is $F_c - F_o$.

Independent Sideband Transmitter

Figure 6-19 shows a block diagram for an independent sideband transmitter with three stages of modulation. The transmitter uses the filter method to produce two independent single-sideband channels (channel A and channel B). The two channels are combined; then a pilot carrier is reinserted. The composite ISB reduced carrier waveform is up-converted to RF with two additional stages of frequency translation. There are two 0-to 5-kHz information signals (channels A and B) that originate from two independent sources. The channel A information spectrum modulates a 100-kHz LF carrier in balanced modulator A. The output from balanced modulator A passes through BPF A, which is tuned to the lower sideband (95 to 100 kHz). The channel B information spectrum modulates the same 100-kHz LF carrier in balanced modulator B. The output from balanced modulator B passes through BPF B, which is tuned to the upper sideband (100 to 105 kHz). The two SSB spectrums are combined in a hybrid network to form a composite ISB suppressed carrier spectrum (95 to 105 kHz) The LF carrier (100 kHz) is reinserted in the linear summer to form an ISB reduced carrier waveform. The ISB spectrum is mixed with a 2.7-MHz MF carrier in balanced modulator 3. The output from balanced modulator 3 passes through BPF 3 to produce an ISB reduced carrier spectrum that extends from 2.795 to 2.805 MHz with a reduced 2.8-MHz pilot carrier. Balanced modulator 4, BPF 4, and the HF carrier translate the MF spectrum to an RF ISB spectrum that extends from 27.795 to 27.8 MHz (channel A) and 27.8 to 27.805 MHz (channel B) with a 27.8-MHz reduced carrier.

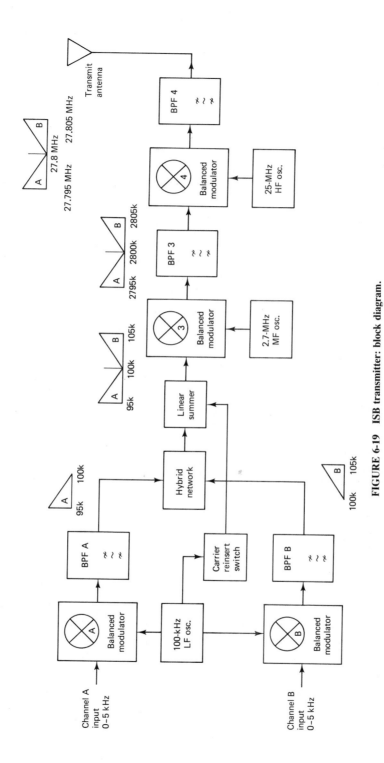

FIGURE 6-19 ISB transmitter: block diagram.

SINGLE-SIDEBAND RECEIVERS

Single-Sideband BFO Receiver

Figure 6-20 shows the block diagram for a simple noncoherent single-sideband receiver. The received RF is selected, amplified, then mixed down to IF for further amplification and band reduction. The output from the IF stage is heterodyned (beat) with the output from a beat frequency oscillator (BFO). The difference between the IF carrier frequency and the BFO frequency is the original information signal spectrum. Demodulation is accomplished simply by mixing and filtering the received signal with locally generated carriers. The receiver is noncoherent because the RF local oscillator and BFO frequencies are not synchronized to the transmitter local oscillators. Consequently, any difference between the transmit and receive carrier frequencies produces a frequency offset error in the demodulated information spectrum. For example, if the receive local oscillator is 100 Hz above its designated frequency and the BFO is 50 Hz above its designated frequency, the restored information is offset 150 Hz from its original input frequency spectrum. Fifty hertz or more offset is distinguishable by a normal listener as a tonal variation.

      Mathematical analysis.   The RF mixer and second detector shown in Figure 6-20 are product detectors (i.e., balanced modulators). Like the balanced modulators in the transmitter, their outputs are the product of their inputs. Essentially, the only difference between a product modulator and a product detector is that in a product modulator the input is tuned to a low-frequency modulating signal and the output is tuned to a high-frequency modulated signal. With a product detector, the input is tuned to a high-frequency modulated signal and the output is tuned to a low-frequency difference signal. With both the modulator and the detector, the single frequency carrier is the switching signal. In a receiver, the received input signal, which is a suppressed RF carrier and

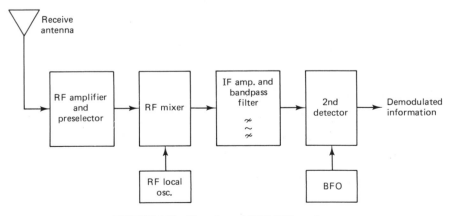

FIGURE 6-20   **Noncoherent BFO SSB receiver.**

one sideband, is multiplied (mixed) with the HF local oscillator frequency to generate an SSB IF difference signal at its output. The output from the second product detector is the sum and difference frequencies between the SSB IF and the beat frequency. The difference frequency band is the original input information spectrum. Mathematically, the output from the second product detector is

$$\text{output} = [\cos{(\omega_c - \omega_a)t}] \times (\sin{\omega_b t})$$

where

$$\omega_c = \text{suppressed IF carrier frequency}$$
$$\omega_a = \text{modulating signal frequencies}$$
$$\omega_c - \omega_a = \text{IF sideband frequencies}$$
$$\omega_b = \text{BFO frequency}$$

and

$$\omega_c = \omega_b$$

The trigonometric identity for the product of a cosine and sine wave of different frequencies is

$$(\cos A) \times (\sin B) = \tfrac{1}{2} \sin{(A + B)} + \tfrac{1}{2} \sin{(A - B)}$$

Therefore, the output is

$$\text{output} = \cos{(\omega_c - \omega_a)t} \times \sin{\omega_b t}$$
$$= \tfrac{1}{2} \sin{(\omega_c - \omega_a + \omega_b)t} + \tfrac{1}{2} \sin{(\omega_c - \omega_a - \omega_b)t}$$
$$\text{output} = \underbrace{\tfrac{1}{2} \sin{(2\omega_c + \omega_a)t}}_{\text{filtered off}} + \underbrace{\tfrac{1}{2} \sin{\omega_a t}}_{\substack{\text{original input} \\ \text{spectrum}}}$$

EXAMPLE 6-1

For the BFO receiver shown in Figure 6-20, an RF received signal of 30 to 30.005 MHz, an RF local oscillator frequency of 20 MHz, an IF frequency band of 10 to 10.005 MHz, and a BFO frequency of 10 MHz, determine:
(a) The demodulated first IF frequency band and demodulated information spectrum.
(b) The demodulated information spectrum when the RF local oscillator drifts down 0.001%.

Solution   (a) The output from the RF mixer is

$$\text{RF mixer output} = F_{rf} - F_{lo}$$
$$F_{ip} = (30 \text{ to } 30.005 \text{ MHz}) - 20 \text{ MHz} = 10 \text{ to } 10.005$$

The demodulated information spectrum is

$$\text{demodulated output} = F_{if} - F_b$$
$$= (10 \text{ to } 10.005 \text{ MHz}) - 10 \text{ MHz} = 0 \text{ to } 5 \text{ kHz}$$

(b) A 0.001% drift would cause the corresponding decrease in the RF local oscillator frequency:

$$\text{Rf local oscillator drift} = (0.00001)(30 \text{ MHz}) = 300 \text{ Hz}$$

Therefore, the output from the RF mixer is

$$\begin{aligned} \text{RF mixer output} &= F_{rf} - F_{lo} \\ &= (30 \text{ to } 30.005 \text{ MHz}) - 19.9997 \text{ MHz} = 10.0003 \text{ to } 10.0053 \text{ MHz} \end{aligned}$$

and

$$\begin{aligned} \text{demodulated output} &= F_{if} - F_b f_o \\ &= (10.0003 \text{ to } 10.0053 \text{ MHz}) - 10 \text{ MHz} = 300 \text{ to } 5300 \text{ Hz} \end{aligned}$$

The 0.001% drift in the RF local oscillator frequency caused a corresponding 300-Hz shift or offset in the demodulated information spectrum.

Coherent SSB BFO Receiver

Figure 6-21 shows a block diagram for a *coherent SSB BFO receiver*. This receiver is identical to the receiver shown in Figure 6-19 except that the carrier and BFO frequencies are synchronized to the transmit carrier oscillators. The carrier recovery circuit is a narrowband PLL that tracks and removes the suppressed carrier from the composite SSBRC received spectrum and uses the recovered pilot to regenerate coherent receive carrier frequencies in the synthesizer. The synthesizer circuit generates the RF local oscillator and BFO frequencies. The carrier recovery circuit tracks the received carrier pilot. Therefore, minor changes in the transmit carrier frequencies are compensated for in the receiver and frequency offset error is eliminated. If the coherent receiver shown

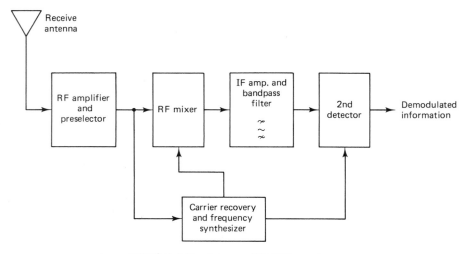

FIGURE 6-21   Coherent SSB BFO receiver.

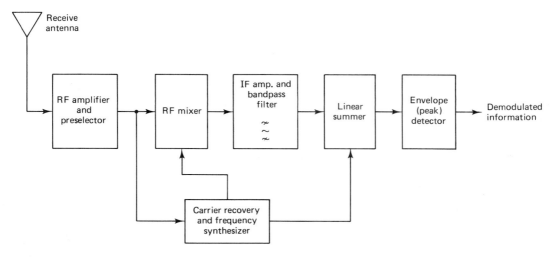

FIGURE 6-22 Single-sideband envelope detection receiver.

in Figure 6-21 had been used in Example 6-1, the receiver oscillators would have been locked onto the received pilot carrier and unable to drift independently.

Single-Sideband Envelope Detection Receiver

Figure 6-22 shows the block diagram for a single-sideband receiver that uses synchronous carriers and envelope detection to demodulate the received signals. The reduced carrier pilot is detected, separated from the composite receive spectrum, and regenerated in the carrier recovery circuit. The regenerated pilot is divided and used as the stable frequency source for a frequency synthesizer, which supplies the receiver with frequency coherent carriers (the receiver carrier oscillators are synchronized to the transmit carrier oscillators). The received RF is mixed down to IF in the first product detector. A regenerated IF carrier is added to the IF spectrum in the linear summer, which produces a SSB full carrier envelope. The SSBFC waveform is demodulated with a conventional peak detector to produce the original input signal spectrum.

Multichannel Pilot Carrier SSB Receiver

Figure 6-23 shows a block diagram for a multichannel pilot carrier SSB receiver that uses a PLL carrier recovery circuit and a frequency synthesizer. The RF input range extends from 4 to 30 MHz, and the VCO natural frequency is coarsely adjusted with an external channel selector switch over a frequency range of 6 to 32 MHz. The VCO frequency tracks above the incoming RF by 2 MHz, which is the first IF. A 1.8-MHz beat frequency sets the second IF to 200 kHz.

Circuit operation. The VCO frequency is coarsely set with the channel selector switch. The output frequency from the VCO mixes with the incoming RF in the first

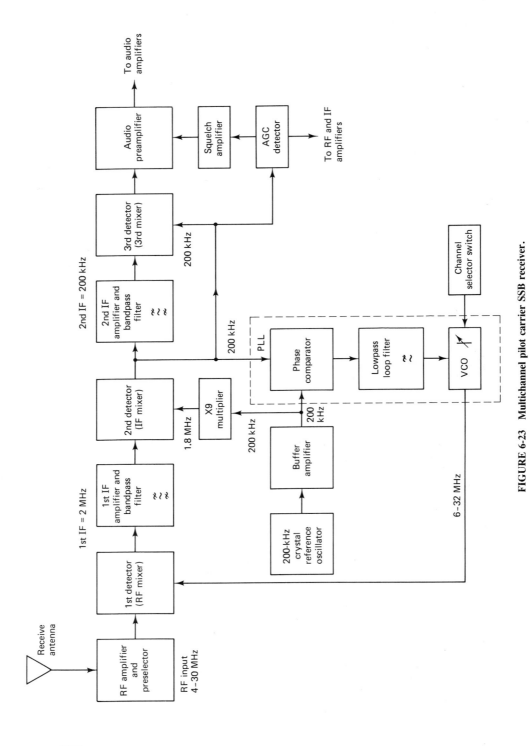

FIGURE 6-23 Multichannel pilot carrier SSB receiver.

258

detector to produce a first IF difference frequency of 2 MHz. The first IF mixes with the 1.8-MHz beat frequency to produce a 200-kHz second IF. The PLL locks onto the 200-kHz pilot and produces a dc correction voltage that fine tunes the VCO. The second IF is beat down to audio in the third detector, which is passed on to the audio preamplifier for processing. The AGC detector produces an AGC voltage that is proportional to the amplitude of the 200-kHz pilot. The AGC voltage is fed back to the RF and/or IF amplifiers to adjust their gains proportionate to the received pilot level and to the squelch circuit to turn the audio preamplifier off in the absence of a received pilot. The PLL compares the 200-kHz pilot to a stable crystal-controlled reference. Consequently, although the receiver carrier supply is not directly synchronized to the transmit oscillators, the first and second IFs are, thus eliminating any frequency offset in the demodulated audio spectrum.

Single-Sideband Measurements

As mentioned previously, single-sideband transmitters are rated in peak envelope power (PEP) and peak envelope volts (PEV) rather than simply rms power and voltage. For a single frequency-modulating signal, the modulated output signal with SSB is not an envelope as with conventional AM, but rather, a continuous single frequency tone. A single frequency is not representative of a typical modulating signal spectrum. Therefore, for test purposes, a *two-frequency* test signal is used for the modulating signal where the two tones have equal amplitudes. Figure 6-24a shows the envelope produced in a

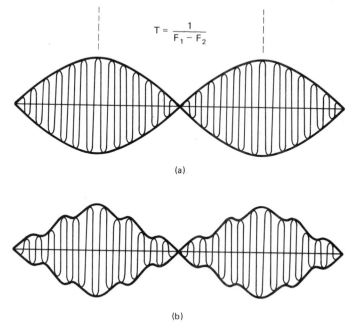

$$T = \frac{1}{F_1 - F_2}$$

(a)

(b)

FIGURE 6-24   Two-tone SSB test signal: (a) without reinserted carrier; (b) with reinserted carrier.

SSB modulator for a two-tone test signal. The envelope is the vector sum of the two equal amplitude side frequencies and is similar to a conventional AM envelope except that the repetition rate is equal to the difference between the two tone frequencies. Figure 6-24b shows the envelope for a two-tone test signal when a low-amplitude pilot carrier is added. The envelope has basically the same shape except with the addition of a low-amplitude sine-wave ripple at the carrier frequency.

A two-tone SSB envelope is an important consideration because it is from this envelope that the output power for a SSB transmitter is determined. The PEP for a SSB transmitter is analogous to the total output power from a conventional AM transmitter. The rated PEP is the rms output power measured at the peak of the envelope when the input is a two-tone test signal and the two tones are equal in amplitude. With such an output signal, the actual power dissipated in the load is equal to half the PEP. Therefore, the voltage developed across the load is

$$e_{total} = \left(e_1^2 + e_2^2 \right)^2$$

where e_1 and e_2 are the rms voltages of the two test tones. Therefore

$$PEP = \frac{(e_1^2 + e_2^2)^2}{R} \quad W$$

and since $e_1 = e_2$.

$$PEP = \frac{(2e)^2}{R}$$

$$= \frac{4e^2}{R} \quad W \tag{6-1}$$

However, the average power dissipated in the load is equal to the sum of the powers of the two tones:

$$P_{ave} = \frac{e_1^2}{R} + \frac{e_2^2}{R} = \frac{2e^2}{R} \quad W \tag{6-2}$$

and

$$P_{ave} = \frac{PEP}{2} \quad W \tag{6-3}$$

Two equal-amplitude test tones are used for the test signal for the following reasons:

1. One tone produces a continuous single-frequency output that does not produce intermodulation.
2. A single-frequency output signal is not analogous to normal conversation.
3. More than two tones makes analysis impractical.
4. Two tones of equal amplitude place a more demanding requirement on the transmitter than is likely to occur during normal operation.

EXAMPLE 6-2

For a two-tone test signal of 1.5 and 3 kHz and a carrier frequency of 100 kHz:
(a) Determine the output frequency spectrum.
(b) For $e_1 = e_2 = 10$ V and a load resistance of 50 Ω, determine the PEP and average output power.

Solution   (a) The output spectrum contains the two upper side frequencies

$$USF_1 = 100 \text{ kHz} + 1.5 \text{ kHz} = 101.5 \text{ kHz}$$

$$USF_2 = 100 \text{ kHz} + 3 \text{ kHz} = 103 \text{ kHz}$$

(b) Substituting into Equation 6-1 yields

$$PEP = \frac{4(3.535)^2}{50} = 1 \text{ W}$$

Substituting into Equation 6-2 gives us

$$P_{ave} = \frac{2(3.535)^2}{50} = \tfrac{1}{2} \text{ W}$$

QUESTIONS

6-1. Describe AM SSBFC. Compare SSBFC to conventional AM.

6-2. Describe AM SSBSC. Compare SSBSC to conventional AM.

6-3. Describe AM SSBRC. Compare SSBRC to conventional AM.

6-4. What is a pilot carrier?

6-5. What is an exalted carrier?

6-6. Describe AM ISB. Compare ISB to conventional AM.

6-7. Describe AM VSB. Compare VSB to conventional AM.

6-8. Define *peak envelope power*.

6-9. Draw the schematic diagram and describe the operation of a balanced ring modulator.

6-10. What is a product modulator?

6-11. Draw the schematic diagram and describe the operation of a FET push-pull balanced modulator.

6-12. Draw the schematic diagram and describe the operation of a balanced bridge modulator.

6-13. What are the advantages of an LIC integrated-circuit balanced modulator over a discrete balanced modulator?

6-14. Draw the block diagram for an SSB transmitter using the filter method. Describe its operation.

6-15. Contrast crystal, ceramic, and mechanical filters.

6-16. Draw the schematic diagram for an SSB transmitter using the phase shift method. Describe its operation.

6-17. Draw the schematic diagram for an SSB transmitter using the "third method." Describe its operation.

6-18. Draw the block diagram for an independent sideband transmitter. Describe its operation.

6-19. What is a product detector?

6-20. What is the difference between a product detector and a product modulator?

6-21. What is the difference between a noncoherent and a coherent receiver?

6-22. Draw the block diagram for a noncoherent single-sideband BFO receiver. Describe its operation.

6-23. Draw the block diagram for a coherent single-sideband BFO receiver. Describe its operation.

6-24. Draw the block diagram for an envelope detection receiver. Describe its operation.

6-25. Draw the block diagram for a coherent multichannel pilot carrier SSB receiver. Describe its operation.

6-26. Why is a two-tone test signal used for making PEP measurements?

PROBLEMS

6-1. For the balanced ring modulator shown in Figure 6-6a, carrier input frequency $F_c = 400$ kHz, and modulating signal frequency spectrum $F_a = 0$ to 4 kHz; determine:
(a) The output frequency range.
(b) The output frequency for a single frequency input $F_a = 2.8$ kHz.

6-2. For the LIC balanced modulator circuit shown in Figure 6-12, carrier input frequency $F_c = 200$ kHz, and modulating signal frequency spectrum $F_a = 0$ to 3 kHz; determine:
(a) The output frequency range.
(b) The output frequency for a single frequency input $F_a = 1.2$ kHz.

6-3. For the SSB transmitter shown in Figure 6-13, LF carrier frequency $F_{1f} = 100$ kHz, MF carrier frequency $F_{mf} = 4$ MHz, HF carrier frequency $F_{hf} = 30$ MHz, and audio input frequency spectrum $F_a = 0$ to 5 kHz:
(a) Sketch the frequency spectrums for the following points: balanced modulator 1 out, BPF 1 out, summer out, balanced modulator 2 out, BPF 2 out, balanced modulator 3 out, and BPF 3 out.
(b) For a single frequency input $F_a = 1.5$ kHz, determine the translated frequency for the following points: BPF 1 out, BPF 2 out, and BPF 3 out.

6-4. Repeat Problem 6-3 except change the LF carrier frequency to 500 kHz. Which transmitter has the more stringent filtering requirements?

6-5. For the SSB transmitter shown in Figure 6-14a, audio input frequency spectrum $F_a = 0$ to 3 kHz and HF carrier frequency $F_{hf} = 28$ MHz:
(a) Sketch the output frequency spectrum.
(b) For a single-frequency input $F_a = 2.2$ kHz, determine the output frequency.

6-6. Repeat Problem 6-5 except change the audio input frequency spectrum to $F_a = 300$ to 5000 Hz.

6-7. For the SSB transmitter shown in Figure 6-17, carrier input frequency $F_c = 500$ kHz, and input frequency spectrum $F_a = 0$ to 4 kHz:
(a) Sketch the frequency spectrum at the output of the linear summer.
(b) For a single audio input frequency $F_a = 3$ kHz, determine the output frequency.

6-8. Repeat Problem 6-7 except change the carrier input frequency to 400 kHz and the input frequency spectrum to F_a = 300 to 5000 Hz.

6-9. For the ISB transmitter shown in Figure 6-19, channel A input frequency spectrum F_a = 0 to 4 kHz, channel B input spectrum F_b = 0 to 4 kHz, LF carrier frequency F_{1f} = 200 kHz, MF carrier frequency F_{mf} = 4 MHz, and HF carrier frequency F_{hf} = 32 MHz:

(a) Sketch the frequency spectrums for the following points: balanced modulator A out, BPF A out, balanced modulator B out, BPF B out, hybrid network out, linear summer out, balanced modulator 3 out, BPF 3 out, balanced modulator 4 out, and BPF 4 out.

(b) For A-channel input frequency F_a = 2.5 kHz and B-channel input frequency F_b = 3 kHz, determine the frequency components at the following points: BPF A out, BPF B out, BPF 3 out, and BPF 4 out.

6-10. Repeat Problem 6-9 except change the channel A input frequency spectrum to 0 to 10 kHz and the channel B input frequency spectrum to 0 to 6 kHz.

6-11. For the SSB receiver shown in Figure 6-20, RF input frequency F_{rf} = 35.602 MHz, RF local oscillator frequency F_{lo} = 25 MHz, and a 2-kHz modulating signal frequency, determine the IF frequency and the BFO frequency.

6-12. For the multichannel pilot carrier SSB receiver shown in Figure 6-23, crystal oscillator frequency F_{co} = 300 kHz, first IF frequency F_{lif} = 3.3 MHz, RF input frequency F_{rf} = 23.303 MHz, and modulating signal frequency F_a = 3 kHz, determine the following: VCO output frequency, multiplication factor, second IF frequency.

6-13. For a two-tone test signal of 2 and 3 kHz and a carrier frequency of 200 kHz:

(a) Determine the output frequency spectrum.

(b) For $e_1 = e_2$ = 12 V_p and a load resistor R_L = 50 Ω, determine the PEP and average power.

Chapter 7

ANGLE MODULATION TRANSMISSION

INTRODUCTION

As stated previously, there are three properties of an analog signal that can be varied: its amplitude, its frequency, or its phase. Chapters 3, 4, and 6 dealt with amplitude modulation. This chapter (as well as Chapter 8) deals with both frequency modulation (FM) and phase modulation (PM). FM and PM are both forms of *angle* modulation. Unfortunately, both forms of angle modulation are often referred to simply as FM when, actually, there is a distinct (although subtle) difference between the two. Angle modulation was first introduced in 1931 as an alternative to amplitude modulation. It was suggested that an angle-modulated wave was less susceptible to noise than AM and, consequently, could improve the performance of radio communications. Major E. H. Armstrong developed the first successful FM radio system in 1936, and in 1939 the first regularly scheduled broadcasting of FM signals began in Alpine, New Jersey.

ANGLE MODULATION

Angle modulation results whenever the phase angle (θ) of a sinusoidal wave is varied with respect to time. An angle-modulated wave is expressed mathematically as

$$M(t) = V_c \cos \left[\omega_c t + \theta(t) \right] \qquad (7\text{-}1)$$

where

$$M(t) = \text{angle-modulated carrier}$$

$$V_c = \text{peak carrier amplitude (V)}$$

ω_c = carrier frequency, $2\pi F_c$

$\theta(t)$ = angle modulation (rad)

With angle modulation it is necessary that $\theta(t)$ be a prescribed function of the modulating signal. Therefore, if $V(t)$ is the modulating signal, the angle modulation is expressed mathematically as

$$\theta(t) = F[V(t)] \qquad (7\text{-}2)$$

where $V(t)$ is the modulating signal = $V_a \sin \omega_a t$.

In essence, the difference between FM and PM lies in which property (the frequency or the phase) is varied directly by the modulating signal. Whenever the frequency of a carrier is varied, the phase is also varied, and vice versa. Therefore, FM and PM must both occur whenever either form of angle modulation is performed. If the frequency of the carrier is varied directly in accordance with the modulating signal, FM results. If the phase of the carrier is varied directly in accordance with the modulating signal, PM results. Therefore, direct FM is indirect PM and direct PM is indirect FM.

Figure 7-1 shows the waveform for a sinusoidal carrier where the frequency is changing in respect to time. Whenever the period (T) of a sine wave is changed, both its frequency and phase change, and if the changes are continuous, the wave is no longer a single frequency. It can be shown that the resultant waveform comprises the original carrier frequency (sometimes called the *carrier rest frequency*) and an infinite number of side frequencies. The change in frequency is called the frequency deviation (ΔF) and the change in phase is called the phase deviation ($\Delta\theta$). Frequency deviation is the relative displacement of the carrier frequency, and phase deviation is the relative *angular displacement* of the carrier in respect to a reference phase.

Figure 7-2 shows a sinusoidal carrier in which the frequency (F) is changed (*deviated*) over a period of time (t). After t seconds, the frequency has changed ΔF hertz and is now $F - \Delta F$. The phase has also changed $\Delta\theta$ radians.

Mathematical Analysis

The difference between FM and PM is more easily understood by defining the following four terms with reference to Equation 7-1: instantaneous phase, instantaneous phase deviation, instantaneous frequency, and instantaneous frequency deviation.

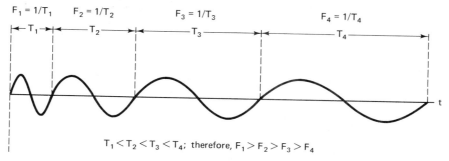

$T_1 < T_2 < T_3 < T_4$; therefore, $F_1 > F_2 > F_3 > F_4$

FIGURE 7-1 Frequency changing with time.

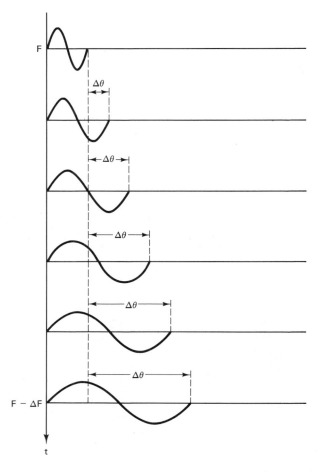

FIGURE 7-2 **Phase changing with frequency.**

Instantaneous phase. The *instantaneous phase* is the precise phase of the carrier at a given time and is expressed mathematically as

$$\text{instantaneous phase} = \omega_c t + \theta(t) \qquad \text{rad} \tag{7-3}$$

where

$$\omega_c t = \left(2\pi \, \frac{\text{rad}}{\text{cycle}} \right) \left(F \, \frac{\text{cycles}}{\text{second}} \right) (t \text{ seconds}) = 2\pi F t \qquad \text{rad}$$

$$\theta(t) = \text{rad}$$

Instantaneous phase deviation. The *instantaneous phase deviation* is the instantaneous change in the phase of the carrier at a given time and is expressed mathematically as

$$\text{instantaneous phase deviation} = \theta(t) \qquad \text{rad} \tag{7-4}$$

Instantaneous frequency. The *instantaneous frequency* of an angle-modulated carrier is the precise frequency of the carrier at a given time and is defined as the first time derivative of the instantaneous phase. In terms of Equation 7-3, the instantaneous frequency is expressed mathematically as

$$\begin{array}{l} \text{instantaneous} \\ \text{frequency} \end{array} = \frac{d}{dt}\,[\omega_c t + \theta(t)] = \omega_c + \theta'(t) \qquad \text{rad/s} \qquad (7\text{-}5)$$

where

$$\omega_c = \left(2\pi\,\frac{\text{rad}}{\text{cycle}}\right)\left(F\,\frac{\text{cycles}}{\text{second}}\right) = 2\pi F \qquad \text{rad/s}$$

$$\theta'(t) = \text{rad/s}$$

Instantaneous frequency deviation. The *instantaneous frequency deviation* is the instantaneous change in the frequency of the carrier and is defined as the first time derivative of the instantaneous phase deviation. Therefore, the instantaneous phase deviation is the first integral of the instantaneous frequency deviation. In terms of Equation 7-4, the instantaneous frequency deviation is expressed mathematically as*

$$\begin{array}{l} \text{instantaneous} \\ \text{frequency deviation} \end{array} = \theta'(t) \qquad \text{rad/s} \qquad (7\text{-}6)$$

For a modulating signal, $V(t)$, the phase and frequency modulation are

$$\text{phase modulation} = \theta(t) = KV(t) \qquad (7\text{-}7)$$

$$\text{frequency modulation} = \theta'(t) = K_1 V(t) \qquad (7\text{-}8)$$

where K and K_1 are constants which are, in fact, equal to the *deviation sensitivities* of the modulators. The deviation sensitivity is the input-versus-output transfer function for the modulator. The deviation sensitivity for a PM modulator is

$$K = \text{rad/V}$$

and for an FM modulator

$$K_1 = 2\pi F \text{ rad/V}$$

or,

$$= \frac{2\pi F \text{ rad/V}}{2\pi \text{ rad}} = \text{Hz/V}$$

Phase modulation is the first integral of the frequency modulation. Therefore, from Equations 7-7 and 7-8,

$$\text{PM} = \theta(t) = \int \theta'(t) = \int K_1 V(t) = K_1 \int V(t) \qquad (7\text{-}9)$$

---

* Note that a prime is used to denote the first derivative.

For a modulating signal $V_a \cos \omega_a t$ and substituting into Equation 7-1 yields

$$\text{phase modulation} = V_c \cos \omega_c t + KV_a \cos \omega_a t)$$

$$\text{frequency modulation} = V_c \cos \left(\omega_c t + \frac{K_1 V_a}{\omega_a} \sin \omega_a t \right)$$

The preceding mathematical relationships are summarized in Table 7-1. Also, the FM and PM waves that result when the modulating signal is a single frequency (sinusoidal) are shown.

FM and PM Waveforms

Figure 7-3 illustrates both frequency and phase modulation of a sinusoidal carrier by a single frequency-modulating signal. It can be seen that the FM and PM waveforms are identical except for their time relationship (phase). Thus it is impossible to distinguish an FM waveform from a PM waveform without knowing the characteristics of the modulating signal. Figure 7-3a shows the unmodulated carrier and Figure 7-3b the modulating signal (both sine waves). Figure 7-3e shows the first derivative of the modulating signal (a cosine wave). Figure 7-3c shows the frequency-modulated wave whose instantaneous frequency is proportional to the modulating signal. Note that the frequency deviation is maximum at the positive and negative peaks of the modulating signal and minimum at the *zero crossings* (the frequency deviation is proportional to the modulating signal—more specifically, to the amplitude of the modulating signal). Also, note in Figure 7-3d that the frequency deviation is maximum at the zero crossings of the modulating signal and minimum at the positive and negative peaks (the frequency deviation is proportional to the slope of the modulating signal). Therefore, for the phase-modulated waveform, the frequency deviation is proportional to the waveform shown in Figure 7-3e, which is a cosine wave (the first derivative of the modulating signal). Note that the amplitudes of both the FM and PM waveforms remain constant. Therefore, the following can be concluded:

1. With frequency modulation, the instantaneous frequency is proportional to the modulating signal and the instantaneous phase is proportional to the first time integral of the modulating signal.

TABLE 7-1   EQUATIONS FOR PHASE- AND FREQUENCY-MODULATED CARRIERS

Type of modulation	Modulating signal	Angle-modulated wave, $M(t)$
(a) Phase	$V(t)$	$V_c \cos [\omega_c t + KV(t)]$
(b) Frequency	$V(t)$	$V_c \cos [\omega_c t + K_1 \int V(t)\, dt]$
(c) Phase	$V_a \cos \omega_a t$	$V_c \cos (\omega_c t + KV_a \cos \omega_a t)$
(d) Frequency	$-V_a \sin \omega_a t$	$V_c \cos \left(\omega_c t + \dfrac{K_1 V_a}{\omega_a} \cos \omega_a t \right)$
(e) Frequency	$V_a \cos \omega_a t$	$V_c \cos \left(\omega_c t + \dfrac{K_1 V_a}{\omega_a} \sin \omega_a t \right)$

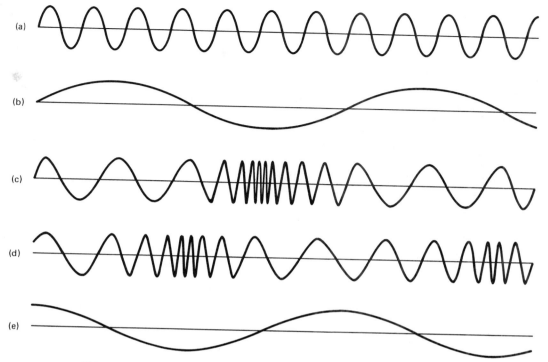

FIGURE 7-3 **Phase and frequency modulation of a sine-wave carrier by a sine-wave signal:**
(a) unmodulated carrier; (b) modulating signal; (c) frequency-modulated wave;
(d) phase-modulated wave; (e) first derivative of the modulating signal.

2. With phase modulation, the instantaneous phase is proportional to the modulating signal and the instantaneous frequency is proportional to the first time derivative (waveform slope) of the modulating signal.

Frequency deviation. *Frequency deviation* (ΔF) is the change in frequency that occurs in the carrier when it is acted on by a modulating signal. From Figure 7-3 it can be seen that ΔF is proportional to the amplitude of the modulating signal V_a, and the rate at which the frequency change occurs is equal to the modulating signal frequency F_a. Frequency deviation is typically given as a peak frequency shift in hertz. The peak-to-peak frequency deviation is sometimes called *carrier swing*. Frequency deviation is a function of the deviation sensitivity of the modulator and the modulating signal amplitude. Mathematically, ΔF is

$$\Delta F = K_1 V_a \qquad 2\pi F \; rad \qquad (7\text{-}10a)$$

or

$$\Delta F = \frac{K_1 V_a}{2\pi \; rad} \qquad F \; Hz \qquad (7\text{-}10b)$$

where

$$V_a = \text{peak modulating signal amplitude (V)}$$

$$K_1 = \text{deviation sensitivity } (2\pi F \text{ rad/V})$$

If the deviation sensitivity is given in Hz/V,

$$\Delta F = K_1 V_a = \text{Hz} \tag{7-10c}$$

Modulation index. Comparing the expressions for the angle-modulated wave types (c), (d), and (e) of Table 7-1 shows that the formula for either a frequency- or a phase-modulated wave with a sinusoidal modulating signal can be written in the general form

$$M(t) = V_c \cos [\omega_c t + \theta(t)]$$

or

$$M(t) = V_c \cos (\omega_c t + m \cos \omega_a t) \tag{7-11}$$

where

$$\theta(t) = m \cos \omega_a t = \text{instantaneous phase deviation}$$

$$m = \text{modulation index}$$

and

$$m = KV_a \text{ for phase modulation} \tag{7-12}$$

$$m = \frac{K_1 V_a}{\omega_a} \text{ for frequency modulation} \tag{7-13}$$

where

$$V_a = \text{peak amplitude of the modulating signal}$$

Therefore for PM

$$m = (\text{rad/V})(V_a) = \text{rad}$$

and for FM

$$m = \frac{(2\pi F \text{ rad/V})(V_a)}{2\pi F_a \text{rad}} = \frac{\Delta F}{F_a} \text{ (unitless)}$$

$\theta(t)$ is the instantaneous phase deviation, and from Equation 7-11 it can be seen that $\theta(t)$ is a function of $\cos \omega_a t$. Therefore, the phase deviation varies at a rate equal to ω_a (the modulating signal frequency) and reaches a maximum peak value of KV_a for PM and $K_1 V_a/\omega_a$ for FM. Therefore, for PM, m is the peak phase deviation in radians and is called the *modulation index*. For FM, the modulation index is a unitless ratio and is used to describe the depth of modulation. From the preceding relationships it can be seen that for phase modulation m is independent of the frequency of the modulating signal. However, for frequency modulation, m is inversely proportional to the frequency of the modulating signal. Also, for both phase and frequency modulation, the modulation index is directly proportional to the deviation sensitivity (K or K_1) and the modulating signal amplitude (V_a).

With FM it is more common to express the modulation index as the peak frequency deviation divided by the modulating signal frequency. Mathematically, modulation index is

$$m = \frac{\Delta F}{F_a}$$

(7-14)

where

$K_1 =$ deviation sensitivity (Hz/V)

$\Delta F = K_1 V_a =$ peak frequency deviation (Hz)

$F_a =$ modulating signal frequency (Hz)

EXAMPLE 7-1

(a) Determine the peak frequency deviation (ΔF) and modulation index (m) for an FM modulator with a deviation sensitivity $K_1 = 5$ kHz/V and a modulating signal $V(t) = 2 \cos (2\pi 2000t)$.

(b) Determine the peak phase deviation (m) for a PM modulator with a deviation sensitivity $K = 2.5$ rad/V and a modulating signal $V(t) = 2 \cos (2\pi 2000t)$.

Solution   (a) From Equation 7-10c,

$$\Delta F = K_1 V_a = \left(\frac{5000 \text{ Hz}}{\text{V}}\right)(2 \text{ V}) = 10 \text{ kHz}$$

From Equation 7-14,

$$m = \frac{\Delta F}{F_a} = \frac{10 \text{ kHz}}{2 \text{ kHz}} = 5$$

(b) From Equation 7-12,

$$m = KV_a = \left(\frac{2.5 \text{ rad}}{\text{V}}\right)(2 \text{ V}) = 5 \text{ rad}$$

In Example 7-1 the modulation index for the frequency-modulated carrier is equal to the peak phase deviation of the phase-modulated carrier (5). If the amplitude of the modulating signal is changed, both the FM modulation index and the peak phase deviation will change proportionally. If the frequency of the modulating signal changes, the FM modulation index changes inversely proportional. However, the phase deviation is unaffected by changes in the modulating signal frequency. Therefore, under identical conditions, FM and PM are indistinguishable for a single modulating frequency; however, when the frequency changes, the PM modulation index remains constant, whereas the FM modulation index increases as the modulating frequency is reduced, and vice versa.

Percent modulation.   With FM, *percent modulation* is simply the ratio of the frequency deviation actually produced to the maximum frequency deviation allowed by law stated in percent form. Mathematically, percent modulation is

$$\% \text{ modulation} = \frac{\Delta F \text{ (actual)}}{\Delta F \text{ (maximum)}} \times 100 \qquad (7\text{-}15)$$

For example, in the United States the FCC (in Canada the DDC), limits the frequency deviation for FM broadcast band transmitters to ± 75 kHz. If a given modulating signal produces ± 50 kHz of frequency deviation, the percent modulation is

$$\% \text{ modulation} = \frac{50 \text{ kHz}}{75 \text{ kHz}} \times 100 = 67\%$$

Phase and Frequency Modulators and Demodulators

A *phase modulator* is a circuit in which the carrier is varied in such a way that its instantaneous phase is proportional to the modulating signal. The unmodulated carrier is a single-frequency sinusoid and is commonly called the *rest* frequency. A *frequency modulator* (often called a *frequency deviator*) is a circuit in which the carrier is varied in such a way that its instantaneous phase is proportional to the integral of the modulating wave. Therefore, with a frequency modulator if the modulating wave $V(t)$ is differentiated prior to being applied to the modulator, the instantaneous phase deviation is proportional to the integral of $V'(t)$ or, in other words, proportional to $V(t)$ [$\int V'(t) = V(t)$]. Similarly, an FM modulator that is preceded by a differentiator produces an output wave in which the phase deviation is proportional to the modulating wave and is, therefore, equivalent to a phase modulator. Several other equivalences are possible. For example, a frequency demodulator followed by an integrator is equivalent to a phase demodulator. Four common equivalences are listed below and illustrated in Figure 7-4.

1. PM modulator = differentiator + FM modulator
2. PM demodulator = FM demodulator + integrator
3. FM modulator = integrator + PM modulator
4. FM demodulator = PM demodulator + differentiator

Frequency Analysis of FM and PM Signals

With angle modulation, the frequency components of the modulated wave are much more complexly related to the frequency components of the modulating signal than with amplitude modulation. In a frequency or a phase modulator, a single-frequency modulating signal produces an infinite number of side frequencies, where each side frequency is displaced from the carrier by an integral multiple of the modulating signal

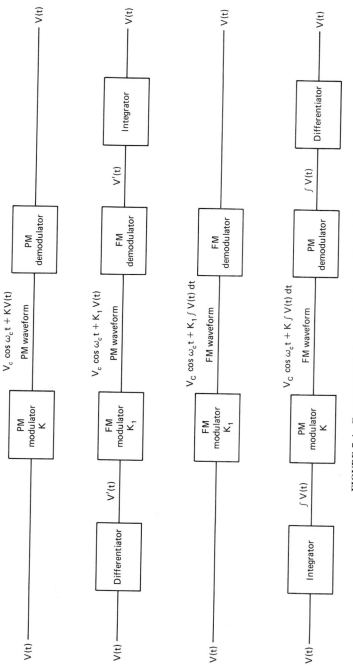

FIGURE 7-4 Frequency and phase modulation and demodulation.

frequency. However, most of the side frequencies are negligibly small in amplitude and can be ignored when using a variety of criteria.

Modulation by a single-frequency sinusoid. Frequency analysis of an angle-modulated wave with a single-frequency sinusoid produces a peak phase deviation of m radians, where m is the modulation index. Again, from Equation 7-11 and for a modulating frequency equal to ω_a, $M(t)$ is written as

$$M(t) = V_c \cos (\omega_c t + m \cos \omega_a t)$$

From Equation 7-11, the individual frequency components that make up the modulated wave are not obvious. However, *Bessel function identities* are available that may be applied directly. One such identity is

$$\cos (\alpha + m \cos \beta) = \sum_{n=-\infty}^{\infty} J_n(m) \cos \left(\alpha + n\beta + \frac{n\pi}{2} \right) \qquad (7\text{-}16)$$

$J_n(m)$ is the Bessel function of the first kind of nth order and of argument m. If Identity 7-16 is applied to Equation 7-11, $M(t)$ may be rewritten as

$$M(t) = V_c \sum_{n=-\infty}^{\infty} J_n(m) \cos \left(\omega_c t + n\omega_a t + \frac{n\pi}{2} \right) \qquad (7\text{-}17)$$

Expanding Equation 7-17 for the first four terms yields

$$\begin{aligned} M(t) = V_c \{ &J_0(m) \cos \omega_c t + J_1(m) \cos \left[(\omega_c + \omega_a)t + \frac{\pi}{2} \right] \\ &+ J_1(m) \cos \left[(\omega_c - \omega_a)t + \frac{\pi}{2} \right] - J_2(m) \cos \left[(\omega_c + 2\omega_a)t \right] \\ &- J_2(m) \cos \left[(\omega_c - 2\omega_a)t \right] + \cdots \end{aligned} \qquad (7\text{-}18)$$

Equations 7-17 and 7-18 show that a single-frequency modulating signal produces an infinite number of sets of side frequencies, each displaced from the carrier by integral multiples of the modulating frequency. A sideband set includes an upper and a lower side frequency ($\pm F_a$, $\pm 2F_a$, $\pm 3F_a$, etc.). The successive sets of sidebands are called the first-order sidebands, second-order sidebands, and so on; and their magnitudes are determined by the coefficients $J_1(m)$, $J_2(m)$, and so on, respectively. Table 7-2 shows the Bessel functions of the first kind for several values of m. The values shown for J_n are relative to the amplitude of the unmodulated carrier. For example, $J_2 = 0.35$ indicates that the amplitude of the second set of side frequencies is equal to 35% of the unmodulated carrier amplitude ($0.35V_c$). It can be seen that the amplitude of the higher-order sidebands rapidly becomes insignificant as the modulation index decreases below unity. For larger values of m, the value of $J_n(m)$ starts to decrease rapidly as soon as $n = m$. As the modulation index increases from zero, the magnitude of the carrier $J_0(m)$ decreases.

TABLE 7-2 BESSEL FUNCTIONS OF THE FIRST KIND, $J_n(m)$

m_f	J_0	J_1	J_2	J_3	J_4	J_5	J_6	J_7	J_8	J_9	J_{10}	J_{11}	J_{12}	J_{13}	J_{14}
0.00	1.00	—	—	—	—	—	—	—	—	—	—	—	—	—	—
0.25	0.98	0.12	—	—	—	—	—	—	—	—	—	—	—	—	—
0.5	0.94	0.24	0.03	—	—	—	—	—	—	—	—	—	—	—	—
1.0	0.77	0.44	0.11	0.02	—	—	—	—	—	—	—	—	—	—	—
1.5	0.51	0.56	0.23	0.06	0.01	—	—	—	—	—	—	—	—	—	—
2.0	0.22	0.58	0.35	0.13	0.03	—	—	—	—	—	—	—	—	—	—
2.4	0	0.52	0.43	0.20	0.06	0.02	0.01	—	—	—	—	—	—	—	—
2.5	−0.05	0.50	0.45	0.22	0.07	0.02	0.01	—	—	—	—	—	—	—	—
3.0	−0.26	0.34	0.49	0.31	0.13	0.04	0.01	—	—	—	—	—	—	—	—
4.0	−0.40	−0.07	0.36	0.43	0.28	0.13	0.05	0.02	—	—	—	—	—	—	—
5.0	−0.18	−0.33	0.05	0.36	0.39	0.26	0.13	0.05	0.02	—	—	—	—	—	—
6.0	0.15	−0.28	−0.24	0.11	0.36	0.36	0.25	0.13	0.06	0.02	—	—	—	—	—
7.0	0.30	0.00	−0.30	−0.17	0.16	0.35	0.34	0.23	0.13	0.06	0.02	—	—	—	—
8.0	0.17	0.23	−0.11	−0.29	−0.10	0.19	0.34	0.32	0.22	0.13	0.06	0.03	—	—	—
9.0	−0.09	0.25	0.14	−0.18	−0.27	−0.06	0.20	0.33	0.31	0.21	0.12	0.06	0.03	0.01	—
10.0	−0.25	0.05	0.25	0.06	−0.22	−0.23	−0.01	0.22	0.32	0.29	0.21	0.12	0.06	0.03	0.01

For m equal to approximately 2.4, $J_0(m) = 0$ and the carrier vanishes (this is called the *first carrier null*). This property is often used to determine the modulation index or set the deviation for an FM modulator. The carrier reappears as m is increased beyond 2.4. When m reaches 5.4, the carrier once again vanishes (this is called the *second carrier null*). The carrier goes to zero at periodic intervals as the index of modulation is further increased. Figure 7-5 shows the curves for the relative amplitudes of the carrier and several sets of side frequencies for values of m up to 10. It can be seen that the amplitude of both the carrier and the side frequencies vary at a periodic rate that is a damped sine wave. The negative values for $J(m)$ simply indicate the relative phase of that side-frequency set.

In Table 7-2 only the significant side frequencies are listed. A side frequency is not considered significant unless it has an amplitude equal to or greater than 1% of the unmodulated carrier amplitude ($J_n \geq 0.01$). From Table 7-2 it can be seen that as m increases, the number of significant sidebands also increases. Therefore, the bandwidth of an angle-modulated wave is directly proportional to the modulation index.

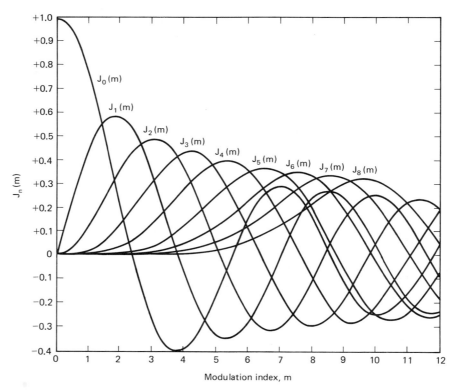

FIGURE 7-5 $J_n(m)$ versus m.

EXAMPLE 7-2

For an FM modulator with a modulation index $m = 1$, a modulating signal $V(t) = V_a \sin (2\pi 1000t)$ and an unmodulated carrier $v_c = 10 \sin (2\pi 500kt)$:
   (a) Determine the number of sets of significant side frequencies.
   (b) Determine their amplitudes.
   (c) Draw the frequency spectrum.

Solution (a) From Table 7-2 a modulation index of 1 yields a reduced carrier component and three sets of side frequencies.
   (b) Their amplitudes are

$$J_0 = 0.77(10) = 7.7 \text{ V}_p$$

$$J_1 = 0.44(10) = 4.4 \text{ V}_p$$

$$J_2 = 0.11(10) = 1.1 \text{ V}_p$$

$$J_3 = 0.02(10) = 0.2 \text{ V}_p$$

   (c) The frequency spectrum is shown in Figure 7-6.

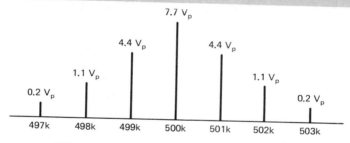

FIGURE 7-6 Frequency spectrum for Example 7-2.

If the FM modulator in Example 7-2 were replaced with a PM modulator and the same carrier and modulating signal frequencies were used, a peak phase deviation $m = 1$ rad produces exactly the same frequency spectrum.

Bandwidth Requirements for Angle-Modulated Waves

In 1922, J. R. Carson mathematically proved that frequency modulation cannot be accommodated in a narrower bandwidth than AM. From the preceding discussion and Example 7-2, it can be seen that the bandwidth of an angle-modulated wave is a function of the modulating signal frequency and the modulation index. With angle modulation, multiple sets of sidebands are produced and, consequently, the bandwidth can be significantly wider than that of an AM wave with the same modulating signal. The modulator output waveform in Example 7-2 requires 6 kHz of bandwidth to pass the carrier and the significant side frequencies. A conventional AM modulator would require only 2 kHz of bandwidth, and a single-sideband system only 1 kHz.

Angle-modulated waveforms are generally classified as either *low*, *medium*, or *high* index. For the low-index case, the peak phase deviation (modulation index) is less than 1 rad, and the high-index case is when the peak phase deviation is greater than 10 rad. Modulation indices between 1 and 10 are classified as medium index. From Table 7-2 it can be seen that with low-index angle modulation, most of the signal information is carried by the first set of sidebands and the minimum bandwidth required is approximately equal to twice the highest modulating signal frequency. For a high-index signal, a method of determining the bandwidth called the quasi-stationary approach is used. With this approach, it is assumed that the modulating signal is changing very slowly. For example, for an FM modulator with a deviation sensitivity $K_1 = 2\pi 2000$ rad per volt-second and a 1-V_p modulating signal, the peak frequency deviation $\Delta F = 2000$ Hz. If the rate of change of the frequency of the modulating signal is very slow, the bandwidth is determined by the peak-to-peak frequency deviation. Therefore, for large indices, the minimum bandwidth required is equal to the peak-to-peak frequency deviation or twice the peak frequency deviation.

Thus for low-index modulation, the minimum bandwidth is approximated by

$$B = 2F_a \tag{7-19}$$

and for high-index modulation, the minimum bandwidth is approximated by

$$B = 2(\Delta F) \tag{7-20}$$

The actual bandwidth required to pass all of the significant sidebands for an angle-modulated wave is equal to two times the product of the highest modulating signal and the number of significant sidebands determined from the table of Bessel functions. Mathematically, the rule for determining the bandwidth for an angle-modulated wave using the Bessel table is

$$B = 2(n \times F_a) \tag{7-21}$$

where

      n = number of significant sidebands
      F_a = highest modulating signal frequency

In an unpublished memorandum dated August 28, 1939, Carson established a general rule to estimate the bandwidth for all angle-modulated systems regardless of the modulation index. This rule is called Carson's rule. Simply stated, Carson's rule approximates the minimum bandwidth of an angle-modulated wave as twice the sum of the peak frequency deviation and the highest modulating signal frequency. Mathematically, Carson's rule is

$$B = 2(\Delta F + F_a) \tag{7-22}$$

where

ΔF = peak frequency modulation
F_a = highest modulating signal frequency

Carson's rule is an approximation and gives transmission bandwidths that are narrower than the bandwidths actually determined using the Bessel table and Equation 7-21. The actual bandwidth required is a function of the modulating signal waveform and the quality of transmission desired.

EXAMPLE 7-3

For an FM modulator with a peak frequency deviation ΔF = 10 kHz, a modulating signal frequency F_a = 10 kHz, and a 500-kHz carrier:
  (a) Determine the actual minimum bandwidth from the Bessel function table.
  (b) Determine the approximate minimum bandwidth using Carson's rule.
  (c) Plot the output frequency spectrum.

Solution   (a) Substituting into Equation 7-14 yields

$$m = \frac{\Delta F}{F_a} = \frac{10,000}{10,000} = 1$$

From Table 7-2 a modulation index of 1 yields three sets of significant sidebands. Substituting into Equation 7-21, the minimum bandwidth is

$$B = 2(3 \times 10,000) = 60 \text{ kHz}$$

(b) Substituting into Equation 7-22, the minimum bandwidth is

$$B = 2(10 \text{ kHz} + 10 \text{ kHz}) = 40 \text{ kHz}$$

(c) The output frequency spectrum is shown in Figure 7-7.

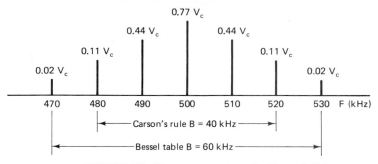

FIGURE 7-7   Frequency spectrum for Example 7-3.

From Example 7-3 it can be seen that the minimum bandwidth determined from Carson's rule is less than the actual minimum bandwidth required to pass all of the significant sideband sets as defined in the Bessel table. Therefore, a system that was designed using Carson's rule would have a poorer performance than a system designed using the Bessel table. For modulation indices above 5, Carson's rule is a close approximation to the actual minimum bandwidth required.

Deviation ratio.   For a given FM system, the minimum bandwidth is greatest when the maximum frequency deviation is produced by the maximum modulating signal frequency (i.e., the highest modulating frequency occurs with the maximum amplitude allowed). By definition, *deviation ratio* is the *worst-case* modulation index and is equal to the maximum peak frequency deviation divided by the maximum modulating signal frequency. The worst-case modulation index produces the widest FM output frequency spectrum. Mathematically, the deviation ratio is

$$\text{DR} = \frac{\Delta F_{\text{max}}}{F_{a(\text{max})}} \tag{7-23}$$

where

$$\text{DR} = \text{deviation ratio}$$
$$F_{a(\text{max})} = \text{maximum modulating signal frequency (Hz)}$$
$$\Delta F_{\text{max}} = \text{maximum peak frequency deviation (Hz)}$$

For example, for the commercial FM broadcast band, the maximum frequency deviation set by the FCC is 75 kHz and the maximum modulating signal frequency is 15 kHz. Therefore, the deviation ratio for FM broadcasting is 75 kHz/15 kHz = 5. This does not mean that whenever a modulation index of 5 occurs, the widest bandwidth also occurs. It means that whenever a modulation index of 5 occurs for the maximum modulating signal frequency, the widest bandwidth occurs.

EXAMPLE 7-4

(a) Determine the deviation ratio and bandwidth for the worst-case (widest-bandwidth) modulation index for an FM broadcast-band transmitter.
(b) Determine the deviation ratio and maximum bandwidth for an equal modulation index with only half the peak frequency deviation and modulating signal frequency.

Solution   (a) $\text{DR} = \dfrac{75 \text{ kHz}}{15 \text{ kHz}} = 5$

From Table 7-2 a modulation index of 5 produces eight sets of significant sidebands. Substituting into Equation 7-21 yields

$$B = 2(8 \times 15,000) = 240 \text{ kHz}$$

(b) For a 37.5-kHz frequency deviation and a modulating signal frequency $F_a = 7.5$ kHz, the modulation coefficient is

$$\frac{37.5 \text{ kHz}}{7.5 \text{ kHz}} = 5$$

and the bandwidth is

$$B = 2(8 \times 7500) = 120 \text{ kHz}$$

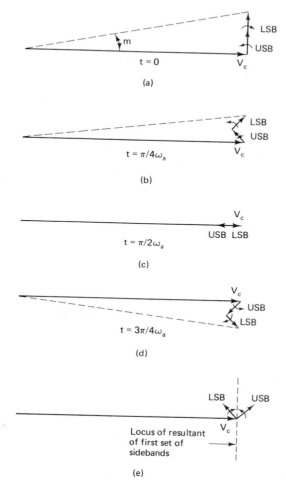

FIGURE 7-8 Angle modulation phasor representation, low modulation index.

It would seem that a higher modulation index (i.e., more sideband sets) with a lower modulating frequency would generate a wider bandwidth because there are considerably more sideband sets generated, but remember, the sidebands are closer together. For example, a 1-kHz modulating signal that produces 10 kHz of frequency deviation has a modulation index $m = 10$ and produces 14 sets of sidebands. However, the sidebands are displaced from each other by only 1 kHz, and therefore the total bandwidth $B = 2(14 \times 1000) = 28,000$ Hz.

Phasor Representation of Angle Modulation

As with AM, an angle-modulated wave can be shown with phasors. The phasor diagram for a low-index-angle modulated wave with a single-frequency modulating signal is shown in Figure 7-8. For this special case ($m < 1$) only the first set of sideband pairs is considered and the phasor diagram is very similar to that of an AM wave except for a phase reversal of one of the side frequencies. The resultant vector has an amplitude that is close to unity at all times and a peak phase deviation of m radians. It is important to note that if the higher-order terms were included, the vector would have no amplitude variations. The dashed line shown in Figure 7-8e is the locus of the resultant formed by the carrier and the first set of side frequencies.

Figure 7-9 shows the phasor diagram for a high-index-angle modulated wave with five sets of side frequencies (for simplicity, the vectors for only two sets are shown). The resultant vector is the sum of the carrier component and the significant side frequencies with their magnitudes adjusted according to the Bessel table. Each side frequency is shifted an additional 90° from the preceding one. The locus of the resultant five-component approximation is curved and closely follows the signal locus. By definition, the locus is a segment of the circle with radius equal to the amplitude of the unmodulated carrier. It should be noted that the resultant signal amplitude and, consequently, the signal power remains constant.

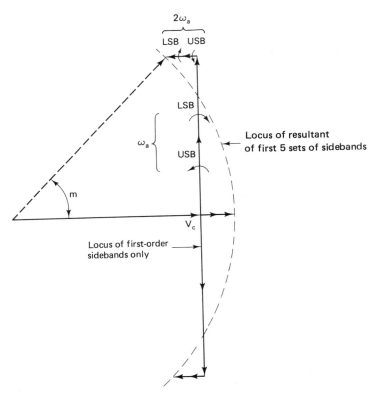

FIGURE 7-9   **Angle modulation phasor representation, high modulation index.**

Average Power of an Angle-Modulated Wave

The total power in an angle-modulated wave is equal to the power of the unmodulated carrier. Therefore, with angle modulation (unlike AM) the power that was originally in the unmodulated carrier is redistributed among the carrier and its sidebands. The rms power of an angle-modulated wave is independent of the modulating signal and equal to the rms power of the unmodulated carrier. Mathematically, the rms power in the unmodulated carrier is simply

$$P_c = \frac{V_c^2}{2R} \text{ Wrms} \qquad (7\text{-}24a)$$

where

P_c = carrier power (watts rms)
V_c = unmodulated carrier voltage (volts peak)
R = load resistance

and peak power is

$$P_c = \frac{V_c^2}{R} \qquad (7\text{-}24b)$$

and the total rms power in an angle-modulated carrier is

$$P_t = \frac{M^2(t)}{R} \qquad (7\text{-}25a)$$

or

$$P_t = \frac{V_c^2}{R} \cos^2[\omega_c t + \theta(t)] \qquad (7\text{-}25b)$$

$$= \frac{V_c^2}{R} \left\{ \frac{1}{2} + \frac{1}{2} \cos[2\omega_c t + 2\theta(t)] \right\} \qquad (7\text{-}25c)$$

In Equation 7-25c, the second term consists of an infinite number of sinusoidal side-frequency components about a frequency equal to twice the unmodulated carrier frequency ($2F_c$). Consequently, the average value of the second term is zero and the rms power of the modulated wave is

$$P = \frac{V_c^2}{2R} \text{ Wrms} \qquad (7\text{-}26)$$

Note that Equations 7-24 and 7-26 are identical, so that the average power of the modulated carrier is equal to the power of the unmodulated carrier. The modulated carrier power is the sum of the powers of the carrier and the side frequencies. Therefore, the total modulated carrier power is

$$P_t = P_c + P_1 + P_2 + P_3 + P_n \tag{7-27a}$$

$$= \frac{V_c^2}{R} + \frac{2(V_1)^2}{R} + \frac{2(V_2)^2}{R} + \frac{2(V_3)^2}{R} + \frac{2(V_n)^2}{R} \tag{7-27b}$$

where

P_c = carrier power
P_1 = power in first set of sidebands
P_2 = power in second set of sidebands
P_3 = power in third set of sidebands
P_n = power in nth set of sidebands

EXAMPLE 7-5

(a) Determine the peak unmodulated carrier power for the FM modulator and conditions given in Example 7-2 (assume a load resistance $R_L = 50 \ \Omega$).
(b) Determine total peak angle-modulated wave power.

Solution   (a) Substituting into Equation 7-24 yields

$$P_c = \frac{10^2}{50} = 2 \ \text{W peak}$$

(b) Substituting into Equation 7-27b gives us

$$P_t = \frac{7.7^2}{50} + \frac{2(4.4)^2}{50} + \frac{2(1.1)^2}{50} + \frac{2(0.2)^2}{50} + \cdots$$

$$= 1.1858 + 0.7744 + 0.0484 + 0.0016$$

$$= 2.0102 \ \text{W peak}$$

The results of (a) and (b) are not exactly equal because the values given in the Bessel table have been rounded off. However, the results are close enough to illustrate that the power in the modulated wave is equal to the power of the unmodulated carrier.

Noise and Angle Modulation

When random white noise with a constant spectral density is added to an FM signal, it produces an unwanted deviation of the carrier frequency. The magnitude of the unwanted modulation depends on the relative amplitude of the noise in respect to the carrier. When this unwanted carrier deviation is demodulated, it becomes noise if it has frequency components that fall into the modulating signal spectrum. The spectral shape of the demodulated noise depends on whether an FM or a PM demodulator is used. The noise voltage at the output of a PM demodulator is constant with frequency, whereas the noise voltage at the output of an FM demodulator increases linearly with frequency. This is commonly called the FM *noise triangle* and is shown in Figure 7-10. From Figure 7-10 it can be seen that the demodulated noise voltage is inherently higher for the higher-modulating signal frequencies.

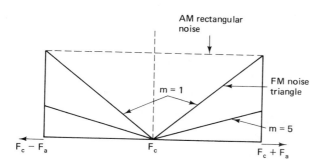

AM rectangular
noise

FM noise
triangle

m = 1

m = 5

$F_c - F_a$ F_c $F_c + F_a$

FIGURE 7-10 FM noise triangle.

PM due to an interfering sinusoid. Figure 7-11a shows phase modulation caused by a single-frequency noise signal. The noise component V_n is separated in frequency from the signal component V_c by frequency ω_n. This is shown in Figure 7-11b. Assuming that $V_c > V_n$, the peak phase deviation due to an interfering single-frequency sinusoid is

$$\Delta\theta(\text{peak}) = \frac{V_n}{V_c} \qquad \text{rad} \tag{7-28}$$

Figure 7-11c shows the result of *limiting* the amplitude of the composite FM signal on noise. (Limiting is commonly used in angle-modulation receivers and is explained in Chapter 8.) It can be seen that the single-frequency noise signal has been transposed into a noise sideband pair each with amplitude $V_n/2$. These sidebands are coherent; therefore, the peak phase deviation is still V_n/V_c radians. However, the unwanted amplitude variations have been removed, which reduces the total power but does not reduce the interference in the demodulated signal due to the unwanted phase deviation. Because the phase modulation is sinusoidal, the rms phase deviation due to the interfering noise signal is the peak phase deviation divided by $\sqrt{2}$. Mathematically, the rms phase deviation is

$$\Delta\theta(\text{rms}) = \frac{V_n}{V_c \sqrt{2}} \tag{7-29}$$

FM due to an interfering sinusoid. As stated previously and from Equation 7-5, the instantaneous frequency deviation $\Delta F(t)$ is the first time derivative of the instantaneous phase deviation $\theta(t)$. When the carrier component is much larger than the interfering noise voltage, the instantaneous phase deviation is approximately

$$\theta(t) = \frac{V_n}{V_c} \sin(\omega_n t + \theta_n) \qquad \text{rad} \tag{7-30}$$

and taking the first derivative, we obtain

$$\Delta F(t) = \frac{V_n}{V_c} \omega_n (\cos \omega_n t + \theta_n) \qquad \text{rad/s} \tag{7-31}$$

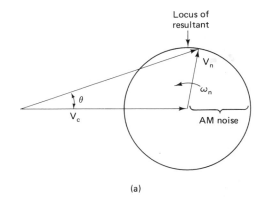

(a)

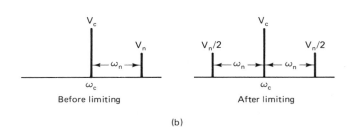

Before limiting   After limiting

(b)

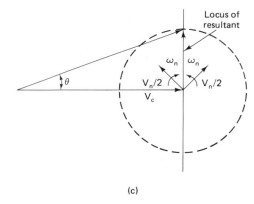

(c)

FIGURE 7-11   Interfering sinusoid of noise: (a) before limiting; (b) frequency spectrum; (c) after limiting.

Therefore, peak frequency deviation is

$$\Delta F(\text{peak}) = \frac{V_n}{V_c} \omega_n \qquad \text{rad/s} \qquad (7\text{-}32\text{a})$$

$$= \frac{V_n}{V_c} F_n \qquad \text{Hz} \qquad (7\text{-}32\text{b})$$

and the rms frequency deviation is

$$\Delta F(\text{rms}) = \frac{V_n}{V_c \sqrt{2}} \omega_n \qquad \text{rad} \qquad (7\text{-}33\text{a})$$

$$= \frac{V_n}{V_c \sqrt{2}} F_n \qquad \text{Hz} \qquad (7\text{-}33\text{b})$$

Rearranging Equation 7-14, it can be seen that the peak frequency deviation (ΔF) is a function of the modulating frequency and the modulation index. Therefore, for a noise modulating frequency F_n, the peak frequency deviation is

$$\Delta F(\text{peak}) = mF_n \qquad (7\text{-}34)$$

where m = peak phase deviation

From Equation 7-34 it can be seen that the further the noise frequency is displaced from the carrier frequency, the larger the frequency deviation. Therefore, noise frequencies that produce components at the high end of the modulating signal spectrum produce more frequency deviation for the same phase deviation than frequencies that fall at the low end. FM demodulators generate an output voltage that is proportional to the frequency deviation and of frequency equal to the difference between the carrier and the interfering signal. Therefore, high-frequency noise components produce more demodulated noise than do low-frequency components.

The signal-to-noise ratio at the output of an FM demodulator due to unwanted frequency deviation from an interfering sinusoid is the ratio of the peak frequency deviation due to the information signal to the peak frequency deviation due to the interfering signal.

$$\frac{S}{N} = \frac{\Delta F(\text{signal})}{\Delta F(\text{noise})} \qquad (7\text{-}35)$$

EXAMPLE 7-6

For an angle-modulated carrier $V_c = 6 \cos(2\pi 110 \text{ MHz}t)$ with 75 kHz of frequency deviation due to the information signal and a single-frequency interfering signal $V_n = 0.3 \cos(2\pi 109.985 \text{ MHz}t)$.
    (a) Determine the frequency of the demodulated interference signal.
    (b) Determine the peak and rms phase and frequency deviations due to the interfering signal.
    (c) Determine the signal-to-noise ratio at the output of the demodulator.

Solution   (a) The frequency of the noise interference is

$$F_c - F_n = 110 \text{ MHz} = 109.985 \text{ MHz} = 15 \text{ kHz}$$

   (b) Substituting into Equation 7-28 yields

$$\theta(\text{peak}) = \frac{0.3}{6} = 0.05 \text{ rad}$$

and from Equation 7-29 the rms phase deviation is

$$\theta(\text{rms}) = \frac{0.3}{6\sqrt{2}} = 0.03535 \text{ rad}$$

Substituting into Equation 7-32b yields

$$\Delta F(\text{peak}) = \left(\frac{0.3}{6}\right) 15 \text{ kHz} = 750 \text{ Hz}$$

Substituting into Equation 7-33b gives us

$$\Delta F(\text{rms}) = \left(\frac{0.3}{6\sqrt{2}}\right) 15 \text{ kHz} = 530 \text{ Hz}$$

(c) The S/N ratio of the interfering tone and the carrier is the ratio of the peak carrier amplitude to the peak noise amplitude, or

$$\frac{6}{0.3} = 20$$

The S/N ratio after demodulation is found by substituting into Equation 7-35:

$$\frac{S}{N} = \frac{75 \text{ kHz}}{750 \text{ Hz}} = 100$$

There is an S/N improvement of $100/20 = 5$ or $10 \log 5 = 7$ dB.

Angle modulation due to random noise.　　Random noise comprises an infinite number of frequencies, of equal amplitude, and of arbitrary phase. It is convenient to analyze system noise on a per-hertz basis and to consider a band of noise N hertz wide to be made up of N uniformly spaced sinusoids. If the total peak noise voltage and, consequently, the total noise power is assumed to be relatively small compared to the carrier, the phase modulation is approximated as

$$\theta(t) = \sum_{N=-1}^{N} \frac{V_n}{V_c} \sin(\omega_n t + \theta_n) \tag{7-36}$$

From Equation 7-36 it can be seen that the total phase modulation due to random noise is equal to the summation of the phase modulation components that are produced by each individual noise component. If these individual noise components are large compared to the carrier, intermodulation distortion occurs. Because phase relationships that exist between the noise sinusoids are not known, the peak phase modulation they produce cannot be defined. Therefore, the total rms voltage is

$$\text{total rms voltage} = \sum_{N=1}^{N} \frac{V_n^2}{2} \tag{7-37a}$$

$$= \frac{V_n}{2} N \tag{7-37b}$$

where $V_n/2$ is the rms amplitude of each sinusoid. The total rms phase deviation is

$$\theta(t)_{total} = \frac{(V_n/\sqrt{2})N}{V_c} \quad \text{rad} \tag{7-38a}$$

$$= \frac{V_n N}{V_c \sqrt{2}} \quad \text{rad} \tag{7-38b}$$

EXAMPLE 7-7

Determine the total rms phase deviation produced by a 10-kHz band of random noise with a peak voltage $V_n = 0.1 \ \mu V$ and a carrier $V_c = 1 \ \sin(2\pi 10 \ \text{MHz} \ t)$.

Solution Substituting into Equation 7-38b gives us

$$\theta(t)_{total} = \frac{(0.1 \ \mu V)(10,000)}{\sqrt{2}} = 0.0007 \ \text{rad}$$

Preemphasis and Deemphasis

The noise triangle shown in Figure 7-10 showed that, with FM, noise at the higher modulating frequencies is inherently greater in amplitude than noise at the lower frequencies. This includes both single-frequency interference and random noise. Therefore, assuming that the amplitudes of all of the information signals are equal, a nonuniform *S/N* is evident and the higher modulating frequencies suffer a greater *S/N* degradation. This is shown in Figure 7-12a. It can be seen that the *S/N* is lower at the high-frequency ends of the triangle. *Preemphasis* is artificially boosting the amplitude of the high-frequency modulating signals prior to modulation, and *deemphasis* is simply the opposite action and is performed in the receiver after demodulation to restore the original modulating signal voltage spectrum. Figure 7-12b shows the effects on the signal-to-noise ratio using pre- and deemphasis. The figure shows that pre- and deemphasis produces a uniform *S/N* ratio throughout the modulating signal spectrum.

A preemphasis network is a high-pass filter (i.e., a differentiator) and a deemphasis network is a low-pass filter (i.e., an integrator). Figure 7-13a shows the schematic diagrams for an active preemphasis network and a passive deemphasis network. Their corresponding frequency response curves are shown in Figure 7-13b. A preemphasis network provides a constant increase in the amplitude of the modulating signal with increase in modulating frequency. With FM, approximately 12 dB of improvement in noise performance is achieved using pre- and deemphasis. The break frequency (where pre- and deemphasis begins) is determined by the *RC* or *L/R* time constant of the network. Mathematically, the break frequency is

$$F = \frac{1}{2\pi RC} \quad \text{or} \quad \frac{1}{2\pi L/R} \tag{7-39}$$

The networks shown in Figure 7-13 are for the FM broadcast band, which uses a 75-μs time constant; therefore, the break frequency is approximately 2.12 kHz. The

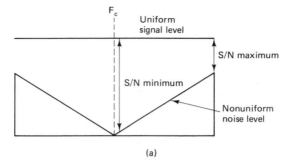

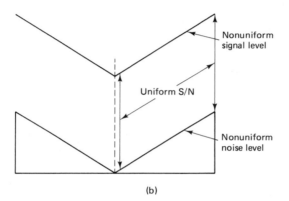

FIGURE 7-12   FM signal-to-noise:
(a) without preemphasis; (b) with
preemphasis.

FM transmission of the audio portion of commercial television broadcasting uses a 50-μs time constant. Because the noise at the output of a PM demodulator is constant with frequency, phase modulators do not require preemphasis networks.

FREQUENCY MODULATION TRANSMISSION

Direct FM

Direct FM is angle modulation in which the frequency of the carrier is varied (deviated) directly by the modulating signal. With direct FM, the instantaneous frequency deviation is directly proportional to the amplitude of the modulating signal. Figure 7-14 shows a schematic diagram for a simple (although highly impractical) direct FM generator. The tank circuit (L and C_m) is the frequency-determining section for a standard LC oscillator. The capacitor microphone is a *transducer* that converts *acoustical energy* to *mechanical energy* which is used to vary the distance between the plates of C_m and, consequently, change its capacitance. As C_m is varied, the resonant frequency is varied. Consequently, the oscillator output frequency varies directly with the input from the external sound

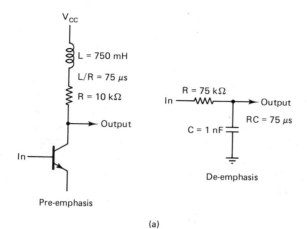

Pre-emphasis

De-emphasis

(a)

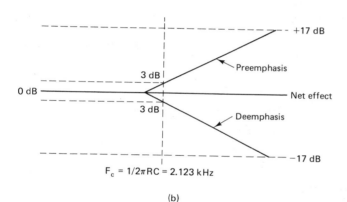

$F_c = 1/2\pi RC = 2.123 \, kHz$

(b)

FIGURE 7-13 Preemphasis and deemphasis: (a) schematic diagrams; (b) attenuation curves.

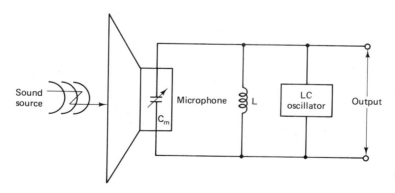

FIGURE 7-14 Simple direct FM modulator.

source. This is direct FM generation because the oscillator frequency is directly changed (deviated) by the modulating signal and the magnitude of the frequency change is proportional to the amplitude of the modulating signal.

Varactor diode modulators. Figure 7-15 shows the schematic diagram for a more practical direct FM generator that uses a varactor diode to deviate the frequency of a crystal oscillator. R_1 and R_2 develop a dc bias that reverse biases varactor diode VD_1 and determines the center or rest frequency for the oscillator. The external audio modulating signal voltage adds to and subtracts from the dc bias which changes the varactor's capacitance and thus the oscillator's frequency. Positive alternations of V_a increase the reverse bias across VD_1, which decreases its capacitance and increases the frequency of oscillation. Negative alternations of V_a decrease the reverse bias across VD_1, increase its capacitance, and decrease the frequency of oscillation. Varactor diode FM modulators are extremely popular because they are simple to use, reliable, and have the stability of a crystal oscillator. However, because a crystal is used, the peak frequency deviation is limited to relatively small values. Consequently, they are used primarily for low-index applications, such as two-way mobile FM radio.

Figure 7-16 shows a simplified schematic diagram for a voltage-controlled oscillator (VCO) FM generator. Again, a varactor diode is used to transform changes in the modulating signal to changes in frequency. The center or rest frequency for the modulator is

$$F_c = \frac{1}{2\pi \sqrt{LC}} \tag{7-40}$$

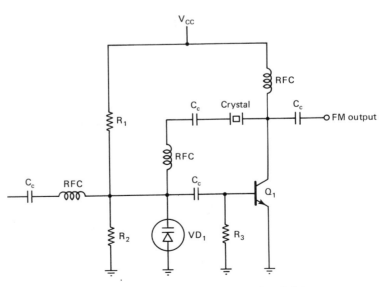

FIGURE 7-15   Varactor diode direct FM modulator.

where

$$L = \text{primary inductance}$$
$$C = \text{varactor diode capacitance}$$

With a modulating signal the frequency is

$$F_m = \frac{1}{2\pi \sqrt{L(C + \Delta C)}} \tag{7-41}$$

where ΔC is the change in varactor diode capacitance due to the modulating signal. The change in frequency with modulation (ΔF) is

$$\Delta F = F_c - F_m \tag{7-42}$$

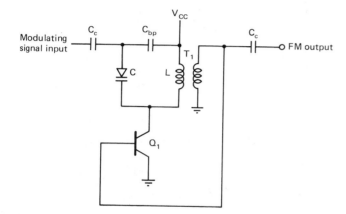

FIGURE 7-16   Varactor diode VCO FM modulator.

Reactance modulator.   Figure 7-17a shows a schematic diagram for a *reactance modulator* using a JFET as the active device. This circuit configuration is called a reactance modulator because the JFET looks like a variable-reactance load to the *LC* tank circuit. The modulating signal varies the reactance of Q_1, which causes a corresponding change in the resonant frequency of the oscillator tank circuit.

Figure 7-17b shows the ac equivalent circuit. R_1, R_2, R_3, R_4, R_E, and R_C provide the dc bias for Q_1. However, R_E is bypassed by C_E and is, therefore, omitted from the ac equivalent circuit. Circuit operation is as follows. Assuming an ideal JFET ($i_g = 0$)

$$v_g = i_R \tag{7-43}$$

where

$$i = \frac{v}{R - jX_c} \tag{7-44}$$

Therefore,

$$v_g = \frac{v}{R - jX_c} R \tag{7-45}$$

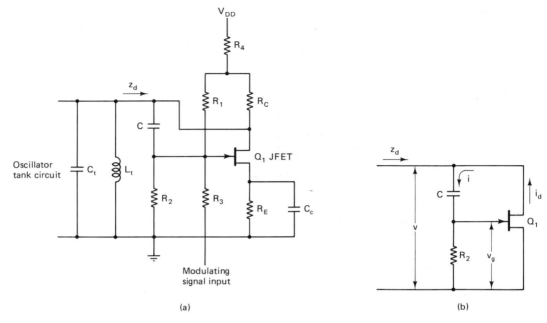

FIGURE 7-17   JFET reactance modulator: (a) schematic diagram; (b) ac equivalent circuit.

and the JFET drain current i_d is

$$i_d = g_m v_g = g_m \frac{v}{R - jX_c} R \qquad (7\text{-}46)$$

where g_m is the JFET transconductance, and the impedance between the drain and ground is

$$z_d = \frac{e}{id} = g_m \frac{v/v}{R - jX_c} R \qquad (7\text{-}47\text{a})$$

Rearranging Equation 7-47a yields

$$z_d = \frac{R - jX_c}{g_m R} = \frac{1}{g_m} - \frac{jX_c}{g_m R} \qquad (7\text{-}47\text{b})$$

Assuming that $R << X_c$,

$$z_d = -j\frac{X_c}{g_m R} = \frac{-j}{2\pi F g_m R C} \qquad (7\text{-}47\text{c})$$

$g_m RC$ is equivalent to a variable capacitance and is inversely proportional to resistance (R), the angular frequency of the modulating signal ($2\pi F$), and the transconductance (g_m) of Q_1, which varies with the gate-to-source voltage. When a modulating signal is applied to R_3, the gate-to-source voltage is varied accordingly, causing a proportional change in g_m. As a result, the equivalent circuit impedance (z_d) is a function of the modulating signal. Therefore, the resonant frequency of the oscillator tank circuit is a function of the amplitude of the modulating signal and the rate at which it deviates is equal to $2\pi F$. Interchanging R_2 and C causes the variable reactance to be inductive rather than capacitive but does not affect the output FM waveform. The maximum frequency deviation for a reactance modulator is approximately 5 kHz.

Linear integrated-circuit direct FM generator. *Linear integrated-circuit* (LIC) *FM modulators* generate a high-quality output waveform that is stable, accurate, and directly proportional to the input modulating signal. Figure 7-18 shows a simplified schematic diagram for a LIC direct FM generator. The VCO center frequency is determined by the external resistor and capacitor (R and C). The input modulating signal deviates the center frequency, which produces an FM output waveform. The analog multiplier and sine shaper convert the VCO square-wave output to a sine wave, and the unity-gain amplifier provides a buffered output. The modulator output frequency is

$$FM = (F_o + \Delta F)N \tag{7-48}$$

where $\Delta F = V_a K$.

Figure 7-19 shows the specification sheets for the XR-2206 *monolithic function generator*. The 2206 can generate either a sine or a triangular FM output waveform. The 2206 can be used for either sweep frequency operation, frequency shift keying (FSK), or direct FM generation.

Indirect FM

Indirect FM is angle modulation in which the frequency of the carrier is changed indirectly by the modulating signal. Indirect FM is accomplished by directly deviating the phase of the carrier and is therefore a form of phase modulation. The instantaneous phase of an indirect FM carrier is directly proportional to the modulating signal amplitude. Therefore, the instantaneous frequency is proportional to the integral of the modulating signal.

Figure 7-20 shows a schematic diagram for an indirect FM modulator (henceforth referred to as a phase modulator). The phase modulator comprises a varactor diode VD_1 in series with an inductive network (tunable coil L_1 and resistor R_1). The combined series–parallel network appears as a series resonant circuit to the output frequency from the crystal oscillator. A modulating signal applied to VD_1 changes its capacitance and,

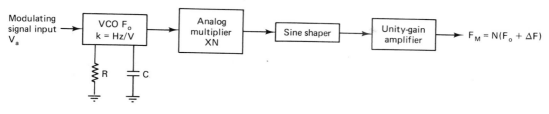

FIGURE 7-18 LIC direct FM generator: simplified schematic diagram.

Monolithic Function Generator

GENERAL DESCRIPTION

The XR-2206 is a monolithic function generator integrated circuit capable of producing high quality sine, square, triangle, ramp, and pulse waveforms of high-stability and accuracy. The output waveforms can be both amplitude and frequency modulated by an external voltage. Frequency of operation can be selected externally over a range of 0.01 Hz to more than 1 MHz.

The circuit is ideally suited for communications, instrumentation, and function generator applications requiring sinusoidal tone, AM, FM, or FSK generation. It has a typical drift specification of 20 ppm/°C. The oscillator frequency can be linearly swept over a 2000:1 frequency range, with an external control voltage, having a very small affect on distortion.

FEATURES

Low-Sine Wave Distortion	.5%, Typical
Excellent Temperature Stability	20 ppm/°C, Typical
Wide Sweep Range	2000:1, Typical
Low-Supply Sensitivity	0.01%/V, Typical
Linear Amplitude Modulation	
TTL Compatible FSK Controls	
Wide Supply Range	10V to 26V
Adjustable Duty Cycle	1% to 99%

APPLICATIONS

Waveform Generation
Sweep Generation
AM/FM Generation
V/F Conversion
FSK Generation
Phase-Locked Loops (VCO)

ABSOLUTE MAXIMUM RATINGS

Power Supply	26V
Power Dissipation	750 mW
Derate Above 25°C	5 mW/°C
Total Timing Current	6 mA
Storage Temperature	−65°C to +150°C

FUNCTIONAL BLOCK DIAGRAM

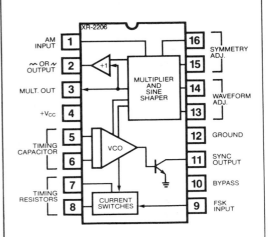

ORDERING INFORMATION

Part Number	Package	Operating Temperature
XR-2206M	Ceramic	−55°C to +125°C
XR-2206N	Ceramic	0°C to +70°C
XR-2206P	Plastic	0°C to +70°C
XR-2206CN	Ceramic	0°C to +70°C
XR-2206CP	Plastic	0°C to +70°C

SYSTEM DESCRIPTION

The XR-2206 is comprised of four functional blocks; a voltage-controlled oscillator (VCO), an analog multiplier and sine-shaper; a unity gain buffer amplifier; and a set of current switches.

The VCO actually produces an output frequency proportional to an input current, which is produced by a resistor from the timing terminals to ground. The current switches route one of the timing pins current to the VCO controlled by an FSK input pin, to produce an output frequency. With two timing pins, two discrete output frequencies can be independently produced for FSK Generation Applications.

EXAR Integrated Systems, Inc., 750 Palomar Avenue, Sunnyvale, CA 94086 * (408) 732-7970 * TWX 910-339-9233

FIGURE 7-19 Specification sheet for XR-2206 monolithic function generator. (Courtesy of EXAR Corporation.)

XR-2206

ELECTRICAL CHARACTERISTICS

Test Conditions: Test Circuit of Figure 1, $V^+ = 12V$, $T_A = 25°$, $C = 0.01 \mu F$, $R_1 = 100 k\Omega$, $R_2 = 10 k\Omega$, $R_3 = 25 k\Omega$ unless otherwise specified. S_1 open for triangle, closed for sine wave.

PARAMETER	XR-2206M			XR-2206C			UNIT	CONDITIONS
	MIN.	TYP.	MAX.	MIN.	TYP.	MAX.		
GENERAL CHARACTERISTCS								
Single Supply Voltage	10		26	10		26	V	
Split-Supply Voltage	±5		±13	±5		±13	V	
Supply Current		12	17		14	20	mA	$R_1 \geqslant 10 k\Omega$
OSCILLATOR SECTION								
Max. Operating Frequency	0.5	1		0.5	1		MHz	$C = 1000 pF, R_1 = 1 k\Omega$
Lowest Practical Frequency		0.01			0.01		Hz	$C = 50 \mu F, R_1 = 2 M\Omega$
Frequency Accuracy		±1	±4		±2		% of f_O	$f_O = 1/R_1 C$
Temperature Stability		±10	±50		±20		ppm/°C	$0°C \leqslant T_A \leqslant 75°C,$
								$R_1 = R_2 = 20 k\Omega$
Supply Sensitivity		0.01	0.1		0.01		%/V	$V_{LOW} = 10V, V_{HIGH} = 20V,$
								$R_1 = R_2 = 20 k\Omega$
Sweep Range	1000:1	2000:1			2000:1		$f_H = f_L$	$f_H @ R_1 = 1 k\Omega$
								$f_L @ R_1 = 2 M\Omega$
Sweep Linearity								
10:1 Sweep		2			2		%	$f_L = 1 kHz, f_H = 10 kHz$
1000:1 Sweep		8			8		%	$f_L = 100 Hz, f_H = 100 kHz$
FM Distortion		0.1			0.1		%	±10% Deviation
Recommended Timing								
Components								
Timing Capacitor: C	0.001		100	0.001		100	μF	See Figure 4.
Timing Resistors: R_1 & R_2	1		2000	1		2000	$k\Omega$	
Triangle Sine Wave Output								See Note 1, Figure 2.
Triangle Amplitude		160			160		mV/kΩ	Figure 1, S_1 Open
Sine Wave Amplitude	40	60	80		60		mV/kΩ	Figure 1, S_1 Closed
Max. Output Swing		6			6		Vp-p	
Output Impedance		600			600		Ω	
Triangle Linearity		1			1		%	
Amplitude Stability		0.5			0.5		dB	For 1000:1 Sweep
Sine Wave Amplitude Stability		4800			4800		ppm/°C	See Note 2.
Sine Wave Distortion								
Without Adjustment		2.5			2.5		%	$R_1 = 30 k\Omega$
With Adjustment		0.4	1.0		0.5	1.5	%	See Figures 6 and 7.
Amplitude Modulation								
Input Impedance	50	100		50	100		$k\Omega$	
Modulation Range		100			100		%	
Carrier Suppression		55			55		dB	
Linearity		2			2		%	For 95% modulation
Square-Wave Output								
Amplitude		12			12		Vp-p	Measured at Pin 11.
Rise Time		250			250		nsec	$C_L = 10 pF$
Fall Time		50			50		nsec	$C_L = 10 pF$
Saturation Voltage		0.2	0.4		0.2	0.6	V	$I_L = 2 mA$
Leakage Current		0.1	20		0.1	100	μA	$V_{11} = 26V$
FSK Keying Level (Pin 9)	0.8	1.4	2.4	0.8	1.4	2.4	V	See section on circuit controls
Reference Bypass Voltage	2.9	3.1	3.3	2.5	3	3.5	V	Measured at Pin 10.

Note 1: Output amplitude is directly proportional to the resistance, R_3, on Pin 3. See Figure 2.

Note 2: For maximum amplitude stability, R_3 should be a positive temperature coefficient resistor.

FIGURE 7-19 *(continued)*

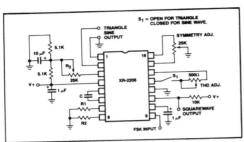

Figure 1: Basic Test Circuit.

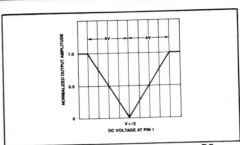

Figure 5: Normalized Output Amplitude versus DC Bias at AM Input (Pin 1).

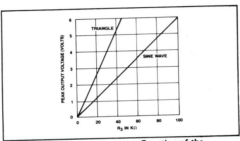

Figure 2: Output Amplitude as a Function of the Resistor, R₃, at Pin 3.

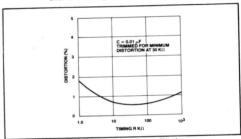

Figure 6: Trimmed Distortion versus Timing Resistor.

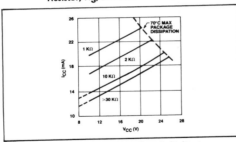

Figure 3: Supply Current versus Supply Voltage, Timing, R.

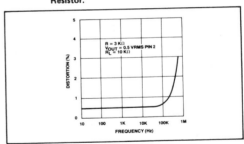

Figure 7: Sine Wave Distortion versus Operating Frequency with Timing Capacitors Varied.

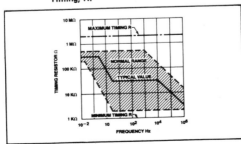

Figure 4: R versus Oscillation Frequency.

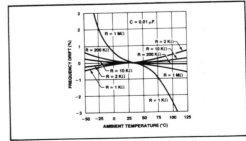

Figure 8: Frequency Drift versus Temperature.

FIGURE 7-19 (*continued*)

XR-2206

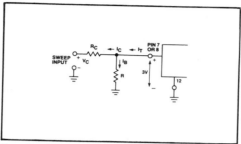

Figure 9: Circuit Connection for Frequency Sweep.

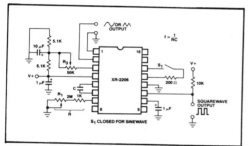

Figure 10: Circuit for Sine Wave Generation without External Adjustment. (See Figure 2 for Choice of R_3.)

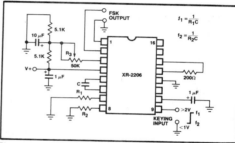

Figure 12: Sinusoidal FSK Generator.

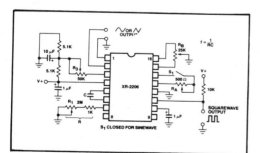

Figure 11: Circuit for Sine Wave Generation with Minimum Harmonic Distortion. (R_3 Determines Output Swing — See Figure 2.)

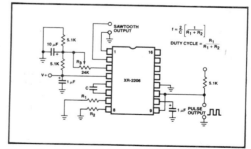

Figure 13: Circuit for Pulse and Ramp Generation.

FIGURE 7-19 (*continued*)

Frequency-Shift Keying:

The XR-2206 can be operated with two separate timing resistors, R_1 and R_2, connected to the timing Pin 7 and 8, respectively, as shown in Figure 12. Depending on the polarity of the logic signal at Pin 9, either one or the other of these timing resistors is activated. If Pin 9 is open-circuited or connected to a bias voltage $\geq 2V$, only R_1 is activated. Similarly, if the voltage level at Pin 9 is $\leq 1V$, only R_2 is activated. Thus, the output frequency can be keyed between two levels, f_1 and f_2, as:

$$f_1 = 1/R_1C \text{ and } f_2 = 1/R_2C$$

For split-supply operation, the keying voltage at Pin 9 is referenced to V^-.

Output DC Level Control:

The dc level at the output (Pin 2) is approximately the same as the dc bias at Pin 3. In Figures 10, 11 and 12, Pin 3 is biased midway between V^+ and ground, to give an output dc level of $\approx V^+/2$.

APPLICATIONS INFORMATION

Sine Wave Generation

Without External Adjustment:

Figure 10 shows the circuit connection for generating a sinusoidal output from the XR-2206. The potentiometer, R_1 at Pin 7, provides the desired frequency tuning. The maximum output swing is greater than $V^+/2$, and the typical distortion (THD) is <2.5%. If lower sine wave distortion is desired, additional adjustments can be provided as described in the following section.

The circuit of Figure 10 can be converted to split-supply operation, simply by replacing all ground connections with V^-. For split-supply operation, R_3 can be directly connected to ground.

With External Adjustment:

The harmonic content of sinusoidal output can be reduced to $\approx 0.5\%$ by additional adjustments as shown in Figure 11. The potentiometer, R_A, adjusts the sine-shaping resistor, and R_B provides the fine adjustment for the waveform symmetry. The adjustment procedure is as follows:

1. Set R_B at midpoint, and adjust R_A for minimum distortion.

2. With R_A set as above, adjust R_B to further reduce distortion.

Triangle Wave Generation

The circuits of Figures 10 and 11 can be converted to triangle wave generation, by simply open-circuiting Pin 13 and 14 (i.e., S_1 open). Amplitude of the triangle is approximately twice the sine wave output.

FSK Generation

Figure 12 shows the circuit connection for sinusoidal FSK signal operation. Mark and space frequencies can be independently adjusted, by the choice of timing resistors, R_1 and R_2; the output is phase-continuous during transitions. The keying signal is applied to Pin 9. The circuit can be converted to split-supply operation by simply replacing ground with V^-.

Pulse and Ramp Generation

Figure 13 shows the circuit for pulse and ramp waveform generation. In this mode of operation, the FSK keying terminal (Pin 9) is shorted to the square-wave output (Pin 11), and the circuit automatically frequency-shift keys itself between two separate frequencies during the positive-going and negative-going output waveforms. The pulse width and duty cycle can be adjusted from 1% to 99%, by the choice of R_1 and R_2. The values of R_1 and R_2 should be in the range of 1 kΩ to 2 MΩ.

FIGURE 7-19 (*continued*)

XR-2206

PRINCIPLES OF OPERATION

Description of Controls

Frequency of Operation:

The frequency of oscillation, f_o, is determined by the external timing capacitor, C, across Pin 5 and 6, and by the timing resistor, R, connected to either Pin 7 or 8. The frequency is given as:

$$f_o = \frac{1}{RC} \ \ Hz$$

and can be adjusted by varying either R or C. The recommended values of R, for a given frequency range, are shown in Figure 4. Temperature stability is optimum for $4 \ k\Omega < R < 200 \ k\Omega$. Recommended values of C are from 1000 pF to 100 μF.

Frequency Sweep and Modulation:

Frequency of oscillation is proportional to the total timing current, I_T, drawn from Pin 7 or 8:

$$f = \frac{320 I_T \ (mA)}{C \ (\mu F)} \ \ Hz$$

Timing terminals (Pin 7 or 8) are low-impedance points, and are internally biased at +3V, with respect to Pin 12. Frequency varies linearly with I_T, over a wide range of current values, from 1 μA to 3 mA. The frequency can be controlled by applying a control voltage, V_C, to the activated timing pin as shown in Figure 9. The frequency of oscillation is related to V_C as:

$$f = \frac{1}{RC} \ \ 1 + \frac{R}{R_C} (1 - \frac{V_C}{3}) \ \ \ Hz$$

where V_C is in volts. The voltage-to-frequency conversion gain, K, is given as:

$$K = \partial f / \partial V_C = - \ \frac{0.32}{R_C C} \ Hz/V$$

CAUTION: For safe operation of the circuit, I_T should be limited to $\leqslant 3$ mA.

Output Amplitude:

Maximum output amplitude is inversely proportional to the external resistor, R_3, connected to Pin 3 (see Figure 2). For sine wave output, amplitude is approximately 60 mV peak per $k\Omega$ of R_3; for triangle, the peak amplitude is approximately 160 mV peak per $k\Omega$ of R_3. Thus, for example, $R_3 = 50 \ k\Omega$ would produce approximately $\pm 3V$ sinusoidal output amplitude.

Amplitude Modulation:

Output amplitude can be modulated by applying a dc bias and a modulating signal to Pin 1. The internal impedance at Pin 1 is approximately 100 $k\Omega$. Output amplitude varies linearly with the applied voltage at Pin 1, for values of dc bias at this pin, within ± 4 volts of $V^+/2$ as shown in Figure 5. As this bias level approaches $V^+/2$, the phase of the output signal is reversed, and the amplitude goes through zero. This property is suitable for phase-shift keying and suppressed-carrier AM generation. Total dynamic range of amplitude modulation is approximately 55 dB.

CAUTION: AM control must be used in conjunction with a well-regulated supply, since the output amplitude now becomes a function of V^+.

EQUIVALENT SCHEMATIC DIAGRAM

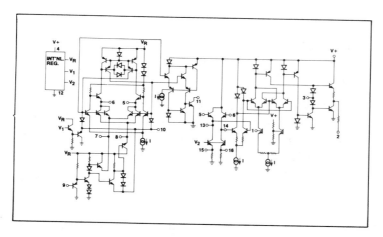

FIGURE 7-19 *(continued)*

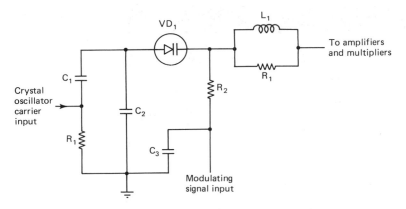

FIGURE 7-20   Indirect FM modulator schematic diagram.

consequently, the phase angle of the impedance seen by the RF carrier varies, which results in a corresponding phase shift in the carrier. The phase shift is directly proportional to the amplitude of the modulating signal. An advantage of indirect FM is that a buffered crystal oscillator is used for the carrier source. Consequently, indirect FM transmitters are more frequency stable than their direct counterparts. A disadvantage is that the capacitance-versus-voltage characteristic of a varactor diode is nonlinear. In fact, it closely resembles a square-root function. Consequently, to minimize distortion of the PM waveform, the amplitude of the modulating signal must be kept small, which, of course, limits the phase deviation (modulation index) to small values.

FM Communication Systems

Before FM transmitters are examined, it is helpful to have a basic understanding of the services allocated by the FCC for FM transmission, their specifications, and limitations.

       Commercial FM broadcast band.   The commercial FM broadcast band is a wideband FM communications system that comprises one hundred 200-kHz channels that occupy the frequency band from 88 to 108 MHz. As stated previously, the maximum frequency deviation for a commercial FM broadcast-band transmitter is 75 kHz, with a maximum modulating signal frequency of 15 kHz. These requirements yield a deviation ratio of 5, which produces eight sets of significant sidebands. The sidebands are separated by 15 kHz, which produces a bandwidth of 240 kHz. It appears that *adjacent channel interference* would be a problem because the 240-kHz bandwidth determined using Bessel approximations exceeds the allocated 200-kHz bandwidth. However, from the Bessel table, it can be seen that the voltage components for the two highest sideband sets are 0.05 and 0.02, respectively, which are considered insignificant. Also, using Carson's approximation, the bandwidth is 2(75 kHz + 15 kHz), or only 180 kHz. The FCC gets around this problem by leaving every other channel assignment vacant in a given metropolitan area, and it is unlikely that any area would have the need for more than 50 FM broadcast stations. Figure 7-21 shows the frequency spectrum allocated by the FCC for commercial FM broadcasting and the worst-case frequency spectrum for a single station's transmission.

Two-way FM mobile communications. The FCC has allocated several frequency bands for the purpose of FM mobile communications, which includes such services as mobile telephone, local law enforcement, civil defense, marine, and fire department communications systems. The primary purpose for two-way FM radio is voice communications. Therefore, the bandwidth allocated for two-way FM is typically only 10 to 30 kHz. The maximum frequency deviation is approximately 5 kHz, and the highest modulating signal frequency is typically 3 kHz. Two-way mobile FM radio is *narrowband FM* (NBFM). That is, the modulation indices are low enough so that only one or two pairs of significant sidebands are generated and the bandwidth resembles that of a conventional AM system. NBFM is sometimes called *low-index FM* because low deviation ratios are used.

Several frequency bands used for two-way FM mobile radio communications are: 25 to 50 MHz, 152 to 162 MHz, 450 to 470 MHz, and 806 to 947 MHz. Cellular radio is a mobile FM telephone communications system and is described in Chapter 8.

Television broadcasting. The sound portion of commercial television broadcasting uses frequency modulation. A 4.5-MHz *subcarrier* is frequency modulated, then up-converted to either VHF or UHF for transmission. The FM subcarrier is transmitted 4.5 MHz above the picture carrier. A more detailed description of television broadcasting is given in Chapter 12. The maximum frequency deviation allowed by the FCC for television broadcasting is 25 kHz, and the maximum modulating signal frequency is 15 kHz. Therefore, the deviation ratio DR = 25 kΩ/15 kΩ or 1.67 and the maximum

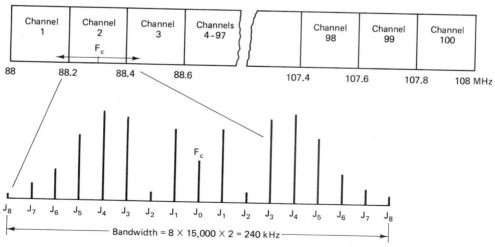

FIGURE 7-21 Commercial FM broadcasting frequency spectrum.

bandwidth using Carson's rule is $B = 2(25 \text{ kHz} + 15 \text{ kHz}) = 80 \text{ kHz}$. Figure 7-22 shows the output frequency spectrum for channel 10 of the commercial television broadcast band.

Microwave and satellite radio communications. FM microwave and satellite systems are gradually being replaced with higher-capacity digital systems. However, due to the rapid increase in the demand for long-distance telecommunications, the existing FM systems will be with us for some time. With both microwave and satellite FM radio, an intermediate frequency (typically between 60 and 80 MHz) is frequency modulated by a wideband signal spectrum, then up-converted in frequency to the gigahertz range for transmission. The modulating signal frequency spectrum can extend from a few kilohertz up to several megahertz. Consequently, low-index NBFM must be used to minimize the transmission bandwidth. Keep in mind that narrowband FM simply means that low modulation indices are used and generally only one set of significant sidebands is generated. However, the bandwidth can still be quite wide, depending on the bandwidth of the modulating signal spectrum.

Direct FM Transmitter

Crosby direct FM transmitter. Figure 7-23 shows the block diagram for a *Crosby direct FM transmitter* with an *automatic frequency control* (AFC) loop. The frequency modulator can be either a reactance modulator or a VCO. The carrier frequency is the unmodulated output frequency from the master oscillator (F_c). It is important to note that the carrier source is not a crystal, and if it were not for the AFC loop, this type of transmitter could not meet the frequency-stability requirements set by the FCC for commercial FM broadcasting. For the transmitter shown in Figure 7-23, the center frequency of the master oscillator is 5.1 MHz, which is multiplied by 18 in three steps ($3 \times 2 \times 3$) to produce a transmit carrier frequency $F_t = 91.8$ MHz. At this time there are three aspects of frequency conversion that should be noted. First, when the frequency of a frequency-modulated carrier is multiplied, its frequency and phase deviation are multiplied as well. Second, the rate at which the carrier is deviated (i.e., the

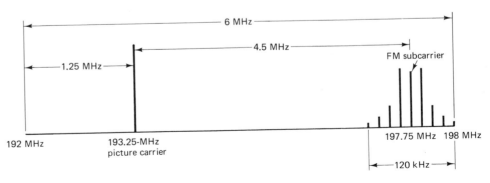

FIGURE 7-22   RF spectrum channel 10, commercial TV broadcasting.

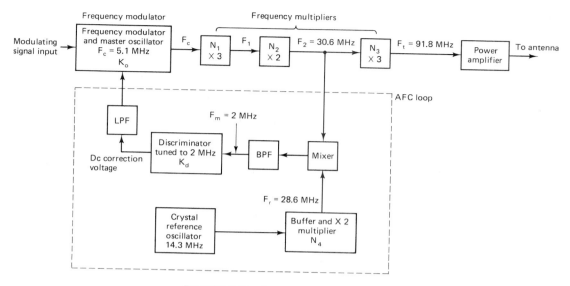

FIGURE 7-23 **Crosby direct FM transmitter.**

modulating signal frequency) is unaffected by the multiplication process. Therefore, the modulation index is also multiplied, Third, when a frequency-modulated carrier is heterodyned with another frequency in a mixer, the carrier can be either up- or down-converted, depending on the passband of the output filter. However, the frequency deviation, phase deviation, and rate of change are unaffected by the heterodyning process. Therefore, for the transmitter shown in Figure 7-23, the frequency and phase deviation at the output of the modulator are also multiplied by 18. To achieve 75 kHz of frequency deviation at the antenna, the deviation at the output of the modulator is

$$\Delta F = \frac{75 \text{ kHz}}{18} = 4166.7 \text{ Hz}$$

and the modulation index is

$$m = \frac{4166.7 \text{ Hz}}{F_a}$$

For a maximum modulating signal frequency $F_a = 15$ kHz,

$$m = \frac{4166.7 \text{ Hz}}{15 \text{ kHz}} = 0.2778$$

Thus the modulation index at the antenna is

$$m = 0.2778(18) = 5$$

which is the deviation ratio for commercial FM broadcasting.

AFC loop. The purpose of the *AFC loop* is to achieve near-crystal stability of the transmit carrier frequency without using a crystal as the master oscillator. With AFC, the carrier is beat against a crystal source in a mixer, down-converted in frequency, then fed back to the input of a *frequency discriminator*. A discriminator is a frequency-selective device whose output is a dc voltage that is proportional to the difference between the input frequency and its resonant frequency (discriminator operation is explained in Chapter 8). For the transmitter shown in Figure 7-23, the output from the doubler $F_2 = 30.6$ MHz, which is mixed with a crystal-controlled reference frequency $F_r = 28.6$ MHz to produce a difference frequency $F_m = 2$ MHz. The discriminator is a relatively high-Q (narrow band) tuned circuit that reacts only to frequencies near 2 MHz. Therefore, the discriminator responds to long-term, low-frequency changes in the carrier center frequency due to master oscillator drift and does not respond to the frequency deviation caused by the modulating signal. If the discriminator responded to the frequency deviation produced by the modulating signal, the feedback loop would cancel the deviation and thus remove the modulation from the FM wave (this is called ''wipe off''). The dc correction voltage is added to the modulating signal to automatically adjust the master oscillator's center frequency to compensate for the low-frequency drift.

EXAMPLE 7-8

Use the transmitter model shown in Figure 7-23 to answer the following questions. For a total frequency multiplication of ×20 and a transmit carrier frequency $F_t = 88.8$ MHz, determine:

    (a) The master oscillator center frequency.
    (b) The maximum frequency deviation at the output of the modulator.
    (c) The deviation ratio at the output of the modulator for a maximum modulating frequency $F_a = 15$ kHz.
    (d) The deviation ratio at the antenna.

Solution   (a) $F_c = \dfrac{F_t}{N_1 N_2 N_3} = \dfrac{88.8 \text{ MHz}}{20} = 4.43$ MHz

   (b) $\Delta F = \dfrac{\Delta F_t}{N_1 N_2 N_3} = \dfrac{75 \text{ kHz}}{20} = 3750$ Hz

   (c) $\text{DR} = \dfrac{\Delta F(\text{maximum})}{F_a(\text{maximum})} = \dfrac{3750 \text{ Hz}}{15 \text{ kHz}} = 0.25$

   (d) $\text{DR} = 0.25 \times 20 = 5$

Automatic frequency control. Because the Crosby transmitter uses either a VCO or a reactance oscillator to generate the carrier frequency, it is more susceptible to *frequency drift* due to temperature change, power supply fluctuations, and so on, than if it were a crystal oscillator. As stated in Chapter 2, the stability of an oscillator is often given in *parts per million* (ppm) per degree celsius. For example, for the transmitter shown in Figure 7-23, an oscillator stability of ±40 ppm could produce ±204 Hz (5.1

MHz $\times$ 40/Hz/million) of frequency drift per degree celsius at the output of the master oscillator. This corresponds to ± 3672 Hz drift at the antenna (18×204), which far exceeds the ± 2 kHz maximum set by the FCC for commercial FM broadcasting. Although an AFC circuit does not totally eliminate frequency drift, it can substantially reduce it. Assuming a rock-stable crystal reference oscillator and a perfectly tuned discriminator, the frequency drift at the output of the second multiplier without feedback is

$$\text{open-loop drift} = dF_{ol} = N_1 N_2 dF_c \qquad (7\text{-}49)$$

and the closed-loop drift is

$$\text{closed-loop drift} = dF_{cl} = dF_{ol} - N_1 N_2 k_d k_o dF_{cl} \qquad (7\text{-}50)$$

Therefore,

$$dF_{cl} + N_1 N_2 k_d k_o dF_{cl} = dF_{ol}$$

and

$$dF_{cl}(1 + N_1 N_2 k_d k_o) = dF_{ol}$$

Thus

$$dF_{cl} = \frac{dF_{ol}}{1 + N_1 N_2 k_d k_o} \qquad (7\text{-}51)$$

where

$\quad K_d = $ discriminator transfer function (V/Hz)

$\quad K_o = $ master oscillator transfer function (Hz/V)

From Equation 7-51 it can be seen that the frequency drift at the output of the second multiplier and consequently at the input to the discriminator is reduced by a factor of $1 + N_1 N_2 k_d k_o$ when the AFC loop is closed. The carrier frequency drift is multiplied by the AFC loop gain and fed back to the master oscillator as a dc correction voltage. Of course, the total frequency error cannot be canceled because then there would not be any error voltage to feed back.

EXAMPLE 7-9

Use the transmitter block diagram and values given in Figure 7-23 to answer the following questions. Determine the reduction in frequency drift at the antenna for a transmitter without AFC compared to a transmitter with AFC. Use a VCO stability = 200 ppm, K_o = 10 kHz/V, and K_d = 2 V/kHz.

Solution   With the feedback loop open, the master oscillator output frequency is

$$F_c = 5.1 \text{ MHz} + (200 \text{ ppm})(5.1 \text{ MHz}) = 5{,}101{,}020 \text{ Hz}$$

and the frequency at the output of the second multiplier is

$$F_2 = N_1 N_2 F_c = (5{,}101{,}020)(6) = 30{,}606{,}120 \text{ Hz}$$

Thus the frequency drift is

$$dF_2 = 30,606,120 - 30,600,000 = 6120 \text{ Hz}$$

Therefore, the antenna transmit frequency is

$$F_t = (30,606,120)(3) = 91.81836 \text{ MHz}$$

which is 18.36 kHz above the assigned frequency and well out of limits.

With the feedback loop closed, the frequency drift at the output of the second multiplier is reduced by a factor of $1 + N_1 N_2 K_o K_d$ or

$$1 + (2)(3)\left(\frac{10 \text{ kHz}}{\text{V}}\right)\left(\frac{2 \text{ V}}{\text{kHz}}\right) = 121$$

Therefore,

$$dF_2 = \frac{6120}{121} = 51 \text{ Hz}$$

Thus

$$F_2 = 30,600,051$$

and the antenna transmit frequency is

$$F_t = 30,600,051 \times 3 = 91,800,153 \text{ Hz}$$

The frequency drift at the antenna has been reduced from 18,360 Hz to 153 Hz, which is now well within the ± 2 kHz FCC requirements.

The preceding discussion and Example 7-9 assumed a perfectly stable crystal reference oscillator and a perfectly tuned discriminator. In actuality, both the discriminator and the reference crystal are subject to drift, and the worst-case situation is when they both drift in the same direction as the VCO. The drift characteristics for a typical discriminator are on the order of 100 ppm. Perhaps now you can see why the output frequency from the second multiplier was mixed down to a relatively low frequency (2 MHz) prior to being fed to the discriminator. For a discriminator tuned to 2 MHz with a stability of 100 ppm, the maximum discriminator drift is

$$dF_d = 100 \text{ ppm} \times 2 \text{ MHz} = 200 \text{ Hz}$$

If the 30.6-MHz signal were fed directly into the discriminator, the maximum drift would be

$$dF_d = 100 \text{ ppm} \times 30.6 \text{ MHz} = 3060 \text{ Hz}$$

Frequency drift due to discriminator instability is multiplied by the AFC open-loop gain. Therefore, the change in the second multiplier output frequency due to discriminator drift is

$$dF_2 = dF_d N_1 N_2 K_o K_d \qquad (7\text{-}52)$$

Similarly, the crystal reference oscillator can drift and also contribute to the total frequency change at the output of the second multiplier. The drift due to crystal instability is multiplied by 2 before entering the mixer; therefore,

$$dF_2 = N_4 dF_o N_1 N_2 K_o K_d \qquad (7\text{-}53)$$

and the maximum open-loop frequency drift at the output of the second multiplier is

$$dF_2(\text{total}) = N_1 N_2 K_o K_d(dF_c + dF_d + N_4 dF_o) \qquad (7\text{-}54)$$

Phase-locked-loop FM transmitter. Figure 7-24 shows a *wideband* FM transmitter that uses a phase-locked loop to achieve crystal stability from a VCO master oscillator and, at the same time, generate a high-index, wideband FM signal at its output. The output frequency from the VCO is divided by N and fed back to the PLL phase comparator, where it is compared to a stable crystal reference frequency. The phase comparator generates a dc voltage that is proportional to the difference in the two frequencies. The dc correction voltage is added to the modulating signal and applied to the VCO input. The correction voltage adjusts the VCO center frequency to its proper value. Again, the LPF prevents changes in the VCO output frequency due to the modulating signal from being converted to dc, fed back to the VCO, and wiping out the modulation. The LPF also prevents the loop from locking onto a sideband frequency.

PM from FM. As stated previously and shown in Figure 7-4, an FM modulator preceded by a differentiator generates a PM waveform. If the transmitters shown in Figures 7-23 and 7-24 are preceded by a preemphasis network, which is a differentiator (high-pass filter), an interesting situation occurs. For a 75-μs time constant, the amplitude of frequencies above 2.12 kHz are emphasized through differentiation. Therefore, for

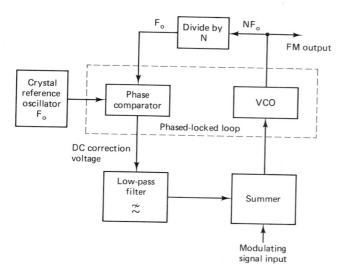

FIGURE 7-24 **Phase-locked-loop FM transmitter.**

modulating frequencies below 2.12 kHz, the output waveform is proportional to the modulating signal, and for frequencies above 2.12 kHz, the output waveform is proportional to the derivative of the input signal. In other words, frequency modulation occurs for frequencies below 2.12 kHz and phase modulation occurs for frequencies above 2.12 kHz. Because the gain of a differentiator increases with frequency above the break frequency (2.12 kHz), and since the frequency deviation is proportional to the modulating signal amplitude, the frequency deviation also increases with frequency above 2.12 kHz. From Equation 7-15 it can be seen that if ΔF and F_a increase proportionately, the modulation index remains constant, which is, of course, a characteristic of PM.

Indirect FM Transmitter

Armstrong indirect FM transmitter.   With indirect FM, the phase of the carrier is deviated directly by the modulating signal, which indirectly deviates the carrier frequency. Figure 7-25 shows the block diagram for a *wideband Armstrong indirect FM transmitter*. The carrier source is a crystal oscillator. Therefore, the stability requirements set by the FCC can be achieved without the use of AFC.

With an Armstrong transmitter, a relatively low-frequency subcarrier is phase shifted 90° and fed to a balanced modulator, where it is mixed with the input modulating signal (F_a). The output from the balanced modulator is a double-sideband suppressed

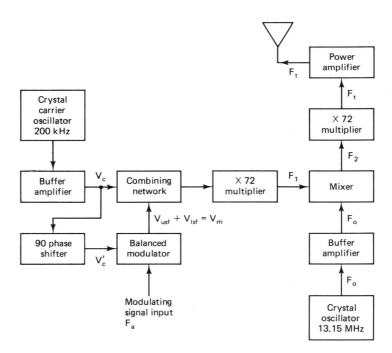

FIGURE 7-25   Armstrong indirect FM transmitter.

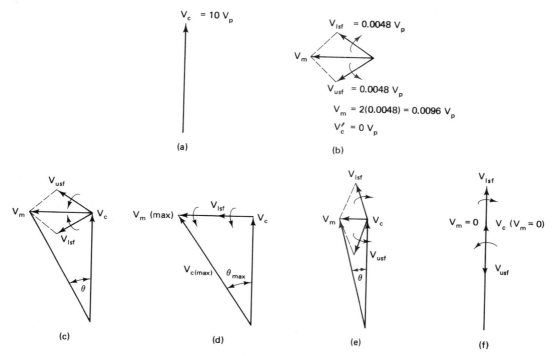

FIGURE 7-26   Phasor addition of V_c, V_{usf}, and V_{lsf}: (a) carrier phasor; (b) sideband phasors; (c) – (f) progressive phasor addition. Part (d) shows the peak phase shift.

carrier waveform which is combined with the original carrier to produce a low-index phase-modulated waveform. Figure 7-26a shows the phasor for the original carrier (V_c), and Figure 7-26b shows the phasors for the sideband components of the suppressed carrier waveform (V_{usf} and V_{lsf}). Because the suppressed carrier (V_c') is 90° out of phase with V_c, the upper and lower sidebands combine to produce a component (V_m) that is always in quadrature (at right angles) with V_c. Figure 7-26c through f show the progressive phasor addition of V_c, V_{usf}, and V_{lsf}. It can be seen that the output from the combining network is a phase-modulated wave where the deviation rate is equal to the modulating signal frequency, F_a, and the peak phase deviation (modulation index) is

$$\theta = m = \arctan \frac{V_m}{V_c} \qquad (7\text{-}55a)$$

For very small angles, the tangent is approximately equal to the angle; therefore,

$$\theta = m \approx \frac{V_m}{V_c} \qquad (7\text{-}55b)$$

EXAMPLE 7-10

For the phase-shifted carrier (V'_c), upper side frequency (V_{usf}), and lower side frequency (V_{lsb}) components shown in Figure 7-26; determine:

     (a) The peak carrier phase shift in both radians and degrees.

     (b) The frequency deviation for modulating signal frequency $F_a = 15$ kHz.

Solution   (a) The peak phase deviation is equal to the modulation index; therefore, from Equation 7-55a,

$$m = \theta = \arctan \frac{V_m}{V_c} \qquad (7\text{-}55a)$$

$$= \arctan \frac{0.0096}{10} = 0.00553°$$

$$= (0.00553°) \frac{\pi r}{180°} = 0.000965 \text{ rad}$$

     (b) Rearranging Equation 7-14 gives us

$$\Delta F = mF_a = (0.000965)(15,000) = 14.47 \text{ Hz}$$

From the phasor diagrams shown in Figure 7-26, it can be seen that the carrier amplitude is varied, which produces unwanted amplitude modulation in the output waveform, and $V_c(\max)$ occurs when V_{usf} and V_{lsf} are in phase with each other and with V_c. The maximum phase deviation that can be produced with this type of modulator is approximately 1.67 mrad. Therefore, from Equation 7-14 and for a maximum modulating signal frequency $F_a = 15$ kHz, the maximum frequency deviation is

$$\Delta F_{\max} = (0.00167)(15 \text{ kHz}) = 25 \text{ Hz}$$

From the preceding discussion it is evident that the modulation index at the output of the combining network is insufficient to produce a wideband FM spectrum and therefore must be multiplied considerably before being transmitted. For the transmitter shown in Figure 7-25, a 200-kHz phase-modulated carrier with a peak phase deviation $\theta = 0.000965$ rad and a frequency deviation $\Delta F = 14.17$ Hz is produced at the output of the combining network. To achieve 75-kHz frequency deviation, the waveform must be multiplied by 5184 (75,000/14.47). However, this would produce a transmit carrier frequency $F_t = 5184 \times 200$ kHz $= 1036.8$ MHz, which is well beyond the commercial FM broadcast band limits. It is apparent that multiplication by itself is inadequate. Therefore, a combination of multiplying and mixing is used to develop the desired transmit carrier frequency with 75-kHz frequency deviation. The waveform at the output of the combining network is multiplied by 72, producing the following output signal:

$$F_1 = 72 \times 200 \text{ kHz} = 14.4 \text{ MHz}$$

$$m = 72 \times 0.000965 = 0.06948 \text{ rad}$$

$$\Delta F = 72 \times 14.47 = 1042 \text{ Hz}$$

The output from the first multiplier is mixed with a 13.15-MHz crystal controlled frequency (F_o) to produce a difference signal (F_2) with the following characteristics:

$$F_2 = 14.4 \text{ MHz} - 13.15 \text{ MHz} = 1.25 \text{ MHz}$$

$$m = 0.06948 \text{ rad}$$

$$\Delta F = 1042 \text{ Hz}$$

Note that only the carrier frequency is affected by the heterodyning process. The output from the mixer is once again multiplied by 72, to produce a transmit signal with the following characteristics:

$$F_t = 1.25 \text{ MHz} \times 72 = 90 \text{ MHz}$$

$$m = 0.06948 \times 72 = 5.0 \text{ rad}$$

$$\Delta F = 1042 \text{ Hz} \times 72 = 75 \text{ kHz}$$

In the preceding example, the carrier was multiplied by a factor of 450, while the frequency deviation and modulation index were multiplied by a factor of 5184.

With the Armstrong transmitter, the phase of the carrier is directly modulated in the combining network, producing indirect frequency modulation. The magnitude of the phase deviation is directly proportional to the amplitude of the modulating signal but independent of its frequency. Therefore, for changes in the modulating signal frequency, the modulation index remains constant and the frequency deviation changes proportionately. For example, for the transmitter show in Figure 7-25, if the modulating signal amplitude is held constant while its frequency is decreased to 5 kHz, the modulation index remains at 5 while the frequency deviation is reduced to $\Delta F = 5 \times 5 \text{ kHz} = 25$ kHz.

FM from PM.   As stated previously and shown in Figure 7-4, a PM modulator preceded by an integrator generates an FM waveform. If the PM transmitter shown in Figure 7-25 is preceded by a low-pass filter (which is an integrator) FM results. The low-pass filter is simply a $1/F$ filter, which is commonly called a *predistorter* or a *frequency correction* network.

FM Versus PM

From a purely theoretical viewpoint, the difference between FM and PM is quite simple: the modulation index for FM is defined differently than for PM. With PM, the phase deviation is directly proportional to the amplitude of the modulating signal and independent of its frequency. With FM, the frequency deviation is directly proportional to the amplitude of the modulating signal and inversely proportional to its frequency.

Considering FM as a form of phase modulation, the larger the frequency deviation, the larger the phase deviation. Therefore, the latter depends, at least to a certain extent, on the amplitude of the modulating signal, just as with PM. With PM, the modulation index is proportional to the amplitude of the modulating signal voltage only, whereas with FM the modulation index is also inversely proportional to the modulating signal frequency. If FM transmissions are received on a PM receiver, the bass frequencies have considerably more phase deviation than a PM modulator would have given them. Since the output from a PM demodulator is proportional to the phase deviation (modulation index), the signal appears excessively bass-boosted. Alternatively (and this is the more practical situation), PM demodulated by an FM receiver appears lacking in bass frequencies.

QUESTIONS

7-1. Define *angle modulation*.

7-2. Define *direct FM*; *indirect FM*.

7-3. Define *direct PM*; *indirect PM*.

7-4. Define *frequency deviation*; *phase deviation*.

7-5. Define *instantaneous phase*; *instantaneous phase deviation*; *instantaneous frequency*; *instantaneous frequency deviation*.

7-6. Define *deviation sensitivity* for a frequency modulator and for a phase modulator.

7-7. Describe the relationship between the instantaneous carrier frequency and the modulating signal for FM.

7-8. Describe the relationship between the instantaneous carrier phase and the modulating signal for PM.

7-9. Describe the relationship between frequency deviation and the modulating signal.

7-10. Define *carrier swing*.

7-11. Define *modulation index* for FM and for PM.

7-12. Describe the relationship between modulation index and the modulating signal for FM; for PM.

7-13. Define *percent modulation*.

7-14. Describe the difference between a frequency and a phase modulator.

7-15. How can a frequency modulator be converted to a phase modulator; a phase modulator to a frequency modulator?

7-16. How many sets of side frequencies are produced when a carrier is frequency modulated by a single input frequency?

7-17. What are the requirements for a side frequency to be significant?

7-18. Define a *low*, a *medium*, and a *high modulation index*.

7-19. Describe the significance of the Bessel table.

7-20. State Carson's general rule for determining the bandwidth for an angle-modulated wave.

7-21. Define *deviation ratio*.

7-22. Describe the relationship between the power in the unmodulated carrier and the power in the modulated wave for FM.

7-23. Describe the FM noise triangle.

7-24. What effect does limiting have on the composite FM waveform?

7-25. Define *preemphasis*; *deemphasis*.

7-26. Describe a preemphasis network; a deemphasis network.

7-27. Describe the operation of a varactor diode FM generator.

7-28. Describe the operation of a reactance FM modulator.

7-29. Describe the operation of a linear integrated-circuit FM modulator.

7-30. List four communications services that use frequency modulation.

7-31. Draw the block diagram for a Crosby direct FM transmitter and describe its operation.

7-32. What is the purpose of an AFC loop? Why is one required for the Crosby system?

7-33. Draw the block diagram for a phase-locked-loop FM transmitter and describe its operation.

7-34. Draw the block diagram for an Armstrong indirect FM transmitter and describe its operation.

7-35. Contrast FM and PM.

PROBLEMS

7-1. If a frequency modulator produces 5 kHz of frequency deviation for a 10-Vp modulating signal, determine the deviation sensitivity.

7-2. If a phase modulator produces 2 rad of frequency deviation for a 5-Vp modulating signal, determine the deviation sensitivity.

7-3. Determine (a) the peak frequency deviation, (b) the carrier swing, and (c) the modulation index for an FM modulator with deviation sensitivity $K_1 = 4$ kHz/V and modulating signal $v_a = 10 \sin (2\pi 2000t)$.

7-4. Determine the peak phase deviation for a PM modulator with deviation sensitivity $K = 1.5$ rad/V and modulating signal $v_a = 2 \sin (2\pi 1000t)$.

7-5. Determine the percent modulation for a television broadcast station with a maximum frequency deviation $\Delta F = 50$ kHz when the modulating signal produces 40-kHz frequency deviation at the antenna.

7-6. From the Bessel table determine the number of sets of side frequencies produced for the following modulation indices: 0.5, 1.0, 2.0, 5.0, and 10.0.

7-7. For an FM modulator with modulation index $m = 2$, modulating signal $V(t) = v_a \sin (2\pi 2000t)$, and unmodulated carrier $v_c = 8 \sin (2\pi 800kt)$:
(a) Determine the number of sets of significant sidebands.
(b) Determine their amplitudes.
(c) Draw the frequency spectrum.

7-8. For an FM transmitter with 60-kHz carrier swing, determine the frequency deviation. If the amplitude of the modulating signal decreases by a factor of 2, determine the new frequency deviation.

7-9. For a given input signal, an FM broadcast transmitter has a frequency deviation $\Delta F = 20$ kHz. Determine the frequency deviation if the amplitude of the modulating signal increases by a factor of 2.5.

7-10. An FM transmitter has a rest frequency $F_c = 96$ MHz and a deviation sensitivity $K_1 = 4$ kHz/V. Determine the frequency deviation for a modulating signal $V(t) = 8 \sin (2\pi 2000t)$. Determine the modulation index.

7-11. Determine the deviation ratio and worst-case bandwidth for an FM system with a maximum frequency deviation $\Delta F = 25$ kHz and maximum modulating signal frequency $F_a = 12.5$ kHz.

7-12. For an FM modulator with 40-kHz frequency deviation and a modulating signal frequency $F_a = 10$ kHz, determine the bandwidth using both the Bessel table and Carson's rule.

7-13. For an FM modulator with an unmodulated carrier voltage $V_c = 20$ V_p, a modulation index $m = 1$, and a load resistance $RL = 10$ Ω; determine the power in the modulated carrier and each significant side frequency, and sketch the power spectrum for the modulated wave.

7-14. For an angle-modulated carrier $v_c = 2 \cos (2\pi 200\text{MHz } t)$ with 50 kHz of frequency deviation due to the modulating signal and a single-frequency interfering signal $v_n = 0.5 \cos (2\pi 200.01$ MHz), determine:
  (a) The frequency of the demodulated interference signal.
  (b) The peak and rms phase and frequency deviation due to the interfering signal.
  (c) The S/N ratio at the output of the demodulator.

7-15. Determine the total rms phase deviation produced by a 5-kHz band of random noise with a peak voltage $V_n = 0.08$ V and a carrier $v_c = 1.5 \sin (2\pi 40\text{MHz } t)$.

7-16. For a Crosby direct FM transmitter similar to the one shown in Figure 7-23 with the following parameters, determine:
  (a) The frequency deviation at the output of the VCO and the power amplifier.
  (b) The modulation index at the same two points.
  (c) The bandwidth at the output of the power amplifier.

$$N_1 = \times 3$$

$$N_2 = \times 3$$

$$N_3 = \times 2$$

Crystal reference oscillator frequency $= 13$ MHz

Reference multiplier $= \times 3$

VCO deviation sensitivity $K_1 = 450$ Hz/V

Modulating signal $v_a = 3 \sin (2\pi 5kt)$

VCO rest frequency $F_c = 4.5$ MHz

Discriminator resonant frequency $F_d = 1.5$ MHz

7-17. For an Armstrong indirect FM transmitter similar to the one shown in Figure 7-25 with the following parameters, determine:
  (a) The modulation index at the output of the combining network and the power amplifier.
  (b) The frequency deviation at the same two points.
  (c) The transmit carrier frequency.

Crystal carrier oscillator $= 210$ kHz

Crystal reference oscillator $= 10.2$ MHz

Sideband voltage $V_m = 0.018$ V_p

Carrier input to combiner $V_c = 5$ V_p

First multiplier $= \times 40$

Second multiplier $= \times 50$

Modulating signal frequency $F_a = 2$ kHz

Chapter 8

ANGLE MODULATION RECEIVERS AND SYSTEMS

INTRODUCTION

The receivers used for receiving angle-modulated signals are essentially identical to those that are used for conventional AM or SSB reception, except for the method used to extract the audio information from the composite IF waveform. In FM receivers, the output of the audio detector is proportional to the frequency deviation at its input. With PM receivers, the output of the audio detector is proportional to the phase deviation at its input. Because frequency and phase modulation both occur with either angle modulation system, FM signals can be demodulated by a PM receiver, and vice versa. Therefore, the circuits used to demodulate FM and PM signals are both described under the heading "FM receivers."

In conventional AM, the modulating signal is impressed onto the carrier in the form of amplitude variations. However, noise introduced in the system also produces changes in the amplitude of the envelope. Therefore, the noise cannot be removed from the composite waveform without also removing a portion of the information. With angle modulation, the information is impressed onto the carrier in the form of frequency or phase variations. Therefore, with angle modulation receivers, amplitude variations caused by noise can be removed from the composite waveform simply by *limiting* (*clipping*) the peaks of the envelope prior to detection. With angle modulation, an improvement in the signal-to-noise ratio is achieved during the demodulation process; thus system performance in the presence of noise can be improved by limiting. In essence, this is the major advantage of angle modulation over conventional AM.

FM RECEIVERS

Figure 8-1 shows a simplified block diagram for a double-conversion superheterodyne FM receiver. As you can see, it is very similar to a conventional AM receiver. The RF, mixer, and IF stages are identical to those used in AM receivers, although FM receivers generally have more IF amplification. Also, due to the inherent noise suppression characteristics of FM receivers, RF amplifiers are often not required. However the audio detector stage in an FM receiver is quite different from those used in AM receivers. The envelope (peak) detector used in conventional AM receivers is replaced by a *limiter*, *frequency discriminator*, and *deemphasis network*. The limiter and deemphasis network contribute to the improvement in the *S/N* ratio that is achieved in the audio demodulator stage. For FM broadcast band receivers, the first IF is a relatively high frequency (generally, 10.7 MHz) for good image frequency rejection, and the second IF is a relatively low frequency (generally, 455 kHz) to reduce the possibility that the amplifiers will oscillate.

FM Demodulators

FM demodulators are frequency-dependent circuits that produce an output voltage that is directly proportional to the instantaneous frequency at its input ($V_o = \Delta FK$, where $K = $ V/Hz and is the transfer function for the demodulator, and ΔF is the difference between the input frequency and the center frequency of the demodulator). There are several circuits that are used for demodulating FM signals. The most common are the *slope detector*, *Foster–Seeley discriminator*, *ratio detector*, *PLL demodulator*, and *quadrature detector*. The slope detector, Foster–Seeley discriminator, and ratio detector are all forms of *tuned circuit frequency discriminators*. Tuned circuit frequency discriminators convert FM to AM, then demodulate the AM waveform with a conventional peak detector. Also, most frequency discriminators require a 180° phase inverter, an adder circuit, and one or more frequency-dependent circuits.

Slope detector.   Figure 8-2a shows the schematic diagram for a *single-ended slope detector*, which is the simplest form of tuned circuit frequency discriminator. The single-ended slope detector has the most nonlinear voltage-versus-frequency characteristics and is, therefore, seldom used. However, its circuit operation is basic to all tuned circuit frequency discriminators.

In Figure 8-2a, tuned circuit (L_a and C_a) produces an output voltage that is proportional to the input frequency. The maximum output voltage occurs at its resonant frequency (F_o), and its output rolls off proportionately as the input frequency is deviated above and below F_o. The circuit is designed so that the IF center frequency (F_c) falls in the center of the most linear portion of the voltage-versus-frequency curve (Figure 8-2b). As the IF deviates above F_c, the output voltage increases and, as the IF deviates below F_c, the output voltage decreases. Therefore, the tuned circuit converts frequency variations to amplitude variations (i.e., FM-to-AM conversion). D_i, C_i, and R_i make up a simple peak detector which converts the amplitude variations to an output voltage

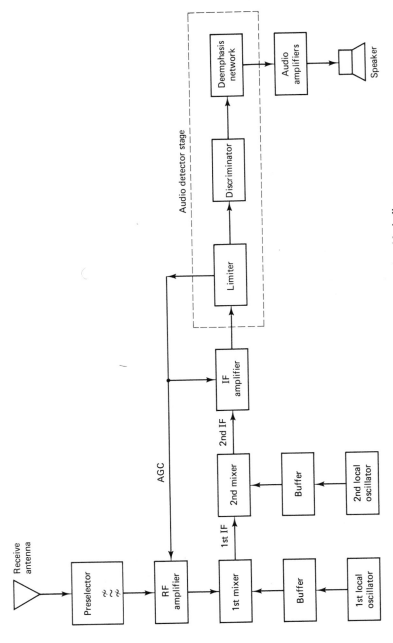

FIGURE 8-1 Double-conversion FM receiver block diagram.

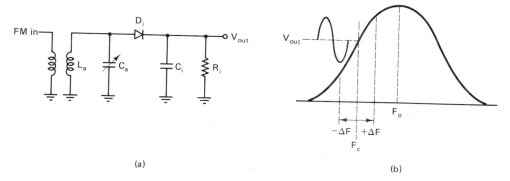

FIGURE 8-2 Slope detector: (a) schematic diagram; (b) voltage-versus-frequency curve.

that varies at a rate equal to that of the input frequency variations and whose amplitude is proportional to the magnitude of the frequency changes.

Balanced slope detector. Figure 8-3a shows the schematic diagram for a *balanced slope detector*. A single-ended slope detector is a tuned circuit frequency discriminator, and a balanced slope detector is simply two slope detectors connected in parallel and fed 180° out of phase. The phase inversion is accomplished by center tapping the tuned secondary windings of transformer T_1. In Figure 8-3a, tuned circuits (L_a, C_a, and L_b, C_b) perform the FM-to-AM conversion, and balanced peak detectors (D_1, C_1, R_1 and D_2, C_2, R_2) remove the information from the AM waveform. The top tuned circuit (L_a and C_a) is tuned to a frequency (F_a) which is above the IF center frequency (F_o) by approximately 1.33 × ΔF (for the FM broadcast band this is approximately 1.33 × 75 kHz = 100 kHz). The lower tuned circuit (L_b and C_b) is tuned to a frequency (F_b) that is below the IF center frequency by an equal amount. Circuit operation is quite simple. The output voltage from each tuned circuit is proportional to the input frequency and each output is rectified by its respective peak detector. Therefore, the closer the input frequency is to the tank circuit resonant frequency, the greater the tank circuit output voltage. The IF center frequency falls exactly halfway between the resonant frequencies of the two tuned circuits. Therefore, at the IF center frequency, the output voltages from the two tank circuits are equal in amplitude but opposite in polarity. Consequently, the rectified output voltages across R_1 and R_2, when added, produce a differential output voltage $V_{out} = 0$ V. When the IF is deviated above resonance, the top tuned circuit produces a higher output voltage than the lower tank circuit and V_{out} goes positive. When the IF is deviated below resonance, the output voltage from the lower tank circuit is larger and V_{out} goes negative. The output voltage-versus-frequency response curve is shown in Figure 8-3b. Although the slope detector is probably the simplest FM detector, it has several disadvantages, which include poor linearity, difficulty in tuning, and lack of provision for limiting. Because limiting is not provided, a slope detector will demodulate amplitude as well as frequency variations and consequently, must be preceded by a limiter stage. A balanced slope detector is aligned by injecting

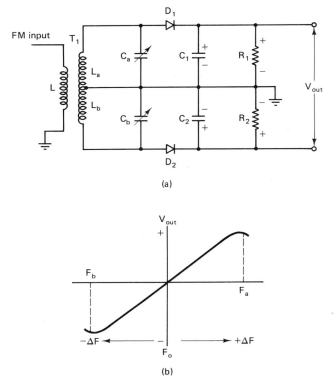

(a)

(b)

FIGURE 8-3 Balanced slope detector:
(a) schematic diagram; (b) voltage-versus-
frequency response curve.

an unmodulated IF center frequency and tuning C_a and C_b for zero volts out. Then F_a and F_b are alternately injected while C_a and C_b are tuned for maximum and equal output voltages with opposite polarities.

Foster–Seeley discriminator.   A *Foster–Seeley discriminator* (sometimes called a *phase-shift discriminator*) is a tuned circuit frequency discriminator whose operation is very similar to that of the balanced slope detector. The schematic diagram for a Foster–Seeley discriminator is shown in Figure 8-4a. The capacitance values for C_c, C_1, and C_2 are chosen such that they are short circuits for the IF center frequency. Therefore, the right side of L_3 is ac ground and the IF signal (V_{in}) is fed directly (in phase) across L_3 (VL_3). The incoming IF is inverted 180° by transformer T_1 and divided equally between L_a and L_b. At the resonant frequency of the secondary tank circuit (the IF center frequency), the secondary current (I_s) is in phase with the total secondary voltage (V_s) and 180° out of phase with VL_3. Also, due to loose coupling, the primary of T_1 is essentially an inductor, so that the primary current I_p is 90° out of phase with V_{in}, and since magnetic induction depends on primary current, the voltage induced in the secondary is 90° out of phase with V_{in} (VL_3). Therefore, VL_a and VL_b are 180° out of phase with each other and in quadrature or 90° out of phase with VL_3. The voltage

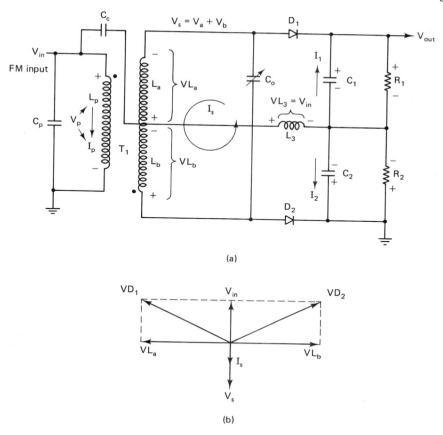

(a)

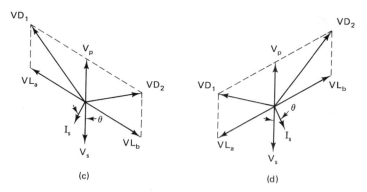

(b)

(c)   (d)

FIGURE 8-4 Foster–Seeley discriminator: (a) schematic diagram; (b) vector diagram, $F_{in} = F_o$; (c) vector diagram, $F_{in} > F_o$; (d) vector diagram, $F_{in} < F_o$.

across the top diode VD_1 is the vector sum of VL_3 and VL_a, and the voltage across the bottom diode VD_2 is the vector sum of VL_3 and VL_b. The corresponding vector diagrams are shown in Figure 8-4b. It can be seen that the voltages across D_1 and D_2 are equal. Therefore, at resonance, I_1 and I_2 are equal and C_1 and C_2 charge to equal voltages with opposite polarities. Consequently, $V_{out} = VC_1 - VC_2 = 0$ V. When the IF is deviated above resonance ($X_L > X_C$) and the secondary tank circuit impedance becomes inductive and the secondary current lags the secondary voltage by some angle θ, which is proportional to the frequency deviation. The corresponding phasor diagram is shown in Figure 8-4c. The figure shows that the vector sum of the voltage across D_1 is greater than the voltage across D_2. Consequently, C_1 charges while C_2 discharges and V_{out} goes positive. When the IF is deviated below resonance ($X_L < X_C$), the secondary current leads the secondary voltage by some angle θ, which is again proportional to the change in frequency. The corresponding phasors are shown in Figure 8-4d. It can be seen that the vector sum of the voltage across D_1 is now less than the voltage across D_2. Consequently, C_1 discharges while C_2 charges and V_{out} goes negative. A Foster–Seeley discriminator is tuned by injecting an unmodulated IF center frequency and tuning C_o for zero volts out.

The preceding discussion and Figure 8-4 show that the output voltage from a Foster–Seeley discriminator is directly proportional to the magnitude and the direction of the frequency deviation. Figure 8-5 shows a typical voltage-versus-frequency response curve for a Foster–Seeley discriminator. For obvious reasons, it is often called an *S-curve*. It can be seen that the output voltage-versus-frequency deviation curve is more linear than that of a slope detector, and because there is only one tank circuit, it is easier to tune. For distortionless demodulation, the frequency deviation should be restricted to the linear portion of the secondary tuned circuit frequency response curve. Like the slope detector, a Foster–Seeley discriminator responds to amplitude variations and therefore must be preceded by a limiter.

Ratio detector. The *ratio detector* has one major advantage over the slope detector and the Foster–Seeley discriminator; a ratio detector is relatively immune to amplitude variations in the input signal. Figure 8-6a shows the schematic diagram for a ratio detector. Like the Foster–Seeley discriminator, a ratio detector has a single tuned circuit

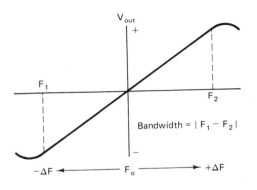

Bandwidth = $|F_1 - F_2|$

FIGURE 8-5 Discriminator voltage-versus-frequency response curve.

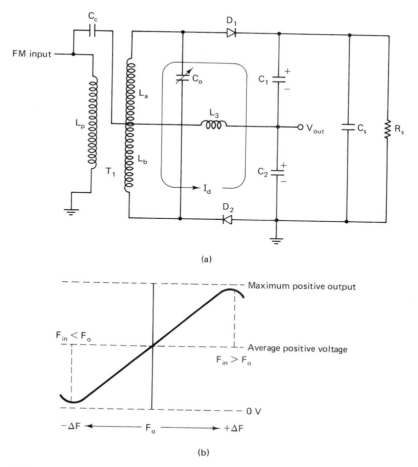

(a)

(b)

FIGURE 8-6 Ratio detector: (a) schematic diagram; (b) voltage-versus-frequency response curve.

in the transformer secondary. Therefore, the operation of a ratio detector is very similar to that of the Foster–Seeley discriminator. In fact, the voltage vectors for D_1 and D_2 are identical to those for the Foster–Seeley discriminator shown in Figure 8-4. However, with the ratio detector, one diode is reversed (D_2) and current (I_d) can flow around the outermost loop of the circuit. Therefore, after several cycles of the input signal, shunt capacitor C_s charges to approximately the peak voltage across the secondary windings of T_1. The reactance of C_s is low and R_s simply provides a dc path for diode current. Therefore, the time constant for R_s and C_s is sufficiently long so that rapid changes in the amplitude of the input signal due to thermal noise or other interfering signals are shorted to ground and therefore have no effect on the average voltage across C_s. Consequently, C_1 and C_2 charge and discharge proportional to frequency changes in the input signal and are relatively immune to amplitude variations. Also, the output from a ratio detector is taken with respect to ground, and for the diode polarities shown in Figure 8-6a, the average output voltage is positive. At resonance, the output voltage is divided equally between C_1 and C_2 and redistributed as the input frequency is deviated above and below resonance. Therefore, changes in V_{out} are due to the changing ratio of the voltages across C_1 and C_2, while the total voltage is clamped by C_s.

Figure 8-6b shows the output response curve for the ratio detector shown in Figure 8-6a. It can be seen that at resonance V_{out} is not equal to 0 V, but rather, to one-half of the voltage across the secondary windings of T_1. Because a ratio detector is relatively immune to amplitude variations, it is often selected over a discriminator. However, a discriminator produces a more linear output voltage-versus-frequency response curve.

PLL FM demodulator.   Since the development of the LSI linear integrated circuit, FM demodulation can be accomplished quite simply with a phase-locked loop. Although the operation of a PLL is quite involved, the operation of a *PLL FM demodulator* is probably the simplest and easiest to understand. A PLL frequency demodulator requires no tuned circuits and automatically compensates for carrier shift due to instability of the transmit oscillator.

In Chapter 5 a detailed description of PLL operation was given. It was shown that after frequency lock had occurred, the VCO tracked frequency changes in the input signal by maintaining a phase error at the input to the phase comparator. Therefore, if the PLL input is a deviated FM signal and the VCO natural frequency is equal to the IF center frequency, the dc correction voltage fed back to the input of the VCO is proportional to the frequency deviation and is thus the demodulated information signal. If the IF is sufficiently limited prior to the PLL and the loop is properly compensated, the PLL loop gain is constant and equal to K_v. Therefore, the demodulated signal can be taken directly from the output of the internal buffer and is mathematically given as

$$V_{out} = \Delta F k_d K_a \qquad (8\text{-}1)$$

Figure 8-7 shows a schematic diagram for an FM demodulator using the XR-2212 PLL, and Figure 8-8 shows the manufacturer's specification sheets for the XR-2212. R_o and C_o are coarse adjustments for setting the VCO's free-running frequency. R_x is for fine tuning, and R_F and R_C set the internal op-amp voltage gain (K_a). The PLL closed-loop frequency response should be compensated to allow unattentuated demodulation of the entire information signal bandwidth. The PLL op-amp buffer provides voltage gain and current drive stability.

Quadrature FM demodulator.   A *quadrature FM demodulator* (sometimes called a *coincidence detector*) extracts the original information signal from the composite IF waveform by multiplying the two quadrature (90° out of phase) signals. A quadrature detector uses a 90° phase shifter, a single tuned circuit, and a product detector to demodulate FM signals. The 90° phase shifter produces a signal that is in quadrature with the received IF. The tuned circuit converts frequency variations to phase variations, and the product detector multiplies the received IF by the phase-shifted signal.

Figure 8-9 shows a simplified schematic diagram for an FM quadrature detector.

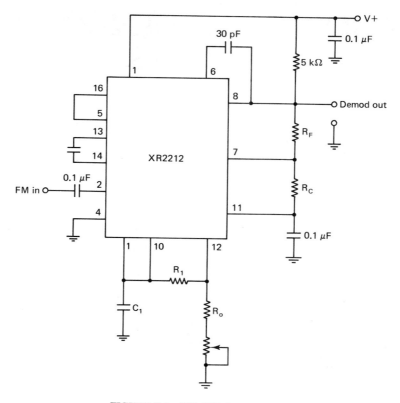

FIGURE 8-7   PLL FM demodulator.

C_i is a high reactance capacitor that, when placed in series with tank circuit (R_o, L_o, and C_o), produces a 90° phase shift at the IF center frequency. The tank circuit is tuned to the IF center frequency and produces an additional phase shift that is proportional to frequency deviation. The IF input signal (V_i) is multiplied by the quadrature signal (V_o) in the product detector and produces an output signal that is proportional to the frequency deviation. At resonance, the tank circuit impedance is resistive. However, frequency variations in the IF produce an additional positive or negative phase shift. Therefore, the product detector output voltage is proportional to the phase difference between the two input signals and is expressed mathematically as

$$V_{out} \propto V_o V_i$$
$$\propto (V_o \cos \omega_o t) \times (V_i \sin \omega_i t + \phi)$$

Substituting into the trigonometric identity for the product of a sine and a cosine wave of equal frequency gives us

$$V_{out} \propto V_i \sin (2\omega_i t + \phi) + V_o \sin \phi$$

Precision Phase-Locked Loop

GENERAL DESCRIPTION

The XR-2212 is an ultra-stable monolithic phase-locked loop (PLL) system especially designed for data communications and control system applications. Its on board reference and uncommitted operational amplifier, together with a typical temperature stability of better than 20 ppm/°C, make it ideally suited for frequency synthesis, FM detection, and tracking filter applications. The wide input dynamic range, large operating voltage range, large frequency range, and ECL, DTL, and TTL compatibility contribute to the usefulness and wide applicability of this device.

FEATURES

Quadrature VCO Outputs
Wide Frequency Range   0.01 Hz to 300 kHz
Wide Supply Voltage Range   4.5V to 20V
DTL/TTL/ECL Logic Compatibility
Wide Dynamic Range   2 mV to 3 Vrms
Adjustable Tracking Range ($\pm 1\%$ to $\pm 80\%$)
Excellent Temp. Stability   20 ppm/°C, Typ.

APPLICATIONS

Frequency Synthesis
Data Synchronization
FM Detection
Tracking Filters
FSK Demodulation

ABSOLUTE MAXIMUM RATINGS

Power Supply   18V
Input Signal Level   3 Vrms
Power Dissipation
Ceramic Package: 750 mW
   Derate Above $T_A = +25°C$ 6 mW/°C
Plastic Package: 625 mW
   Derate Above $T_A = +25°C$ 5 mW/°C

FUNCTIONAL BLOCK DIAGRAM

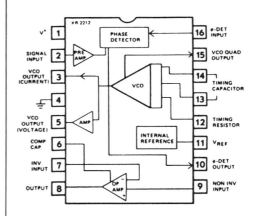

ORDERING INFORMATION

Part Number	Package	Operating Temperature
XR-2212M	Ceramic	−55°C to +125°C
XR2212CN	Ceramic	0°C to +70°C
XR-2212CP	Plastic	0°C to +70°C
XR-2212N	Ceramic	−40°C to +85°C
XR-2212P	Plastic	−40°C to +85°C

SYSTEM DESCRIPTION

The XR-2212 is a complete PLL system with buffered inputs and outputs, an internal reference, and an uncommited op amp. Two VCO outputs are pinned out; one sources current, the other sources voltage. This enables operation as a frequency synthesizer using an external programmable divider. The op amp section can be used as an audio preamplifier for FM detection or as a high speed sense amplifier (comparator) for FSK demodulation. The center frequency, bandwidth, and tracking range of the PLL are controlled independantly by external components. The PLL output is directly compatible with MOS, DTL, ECL, and TTL logic families as well as microprocessor peripheral systems.

The precision PLL system operates over a supply voltage range of 4.5 V to 20 V, a frequency range of 0.01 Hz to 300 kHz, and accepts input signals in the range of 2 mV to 3 Vrms. Temperature stability of the VCO is typically better than 20 ppm/°C.

 Corporation, 750 Palomar Avenue, Sunnyvale, CA 94086 • (408) 732-7970 • TWX 910-339-9233

FIGURE 8-8   Specification sheet for XR-2212 PLL. (Courtesy of EXAR Corporation.)

XR-2212

ELECTRICAL CHARACTERISTICS

Test Conditions: $V^+ = +12V$, $T_A = +25°C$, $R_0 = 30$ kΩ, $C_0 = 0.033$ μF, unless otherwise specified. See Figure 2 for component designation.

PARAMETERS	XR-2212/2212M			XR-2212C			UNITS	CONDITIONS
	MIN	TYP	MAX	MIN	TYP	MAX		
GENERAL								
Supply Voltage	4.5		15	4.5		15	V	
Supply Current		6	10		6	12	mA	$R_0 \geq 10$ KΩ. See Fig. 4
OSCILLATOR SECTION								
Frequency Accuracy		± 1	± 3		± 1		%	Deviation from $f_0 = 1/R_0C_0$
Frequency Stability								$R_1 = \infty$
Temperature		± 20	± 50		± 20		ppm/°C	See Fig. 8.
Power Supply		0.05	0.5		0.05		%/V	$V^+ = 12 \pm 1$ V. See Fig. 7.
		.2			.2		%/V	$V^+ = 5 \pm 0.5$ V.
Upper Frequency Limit	100	300			300		kHz	See Fig. 7. $R_0 = 8.2$ KΩ, $C_0 = 400$ pF
Lowest Practical								
Operating Frequency			0.01		0.01		Hz	$R_0 = 2$ MΩ, $C_0 = 50$ μF
Timing Resistor, R_0								See Fig. 5.
Operating Range	5		2000	5		2000	KΩ	
Recommended Range	15		100	15		100	KΩ	See Fig. 7 and 8.
OSCILLATOR OUTPUTS								
Voltage Output								Measured at Pin 5.
Positive Swing, V_{OH}		11			11		V	
Negative Swing, V_{OL}	.8	4			.5		V	
Current Sink Capability		1			1		mA	
Current Output								Measured at Pin 3.
Peak Current Swing	100	150			150		μA	
Output Impedance		1			1		MΩ	
Quadrature Output								Measured at Pin 15.
Output Swing		0.6			0.6		V	
DC Level		0.3			0.3		V	Referenced to Pin 11.
Output Impedance		3			3		KΩ	
LOOP PHASE DETECTOR SECTION								Measured at Pin 10.
Peak Output Current	± 150	± 200	± 300	± 100	± 200	± 300	μA	
Output Offset Current		± 1			± 2		μA	
Output Impedance		1			1		MΩ	
Maximum Swing	± 4	± 5		± 4	± 5		V	Referenced to Pin 11.
INPUT PREAMP SECTION								Measured at Pin 2.
Input Impedance		20			20		KΩ	
Input Signal to Cause Limiting		2	10		2		mVrms	
OP AMP SECTION								
Voltage Gain	55	70		55	70		dB	$R_L = 5.1$ KΩ, $R_F = \infty$
Input Bias Current		0.1	1		0.1	1	μA	
Offset Voltage		± 5	± 20		± 5	± 20	mV	
Slew Rate		2			2		V/μsec	
INTERNAL REFERENCE								Measured at Pin 11.
Voltage Level	4.9	5.3	5.7	4.75	5.3	5.85	V	
Output Impedance		100			100		Ω	

FIGURE 8-8 (*continued*)

329

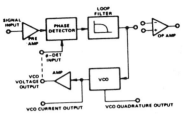

Figure 1. Functional Block Diagram of XR-2212 Precision PLL System

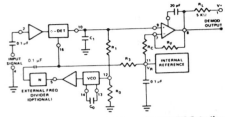

Figure 2. Generalized Circuit Connection for FM Detection, Signal Tracking or Frequency Synthesis

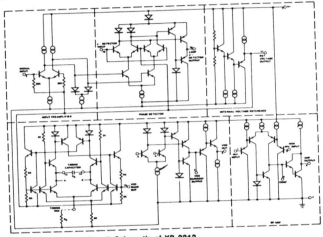

Figure 3. Simplified Circuit Schematic of XR-2212

TYPICAL CHARACTERISTICS

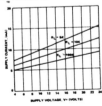

Figure 4. Typical Supply Current vs V+ (Logic Outputs Open Circuited)

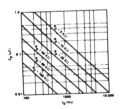

Figure 5. VCO Frequency vs Timing Resistor

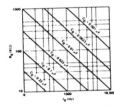

Figure 6. VCO Frequency vs Timing Capacitor

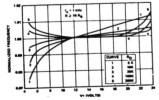

Figure 7. Typical f_0 vs Power Supply Characteristics

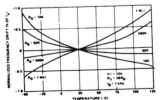

Figure 8. Typical Center Frequency Drift vs Temperature

FIGURE 8-8 (*continued*)

XR-2212

DESCRIPTION OF CIRCUIT CONTROLS

Signal Input (Pin 2): Signal is ac coupled to this terminal. The internal impedance at Pin 2 is 20 KΩ. Recommended input signal level is in the range of 10 mV to 5V peak-to-peak.

VCO Current Output (Pin 3): This is a high impedance (MΩ) current output terminal which can provide ± 100 μA drive capability with a voltage swing equal to V$^+$. This output can directly interface with CMOS or NMOS logic families.

VCO Voltage Output (Pin 5): This terminal provides a low-impedance ($\approx 50\Omega$) buffered output for the VCO. It can directly interface with low-power Schottley TTL. For interfacing with standard TTL circuits, a 750Ω pull-down resistor from pin 5 to ground is required. For operation of the PLL without an external divider, pin 5 can be dc coupled to pin 16.

Op Amp Compensation (Pin 6): The op amp section is frequency compensated by connecting an external capacitor from pin 6 to the amplifier output (pin 8). For unity-gain compensation a 20 pF capacitor is recommended.

Op Amp Inputs (Pins 7 and 9): These are the inverting and the non-inverting inputs for the op amp section. The common-mode range of the op amp inputs is from +1V to (V$^+$ $-$ 1.5) volts.

Op Amp Output (Pin 8): The op amp output is an open-collector type gain stage and requires a pull-up resistor, R$_L$, to V$^+$ for proper operation. For most applications, the recommended value of R$_L$ is in 5 kΩ to 10 kΩ range.

Phase Detector Output (Pin 10): This terminal provides a high-impedance output for the loop phase-detector. The PLL loop filter is formed by R$_1$ and C$_1$ connected to Pin 10 (see Figure 2). With no input signal, or with no phase-error within the PLL, the dc level at Pin 10 is very nearly equal to V$_R$. The peak voltage swing available at the phase detector output is equal to $\pm$V$_R$.

Reference Voltage, V$_R$ (Pin 11): This pin is internally biased at the reference voltage level, V$_R$: V$_R$ = V$^+$/2 $-$ 650 mV. The dc voltage level at this pin forms an internal reference for the voltage levels at pins 10, 12 and 16. Pin 1 *must* be bypassed to ground with a 0.1 μF capacitor, for proper operation of the circuit.

VCO Control Input (Pin 12): VCO free-running frequency is determined by external timing resistor, R$_0$, connected from this terminal to ground. For optimum temperature stability, R$_0$ must be in the range of 10 KΩ to 100 KΩ (see Figure 8).

VCO Frequency Adjustment: VCO can be fine-tuned by connecting a potentiometer, R$_X$, in series with R$_0$ at Pin 12 (see Figure 10).

This terminal is a low-impedance point, and is internally biased at a dc level equal to V$_R$. The maximum timing current drawn from Pin 12 must be limited to ≤ 3 mA for proper operation of the circuit.

VCO Timing Capacitor (Pins 13 and 14): VCO frequency is inversely proportional to the external timing capacitor, C$_0$, connected across these terminals (see Figure 5). C$_0$ must be nonpolar, and in the range of 200 pF to 10 μF.

VCO Quadrature Output (Pin 15): The low-level (≈ 0.6 Vpp) output at this pin is at quadrature phase (i.e. 90° phase-offset) with the other VCO outputs at pins 3 and 5. The dc level at pin 15 is approximately 300 mV above V$_R$. The quadrature output can be used with an external multiplier as a "lock detect" circuit. In order not to degrade oscillator performance, the output at pin 15 must be buffered with an external high-impedance low-capacitance amplifier. When not in use, pin 15 should be left open-circuited.

Phase Detector Input (Pin 16): Voltage output of the VCO (pin 5) or the output of an external frequency divider is connected to this pin. The dc level of the sensing threshold for the phase detector is referenced to V$_R$. If the signal is capacitively coupled to pin 16, then this pin must be biased from pin 11, through an external resistor, R$_B$ (R$_B$ $\approx$ 10 KΩ). The peak voltage swing applied to pin 16 *must not* exceed (V$^+$ $-$ 1.5) volts.

PHASE-LOCKED LOOP PARAMETERS:

Transfer Characteristics:

Figure 9 shows the basic frequency to voltage characteristics of XR-2212. With no input signal present, filtered phase detector output voltage is approximately equal to the internal reference voltage, V$_R$, at pin 11. The PLL can track an input signal over its tracking bandwidth, shown in the figure. The frequencies f$_{TL}$ and f$_{TH}$ represent the lower and the upper edge of the tracking range, f$_0$ represents the VCO center frequency.

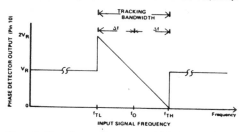

Figure 9. Phase Detector Output Voltage (Pin 10) as a Function of Input Signal Frequency. Note: Output Voltage is Referenced to Internal Reference Voltage V$_R$ at Pin 11

Design Equations:

(See Figure 2 and Figure 9 for definition of components.)

1. VCO Center Frequency, f$_0$:   f$_0$ = 1/R$_0$C$_0$ Hz

FIGURE 8-8 (*continued*)

2. Internal Reference Voltage, V_R (measured at Pin 11)

$$V_R = V + /2 - 650 \text{ mV}$$

3. Loop Low-Pass Filter Time Constant, τ: $\quad \tau = R_1 C_1$

4. Loop Damping, ζ: $\quad \zeta = 1/4 \sqrt{\dfrac{NC_0}{C_1}}$

where N is the external frequency divider modular (See 2). If no divider is used, $N = 1$.

5. Loop Tracking Bandwidth, $\pm \Delta f / f_0$: $\quad \Delta f / f_0 = R_0 / R_1$

6. Phase Detector Conversion Gain, K_ϕ: (K_ϕ is the differential dc voltage across Pins 10 and 11, per unit of phase error at phase-detector input) $\quad K_\phi = -2V_R / \pi$ volts/radian

7. VCO Conversion Gain, K_0: (K_0 is the amount of change in VCO frequency, per unit of dc voltage change at Pin 10. It is the reciprocal of the slope of conversion characteristics shown in Figure 9). $K_0 = -1/V_R C_0 R_1$ Hz/volt

8. Total Loop Gain, K_T:

$$K_T = 2\pi K_\phi K_0 = 4/C_0 R_1 \text{ rad/sec/volt}$$

9. Peak Phase-Detector Current, I_A; available at pin 10.

$$I_A = V_R \text{ (volts)}/25 \text{ mA}$$

APPLICATION INFORMATION

FM DEMODULATION:

XR-2212 can be used as a linear FM demodulator for both narrow-band and wide-band FM signals. The generalized circuit connection for this application is shown in Figure 10, where the VCO output (pin 5) is directly connected to the phase detector input (pin 16). The demodulated signal is obtained at phase detector output (pin 10). In the circuit connection of Figure 10, the op amp section of XR-2212 is used as a buffer amplifier to provide both additional voltage amplification as well as current drive capability. Thus, the demodulated output signal available at the op amp output (pin 8) is fully buffered from the rest of the circuit.

In the circuit of Figure 10, $R_0 C_0$ set the VCO center frequency, R_1 sets the tracking bandwidth, C_1 sets the low-pass filter time constant. Op amp feedback resistors R_F and R_C set the voltage gain of the amplifier section.

Design Instructions:

The circuit of Figure 10 can be tailored to any FM demodulation application by a choice of the external components R_0, R_1, R_C, R_F, C_0 and C_1. For a given FM center frequency and frequency deviation, the choice of these components can be calculated as follows, using the design equations and definitions given on page 1-34, 1-35 and 1-36.

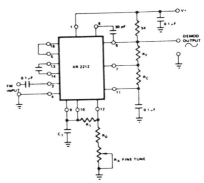

Figure 10. Circuit Connection for FM Demodulation

a) Choose VCO center frequency f_0 to be the same as FM carrier frequency.

b) Choose value of timing resistor R_0, to be in the range of 10 KΩ to 100 KΩ. This choice is arbitrary. The recommended value is $R_0 \cong 20$ KΩ. The final value of R_0 is normally fine-tuned with the series potentiometer, R_X.

c) Calculate value of C_0 from design equation (1) or from Figure 6:

$$C_0 = 1/R_0 f_0$$

d) Choose R_1 to determine the tracking bandwidth, Δf (see design equation 5). The tracking bandwidth, Δf, should be set significantly wider than the maximum input FM signal deviation, Δf_{SM}. Assuming the tracking bandwidth to be "N" times larger than Δf_{SM}, one can re-unite design equation 5 as:

$$\frac{\Delta f}{f_0} = \frac{R_0}{R_1} = N \frac{\Delta f_{SM}}{f_0}$$

Table I lists recommended values of N, for various values of the maximum deviation of the input FM signal.

% Deviation of FM Signal ($\Delta f_{SM}/f_0$)	Recommended value of Bandwidth Ratio, N ($N = \Delta f/\Delta f_{SM}$)
1% or less	10
1 to 3%	5
1 to 5%	4
5 to 10%	3
10 to 30%	2
30 to 50%	1.5

TABLE I

Recommended values of bandwidth ratio, N, for various values of FM signal frequency deviation. (Note: N is the ratio of tracking bandwidth Δf to max. signal frequency deviation, Δf_{SM}).

FIGURE 8-8 (continued)

e) Calculate C_1 to set loop damping (see design equation 4). Normally, $\zeta = 1/2$ is recommended. Then, $C_1 = C_0/4$ for $\zeta = 1/2$.

f) Calculate R_C and R_F to set peak output signal amplitude. Output signal amplitude, V_{out}, is given as:

$$V_{out} = \left(\frac{\Delta f_{SM}}{f_0}\right)\left(v_R\right)\left(\frac{R_1}{R_0}\right)\left[\frac{R_C + R_F}{R_C}\right]$$

In most applications, $R_F = 100\ K\Omega$ is recommended; then R_C, can be calculated from the above equation to give desired output swing. The output amplifier can also be used as a unity-gain voltage follower, by open circuiting R_C (i.e., $R_C = \infty$).

Note: All calculated component values except R_0 can be rounded-off to the nearest standard value, and R_0 can be varied to fine-tune center frequency, through a series potentiometer, R_X. (See Figure 10.)

Design Example:

Demodulator for FM signal with 67 kHz carrier frequency with ±5 kHz frequency deviation. Supply voltage is + 12V and required peak output swing is ±4 volts.

Step a) f_0 is chosen as 67 kHz.

Step b) Choose $R_0 = 20\ K\Omega$ (18 KΩ fixed resistor in series with 5 KΩ potentiometer).

Step c) Calculate C_0; from design Eq. (1).

$C_0 = 746$ pF

Step d) Calculate R_1. For given FM deviation, $\Delta f_{SM}/f_0 = 0.0746$, and N = 3 from Table I.

Then:

$R_0/R_1 = (3)(0.0746) = 0.224$

or:

$R_1 = 89.3\ K\Omega$.

Step e) Calculate $C_1 = (C_0/4) = 186$ pF.

Step f) Calculate R_C and R_F to get ±4 volts peak output swing: Let $R_F = 100\ K\Omega$. Then,

$R_C = 80.6\ K\Omega$.

Note: All values except R_0 can be rounded-off to nearest standard value.

FREQUENCY SYNTHESIS

Figure 11 shows the generalized circuit connection for frequency synthesis. In this application an external frequency divider is connected between the VCO output (pin 5) and the phase-detector input (pin 16). When the circuit is in lock, the two signals going into the phase-detector are at the same frequency, or $f_S = f_1/N$ where

N is the modulus of the external frequency divider. Conversely, the VCO output frequency, f_1 is equal to Nf_S.

In the circuit configuration of Figure 11, the external timing components, R_0 and C_0, set the VCO free-running frequency; R_1 sets the tracking bandwidth and C1 sets the loop damping, i.e., the low-pass filter time constant (see design equations).

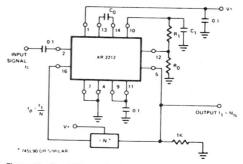

Figure 11. Circuit Connection for Frequency Synthesizer

The total tracking range of the PLL (see Figure 9), should be chosen to accommodate the lowest and the highest frequency, f_{max} and f_{min}, to be synthesized. A recommended choice for most applications is to choose a tracking half-bandwidth Δf, such that:

$$\Delta f \approx f_{max} - f_{min}.$$

If a fixed output frequency is desired, i.e. N and f_S are fixed, then a ±10% tracking bandwidth is recommended. Excessively large tracking bandwidth may cause the PLL to lock on the harmonics of the input signals; and the small tracking range increases the "lock-up" or acquisition time.

If a variable input frequency and a variable counter modulus N is used, then the maximum and the minimum values of output frequency will be:

$$f_{max} = N_{max}(f_S)_{max} \text{ and } f_{min} = N_{min}(f_S)_{min}.$$

Design Instructions:

For a given performance requirement, the circuit of Figure 11 can be optimized as follows:

a) Choose center frequency, f_0, to be equal to the output frequency to be synthesized. If a range of output frequencies is desired, set f_0 to be at mid-point of the desired range.

b) Choose timing resistor R_0 to be in the range of 15 KΩ to 100 KΩ. This choice is arbitrary. R_0 can be fine tuned with a series potentiometer, R_X.

c) Choose timing capacitor, C_0 from Figure 6 or Equation 1.

FIGURE 8-8 (*continued*)

d) Calculate R_1 to set tracking bandwidth (see Figure 9, and design equation 5). If a range of output frequencies are desired, set R_1 to get:

$$\Delta f = f_{max} - f_{min}.$$

If a single fixed output frequency is desired, set R_1 to get:

$$\Delta f = 0.1 \, f_0.$$

e) Calculate C_1 to obtain desired loop damping. (See design equation 4). For most applications, $\zeta = 1/2$ is recommended, thus:

$$C_1 = NC_0/4$$

Note: All component values except R_0 can be rounded-off to nearest standard value.

FIGURE 8-8 (*continued*)

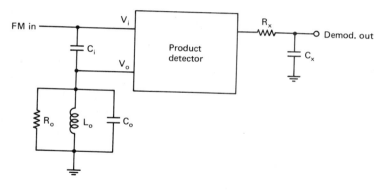

FIGURE 8-9 Quadrature FM demodulator.

The second harmonic ($2\omega_i$) is filtered off, leaving

$$V_{\text{out}} \propto V_o \sin \phi$$

where $\phi = \tan^{-1} \rho Q$ and

$$\rho \approx 2\pi F/F_o \text{ (fractional frequency deviation)}$$
$$Q = \text{tank circuit quality factor}$$

Amplitude Limiters and FM Thresholding

The vast majority of terrestrial FM communications systems use conventional noncoherent demodulation because most standard frequency discriminators use envelope detection to remove the intelligence from the FM waveform. Unfortunately, envelope detectors (including ratio detectors) will demodulate incidental amplitude variations in the IF. Transmission noise and interference add to the signal and thus produce unwanted amplitude variations. Also, FM is generally accompanied by small amounts of residual amplitude modulation. In the receiver, the unwanted AM and random noise interference are demodulated along with the signal and produce unwanted distortion in the recovered information signal. The noise is more prevalent at the peaks of the FM envelope and relatively insignificant during the zero crossings. A limiter is a circuit that produces a constant-amplitude output for all input signals above a prescribed minimum input level; which is often called either the *threshold*, *quieting*, or *capture* level. Limiters are required in most FM receivers because many of the demodulators that were discussed earlier in this chapter will demodulate AM as well as FM. With amplitude limiters, the *S/N* ratio at the output of the demodulator (*postdetection S/N*) can be improved by as much as 20 or more dB over the input (*predetection*) *S/N*.

Essentially, an amplitude limiter is an additional IF amplifier that is overdriven. Limiting begins when the IF signal is sufficiently large that it drives the amplifier both into saturation and cutoff. Figure 8-10 shows the input and output waveforms for a typical limiter. In Figure 8-10b, it can be seen that for IF signals that are below threshold, the noise is not reduced, and for IF signals above threshold, there is a large reduction

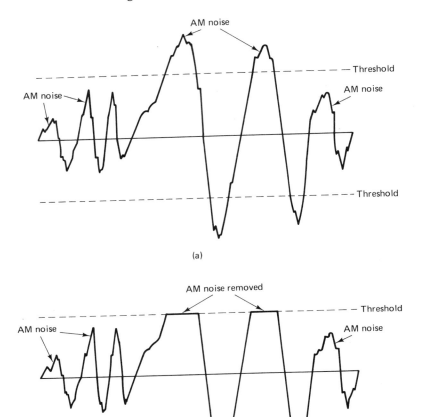

FIGURE 8-10   Amplitude limiter input and output waveforms: (a) input waveform; (b) output waveform.

in the noise level. The purpose of the limiter is to remove all amplitude variations from the IF signal.

Figure 8-11a shows the output from a limiter when the noise is greater than the signal (i.e., the noise has captured the limiter). The irregular widths of the serrations are caused by noise impulses saturating the limiter. Figure 8-11b shows the limiter output when the signal is sufficiently greater than the noise (i.e., the signal has captured the limiter). The sinusoidal signal peaks have the limiter so far into saturation that the weaker noise is totally masked. The improvement in the *S/N* ratio is called *FM thresholding*, *FM quieting*, or the *FM capture effect*. There are three criteria that must be satisfied before FM thresholding can occur. They are:

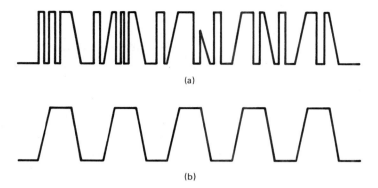

FIGURE 8-11   Limiter output: (a) captured by noise; (b) captured by signal.

1. The predetection *S/N* must be 10 dB or greater.
2. The IF signal must be sufficiently amplified to overdrive the limiter.
3. The signal must have a modulation index $m \geq 1$.

Figure 8-12 shows typical FM thresholding curves for a low modulation index signal ($m = 1$) and a high modulation index signal ($m = 4$). The output voltage from

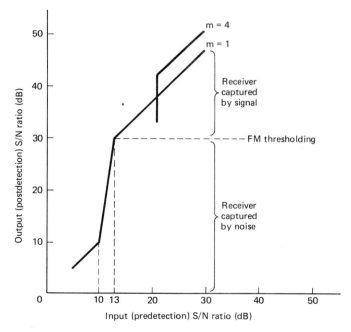

FIGURE 8-12   FM thresholding.

an FM detector is proportional to m^2. Therefore, doubling m increases the S/N ratio by 6 dB. The quieting ratio for $m = 1$ is an input $S/N = 13$ dB, and 22 dB for $m = 4$. For S/N ratios below threshold, the receiver is said to be captured by the noise, and for S/N ratios above threshold, the receiver is said to be captured by the signal. Figure 8-12 shows that IF signals at the input to the limiter with 13 or more dB of S/N undergo 17 dB of S/N improvement.

Limiter circuits. Figure 8-13a shows a schematic diagram for a single-stage limiter with a built-in output filter. This configuration is commonly called a *bandpass limiter/amplifier* (BPL). A BPL is essentially a class A biased tuned IF amplifier, and for limiting and FM quieting to occur, requires an IF input signal sufficient to drive it into saturation and cutoff. The output tank circuit is tuned to the IF center frequency. Filtering removes the harmonic and intermodulation distortion present in the rectangular pulses due to hard limiting. This is shown in Figure 8-14. If resistor R_2 were removed entirely, the amplifier would be biased for class C operation, which is also appropriate for this type of circuit. Figure 8-13b shows limiter action for the circuit shown in Figure 8-13a. It can be seen that for small signals (below the threshold voltage) no

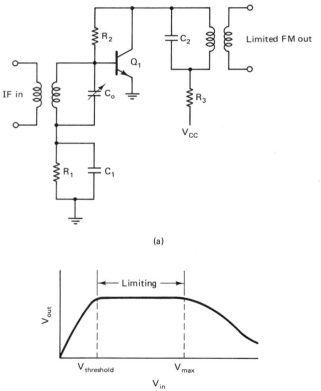

(a)

(b)

FIGURE 8-13 Single-stage tuned limiter: (a) schematic diagram; (b) limiter action.

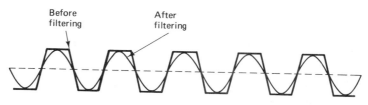

Before filtering   After filtering

FIGURE 8-14   Filtered limiter output.

limiting occurs. When V_{in} reaches $V_{threshold}$, limiting begins, and for input amplitudes above V_{max}, there is actually a decrease in V_{out} with V_{in}. This is because with high input drive levels, the collector current pulses are sufficiently narrow that they actually develop less tank circuit power. The problem of overdriving the limiter can be rectified by incorporating AGC into the circuit.

When two limiter stages are used, it is called *double* limiting; three stages, *triple* limiting; and so on. Figure 8-15 shows a three-stage *cascoded limiter* without a built-in filter. This type of limiter must be followed by either a ceramic or crystal filter to remove the nonlinear distortion. The limiter shown has three *RC*-coupled limiter stages that are dc series connected to reduce the current drain. Cascoded amplifiers combine several of the advantages of common-emitter and common-gate amplifiers. Cascoding amplifiers also decreases the thresholding level and thus improves the quieting capabilities of the stage. The effects of double and triple limiting are shown in Figure 8-16.

EXAMPLE 8-1

For an FM receiver with a 200-kHz bandwidth, a noise figure = 8 dB, and an input noise temperature T = 100 K; determine the minimum receive carrier power to achieve a postdetection S/N = 37 dB. Use the receiver block diagram shown in Figure 8-1 as the receiver model and the FM thresholding curve shown in Figure 8-12.

Solution   From Figure 8-12 it can be seen that to achieve a post detection S/N = 37 dB, the S/N at the input to the limiter must be at least

$$37 \text{ dB} - 17 \text{ dB} = 20 \text{ dB}$$

Therefore, for a receiver noise figure = 8 dB, the S/N ratio at the receiver input must be at least

$$20 \text{ dB} + 8 \text{ dB} = 28 \text{ dB}$$

The receiver input noise power, N, is

$$10 \log \frac{KTB}{0.001} = 10 \log \frac{(1.38 \times 10^{-23})(100)(200,000)}{0.001}$$

$$N = -125.6 \text{ dBm}$$

Therefore, the minimum receive carrier power is

$$-125.6 \text{ dBm} + 28 \text{ dB} = -97.6 \text{ dBm}$$

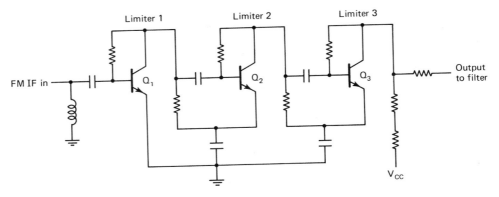

FIGURE 8-15   Three-stage cascoded limiter.

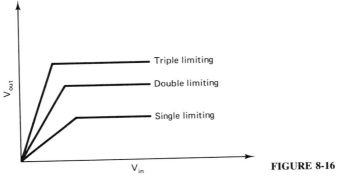

FIGURE 8-16   Limiter response curves.

FM SYSTEMS

FM Stereo

Until 1961, all commercial FM broadcast-band transmissions were *monophonic*. That is, a single 50-Hz to 15-kHz audio channel made up the entire voice and music information spectrum. This single audio channel modulated a high-frequency carrier and was transmitted through a 200-kHz-bandwidth FM channel. With mono transmission, each speaker assembly at the receiver reproduces exactly the same information. It is possible to separate the information in the frequency domain with special speakers, such as *woofers* for low frequencies and *tweeters* for high frequencies. However, it is impossible to separate monophonic sound *spatially*. The entire information signal sounds as though it is coming from the same direction (i.e., from a *point source*—there is no directivity to the sound). In 1961, the FCC authorized *stereophonic* transmission for the commercial FM broadcast band. With stereophonic transmission, the information signal is spatially divided into two 50-Hz to 15-kHz audio channels (a left and a right). Music that originated on the left side is reproduced only on the left speaker, and music that originated on the right side is reproduced only on the right speaker. Therefore, with stereophonic transmission, it is possible to reproduce music with a unique directivity and spatial dimension that before was possible only with live entertainment (i.e., from an *extended* source). Also, with stereo transmission, it is possible to separate music or sound by *tonal quality*, such as percussion, strings, horns, and so on.

A primary concern of the FCC before authorizing stereophonic transmission was its compatibility with monophonic receivers. Stereo transmission was not to affect mono reception. Also, monophonic receivers must be able to receive stereo transmission as monaural without any perceptible degradation in program quality. In addition, stereophonic receivers were to receive stereo programming with nearly perfect separation (40 dB or more) between the left and right channels.

The original FM audio spectrum is shown in Figure 8-17a. The audio channel extended from 50 Hz to 15 kHz. In 1955, the FCC approved *subcarrier* transmission under the Subsidiary Communications Authorization (SCA). SCA is used to broadcast uninterrupted music to private subscribers, such as department stores, restaurants, medical offices, and so on, equipped with special SCA receivers. Originally, the SCA subcarrier ranged from 25 to 75 kHz, but has since been standardized at 67 kHz. The subcarrier and its associated sidebands become part of the total signal that modulates the main

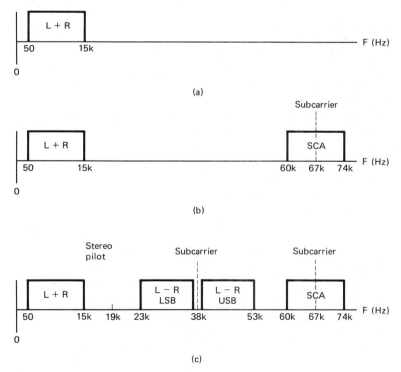

FIGURE 8-17   FM baseband spectrum: (a) prior to 1955; (b) prior to 1961; (c) since 1961.

carrier. At the receiver, the subcarrier is demodulated along with the primary channel but, of course, cannot be heard because of its high frequency. The process of stacking two or more independent channels in the frequency domain then modulating a single carrier is called frequency-division multiplexing (Chapter 1). With FM stereophonic broadcasting, three voice or music channels are frequency-division multiplexed onto a single FM carrier. Figure 8-17b shows the FM baseband spectrum prior to 1961 (baseband comprises the total modulating signal spectrum). The primary audio channel remained at 50 Hz to 15 kHz, while an additional SCA audio channel is translated to the 60- to 74-kHz passband. The SCA subcarrier may be AM single- or double-sideband transmission, or FM with a maximum modulating signal frequency of 7 kHz. However, the SCA modulation of the main carrier is low-index narrowband FM and, consequently, is a much lower-quality transmission than the primary FM channel. The total frequency deviation remained at 75 kHz with 90% (67.5 kHz) reserved for the primary channel and 10% (7.5 kHz) reserved for SCA.

Figure 8-17c shows the FM baseband spectrum since 1961. It comprises the original 50-Hz to 15-kHz audio channel plus two additional audio channels frequency-division multiplexed into a composite baseband signal. The three channels are (1) the left (L) plus the right (R) audio channels (i.e., the L + R stereo channel), (2) the left plus the inverted right audio channels (i.e., the L − R stereo channel), and (3) the SCA subcarrier and its associated sidebands. The L + R stereo channel occupies the 0- to 15-kHz passband, and the 23- to 53-kHz passband is used with stereophonic transmission for carrying the L − R stereo channel. The L − R signal amplitude modulates a 38-kHz subcarrier and produces a double-sideband suppressed carrier signal that occupies the 23- to 53-kHz passband. SCA transmissions occupy the 60- to 74-kHz spectrum. The information contained in the L + R and L − R stereo channels is identical except for their phase. With this scheme, mono receivers can demodulate the total baseband spectrum, but only the 50-Hz to 15-kHz L + R stereo channel is amplified and fed to the speakers. Stereophonic receivers must provide additional demodulation of the 23- to 53-kHz L − R stereo channel, then separate the left and right audio information and feed them to their respective speakers. Again, the SCA subcarrier is demodulated by all FM receivers, although only those with special SCA equipment demodulate the subcarrier to audio.

With stereo transmission, the maximum frequency deviation is still 75 kHz; 7.5 kHz (10%) is reserved for SCA transmission and another 7.5 kHz (10%) is reserved for the 19 kHz stereo pilot, which is explained in the next section. This leaves 60 kHz of frequency deviation for the actual stereophonic transmission of the L + R and L − R stereo channels. However, the L + R and L − R channels are not necessarily limited to 30 kHz deviation each. A rather simple but unique technique is used to *interleave* the two channels such that at times either the L + R or the L − R channel may deviate the main carrier 60 kHz by themselves. However, the total deviation will never exceed 60 kHz. This interleaving technique is explained later in the chapter.

FM stereo generation.   Figure 8-18 shows a simplified block diagram for a stereo FM transmitter. The L and R audio channels are combined in a matrix network

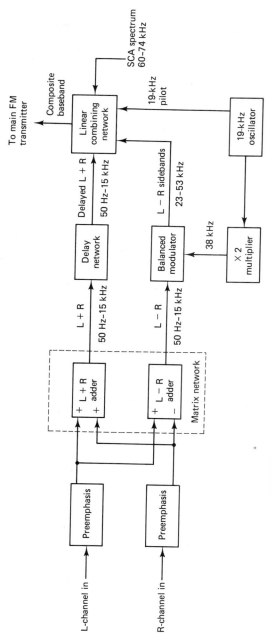

FIGURE 8-18 Stereo FM transmitter using frequency-division multiplexing.

to produce the L + R and L − R stereo channels. The L − R audio channel modulates a 38-kHz subcarrier and produces a 23- to 53-kHz L − R stereo channel. Because there is a time delay introduced in the L − R signal path as it propagates through the balanced modulator, the L + R channel must be artificially delayed somewhat to maintain phase integrity with the L − R channel for demodulation purposes. Also, for demodulation purposes, a 19-kHz subharmonic of the 38-kHz suppressed carrier is added to the baseband signal. The 19-kHz pilot is transmitted rather than the 38-kHz carrier because it is considerably more difficult to recover the 38-kHz carrier in the receiver. The composite baseband signal is fed to the FM transmitter, where it modulates the main carrier.

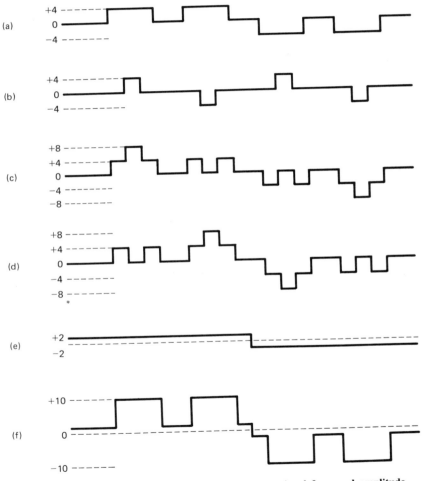

FIGURE 8-19 Development of the composite stereo signal for equal amplitude L and R signals: (a) L audio signal; (b) R audio signal; (c) L + R stereo channel; (d) L − R stereo channel; (e) SCA + 19-kHz pilot; (f) composite baseband waveform.

L + R and L − R channel interleaving. Figure 8-19 shows the development of the composite stereo signal for equal-amplitude L and R channel signals. For illustration purposes, rectangular waveforms are shown. Table 8-1 is a tabular summary of the individual and total signal voltages for Figure 8-19. Note that the L − R audio channel does not appear in the composite waveform. The L − R channel modulates the 38-kHz subcarrier to form the L − R sidebands, which are part of the composite spectrum.

For the FM modulator in this example, it is assumed that 10 V of baseband signal will produce 75 kHz of frequency deviation of the main carrier, and the SCA and 19-kHz pilot polarities shown are for maximum frequency deviation. The L and R channels are each limited to a maximum value of 4 V; 1 V is for SCA, and 1 V is for the 19-kHz stereo pilot. Therefore, 8 V is left for the L + R and L − R stereo channels. Figure 8-19 shows the L, R, L + R, L − R channels, the SCA and 19-kHz pilot, and the composite stereo waveform. It can be seen that the L + R and L − R stereo channels interleave and never produce more than 8 V of total amplitude and therefore never produce more than 60 kHz of frequency deviation. The total composite baseband never exceeds 10 V (75 kHz deviation).

Figure 8-20 shows the development of the composite stereo waveform for unequal values of the L and R signals. Again, it can be seen that the composite stereo waveform never exceeds 10 V or 75 kHz of frequency deviation. For the first set of waveforms, it appears that the sum of the L + R and L − R waveforms completely cancels. Actually, this is not true; it only appears that way because rectangular waveforms are used in this example.

FM stereo reception. FM stereo receivers are identical to standard FM receivers up to the output of the audio detector stage. The output of the discriminator is the total baseband spectrum that was shown in Figure 8-17c.

Figure 8-21 shows a simplified block diagram for an FM receiver giving both mono and stereophonic audio outputs. In the mono signal processing, the L + R stereo channel, which contains all of the original information from both the L and R audio channels, is simply filtered, amplified, then fed to both the L and R speakers. In the stereo portion of the signal processor, the baseband signal is fed to a stereo demodulator where the L and R channels are separated and fed to their respective speakers. The

TABLE 8-1 COMPOSITE FM VOLTAGES

L	R	L + R	L − R	SCA and pilot	Total
0	0	0	0	2	2
4	0	4	4	2	10
0	4	4	−4	2	2
4	4	8	0	2	10
4	−4	0	8	2	10
−4	4	0	−8	−2	−10
−4	−4	−8	0	−2	−10

(a)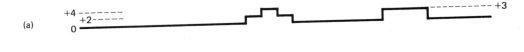

(b)

(c)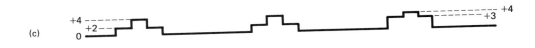

(d)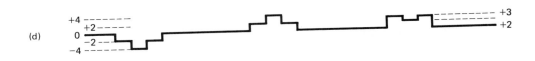

(e)

(f)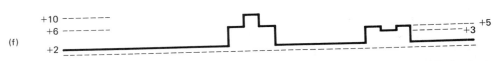

FIGURE 8-20   Development of the composite stereo signal for unequal amplitude L and R signals:
(a) L audio signal; (b) R audio signal; (c) L + R stereo channel; (d) L− R stereo channel;
(e) SCA + 19-kHz pilot; (f) composite baseband waveform.

L + R and L − R stereo channels and the 19-kHz pilot are separated from the composite baseband signal with filters. The 19-kHz pilot is filtered off with a high-Q bandpass filter, multiplied by 2, amplified, and then fed to the L − R demodulator. The L + R stereo channel is filtered off by a low-pass filter with an upper cutoff frequency of 15 kHz. The L − R double-sideband signal is separated with a broadly tuned bandpass

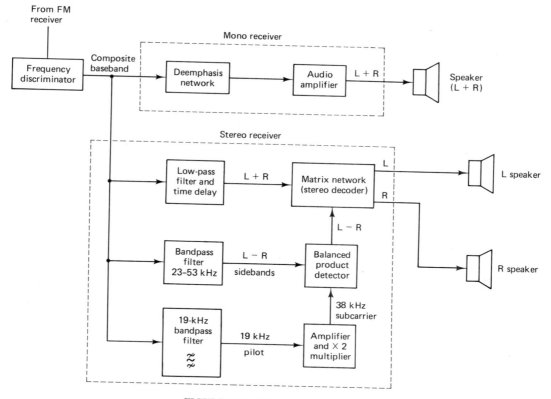

FIGURE 8-21   FM stereo and mono receiver.

filter, then mixed with the recovered 38-kHz carrier in a balanced modulator to produce the L − R audio information. The matrix network combines the L + R and L − R signals in such a way to separate the L and R audio information signals, which are fed to their respective deemphasis networks and speakers.

Figure 8-22 shows the block diagram for the stereo matrix decoder. The L − R audio channel is added directly to the L + R audio channel. The output from the adder is

$$
\begin{array}{r}
L + R \\
+ \underline{(L - R)} \\
2L
\end{array}
$$

The L − R audio channel is inverted, then added to the L + R audio channel. The output from the adder is

$$
\begin{array}{r}
L + R \\
- \underline{(L - R)} \\
2R
\end{array}
$$

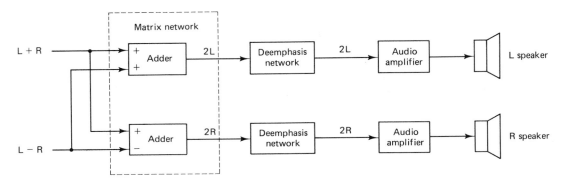

FIGURE 8-22   Stereo matrix network decoder.

Large-scale integration stereo demodulator.   Figure 8-23 shows the specification sheet for an XR-1310 stereo demodulator. The XR-1310 is a monolithic FM stereo demodulator which uses phase locked techniques to derive the right and left audio channels from the composite stereo signal. The XR-1310 uses no external LC tank circuits for tuning, and alignment is accomplished with a single potentiometer. The XR-1310 features simple noncritical tuning, excellent channel separation, low distortion, and a wide dynamic range.

Two-Way FM Radio Communications

Two-way FM radio communications is used extensively for *public safety mobile* communications, such as police and fire department calls and emergency medical services. There are three primary frequency bands allocated by the FCC for two-way FM radio communications: 132 to 174 MHz, 450 to 470 MHz, and 806 to 947 MHz. The maximum frequency deviation for two-way FM transmitters is typically 5 kHz, and the maximum modulating signal frequency is 3 kHz. These values give a deviation ratio of 1.67 and a maximum Bessel bandwidth of approximately 24 kHz. However, the allocated FCC channel spacing is 30 kHz. Two-way FM radio is half duplex, which supports two-way communications but not simultaneously; only one party can transmit at a time. Transmissions are initiated by closing a *push-to-talk* (PTT) switch, which turns on the transmitter and shuts off the receiver. During idle conditions, the transmitter is shut off and the receiver is turned on to allow monitoring of the radio channel for transmissions from other parties.

Two-way FM transmitter.   The simplified block diagram for a *modular integrated circuit* two-way FM transmitter is shown in Figure 8-24. This is actually a direct PM transmitter. PM is generally used because direct FM transmitters do not have the stability necessary to meet FC standards without using AFC. The transmitter shown is a four-channel unit that operates in the 150- to 174-MHz band. The channel selector switch applies power to one of four crystal oscillator modules that operate at

STEREO DEMODULATOR

GENERAL DESCRIPTION

The XR-1310 is a unique FM stereo demodulator which uses phase-locked techniques to derive the right and left audio channels from the composite signal. Using a phase-locked loop to regenerate the 38 kHz subcarrier, it requires no external L-C tanks for tuning. Alignment is accomplished with a single potentiometer.

FEATURES

Requires No Inductors
Low External Part Count
Simple, Noncritical Tuning by Single Potentiometer Adjustment
Internal Stereo/Monaural Switch with 100 mA Lamp Driving Capability
Wide Dynamic Range: 600 mV (RMS) Maximum Composite Input Signal
Wide Supply Voltage Range: 8 to 14 Volts
Excellent Channel Separation
Low Distortion
Excellent SCA Rejection

ORDERING INFORMATION

Part Number	Package	Operating Temperature
XR-1310CP	Plastic	−40°C to +85°C

FUNCTIONAL BLOCK DIAGRAM   March 1982

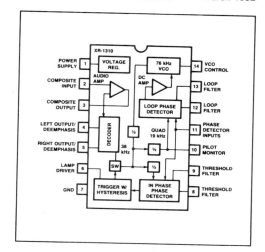

APPLICATIONS

FM Stereo Demodulation

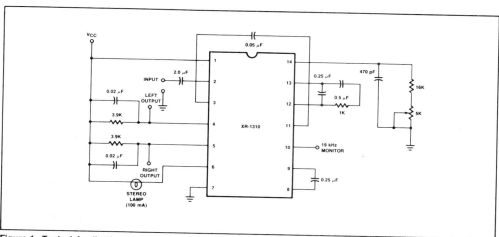

Figure 1. Typical Application

EXAR INTEGRATED SYSTEMS, INC.
750 Palomar Avenue, Sunnyvale, CA 94088   (408) 732-7970   TWX 910-339-9233   3-82 Rev. 3

FIGURE 8-23   XR-1310 stereo demodulator. (Courtesy of EXAR Corporation.)

ELECTRICAL CHARACTERISTICS

Test Conditions: Unless otherwise noted; $V_{CC}* = +12$ Vdc, $T_A = +25°C$, 560 mV (RMS) (2.8 Vp-p) standard multiplex composite signal with L or R channel only modulated at 1.0 kHz and with 100 mV (RMS) (10 % pilot level), using circuit of Figure 1.

PARAMETERS	MIN.	TYP.	MAX.	UNIT
Maximum Standard Composite Input Signal (0.5 % THD)	2.8			Vp-p
Maximum Monaural Input Signal (1.0 % THD)	2.8			Vp-p
Input Impedance		50		kΩ
Stereo Channel Separation (50 Hz — 15 kHz)	30	40		dB
Audio Output Voltage (desired channel)		485		mV (RMS)
Monaural Channel Balance (pilot tone "off")			1.5	dB
Total Harmonic Distortion		0.3		%
Ultrasonic Frequency Rejection 19 kHz		34.4		dB
38 kHz		45		
Inherent SCA Rejection (f = 67 kHz; 9.0 kHz beat note measured with 1.0 kHz modulation "off")		80		dB
Stereo Switch Level (19 kHz input for lamp "on")	13		20	mV (RMS)
Hysteresis		6		dB
Capture Range (permissable tuning error of internal oscillator, reference circuit values of Figure 1)		±3.5		%
Operating Supply Voltage (loads reduced to 2.7 kΩ for 8.0-volt operation)	8.0		14	Vdc
Current Drain (lamp "off")		13		mAdc

*Symbols conform to JEDEC Engineering Bulletin No. 1 when applicable.

ABSOLUTE MAXIMUM RATINGS
($T_A = +25°C$ unless otherwise noted)

Power Supply Voltage	14 V	
Lamp Current (nominal rating, 12 V lamp)	75 mA	

Power Dissipation (package limitation)	625 mW
Derate above $T_A = +25°C$	5.0 mW/°C
Operating Temperature Range (Ambient)	−40 to +85°C
Storage Temperature Range	−65 to +150°C

FIGURE 8-23 (*continued*)

frequencies between 12.5 and 14.5 MHz, depending on the specific channel assignment. The oscillator frequency is temperature compensated by the compensation module to ensure a stability of ±0.0002%. The phase modulator uses a varactor diode which is modulated by the audio signal at the output of the audio limiter. The audio signal amplitude is limited to ensure that the transmitter is not overdeviated. The modulated IF carrier is amplified, then multiplied by 12 to produce the desired RF carrier frequency. The RF is further amplified and filtered prior to transmission. The electronic push-to-talk module ensures that power is applied to the transmitter only when a transmission is in progress and that the receiver is on only when the transmitter is off. An electronic PTT is used rather than a simple mechanical switch, to reduce the static noise associated with *contact bounce* in mechanical switches. Keying the PTT applies dc power to the selected transmit oscillator module and the RF power amplifiers. Figure 8-25 shows the schematic diagram for a typical PTT module. Keying the PTT switch grounds the base of Q_1, causing it to conduct and turn off Q_2. With Q_2 off, V_{CC} is applied to the

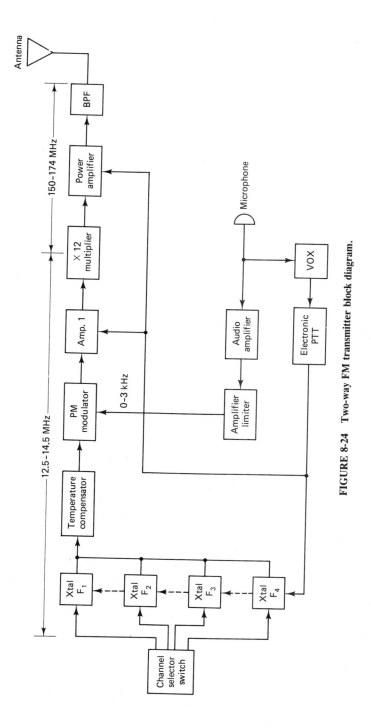

FIGURE 8-24 Two-way FM transmitter block diagram.

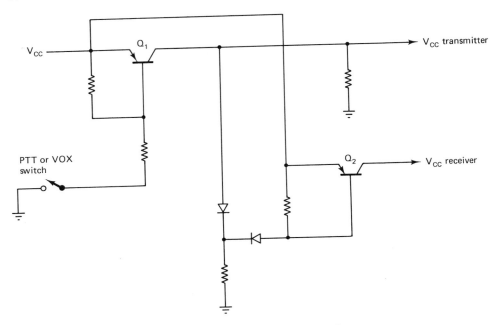

FIGURE 8-25 Electronic PTT schematic diagram.

transmitter and removed from the receiver. With the PTT switch released, Q_1 shuts off, removing V_{CC} from the transmitter and Q_2 is turned on applying V_{CC} to the receiver. Transmitters equipped with VOX (*voice-operated transmitter*) are automatically keyed each time the operator speaks into the microphone, regardless of whether or not the PTT button is depressed. Transmitters equipped with VOX require an external microphone. The schematic diagram for a typical VOX module is shown in Figure 8-26. Audio signal power in the 400- to 600-Hz passband is filtered and amplified by Q_1, Q_2, and Q_3. The output from Q_3 is rectified and used to turn on Q_4, which places a ground on the PTT circuit, enabling the transmitter and disabling the receiver. With no audio input signal, Q_4 is off and the PTT pin is open, disabling the transmitter and enabling the receiver.

Two-way FM receiver.   The block diagram for a typical two-way FM receiver is shown in Figure 8-27. Again, this is a four-channel integrated-circuit modular receiver with four separate crystal oscillator modules. Whenever the receiver is on, one of the four oscillator modules is activated, depending on the position of the channel selector switch. The oscillator frequency is again temperature compensated, then multiplied by 9. The output from the multiplier is applied to the mixer, where it beats with the incoming RF to produce a 20-MHz IF. This receiver uses low-side injection and the crystal oscillator frequency is determined as follows:

$$\text{crystal frequency} = \frac{\text{RF frequency} - 20 \text{ MHz}}{9}$$

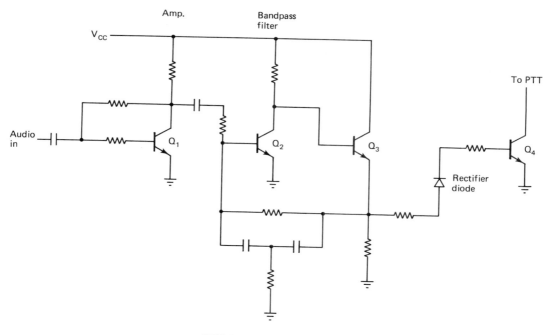

FIGURE 8-26   VOX schematic diagram.

The IF signal is filtered, amplified, limited, and then applied to the frequency discriminator for demodulation. The output from the discriminator is amplified, then applied to the speaker. A typical noise amplifier/squelch circuit is shown in Figure 8-28. The squelch circuit is keyed by out-of-band noise at the output of the audio amplifier. With no receive RF, the AGC causes the gain of the IF amplifiers to increase, which increases the receiver noise in the 3- to 5-kHz band. Whenever excessive noise is present, the audio amplifier is turned off and the receiver is quieted. The input bandpass filter passes the 3- to 5-kHz noise signal, which is amplified and rectified. The rectified output voltage determines the off/on condition of squelch switch, Q_3. When Q_3 is on, V_{CC} is applied to the receive audio amplifier. When Q_3 is off, V_{CC} is removed from the audio amplifier, quieting the receiver. R_x is a squelch sensitivity adjustment.

Mobile Telephone Service

Mobile telephone service is best described by explaining the differences between it and two-way mobile radio. As stated previously, mobile radio is half duplex and all transmissions (unless scrambled) can be heard by any listener tuned to that channel. Mobile telephone is full duplex and operates much the same as the wireline telephone service provided by local telephone companies. Mobile telephone permits two-way simultaneous transmission and, for privacy, each mobile unit is assigned a unique telephone number.

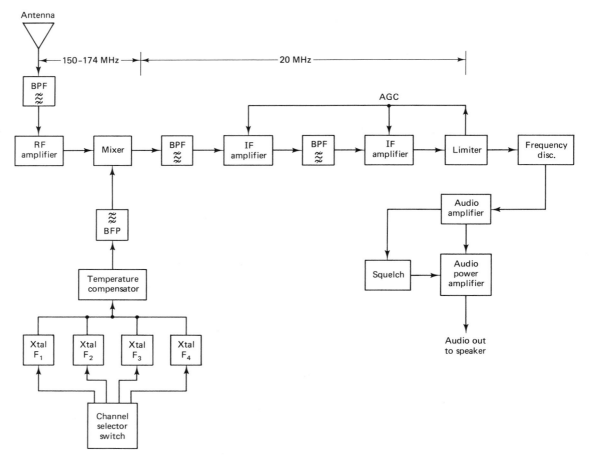

FIGURE 8-27   Two-way FM receiver block diagram.

Historical perspective.   The first mobile telephone system in the United States was established in 1946 in St. Louis, Missouri. The FCC allocated six 60-kHz mobile telephone channels in the 150-MHz frequency range. In 1947, a public mobile telephone system was established along the highway between New York City and Boston that operated in the 35- to 40-MHz frequency range. In 1949, the FCC authorized six additional mobile channels to *radio common carriers*, which are defined as companies that do not provide public wireline telephone service but do interconnect to the public telephone network and provide equivalent nonwireline telephone service. The FCC later increased the number of channels from 6 to 11 by reducing the bandwidth to 30 kHz and spacing the new channels between the old ones. In 1950, the FCC added 12 new channels in the 450-MHz band. Until 1964, mobile telephone systems operated only in the manual mode; a special mobile telephone operator handled every call to and from each *mobile unit*. In 1964, *automatic channel selection systems* were placed in service

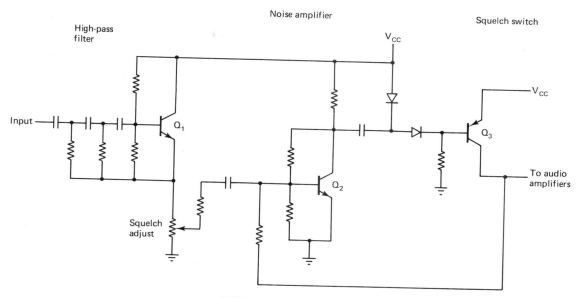

FIGURE 8-28 Squelch circuit.

for mobile telephone systems. This eliminated the need for push-to-talk operation and allowed customers to direct dial their calls without the aid of an operator. *Automatic call completion* was extended to the 450-MHz band in 1969 and *improved mobile telephone systems* (IMTS) became the United States standard mobile telephone service. Presently, there are more than 160,000 mobile telephone service (MTS) subscribers nationwide. MTS uses FM radio channels to establish communication links between mobile telephone and central base station transceivers, which are linked to the local telephone exchange via normal metallic telephone lines. Most MTS systems serve an area approximately 40 mi in diameter and each channel operates similar to a party line. Each channel may be assigned to several subscribers, but only one at a time. If the subscriber's preassigned channel is busy, the subscriber must wait until it is idle before he or she can either place or receive a call.

The growing demand for the overcrowded mobile telephone frequency spectrum prompted the FCC to issue Docket 18262, which inquired into a means for providing a higher spectrum efficiency. In 1971, AT&T submitted a proposal on the technical feasibility of providing efficient use of the mobile telephone frequency spectrum. AT&T's report, entitled *High Capacity Mobile Phone Service*, outlined the principles of cellular radio.

Cellular radio.   *Cellular radio* corrects many of the problems of traditional two-way mobile telephone service and creates a totally new environment for both mobile radio and traditional wireline telephone service. The key concepts of cellular radio were uncovered by researchers at Bell Telephone Laboratories in 1947. It was determined that by subdividing a relatively large geographical area into smaller sections called *cells*, a concept called *frequency reuse* could be employed to increase dramatically the capacity of a mobile telephone channel (reuse is explained later in the chapter). In addition, integrated-circuit technology and microprocessors have recently enabled complex radio and logic circuits to be used in electronic switching machines to store programs that provide faster and more efficient call processing.

In 1974, the FCC allocated an additional 40 MHz of bandwidth for cellular radio service (825 to 845 MHz and 870 to 890 MHz). These frequency bands were formerly allocated to UHF television channels 70 to 83. In 1975, AT&T was granted the first license to operate a developmental cellular radio service in Chicago and AT&T subsequently formed the *Advanced Mobile Phone Service* (AMPS). The following year *American Radio Telephone Service* (ARTS) was granted authorization from the FCC to install a second developmental system in the Baltimore/Washington, D.C., area.

The basic cellular radio concept is quite simple. The FCC defined geographic cellular radio coverage areas based on 1980 census figures. With the cellular concept, each area is further divided into *hexagonal cells* that fit together to form a honeycomb pattern. The hexagon shape was chosen because it provides the most effective transmission by approximating a circular pattern while eliminating gaps present between adjacent circles. The number of cells per system is not defined by the FCC and has been left to the provider to establish in accordance with anticipated traffic patterns. Each geographic mobile service area is allocated 666 cellular radio channels. Each transceiver within a covered area has a fixed subset of the 666 available radio channels based on expected traffic flow.

Figure 8-29 shows a simplified cellular telephone system which includes all of the basic components necessary for cellular radio communications. There is a radio-frequency transceiver located at the physical center of each cell. Cellular radio uses several moderately powered transceivers over a relatively wide service area, as opposed to MTS, which uses a single high-powered transceiver at high elevation. Also, each cell contains a computerized cell-site controller to handle all cell-site control functions. All cell sites are connected to an electronic switching center through a leased four-wire metallic telephone line. The electronic switching center also provides access to the public switched telephone network. Generally, each cell can accommodate up to 70 different channels simultaneously. Within a cell, each channel can support only one mobile telephone user at a time. Channels are dynamically assigned and dedicated to a single user for the duration of the call, and any user may be assigned to any channel. Called *frequency reuse*, this allows a cellular telephone system in a single area to handle considerably more than the 666 available channels. Thus cellular radio makes more efficient use of the available frequency spectrum than does traditional MTS service.

As a car moves away from the transceiver in the center of a cell, the received signal begins to decrease. When the signal strength is reduced to a predetermined level, the electronic switching center locates the cell in the *honeycomb* that is receiving the strongest signal from the mobile unit and transfers the mobile unit to the transceiver in the new cell. The transfer includes converting the call to an available frequency within the new cell's allocated channel subset. This transfer is called a *handoff* and is completely

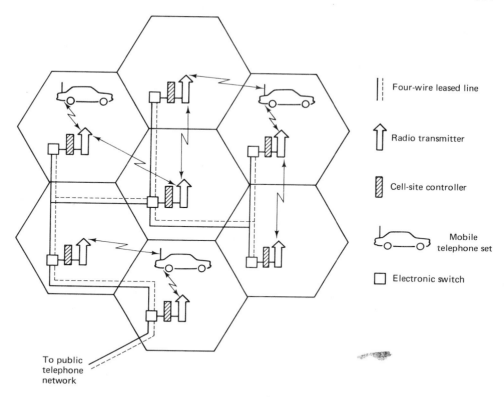

Four-wire leased line

Radio transmitter

Cell-site controller

Mobile
telephone set

Electronic switch

To public
telephone
network

FIGURE 8-29   Typical cellular telephone system.

transparent to the subscriber (i.e., the subscriber does not know that his facility has been switched). The transfer takes approximately 0.2 s, which is imperceptible to voice telephone users. However, this delay may be disruptive to data communications.

There are five primary components of a cellular radio system. They are the *electronic switching center*, a *controller*, a *radio transceiver*, *system interconnections*, and *mobile telephone units*.

The electronic switching center. The electronic switching center is a digital telephone exchange and is the heart of the system. The switch performs two essential functions: (1) it controls the switching between the public telephone network and the cell sites for all wireline-to-mobile, mobile-to-wireline, and mobile-to-mobile calls; and (2) it processes data received from the cell-site controllers concerning mobile unit status, diagnostic data, and bill compiling information.

The cell site. Each cell contains one cell-site controller that operates under the direction of the switching center. The cell-site controller manages each of the radio channels at the site, supervises calls, turns the radio transmitter and receiver on and off, injects data onto the control and user channels, and performs diagnostic tests on the cell-site equipment.

System interconnections. Four-wire leased telephone lines are generally used to connect the switching centers to each of the cell sites. There is one dedicated four-wire trunk for each of the cell's user channels. Also, there must be at least one four-wire circuit to connect the switch to the cell-site controller as a control channel.

Mobile units. A mobile telephone unit consists of a control unit, a radio transceiver, a logic unit, and a mobile antenna. The control unit houses all of the user interfaces, including a handset. The transceiver uses a frequency synthesizer to tune into any designated cellular system channel. The logic unit interprets subscriber actions and system commands and manages the transceiver and control units.

QUESTIONS

8-1. Describe the differences between an AM receiver and an FM receiver.

8-2. Draw the schematic diagram for a single-ended slope detector and describe its operation.

8-3. Draw the schematic diagram for a doubled-ended slope detector and describe its operation.

8-4. Draw the schematic diagram for a Foster–Seeley discriminator and describe its operation.

8-5. Draw the schematic diagram for a ratio detector and describe its operation.

8-6. Describe the operation of a PLL FM demodulator.

8-7. Draw the schematic diagram for a quadrature FM demodulator and describe its operation.

8-8. Contrast the advantages and disadvantages of the FM demodulators described in Questions 8-2 through 8-7.

8-9. What is the purpose of a limiter in an FM receiver?

8-10. Describe FM thresholding.

8-11. Describe the operation of an FM stereo transmitter; an FM stereo receiver.

8-12. Draw the block diagram for a two-way FM transmitter and describe its operation.

8-13. Draw the block diagram for a two-way FM receiver and describe its operation.

8-14. Describe the operation of an electronic PTT circuit.

8-15. Describe the operation of a cellular radio system.

PROBLEMS

8-1. Determine the minimum input S/N ratio required for a receiver with 15 dB of FM improvement, a noise figure $F = 4$ dB, and a desired postdetection signal-to-noise ratio $= 33$ dB.

8-2. For an FM receiver with a 100 kHz bandwidth, a noise figure $F = 6$ dB, and an input noise temperature $T = 200°C$; determine the minimum receive carrier power to achieve a postdetection $S/N = 40$ dB. Use the receiver block diagram shown in Figure 8-1 as the receiver model and the FM thresholding curve shown in Figure 8-12.

8-3. For an FM receiver tuned to 92.75 MHz using high-side injection and a first IF of 10.7 MHz, determine the image frequency. The local oscillator frequency.

8-4. For an FM receiver with an input frequency deviation $\Delta F = 40$ kHz and a transfer ratio $K = 0.01$ V/kHz, determine V_{out}.

8-5. For the balanced slope detector shown in Figure 8-3a, a center frequency $F_c = 20.4$ MHz, and a maximum input frequency deviation $\Delta F = 50$ kHz, determine the upper and lower cutoff frequencies for the tuned circuit.

8-6. For the Foster–Seeley discriminator shown in Figure 8-4, $VC_1 = 1.2$ V and $VC_2 = 0.8$ V, determine V_{out}.

8-7. For the ratio detector shown in Figure 8-6, $VC_1 = 1.2$ V and $VC_2 = 0.8$ V, determine V_{out}.

8-8. For an FM demodulator with an FM improvement factor equal to 23 dB and an input (predetection) signal-to-noise ratio $S_i/N_i = 26$ dB, determine the postdetection S/N.

8-9. From Figure 8-12, determine the approximate FM improvement factor for an input $S/N = 10.5$ dB and $m = 1$.

Chapter 9

TRANSMISSION LINES

INTRODUCTION

A *transmission line* is a *metallic conductor system* that is used to transfer electrical energy from one point to another. More specifically, a transmission line is two or more conductors separated by an insulator, such as a pair of wires or a system of wire pairs. A transmission line can be as short as a few inches or it can span several thousand miles. Transmission lines can be used to propagate dc or low-frequency ac (such as 60 cycle power and audio signals); they can also be used to propagate very high frequencies (such as intermediate and radio-frequency signals). When propagating low-frequency signals, transmission-line behavior is rather simple and quite predictable. However, when propagating high-frequency signals, the characteristics of transmission lines become more involved and their behavior somewhat peculiar to a student of lumped constant circuits and systems.

TRANSVERSE ELECTROMAGNETIC WAVES

Propagation of electrical power along a transmission line occurs in the form of *transverse electromagnetic* (TEM) *waves*. A wave is an *oscillatory motion*. The vibration of a particle excites similar vibrations in nearby particles. A TEM wave propagates primarily in the nonconductor (dielectric) that separates the two conductors of a transmission line. Therefore, a wave travels or propagates itself through a medium. A transverse wave is a wave in which the direction of displacement is perpendicular to the direction of propagation. A surface wave of water is a transverse wave. A wave in which the displacement is in the direction of propagation is called a *longitudinal wave*. Sound waves are longitudinal. An electromagnetic (EM) wave is a wave produced by the acceleration of an electric charge. In a conductor, current and voltage are always accompanied by an electric (*E*) and a magnetic (*H*) field in the adjoining region of space. Figure 9-1a shows the spatial relationships between the *E* and *H* fields of an electromagnetic wave. Figure 9-1b shows the cross-sectional views of the *E* and *H* fields that

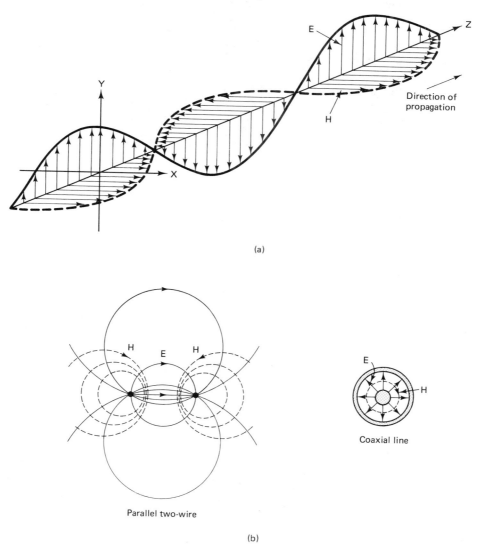

(a)

Parallel two-wire

Coaxial line

(b)

FIGURE 9-1 (a) Spatial and (b) cross-sectional views showing the relative displacement of the *E* and *H* fields on a transmission line.

surround a parallel two-wire and a coaxial line. It can be seen that the E and H fields are perpendicular to each other (at 90° angles) at all points. This is referred to as *space quadrature*. Electromagnetic waves that travel along a transmission line from the source toward the load are called *incident waves*, and those that travel from the load back toward the source are called *reflected waves*.

Characteristics of Electromagnetic Waves

Wave velocity. Waves travel at various speeds, depending on the type of wave and the characteristics of the propagation medium. Sound waves travel at approximately 1100 f/s in the normal atmosphere. Electromagnetic waves travel much faster. In free space (a vacuum), TEM waves travel at the speed of light, $c = 186,283$ statute mi/s or 299,793 m/s, rounded off to 186,000 mi/s and 3×10^8 m/s. However, in air (such as the earth's atmosphere), TEM waves travel slightly more slowly, and along a transmission line, electromagnetic waves travel considerably more slowly.

Frequency and wavelength. The oscillations of an electromagnetic wave are periodic and repetitious. Therefore, they are characterized by a frequency. The rate at which the periodic wave repeats is its frequency. The distance of one cycle occurring in space is called the *wavelength* and is determined from the following fundamental equation:

$$\text{distance} = \text{velocity} \times \text{time} \qquad (9\text{-}1)$$

If the time for one cycle is substituted into Equation 9-1, we get the length of one cycle, which is called the wavelength and whose symbol is the Greek lowercase letter lambda (λ).

$$\lambda = \text{velocity} \times \text{period}$$
$$= v \times T$$

and since $T = 1/F$,

$$\lambda = \frac{v}{F} \qquad (9\text{-}2)$$

for free-space propagation, $v = c$; therefore, the length of one cycle is

$$\lambda = \frac{c}{F} = \frac{3 \times 10^8 \text{ m/s}}{F \text{ cycles/s}} = \frac{\text{meters}}{\text{cycle}} \qquad (9\text{-}3)$$

Figure 9-2 shows a graph of the displacement and velocity of a longitudinal wave as it propagates along a transmission line from a source to a load. The horizontal (x) axis is distance and the vertical (y) axis is displacement. One wavelength is the distance covered by one cycle of the wave. It can be seen that the wave moves to the right or propagates down the line with time. If a voltmeter is placed at any stationary point on the line, the voltage measured will fluctuate from 0 to maximum positive, back to zero, to maximum negative, back to zero again, and then the cycle repeats.

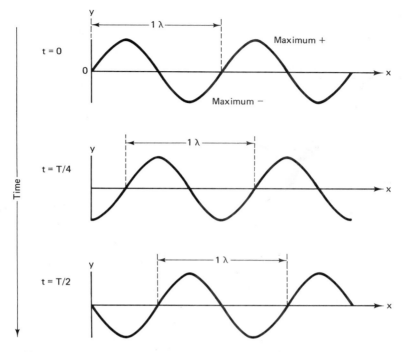

FIGURE 9-2 Displacement and velocity of a longitudinal wave as it propagates down a transmission line.

TYPES OF TRANSMISSION LINES

Transmission lines can be generally classified as balanced or unbalanced. With a *balanced transmission line*, both conductors carry current and the current in each wire is 180° out of phase with the current in the other wire. With an *unbalanced line*, one wire is at ground potential, while the other wire carries all of the current. Both conductors in a balanced line carry current and the signal currents are equal magnitude with respect to electrical ground but travel in opposite directions. Currents that flow in opposite directions in a balanced wire pair are called *metallic circuit currents*. Currents that flow in the same direction are called *longitudinal currents*. A balanced pair has the advantage that most noise interference is induced equally in both wires, producing longitudinal currents that cancel in the load. Figure 9-3 shows the results of metallic and longitudinal currents on a balanced transmission line. It can be seen that longitudinal currents (generally produced by static interference) cancel in the load. Balanced transmission lines can be connected to unbalanced loads, and vice versa, with special transformers called *baluns* (*bal*anced to *un*balanced).

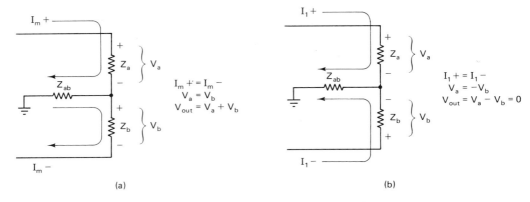

FIGURE 9-3 **Results of metallic and longitudinal currents on a balanced transmission line:
(a) metallic currents due to signal voltages; (b) longitudinal currents due to noise voltages.**

Parallel-Conductor Transmission Lines

Open-wire transmission line.   An *open-wire transmission line* is a *two-wire parallel conductor* and is shown in Figure 9-4a. It consists simply of two parallel wires, closely spaced and separated by air. Nonconductive spacers are placed at periodic intervals for support and to keep the distance between the conductors constant. The distance between the two conductors is generally between 2 and 6 in. The dielectric is simply the air between and around the two conductors in which the TEM wave propagates. The only real advantage of this type of transmission line is its simple construction. Because there is no shielding, radiation losses are high and it is susceptible to noise pickup. These are the primary disadvantages of an open-wire transmission line. Therefore, open-wire transmission lines are normally operated in the balanced mode.

Twin-lead.   *Twin-lead* is another form of two-wire parallel conductor transmission line and is shown in Figure 9-4b. Twin-lead is often called *ribbon cable*. Twin-lead is essentially the same as an open-wire transmission line except that the spacers between the two conductors are replaced with a continuous solid dielectric. This assures uniform spacing along the entire cable, which is a desirable characteristic for reasons that are explained later in the chapter. Typically, the distance between the two conductors is $\frac{5}{16}$ in. Common dielectric materials are Teflon and polyethylene.

Twisted-pair cable.   A *twisted-pair cable* is formed by twisting together two insulated conductors. Pairs are often stranded in *units*, and the units are then cabled into *cores*. The cores are covered with various types of *sheaths*, depending on their intended use. Neighboring pairs are twisted with different *pitch* (twist length) to reduce interference between pairs due to mutual induction. The *primary constants* of twisted-pair cable are its electrical parameters (resistance, inductance, capacitance, and conductance), which are subject to variations with the physical environment such as temperature,

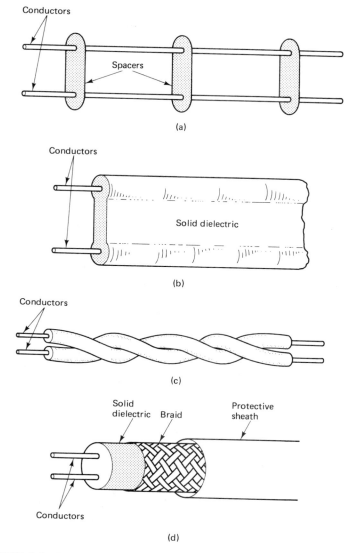

FIGURE 9-4 Transmission lines: (a) open-wire; (b) twin-lead; (c) twisted pair; (d) shielded pair.

moisture, and mechanical stress and are dependent on manufacturing deviations. A twisted-pair cable is shown in Figure 9-4c.

Shielded cable pair. To reduce radiation losses and interference, parallel two-wire transmission lines are often enclosed in a conductive metal *braid*. The braid is connected to ground and acts like a shield. The braid also prevents signals from radiating

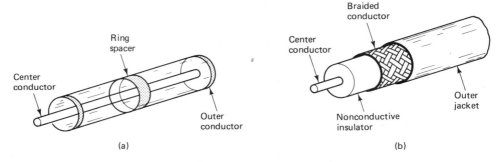

FIGURE 9-5   Concentric or coaxial transmission lines; (a) rigid air filled; (b) solid flexible line.

beyond its boundaries and keeps electromagnetic interference from reaching the signal conductors. A shielded parallel wire pair is shown in Figure 9-4d. It consists of two parallel wire conductors separated by a solid dielectric material. The entire structure is enclosed in a braided conductive tube, then covered with a protective plastic coating.

Concentric or Coaxial Transmission Lines

Parallel-conductor transmission lines are suitable for low-frequency applications. However, at high frequencies, their radiation and dielectric losses, as well as their susceptibility to external interference, are excessive. Therefore, *coaxial conductors* are used extensively for high-frequency applications, to reduce losses and to isolate transmission paths. The basic coaxial cable consists of a center conductor surrounded by a *concentric* (uniform distance from the center) *outer conductor*. At relatively high operating frequencies, the coaxial outer conductor provides excellent shielding against external interference. However, at lower frequencies, the shielding is ineffective. Also, a coaxial cable's outer conductor is generally grounded, which limits its use to unbalanced applications.

Essentially, there are two types of coaxial cables: *rigid air filled* or *solid flexible* lines. Figure 9-5a shows a rigid air coaxial line. It can be see that the center conductor is surrounded coaxially by a tubular outer conductor and the insulating material is air. The outer conductor is physically isolated and separated from the center conductor by a spacer, which is generally made of Pyrex, polystyrene, or some other nonconductive material. Figure 9-5b shows a solid flexible coaxial cable. The outer conductor is braided, flexible, and coaxial to the center conductor. The insulating material is a solid nonconductive polyethylene material that provides both support and electrical isolation between the inner and outer conductors. The inner conductor is a flexible copper wire that can either be solid or hollow.

Rigid air-filled coaxial cables are relatively expensive to manufacture, and to minimize losses, the air insulator must be relatively free of moisture. Solid coaxial cables have lower losses, are easier to construct, and are easier to install and maintain. Both types of coaxial cables are relatively immune to external radiation, radiate little themselves, and can operate at higher frequencies than can their parallel-wire counterparts.

The basic disadvantages of coaxial transmission lines is that they are expensive and must be used in the unbalanced mode.

TRANSMISSION-LINE EQUIVALENT CIRCUIT

Uniformly Distributed Lines

The characteristics of a transmission line are determined by its electrical properties, such as wire conductivity and insulator dielectric constant, and its physical properties; such as wire diameter and conductor spacing. These properties, in turn, determine the primary electrical constants: series dc resistance (R), series inductance (L), shunt capacitance (C), and shunt conductance (G). Resistance and inductance occur along the line, whereas capacitance and conductance occur between the two conductors. The primary constants are uniformly distributed throughout the length of the line and are therefore commonly called distributed parameters. To simplify analysis, distributed parameters are commonly *lumped* together per a given unit length to form an artificial electrical model of the line. For example, series resistance is generally given in ohms per mile or kilometer.

Figure 9-6 shows the electrical equivalent circuit for a metallic two-wire transmission line showing the relative placement of the various lumped parameters. The conductance between the two wires is shown in reciprocal form and given as a shunt leakage resistance (R_s).

Transmission Characteristics

The transmission characteristics of a transmission line are called secondary constants and are determined from the four primary constants. The secondary constants are characteristic impedance and propagation constant.

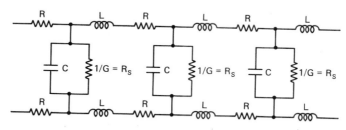

C = capacitance — two conductors separated
    by an insulator
R = resistance — opposition to current flow
L = current-carrying conductor that enables
    the setting up of a magnetic field
1/G = leakage resistance of dielectric
R_s = shunt leakage resistance

FIGURE 9-6   Two-wire parallel transmission line — electrical equivalent circuit.

Characteristic impedance.   For maximum power transfer from the source to the load (i.e., no reflected energy), a transmission line must be terminated in a purely resistive load equal to the *characteristic impedance* of the line. The characteristic impedance (Z_o) of a transmission line is a complex ac quantity which is expressed in ohms, is totally independent of both length and frequency, and cannot be measured directly. Characteristic impedance (which is sometimes called *surge impedance*) is defined as the impedance seen looking into an infinitely long line or the impedance seen looking into a finite length of line which is terminated in a purely resistive load equal to the characteristic impedance of the line. A transmission line stores energy in its distributed inductance and capacitance. If the line is infinitely long, it can store energy indefinitely; energy from the source is entering the line and none is returned. Therefore, the line acts like a resistor that dissipates all of the energy. An infinite line can be simulated if a finite line is terminated in a purely resistive load equal to Z_o; all of the energy that enters the line from the source is dissipated in the load (this assumes a totally lossless line).

Figure 9–7 shows a single section of a transmission line terminated in a load Z_L that is equal to Z_o. The impedance seen looking into a line of n such sections is determined from the following expression:

$$Z_o^2 = Z_1 Z_2 + \frac{ZL^2}{n} \tag{9-4}$$

where n is the number of sections. For an infinite number of sections, ZL^2/n approaches 0 if

$$\lim \frac{ZL^2}{n}\bigg|_{n \to \infty} = 0$$

then

$$Z_o = \sqrt{Z_1 Z_2}$$

where

$$Z_1 = R + j\omega L$$

$$\frac{1}{Z_2} = \frac{1}{R_s} + \frac{1}{1/j\omega C}$$

$$\frac{1}{Z_2} = G + j\omega C$$

$$Z_2 = \frac{1}{G} + \frac{1}{j\omega C} = \frac{1}{G + j\omega C}$$

Therefore,

$$Z_o = \sqrt{(R + j\omega L)\frac{1}{G + j\omega C}}$$

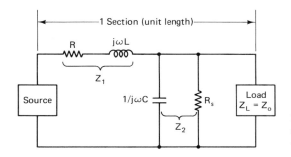

FIGURE 9-7 Equivalent circuit for a single section of transmission line terminated in a load equal to Z_o.

or

$$Z_o = \sqrt{\frac{R + j\omega L}{G + j\omega C}} \tag{9-5}$$

For extremely low frequencies, the resistances dominate and Equation 9-5 simplifies to

$$Z_o = \sqrt{\frac{R}{G}} \tag{9-6}$$

For extremely high frequencies, the inductance and capacitance dominate and Equation 9-5 simplifies to

$$Z_o = \sqrt{\frac{j\omega L}{j\omega C}} = \sqrt{\frac{L}{C}} \tag{9-7}$$

From Equation 9-7 it can be seen that for high frequencies the characteristic impedance of a transmission line approaches a constant, is independent of both frequency and length, and is determined solely by the distributed inductance and capacitance. It can also be seen that the phase angle is 0°. Therefore, Z_o looks purely resistive and all of the incident energy is absorbed by the line.

From a purely resistive approach, it can easily be seen that the impedance seen looking into a transmission line made up of an infinite number of sections approaches the characteristic impedance. This is shown in Figure 9-8. Again, for simplicity, only

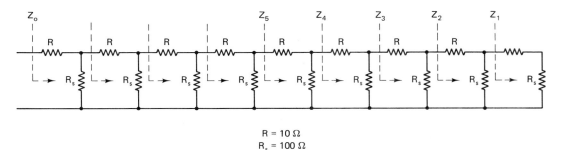

R = 10 Ω
R_s = 100 Ω

FIGURE 9-8 Characteristic impedance of a transmission line of infinite sections or terminated in load equal to Z_o.

the series resistance R and the shunt resistance R_s are considered. The impedance seen looking into the last section of the line is simply the sum of R and R_s. Mathematically, Z_1 is

$$Z_1 = R + R_s = 10 + 100 = 110$$

Adding a second section, Z_2, gives

$$Z_2 = R + \frac{R_s Z_1}{R_s + Z_1} = 10 + \frac{10 \times 110}{10 + 110} = 10 + 52.38 = 62.38$$

and a third section, Z_3, is

$$Z_3 = R + \frac{R_s Z_2}{R_s + Z_2}$$

$$= 10 + \frac{10 \times 62.38}{10 + 62.38} = 10 + 38.42 = 48.32$$

A fourth section, Z_4, is

$$Z_4 = 10 + \frac{10 \times 48.32}{10 + 48.32} = 10 + 32.62 = 42.62$$

It can be seen that after each additional section, the total impedance seen looking into the line decreases from its previous value. However, each time the magnitude of the decrease is less than the previous value. If the process shown above were continued, the impedance seen looking into the line decreases asymptotically toward 37 Ω, which is the characteristic impedance of the line.

If the transmission line shown in Figure 9-8 were terminated in a load resistance $Z_L = 37$ Ω, the impedance seen looking into any number of sections would equal 37 Ω, the characteristic impedance. For a single section of line, Z_o is

$$Z_o = Z_1 = R + \frac{R_s \times Z_L}{R_s + Z_L} = 10 + \frac{100 \times 37}{100 + 37} = 10 + \frac{3700}{137} = 37 \ \Omega$$

Adding a second section, Z_2, is

$$Z_o = Z_2 = R + \frac{R_s \times Z_1}{R_s + Z_1} = 10 + \frac{100 \times 37}{100 + 37} = 10 + \frac{3700}{137} = 37 \ \Omega$$

Therefore, if this line were terminated into a load resistance $Z_L = 37$ Ω, $Z_o = 37$ Ω no matter how many sections are included.

The characteristic impedance of a transmission line can also be determined using Ohm's law. When a source is connected to an infinitely long line and a voltage is applied, a current flows. Even though the load is open, the circuit is complete through the distributed constants of the line. The characteristic impedance is simply the ratio of source voltage (E_o) to line current (I_o). Mathematically, Z_o is

$$Z_o = \frac{E_o}{I_o} \tag{9-8}$$

The characteristic impedance of a two-wire parallel transmission line with an air dielectric can be determined from its physical dimensions (see Figure 9-9a) and the formula

$$Z_o = 276 \log \frac{D}{r} \tag{9-9}$$

where

D = distance between the centers of the two conductors
r = radius of the conductor

and $D \gg r$.

EXAMPLE 9-1

Determine the characteristic impedance for an air dielectric two-wire parallel transmission line with a D/r ratio = 12.22.

Solution   Substituting into Equation 9-9, we obtain

$$Z_o = 276 \log 12.22 = 300 \ \Omega$$

The characteristic impedance of a concentric coaxial cable can also be determined from its physical dimensions (see Figure 9-9b) and the formula

$$Z_o = \frac{138}{\sqrt{\epsilon_r}} \log \frac{D}{d} \tag{9-10}$$

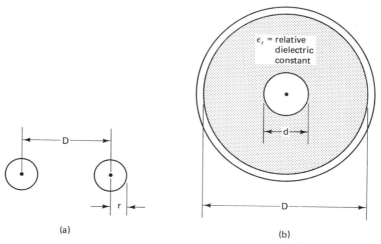

FIGURE 9-9   Physical dimensions of transmission lines: (a) two-wire parallel transmission line; (b) coaxial cable transmission line.

where

D = inside diameter of the outer conductor
d = outside diameter of the inner conductor
ϵ_r = relative dielectric constant of the insulating material

---

EXAMPLE 9-2

Determine the characteristic impedance for a RG-59A coaxial cable with the following specifications: $L = 0.118$ μH/ft, $C = 21$ pF/ft, $d = 0.25$ in., $D = 0.87$ in., and $\epsilon = 1$.

Solutions   Substituting into Equation 9-8 yields

$$Z_o = \sqrt{\frac{L}{C}} = \sqrt{\frac{0.118 \times 10^{-6}\,\text{H/ft}}{21 \times 10^{-12}\,\text{F/ft}}} = 75\,\Omega$$

Substituting into Equation 9-10 gives us

$$Z_o = \frac{138}{\sqrt{l}} \log \frac{0.87\,\text{in.}}{0.25\,\text{in.}} = 75\,\Omega$$

Transmission lines can be summarized thus far as follows:

1. The input impedance of an infinitely long line at radio frequencies is resistive and equal to Z_o.
2. Electomagnetic waves travel down the line without reflections; such a line is called nonresonant.
3. The ratio of voltage to current at any point along the line is equal to Z_o.
4. The incident voltage and current at any point along the line are in phase.
5. Line losses on a nonresonant line are minimum per unit length.
6. Any transmission line that is terminated in a purely resistive load equal to Z_o acts like an infinite line.
   a. $Z_i = Z_o$.
   b. There are no reflected waves.
   c. V and I are in phase.
   d. There is maximum transfer of power from source to load.

Propagation constant. *Propagation constant* (sometimes called *propagation coefficient*) is used to express the attenuation (signal loss) and the phase shift per unit length of a transmission line. As a wave propagates down a transmission line, its amplitude decreases with distance traveled. The propagation constant is used to determine the reduction in voltage or current with distance as a TEM wave propagates down a transmission line. For an infinitely long line, all of the incident power is dissipated in the resistance of the wire as the wave propagates down the line. Therefore, with an infinitely long line or a line that looks infinitely long, such as a finite line terminated in a matched load ($Z_o = Z_L$), no energy is returned or reflected back toward the source. Mathematically, the propagation constant is

$$\gamma = \alpha + j\beta \qquad (9\text{-}11\text{a})$$

where

γ = propagation constant
α = attenuation coefficient (nepers per unit length)
β = phase shift coefficient (radians per unit length)

and since the propagation constant is a complex quantity,

$$\gamma = \sqrt{(R + j\omega L)(G + j\omega C)} \qquad (9\text{-}11\text{b})$$

Since a phase shift of 2π rad occurs over a distance of one wavelength, then

$$\beta = \frac{2\pi}{\lambda} \qquad (9\text{-}12)$$

At intermediate and radio frequencies, $\omega L > R$ and $\omega C > G$; thus

$$\alpha = \frac{R}{2Z_o} + \frac{GZ_o}{2} \qquad (9\text{-}13)$$

and

$$\beta = \omega\sqrt{LC} \qquad (9\text{-}14)$$

The current and voltage distribution along a transmission line that is terminated in a load equal to its characteristic impedance (i.e., a matched line) are determined from the formula

$$I = I_s e^{-l\gamma} \qquad (9\text{-}15)$$

$$V = V_s e^{-l\gamma} \qquad (9\text{-}16)$$

where

I_s = current at the source end of the line
V_s = voltage at the source end of the line
γ = propagation constant
l = length of the line at which the current or voltage is determined

For a matched load ($Z_L = Z_o$), the loss in signal voltage or current and phase shift for a given length of cable is equal to γl.

TRANSMISSION-LINE WAVE PROPAGATION

As stated previously, electromagnetic waves travel at the speed of light when propagating through a vacuum, and nearly at the speed of light when propagating through air. However, in metallic transmission lines where the conductor is generally copper and the dielectric materials vary considerably with cable type, an electromagnetic wave travels much more slowly.

Velocity Factor

Velocity factor (sometimes called *velocity constant*) is defined simply as the ratio of the actual velocity of propagation through a given medium to the velocity of propagation through free space. Mathematically, the velocity factor is

$$V_f = \frac{V_p}{c} \qquad (9\text{-}17)$$

where

V_f = velocity factor
V_p = actual velocity of propagation
c = velocity of propagation through free space, $c = 3 \times 10^8$ m/s

and

$$V_f \times c = V_p$$

The velocity at which an electromagnetic wave travels through a transmission line is dependent on the dielectric constant of the insulating material separating the two conductors. The velocity factor is closely approximated with the formula

$$V_f = \frac{1}{\sqrt{\epsilon_r}} \qquad (9\text{-}18)$$

where ϵ_r is the dielectric constant of a given material relative to the dielectric constant of a vacuum (ϵ/ϵ_o).

Dielectric constant is simply the *permittivity* of a material. The relative dielectric constant of air is 1.0006. However, the relative dielectric constant of materials commonly used in transmission lines range from 1.2 to 2.8, giving velocity factors from 0.6 to 0.9. The velocity factors for several common transmission-line configurations are given in Table 9-1, and the relative dielectric constants for several insulating materials are listed in Table 9-2.

The dielectric constant of a material is dependent on the primary constants inductance and capacitance. Inductors store magnetic energy and capacitors store electric

TABLE 9-1   VELOCITY FACTORS

Material	Velocity factor
Air	0.95–0.975
Rubber	0.56–0.65
Polyethylene	0.66
Teflon	0.70
Teflon foam	0.82
Teflon pins	0.81
Teflon spiral	0.81

TABLE 9-2   RELATIVE DIELECTRIC CONSTANTS

Material	Relative dielectric constant
Vacuum	1.0
Air	1.0006
Teflon	2.0
Paper, paraffined	2.5
Rubber	3.0
Mica	5.0
Glass	7.5

energy. It takes a finite amount of time for an inductor or a capacitor to take on or give off energy. Therefore, the velocity at which an electromagnetic wave propagates along a transmission line varies with the inductance and capacitance of the cable. It can be shown that time $T = \sqrt{LC}$. Therefore, inductance, capacitance, and velocity of propagation are mathematically related by the formula

$$\text{velocity} \times \text{time} = \text{distance}$$

Therefore,

$$V_p = \frac{\text{distance}}{\text{time}} = \frac{D}{T} \qquad (9\text{-}19)$$

Substituting for time yields

$$V_p = \frac{D}{\sqrt{LC}} \qquad (9\text{-}20)$$

If distance is normalized to 1 m, the velocity of propagation for a lossless line is

$$V_p = \frac{1 \text{ m}}{\sqrt{LC}} = \frac{1}{\sqrt{LC}} \qquad \text{m/s} \qquad (9\text{-}21)$$

EXAMPLE 9-3

For a given length of RG8A/U coaxial cable with a distributed capacitance $C = 96.6$ pF/m, a distributed inductance $L = 241.56$ nH/m, and a relative dielectric constant $\epsilon_r = 2.3$; determine the velocity of propagation and the velocity factor.

Solution   From Equation 9-21,

$$V_p = \frac{1}{\sqrt{96.6 \times 10^{-12} \times 241.56 \times 10^{-6}}}$$

$$= 2.07 \times 10^8 \text{ m/s}$$

From Equation 9-17,

$$V_f = \frac{2.07 \times 10^8 \text{ m/s}}{3 \times 10^8 \text{ m/s}} = 0.69$$

From Equation 9-18,

$$V_p = \frac{1}{\sqrt{2.3}} = 0.66$$

Because wavelength is directly proportional to velocity and the velocity of propagation of a TEM wave varies with dielectric constant, the wavelength of a TEM wave also varies with dielectric constant. Therefore, for transmission media other than free space, Equation 9-3 can be rewritten as

$$\lambda = \frac{V_p}{F} = \frac{cV_f}{F} = \frac{c}{F\sqrt{\epsilon_r}} \tag{9-22}$$

Electrical Length of a Transmission Line

The length of a transmission line relative to the length of the wave propagating down it is an important consideration when analyzing transmission line behavior. At low frequencies (i.e., long wavelengths), the voltage along the line remains relatively constant. However, for high frequencies, several wavelengths of the signal may be present on the line at the same time. Therefore, the voltage along the line may vary appreciably. Consequently, the length of a transmission line is often given in wavelengths rather than in linear dimensions. Transmission-line phenomena apply to long lines. Generally, a transmission line is defined as long if its length exceeds one-sixteenth of a wavelength; otherwise, it is considered short. A given length of transmission line may appear short at one frequency and long at another frequency. For example, a 10-m length of transmission line at 1000 Hz is short (λ = 300,000 m; 10 m is only a small fraction of a wavelength). However, the same line at 6 HGz is long (λ = 5 cm; the line is 200 wavelengths long). It will be apparent later in this chapter, in Chapter 10, and in Appendix A that electrical length is used extensively for transmission line calculations and antenna design.

TRANSMISSION-LINE LOSSES

For analysis purposes, transmission lines are often considered totally lossless. In reality, however, there are several ways in which power is lost in a transmission line. They are conductor loss, radiation loss, dielectric heating loss, coupling loss, and corona.

Conductor Loss

Because current flows through a transmission line and the transmission line has a finite resistance, there is an inherent and unavoidable power loss. This is sometimes called *conductor* or *conductor heating loss* and is simply an I^2R loss. Because resistance is distributed throughout a transmission line, conductor loss is directly proportional to line length. Also, because power dissipation is directly proportional to current, conductor loss is inversely proportional to characteristic impedance. To reduce conductor loss, simply shorten the transmission line or use a larger-diameter wire (keep in mind that changing the wire diameter also changes the characteristic impedance and consequently the current).

Conductor loss is somewhat dependent on frequency. This is because of an action called the *skin effect*. When current flows through an isolated round wire, the magnetic flux associated with it is in the form of concentric circles. This is shown in Figure 9-10. It can be seen that the flux density near the center of the conductor is greater than it is near the surface. Consequently, the lines of flux near the center of the conductor encircle the current and reduce the mobility of the encircled electrons. This is a form of self-induction and causes the inductance near the center of the conductor to be greater than at the surface. Therefore, at radio frequencies, most of the current flows along the surface (outer skin) rather than near the center of the conductor. This is equivalent

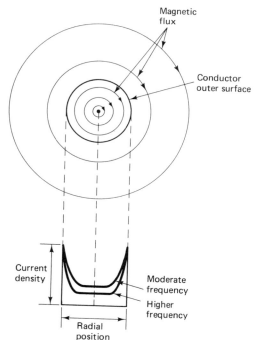

FIGURE 9-10   Isolated round conductor showing magnetic lines of flux, current distributions, and the skin effect.

to reducing the cross-sectional area of the conductor and increasing the opposition to current flow (i.e., resistance). The additional opposition has a 0° phase angle and is therefore a resistance and not a reactance. Therefore, the ac resistance of the conductor is directly proportional to frequency. The ratio of the ac resistance to the dc resistance of a conductor is called the resistance ratio. Above approximately 100 MHz, the center of a conductor can be completely removed and have absolutely no effect on the total conductor loss or EM wave propagation.

Conductor loss in transmission lines varies from as low as a fraction of a decibel per 100 m for rigid air dielectric coaxial cable to as high as 200 dB per 100 m for solid dielectric flexible line.

Radiation Loss

If the separation between conductors in a transmission line is an appreciable fraction of a wavelength, the electrostatic and electromagnetic fields that surround the conductor cause the line to act like an antenna and transfer energy to nearby conductors. The amount of energy radiated depends on the dielectric material, the conductor spacing, and the length of the line. *Radiation losses* are reduced by properly shielding the cable. Therefore, coaxial cables have less radiation loss than do two-wire parallel lines. Radiation loss is also directly proportional to frequency.

Dielectric Heating Loss

A difference of potential between the two conductors of a transmission line causes *dielectric heating*. Heat is a form of energy and must be taken from the energy propagating down the line. For air dielectric lines, the heating loss is negligible. However, for solid lines, dielectric heating loss increases with frequency.

Coupling Loss

Coupling loss occurs whenever a connection is made to or from a transmission line or when two separate pieces of transmission line are connected together. Mechanical connections are discontinuities (places where dissimilar materials meet). Discontinuities tend to heat up, radiate energy, and dissipate power.

Corona

Corona is arcing that occurs between the two conductors of a transmission line when the difference of potential between them exceeds the *breakdown* voltage of the dielectric insulator. Generally, once corona has occurred, the transmission line is destroyed.

INCIDENT AND REFLECTED WAVES

An ordinary transmission line is bidirectional; power can propagate equally well in both directions. Voltage that propagates from the source toward the load is called *incident voltage*, and voltage that propagates from the load toward the source is called *reflected*

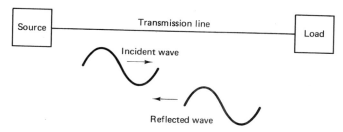

FIGURE 9-11 Source, load, transmission line, and their corresponding incident and reflected waves.

voltage. Similarly, there are incident and reflected currents. Consequently, incident power is power that propagates toward the load, and reflected power is power that propagates toward the source. Incident voltage and current are always in phase. For an infinitely long line, all of the incident power is absorbed by the line and there is no reflected power. Also, if the line is terminated in a purely resistive load equal to the characteristic impedance of the line, the load absorbs all of the incident power (this assumes a lossless line). For a more practical definition, reflected power is the portion of the incident power that was not absorbed by the load. Therefore, the reflected power can never exceed the incident power.

Resonant and Nonresonant Lines

A line with no reflected power is called a *flat* or *nonresonant line*. On a flat line, the voltage and current are constant throughout its length, assuming no losses. When the load is either a short or an open circuit, all of the incident power is reflected back toward the source. If the source were replaced with an open or a short and the line were lossless, energy present on the line would reflect back and forth (oscillate) between the load and source ends similar to the power in a tank circuit. This is called a *resonant line*. In a resonant line, the energy is alternately transferred between the magnetic and electric fields of the distributed inductance and capacitance. Figure 9-11 shows a source, transmission line, and load with their corresponding incident and reflected waves.

Reflection Coefficient

The reflection coefficient (sometimes called the *coefficient of reflection*) is a vector quantity that represents the ratio of incident voltage to reflected voltage or incident current to reflected current. Mathematically, the reflection coefficient is

$$\Gamma = \frac{E_r}{E_i} \qquad \text{or} \qquad \frac{I_r}{I_i} \qquad (9\text{-}23)$$

where

Γ = reflection coefficient
E_i = incident voltage

E_r = reflected voltage
I_i = incident current
I_r = reflected current

From Equation 9-23 it can be seen that the maximum and worst-case value for Γ is 1 ($E_r = E_i$) and the minimum value and ideal condition is when $\Gamma = 0$ ($E_r = 0$).

STANDING WAVES

When $Z_o = Z_L$, all of the incident power is absorbed by the load. This is called a *matched line*. When $Z_o \neq Z_L$, some of the incident power is absorbed by the load and some is returned (reflected) to the source. This is called an *unmatched* or *mismatched line*. With a mismatched line, there are two electromagnetic waves, traveling in opposite directions, present on the line at the same time (these waves are in fact called *traveling waves*). The two traveling waves set up an interference pattern known as a *standing wave*. This is shown in Figure 9-12. As the incident and reflected waves pass each other, a stationary voltage and current are produced on the line. These stationary waves are called standing waves because they appear to remain in a fixed position on the line, varying only in amplitude. The standing wave has minima (nodes) and maxima (antinodes), which are separated by half of a wavelength of the traveling waves.

Standing-Wave Ratio

The *standing-wave ratio* (SWR) is defined as the ratio of the maximum voltage to the minimum voltage or the maximum current to the minimum current of a standing wave on a transmission line. SWR is often called the *voltage standing-wave ratio* (VSWR).

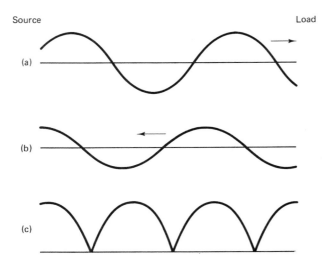

FIGURE 9-12 Developing a standing wave on a transmission line: (a) incident wave; (b) reflected wave; (c) standing wave.

Essentially, SWR is a measure of the mismatch between the load impedance and the characteristic impedance of the transmission line. Mathematically, SWR is

$$\text{SWR} = \frac{V_{\max}}{V_{\min}} \tag{9-24}$$

The voltage maxima ($V_{\max}$) occur when the incident and reflected waves are in phase (i.e., their maximum peaks pass the same point on the line with the same polarity), and the voltage minima ($V_{\min}$) occur when the incident and reflected waves are 180° out of phase. Mathematically, $V_{\max}$ and $V_{\min}$ are

$$V_{\max} = E_i + E_r \tag{9-25}$$
$$V_{\min} = E_i - E_r \tag{9-26}$$

Therefore, Equation 9-24 can be rewritten as

$$\text{SWR} = \frac{V_{\max}}{V_{\min}} = \frac{E_i + E_r}{E_i - E_r} \tag{9-27}$$

From Equation 9-27 it can be seen that when the incident and reflected waves are equal in amplitude (a total mismatch), SWR = infinity. This, of course, is the worst-case condition. Also, from Equation 9-27, it can be seen that when there is no reflected wave ($E_r = 0$), SWR = E_i/E_i or 1. This condition occurs when $Z_o = Z_L$ and is the ideal situation.

The standing-wave ratio can also be written in terms of Γ. Rearranging Equation 9-23 yields

$$\Gamma E_i = E_r$$

Substituting into Equation 9-27 gives us

$$\text{SWR} = \frac{E_i + E_i \Gamma}{E_i - E_i \Gamma}$$

Factoring out E_i yields

$$\text{SWR} = \frac{E_i(1 + \Gamma)}{E_i(1 - \Gamma)} = \frac{1 + \Gamma}{1 - \Gamma} \tag{9-28}$$

Cross-multiplying gives

$$\text{SWR}(1 - \Gamma) = 1 + \Gamma$$
$$\text{SWR} - \text{SWR }\Gamma = 1 + \Gamma$$
$$\text{SWR} = 1 + \Gamma + \text{SWR}$$
$$\text{SWR} - 1 = \Gamma(1 + \text{SWR}) \tag{9-29}$$

$$\Gamma = \frac{\text{SWR} - 1}{\text{SWR} + 1} \tag{9-30}$$

EXAMPLE 9-4

For a transmission line with incident voltage $E_i = 5$ V_p and reflected voltage $E_r = 3$ V_p, determine:

    (a) The reflection coefficient.

    (b) The SWR.

Solution   (a) Substituting into Equation 9-23 yields

$$\Gamma = \frac{E_r}{E_i} = \frac{3}{5} = 0.6$$

(b) Substituting into Equation 9-27 gives us

$$\text{SWR} = \frac{E_i + E_r}{E_i - E_r} = \frac{5 + 3}{5 - 3} = \frac{8}{2} = 4$$

Substituting into Equation 9-30, we obtain

$$\Gamma = \frac{4 - 1}{4 + 1} = \frac{3}{5} = 0.6$$

When the load is purely resistive, SWR can also be expressed as a ratio of the characteristic impedance to the load impedance, or vice versa. Mathematically, SWR is

$$\text{SWR} = \frac{Z_o}{Z_L} \qquad \text{or} \qquad \frac{Z_L}{Z_o} \quad \text{(whichever gives an SWR greater than 1)} \qquad (9\text{-}31)$$

The numerator and denominator for Equation 9-31 are chosen such that the SWR is always a number greater than 1, to avoid confusion and comply with the convention established in Equation 9-27. From Equation 9-31 it can be seen that a load resistance $Z_L = 2Z_o$ gives the same SWR as a load resistance $Z_L = Z_o/2$; the degree of mismatch is the same.

The disadvantages of not having a matched (flat) transmission line can be summarized as follows:

1. One hundred percent of the source incident power does not reach the load.
2. The dielectric separating the two conductors can break down and cause corona as a result of the high-voltage standing-wave ratio.
3. Reflections and rereflections cause more power loss.
4. Reflections cause ghost images.
5. Mismatches cause noise interference.

Although it is highly unlikely that a transmission line will be terminated in a load that is either an open or a short circuit, these conditions are examined because they illustrate the worst possible conditions that could occur and produce standing waves that are representative of less severe conditions.

Standing Waves on an Open Line

When incident waves of voltage and current reach an open termination, none of the power is absorbed; it is all reflected back toward the source. The incident voltage wave is reflected in exactly the same manner as if it were to continue down an infinitely long line. However, the incident current is reflected 180° reversed from how it would have continued if the line were not open. As the incident and reflected waves pass, standing waves are produced on the line. Figure 9-13 shows the voltage and current standing waves on a transmission line that is terminated in an open circuit. It can be seen that the voltage standing wave has a maximum value at the open end and a minimum value one quarter wavelength from the open. The current standing wave has a minimum value at the open end and a maximum value one quarter wavelength from the open. It stands to reason that maximum voltage occurs across an open and there is minimum current.

The characteristics of a transmission line terminated in an open can be summarized as follows:

1. The voltage standing wave is reflected back just as if it were to continue (i.e., no phase reversal).
2. The current standing wave is reflected back 180° from how it would have continued.
3. The sum of the incident and reflected current waveforms is minimum at the open.
4. The sum of the incident and reflected voltage waveforms is maximum at the open.

From Figure 9-13 it can also be seen that the voltage and current standing waves repeat every one-half wavelength. The impedance at the open end $Z = V_{max}/I_{min}$ and is maximum. The impedance one-quarter wavelength from the open $Z = V_{min}/I_{max}$ and is minimum. Therefore, one-quarter wavelength from the open an impedance inversion occurs and additional impedance inversions occur each quarter wavelength.

Figure 9-14 shows the development of a voltage standing wave on a transmission line that is terminated in an open circuit. Figure 9-14 shows an incident wave propagating down a transmission line toward the load. The wave is traveling at approximately the speed of light; however, for illustration purposes, the wave has been frozen at quarter-wavelength intervals. In Figure 9-14a it can be seen that the incident wave has not

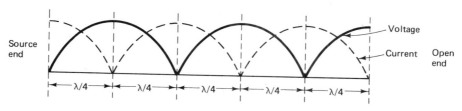

FIGURE 9-13 Voltage and current standing waves on a transmission line that is terminated in an open circuit.

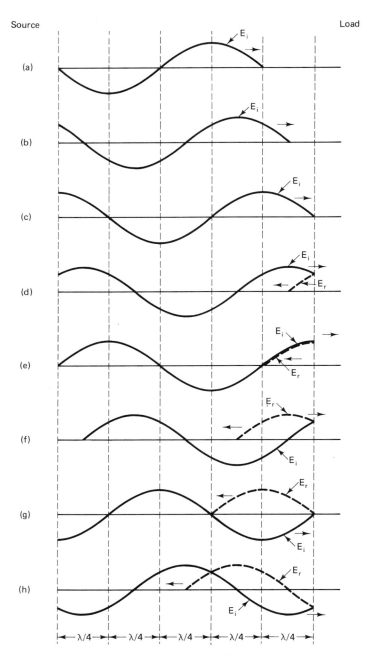

Source
Load

(a)

(b)

(c)

(d)

(e)

(f)

(g)

(h)

|← λ/4 →|← λ/4 →|← λ/4 →|← λ/4 →|← λ/4 →|

FIGURE 9-14   Incident and reflected waves on a transmission line terminated in an open circuit.

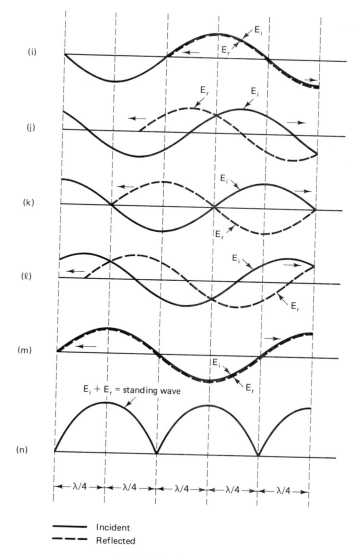

(i)

(j)

(k)

(ℓ)

(m)

(n)

$E_i + E_r$ = standing wave

⟵ λ/4 ⟶⟵ λ/4 ⟶⟵ λ/4 ⟶⟵ λ/4 ⟶⟵ λ/4 ⟶

——— Incident
– – – Reflected

FIGURE 9-14 (*continued*)

reached the open. Figure 9-14b shows the wave one time unit later (for this example, the wave travels one-quarter wavelength per time unit). As you can see, the wave has moved one-quarter wavelength closer to the open. Figure 9-14c shows the wave just as it arrives at the open. Thus far there has been no reflected wave and, consequently, no standing wave. Figure 9-14d shows the incident and reflected waves one time unit after the incident wave has reached the open; the reflected wave is propagating away from the open. Figure 9-14e, f, and g show the incident and reflected waves for the next three time units. In Figure 9-14e it can be seen that the incident and reflected waves are at their maximum positive values at the same time, thus producing a voltage maximum at the open. It can also be seen that one-quarter wavelength from the open the sum of the incident and reflected waves (the standing wave) is always equal to 0 V (a minimum). Figure 9-14h through m show propagation of the incident and reflected waves until the reflected wave reaches the source, and Figure 9-14n shows the resulting standing wave. It can be seen that the standing wave remains stationary (the voltage nodes and antinodes remain at the same points). However, the amplitude of the antinodes vary from maximum positive to zero to maximum negative then repeats. For an open load, all of the incident voltage is reflected ($E_r = E_i$); therefore, $V_{max} = E_i + E_r$ or $2E_i$. A similar illustration can be shown for a current standing wave (however, remember that the current reflects back with a 180° phase inversion).

Standing Waves on a Shorted Line

As with an open line, none of the incident power is absorbed by the load when a transmission line is terminated in a short circuit. However, with a shorted line, the incident voltage and current waves are reflected back in the opposite manner. The voltage wave is reflected 180° reversed from how it would have continued down an infinitely long line, and the current wave is reflected in exactly the same manner as if there were no short.

Figure 9-15 shows the voltage and current standing waves on a transmission line that is terminated in a short circuit. It can be seen that the voltage standing wave has a minimum value at the shorted end and a maximum value one quarter wavelength from the short. The current standing wave has a maximum value at the short and a minimum value one quarter wavelength back. The voltage and current standing waves repeat every quarter wavelength. Therefore, there is an impedance inversion every quarter wavelength. The impedance at the short $Z = V_{min}/I_{max}$ = minimum, and one quarter wavelength back $Z = V_{max}/I_{min}$ = maximum. Again, it stands to reason that a voltage minimum will occur across a short and there is maximum current.

The characteristics of a transmission line terminated in a short can be summarized as follows:

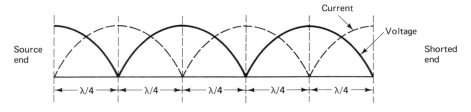

FIGURE 9-15 **Voltage and current standing waves on a transmission line that is terminated in a short circuit.**

1. The voltage standing wave is reflected back 180° reversed from how it would have continued.
2. The current standing wave is reflected back the same as if it had continued.
3. The sum of the incident and reflected current waveforms is maximum at the short.
4. The sum of the incident and reflected voltage waveforms is zero at the short.

For a transmission line terminated in either a short or an open circuit, the reflection coefficient is 1 (the worst case) and the SWR is infinity (also the worst-case condition).

TRANSMISSION-LINE INPUT IMPEDANCE

In the preceding section it was shown that when a transmission line is terminated in either a short or an open circuit, there is an *impedance inversion* every quarter wavelength. For a lossless line, the impedance varies from infinity to zero. However, in a more practical situation where power losses occur, the amplitude of the reflected wave is always less than that of the incident wave. Therefore, the impedance varies from some maximum to some minimum value, or vice versa, depending on whether the line is terminated in a short or an open. The input impedance seen looking into a transmission line that is terminated in a short or an open can be resistive, inductive, or capacitive, depending on the distance from the termination.

Phasor Analysis of Input Impedance—Open Line

Phasor diagrams are generally used to analyze the input impedance of a transmission line because they are relatively simple and give a pictorial representation of the voltage and current phase relationships. Voltage and current phase relations refer to variations in time. Figures 9-13, 9-14, and 9-15 show standing waves of voltage and current plotted versus distance and are therefore not indicative of true phase relationships. The succeeding sections use phasor diagrams to analyze the input impedance of several transmission-line configurations.

Quarter-wavelength transmission line.   Figure 9-16a shows the phasor diagram for the voltage and Figure 9-16b shows the phasor diagram for the current on a quarter-wave section of a transmission line terminated in an open circuit. I_i and V_i are the in-phase incident current and voltage waveforms at the input (source) end of the line at a given instant in time, respectively. Any reflected voltage (E_r) present at the input of the line has traveled one-half wavelength (from the source to the open and back) and is, consequently, 180° behind the incident voltage. Therefore, the total voltage (E_t) at the input end is the sum of E_i and E_r. $E_t = E_i + E_r\underline{/-180°}$ and, assuming a small line loss, $E_t = E_i - E_r$. The reflected current is delayed 90° propagating from the source to the load and another 90° from the load back to the source. Also, the reflected current undergoes an 180° phase reversal at the open. The reflected current has effectively

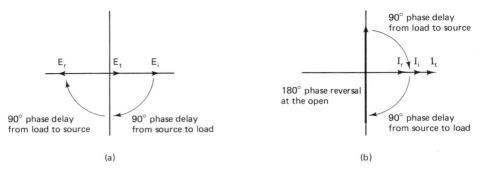

(a)   (b)

FIGURE 9-16 Voltage and current phase relationships for a quarter-wave line terminated in an open circuit: (a) voltage phase relationships; (b) current phase relationships.

been delayed 360°. Therefore, when the reflected current reaches the source end, it is in phase with the incident current and the total current $I_t = I_i + I_r$. By examining Figures 9-16, it can be seen that E_t and I_t are in phase. Therefore, the input impedance seen looking into a transmission line one quarter wavelength long that is terminated in an open circuit $Z_{in} = E_t\underline{/0°}/I_t\underline{/0°} = Z_{in}\underline{/0°}$. Z_{in} has a 0° phase angle, is resistive, and is minimum. Therefore, a quarter-wavelength transmission line terminated in an open circuit is equivalent to a series LC circuit.

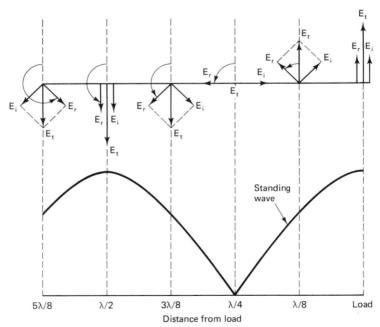

FIGURE 9-17 Vector addition of incident and reflected waves producing a standing wave.

Figure 9-17 shows several voltage phasors for the incident and reflected waves on a transmission line that is terminated in an open circuit and how they produce a voltage standing wave.

Transmission line less than one-quarter wavelength long. Figure 9-18a shows the voltage phasor diagram and Figure 9-18b the current phasor diagram for a transmission line that is less than one-quarter wavelength long and terminated in an open circuit. Again, the incident current (I_i) and voltage (E_i) are in phase. The reflected voltage wave is delayed 45° traveling from the source to the load (a distance of one eighth wavelength) and another 45° traveling from the load back to the source (an additional one eighth wavelength). Therefore, when the reflected wave reaches the source end, it lags the incident wave by 90°. The total voltage at the source end is the vector sum of the incident and reflected waves. Thus $E_t = \sqrt{E_i^2 + E_r^2} = E_t \underline{/-45°}$. The reflected current wave is delayed 45° traveling from the source to the load and another 45° from the load back to the source (a total distance of one-quarter wavelength). In addition, the reflected current wave has undergone an 180° phase reversal at the open prior to being reflected. The reflected current wave has been delayed a total of 270°. Therefore, the reflected wave effectively leads the incident wave by 90°. The total current at the source end is the vector sum of the present and reflected waves. Thus $I_t = \sqrt{I_i^2 + I_r^2} = I_t \underline{/+45°}$. By examining Figure 9-18, it can be seen that E_t lags I_t by 90°. Therefore, $Z_{in} = E_t \underline{/-45°} / I_t \underline{/+45°} = Z_{in} \underline{/-90°}$. Z_{in} has a $-90°$ phase angle and is therefore capacitive. Any transmission line that is less than one-quarter wavelength and terminated in an open circuit is equivalent to a capacitor. The amount of capacitance depends on the exact electrical length of the line.

Transmission line more than one-quarter wavelength long. Figure 9-19a shows the voltage phasor diagram and Figure 9-19b shows the current phasor diagram for a transmission line that is more than one-quarter wavelength long and terminated in an open circuit. For this example, a three-eighths wavelength transmission line is used.

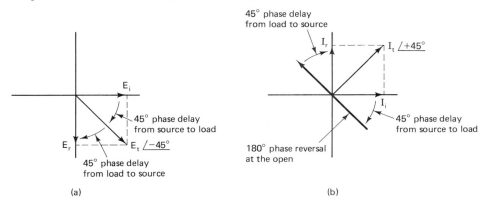

(a)   (b)

FIGURE 9-18 **Voltage and current phase relationships for a transmission line less than one-quarter wavelength terminated in an open circuit: (a) voltage phase relationships; (b) current phase relationships.**

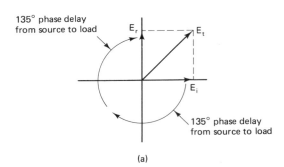
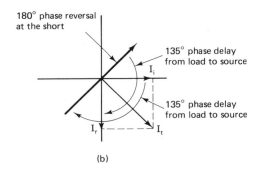

(a)   (b)

FIGURE 9-19 **Voltage and current phase relationships for a transmission more than one-quarter wavelength terminated in an open circuit: (a) voltage phase relationships; (b) current phase relationships.**

The reflected voltage is delayed three-quarters wavelength or $270°$. Therefore, the reflected voltage effectively leads the incident voltage by $90°$. Consequently, the total voltage $E_t = \sqrt{E_i^2 + E_r^2} = E_t/+45°$. The reflected current wave has been delayed $270°$ and undergone an $180°$ phase reversal. Therefore, the reflected current effectively lags the incident current by $90°$. Consequently, the total current $I_t = \sqrt{I_i^2 + I_r^2} = I_t/-45°$. Therefore, $Z_{in} = E_t/+45°/I_t/-45° = Z_{in}/+90°$. Z_{in} has a $+90°$ phase angle and is therefore inductive. A transmission line between one-quarter and one-half wavelength that is terminated in an open circuit is equivalent to an inductor. The amount of inductance depends on the exact electrical length of the line.

Open transmission line as a circuit element.   From the preceding discussion and Figures 9-16 through 9-19, it is obvious that an open transmission line can behave like a resistor, an inductor, or a capacitor, depending on its electrical length. Because standing-wave patterns on an open line repeat every half wavelength, the input impedance also repeats. Figure 9-20 shows the variations in input impedance for an open transmission line of various electrical lengths. It can be seen that an open line is resistive and maximum at the open and at each successive half-wavelength interval, and resistive and minimum one-quarter wavelength from the open and at each successive half-wavelength interval. For electrical lengths less than one-quarter wavelength, the input impedance is capacitive and decreases with length. For electrical lengths between one-quarter and one-half wavelength, the input impedance is inductive and increases with length. The capacitance and inductance patterns also repeat every half wavelength.

Phasor Analysis of Input Impedance—Shorted Line

The following explanations use phasor diagrams to analyze shorted transmission lines in the same manner as with open lines. The difference is that with shorted transmission lines, the voltage waveform is reflected back with an $180°$ phase reversal and the current waveform is reflected back as if there were no short.

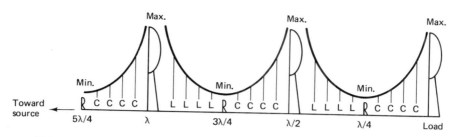

FIGURE 9-20 Input impedance variations for an open-circuited transmission line.

Quarter-wavelength transmission line. The voltage and current phasor diagrams for a quarter-wavelength transmission line terminated in a short circuit are identical to those shown in Figure 9-16, except reversed. The incident and reflected voltages are in phase; therefore, $E_t = E_i + E_r$ and maximum. The incident and reflected currents are 180° out of phase; therefore, $I_t = I_i - I_r$ and minimum. $Z_{in} = E_t \underline{/0°} / I_t \underline{/0°} = Z_{in} \underline{/0°}$ and maximum. Z_{in} has a 0° phase angle, is resistive, and is maximum. Therefore, a quarter-wavelength transmission line terminated in a short circuit is equivalent to a parallel LC circuit.

Transmission line less than one-quarter wavelength long. The voltage and current phasor diagrams for a transmission line less than one-quarter wavelength long and terminated in a short circuit are identical to those shown in Figure 9-18, except reversed. The voltage is reversed 180° at the short and the current is reflected with the same phase as if it had continued. Therefore, the total voltage at the source end of the line lags the current by 90° and the line looks inductive.

Transmission line more than one-quarter wavelength long. The voltage and current phasor diagrams for a transmission line more than one-quarter wavelength long and terminated in a short circuit are identical to those shown in Figure 9-19, except reversed. The total voltage at the source end of the line leads the current by 90° and the line looks capacitive.

Open transmission line as a circuit element. From the preceding discussion it is obvious that a shorted transmission line can behave like a resistor, an inductor, or a capacitor, depending on its electrical length. On a shorted transmission line, standing waves repeat every half wavelength; therefore, the input impedance also repeats. Figure 9-21 shows the variations in input impedance of a shorted transmission line for various electrical lengths. It can be seen that a shorted line is resistive and minimum at the short and at each successive half-wavelength interval, and resistive and maximum one-quarter wavelength from the short and at each successive half-wavelength interval. For electrical lengths less than one-quarter wavelength, the input impedance is inductive and increases with length. For electrical lengths between one-quarter and one-half wavelength, the input impedance is capacitive and decreases with length. The inductance and capacitance patterns also repeat every half wavelength.

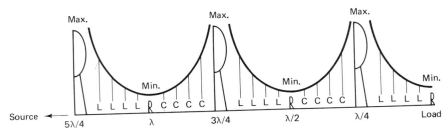

FIGURE 9-21   Input impedance variations for a short-circuited transmission line.

Transmission-line input impedance summary. Figure 9-22 summarizes the transmission-line configurations described in the preceding sections, their input impedance characteristics, and their equivalent LC circuits. It can be seen that both shorted and open sections of transmission lines can behave as resistors, inductors, or capacitors, depending on their electrical length.

Transmission-Line Impedance Matching

Power is transferred most efficiently to a load when there are no reflected waves: that is, when the load is purely resistive and equal to Z_o. Whenever the characteristic impedance of a transmission line and its load are not matched (equal), standing waves are present on the line and maximum power is not transferred to the load. As stated previously, standing waves cause power loss, dielectric breakdown, noise, radiation, and *ghost signals*. Therefore, whenever possible a transmission line should be matched to its load. There are two common transmission-line techniques that are used to match a transmission line to a load with an impedance that is not equal to Z_o. They are quarter-wavelength transformer matching and stub matching.

Quarter-wavelength transformer matching. Quarter-wavelength *transformers* are used to match transmission lines to purely resistive loads whose resistance is not equal to the characteristic impedance of the line. Keep in mind that a quarter-wavelength transformer is not actually a transformer, but rather, a quarter-wavelength section of transmission line that acts like a transformer. In a previous section it was shown that the input impedance to a transmission line varies from some maximum value to some minimum value, or vice versa, every quarter wavelength. Therefore, a transmission line one-quarter wavelength long acts like a *step-up* or *step-down transformer*, depending on whether Z_L is greater than or less than Z_o. A quarter-wavelength transformer is not a broadband impedance-matching device; it is a quarter wavelength at only a single frequency. The impedance transformations for a quarter wavelength transmission line are as follows:

1. $R_L = Z_o$: The quarter-wavelength line acts like a transformer with a 1:1 turns ratio.
2. $R_L > Z_o$: The quarter-wavelength line acts like a step-down transformer.
3. $R_L < Z_o$: The quarter-wavelength line acts like a step-up transformer.

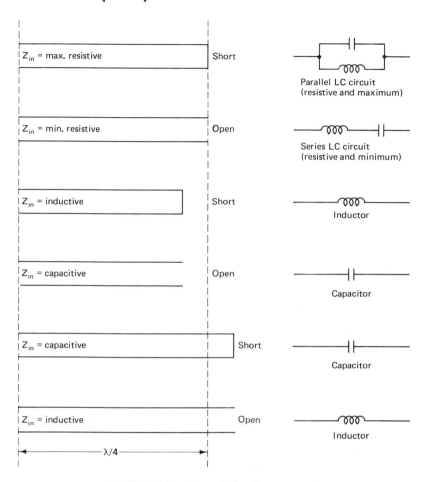

FIGURE 9-22 Transmission-line summary.

Like a transformer, a quarter-wavelength transformer is placed between a transmission line and its load. A quarter-wavelength transformer is simply a length of transmission line one-quarter wavelength long. Figure 9-23 shows how a quarter-wavelength transformer is used to match a transmission line to a purely resistive load. The characteristic impedance of the quarter-wavelength section is determined mathematically from the formula

$$Z_o' = \sqrt{Z_o Z_L} \qquad (9\text{-}32)$$

where

Z_o' = characteristic impedance of quarter-wavelength transformer
Z_o = characteristic impedance of the transmission line that is being matched
Z_L = load impedance

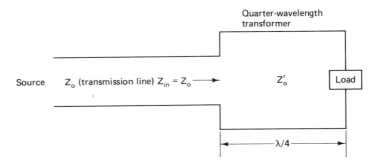

FIGURE 9-23   Quarter-wavelength transformer.

EXAMPLE 9-5

Determine the physical length and characteristic impedance for a quarter-wavelength transformer that is used to match a section of RG-8A/U ($Z_o = 50\ \Omega$) to a 150-Ω resistive load. The frequency of operation is 150 MHz and the velocity factor $V_f = 1$.

Solution The physical length of the transformer is dependent on the wavelength of the signal. Substituting into Equation 9-2 yields

$$\lambda = \frac{c}{F} = \frac{3 \times 10^8\ \text{m/s}}{150\ \text{MHz}} = 2\ \text{m}$$

$$\frac{\lambda}{4} = \frac{2\ \text{m}}{4} = 0.5\ \text{m}$$

The characteristic impedance of the 0.5-m transformer is determined from Equation 9-32:

$$Z_o' = \sqrt{Z_o Z_L} = \sqrt{(50)(150)} = 86.6\ \Omega$$

Stub matching.   When a load is purely inductive or purely capacitive, it absorbs no energy. The reflection coefficient is 1 and the SWR is infinity. When the load is a complex impedance (which is usually the case), it is necessary to remove the reactive component to match the transmission line to the load. Transmission-line *stubs* are commonly used for this purpose. A transmission-line stub is simply a piece of additional transmission line that is placed across the primary line as close to the load as possible. The susceptance of the stub is used to tune out the susceptance of the load. With *stub matching*, either a shorted or an open stub can be used. However, shorted stubs are preferred because open stubs have a tendency to radiate, especially at the higher frequencies.

Figure 9-24 shows how a shorted stub is used to cancel the susceptance of the load and match the load resistance to the characteristic impedance of the transmission line. In previous sections it was shown how a shorted section of transmission line can look resistive, inductive, or capacitive, depending on its electrical length. A transmission line that is one-half wavelength or shorter can be used to tune out the reactive component of a load.

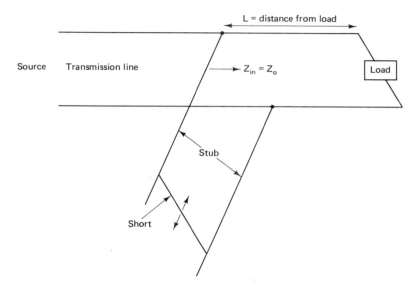

FIGURE 9-24   Shorted stub impedance matching.

The process of matching a load to a transmission line with a shorted stub is as follows.

1. Locate a point as close to the load as possible where the resistive component of the input impedance is equal to the characteristic impedance of the transmission line. $Z_{in} = R + jX$ where $R = Z_o$.

2. Attach the shorted stub to the point on the transmission line identified in step 1.

3. Depending on whether the reactive component of the load is inductive or capacitive, the stub length is adjusted accordingly. $Z_{in} = R + jX_L - jX_C = R$ or $R + jX_C - jX_L = R$.

For a more complete explanation of stub matching using the Smith chart, refer to Appendix A.

QUESTIONS

9-1. Define *transmission line*.

9-2. Describe a transverse electromagnetic wave.

9-3. Define *wave velocity*.

9-4. Define *frequency* and *wavelength* for a transverse electromagnetic wave.

9-5. Describe balanced and unbalanced transmission lines.

9-6. Describe an open-wire transmission line.

9-7. Describe a twin-lead transmission line.

9-8. Describe a twisted-pair transmission line.

9-9. Describe a shielded cable transmission line.

9-10. Describe a concentric transmission line.

9-11. Describe the electrical and physical properties of a transmission line.

9-12. List and describe the four primary constants of a transmission line.

9-13. Define *characteristic impedance* for a transmission line.

9-14. What properties of a transmission line determine its characteristic impedance?

9-15. Define *propagation constant* for a transmission line.

9-16. Define *velocity factor* for a transmission line.

9-17. What properties of a transmission line determine its velocity factor?

9-18. What properties of a transmission line determine its dielectric constant?

9-19. Define *electrical length* for a transmission line.

9-20. List and describe five types of transmission-line losses.

9-21. Describe an incident wave; a reflected wave.

9-22. Describe a resonant transmission line; a nonresonant transmission line.

9-23. Define *reflection coefficient*.

9-24. Describe standing waves; standing-wave ratio.

9-25. Describe the standing waves present on an open transmission line.

9-26. Describe the standing waves present on a shorted transmission line.

9-27. Define *input impedance* for a transmission line.

9-28. Describe the behavior of a transmission line that is terminated in a short circuit that is greater than one-quarter wavelength long; less than one-quarter wavelength.

9-29. Describe the behavior of a transmission line that is terminated in an open circuit that is greater than one-quarter wavelength long; less than one-quarter wavelength long.

9-30. Describe the behavior of an open transmission line as a circuit element.

9-31. Describe the behavior of a shorted transmission line as a circuit element.

9-32. Describe the input impedance characteristics of a quarter-wavelength transmission line.

9-33. Describe the input impedance characteristics of a transmission line that is less than one-quarter wavelength long; greater than one-quarter wavelength long.

9-34. Describe quarter-wavelength transformer matching.

9-35. Describe how stud matching is accomplished.

PROBLEMS

9-1. Determine the wavelengths for electromagnetic waves with the following frequencies: 1 kHz, 100 kHz, 1 MHz, and 1 GHz.

9-2. Determine the frequencies for electromagnetic waves with the following wavelengths: 1 cm, 1 m, 10 m, 100 m, and 1000 m.

9-3. Determine the characteristic impedance for an air-dielectric transmission line with D/r ratio of 8.8.

9-4. Determine the characteristic impedance for a concentric transmission line with D/d ratio of 4.

9-5. Determine the characteristic impedance for a coaxial cable with inductance $L = 0.2$ μH/ft and capacitance $C = 16$ pF/ft.

9-6. For a given length of coaxial cable with distributed capacitance $C = 48.3$ pF/m and distributed inductance $L = 241.56$ nH/m, determine the velocity factor and velocity of propagation.

9-7. Determine the reflection coefficient for a transmission line with incident voltage $E_i = 0.2$ V and reflected voltage $E_r = 0.01$ V.

9-8. Determine the standing-wave ratio for the transmission line described in Question 9-7.

9-9. Determine the SWR for a transmission line with maximum voltage standing-wave amplitude $V_{max} = 6$ V and minimum voltage standing-wave amplitude $V_{min} = 0.5$.

9-10. Determine the SWR for a 50-Ω transmission line that is terminated in a load resistance $Z_L = 75$ Ω.

9-11. Determine thee SWR for a 75-Ω transmission line that is terminated in a load resistance $Z_L = 50$ Ω.

9-12. Determine the characteristic impedance for a quarter-wavelength transformer that is used to match a section of 75-Ω transmission line to a 100-Ω resistive load.

Chapter 10

WAVE PROPAGATION

INTRODUCTION

In Chapter 9 we explained how a metallic wire is used as a transmission medium to propagate electromagnetic waves from one point to another. However, very often in communications systems it is impractical or impossible to interconnect two pieces of equipment with a physical facility such as a wire; for example, across large spans of water, rugged mountains, or desert terrain, or to and from satellite transponders parked 22,000 mi above the earth. Also, when the transmitters and receivers are mobile, as with two-way radio or mobile telephone, metallic facilities are impossible. Therefore, free space or the earth's atmosphere is often used as a transmission medium. Free-space propagation of electromagnetic waves is often called *radio-frequency* (RF) *propagation* or simply *radio propagation*.

To propagate TEM waves through the earth's atmosphere, it is necessary that energy be radiated from the source; then the energy must be *captured* at the receive end. Radiating and capturing energy are antenna functions and are explained in Chapter 11, and the properties of electromagnetic waves were explained in Chapter 9.

RAYS AND WAVEFRONTS

Electromagnetic waves are invisible. Therefore, they must be analyzed by indirect methods using schematic diagrams. The concept of *rays* and *wavefronts* is an aid to illustrating the effects of electromagnetic wave propagation through free space. A ray is a line

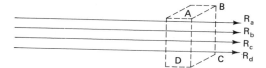

FIGURE 10-1   Plane wave.

drawn along the direction of propagation of an electromagnetic wave. Rays are used to show the relative direction of electromagnetic wave propagation. However, a ray does not necessarily represent the propagation of a single electromagnetic wave. Several rays are shown in Figure 10-1 (R_a, R_b, R_c, etc.). A wavefront shows a surface of constant phase of a wave. A wavefront is formed when points of equal phase on rays propagated from the same source are joined together. Figure 10-1 shows a wavefront with a surface that is perpendicular to the direction of propagation (rectangle *ABCD*). When a surface is plane, its wavefront is perpendicular to the direction of propagation. The closer to the source, the more complicated the wavefront becomes.

Most wavefronts are more complicated than a simple plane wave. Figure 10-2 shows a point source, several rays propagating from it, and the corresponding wavefront. A *point source* is a single location from which rays propagate equally in all directions (i.e., an *isotropic source*). The wavefront generated from a point source is simply a sphere with radius R and its center located at the point of origin of the waves. In free space and a sufficient distance from the source, the rays within a small area of a spherical wavefront are nearly parallel. Therefore, the farther from a source, the closer to plane a wavefront appears.

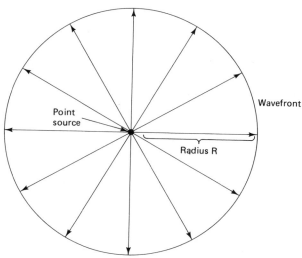

FIGURE 10-2   Wavefront from a point source.

ELECTROMAGNETIC RADIATION

Power Density and Field Intensity

Electromagnetic waves represent the flow of energy in the direction of propagation. The rate at which energy passes through a given surface area in free space is called *power density*. Therefore, power density is energy per unit time per unit of area and is usually given in watts per square meter. *Field intensity* is the intensity of the voltage and magnetic fields of an electromagnetic wave propagating in free space. Voltage intensity is usually given in volts per meter and magnetic intensity in ampere-turns/meter. Mathematically, the rms power density is

$$\mathcal{P} = \mathcal{E}H \text{ watts per meter squared} \tag{10-1}$$

where

$\mathcal{P}$ = power density (W/m^2)
$\mathcal{E}$ = rms voltage intensity (V/m)
H = rms magnetic intensity (At/m)

Characteristic Impedance of Free Space

The electric and magnetic intensities of an electromagnetic wave in free space are related through the characteristic impedance (resistance) of free space. The characteristic impedance of a lossless transmission medium is equal to the square root of the ratio of its magnetic permeability to its electric permittivity. Mathematically, the characteristic impedance of free space (Z_s) is

$$Z_s = \frac{\mu}{\epsilon} \tag{10-2}$$

where

Z_s = characteristic impedance of free space
μ = magnetic permeability of free space (1.26×10^{-6} H/m)
ϵ = electric permittivity of free space (8.85×10^{-12} F/m)

Substituting into Equation 10-2, we have

$$Z_s = \sqrt{\frac{1.26 \times 10^{-6}}{8.85 \times 10^{-12}}} = 377 \ \Omega$$

Therefore, using Ohm's law, we obtain

$$\mathcal{P} = \frac{\mathcal{E}^2}{377} = 377H^2 \qquad \text{watts per meter squared} \tag{10-3}$$

and

$$H = \frac{\mathscr{E}}{377} \quad \text{ampere-turns per meter} \tag{10-4}$$

SPHERICAL WAVEFRONT AND THE INVERSE SQUARE LAW

Spherical Wavefront

Figure 10-3 shows a point source that radiates power at a constant rate uniformly in all directions. Such a source is called an *isotropic radiator*. A true isotropic radiator does not exist. However, it is closely approximated by an *omnidirectional antenna*. An isotropic radiator produces a spherical wavefront with radius R. All points distance R from the source lie on the surface of the sphere and have equal power densities. For example, in Figure 10-3 points A and B are equal distance from the source. Therefore, the power density at points A and B are equal. At any instant of time, the total power radiated, P_r watts, is uniformly distributed over the total surface of the sphere (this assumes a lossless transmission medium). Therefore, the power density at any point on the sphere is the total radiated power divided by the total area of the sphere. Mathematically, the power density at any point on the surface of a spherical wavefront is

$$\mathscr{P} = \frac{P_r}{4\pi R^2} \tag{10-5}$$

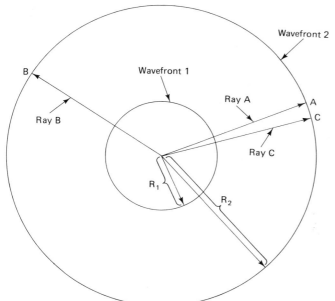

FIGURE 10-3 Spherical wavefront from an isotropic source.

where

P_r = total power radiated

R = radius of the sphere (which is equal to the distance from any point on the surface of the sphere to the source)

$4\pi R^2$ = area of the sphere

and for a distance R_a meters from the source, the power density is

$$\mathcal{P} = \frac{P_r}{4\pi R_a^2}$$

Equating Equations 10-3 and 10-5 gives

$$\frac{P_r}{4\pi R^2} = \frac{\mathcal{E}^2}{377}$$

Therefore,

$$\mathcal{E}^2 = \frac{377 P_r}{4\pi R^2} \qquad \text{and} \qquad \mathcal{E} = \frac{\sqrt{30\, P_r}}{R} \qquad (10\text{-}6)$$

Inverse Square Law

From Equation 10-3 it can be seen that the farther the wavefront moves from the source, the smaller the power density (R_a and R_c move farther apart). The total power distributed over the surface of the sphere remains the same. However, because the area of the sphere increases in direct proportion to the distance from the source squared (i.e., the radius of the sphere squared), the power density is inversely proportional to the square of the distance from the source. This relationship is called the *inverse square law*. Therefore, the power density at any point on the surface of the outer sphere is

$$\mathcal{P}_2 = \frac{P_r}{4\pi R_2^2}$$

and the power density at any point on the inner sphere is

$$\mathcal{P}_1 = \frac{P_r}{4\pi R_1^2}$$

Therefore,

$$\frac{\mathcal{P}_2}{\mathcal{P}_1} = \frac{P_r/4\pi R_2^2}{P_r/4\pi R_1^2} = \frac{R_1^2}{R_2^2} = \left(\frac{R_1}{R_2}\right)^2 \qquad (10\text{-}7)$$

From Equation 10-7 it can be seen that as the distance from the source doubles, the power density decreases by a factor of 2^2, or 4. When deriving the inverse square law of radiation (Equation 10-7) it was assumed that the source radiates isotropically, although it is not necessary. However, it is necessary that the velocity of propagation in all directions be uniform. Such a propagation medium is called an *isotropic medium*.

EXAMPLE 10-1

For an isotropic antenna radiating 100 W of power, determine:
   (a) The power density 1000 m from the source.
   (b) The power density 2000 m from the source.

Solution
   (a) Substituting into Equation 10-5 yields

$$\mathcal{P} = \frac{100}{4\pi 1000^2} = 7.96 \ \mu W/m^2$$

   (b) Again, substituting into Equation 10-5 gives us

$$\mathcal{P} = \frac{100}{4\pi 2000^2} = 1.99 \ \mu W/m^2$$

or substituting into Equation 10-7, we have

$$\frac{\mathcal{P}_2}{\mathcal{P}_1} = \frac{1000^2}{2000^2} = 7.96 \ \mu W/m^2$$

$$\mathcal{P}_2 = 7.96 \ \mu W/m^2 \left(\frac{1000^2}{2000^2}\right) = 1.99 \ \mu W/m^2$$

WAVE ATTENUATION AND ABSORPTION

Attenuation

The inverse square law for radiation mathematically describes the reduction in power density with distance from the source. As a wavefront moves away from the source, the continuous electromagnetic field that is radiated from that source spreads out. That is, the waves move farther away from each other and, consequently, the number of waves per unit area decreases. None of the radiated power is lost or dissipated because the wavefront is moving away from the source; the wave simply spreads out or disperses over a larger area, decreasing the power density. The reduction in power density with distance is equivalent to a power loss and is commonly called *wave attenuation*. Because the attenuation is due to the spherical spreading of the wave, it is sometimes called the *space attenuation* of the wave. Wave attenuation is generally expressed in terms of the common logarithm of the power density ratio (dB loss). Mathematically, wave attenuation (γ_a) is

$$\gamma_a = 10 \log \frac{\mathcal{P}_1}{\mathcal{P}_2} \tag{10-8}$$

The reduction in power density due to the inverse-square law presumes free-space propagation (i.e., a vacuum or nearly a vacuum) and is called wave attenuation. The reduction in power density due to nonfree-space propagation is called *absorption*.

Absorption

The earth's atmosphere is not a vacuum. Rather, it is made up of atoms and molecules of various substances, such as gases, liquids, and solids. Some of these materials are capable of absorbing electromagnetic waves. As an electromagnetic wave propagates through the earth's atmosphere, energy is transferred from the wave to the atoms and molecules of the atmosphere. Wave absorption by the atmosphere is analogous to an I^2R power loss. Once absorbed, the energy is lost forever and causes an attenuation in the voltage and magnetic field intensities and a corresponding reduction in power density.

Absorption of radio frequencies in a normal atmosphere is dependent on frequency and relatively insignificant below approximately 10 GHz. Figure 10-4 shows atmospheric absorption in dB/km due to oxygen and water vapor for radio frequencies above 10 GHz. It can be seen that certain frequencies are affected more or less by absorption, creating peaks and valleys in the curves. Wave attenuation due to absorption does not depend on distance from the radiating source, but rather, the total distance that the wave propagates through the atmosphere. In other words, for a *homogeneous medium* (one with uniform properties throughout), the absorption experienced during the first mile of propagation is the same as for the last mile. Also, abnormal atmospheric conditions such as heavy rain or dense fog absorb more energy than a normal atmosphere. Atmospheric absorption (η) for a wave propagating from R_1 to R_2 is $\gamma(R_2 - R_1)$, where γ is the absorption coefficient. Therefore, wave attenuation depends on the ratio R_2/R_1 and wave absorption depends on the distance between R_1 and R_2. In a more practical situation (i.e., an *inhomogeneous medium*), the absorption coefficient varies considerably with location, thus creating a difficult problem for radio systems engineers.

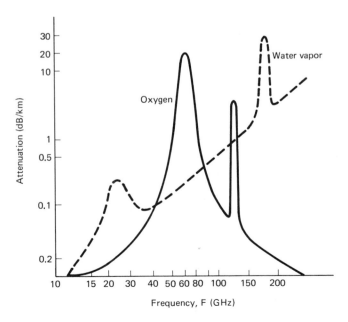

FIGURE 10-4 Atmospheric absorption of electromagnetic waves.

OPTICAL PROPERTIES OF RADIO WAVES

In the earth's atmosphere, ray-wavefront propagation may be altered from free-space behavior by *optical* effects such as *refraction, reflection, diffraction,* and *interference.* Using rather unscientific terminology, refraction can be thought of as *bending,* reflection as *bouncing,* diffraction as *scattering,* and interference as *colliding.* Refraction, reflection, diffraction, and interference are called optical properties because they were first observed in the science of optics, which is the behavior of light waves. Because light waves are high-frequency electromagnetic waves, it stands to reason that optical properties will also apply to radio-wave propagation. Although optical principles can be analyzed completely by application of Maxwell's equations, this is necessarily complex. For most applications, *geometric ray tracing* can be substituted for analysis by Maxwell's equations.

Refraction

Electromagnetic *refraction* is the change in direction of a ray as it passes obliquely from one medium to another with different velocities of propagation. As stated previously, the velocity at which an electromagnetic wave propagates is inversely proportional to the density of the medium in which it is propagating. Therefore, refraction occurs whenever a radio wave passes from one medium into another medium of different density. Figure 10-5 shows refraction of a wavefront at a *plane* boundary between two media with different densities. For this example, medium 1 is less dense than medium 2 (i.e.,

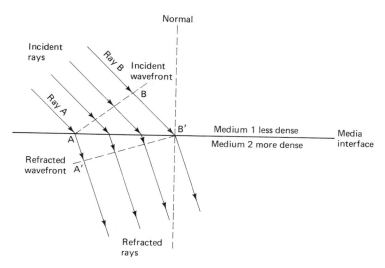

FIGURE 10-5 Refraction at a plane boundary between two media.

$v_1 > v_2$). It can be seen that ray A enters the more dense medium before ray B. Therefore, ray B propagates more slowly than ray A and travels distance B–B' during the same time that ray A travels distance A–A'. Therefore, wavefront ($A'B'$) is *tilted* or bent in a downward direction. Since a ray is defined as being perpendicular to the wavefront at all points, the rays in Figure 10-5 have changed direction at the interface of the two media. Whenever a ray passes from a less dense to a more dense medium, it is effectively bent toward the *normal*. (The normal is simply an imaginary line drawn perpendicular to the interface at the point of incidence.) Conversely, whenever a ray passes from a more dense to a less dense medium, it is effectively bent away from the normal. The *angle of incidence* is the angle formed between the incident wave and the normal, and the *angle of refraction* is the angle formed between the refracted wave and the normal.

The amount of bending or refraction that occurs at the interface of two materials of different densities is quite predictable and depends on the *refractive index* (also called the *index of refraction*) of the two materials. The refractive index is simply the ratio of the velocity of propagation of a light ray in free space to the velocity of propagation of a light ray in a given material. Mathematically, the refractive index is

$$n = \frac{c}{v} \qquad (10\text{-}9)$$

where

$\quad\quad\quad n =$ refractive index
$\quad\quad\quad c =$ speed of light in free space
$\quad\quad\quad v =$ speed of light in a given material

The refractive index is also a function of frequency. However, the variation in most applications is insignificant and is therefore omitted from this discussion. How an electromagnetic wave reacts when it meets the interface of two transmissive materials that have different indexes of refraction can be explained with *Snell's law*. Snell's law simply states:

$$n_1 \sin \theta_1 = n_2 \sin \theta_2 \qquad (10\text{-}10)$$

and

$$\frac{\sin \theta_1}{\sin \theta_2} = \frac{n_2}{n_1}$$

where

$\quad\quad\quad n_1 =$ refractive index of material 1
$\quad\quad\quad n_2 =$ refractive index of material 2
$\quad\quad\quad \theta_1 =$ angle of incidence
$\quad\quad\quad \theta_2 =$ angle of refraction

and since the refractive index of a material is equal to the square root of its dielectric constant,

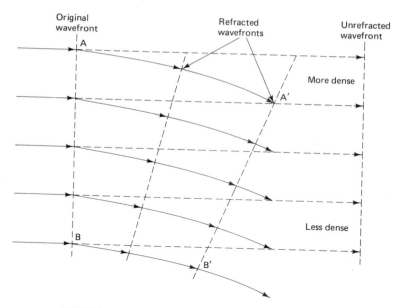

FIGURE 10-6 Wavefront refraction in a gradient medium.

$$\frac{\sin \theta_1}{\sin \theta_2} = \sqrt{\frac{\epsilon_2}{\epsilon_1}} \qquad (10\text{-}11)$$

where

ϵ_1 = dielectric constant of medium 1
ϵ_2 = dielectric constant of medium 2

Refraction also occurs when a wavefront propagates in a medium that has a *density gradient* that is perpendicular to the direction of propagation (i.e., parallel to the wavefront). Figure 10-6 shows wavefront refraction in a transmission medium that has a gradual variation in its refractive index. The medium is more dense near the bottom and less dense at the top. Therefore, rays traveling near the top travel faster than rays near the bottom and consequently, the wavefront tilts downward. The tilting occurs in a gradual fashion as the wave progresses as shown.

Reflection

Reflect means to cast or turn back, and *reflection* is the act of reflecting. Electromagnetic reflection occurs when an incident wave strikes a boundary of two media and some or all of the incident power does not enter the second material. The waves that do not penetrate the second medium are reflected. Figure 10-7 shows electromagnetic wave reflection at a plane boundary between two media. Because all the reflected waves

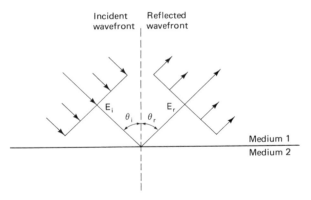

**FIGURE 10-7   Electromagnetic reflection
at a plane boundary of two media.**

remain in medium 1, the velocities of the reflected and incident waves are equal. Consequently, the *angle of reflection* equals the *angle of incidence* ($\theta_i = \theta_r$). However, the
reflected voltage field intensity is less than the incident voltage field intensity. The
ratio of the reflected to the incident voltage intensities is called the *reflection coefficient*,
Γ (sometimes called the *coefficient of reflection*). For a perfect conductor, $\Gamma = 1$. Γ is
used to indicate both the relative amplitude of the incident and reflected fields and also
the phase shift that occurs at the point of reflection. Mathematically, the reflection
coefficient is

$$\Gamma = \frac{E_r e^{j\theta_r}}{E_i e^{j\theta_i}} = \frac{E_r}{E_i} e^{j(\theta_r - \theta_i)} \tag{10-12}$$

where

Γ = reflection coefficient
E_i = incident voltage intensity
E_r = reflected voltage intensity
θ_i = incident phase
θ_r = reflected phase

The ratio of the reflected and incident power densities is Γ. The portion of the
total incident power that is not reflected is called the *power transmission coefficient* (T)
(or simply the *transmission coefficient*). For a perfect conductor, $T = 0$. The *law of
conservation of energy* states that for a perfect reflective surface, the total reflected
power must equal the total incident power. Therefore,

$$T + |\Gamma|^2 = 1 \tag{10-13}$$

For imperfect conductors, both $|\Gamma|^2$ and T are functions of the angle of incidence,
the electric field polarization, and the dielectric constants of the two materials. If medium
2 is not a perfect conductor, some of the incident waves penetrate it and are absorbed.
The absorbed waves set up currents in the resistance of the material and the energy is
converted to heat. The fraction of power that penetrates medium 2 is called the *absorption
coefficient* (or sometimes the *coefficient of absorption*).

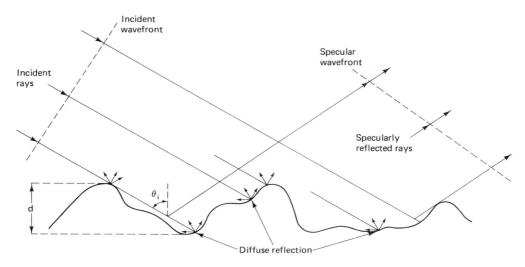

FIGURE 10-8 **Reflection from a semirough surface.**

When the reflecting surface is not plane (i.e., it is curved) the curvature of the reflected wave is different from that of the incident wave. When the wavefront of the incident wave is curved and the reflective surface is plane, the curvature of the reflected wavefront is the same as that of the incident wavefront.

Reflection also occurs when the reflective surface is *irregular* or *rough*. However, such a surface may destroy the shape of the wavefront. When an incident wavefront strikes an irregular surface, it is randomly scattered in many directions. Such a condition is called *diffuse reflection*, whereas reflection from a perfectly smooth surface is called *specular* (mirror-like) *reflection*. Surfaces that fall between smooth and irregular are called *semirough surfaces*. Semirough surfaces cause a combination of diffuse and specular reflection. A semirough surface will not totally destroy the shape of the reflected wavefront. However, there is a reduction in the total power. The *Rayleigh criterion* states that a semirough surface will reflect as if it were a smooth surface whenever the cosine of the angle of incidence is greater than $\lambda/8d$, where d is the depth of the surface irregularity and λ is the wavelength of the incident wave. Reflection from a semirough surface is shown in Figure 10-8. Mathematically, Rayleigh's criterion is

$$\cos \theta_i > \frac{\lambda}{8d} \qquad (10\text{-}14)$$

Diffraction

Diffraction is defined as the modulation or redistribution of energy within a wavefront when it passes near the edge of an *opaque* object. Diffraction is the phenomenon that allows light or radio waves to propagate (*peek*) around corners. The previous discussions of refraction and reflection assumed that the dimensions of the refracting and reflecting

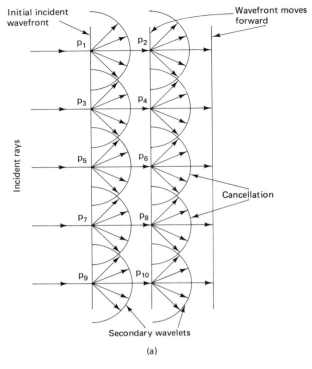

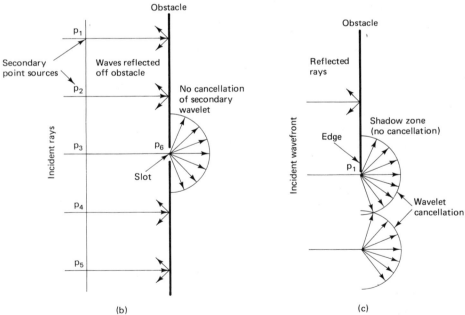

FIGURE 10-9 Electromagnetic wave diffraction: (a) Huygens' principle for a plane wavefront; (b) finite wavefront through a slot; (c) around an edge.

surfaces were large in respect to a wavelength of the signal. However, when a wavefront passes near an obstacle or discontinuity with dimensions comparable in size to a wavelength, simple geometric analysis cannot be used to explain the results and *Huygens' principle* (which is deduced from Maxwell's equations) is necessary.

Huygens' principle states that every point on a given spherical wavefront can be considered as a secondary point source of electromagnetic waves from which other secondary waves (wavelets) are radiated outward. Huygens' principle is illustrated in Figure 10-9. Normal wave propagation considering an infinite plane is shown in Figure 10-9a. Each secondary point source (p_1, p_2, etc.) radiates energy outward in all directions. However, the wavefront continues in its original direction rather than spreading out because cancellation of the secondary wavelets occurs in all directions except straight forward. Therefore, the wavefront remains plane.

When a finite plane wavefront is considered, as in Figure 10-9b, cancellation in random directions is incomplete. Consequently, the wavefront spreads out or *scatters*. This scattering effect is called *diffraction*. Figure 10-9c shows diffraction around the edge of an obstacle. It can be seen that wavelet cancellation occurs only partially. Diffraction occurs around the edge of the obstacle, which allows secondary waves to "sneak" around the corner of the obstacle into what is called the *shadow zone*. This phenomenon can be observed when a door is opened into a dark room. Light rays diffract around the edge of the door and illuminate the area behind the door.

Interference

Interfere means to come into opposition, and *interference* is the act of interfering. Radio-wave interference occurs when two or more electromagnetic waves combine in such a way that system performance is degraded. Refraction, reflection, and diffraction are categorized as geometric optics, which means that their behavior is analyzed primarily in terms of rays and wavefronts. Interference, on the other hand, is subject to the principle of *linear superposition* of electromagnetic waves and occurs whenever two or more waves simultaneously occupy the same point in space. The principle of linear superposition states that the total voltage intensity at a given point in space is the sum of the individual wave vectors. Certain types of propagation media have nonlinear properties; however, in an ordinary medium (such as air or the earth's atmosphere), linear superposition holds true.

Figure 10-10 shows the linear addition of two instantaneous voltage vectors whose phase angles differ by angle θ. It can be seen that the total voltage is not simply the sum of the two vectors, but rather, the phasor addition of the two. With free-space propagation, a phase difference may exist simply because the electromagnetic polariza-

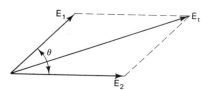

FIGURE 10-10 **Linear addition of two vectors with differing phase angles.**

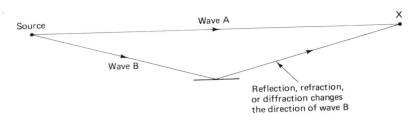

FIGURE 10-11   Electromagnetic wave interference.

tions of two waves differ. Depending on the phase angles of the two vectors, either addition or subtraction can occur. (This implies simply that the result may be more or less than either vector because the two electromagnetic waves can reinforce or cancel.)

Figure 10-11 shows interference between two electromagnetic waves in free space. It can be seen that at point X the two waves occupy the same area of space. However, wave B has traveled a different path than wave A, and therefore their relative phase angles may be different. If the difference in distance traveled is an odd integral multiple of one-half wavelength, reinforcement takes place. If the difference is an even integral multiple of one-half wavelength, total cancellation occurs. More likely the difference in distance falls somewhere between the two and partial cancellation occurs. For frequencies below VHF, the relatively large wavelengths prevent interference from being a significant problem. However, with UHF and above, wave interference can be severe.

PROPAGATION OF WAVES

In radio communications systems, there are several ways in which waves can be propagated, depending on the type of system and the environment. Also, as previously explained, electromagnetic waves travel in straight lines except when the earth and its atmosphere alter their path. There are three ways of propagating electromagnetic waves: ground wave, space wave (which includes both direct and ground-reflected waves), and sky wave propagation.

Figure 10-12 shows the normal modes of propagation between two radio antennas. Each of these modes exists in every radio system; however, some are negligible in certain frequency ranges or over a particular type of terrain. At frequencies below 1.5 MHz, ground waves provide the best coverage. This is because ground losses increase rapidly with frequency. Sky waves are used for high-frequency applications, and space waves are used for very high frequencies and above.

Ground-Wave Propagation

A *ground wave* is an electromagnetic wave that travels along the surface of the earth. Therefore, ground waves are sometimes called *surface waves*. Ground waves must be vertically polarized. This is because the electric field in a horizontally polarized wave would be parallel to the earth's surface, and such waves would be short-circuited by

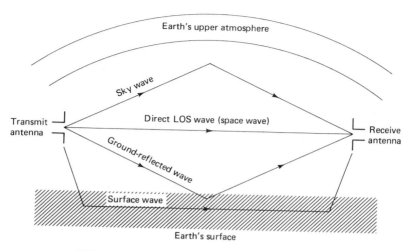

FIGURE 10-12   Normal modes of wave propagation.

the conductivity of the ground. With ground waves, the changing electric field induces voltages in the earth's surface, which cause currents to flow that are very similar to those in a transmission line. The earth's surface also has resistance and dielectric losses. Therefore, ground waves are attenuated as they propagate. Ground waves propagate best over a surface that is a good conductor, such as salt water, and poorly over dry desert areas. Ground-wave losses increase rapidly with frequency. Therefore, ground-wave propagation is generally limited to frequencies below 2 MHz.

Figure 10-13 shows ground-wave propagation. The earth's atmosphere has a *gra-*

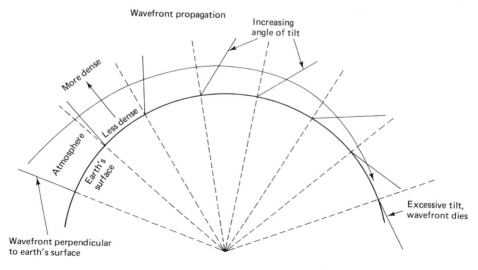

FIGURE 10-13   Ground-wave propagation.

dient density (i.e., the density decreases gradually with distance from the earth's surface), which causes the wavefront to tilt progressively forward. Therefore, the ground wave propagates around the earth, remaining close to its surface, and if enough power is transmitted, the wavefront could propagate beyond the horizon or even around the entire earth's circumference. However, care must be taken when selecting the frequency and the terrain over which the ground wave will propagate, to ensure that the wavefront does not tilt excessively and simply turn over, lie flat on the ground, and cease to propagate.

Ground-wave propagation is commonly used for ship-to-ship and ship-to-shore communications, for radio navigation, and for maritime mobile communications. Ground waves are used at frequencies as low as 15 kHz.

The disadvantages of ground-wave propagation are as follows:

1. Ground waves require a relatively high transmission power.
2. Ground waves are limited to low and very low frequencies (LF and VLF), facilitating large antennas (the reason for this is explained in Chapter 11).
3. Ground losses vary considerably with surface material.

The advantages of ground-wave propagation are as follows:

1. Given enough transmit power, ground waves can be used to communicate between any two locations in the world.
2. Ground waves are relatively unaffected by changing atmospheric conditions.

Space-Wave Propagation

Space-wave propagation includes radiated energy that travels in the lower few miles of the earth's atmosphere. Space waves include both direct and ground-reflected waves (see Figure 10-14). *Direct waves* are waves that travel essentially in a straight line between the transmit and receive antennas. Space-wave propagation with direct waves is commonly called *line-of-sight* (LOS) *transmission*. Therefore, space-wave propagation

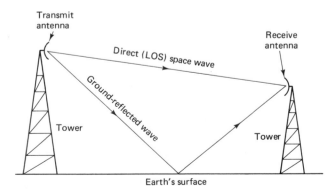

FIGURE 10-14   Space-wave propagation.

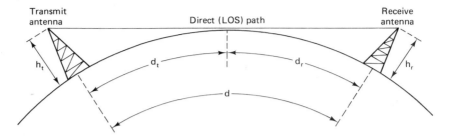

FIGURE 10-15 Space waves and radio horizon.

is limited by the curvature of the earth. Ground-reflected waves are those waves that are reflected by the earth's surface as they propagate between the transmit and receive antennas.

Figure 10-14 shows space-wave propagation between two antennas. It can be seen that the field intensity at the receive antenna depends on the distance between the two antennas (attenuation and absorption) and whether the direct and ground-reflected waves are in phase (interference).

The curvature of the earth presents a horizon to space wave propagation commonly called the *radio horizon*. Due to atmospheric refraction, the radio horizon extends beyond the *optical horizon* for the common *standard atmosphere*. The radio horizon is approximately four-thirds that of the optical horizon. Refraction is caused by the troposphere, due to changes in its density, temperature, water vapor content, and relative conductivity. The radio horizon can be lengthened simply by elevating the transmit or receive antennas (or both) above the earth's surface with towers or by placing them on top of mountains or high buildings.

Figure 10-15 shows the effect of antenna height on the radio horizon. The line-of-sight radio horizon for a single antenna is given as

$$d = \sqrt{2h} \qquad (10\text{-}15)$$

where

d = distance to radio horizon (mi)
h = antenna height above sea level (ft)

Therefore, for a transmit and receive antenna, the distance between the two antennas is

$$d = d_t + d_r$$

or

$$d = \sqrt{2h_t} + \sqrt{2h_r} \qquad (10\text{-}16)$$

where

d = total distance (mi)
d_t = radio horizon for transmit antenna (mi)

d_r = radio horizon for receive antenna (mi)
h_t = transmit antenna height (ft)
h_r = receive antenna height (ft)

or

$$d = 4\sqrt{h_t} + 4\sqrt{h_r} \qquad (10\text{-}17)$$

where d_t and d_r are distance in kilometers and h_t and h_r are height in meters.

From Equations 10-16 and 10-17 it can be seen that the space-wave propagation distance can be extended simply by increasing either the transmit or receive antenna height, or both.

Because the conditions in the earth's lower atmosphere are subject to change, the degree of refraction can vary with time. A special condition called *duct propagation* occurs when the density of the lower atmosphere is such that electromagnetic waves are trapped between it and the earth's surface. The layers of the atmosphere act like a duct and an electromagnetic wave can propagate for great distances around the curvature of the earth within this duct. Duct propagation is shown in Figure 10-16.

Sky-Wave Propagation

Electromagnetic waves that are directed above the horizon level are called *sky waves*. Typically, sky waves are radiated in a direction that produces a relatively large angle with reference to the earth. Sky waves are radiated toward the sky, where they are either reflected or refracted back to earth by the ionosphere. The ionosphere is the region of space located approximately 50 to 400 km (30 to 250 mi) above the earth's surface. The ionosphere is the upper portion of the earth's atmosphere. Therefore, it absorbs large quantities of the sun's radiant energy, which ionizes the air molecules, creating free electrons. When a radio wave passes through the ionosphere, the electric field of the wave exerts a force on the free electrons, causing them to vibrate. The vibrating electrons decrease current, which is equivalent to reducing the dielectric constant. Reducing the dielectric constant increases the velocity of propagation and causes electromagnetic waves to bend away from the regions of high electron density toward regions of low electron density (i.e., increasing refraction). As the wave moves farther from earth, ionization increases. However, there are fewer air molecules to ionize.

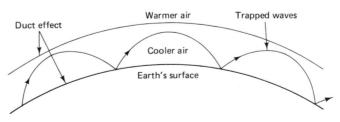

FIGURE 10-16   **Duct propagation.**

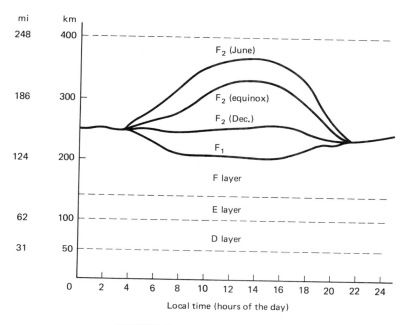

FIGURE 10-17 Ionospheric layers.

Therefore, in the upper atmosphere, there is a higher percentage of ionized molecules than in the lower atmosphere. The higher the ion density, the more refraction. Also, due to the ionosphere's nonuniform composition and its temperature and density variations, it is *stratified*. Essentially, there are three layers that comprise the ionosphere (the D, E, and F layers), which are shown in Figure 10-17. It can be seen that all three layers of the ionosphere vary in location and in *ionization density* with the time of day. They also fluctuate in a cyclic pattern throughout the year, and according to the 11-year *sunspot cycle*. The ionosphere is most dense during times of maximum sunlight (i.e., during the daylight hours and in the summer).

D layer. The *D layer* is the lowest layer of the ionosphere and is located between 30 and 60 mi (50 to 100 km) from the earth's surface. Because it is the layer farthest from the sun, there is very little ionization in this layer. Therefore, the D layer has very little effect on the direction of propagation of radio waves. However, the ions in the D layer can absorb appreciable amounts of electromagnetic energy. The amount of ionization in the D layer depends on the altitude of the sun above the horizon. Therefore, it disappears at night. The D layer reflects VLF and LF waves and absorbs MF and HF waves. (See Table 1-1 for VLF, LF, MF, and HF frequency regions.)

E layer. The *E layer* is located between 60 and 85 mi (100 to 140 km) above the earth's surface. The E layer is sometimes called the *Kennelly–Heaviside layer* after the two scientists who discovered it. The E layer has its maximum density at approximately 70 km at noon, when the sun is at its highest point. Like the D layer, the E layer almost totally disappears at night. The E layer aids MF surface-wave propagation and reflects HF waves somewhat during the daytime. The upper portion of the E layer is sometimes considered separately and called the sporadic E layer because it seems to come and go rather unpredictably. The sporadic E layer is caused by *solar flares* and *sunspot activity*. The sporadic E layer is a thin layer with a very high ionization density. When it appears, there generally is an unexpected improvement in long-distance transmission.

F layer. The *F layer* is actually made up of two layers: the F_1 and F_2 layers. During the daytime, the F_1 layer is located between 85 and 155 mi (140 to 250 km) above the earth's surface, and the F_2 layer is located 85 to 185 mi (140 to 300 km) above the earth's surface during the winter and 155 to 220 mi (250 to 350 km) in the summer. During the night, the F_1 layer combines with the F_2 layer to form a single layer. The F_1 layer absorbs and attenuates some HF waves, although most of the waves pass through to the F_2 layer, where they are refracted back to earth.

PROPAGATION TERMS AND DEFINITIONS

Critical Frequency and Critical Angle

Frequencies above the UHF range are virtually unaffected by the ionosphere because of their extremely short wavelengths. At these frequencies, the distance between ions are appreciably large, and consequently, the electromagnetic waves pass through them with little noticeable effect. Therefore, it stands to reason that there must be an upper frequency limit for sky-wave propagation. *Critical frequency* (F_c) is defined as the highest

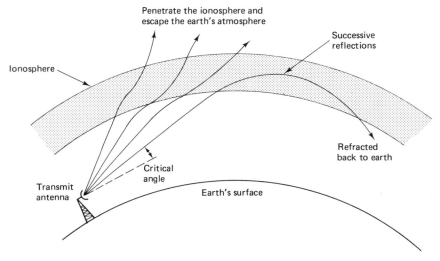

FIGURE 10-18   Critical angle.

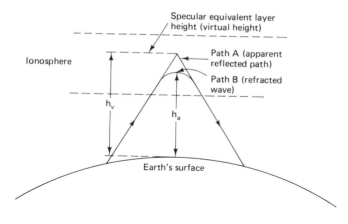

FIGURE 10-19   Virtual and actual height.

frequency that can be propagated directly upward and still be returned to earth by the ionosphere. The critical frequency depends on the ionization density and therefore varies with the time of day and the season. If the vertical angle of radiation is decreased, frequencies at or above the critical frequency can still be refracted back to the earth's surface because they will travel a longer distance in the ionosphere and thus have a longer time to be refracted. Therefore, critical frequency is used only as a point of reference for comparison purposes. However, every frequency has a maximum vertical angle at which it can be propagated and still be refracted back by the ionosphere. This angle is called the *critical angle*. The critical angle θ_c is shown in Figure 10-18.

Virtual Height

Virtual height is the height above the earth's surface from which a refracted wave appears to have been reflected. Figure 10-19 shows a wave that has been radiated from the earth's surface toward the ionosphere. The radiated wave is refracted back to earth and follows path *B*. The actual maximum height that the wave reached is height h_a. However, path *A* shows the projected path that a reflected wave could have taken and still been returned to earth at the same location. The maximum height that this hypothetical reflected wave would have reached is the virtual height (h_v).

Maximum Usable Frequency

The *maximum usable frequency* (MUF) is the highest frequency that can be used for sky-wave propagation between two specific points on the earth's surface. It stands to reason, then, that there are as many values possible for MUF as there are points on earth and frequencies—an infinite number. MUF, like the critical frequency, is a limiting frequency for sky-wave propagation. However, the maximum usable frequency is for a specific angle of incidence (the angle between the incident wave and the normal). Mathematically, MUF is

$$\text{MUF} = \frac{\text{critical frequency}}{\cos \theta} \tag{10-18a}$$

$$= \text{critical frequency} \times \sec \theta \tag{10-18b}$$

where θ is the angle of incidence.

Equations 10-18 are called the *secant law*. The secant law assumes a flat earth and a flat reflecting layer, which of course, can never exist. Therefore, MUF is used only for making preliminary calculations.

Skip Distance

The *skip distance* (t_s) is the minimum distance from a transmit antenna that a sky wave of given frequency (which must be greater than the critical frequency) will be returned to earth. Figure 10-20a shows several waves with different angles of incidence

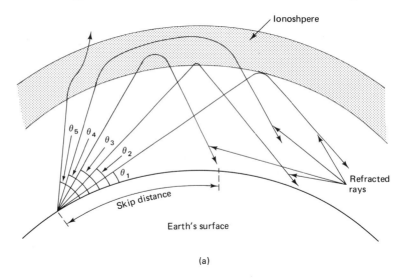

(a)

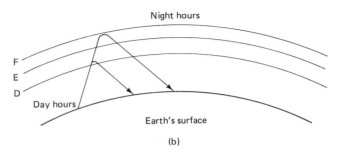

(b)

FIGURE 10-20   Skip distance: (a) skip distance; (b) daytime versus nighttime propagation.

being radiated from the same point on earth. It can be seen that the point where the wave is returned to earth moves closer to the transmitter as the angle of incidence (θ) is increased. Eventually, however, the angle of incidence is sufficiently high that the wave penetrates through the ionosphere and totally escapes the earth's atmosphere.

Figure 10-20b shows the effect on the skip distance of the disappearance of the D and E layers during nighttime. Effectively, the *ceiling* formed by the ionosphere is raised, allowing sky waves to travel higher before being refracted back to earth. This effect explains how far-away radio stations are sometimes heard during the night that cannot be heard during daylight hours. Figure 10-20c shows the effects of *multipath* and *multiple skip* sky-wave propagation.

QUESTIONS

10-1. Describe an electromagnetic ray; a wavefront.

10-2. Describe power density; voltage intensity.

10-3. Describe a spherical wavefront.

10-4. Explain the inverse square law.

10-5. Describe wave attenuation.

10-6. Describe wave absorption.

10-7. Describe refraction. Explain Snell's law for refraction.

10-8. Describe reflection. Explain Snell's law for reflection.

10-9. Describe diffraction. Explain Huygens' principle.

10-10. Describe the composition of a good reflector.

10-11. Describe the atmospheric conditions that cause electromagnetic refraction.

10-12. Define electromagnetic wave interference.

10-13. Describe ground-wave propagation. List its advantages and disadvantages.

10-14. Describe space-wave propagation.

10-15. Explain why the radio horizon is at a greater distance than the optical horizon.

10-16. Describe the various layers of the ionosphere.

10-17. Describe sky-wave propagation.

10-18. Explain why atmospheric conditions vary with time of day, month of year, and so on.

10-19. Define *critical frequency*; *critical angle*.

10-20. Describe virtual height.

10-21. Define *maximum usable frequency*.

10-22. Define *skip distance* and give the reasons that it varies.

PROBLEMS

10-1. Determine the power density for a radiated power of 1000 W at distance of 20 km from an isotropic antenna.

10-2. Determine the power density for Problem 10-1 for a point that is 30 km from the antenna.

10-3. Describe the effects on power density if the distance from a transmit antenna is tripled.

10-4. Determine the radio horizon for a transmit antenna that is 100 ft high and a receiving antenna that is 50 ft high; 100 m and 50 m.

10-5. Determine the maximum usable frequency for a critical frequency of 10 MHz and an angle of incidence of 45°.

10-6. Determine the voltage intensity for the same point in Problem 10-1.

10-7. Determine the voltage intensity for the same point in Problem 10-2.

10-8. For a radiated power $P_r = 10$ kW, determine the voltage intensity at a distance 20 km from the source.

10-9. Determine the change in power density when the distance from the source increases by a factor of 4.

10-10. If the distance from the source is reduced to one-half its value, what affect does this have on the power density?

10-11. The power density at a point from a source is 0.001 μW and the power density at another point is 0.00001 μW; determine the attenuation in decibels.

10-12. For a dielectric ratio $\sqrt{\epsilon_2/\epsilon_1} = 0.8$ and an angle of incidence $\theta_i = 26°$, determine the angle of refraction, θ_r.

10-13. Determine the distance to the radio horizon for an antenna located 40 ft above sea level.

10-14. Determine the distance to the radio horizon for an antenna that is 40 ft above the top of a 4000-ft mountain peak.

10-15. Determine the maximum distance between identical antennas equally distant above sea level for Problem 10-13.

Chapter 11

ANTENNAS

INTRODUCTION

In essence, an *antenna* is a metallic conductor system capable of transmitting and receiving electromagnetic waves. An antenna is used to interface a transmitter to free space or free space to a receiver. An antenna couples energy from the output of a transmitter to the earth's atmosphere or from the earth's atmosphere to the input of a receiver. An antenna is a *passive reciprocal device*: *passive* in that it cannot actually amplify a signal, at least not in the true sense of the word (however, as you will see later in this chapter, an antenna can have *gain*); and *reciprocal* in that the transmit and receive characteristics of an antenna are identical, except where *feed* currents to the antenna element are tapered to modify the transmit pattern.

Basic Antenna Operation

Basic antenna operation is best understood by looking at the voltage standing-wave patterns on a transmission line, which are shown in Figure 11-1a. The transmission line is terminated in an open circuit, which represents an abrupt discontinuity to the incident voltage wave in the form of a phase reversal. The phase reversal causes some of the incident voltage to be radiated, not reflected back toward the source. The radiated energy propagates away from the antenna in the form of transverse electromagnetic waves. The *radiation efficiency* of an open transmission line is extremely low. Radiation efficiency is the ratio of radiated to reflected energy. To radiate more energy, simply spread the conductors farther apart. Such an antenna is called a *dipole* (meaning two poles) and is shown in Figure 11-1b.

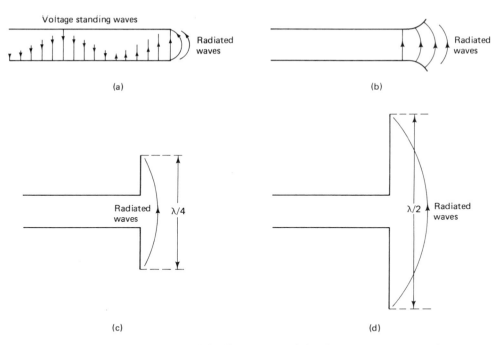

FIGURE 11-1 Radiation from a transmission line: (a) transmission-line radiation; (b) spreading conductors; (c) Marconi antenna; (d) Hertz antenna.

In Figure 11-1c the conductors are spread out in a straight line to a total length of one-quarter wavelength. Such an antenna is called a basic *quarter-wave dipole* or a *Marconi antenna*. A half-wave dipole is called a *Hertz antenna* and is shown in Figure 11-1d.

TERMS AND DEFINITIONS

Radiation Pattern

A *radiation pattern* is a *polar* diagram or graph representing field strengths or power densities at various points in space relative to an antenna. If the radiation pattern is plotted in terms of electric field strength ($\mathscr{E} = V/m$) or power density ($\mathscr{P} = W/m^2$), it is called an *absolute* radiation pattern. If it plots field strength or power density in respect to a reference point, it is called a *relative* radiation pattern. Figure 11-2a shows an absolute radiation pattern for an unspecified antenna. The pattern is plotted on *polar* coordinate paper with the heavy solid line representing points of equal power density ($10~\mu W/m^2$). The circular gradients indicate distance in 2-km steps. It can be seen that maximum radiation is in a direction $90°$ from the reference. The power density 10 km from the antenna in a $90°$ direction is $10~\mu W/m^2$. In a $45°$ direction, the point of equal

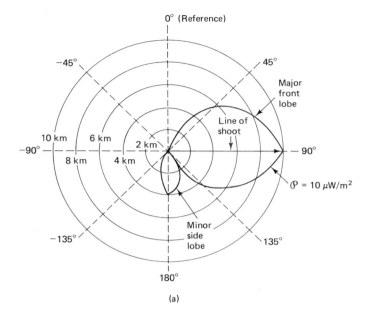

(a)

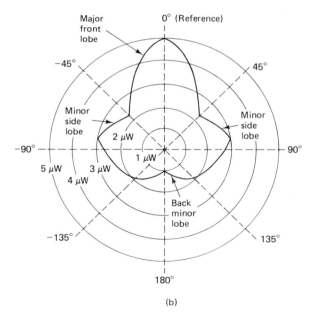

(b)

FIGURE 11-2 Radiation patterns: (a) absolute radiation pattern; (b) relative radiation pattern; (c) relative radiation pattern in decibels; (d) relative radiation pattern for an omnidirectional antenna.

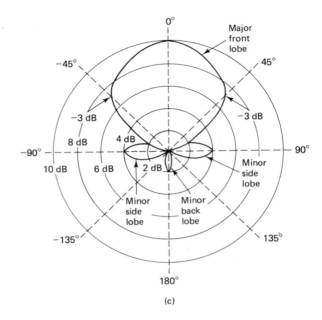

(c)

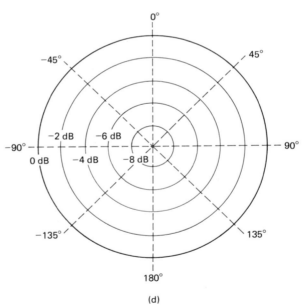

(d)

FIGURE 11-2 (*continued*)

power density is 5 km from the antenna; at 180°, only 4 km; and in a −90° direction, there is essentially no radiation.

In Figure 11-2a the primary beam is in a 90° direction and is called the *major lobe*. There can be more than one major lobe. There is also a *secondary* beam or *minor* lobe in a −180° direction. Normally, minor lobes represent undesired radiation or reception. Because the major lobe propagates and receives the most energy, that lobe is called the *front* lobe (i.e., the front of the antenna). Lobes adjacent to the front lobe are called *side* lobes (the 180° minor lobe is a side lobe), and lobes in a direction exactly opposite the front lobe are called *back* lobes (there is no back lobe shown on this pattern). The ratio of the front lobe to the back lobe is simply called the *front-to-back ratio*, and the ratio of the front lobe to a side lobe is called the *front-to-side ratio*. The line bisecting the major lobe or pointing from the center of the antenna in the direction of maximum radiation is called the *line of shoot*.

Figure 11-2b shows a relative radiation pattern for an unspecified antenna. The heavy solid line represents points of equal distance from the antenna (10 km), and the circular gradients indicate power density in 1 μW/m^2 divisions. It can be seen that maximum radiation (5 μW/m^2) is in the direction of the reference (0°) and the antenna radiates the least power (1 μW/m^2) in a direction 180° from the reference. Consequently, the front-to-back ratio is 5 : 1 = 5. Generally, relative field strength and power density are plotted in decibels (dB), where dB = 20 log (ϵ/ϵ_{max}) or 10 log ($\mathscr{P}/\mathscr{P}_{max}$). Figure 11-2c shows a relative radiation pattern for power density in decibels. In a direction ±45° from the reference, the power density is −3 dB (half power) relative to the power density in the direction of maximum radiation (0°). Figure 11-2d shows a relative radiation pattern for power density for an omnidirectional antenna. As stated previously, an omnidirectional antenna radiates energy equally in all directions; therefore, the radiation pattern is simply a circle (actually, a sphere). Also, with an omnidirectional antenna, there are no front, back, or side lobes because radiation is equal in all directions.

The radiation patterns shown in Figure 11-2 are two-dimensional. However, radiation from an actual antenna is three-dimensional. Therefore, radiation patterns are taken in both the horizontal (from the top) and the vertical (from the side) planes. For the omnidirectional antenna shown in Figure 11-2d, the radiation patterns in the horizontal and vertical planes are circular and equal because the actual radiation pattern for an isotropic radiator is a sphere.

Near and Far Fields

The radiation field that is close to an antenna is not the same as the radiation field that is at a great distance. The term *near field* refers to the field pattern that is close to the antenna, and the term *far field* refers to the field pattern that is at great distance. During one half of a cycle, power is radiated from an antenna where some of the power is stored temporarily in the near field. During the second half of the cycle, power in the near field is returned to the antenna. This action is similar to the way in which an inductor stores and releases energy. Therefore, the near field is sometimes called the *induction field*. Power that reaches the far field continues to radiate outward and is never returned to the antenna. Therefore, the far field is sometimes called the *radiation field*. Radiated power is usually the more important of the two; therefore, antenna radiation patterns are generally given for the far field. The near field is defined as the area within a distance D^2/λ from the antenna where λ is the wavelength and D the antenna diameter in the same units.

Radiation Resistance and Antenna Efficiency

All of the power supplied to an antenna is not radiated. Some of it is converted to heat and dissipated. *Radiation resistance* is somewhat "unreal" in that it cannot be measured directly. Radiation resistance is an ac antenna resistance and is equal to the ratio of the power radiated by the antenna to the square of the current at its feed point. Mathematically, radiation resistance is

$$R_r = \frac{P}{i^2} \qquad (11-1)$$

where

$$R_r = \text{radiation resistance}$$
$$P = \text{rms power radiated by the antenna}$$
$$i = \text{rms antenna current at the feedpoint}$$

Radiation resistance is the resistance which, if it replaced the antenna, would dissipate exactly the same amount of power that the antenna radiates.

Antenna efficiency is the ratio of the power radiated by an antenna to the sum of the power radiated and the power dissipated or the ratio of the power radiated by the antenna to the total input power. Mathematically, antenna efficiency is

$$\eta = \frac{P_r}{P_r + P_d} \times 100 \qquad (11-2)$$

where

$$\eta = \text{antenna efficiency (\%)}$$
$$P_r = \text{power radiated by antenna}$$
$$P_d = \text{power dissipated in antenna}$$

Figure 11-3 shows a simplified electrical equivalent circuit for an antenna. Some of the input power is dissipated in the dc resistances (ground resistance, corona, imperfect dielectrics, eddy currents, etc.) and the remainder is radiated. The total antenna power is the sum of the dissipated and radiated powers. Therefore, in terms of resistance and current, antenna efficiency is

$$\eta = \frac{i^2 R_r}{i^2(R_r + R_{dc})} = \frac{R_r}{R_r + R_{dc}} \qquad (11-3)$$

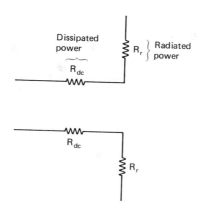

FIGURE 11-3 Simplified equivalent circuit of an antenna.

where

η = antenna efficiency
i = antenna current
R_r = radiation resistance
R_{dc} = dc antenna resistance

Directive Gain and Power Gain

The terms *directive gain* and *power gain* are often misunderstood and, consequently, misused. *Directive gain* is the ratio of the power density radiated in a particular direction to the power density radiated to the same point by a reference antenna, assuming both antennas are radiating the same amount of power. The relative power density radiation pattern for an antenna is actually a directive gain pattern if the power density reference is taken from a standard reference antenna, which is generally an isotropic antenna. The maximum directive gain is called *directivity*. Mathematically, directive gain is

$$\mathscr{D} = \frac{\mathscr{P}}{\mathscr{P}_{ref}}$$

(11-4)

where

$\mathscr{D}$ = directive gain
$\mathscr{P}$ = power density at some point with a given antenna
$\mathscr{P}_{ref}$ = power density at the same point with a reference antenna

Power gain is the same as directive gain except that the total power fed to the antenna is used (i.e., antenna efficiency is taken into account). It is assumed that the given antenna and the reference antenna have the same input power and that the reference antenna is lossless (i.e., η = 100%). Mathematically, power gain (A_p) is

$$A_p = \mathscr{D}\eta$$

(11-5)

If an antenna is lossless, it radiates 100% of the input power and the power gain is equal to the directive gain. The power gain for an antenna is also given in decibels relative to some reference antenna. Therefore, power gain is

$$A_p = 10 \log \frac{\mathcal{P}\eta}{\mathcal{P}_{\text{ref}}} \tag{11-6}$$

For an isotropic reference the directivity of a half-wave dipole is approximately 1.5 (1.76 dB). It is usual to state the power gain in decibels if referred to a λ/2 dipole. However, if referenced to an isotropic radiator, the decibel figure is stated as dBi, or dB/isotropic radiator and is 1.76 dB greater than if a half-wave dipole were used for the reference. It is important to note that the power radiated from an antenna can never exceed the input power. Therefore, the antenna does not actually amplify the input power. An antenna simply concentrates its radiated power in a particular direction. Therefore, points that are located in areas where the radiated power is concentrated realize an apparent gain relative to the power density at the same points had an isotropic antenna been used. If gain is realized in one direction, a corresponding reduction in power density (a loss) must be realized in another direction. The direction in which an antenna is ''pointing'' is always the direction of maximum radiation. Because an antenna is a reciprocal device, its radiation pattern is also its reception pattern. For maximum *captured* power, a receive antenna must be pointing in the direction from which reception is desired. Therefore, receive antennas have directivity and power gain just like transmit antennas.

Effective Isotropic Radiated Power

Effective isotropic radiated power (EIRP) is defined as an equivalent transmit power and is expressed mathematically as

$$\text{EIRP} = P_r A_t \qquad \text{watts} \tag{11-7a}$$

where

$\qquad P_r$ = total radiated power
$\qquad A_t$ = transmit antenna directive gain

or

$$\text{EIRP (dBm)} = 10 \log \frac{P_r}{0.001} + 10 \log A_t \tag{11-7b}$$

Equation 11-7a can be rewritten using antenna input power and power gain as

$$\text{EIRP} = P_{\text{in}} A_p \tag{11-7c}$$

EIRP or simply ERP (effective radiated power) is the equivalent power that an isotropic antenna would have to radiate to achieve the same power density in the chosen direction at a given point as another antenna. For instance, if a given antenna has a power gain of 10, the power density is 10 times greater than it would have been had the antenna been an isotropic radiator. An isotropic antenna would have to radiate 10 times as much power in the desired direction to achieve the same power density. Therefore, the given antenna "effectively" radiates 10 times as much power as an isotropic antenna with the same input power and efficiency.

To determine the power density at a given point, Equation 10-5 is expanded to include the transmit antenna gain and rewritten as

$$\mathcal{P} = \frac{P_r A_t}{4\pi R^2} \tag{11-8a}$$

EXAMPLE 11-1

If a transmit antenna has a directive gain $A_t = 10$ and radiated power $P_r = 100$ W, determine:
(a) The EIRP.
(b) The power density at a point 10 km away.
(c) The power density had an isotropic antenna been used with the same input power and efficiency.

Solution (a) Substituting into Equation 11-7a yields

$$\text{EIRP} = P_r A_t = 100 \text{ W} \times 10 = 1000 \text{ W}$$

(b) Substituting into Equation 11-8a gives us

$$\mathcal{P} = \frac{P_r A_t}{4\pi R^2} = \frac{\text{EIRP}}{4\pi R^2} = \frac{1000 \text{ W}}{4\pi(10,000 \text{ m})^2} = 0.796 \text{ } \mu\text{W/m}^2$$

(c) Substituting into Equation 10-5, we obtain

$$\mathcal{P} = \frac{P_r}{4\pi R^2} = \frac{100}{4\pi(10,000 \text{ m})^2} = 0.0796 \text{ } \mu\text{W/m}^2$$

It can be seen that the power density at a point 10,000 m from the transmit antenna is 10 times greater with the first antenna than with the isotropic radiator. To achieve the same power density, the isotrope would have to radiate 1000 W. Therefore, the first antenna effectively radiates 1000 W.

As stated previously, antennas are reciprocal devices; an antenna has the same power gain and directivity when it is used to receive electromagnetic waves as it has for transmitting electromagnetic waves. Consequently, the power received or captured by an antenna is the product of the power density in the space immediately surrounding the antenna and the antenna directive gain. Therefore, Equation 11-8a can be expanded to

$$\text{captured power} = c = \frac{P_r A_t A_r}{4\pi R^2} \tag{11-8b}$$

In Example 11-1, if an antenna that was identical to the transmit antenna were used to receive the signal, the captured power would be

$$c = \mathscr{P}A_r$$
$$= (0.0796 \ \mu W/m^2)(10) = 0.796 \ \mu W$$

The captured power is not all useful; some of it is dissipated in the receive antenna. The actual useful received power is the product of the received power density, the receive antenna's direct gain, and the receive antenna's efficiency or the receive power density times the receive antenna's power gain.

Antenna Polarization

The *polarization* of an antenna refers simply to the orientation of the electric field radiated from it. An antenna may be *linearly* (generally, either horizontally or vertically, assuming that the antenna elements lie in a horizontal or vertical plane), *elliptically*, or *circularly polarized*. If an antenna radiates a vertically polarized electromagnetic wave, the antenna is defined as vertically polarized; if an antenna radiates a horizontally polarized electromagnetic wave, the antenna is said to be horizontally polarized; if the radiated electric field rotates in an elliptical pattern, it is elliptically polarized; and if the electric field rotates in a circular pattern, it is circularly polarized.

Antenna Beamwidth

Antenna *beamwidth* is simply the angular separation between the two half-power (-3 dB) points on the major lobe of an antenna's plane radiation pattern, usually taken in one of the "principal" planes. The beamwidth for the antenna whose radiation pattern is shown in Figure 11-4 is the angle formed between points A, X, and B (angle θ). Points A and B are the half-power points (the power density at these points is one-half of what it is an equal distance from the antenna in the direction of maximum radiation). Antenna beamwidth is sometimes called -3-dB beamwidth or half-power beamwidth.

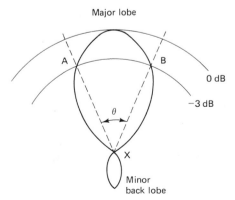

FIGURE 11-4   Antenna beamwidth.

Antenna Bandwidth

Antenna *bandwidth* is vaguely defined as the frequency range over which antenna operation is "satisfactory." This is normally taken between the half-power points, but it sometimes refers to variations in the antenna's input impedance.

Antenna Input Impedance

Radiation from an antenna is a direct result of the flow of RF current. The current flows to the antenna through a transmission line, which is connected to a small gap between the conductors that make up the antenna. The point on the antenna where the transmission line is connected is called the antenna input terminal or simply the *feedpoint*. The feedpoint presents an ac load to the transmission line called the *antenna input impedance*. If the transmitter's output impedance and the antenna's input impedance are equal to the characteristic impedance of the transmission line, there will be no standing waves on the line, and maximum power is transferred to the antenna and radiated.

Antenna input impedance is simply the ratio of the antenna's input voltage to input current. Mathematically, input impedance is

$$Z_{in} = \frac{E_i}{I_i}$$

(11-9)

where

Z_{in} = antenna input impedance
E_i = antenna input voltage
I_i = antenna input current

Antenna input impedance is generally complex. However, if the feedpoint is at a current maximum and there is no reactive component, the input impedance is equal to the sum of the radiation resistance and the dc resistance.

BASIC ANTENNAS

Elementary Doublet

The simplest type of antenna is the *elementary doublet*. The elementary doublet is an electrically short dipole and is often referred to simply as a *short dipole*. "Electrically short" means short compared to one-half wavelength (generally, any dipole that is less than one-tenth wavelength long is considered electrically short). In reality, an elementary doublet cannot be achieved. However, the concept of a short dipole is useful in understanding more practical antennas.

An elementary doublet is a short dipole that has uniform current throughout its length. However, the current is assumed to vary sinusoidally in time and at any instant is

$$i = I \sin (2\pi ft + \theta)$$

where

i = instantaneous current
I = peak amplitude of the RF current
f = frequency (Hz)
t = instantaneous time
θ = phase angle

With the aid of Maxwell's equations, it can be shown that the far (radiation) field is

$$\mathscr{E} = \frac{60\pi Il \sin \theta}{\lambda R} \qquad (11\text{-}10)$$

where

$\mathscr{E}$ = electric field intensity (V rms/m)
I = dipole current (A rms)
l = length of the dipole (m)
R = distance from the dipole (m)
λ = wavelength (m)
θ = phase angle

Plotting Equation 11-10 gives the relative electric field intensity pattern for an elementary dipole, and is shown in Figure 11-5. It can be seen that radiation is maximum at right angles to the dipole and falls off to zero at the ends.

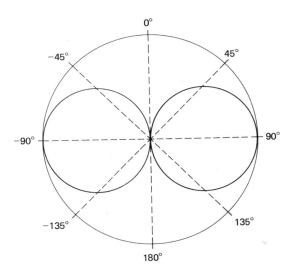

FIGURE 11-5 Relative radiation pattern for an elementary doublet in a plane perpendicular to the dipole axis.

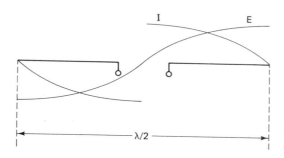

FIGURE 11-6 Idealized voltage and current distributions along a half-wave dipole.

The relative power density pattern can be derived from Equation 11-10 by substituting $\mathscr{P} = \mathscr{E}^2/120\pi$. Mathematically, we have

$$\mathscr{P} = \frac{30\pi I^2 l^2 \sin^2 \theta}{\lambda^2 R^2}$$ (11-11)

Half-wave Dipole

The linear half-wave dipole is one of the most widely used antennas at frequencies above 2 MHz. At frequencies below 2 MHz, the physical length of a half-wavelength antenna is prohibitive. The half-wave dipole is generally referred to as a *Hertz antenna.*

A Hertz antenna is a *resonant* antenna. That is, it is a multiple of quarter-wavelengths long and open circuited at the far end. Standing waves of voltage and current exist along a resonant antenna. Figure 11-6 shows the idealized voltage and current distributions along a half-wave dipole for each half-cycle of the input signal. Each pole of the antenna looks like an open quarter-wavelength section of transmission line. Therefore, there is a voltage maximum and current minimum at the ends and a voltage minimum and current maximum in the middle. Consequently, assuming that the feedpoint is in the center of the antenna, the input impedance is E_{min}/I_{max} and a minimum value. The impedance at the ends of the antenna is E_{max}/I_{min} and a maximum value. Figure 11-7 shows the impedance curve for a center-fed half-wave dipole. The impedance varies from a maximum value at the ends of approximately 2500 Ω to a minimum value at the feedpoint of approximately 73 Ω (of which between 68 and 70 Ω is the radiation resistance).

A wire radiator such as a half-wave dipole can be thought of as an infinite number

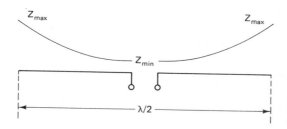

FIGURE 11-7 Impedance curve for a center fed half-wave dipole.

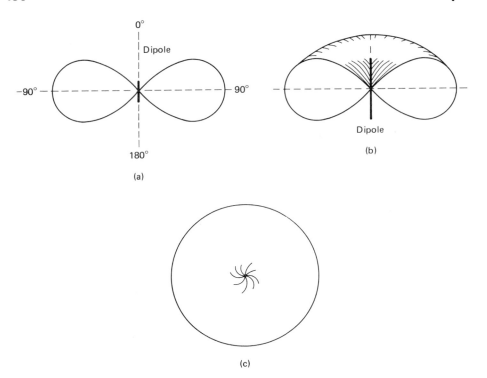

FIGURE 11-8 Half-wave dipole radiation patterns: (a) vertical view of a vertically mounted dipole; (b) cross-sectional view; (c) horizontal view.

of elementary doublets placed end to end. Therefore, the radiation pattern can be obtained by integrating Equation 11-10 over the length of the antenna. The free-space radiation pattern for a half-wave dipole depends on whether the antenna is placed horizontally or vertically in respect to the earth's surface. Figure 11-8a shows the vertical (from the side) radiation pattern for a vertically mounted half-wave dipole. Note that there are two major lobes radiating in opposite directions which are at right angles to the antenna. Also note that the lobes are not circles. Circular lobes are obtained only for the ideal case when the current is constant throughout the antenna's length and this is, of course, unachievable in a practical antenna. Figure 11-8b shows the cross-sectional view. Note that the radiation pattern has a figure-eight pattern and resembles the shape of a doughnut. Maximum radiation is in a plane parallel to the earth's surface. The higher the angle of elevation (θ), the less the radiation, and for $\theta = 90°$ there is no radiation. Figure 11-8c shows the horizontal (from the top) radiation pattern for a vertically mounted half-wave dipole. The pattern is circular because radiation is uniform in all directions perpendicular to the antenna.

Ground effects on a half-wave dipole.   The radiation patterns shown in Figure 11-8 are for free-space conditions. In the earth's atmosphere, wave propagation is affected

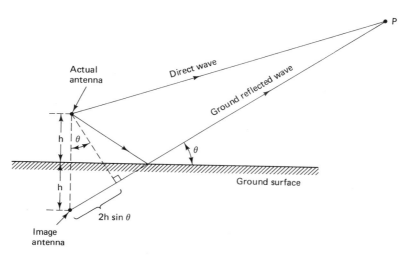

Actual
antenna

Direct wave

Ground reflected wave

P

h θ

θ

Ground surface

h

$2h \sin \theta$

Image
antenna

FIGURE 11-9   Ground effects on a half-wave dipole.

by antenna orientation, atmospheric absorption, and ground effects such as reflection. The effect of ground reflection on the radiation pattern for an ungrounded half-wave dipole is shown in Figure 11-9. The antenna is mounted an appreciable number of wavelengths (height h) above the surface of the earth. The field strength at any given point in space is the sum of the direct and ground reflected waves. The ground reflected wave appears to be radiating from an image antenna distance h below the earth's surface. This apparent antenna is a mirror image of the actual antenna. The ground reflected wave is inverted 180° and travels a distance $2h \sin \theta$ farther than the direct wave to reach the same point in space (point P). The resulting radiation pattern is a summation of the radiations from the actual antenna and the mirror antenna. Note that this is the classical ray-tracing technique used in optics.

Figure 11-10 shows the vertical radiation patterns for a horizontally mounted half-wave dipole one-quarter and one-half wavelength above the ground. For an antenna mounted one-quarter wavelength above the ground, the lower lobe is completely gone and the field strength directly upward is doubled. Figure 11-10a shows the vertical distribution in a plane parallel (and through) the antenna, and Figure 11-10b shows the vertical distribution in a plane at right angles to the antenna. Figure 11-10c shows the vertical radiation pattern for a horizontal dipole one-half wavelength above the ground. The figure shows that the pattern is now broken into two lobes and the direction of maximum radiation (end view) is now at 30° to the horizontal instead of directly upward. There is no component along the ground for horizontal polarization because of the phase shift of the reflected component. Ground-reflected waves have similar effects on all antennas. The best way to eliminate or reduce the effect of ground reflected waves is to mount the antenna far enough above the earth's surface to obtain free-space conditions. However, in many applications, this is impossible. Ground reflections are sometimes desirable to get the desired elevation angle for the major lobe maximum response.

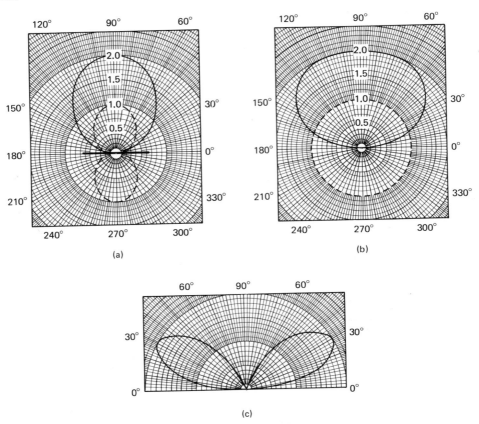

FIGURE 11-10   Vertical radiation pattern for a half-wave dipole.

The height of an ungrounded antenna above the earth's surface also affects the antenna's radiation resistance. This is due to the reflected waves cutting through or intercepting the antenna and altering its current. Depending on the phase of the ground reflected wave, the antenna current can increase or decrease, causing a corresponding increase or decrease in the input impedance.

Grounded Antenna

A *monopole* (single pole) antenna one-quarter wavelength long mounted vertically with the lower end either connected directly to ground or grounded through the antenna coupling network is called a *Marconi antenna*. The characteristics of a Marconi antenna are similar to those of the Hertz antenna because of the ground reflected waves. Figure 11-11 shows the voltage and current standing waves for a quarter-wave grounded antenna. It can be seen that if the Marconi antenna is mounted directly on the earth's surface, the actual antenna and its *image* combine and produce exactly the same standing-wave

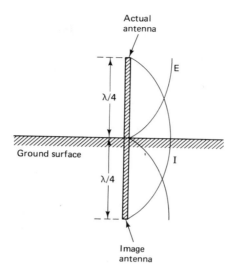

FIGURE 11-11 Voltage and current standing waves for a quarter-wave grounded antenna.

patterns as those of the half-wave ungrounded (Hertz) antenna. Current maxima occur at the grounded ends, which causes high current flow through ground. To reduce power losses, the ground should be a good conductor, such as rich, loamy soil. If the ground is a poor conductor, such as sandy or rocky terrain, an artificial *ground plane* system made of heavy copper wires spread out radially below the antenna may be required. Another way of artificially improving the conductivity of the ground area below the antenna is with a *counterpoise*. A counterpoise is a wire structure placed below the antenna and erected above the ground. The counterpoise should be insulated from earth ground. A counterpoise is a form of capacitive ground system; capacitance is formed between the counterpoise and the earth's surface.

Figure 11-12 shows the radiation pattern for a quarter-wave grounded (Marconi) antenna. It can be seen that the lower half of each lobe is canceled by the ground reflected waves. This is generally of no consequence because radiation in the horizontal direction is increased, thus increasing radiation along the earth's surface (ground waves)

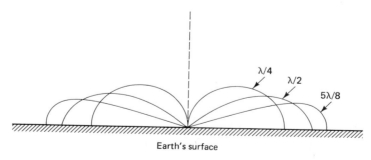

Earth's surface

FIGURE 11-12 Grounded antenna radiation patterns.

and improving area coverage. It can also be seen that increasing the antenna length improves horizontal radiation at the expense of sky-wave propagation. This is also shown in Figure 11-12. Optimum horizontal radiation occurs for an antenna that is approximately five-eighths wavelength long. For a one-wavelength antenna, there is no ground-wave propagation.

A Marconi antenna has the obvious advantage over a Hertz antenna of being only half as long. The disadvantage of a Marconi antenna is that it must be located close to the ground.

Antenna Loading

Thus far we have considered antenna length in terms of wavelengths rather than physical dimensions. By the way, how long is a quarter-wavelength antenna? For a transmit frequency of 1 GHz, one-quarter wavelength is 0.075 m (2.95 in.). However, for a transmit frequency of 1 MHz, one-quarter wavelength is 75 m, and at 100 kHz, one-quarter wavelength is 750 m. It is obvious that the physical dimensions for low-frequency antennas are not practical, especially for mobile radio applications. However, it is possible to increase the electrical length of an antenna by a technique called *loading*. When an antenna is loaded, its physical length remains unchanged although its effective electrical length is increased. There are several techniques used for loading antennas.

Loading coils. Figure 11-13a shows how a coil (inductor) added in series with a dipole antenna effectively increases the antenna's electrical length. Such a coil is appropriately called a *loading coil*. The loading coil effectively cancels out the capacitance component of the antenna input impedance. Thus the antenna looks like a resonant circuit, is resistive, and can now absorb 100% of the incident power. Figure 11-13b shows the current standing-wave patterns on an antenna with a loading coil. The loading coil is generally placed at the bottom of the antenna, allowing the antenna to be easily

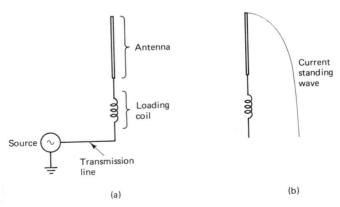

FIGURE 11-13 **Loading coil: (a) antenna with loading coil; (b) current standing wave with loading coil.**

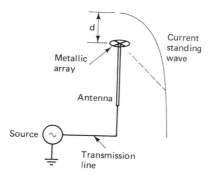

FIGURE 11-14 Antenna top loading.

tuned to resonance. A loading coil effectively increases the radiation resistance of the antenna by approximately 5 Ω. Note also that the current standing wave has a maximum value at the coil, increasing power losses, creating a situation of possible corona, and effectively reducing the radiation efficiency of the antenna.

Top loading. Loading coils have several shortcomings that can be avoided by using a technique called antenna *top loading*. With top loading, a metallic array that resembles a spoked wheel is placed on top of the antenna. The wheel increases the shunt capacitance to ground, reducing the overall antenna capacitance. Antenna top loading is shown in Figure 11-14. Notice that the current standing-wave pattern is pulled up along the antenna as though the antenna length had been increased distance *d*, placing the current maximum at the base. Top loading results in a considerable increase in the radiation resistance and radiation efficiency. It also reduces the voltage of the standing wave at the antenna base. Unfortunately, top loading is awkward for mobile applications.

The current loop of the standing wave can be raised even further (improving the radiation efficiency even more) if a *flat top* is added to the antenna. If a vertical antenna is folded over on top to form an L or T, as shown in Figure 11-15, the current loop will occur nearer the top of the radiator. If the flat-top and vertical portions are each one-quarter wavelength long, the current maximum will occur at the top of the vertical radiator.

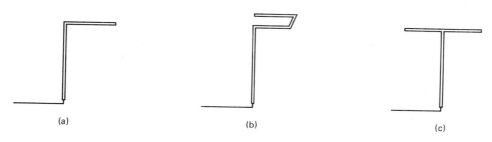

FIGURE 11-15 Flat top antenna loading.

ANTENNA ARRAYS

An antenna *array* is formed when two or more antenna elements are combined to form a single antenna. An antenna element is an individual radiator such as a half- or quarter-wave dipole. The elements are physically placed in such a way that their radiation fields interact with each other, producing a total radiation pattern that is the vector sum of the individual fields. The purpose of an array is to increase the directivity of an antenna and concentrate the radiated power within a smaller geographic area.

In essence, there are two types of antenna elements: *driven* and *parasitic* (non-driven). Driven elements are elements that are directly connected to the transmission line and receive power from or are driven by the source. Parasitic elements are not connected to the transmission line; they receive energy only through mutual induction with a driven element. A parasitic element that is longer than the driven element from which it receives energy is called a *reflector*. A reflector effectively reduces the signal strength in its direction and increases it in the opposite direction. Therefore, it acts like a concave mirror. This action occurs because the wave passing through the parasitic element induces a voltage that is reversed 180° in respect to the wave that induced it. The induced voltage produces an in-phase current and the element radiates (it actually reradiates the energy it just received). The reradiated energy sets up a field that cancels in one direction and reinforces in the other. A parasitic element that is shorter than its associated driven element is called a *director*. A director increases field strength in its direction and reduces it in the opposite direction. Therefore, it acts like a convergent convex lens. This is shown in Figure 11-16.

Radiation directivity can be increased in either the horizontal or the vertical plane, depending on the placement of the elements and whether they are driven or not, and if not, whether they are reflectors or directors.

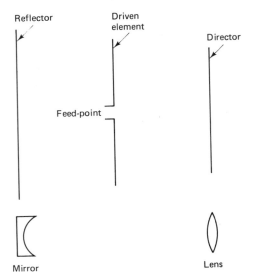

FIGURE 11-16   Antenna array.

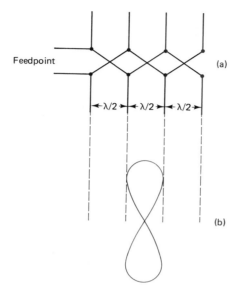

FIGURE 11-17 Broadside antenna: (a) broadside array; (b) radiation pattern.

Broadside Array

A *broadside array* is one of the simplest types of antenna arrays. It is made by simply placing several resonant dipoles of equal size (both length and diameter) in parallel with each other and in a straight line (i.e., collinear). All elements are fed in phase from the same source. As the name implies, a broadside array radiates at right angles to the plane of the array, and radiates very little in the direction of the plane. Figure 11-17a shows a broadside array that is comprised of four driven half-wave elements separated by one-half wavelength. Therefore, the signal that is radiated from element 2 has traveled one-half wavelength farther than the signal radiated from element 1 (i.e., they are radiated 180° out of phase). Crisscrossing the transmission line produces an additional 180° phase shift. Therefore, the currents in all of the elements are in phase and the radiated signals are in phase and additive in a plane at right angles to the plane of the array. Although the horizontal radiation pattern for each element by itself is omnidirectional, when combined their fields produce a highly directive bidirectional radiation pattern (Figure 11-17b). Directivity can be increased even further by increasing the length of the array (i.e., adding more elements).

End-Fire Array

An *end-fire array* is essentially the same element configuration as the broadside array except that the transmission line is not crisscrossed between elements. As a result, the fields are additive in line with the plane of the array. Figure 11-18 shows an end-fire array and its resulting radiation pattern.

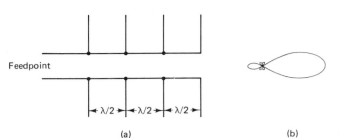

Feedpoint

←λ/2→←λ/2→←λ/2→

(a)   (b)

FIGURE 11-18   End-fire antenna: (a) end-fire array; (b) radiation pattern.

Nonresonant Array—The Rhombic Antenna

The *Rhombic antenna* is a nonresonant antenna that is capable of operating satisfactorily over a relatively wide bandwidth, making it ideally suited for HF transmission (range 3 to 30 MHz). The Rhombic antenna is made up of four nonresonant elements each several wavelengths long. The entire array is terminated in a resistor if unidirectional operation is desired. The most widely used arrangement for the Rhombic antenna resembles a transmission line that has been pinched out in the middle and is shown in Figure 11-19. The antenna is mounted horizontally and placed one-half wavelength or more above the ground. The exact height depends on the precise radiation pattern desired. Each set of elements acts like a transmission line terminated in its characteristic impedance; thus waves are radiated only in the forward direction. The terminating resistor absorbs approximately one-third of the total antenna input power. Therefore, a Rhombic antenna has a maximum efficiency of 67%. Gains of over 40 (16 dB) have been achieved with Rhombic antennas.

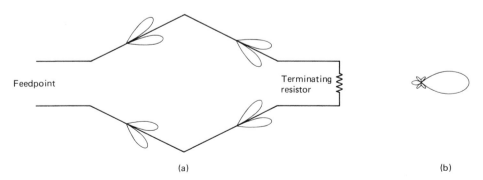

Feedpoint   Terminating
                                                       resistor

(a)   (b)

FIGURE 11-19   Rhombic antenna: (a) Rhombic array; (b) radiation pattern.

SPECIAL-PURPOSE ANTENNAS

Folded Dipole

A two-wire *folded dipole* and its associated voltage standing-wave pattern are shown in Figure 11-20a. The folded dipole is essentially a single antenna made up of two elements. One element is fed directly, while the other is conductively coupled at the ends. Each element is one-half wavelength long. However, because current can flow around corners, there is a full wavelength of current on the antenna. Therefore, for the same input power, the input current will be one-half that of the basic half-wave dipole and the input impedance is four times higher ($4 \times 72 = 288$). The input impedance of a folded dipole is equal to the half-wave impedance (72 Ω) times the number of folded wires squared. For example, if there are three dipoles, as shown in Figure 11-20b, the input impedance is $3^3 \times 72 = 648$ Ω. Another advantage of a folded dipole over a basic half-wave dipole is wider bandwidth. The bandwidth can be increased even further by making the dipole elements larger in diameter (such an antenna is appropriately called a *fat dipole*). However, fat dipoles have slightly different current distributions and input impedance characteristics than thin ones.

Yagi–Uda antenna.   A widely used antenna that commonly uses a folded dipole as the driven element is the *Yagi–Uda antenna*, named after two Japanese scientists who invented and described its operation. (The Yagi–Uda is generally called simply Yagi.) A Yagi antenna is a linear array consisting of a dipole and two or more parasitic elements: one reflector and one or more directors. A simple three-element Yagi is shown in Figure 11-21a. The driven element is a half-wavelength folded dipole. (This element is referred to as the driven element because it is connected to the transmission line.

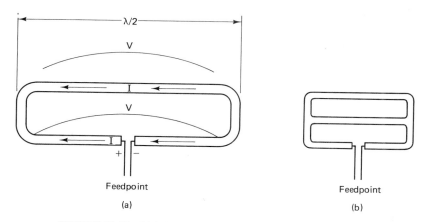

FIGURE 11-20   (a) Folded dipole; (b) three-element folded dipole.

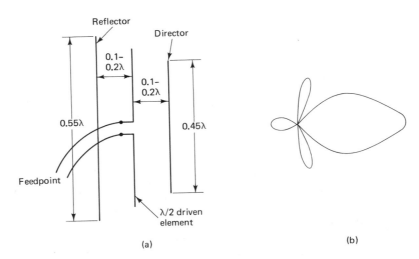

FIGURE 11-21 Yagi–Uda antenna: (a) three-element Yagi; (b) radiation pattern.

However, it is generally used for receiving only.) The reflector is a straight aluminum rod approximately 5% longer than the dipole, and the director is cut approximately 5% shorter than the driven element. The spacing between elements is generally between 0.1 and 0.2 wavelength. Figure 11-12b shows the radiation pattern for a Yagi antenna. The typical directivity for a Yagi is between 7 and 9 dB. The bandwidth of the Yagi can be increased by using more than one folded dipole, each cut to a slightly different length. Therefore, the Yagi antenna is commonly used for VHF television reception because of its wide bandwidth (the VHF TV band extends from 54 to 216 MHz).

Log-Periodic Antenna

A class of frequency-independent antennas called *log-periodics* evolved from the initial work of V. H. Rumsey, J. D. Dyson, R. H. DuHamel, and D. E. Isbell at the University of Illinois in 1957. The primary advantages of log-periodic antennas is the independence of their radiation resistance and radiation pattern to frequency. Log-periodic antennas have bandwidth ratios of 10:1 or greater. The bandwidth ratio is the ratio of the highest to the lowest frequency over which an antenna will satisfactorily operate. The bandwidth ratio is often used rather than simply stating the percentage of the bandwidth to the center frequency. Log-periodics are not simply a type of antenna but rather a class of antenna, because there are many different types, some that are quite unusual. Log-periodic antennas can be unidirectional or bidirectional and have a low-to-moderate directive gain. High gains may also be achieved by using them as an element in a more complicated array.

The physical structure of a log-periodic antenna is repetitive, which results in repetitive behavior in its electrical characteristics. In other words, the design of a log-periodic antenna consists of a basic geometric pattern that repeats, except with a different

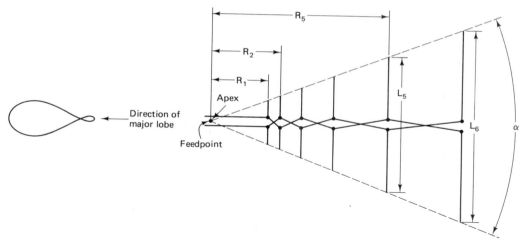

FIGURE 11-22 Log-periodic antenna.

size pattern. A basic log-periodic dipole array is probably the closest that a log-periodic comes to a conventional antenna and is shown in Figure 11-22. It consists of several dipoles of different length and spacing that are fed from a single source at the small end. The transmission line is crisscrossed between the feed points of adjacent pairs of dipoles. The radiation pattern for a basic log-periodic antenna has maximum radiation outward from the small end. The lengths of the dipoles and their spacing are related in such a way that adjacent elements have a constant ratio to each other. Dipole lengths and spacings are related by the formula

$$\frac{R_2}{R_1} = \frac{R_3}{R_2} = \frac{R_4}{R_3} = \frac{1}{\tau} = \frac{L_2}{L_1} = \frac{L_3}{L_2} = \frac{L_4}{L_3} \qquad (11\text{-}12)$$

or

$$\tau = \frac{R_{n+1}}{R_n} = \frac{L_{n+1}}{L_n}$$

where

      R = dipole spacing
      L = dipole length
      τ = design ratio (number < 1)

    The ends of the dipoles lie along a straight line, and the angle where they meet is designated α. For a typical design, $\tau = 0.7$ and $\alpha = 30°$. With the preceding structural stipulations, the antenna input impedance varies repetitively when plotted as a function of frequency, and when plotted against the log of the frequency, varies periodically (hence the name "log-periodic"). A typical plot of the input impedance is shown in Figure 11-23. Although the input impedance varies periodically, the variations are not

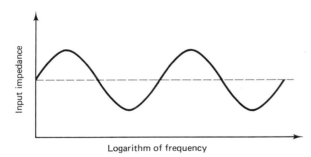

FIGURE 11-23 Log-periodic input impedance versus frequency.

necessarily sinusoidal. Also, the radiation pattern, directivity, power gain, and beamwidth undergo a similar variation with frequency.

The magnitude of a log-frequency period is dependent on the design ratio and, if two successive maxima occur at frequencies F_1 and F_2, they are related by the formula

$$\log F_2 - \log F_1 = \log \frac{F_2}{F_1} = \log \frac{1}{\tau} \qquad (11\text{-}13)$$

Therefore, the measured properties of a log-periodic antenna at frequency F will have identical properties at frequency τF, $\tau^2 F$, $\tau^3 F$, and so on. Log-periodic antennas, like Rhombic antennas, are used mainly for HF and VHF communications. However, log-periodic antennas do not have a terminating resistor and are therefore more efficient. Very often TV antennas advertised as so-called "high-gain" or "high-performance" antennas are log-periodic antennas.

Loop Antenna

The most fundamental *loop antenna* is simply a single-turn coil of wire that is sufficiently shorter than one wavelength and carries RF current. Such a loop is shown in Figure 11-24. If the radius (r) is small compared to a wavelength, current is essentially in phase throughout the loop. A loop can be thought of as many elemental dipoles connected together. Dipoles are straight; therefore, the loop is actually a polygon rather than circular.

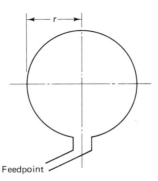

Feedpoint

FIGURE 11-24 Loop antenna.

However, a circle can be approximated if the dipoles are assumed to be sufficiently short. The loop is surrounded by a magnetic field which is at right angles to the wire and the directional pattern is independent of its exact shape. Generally, loops are circular; however, any shape will work. The radiation pattern for a loop antenna is essentially the same as that of a short horizontal dipole.

The radiation resistance for a small loop is

$$R_r = \frac{31,200A^2}{\lambda4} \tag{11-14}$$

where A is the area of the loop. For very low-frequency applications, loops are often made with more than one turn of wire. The radiation resistance of a multiturn loop is simply the radiation resistance for a single turn loop times the number of turns squared. The polarization of a loop antenna, like an elementel dipole, is linear. However, a vertical loop is vertically polarized and a horizontal loop is horizontally polarized.

Small vertically polarized loops are very often used as direction-finding antennas. The direction of the received signal can be found by orienting the loop until a null or zero value is found. This is the direction of the received signal. Loops have an advantage over most other types of antennas in direction finding in that loops are generally much smaller and therefore more easily adapted to mobile communications applications.

QUESTIONS

11-1. Define *antenna*.

11-2. Describe basic antenna operation using standing waves.

11-3. Describe a relative radiation pattern; an absolute radiation pattern.

11-4. Define *front-to-back ratio*.

11-5. Describe an omnidirectional antenna.

11-6. Define *near field*; *far field*.

11-7. Define *radiation resistance*; *antenna efficiency*.

11-8. Define and contrast *directive gain* and *power gain*.

11-9. What is the directivity for an isotropic antenna?

11-10. Define *effective isotropic radiated power*.

11-11. Define *antenna polarization*.

11-12. Define *antenna beamwidth*.

11-13. Define *antenna bandwidth*.

11-14. Define *antenna input impedance*. What factors contribute to an antenna's input impedance?

11-15. Describe the operation of an elementary doublet.

11-16. Describe the operation of a half-wave dipole.

11-17. Describe the effects of ground on a half-wave dipole.

11-18. Describe the operation of a grounded antenna.

11-19. What is meant by *antenna loading*?

11-20. Describe an antenna loading coil.

11-21. Describe antenna top loading.

11-22. Describe an antenna array.

11-23. What is meant by *driven element*; *parasitic element*?

11-24. Describe the radiation pattern for a broadside array; an end-fire array.

11-25. Define *nonresonant antenna*.

11-26. Describe the operation of the Rhombic antenna.

11-27. Describe a folded dipole antenna.

11-28. Describe a Yagi–Uda antenna.

11-29. Describe a log-periodic antenna.

11-30. Describe the operation of a loop antenna.

PROBLEMS

11-1. For an antenna with input power $P_i = 100$ W, rms current $i = 2$ A, and dc resistance $R_{dc} = 2 \, \Omega$; determine:
   (a) The antenna's radiation resistance.
   (b) The antenna's efficiency.
   (c) The power radiated from the antenna, P_r.

11-2. Determine the directivity in decibels for an antenna that produces power density $\mathscr{P} = 2$ μW_m^2 at a point when a reference antenna produces 0.5 μW_m^2 at the same point.

11-3. Determine the power gain in decibels for an antenna with directive gain $\mathscr{D} = 40$ and efficiency $\eta = 65\%$.

11-4. Determine the effective isotropic radiated power for an antenna with power gain $A_p = 43$ dB and radiated power $P_r = 200$ W.

11-5. Determine the effective isotropic radiated power for an antenna with directivity $\mathscr{D} = 33$ dB, efficiency $\eta = 82\%$, and input power $P_i = 100$ W.

11-6. Determine the power density at a point 20 km from an antenna that is radiating 1000 W and has power gain $A_p = 23$ dB.

11-7. Determine the power density at a point 30 km from an antenna that has input power $P_{in} = 40$ W, efficiency $\eta = 75\%$, and directivity $\mathscr{D} = 16$ dB.

11-8. Determine the power captured by a receiving antenna for the following parameters:

> Power radiated $P_r = 50$ W
> Transmit antenna directive gain $A_t = 30$ dB
> Distance between transmit and receive antennas $d = 20$ km
> Receive antenna directive gain $A_r = 26$ dB

11-9. Determine the directivity (in decibels) for an antenna that produces a power density at a point that is 40 times greater than the power density at the same point when the reference antenna is used.

11-10. Determine the effective radiated power for an antenna with directivity $\mathscr{D} = 400$, efficiency $\eta = 0.60$, and input power $P_{in} = 50$ W.

11-11. Determine the efficiency for an antenna with radiation resistance $R_r = 18.8 \ \Omega$, dc resistance $R_{dc} = 0.4 \ \Omega$, and directive gain $\mathcal{D} = 200$.

11-12. Determine the power gain A_p for Problem 11-11.

11-13. Determine the efficiency for an antenna with radiated power $P_r = 44$ W, dissipated power $P_d = 0.8$ W, and directive gain $\mathcal{D} = 400$.

11-14. Determine power gain A_p for Problem 11-13.

Chapter 12

BASIC TELEVISION PRINCIPLES

INTRODUCTION

The word *television* comes from the Greek word *tele* (meaning distant) and the Latin word *vision* (meaning sight). Therefore, television simply means to see from a distance. In its simplest form, television is the process of converting *images* (either stationary or in motion) to electrical signals, then transmitting those signals to a distant receiver, where they are converted back to images that can be perceived with the human eye. Thus television is a system in which images are transmitted from a central location, then received at distant receivers, where they are reproduced in their original form.

HISTORY OF TELEVISION

The idea of transmitting images or pictures was first experimented with in the 1880s when Paul Nipkow, a German scientist, conducted experiments using revolving *disks* placed between a powerful light source and the subject. A spiral row of holes was punched in the disk, which permitted light to scan the subject from top to bottom. After one complete revolution of the disk, the entire subject had been scanned. Light reflected from the subject was directed to a light-sensitive cell, producing current that was proportional in intensity to the reflected light. The fluctuating current operated a neon lamp, which gave off light in exact proportion to that reflected from the subject. A second disk exactly like the one in the transmitter was used in the receiver and the two disks revolved in exact synchronization. The second disk was placed between the neon lamp and the eye of the observer, who thus saw a reproduction of the subject.

The images reproduced with Nipkow's contraption were barely recognizable, although his scanning and synchronization principles are still used today.

In 1925, C. Francis Jenkins in the United States, and John L. Baird in England, using scanning disks connected to vacuum-tube amplifiers and photoelectric cells were able to reproduce images that were recognizable, although still of poor quality. Scientists worked for several years trying to develop effective mechanical scanning disks that with improved mirrors and lens and a more intense light source would improve the quality of the reproduced image. However, in 1933, Radio Corporation of America (RCA) announced a television system, developed by Vladimir K. Zworykin, that used

TABLE 12-1   FCC CHANNEL AND FREQUENCY ASSIGNMENTS

Channel number	Frequency band (MHz)	Channel number	Frequency band (MHz)	Channel number	Frequency band (MHz)
1[a]	44–50	29	560–566	57	728–734
2	54–60	30	566–572	58	734–740
3	60–66	31	572–578	59	740–746
4	66–72	32	578–584	60	746–752
5	76–82	33	584–590	61	752–758
6	82-88	34	590–596	62	758–764
7	174–180	35	596–602	63	764–770
8	180–186	36	602–608	64	770–776
9	186–192	37	608–614	65	776–782
10	192–198	38	614–620	66	782–788
11	198–204	39	620–626	67	788–794
12	204–210	40	626–632	68	794–800
13	210–216	41	632–638	69	800–806
14	470–476	42	638–644	70	806–812
15	476–482	43	644–650	71	812–818
16	482–488	44	650–656	72	818–824
17	488–494	45	656–662	73[a]	824–830
18	494–500	46	662–668	74[a]	830–836
19	500–506	47	668–674	75[a]	836–842
20	506–512	48	674–680	76[a]	842–848
21	512–518	49	680–686	77[a]	848–854
22	518–524	50	686–692	78[a]	854–860
23	524–530	51	692–698	79[a]	860–866
24	530–536	52	698–704	80[a]	866–872
25	536–542	53	704–710	81[a]	872–878
26	542–548	54	710–716	82[a]	878–884
27	548–554	55	716–722	83[a]	884–890
28	554–560	56	722–728		

[a] No longer assigned to television broadcasting.

an electronic scanning technique. Zworykin's system required no mechanical moving parts and is essentially the system used today.

In 1941, commercial broadcasting of *monochrome* (black and white) television signals began in the United States. In 1945, the FCC assigned 13 VHF television channels: 6 low-band channels, 1 to 6 (44 to 88 MHz), and 7 high-band channels, 7 to 13 (174 to 216 MHz). However, in 1948 it was found that channel 1 (44 to 50 MHz) caused interference problems; consequently, this channel was reassigned to mobile radio services. In 1952, UHF channels 14 to 83 (470 to 890 MHz) were assigned by the FCC to provide even more television stations. In 1974, the FCC reassigned to cellular telephone frequency bands at 825 to 845 MHz and 870 to 890 MHz, thus eliminating UHF channels 73 to 83 (however, existing licenses are renewable). Table 12-1 shows a complete list of the FCC channel and frequency assignments used in the United States. In 1947, R. B. Dome of General Electric Corporation proposed the method of *intercarrier* sound transmission for television broadcasting that is used today. In 1949, experiments began with color transmission, and in 1953, the FCC adopted the *National Television Systems Committee* (NTSC) system for color television broadcasting, which is also still used today.

MONOCHROME TELEVISION TRANSMISSION

Television Transmitter

Monochrome television broadcasting involves the transmission of two separate signals: an *aural* (sound) signal and a *video* (picture) signal. Every television transmitter broadcasts two totally separate signals for the picture and sound information. Aural transmission uses frequency modulation and video transmission uses amplitude modulation. Figure 12-1 shows a simplified block diagram for a monochrome television transmitter. It shows two totally separate transmitters (an FM transmitter for the sound information and an AM transmitter for the picture information) whose outputs are combined in a *diplexer bridge* and fed to a single antenna. A diplexer bridge is a network that is used to combine the outputs from two transmitters operating at different frequencies that use the same antenna system. The video information is limited to frequencies below 4 MHz and can originate from either a *camera* (for live transmissions), a video tape or cassette recorder, or a video disk recorder. The video switcher is used to select the desired video information source for broadcasting. The audio information is limited to frequencies below 15 kHz and can originate from either a microphone (again, only for live transmissions), from sound tracks on tape or disk recorders, or from a separate audio cassette or disk recorder. The audio mixer/switcher is used to select the appropriate audio source for broadcasting. Figure 12-1 also shows *horizontal* and *vertical* synchronizing signals which are combined with the picture information prior to modulation. These signals are used in the receivers to synchronize the horizontal and vertical *scanning rates* (synchronization is discussed in detail later in the chapter).

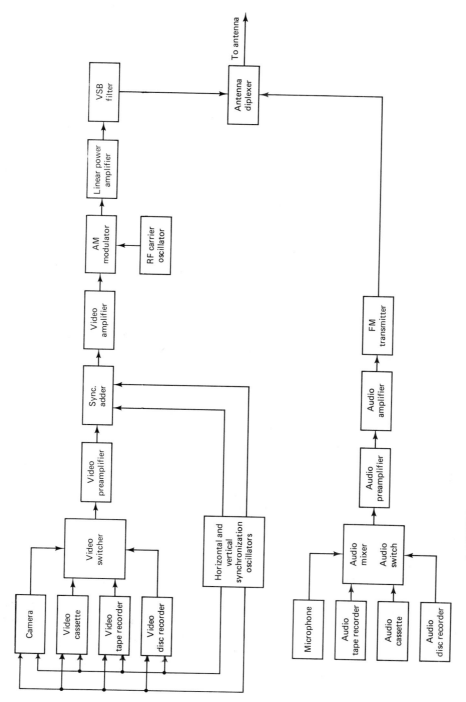

FIGURE 12-1 Simplified block diagram for a monochrome television transmitter.

455

Television Broadcast Standards

Figure 12-2 shows the frequency spectrum for a standard television broadcast channel. Its total bandwidth is 6 MHz. The picture carrier is spaced 1.25 MHz above the lower limit for the channel, and the sound carrier is spaced 0.25 MHz below the upper limit. Therefore, the picture and sound carriers are always 4.5 MHz apart. The color *subcarrier* is located 3.58 MHz above the picture carrier. Commercial television broadcasting uses AM vestigial sideband transmission for the picture information. The lower sideband is 0.75 MHz wide and the upper sideband 4 MHz. Therefore, the low video frequencies (rough outline of the image) are emphasized relative to the high video frequencies (fine detail of the image). The FM sound carrier has a bandwidth of approximately 50 kHz (±25 kHz deviation for 100% modulation). Both amplitude and phase modulation are used to encode the color information onto the 3.58-MHz color subcarrier. The bandwidth and composition of the color spectrum are discussed later in the chapter.

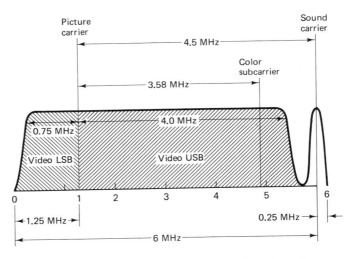

FIGURE 12-2   Standard television broadcast channel.

THE COMPOSITE VIDEO SIGNAL

Composite means made up of disparate or different parts. The composite video signal includes three separate parts. These parts are (1) the luminance signal, (2) the synchronization pulses, and (3) the blanking pulses. These three signals are combined in such a way to form the composite or total video signal.

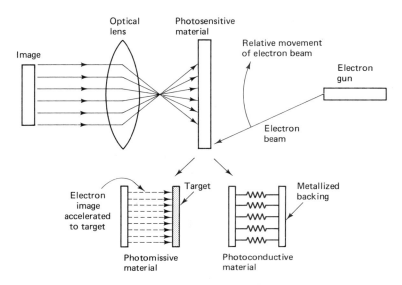

FIGURE 12-3 Simplified diagram of a black-and-white television camera.

The Luminance Signal

The *luminance signal* is the picture information or video signal. This signal originates in the camera and varies in amplitude proportional to the intensity (brightness) of the image. With *negative* transmission, the lower amplitudes correspond to the whitest parts of the image and the higher amplitudes correspond to the darkest. With *positive* transmission, the lower amplitudes correspond to the darkest parts and the higher amplitudes to the whitest. Negative transmission is the FCC standard for modulation of the final picture carrier. However, at intermediate points in both the transmitter and receiver, both negative or positive transmission signals occur and can be observed with a standard oscilloscope.

Figure 12-3 shows a simplified diagram of a black-and-white television camera tube. Light is reflected from an image through an optical lens onto the surface of a photosensitive material. The surface can be made from either a photomissive or photoconductive material and is divided into smaller discrete segments called *picture elements*. A *photomissive* material emits photoelectrons proportional to the intensity of the light striking its surface. A *photoconductive* material has a resistance that is inversely proportional to the intensity of the light striking it. When excited by an electron beam, a picture element outputs a signal that is proportional to the intensity of the light striking it. Therefore, if elements are individually scanned (excited) in sequence, the amplitude of the output signal will vary in accordance with the intensity of the image being scanned.

Figure 12-4 shows the amplitude changes in the luminance signal for a single

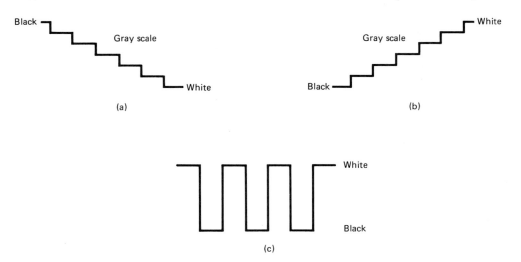

FIGURE 12-4 Luminance signal for a black-and-white camera: (a) gray scale, negative transmission; (b) gray scale, positive transmission; (c) checkerboard pattern, positive transmission.

horizontal scan at the output of a black-and-white camera for a image with varying light intensities. Figure 12-4a shows the amplitude changes for a negative transmission signal as the image changes from black, to a gray scale, to pure white. Figure 12-4b shows the same signal except for positive transmission. A positive transmission luminance signal for an alternating black/white checkerboard pattern is shown in Figure 12-4c. It can be seen that the amplitude of the luminance signal simply alternates from minimum to maximum with the brightness of the image.

Scanning

To produce the complete image, the entire surface of the camera tube must be *scanned*. Figure 12-5 shows a simple scanning sequence. The scanning is done in essentially the same manner in which a page from a book is read. This is called *sequential* horizontal scanning. The entire image is scanned in a sequential series of horizontal lines, one under the other. When the electron beam strikes the back of a picture element, a signal is produced whose amplitude is proportional to the light intensity striking the front of the element. Figure 12-5a shows the scanning beam beginning the active portion of the scan from the upper left corner and moving diagonally to the far right (line *A–B*). This is called the *active* portion of the scan line because this is the time in which the image is converted to electrical signals. Once the beam has reached the far-right side of the photosensitive surface, it immediately returns or retraces to the left side (point *C*). The return time is called horizontal *retrace* or *flyback time*. When the electron scanning beam has reached the bottom-right portion of the photosensitive surface (point *Z*), the beam is returned to top left (point *A*) and the sequence repeats. The return time is called the *vertical retrace time*. While the beam is retracing from the left to the right

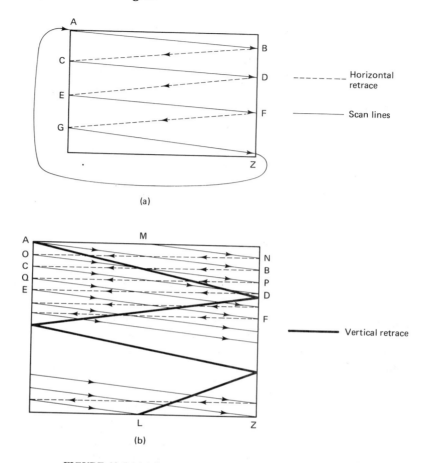

FIGURE 12-5 (a) Sequential scanning; (b) interlaced scanning.

side of the image and from the bottom to the top, it is *shut off* or *blanked*. Consequently, there is no video signal produced during either the horizontal or the vertical retrace times. The active and blanked portions of a single horizontal scan constitute one complete horizontal scan line. The number of horizontal scan lines depends on the detail desired and several other factors that are discussed later in the chapter.

In the United States, a total of 525 horizontal scan lines constitutes one *picture frame*, which is divided into two *fields* of 262.5 horizontal lines each. This scanning technique is called *interlaced scanning* and is shown in Figure 12-5b. Horizontal scanning produces the left-to-right movement of the electron beam and vertical scanning produces the downward movement. The vertical scanning rate is 30 Hz. Therefore, 30 frames per second are produced. Because the human eye can barely perceive a 30-Hz flicker, the frame is divided into two fields. 262.5 horizontal scan lines beginning at the top left (point *A*) and ending at the bottom middle (point *L*) constitute one picture field (the odd field). The second field (the even field) comprises the remaining 262.5 horizontal scan lines interlaced between the scan lines of the first field. The second field begins at the top middle (point *M*) and ends at the bottom right (point *Z*). Between fields, the electron beam retraces from the bottom of the picture back to the top in a zigzag pattern. This is called the *vertical retrace time*. Each field is vertically scanned at a 60-Hz rate. Therefore, although the entire picture changes every $\frac{1}{30}$ s, only half of the picture changes every $\frac{1}{60}$ s. This scanning technique allows 525 lines to be scanned at a 30-Hz rate without producing a noticeable flicker in the picture. To scan 525 horizontal scan lines in $\frac{1}{30}$ s, a 15,750-Hz scanning frequency is required ($30 \times 525 = 15,750$).

Horizontal lines are called *raster lines*, and 525 horizontal lines constitute a *raster*. The raster is the luminance that you see when there is no picture (i.e., when you are tuned to an unassigned channel). A raster simply means that there is horizontal and vertical scanning and brightness, but not necessarily a picture or an image on the screen.

Scanning waveforms.   The scanning beam for the camera and the receiver CRT must move both horizontally and vertically at *uniform* rates. This is called *linear scanning*. Linear scanning is necessary to ensure that picture elements are not "squashed" together or "bunched" to one side or to the top or bottom of the screen. *Magnetic deflection* is generally used to move the electron beam. Magnetic deflection produces essentially the same results as *electrostatic deflection*, which is commonly used in the CRT circuitry of an oscilloscope. Electrostatic deflection cannot be used with large CRTs such as those found in most television sets. With magnetic deflection, a linear rise in current through the deflection coils produces a linear change in magnetic flux. The force from the magnetic flux pulls the scanning beam from the left to right and from the top to bottom of the screen in a continuous, uniform motion. Figure 12-6a shows an ideal horizontal scanning current waveform. The positive slope of the sawtooth moves the beam from left to right in a smooth, constant motion. The negative slope of the waveform produces a magnetic field with the opposite polarity; thus it pulls the beam back to the left side. This is the retrace time. The rate at which the beam moves is proportional to the slope. During retrace it is desirable that the beam move as rapidly as possible. Therefore, the slope during retrace is much steeper than during the active portion of the scan line. In Figure 12-6a it can be seen that the most negative current corresponds to the left side of the screen and the most positive current to the right side. When zero current is flowing, the beam is in the center of the screen. Each scan line takes the time of one cycle of the sawtooth wave. Therefore, the frequency of the sawtooth is equal to the horizontal scan rate, 15,750 Hz, and the time for each scan line is 63.5 μs.

Figure 12-6b shows the vertical scanning waveform. Like the horizontal waveform, it is a sawtooth wave which ensures a uniform movement of the beam in the vertical direction. The bottom of the waveform corresponds to the top of the screen, the top of the waveform corresponds to the bottom of the screen, and zero current again corresponds to the center of the screen. Each vertical cycle corresponds to one complete vertical scan plus the vertical retrace time. Therefore, the vertical sawtooth frequency is 60 Hz.

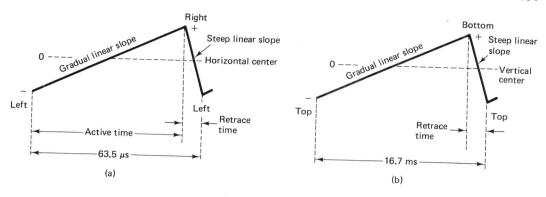

(a)

(b)

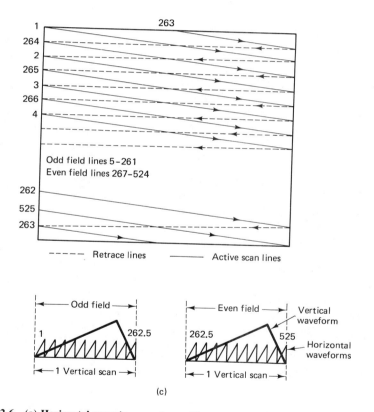

Odd field lines 5–261
Even field lines 267–524

----- Retrace lines   ——— Active scan lines

(c)

FIGURE 12-6 (a) Horizontal scanning waveform; (b) vertical scanning waveform; (c) interlaced scanning waveform.

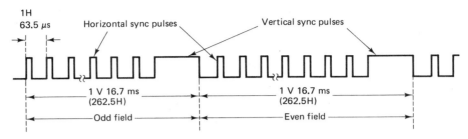

FIGURE 12-7 **Horizontal and vertical sync pulses.**

Figure 12-6c shows the interlaced scanning patterns for the odd and even fields of a standard U.S. television broadcast system. The vertical scanning waveforms are also shown superimposed over the horizontal scanning waveforms.

Synchronizing pulses. To reproduce the original image, the horizontal and vertical scanning rates at the receiver must be equal to those at the camera. Also, the scan lines at the camera and the receiver must begin and end in exact time synchronization. Therefore, a horizontal synchronizing pulse of 15,750 Hz and a vertical synchronizing pulse of 60 Hz are added to the luminance signal at the transmitter. The synchronizing (sync) pulses are stripped off at the receiver and used to synchronize its scanning circuits.

Figure 12-7 shows the horizontal and vertical sync pulses for both the odd and even fields. A new field is scanned once every $\frac{1}{60}$ s. Therefore, the time between vertical sync pulses is $1V$, where $V = \frac{1}{60}$ s or 16.7 ms. The time between horizontal sync pulses is $1H$, where $H = 63.5$ μs. $1V$ is sufficient time to scan 262.5 horizontal lines (1/15,750 = 63.5 μs and 16.7 ms/63.5 μs = 262.5). Each horizontal sync pulse produces one sawtooth horizontal scanning waveform, and each vertical sync pulse produces one sawtooth vertical scanning waveform.

Blanking Pulses

Blanking pulses are video signals that are added to the luminance and synchronizing pulses with the proper amplitude to ensure that the receiver is blacked out during the vertical and horizontal retrace times. The image is not scanned by the camera during retrace, and therefore there is no luminance information transmitted for those times. Blanking pulses are essentially video signals with amplitudes that do not produce any luminance (brightness) on the CRT. Horizontal and vertical sync pulses occur during their respective blanking times.

The Composite Signal

The *composite video signal* includes luminance (brightness) signals, horizontal and vertical sync pulses, and blanking pulses. Figure 12-8 shows the composite video signal for a single horizontal scan line, $1H$ (63.5 μs). (Notice that a positive transmission signal is shown.) The figure shows that the active portion of the scan line occurs during the positive slope of the scanning waveform and horizontal retrace occurs during the

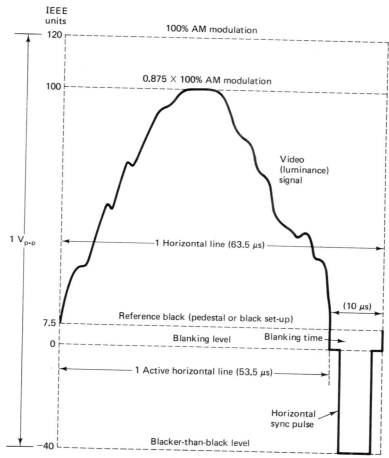

FIGURE 12-8 Composite video signal.

negative slope (i.e., the blanking time). The brightness range for standard television broadcasting is 160 IEEE units peak to peak. 160 IEEE units is generally normalized to 1 V_{p-p}. The exact value of 1 IEEE unit is unimportant; however, the relative value of a video signal in IEEE units determines its brightness. For example, maximum brightness (pure white) is 120 IEEE units, and no brightness is produced for signals below the reference black level (7.5 IEEE units). The *reference* black level is also called the *pedestal* or *black setup* level. The blanking level is 0 IEEE units, which is below the black level or, in other words, *blacker than black*. Sync pulses are negative-going pulses that occupy 25% of the total IEEE range. A sync pulse has a maximum level of 0 IEEE units and a minimum level of −40 IEEE units. Therefore, the entire sync pulse is below black and thus produces no brightness. The brightness range occupies 75% of the total IEEE scale and extends from 0 to 120 IEEE units, with 120 units corresponding to 100% AM modulation of the RF carrier. However, to ensure that

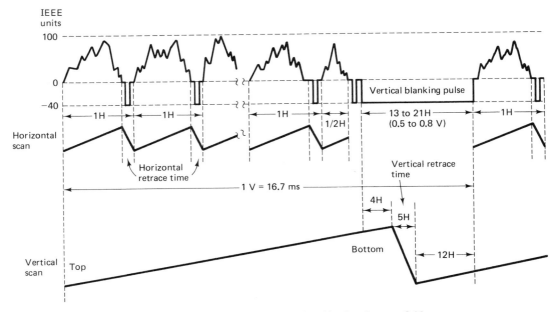

FIGURE 12-9   Composite video for the even field.

overmodulation does not occur, the FCC has established the maximum brightness (pure white) level to be 87.5% or 100 IEEE units ($0.875 \times 160 = 140$ units, $-40 + 140 = 100$ units).

Figure 12-9 shows the composite video signal for the even field, which equals $1V$ or 16.7 ms and is sufficient time for 262.5 horizontal scan lines ($262.5H$). However, the vertical blanking pulse width is between 0.05 and $0.08V$ or 833 to 1333 μs. Therefore, the vertical blanking pulse occupies the time of 13 to 21 horizontal scan lines, which leaves 241.5 to 249.5 active horizontal scan lines. The figure also shows that most of the active scan lines occur during the positive slope of the vertical scanning waveform, and the vertical blanking pulse occurs during the negative slope of the waveform (i.e., the retrace time).

Horizontal blanking time.   Figure 12-10 shows the blanking time for a single horizontal scan line. The total blanking time is approximately $0.16H$ or 9.5 to 11.5 μs. Therefore, the active (visible) time for a horizontal line is approximately $0.84H$ or 52 to 54 μs. Figure 12-10 shows that the sync pulse does not occupy the entire blanking time. The width of the actual sync pulse is approximately $0.08H$ or 4.25 to 5.25 μs. The time between the beginning of the blanking time and the leading edge of the sync pulse is called the *front porch* and is approximately $0.02H$ with a minimum time of 1.27 μs. The time between the trailing edge of the sync pulse and the end of the blanking time is called the *back porch* and is approximately $0.06H$ with a minimum time of 3.81 μs.

Period	Time
Horizontal line	1H 63.5 μs
Horizontal blanking	0.16H 9.5–11.5 μs
Sync pulse	0.08H 4.75 ± 0.5 μs
Front porch	0.02H 1.27 μs minimum
Back porch	0.06H 3.81 μs minimum

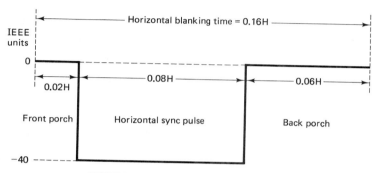

FIGURE 12-10 Horizontal blanking time.

Vertical blanking time.   Figure 12-11 shows the first 10H of a vertical blanking pulse for a negative transmission waveform. The figure shows that the entire blanking pulse is below the level for reference black (i.e., below 7.5 IEEE units). Each vertical blanking pulse begins with six *equalizing* pulses, a vertical sync pulse, and six more equalizing pulses. The equalizing pulses ensure a smooth, synchronized transition be-

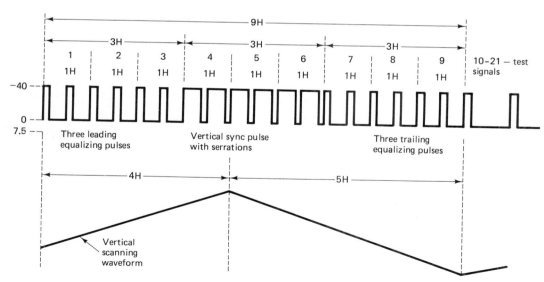

FIGURE 12-11 Vertical blanking pulse.

tween the odd and even fields. Equalizing pulses are explained in more detail in a later section of the chapter. The equalizing pulse rate is 31.5 kHz, which is twice the horizontal scanning rate. Therefore, each equalizing pulse takes $\frac{1}{2}H$ and the 12 pulses occupy a total time of $6H$. The actual vertical sync pulse occupies the time of $3H$. The *serrations* in the vertical sync pulse ensure that the receiver maintains horizontal synchronization during the vertical retrace time. Vertical serrations are explained in more detail in a later section. A total time of nine horizontal scan lines ($9H$) is required to transmit the equalizing and vertical sync pulses. From the figure it can be seen that the first $4H$ occur at the end of a vertical scan (i.e., at the bottom of the CRT). The following $5H$ occur during the retrace time, and all $9H$ occur during the vertical blanking time and are therefore not visible. The exact vertical blanking time is determined by the transmitting station; however, it is generally $21H$. Horizontal lines 10 through 21 of each field are often used to send studio test signals and automatic color and brightness signals.

RF Transmission of the Composite Video

The transmitted AM picture carrier is shown in Figure 12-12 for negative polarity modulation, which is the FCC standard. As stated previously, negative transmission of the RF carrier simply means that changes toward white in the picture decrease the amplitude of the AM picture carrier. The advantage of negative transmission is that noise pulses

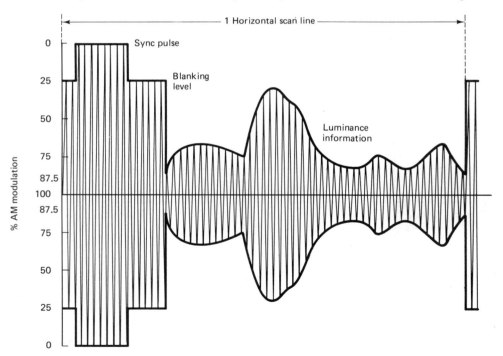

FIGURE 12-12   RF envelope for negative transmission video signal.

in the RF signal increase the carrier toward black, which makes the noise less annoying to the viewer than changes toward white. Also, with negative transmission, brighter images occur more often than darker ones; thus negative transmission uses less power than positive transmission. Notice that the AM envelope shown in Figure 12-12 has the shape of the composite video signal, and the luminance, blanking, and sync signals can easily be identified. Also, note that during the tips of the horizontal sync pulse there is no AM modulation, and the luminance signal never exceeds 87.5% AM modulation.

MONOCHROME TELEVISION RECEPTION

Figure 12-13 shows a block diagram for a monochrome television receiver. The receiver can be separated into five primary sections: RF, IF, video, horizontal and vertical deflection, and sound.

RF Section

A block diagram of an RF section is shown in Figure 12-14. The RF section includes the UHF and VHF antennas, the antenna coupling circuits, the preselectors, an RF amplifier, and a mixer/converter. A Yagi–Uda antenna is used for the VHF channels, and a simple loop antenna is used for the UHF channels.

The purposes of the RF or front-end section are channel selection (i.e., tuning), image frequency rejection, to isolate the local oscillator from the antenna (i.e., preventing the local oscillator signal from radiating), to convert RF signals to IF signals, amplification, and antenna coupling. VHF signals are captured by the antenna, coupled to the receiver input, bandlimited by the preselector, then amplified by the RF amplifier and fed to the mixer/converter. The mixer/converter beats the local oscillator frequency with the RF to produce the difference frequency, which is the IF. Channel selection is accomplished by changing the bandpass characteristics of the preselector and RF amplifier by switching capacitors or inductors in their tuned circuits and, at the same time, changing the local oscillator frequency. The preselector and local oscillator tuning circuits are ganged together. Commercial television receivers use high-side injection (the local oscillator is tuned to the IF above the desired RF channel frequency).

UHF signals are captured by the loop antenna, then immediately mixed down to IF. The UHF mixer is generally a simple diode mixer. When receiving UHF signals, the RF amplifier is simply an additional IF amplifier.

IF Section

The IF section of a television receiver provides most of the receiver's selectivity and gain. The block diagram for a three-stage IF amplifier is shown in Figure 12-15a. The IF section is generally several cascaded high-gain tuned amplifiers. In modern receivers, the IF section processes both the picture and sound IF signals. Such receivers are called

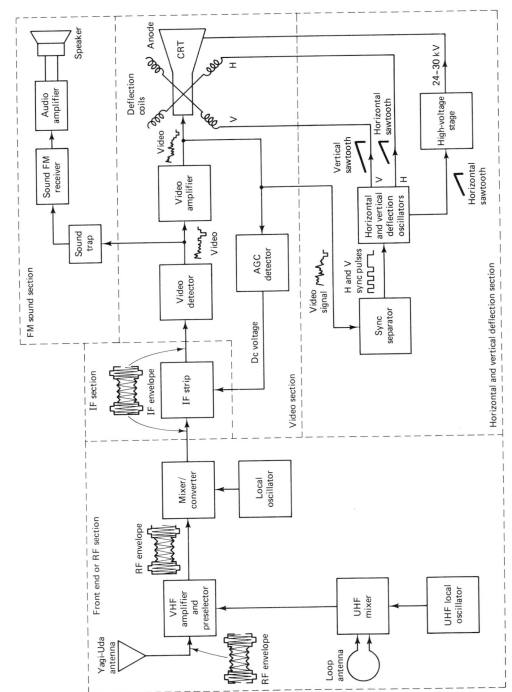

FIGURE 12-13 Block diagram monochrome television receiver.

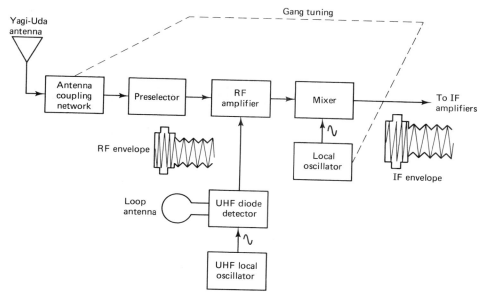

FIGURE 12-14   Block diagram RF section.

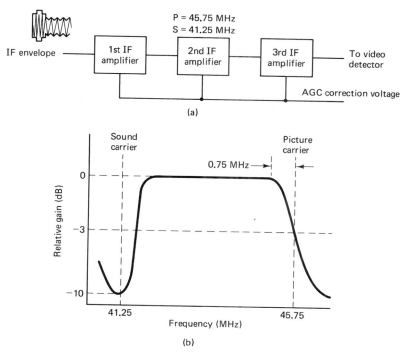

FIGURE 12-15   IF amplifiers: (a) block diagram; (b) frequency response curve.

intercarrier receivers. The standard IFs used in commercial television receivers are 45.75 MHz for the picture and 41.25 MHz for the sound. The IF carriers are separated by 4.5 MHz just as the RF carriers are. IF amplifiers use tuned bandpass filters that bandlimit the signal and prevent adjacent channel interference. A typical IF response curve is shown in Figure 12-15b. Special narrowband bandstop filters called *wavetraps* are used to trap or block the adjacent channel picture and sound carrier frequencies (39.75 MHz and 47.25 MHz, respectively). Wavetraps are also used to attenuate the sound and picture carriers of the selected channel and limit the IF passband to approximately 3 MHz. 3 MHz is used rather than 4 MHz to minimize interference from the color signal, which is explained later in this chapter.

The Video Section

The video section includes a video detector and a series of video amplifiers. A simplified block diagram for a video section is shown in Figure 12-16a. The detector down-converts the picture IF signals to video frequencies and the first sound IF to a second sound IF. The second sound IF is fed to the FM receiver, where the aural information is removed and fed to the audio amplifiers. The video detector is generally a single-diode peak detector. The IF input signal provides the ac voltage necessary to drive the diode into

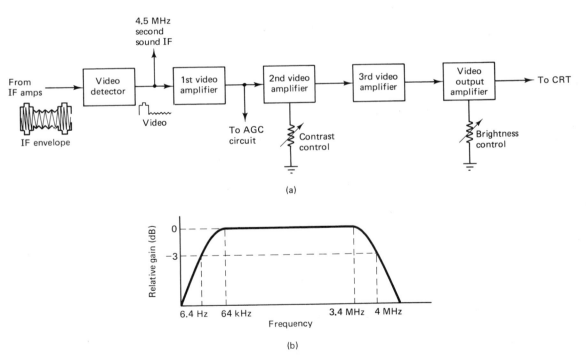

FIGURE 12-16   Video amplifiers: (a) block diagram; (b) frequency response curve.

conduction as a half-wave peak rectifier. The output from the video detector is the composite video signal which is fed to the video amplifiers. The video amplifiers provide the gain necessary for the luminance signal to drive the CRT. Video amplifiers are generally direct-coupled to provide dc restoration of the picture brightness. The contrast and brightness controls are located in the video section, and the AGC takeoff point is generally at the output of the first video amplifier. The brightness control simply allows the viewer to vary the dc bias voltage of the video signal. The contrast control adjusts the gain of the video amplifiers. The picture and sound IF signals mix in the diode detector, which is a nonlinear device, and produce a difference signal of 4.5 MHz, which is the second sound IF. A typical frequency response curve for a video amplifier section is shown in Figure 12-16b.

The Horizontal and Vertical Deflection Circuits

A simplified block diagram showing the vertical and horizontal deflection circuits is shown in Figure 12-17. The deflection section includes a sync separator, horizontal and vertical deflection oscillators, and a high-voltage stage. The horizontal and vertical synchronizing pulses are removed from the composite video signal by the sync separator circuit. The horizontal and vertical sync pulses are then separated with filters and fed to their respective deflection circuits. The deflection circuits convert the sync pulses to sawtooth scanning signals and provide the dc high voltage required for the anode of the CRT.

Sync separator. Figure 12-18a shows the schematic diagram for a single-transistor sync separator, which is a simple clipper circuit. Q_1 is a class C amplifier and the R_1C_1 coupling circuit provides signal bias. The positive portion of the composite video signal (i.e., the sync pulses) forward biases Q_1, causing base current to flow, which charges C_1 to the polarity shown. Between sync pulses, C_1 discharges slightly through R_1. The long R_1C_1 time constant keeps C_1 charged to approximately 90% of the peak positive value. Therefore, once C_1 has charged, the luminance signal drives Q_1 further into cutoff. Thus Q_1 conducts only during the more positive sync pulses. Consequently, the sync pulses are the only portion of the composite video signal that appears at the collector of Q_1. The base–emitter circuit of Q_1 is effectively a diode rectifier. The rectifier operation is shown in Figure 12-18b. Once removed, the horizontal and vertical sync pulses are separated with filters. A high-pass filter (differentiator) detects the 15,750-Hz horizontal sync pulses and a low-pass filter (integrator) detects the 60-Hz vertical sync pulses.

Vertical deflection oscillator. The output from the integrator is a 60-Hz waveform, which is fed to the vertical deflection oscillator. The deflection oscillator produces a 60-Hz linear sawtooth deflection voltage, which produces the vertical scan on the CRT. Figure 12-19 shows a schematic diagram for a transistorized *blocking oscillator*, which is a circuit often used to produce the sawtooth scanning waveform. A blocking oscillator is simply a *triggered oscillator* that produces a sawtooth output waveform

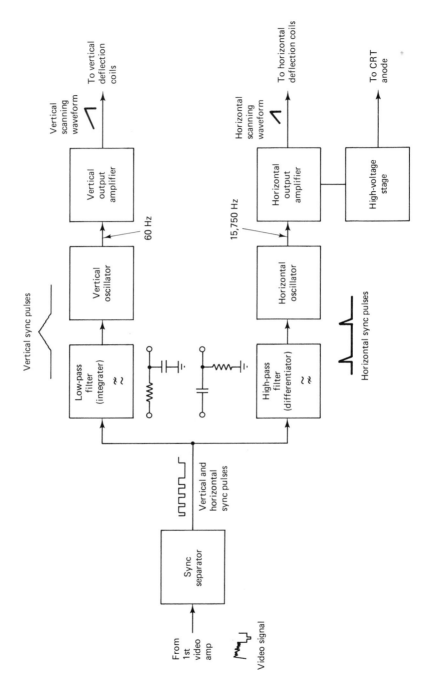

FIGURE 12-17 Horizontal and vertical deflection section.

472

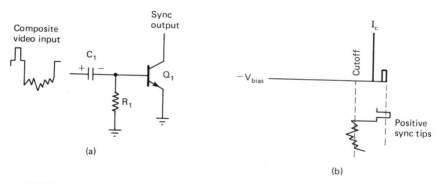

FIGURE 12-18 Sync separator circuit: (a) schematic diagram; (b) bias operation.

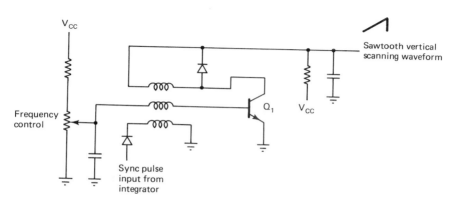

FIGURE 12-19 Vertical blocking oscillator.

that is synchronized to the incoming vertical sync pulse rate. The frequency control sets the *threshold* or *trigger level* for the oscillator. However, the frequency of oscillation is determined by the recovered vertical sync pulses. The output from the vertical oscillator is fed to a vertical output amplifier, which produces the high-voltage sawtooth waveform required to drive the vertical deflection coils.

The integrator (low-pass filter) passes only the vertical sync pulses and produces a 60-Hz trigger pulse for the blocking oscillator. Figure 12-20a shows the operation of the integrator without equalizing pulses. It can be seen that the sync pulses from the odd and even fields begin charging the capacitor from different initial levels; thus the two trigger pulses reach the threshold level for the oscillator at different times (Δt). This causes the vertical oscillator to change frequency and lose synchronization between fields, which is noticeable to the viewer as a slight vertical roll. To prevent loss of synchronization, 12 equalizing pulses are transmitted during the vertical blanking interval, six immediately before and six immediately after the vertical sync pulse. Figure 12-20b shows the waveform produced across the capacitor when the equalizing pulses are

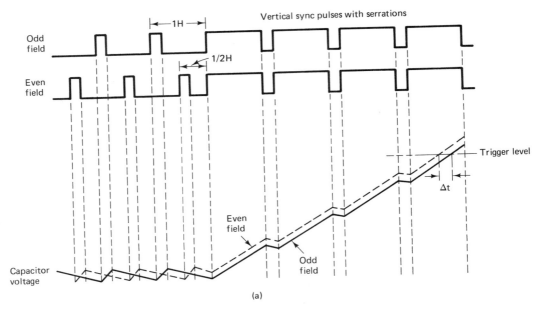

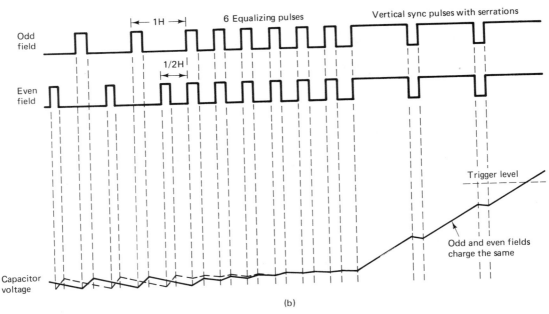

FIGURE 12-20 Integrator operation: (a) without equalizing pulses; (b) with equalizing pulses.

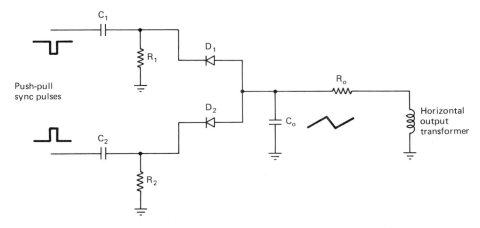

FIGURE 12-21   Push-pull horizontal AFC circuit.

included. The sync pulses from each field begin charging the capacitor at precisely the same time with exactly the same initial voltage; thus the capacitor voltage reaches the threshold level at the same time for each field, preventing the vertical sync oscillator from losing synchronization during the transition between the odd and even fields.

       Horizontal deflection oscillator.   Deflection oscillators, such as the one shown in Figure 12-20, are highly susceptible to noise. Noise pulses can be mistaken for synchronizing pulses and trigger the oscillator at the wrong time, thus changing the horizontal scanning rate. To improve noise immunity, automatic frequency control (AFC) circuits are often used for the horizontal deflection oscillator in television receivers. Figure 12-21 shows the schematic diagram for a push-pull sync discriminator commonly used for horizontal AFC. A phase splitter generates two 180° out-of-phase sync pulses which are required for push-pull operation. The dual-diode sync discriminator produces a horizontal sawtooth waveform across output capacitor C_o. Consequently, the sawtooth frequency is synchronized to the recovered horizontal sync pulses. The output from the AFC circuit is fed to the horizontal deflection circuit, where it provides horizontal scanning for the CRT. The AFC output is also fed to the receiver high-voltage section, where the anode voltage for the CRT is produced.

COLOR TELEVISION TRANSMISSION AND RECEPTION

Color Television Transmitter

In essence, a color television transmitter is identical to the black-and-white transmitter shown in Figure 12-1, except that a color camera is used to produce the video signal. With color broadcasting, all of the colors are produced by mixing different amounts of the three *primary colors*: red, blue, and green. A color camera is actually three cameras

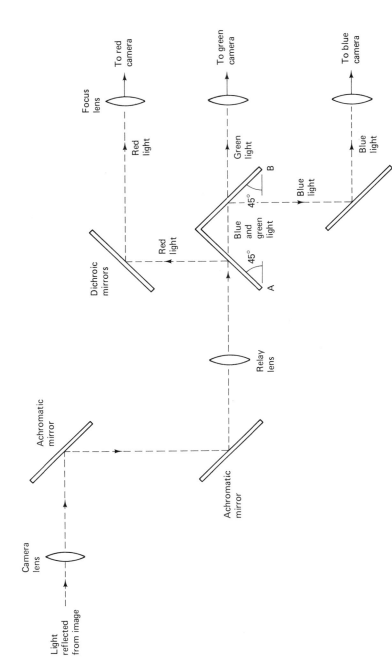

FIGURE 12-22 Mirror configuration used in color television camera to separate R, G, and B video signals.

in one, each with separate video output signals. When an image is scanned, separate camera tubes are used for each of the primary colors. The red camera produces the R video signal, the green camera produces the G video signal, and the blue camera produces the B video signal. The R, G, and B video signals are combined in an encoder to produce the composite color signal, which when combined with the luminance signal, amplitude modulates the RF carrier.

Color Camera

Figure 12-22 shows a configuration of mirrors that can be used to split an image into the three primary colors. The *chromatic* mirrors reflect light of all colors. The *dichroic* mirrors are coated to reflect light of only one frequency (color) and allow all other frequencies (colors) to pass through. Light reflected from the image passes through a single camera lens, is reflected by the two achromatic mirrors, and passes through the relay lens. Dichroic mirrors A and B are mounted on opposing 45° angles. Mirror A reflects red light, while blue and green light pass straight through to mirror B. Mirror B reflects blue light and allows green light to pass through. Consequently, the image is separated into red, green, and blue light frequencies. Once separated, the three color frequency signals modulate their respective camera tubes and produce the R, G, and B video signals.

Color Encoding

Figure 12-23 shows a simplified block diagram for a color television transmitter. The R, G, and B video signals are combined in specific proportions in the *color matrix* to produce the brightness (luminance) or Y video signal and the I and Q chrominance (color) video signals. The luminance signal corresponds to a monochrome video signal. The I and Q color signals amplitude modulate a 3.58-MHz color subcarrier to produce the total color signal, C. The I signal modulates the subcarrier directly in the I balanced modulator, while the Q signal modulates a quadrature (90° out of phase) subcarrier in the Q balanced modulator. The I and Q modulated signals are linearly combined to produce a quadrature amplitude modulation (QAM) signal, C, which is a combination of both phase and amplitude modulation. The C signal is combined with the Y signal to produce the total composite video signal (T).

The luminance signal. The Y or *luminance signal* is formed by combining 30% of the R video signal, 59% of the G video signal, and 11% of the B video signal. Mathematically, Y is expressed as

$$Y = 0.30R + 0.59G + 0.11B \qquad (12-1)$$

The percentages shown in Equation 12-1 correspond to the relative brightness of the three primary colors. Consequently, a scene reproduced in black and white by the

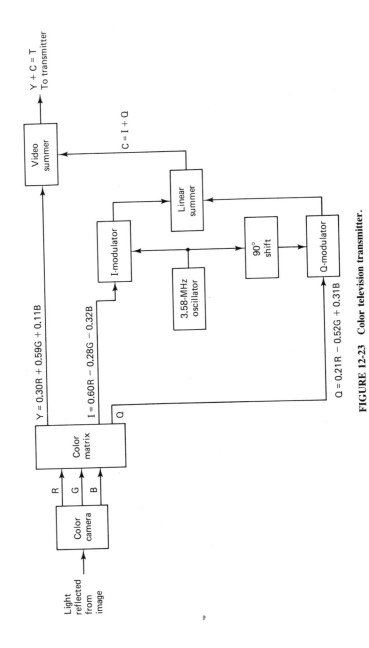

FIGURE 12-23 Color television transmitter.

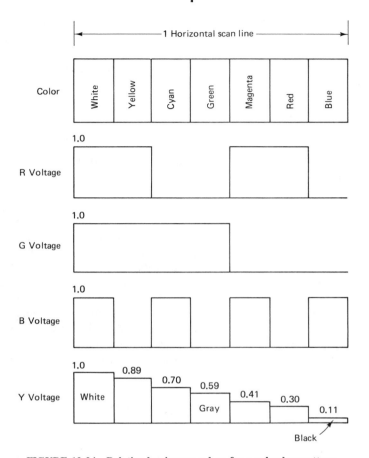

FIGURE 12-24 Relative luminance values for a color bar pattern.

Y signal has exactly the same brightness as the original image. Figure 12-24 shows how the Y signal voltage is formed from several values of R, G, and B.

The Y signal has a maximum relative amplitude of unity or 1, which is 100% white. For maximum values of R, G, and B ($1V$ each), the value for brightness is determined from Equation 12-1 as follows:

$$Y = 0.30(1) + 0.59(1) + 0.11(1) = 1.00$$

The voltage values for Y shown in Figure 12-24 are the relative luminance values for each of the colors. If only the Y signal is used to reproduce the pattern in a receiver, it would appear on the CRT as seven monochrome bars shaded from white on the left to gray in the middle and black at the right. The Y signal is transmitted with a bandwidth of 0 to 4 MHz. However, most receivers bandlimit the Y signal to 3.2 MHz to minimize interference with the 3.58-MHz color signal. The I signal is transmitted with a bandwidth of 1.5 MHz, while the Q signal is transmitted with a bandwidth of 0.5 MHz. However, most receivers limit both the I and Q signals to 0.5-MHz bandwidth.

The chrominance signal.   The *chrominance* or *C signal* is a combination of the I and Q color signals. The I or in-phase color signal is produced by combining 60% of the R video signal, 28% of the inverted G video signal, and 32% of the inverted B video signal. Mathematically, I is expressed as

$$I = 0.60R - 0.28G - 0.32B \qquad (12\text{-}2)$$

The Q or in quadrature color signal is produced by combining 21% of the R video signal, 52% of the inverted G video signal, and 31% of the B video signal. Mathematically, Q is expressed as

$$Q = 0.21R - 0.52G + 0.31B \qquad (12\text{-}3)$$

The I and Q signals are combined to produce the C signal, and since the I and Q signals are in quadrature, the C signal is the phasor sum of the two (i.e., $C = \sqrt{I^2 + Q^2}$). Therefore, the amplitude and phase of the C signal is dependent on the amplitudes of the I and Q signals, which are in turn proportional to the R, G, and B video signals. Figure 12-25 shows the *color wheel* for television broadcasting. The R-Y and B-Y signals are used in most color television receivers for demodulating the R, G, and B video signals and are explained later in the chapter. In the receiver, the C signal reproduces colors proportionate to the amplitudes of the I and Q signals. The hue (color tone) is determined by the phase of the C signal, and the depth or saturation is proportional to

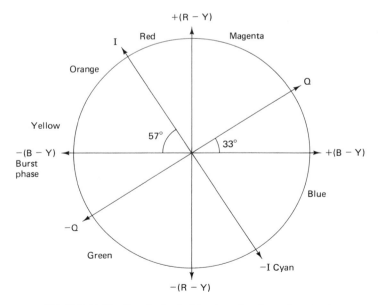

FIGURE 12-25   **Standard television broadcasting color wheel.**

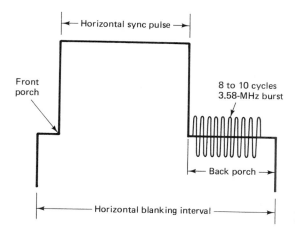

FIGURE 12-26 Horizontal blanking interval and 3.58-MHz burst.

the magnitude of the *C* signal. The outside of the circle corresponds to a relative value of 1.0.

The color burst.   The phase of the 3.58-MHz color subcarrier is the reference phase for color demodulation. Therefore, the color subcarrier must be transmitted together with the composite video so that a receiver can reconstruct the subcarrier with the proper frequency and reference phase and thus determine the phase (color) of the received signal. Eight to ten cycles of the 3.58-MHz subcarrier are inserted on the back porch of each horizontal blanking pulse. This is referred to as the *color burst*. In the receiver, the burst is removed and used to synchronize a local 3.58-MHz color oscillator. The color burst is shown in Figure 12-26. Figure 12-27 shows the composite RF frequency spectrum for color television broadcasting.

Scanning frequencies for color transmission.   The frequency of the color subcarrier is determined by harmonic relations among the color subcarrier and the horizontal and vertical scanning rates. The exact value for the color subcarrier is 3.579545 MHz.

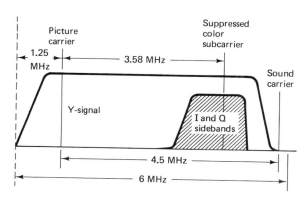

FIGURE 12-27 Composite RF frequency spectrum for color television broadcasting.

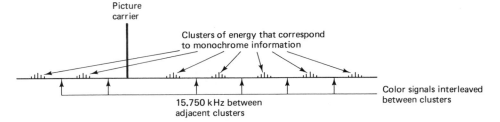

FIGURE 12-28   Frequency interleaving of color and luminance signals.

The sound subcarrier (4.5 MHz) is the 286th harmonic of the horizontal line frequency. Therefore, the horizontal line rate (F_H) for color transmission is not exactly 15.750 kHz. Mathematically, F_H is

$$F_H = \frac{4.5 \text{ MHz}}{286} = 15{,}734.26 \text{ Hz}$$

The exact value of the vertical scan rate (F_V) is

$$F_V = \frac{15{,}734.26}{262.5} = 59.94 \text{ Hz}$$

The color subcarrier frequency (C) is chosen as the 455th harmonic of one-half of the horizontal scan rate. Therefore,

$$C = \frac{15{,}734.26}{2} \times 455 = 3.579545 \text{ MHz}$$

Frequency interlacing.   The Y portion of the video signal produces clusters of energy at 15.73426-kHz intervals throughout the 4-MHz video bandwidth. By producing color signals around a 3.579545-MHz color subcarrier, the color energy is *clustered* within the void intervals between the black-and-white information. This is called frequency *interlacing* or sometimes frequency *interleaving* and is a form of *multiplexing* (i.e., the color and black-and-white information is frequency-division multiplexed into the total video spectrum). Figure 12-28 shows the spectrum for frequency interlacing.

Color Television Receivers

A color television receiver is essentially the same as a black-and-white receiver except for the picture tube and the addition of the color decoding circuits. Figure 12-29 shows the simplified block diagram for the color circuits in a color television receiver.

The composite video signal is fed to the *chroma* bandpass amplifier, which is tuned to the 3.58-MHz subcarrier and has a bandpass of 0.5 MHz. Therefore, only the C signal is amplified and passed on to the B-Y and R-Y demodulators. The 3.58-MHz color burst is separated from the horizontal blanking pulse by keying on the burst separator only during the horizontal flyback time. A synchronous 3.58-MHz color subcarrier is

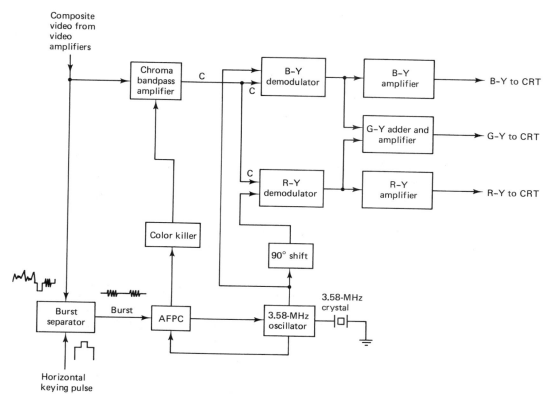

FIGURE 12-29   Color demodulator circuits.

reproduced in the color AFC circuit, which consists of a 3.58-MHz color oscillator and a color AFPC (automatic frequency and phase control) circuit. The *color killer* shuts off the chroma amplifier during monochrome reception (no colors are better than wrong colors). The *C* signal is demodulated in the B-*Y* and R-*Y* demodulators by mixing it with the phase coherent 3.58-MHz subcarrier. The B-*Y* and R-*Y* signals produce the R and B video signal by combining them with the *Y* signal in the following manner:

$$B - Y + Y = B$$
$$R - Y + Y = R$$

The *G* video signal is produced by combining the B-*Y* and R-*Y* signals with the proper proportions.

QUESTIONS

12-1. Briefly describe the meaning of the word *television*.

12-2. Describe a diplexer bridge.

12-3. What components make up the composite video signal?

12-4. Describe negative transmission; positive transmission. Which is the FCC standard for RF transmission?

12-5. Name and briefly describe two types of photosensitive materials used in television cameras.

12-6. Describe sequential horizontal scanning.

12-7. What is meant by the *active portion* of a horizontal line?

12-8. What is meant by *retrace time*?

12-9. Describe interlaced scanning.

12-10. Define *raster lines*; *raster*.

12-11. Why should the horizontal and vertical scanning waveforms be linear?

12-12. Why are horizontal and vertical synchronizing pulses included with the composite video signal?

12-13. Describe a blanking pulse.

12-14. Describe the IEEE scale used for television broadcasting.

12-15. What is meant by *reference black*; *black setup*; *pedestal*; *blacker than black*?

12-16. Describe the horizontal blanking time.

12-17. Describe the vertical blanking time.

12-18. Why are equalizing pulses transmitted during the vertical blanking interval?

12-19. Why are there serrations in the vertical sync pulses?

12-20. Draw the block diagram for a monochrome television receiver and describe its basic operation and the primary purpose of each section.

12-21. Describe the operation of a vertical blocking oscillator.

12-22. Describe the operation of a horizontal AFC circuit.

12-23. What are the three primary colors for television broadcasting?

12-24. Describe the basic operation of a color television camera.

12-25. Describe the Y or luminance signal.

12-26. Describe the I, Q, and C signals.

12-27. What is the color burst? How is it transmitted? What is its purpose?

12-28. Explain why the scanning frequencies used with color television are slightly different than those used for black-and-white transmission.

12-29. Describe frequency interlacing.

12-30. Draw the block diagram of the color decoding circuits in a color television receiver and briefly describe the decoding operation.

PROBLEMS

12-1. Draw the RF frequency spectrum for channel 8. Include the picture, sound, and the color frequencies and their respective bandwidths.

12-2. How long is $0.8H$? $0.8V$?

12-3. How many horizontal scan lines occur during $0.12V$?

12-4. If the back porch of the horizontal blanking pulse is $0.06H$, what is the maximum number of cycles of the color burst signal that can be transmitted?

12-5. If the vertical blanking interval is $20H$ long, how long is this?

12-6. What is the color subcarrier frequency for channel 55; channel 66?

12-7. What is the sound subcarrier frequency for channel 55; Channel 66?

12-8. Determine the value for Y for the following R, G, and B signals: $R = 0.8V$, $G = 0.6V$, and $G = 0.2V$.

12-9. Determine I and Q for the R, G, and B values given in Problem 12-8.

12-10. Determine the C signal for Problem 12-9.

Chapter 13

DIGITAL COMMUNICATIONS

INTRODUCTION

During the past several years, traditional analog communications systems that use conventional amplitude modulation (AM), frequency modulation (FM), or phase modulation (PM) have gradually been replaced with more modern digital communications systems. Digital communications systems offer several outstanding advantages over traditional analog systems: ease of processing, ease of multiplexing, and noise immunity.

The term *digital communications* covers a broad area of communications techniques, including digital transmission and digital radio. Digital transmission is the transmittal of digital pulses between two points in a communications system. Digital radio is the transmittal of digitally modulated analog carriers between two points in a communications system. Digital transmission systems require a physical facility between the transmitter and receiver, such as a metallic wire pair, a coaxial cable, or a fiber optic cable. In digital radio systems, the transmission medium is free space or the earth's atmosphere.

Figure 13-1 shows simplified block diagrams of both a digital transmission system and a digital radio system. In a digital transmission system, the original source information may be in digital or analog form. If it is in analog form, it must be converted to digital pulses prior to transmission and converted back to analog form at the receive end. In a digital radio system, the modulating input signal and the demodulated output signal are digital pulses. The digital pulses could originate from a digital transmission system, from a digital source such as a mainframe computer, or from the binary encoding of an analog signal.

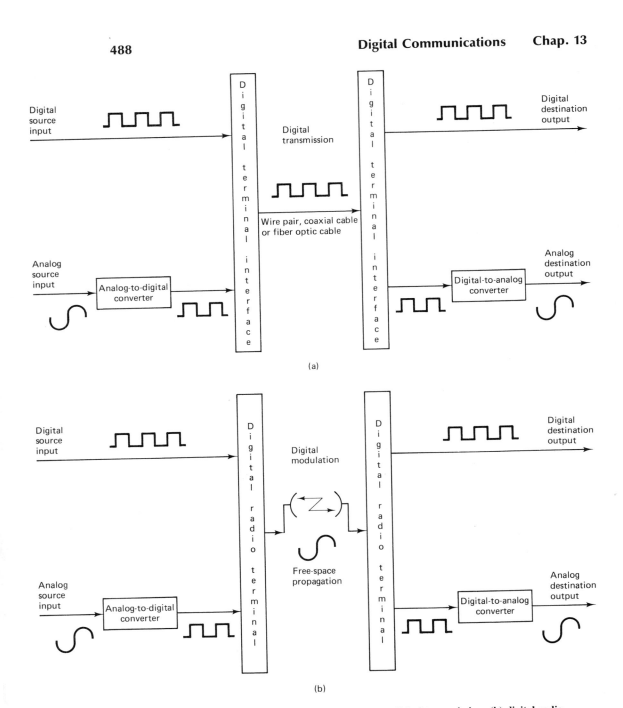

FIGURE 13-1 **Digital communications systems: (a) digital transmission; (b) digital radio.**

INFORMATION CAPACITY

The *information capacity* of a communications system represents the number of independent symbols that can be carried through the system in a given unit of time. The most basic symbol is the *binary digit* (bit). Therefore, it is often convenient to express the information capacity of a system in *bits per second* (bps). In 1928, R. Hartley of Bell Telephone Laboratories developed a useful relationship among bandwidth, transmission time, and information capacity. Simply stated, *Hartley's law* is

$$C \propto B \times T \qquad (13\text{-}1)$$

where

C = information capacity
B = bandwidth
T = transmission time

From Equation 13-1 it can be seen that the information capacity is a linear function and is directly proportional to both the system bandwidth and the transmission time. If either the bandwidth or the transmission time is changed, the information capacity will change proportionally.

In 1948, C. E. Shannon (also of Bell Telephone Laboratories) published a paper in the *Bell System Technical Journal* relating the information capacity of a communications channel to bandwidth, transmission time, and signal-to-noise ratio. Mathematically stated, the *Shannon limit for information capacity* is

$$C = B \log_2 \left(1 + \frac{S}{N} \right) \qquad (13\text{-}2)$$

where

C = information capacity (bps)
B = bandwidth
$\dfrac{S}{N}$ = signal power-to-noise power ratio

For a standard voice band communications channel with a signal-to-noise ratio of 1000 (30 dB) and a bandwidth of 2.7 kHz, the Shannon limit for information capacity is

$$C = 2700 \log_2 (1 + 1000)$$

$$= 26.9 \text{ kbps}$$

Shannon's formula is often misunderstood. The results of the preceding example indicate that 26.9 kbps can be transferred through a 2.7-kHz channel. This may be true, but it cannot be done with a binary system. To achieve an information transmission rate of 26.9 kbps through a 2.7-kHz channel, each symbol transmitted must contain more than one bit of information. Therefore, to achieve the Shannon limit for information capacity, digital transmission systems that have more than two output conditions must be used. Several such systems are described in the following chapters. These systems include both analog and digital modulation techniques and the transmission of both digital and analog signals.

DIGITAL RADIO

The property that distinguishes a digital radio system from a conventional AM, FM, or PM radio system is that in a digital radio system the modulating and demodulated signals are digital pulses rather than analog waveforms. Digital radio uses analog carriers just as conventional systems do. Essentially, there are three digital modulation techniques that are commonly used in digital radio systems: *frequency shift keying* (FSK), *phase shift keying* (PSK), and *quadrature amplitude modulation* (QAM).

FREQUENCY SHIFT KEYING

Frequency shift keying (FSK) is a relatively simple, low-performance form of digital modulation. FSK is a constant-envelope form of angle modulation similar to conventional frequency modulation except that the modulating signal is a binary pulse stream that varies between two discrete voltage levels rather than a continuously changing waveform.

FSK Transmitter

With binary FSK, the center or carrier frequency is shifted (deviated) by the binary input data. Consequently, the output of an FSK modulator is a step function in the frequency domain. As the binary input signal changes from a logic 0 to a logic 1, and vice versa, the FSK output shifts between two frequencies: a *mark* or *logic 1 frequency* and a *space* or *logic 0 frequency*. With FSK, there is a change in the output frequency each time the logic condition of the binary input signal changes. Consequently, the output rate of change is equal to the input rate of change. In digital modulation, the rate of change at the input to the modulator is called the *bit rate* and has the units of bits per second (bps). The rate of change at the output of the modulator is called *baud* or *baud rate* and is equal to the reciprocal of the time of one output signaling element. In FSK, the input and output rates of change are equal; therefore, the bit rate and baud rate are equal. A simple binary FSK transmitter is shown in Figure 13-2.

Bandwidth Considerations of FSK

As with all electronic communications systems, bandwidth is one of the primary considerations when designing an FSK transmitter. FSK is similar to conventional frequency modulation and so can be described in a similar manner.

Figure 13-3 shows an FSK modulator. An FSK modulator is a type of FM

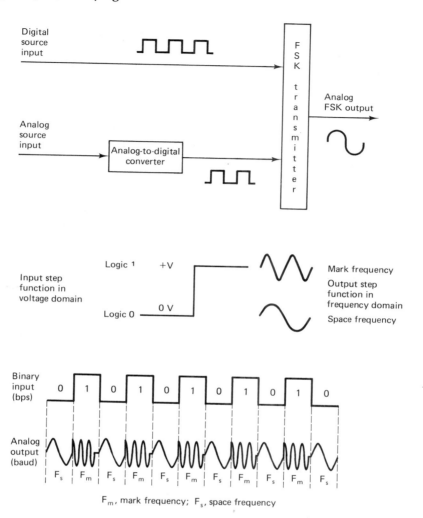

F_m, mark frequency; F_s, space frequency

FIGURE 13-2 Binary FSK transmitter.

transmitter and is very often a *voltage-controlled oscillator* (VCO). It can be seen that the fastest input rate of change occurs when the binary input is a series of alternating 1's and 0's: namely, a square wave. The *fundamental frequency* of a binary square wave is equal to one-half of the bit rate. Consequently, if only the fundamental frequency of the input is considered, the highest modulating frequency to the FSK modulator is equal to one-half of the input bit rate.

The rest frequency of the VCO is chosen such that it falls halfway between the mark and space frequencies. A logic 1 condition at the input shifts the VCO from its rest frequency to the mark frequency, and a logic 0 condition at the input shifts the

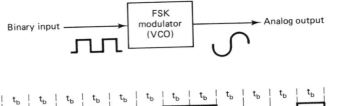

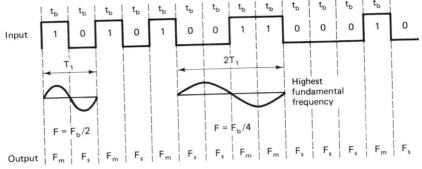

FIGURE 13-3   FSK modulator. t_b, Time of one bit = 1/bps; F_m, mark frequency; F_s, space frequency; T1, period of shortest cycle; 1/T1, fundamental frequency of binary square wave; F_b, input bit rate (bps).

VCO from its rest frequency to the space frequency. Consequently, as the input binary signal changes from a logic 1 to a logic 0, and vice versa, the VCO output frequency *shifts* or *deviates* back and forth between the mark and space frequencies. Because FSK is a form of frequency modulation, the formula for *modulation index* used in FM is also valid for FSK. Modulation index is given as

$$MI = \frac{\Delta F}{F_a}$$

where

  MI = modulation index
  ΔF = frequency deviation (Hz)
  F_a = modulating frequency (Hz)

The worst-case modulation index is the modulation index that yields the widest output bandwidth, called the *deviation ratio*. The worst-case or widest bandwidth occurs when both the frequency deviation and the modulating frequencies are at their maximum values.

In an FSK modulator, ΔF is the peak frequency deviation of the carrier and is equal to the difference between the rest frequency and either the mark or space frequency (or half the difference between the mark and space frequencies). The peak frequency deviation depends on the amplitude of the modulating signal. In a binary digital signal, all logic 1's have the same voltage and all logic 0's have the same voltage; consequently, the frequency deviation is constant and always at its maximum value. F_a is equal to the fundamental frequency of the binary input which under the worst-case condition (alternating 1's and 0's) is equal to one-half of the bit rate. Consequently, for FSK,

$$\text{MI} = \frac{\left|\dfrac{F_m - F_s}{2}\right|}{\dfrac{F_b}{2}} = \frac{|F_m - F_s|}{F_b} \qquad (13\text{-}3)$$

where

$$\frac{|F_m - F_s|}{2} = \text{peak frequency deviation}$$

$$\frac{F_b}{2} = \text{fundamental frequency of the binary input signal}$$

With conventional FM, the bandwidth is directly proportional to the modulation index. Consequently, in FSK the modulation index is generally kept below 1.0, thus producing a relatively narrowband FM output spectrum. The minimum bandwidth required to propagate a signal is called the *minimum Nyquist bandwidth* (F_n). When modulation is used and a double-sided output spectrum is generated, the minimum bandwidth is called the *minimum double-sided Nyquist bandwidth* or the *minimum IF bandwidth*.

EXAMPLE 13-1

For an FSK modulator with space, rest, and mark frequencies of 60, 70, and 80 MHz, respectively and an input bit rate of 20 Mbps, determine the output baud and the minimum required bandwidth.

Solution   Substituting into Equation 13-3, we have

$$\text{MI} = \frac{|F_m - F_s|}{F_b} = \frac{|80\ \text{MHz} - 60\ \text{MHz}|}{20\ \text{Mbps}}$$

$$= \frac{20\ \text{MHz}}{20\ \text{Mbps}} = 1.0$$

From the Bessel chart (Table 13-1), a modulation index of 1.0 yields three sets of significant side frequencies. Each side frequency is separated from the center frequency or an adjacent side frequency by a value equal to the modulating frequency, which in this example is 10 MHz ($F_b/2$). The output spectrum for this modulator is shown in Figure 13-4. It can be seen that the minimum double-sided Nyquist bandwidth is 60 MHz and the baud rate is 20 megabaud, the same as the bit rate.

Because FSK is a form of narrowband frequency modulation, the minimum bandwidth is dependent on the modulation index. For a modulation index between 0.5 and 1, either two or three sets of significant side frequencies are generated. Thus the minimum bandwidth is two to three times the input bit rate.

TABLE 13-1 BESSEL FUNCTION CHART

MI	J_0	J_1	J_2	J_3	J_4
0.0	1.00				
0.25	0.98	0.12			
0.5	0.94	0.24	0.03		
1.0	0.77	0.44	0.11	0.02	
1.5	0.51	0.56	0.23	0.06	0.01
2.0	0.22	0.58	0.35	0.13	0.03

FSK Receiver

The most common circuit used for demodulating FSK signals is the *phase-locked loop* (PLL), which is shown in Figure 13-5. A PLL-FSK demodulator works very much like a PLL-FM demodulator. As the input to the PLL shifts between the mark and space frequencies, the *dc error voltage* at the output of the phase comparator follows the frequency shift. Because there are only two input frequencies (mark and space), there are also only two output error voltages. One represents a logic 1 and the other a logic 0. Therefore, the output is a two-level (binary) representation of the FSK input. Generally, the natural frequency of the PLL is made equal to the center frequency of the FSK modulator. As a result, the changes in the dc error voltage follow the changes in the analog input frequency and are symmetrical around 0 V dc.

FSK has a poorer error performance than PSK or QAM and, consequently, is seldom used for high-performance digital radio systems. Its use is restricted to low-performance, low-cost, asynchronous data modems that are used for data communications over analog, voice band telephone lines (see Chapter 14).

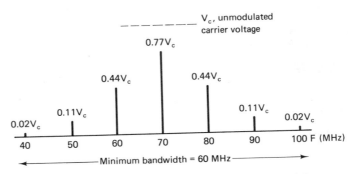

FIGURE 13-4 FSK output spectrum for Example 13-1.

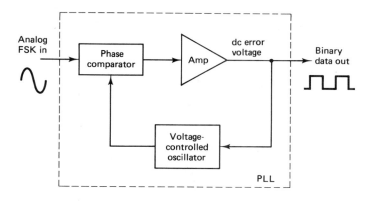

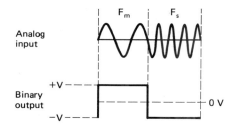

FIGURE 13-5   PLL-FSK demodulator.

Minimum Shift-Keying FSK

Minimum shift-keying FSK (MSK) is a form of *continuous-phase* frequency shift keying (CPFSK). Essentially, MSK is binary FSK except that the mark and space frequencies are *synchronized* with the input binary bit rate. Synchronous simply means that there is a precise time relationship between the two; it does not mean they are equal. With MSK, the mark and space frequencies are selected such that they are separated from the center frequency by an exact odd multiple of one-half of the bit rate $[F_m$ and $F_s = n(F_b/2)$, where $n =$ any odd whole integer]. This ensures that there is a smooth phase transition in the analog output signal when it changes from a mark to a space frequency, or vice versa. Figure 13-6 shows a *noncontinuous* FSK waveform. It can be seen that when the input changes from a logic 1 to a logic 0, and vice versa, there is an abrupt phase discontinuity in the analog output signal. When this occurs, the demodulator has trouble following the frequency shift; consequently, an error may occur.

Figure 13-7 shows a continuous-phase MSK waveform. Notice that when the output frequency changes, it is a smooth, continuous transition. Each transition occurs at a zero crossing; consequently, there are no phase discontinuities. MSK has a better bit-error performance than conventional FSK for a given signal-to-noise ratio. The disad-

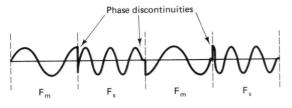

FIGURE 13-6   Noncontinuous FSK waveform.

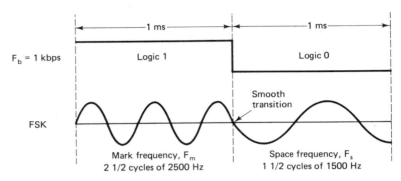

FIGURE 13-7   Continuous-phase MSK waveform.

vantage of MSK is that it requires synchronizing circuits and is therefore more expensive to implement.

PHASE SHIFT KEYING

Phase shift keying (PSK) is another form of angle-modulated, constant-envelope digital modulation. PSK is similar to conventional phase modulation except that with PSK the input signal is a binary digital signal and a limited number of output phases are possible.

BINARY PHASE SHIFT KEYING

With *binary phase shift keying* (BPSK), two output phases are possible for a single carrier frequency ("binary" meaning "2"). One output phase represents a logic 1 and the other a logic 0. As the input digital signal changes, the phase of the output carrier shifts between two angles that are 180° out of phase. Another name for BPSK is *phase reversal keying* (PRK).

BPSK Transmitter

Figure 13-8 shows a simplified block diagram of a BPSK modulator. The balanced modulator acts like a phase reversing switch. Depending on the logic condition of the digital input, the carrier is transferred to the output either in phase or 180° out of phase with the reference carrier oscillator.

Figure 13-9a shows the schematic diagram of a balanced ring modulator. The balanced modulator has two inputs: a carrier that is in phase with the reference oscillator and the binary digital data. For the balanced modulator to operate properly, the digital input voltage must be much greater than the peak carrier voltage. This ensures that the digital input controls the on/off state of diodes D1–D4. If the binary input is a logic 1 (positive voltage), diodes D1 and D2 are forward biased and "on," while diodes D3 and D4 are reverse biased and "off" (Figure 13-9b). With the polarities shown, the carrier voltage is developed across transformer T2 in phase with the carrier voltage across T1. Consequently, the output signal is in phase with the reference oscillator.

If the binary input is a logic 0 (negative voltage), diodes D1 and D2 are reverse biased and "off," while diodes D3 and D4 are forward biased and "on" (Figure 13-9c). As a result, the carrier voltage is developed across transformer T2 180° out of phase with the carrier voltage across T1. Consequently, the output signal is 180° out of phase with the reference oscillator. Figure 13-10 shows the truth table, phasor diagram, and constellation diagram for a BPSK modulator. A *constellation diagram*, which is sometimes called a *signal state-space diagram*, is similar to a phasor diagram except that the entire phasor is not drawn. In a constellation diagram, only the relative positions of the peaks of the phasors are shown.

Bandwidth Considerations of BPSK

A balanced modulator is a *product modulator*; the output signal is the product of the two input signals. In a BPSK modulator, the carrier input signal is multiplied by the binary data. If $+1$ V is assigned to a logic 1 and -1 V is assigned to a logic 0, the input carrier ($\sin \omega_c t$) is multiplied by either a $+$ or $-$ 1. Consequently, the output signal is either $+1 \sin \omega_c t$ or $-1 \sin \omega_c t$; the first represents a signal that is *in phase* with the reference oscillator, the latter a signal that is 180° out of phase with the reference

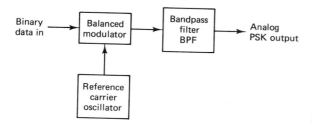

FIGURE 13-8   BPSK modulator.

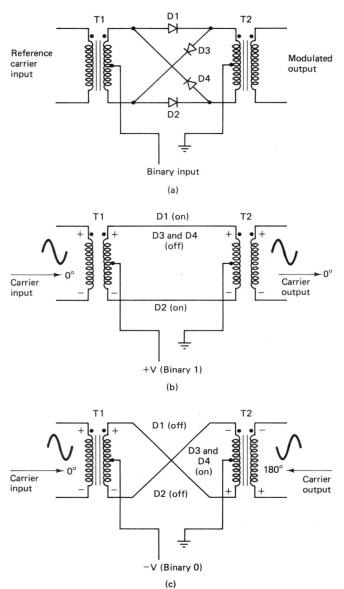

FIGURE 13-9 (a) Balanced ring modulator; (b) logic 1 input; (c) logic 0 input.

oscillator. Each time the input logic condition changes, the output phase changes. Consequently, for BPSK, the output rate of change (baud) is equal to the input rate of change (bps), and the widest output bandwidth occurs when the input binary data are an alternating 1/0 sequence. The fundamental frequency (F_a) of an alternating 1/0 bit sequence is

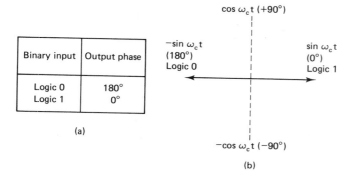

Binary input	Output phase
Logic 0	180°
Logic 1	0°

(a)

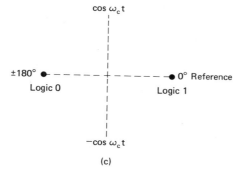

(c)

FIGURE 13-10 BPSK modulator: (a) truth table; (b) phasor diagram; (c) constellation diagram.

equal to one-half of the bit rate ($F_b/2$). Mathematically, the output phase of a BPSK modulator is

$$\theta \quad = \quad \underbrace{(\sin \omega_a t)}_{\substack{\text{fundamental frequency} \\ \text{of the binary} \\ \text{modulating signal}}} \quad \times \quad \underbrace{(\sin \omega_c t)}_{\text{carrier}}$$

or

$$\tfrac{1}{2} \cos (\omega_c t - \omega_a t) - \tfrac{1}{2} \cos (\omega_c t + \omega_a t)$$

Consequently, the minimum double-sided Nyquist bandwidth (F_n) is

$$
\begin{array}{ccc}
\omega_c t + \omega_a t & & \omega_c t + \omega_a t \\
\underline{- (\omega_c t - \omega_a t)} & \text{or} & \underline{- \omega_c t + \omega_a t} \\
& & 2\omega_a t
\end{array}
$$

and because $\omega_a t = F_b/2$,

$$F_n = 2 \left(\frac{F_b}{2} \right) = F_b$$

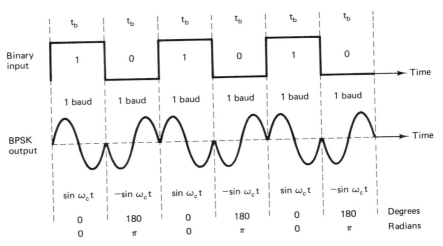

FIGURE 13-11   Output phase versus time relationship for a BPSK modulator.

Figure 13-11 shows the output phase versus time relationship for a BPSK waveform. It can be seen that the output spectrum from a BPSK modulator is simply a double-sideband suppresssed carrier signal where the upper and lower side frequencies are separated from the carrier frequency by a value equal to one-half of the bit rate. Consequently, the minimum bandwidth (F_n) required to pass the worst-case BPSK output signal is equal to the input bit rate.

EXAMPLE 13-2

For a BPSK modulator with a carrier frequency of 70 MHz and an input bit rate of 10 Mbps, determine the maximum and minimum upper and lower side frequencies, draw the output spectrum, determine the minimum Nyquist bandwidth, and calculate the baud.

Solution   Substituting into the equation for the output of a balanced modulator yields

$$\text{output} = (\sin \omega_a t)\,(\sin \omega_c t)$$

$$= [\sin 2\pi(5 \text{ MHz})t)]\|[\sin 2\pi(70 \text{ MHz})t]$$

$$= \tfrac{1}{2}\cos 2\pi(70 \text{ MHz} - 5 \text{ MHz})t - \tfrac{1}{2}\cos 2\pi(70 \text{ MHz} + 5 \text{ MHz})t$$

$$\underbrace{\phantom{= \tfrac{1}{2}\cos 2\pi(70 \text{ MHz} - 5 \text{ MHz})t}}_{\text{lower side frequency}} \qquad \underbrace{\phantom{\tfrac{1}{2}\cos 2\pi(70 \text{ MHz} + 5 \text{ MHz})t}}_{\text{upper side frequency}}$$

Minimum lower side frequency (LSF):

$$\text{LSF} = 70 \text{ MHz} - 5 \text{ MHz} = 65 \text{ MHz}$$

Maximum upper side frequency (USF):

$$\text{USF} = 70 \text{ MHz} + 5 \text{ MHz} = 75 \text{ MHz}$$

Therefore, the output spectrum for the worst-case binary input conditions is as follows:

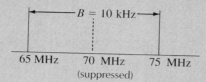

The minimum Nyquist bandwidth (F_n) is

$$F_n = 75 \text{ MHz} - 65 \text{ MHz} = 10 \text{ MHz}$$

and the baud $= F_b$ or 10 megabaud.

BPSK Receiver

Figure 13–12 shows the block diagram of a BPSK receiver. The input signal may be $+\sin \omega_c t$ or $-\sin \omega_c t$. The coherent carrier recovery circuit detects and regenerates a carrier signal that is both frequency and phase coherent with the original transmit carrier. The balanced modulator is a product detector; the output is the product of the two inputs (the BPSK signal and the recovered carrier). The low-pass filter (LPF) separates the recovered binary data from the complex demodulated spectrum. Mathematically, the demodulation process is as follows.

For a BPSK input signal of $+\sin \omega_c t$ (logic 1), the output of the balanced modulator is

$$\text{output} = (\sin \omega_c t)(\sin \omega_c t) = \sin^2 \omega_c t$$

or

$$\text{(filtered out)}$$
$$\sin^2 \omega_c t = \tfrac{1}{2}(1 - \cos 2\omega_c t) = \tfrac{1}{2} - \tfrac{1}{2}\cos 2\omega_c t$$

$$\text{output} = +\tfrac{1}{2} \text{ V dc} = \text{logic 1}$$

It can be seen that the output of the balanced modulator contains a positive dc voltage ($+\tfrac{1}{2}$ V) and a cosine wave at twice the carrier frequency ($2\omega_c t$). The LPF has a cutoff frequency much lower than $2\omega_c t$ and thus blocks the second harmonic of the

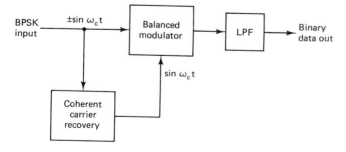

FIGURE 13-12   BPSK receiver.

carrier and passes only the positive dc component. A positive dc voltage represents a demodulated logic 1.

For a BPSK input signal of $-\sin \omega_c t$ (logic 0), the output of the balanced modulator is

$$\text{output} = (-\sin \omega_c t)(\sin \omega_c t) = -\sin^2 \omega_c t$$

or   (filtered out)
$$-\sin^2 \omega_c t = -\tfrac{1}{2}(1 - \cos 2\omega_c t) = -\tfrac{1}{2} + \tfrac{1}{2} \cos 2\omega_c t$$

$$\text{output} = -\tfrac{1}{2} \text{ V dc} = \text{logic } 0$$

The output of the balanced modulator contains a negative dc voltage $(-\tfrac{1}{2} \text{ V})$ and a cosine wave at twice the carrier frequency $(2\omega_c t)$. Again, the LPF blocks the second harmonic of the carrier and passes only the negative dc component. A negative dc voltage represents a demodulated logic 0.

M-ary Encoding

M-ary is a term derived from the word "binary." *M* is simply a digit that represents the number of conditions possible. The two digital modulation techniques discussed thus far (binary FSK and BPSK) are binary systems; there are only two possible output conditions. One represents a logic 1 and the other a logic 0; thus they are *M*-ary systems where $M = 2$. With digital modulation, very often it is advantageous to encode at a level higher than binary. For example, a PSK system with four possible output phases is an *M*-ary system where $M = 4$. If there were eight possible output phases, $M = 8$, and so on. Mathematically,

$$N = \log_2 M$$

where

N = number of bits
M = number of output conditions possible with n bits

For example, if 2 bits were allowed to enter a modulator before the output were allowed to change,

$$2 = \log_2 M \quad \text{and} \quad 2^2 = M \quad \text{thus } M = 4$$

An $M = 4$ indicates that with 2 bits, four different output conditions are possible. For $N = 3$, $M = 2^3$ or 8, and so on.

QUATERNARY PHASE SHIFT KEYING

Quaternary phase shift keying (QPSK), or *quadrature PSK* as it is sometimes called, is another form of angle-modulated, constant-envelope digital modulation. QPSK is an

M-ary encoding technique where *M* = 4 (hence the name "quaternary," meaning "4"). With QPSK four output phases are possible for a single carrier frequency. Because there are four different output phases, there must be four different input conditions. Because the digital input to a QPSK modulator is a binary (base 2) signal, to produce four different input conditions it takes more than a single input bit. With 2 bits, there are four possible conditions: 00, 01, 10, and 11. Therefore, with QPSK, the binary input data are combined into groups of 2 bits called *dibits*. Each dibit code generates one of the four possible output phases. Therefore, for each 2-bit dibit clocked into the modulator, a single output change occurs. Therefore, the rate of change at the output (baud rate) is one-half of the input bit rate.

QPSK Transmitter

A block diagram of a QPSK modulator is shown in Figure 13-13. Two bits (a dibit) are clocked into the bit spliter. After both bits have been serially inputted, they are simultaneously parallel outputted. One bit is directed to the I channel and the other to the Q channel. The I bit modulates a carrier that is in phase with the reference oscillator (hence the name "I" for "in phase" channel), and the Q bit modulates a carrier that is 90° out of phase or in quadrature with the reference carrier (hence the name "Q" for "quadrature" channel).

It can be seen that once a dibit has been split into the I and Q channels, the operation is the same as in a BPSK modulator. Essentially, a QPSK modulator is two

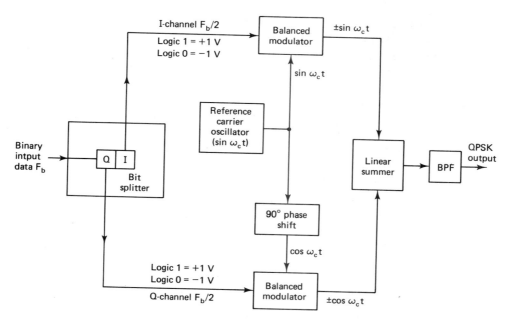

FIGURE 13-13   QPSK modulator.

BPSK modulators combined in parallel. Again, for a logic $1 = +1$ V and a logic $0 = -1$ V, two phases are possible at the output of the I balanced modulator ($+\sin \omega_c t$ and $-\sin \omega_c t$), and two phases are possible at the output of the Q balanced modulator ($+\cos \omega_c t$ and $-\cos \omega_c t$). When the linear summer combines the two quadrature (90° out of phase) signals, there are four possible resultant phases: $+\sin \omega_c t + \cos \omega_c t$, $+\sin \omega_c t - \cos \omega_c t$, $-\sin \omega_c t + \cos \omega_c t$, and $-\sin \omega_c t - \cos \omega_c t$.

EXAMPLE 13-3

For the QPSK modulator shown in Figure 13-13, construct the truth table, phasor diagram, and constellation diagram.

Solution   For a binary data input of Q = 0 and I = 0, the two inputs to the I balanced modulator are -1 and $\sin \omega_c t$, and the two inputs to the Q balanced modulator are -1 and $\cos \omega_c t$. Consequently, the outputs are

$$\text{I balanced modulator} = (-1)(\sin \omega_c t) = -1 \sin \omega_c t$$

$$\text{Q balanced modulator} = (-1)(\cos \omega_c t) = -1 \cos \omega_c t$$

and the output of the linear summer is

$$-1 \cos \omega_c t - 1 \sin \omega_c t = 1.414 \sin \omega_c t - 135°$$

For the remaining dibit codes (01, 10, and 11), the procedure is the same. The results are shown in Figure 13-14.

In Figure 13-14b it can be seen that with QPSK each of the four possible output phasors has exactly the same amplitude. Therefore, the binary information must be encoded entirely in the phase of the output signal. This is the most important characteristic of PSK that distinguishes it from QAM, which is explained later in this chapter. Also, from Figure 13-14b it can be seen that the angular separation between any two adjacent phasors in QPSK is 90°. Therefore, a QPSK signal can undergo almost a $+45°$ or $-45°$ shift in phase during transmission and still retain the correct encoded information when demodulated at the receiver. Figure 13-15 shows the output phase versus time relationship for a QPSK modulator.

Bandwidth Considerations of QPSK

With QPSK, since the input data are divided into two channels, the bit rate in either the I or the Q channel is equal to one-half of the input data ($F_b/2$). (Essentially, the bit splitter stretches the I and Q bits to twice their input bit length.) Consequently, the highest fundamental frequency present at the data input to the I or the Q balanced modulator is equal to one-fourth of the input data rate (one-half of $F_b/2 = F_b/4$). As a result, the output of the I and Q balanced modulators requires a minimum double-sided Nyquist bandwidth equal to one-half of the incoming bit rate ($F_n =$ twice $F_b/4 = F_b/2$). Thus with QPSK, a bandwidth compression is realized (the minimum bandwidth is less than the incoming bit rate). Also, since the QPSK output signal does not change

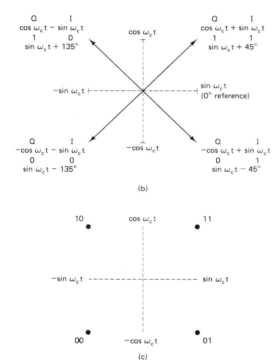

Binary input		QPSK output phase
Q	I	
0	0	−135°
0	1	−45°
1	0	+135°
1	1	+45°

(a)

(b)

(c)

FIGURE 13-14 QPSK modulator: (a) truth table; (b) phasor diagram; (c) constellation diagram.

phase until 2 bits (a dibit) have been clocked into the bit splitter, the fastest output rate of change (baud) is also equal to one-half of the input bit rate. As with BPSK, the minimum bandwidth and the baud are equal. This relationship is shown in Figure 13-16.

In Figure 13-16 it can be seen that the worst-case input condition to the I or Q balanced modulator is an alternating 1/0 pattern, which occurs when the binary input data has a 1100 repetitive pattern. One cycle of the fastest binary transition (a 1/0 sequence) in the I or Q channel takes the same time as 4 input data bits. Consequently, the highest fundamental frequency at the input and fastest rate of change at the output of the balanced modulators is equal to one-fourth of the binary input bit rate.

The output of the balanced modulators can be expressed mathematically as

$$\theta = (\sin \omega_a t)(\sin \omega_c t)$$

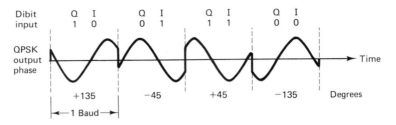

FIGURE 13-15   Output phase versus time relationship for a QPSK modulator.

where

$$\underbrace{\omega_a t = 2\pi \frac{F_b t}{4}}_{\substack{\text{modulating} \\ \text{signal}}} \quad \text{and} \quad \underbrace{\omega_c t = 2\pi F_c t}_{\text{carrier}}$$

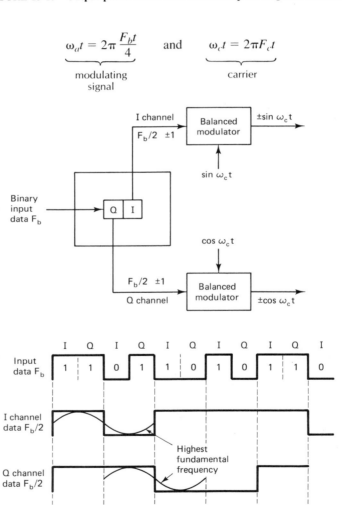

FIGURE 13-16   Bandwidth considerations of a QPSK modulator.

Thus

$$\theta = \left(\sin 2\pi \, \frac{F_b t}{4}\right)(\sin 2\pi \, F_c t)$$

$$= \frac{1}{2} \cos 2\pi \left(F_c - \frac{F_b}{4}\right) t - \frac{1}{2} \cos 2\pi \left(F_c + \frac{F_b}{4}\right) t$$

The output frequency spectrum extends from $F_c + F_b/4$ to $F_c - F_b/4$ and the minimum bandwidth (F_N) is

$$\left(F_c + \frac{F_b}{4}\right) - \left(F_c - \frac{F_b}{4}\right) = \frac{2F_b}{4} = \frac{F_b}{2}$$

EXAMPLE 13-4

For a QPSK modulator with an input data rate (F_b) equal to 10 Mbps and a carrier frequency of 70 MHz, determine the minimum double-sided Nyquist bandwidth (F_N) and the baud. Also, compare the results with those achieved with the BPSK modulator in Example 13-2. Use the QPSK block diagram shown in Figure 13-13 as the modulator model.

Solution The bit rate in both the I and Q channels is equal to one-half of the transmission bit rate or

$$F_{bQ} = F_{bI} = \frac{F_b}{2} = \frac{10 \text{ Mbps}}{2} = 5 \text{ Mbps}$$

The highest fundamental frequency presented to either balanced modulator is

$$F_a = \frac{F_{bQ}}{2} \quad \text{or} \quad \frac{F_{bI}}{2} = \frac{5 \text{ Mbps}}{2} = 2.5 \text{ MHz}$$

The output wave from each balanced modulator is

$$(\sin 2\pi F_a t)(\sin 2\pi F_c t)$$

$$\tfrac{1}{2} \cos 2\pi (F_c - F_a)t - \tfrac{1}{2} \cos 2\pi (F_c + F_a)t$$

$$\tfrac{1}{2} \cos 2\pi [(70 - 2.5) \text{ MHz}]t - \tfrac{1}{2} \cos 2\pi [(70 + 2.5) \text{ MHz}]t$$

$$\tfrac{1}{2} \cos 2\pi (67.5 \text{ MHz})t - \tfrac{1}{2} \cos 2\pi (72.5 \text{ MHz})t$$

The minimum Nyquist bandwidth is

$$F_N = (72.5 - 67.5) \text{ MHz} = 5 \text{ MHz}$$

The baud equals the bandwidth; thus

$$\text{baud} = 5 \text{ megabaud}$$

The output spectrum is as follows:

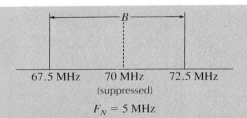

$F_N = 5$ MHz

It can be seen that for the same input bit rate the minimum bandwidth required to pass the output of the QPSK modulator is equal to one-half of that required for the BPSK modulator in Example 13-2. Also, the baud rate for the QPSK modulator is one-half that of the BPSK modulator.

QPSK Receiver

The block diagram of a QPSK receiver is shown in Figure 13-17. The power splitter directs the input QPSK signal to the I and Q product detectors and the carrier recovery circuit. The carrier recovery circuit reproduces the original transmit carrier oscillator signal. The recovered carrier must be frequency and phase coherent with the transmit reference carrier. The QPSK signal is demodulated in the I and Q product detectors, which generate the original I and Q data bits. The outputs of the product detectors are fed to the bit combining circuit, where they are converted from parallel I and Q data channels to a single binary output data stream.

The incoming QPSK signal may be any one of the four possible output phases shown in Figure 13-14. To illustrate the demodulation process, let the incoming QPSK signal be $-\sin \omega_c t + \cos \omega_c t$. Mathematically, the demodulation process is as follows.

The receive QPSK signal ($-\sin \omega_c t + \cos \omega_c t$) is one of the inputs to the I product detector. The other input is the recovered carrier ($\sin \omega_c t$). The output of the I product detector is

$$I = \underbrace{(-\sin \omega_c t + \cos \omega_c t)}_{\text{QPSK input signal}} \underbrace{(\sin \omega_c t)}_{\text{carrier}}$$

$$= (-\sin \omega_c t)(\sin \omega_c t) + (\cos \omega_c t)(\sin \omega_c t)$$

$$= -\sin^2 \omega_c t + (\cos \omega_c t)(\sin \omega_c t)$$

$$= -\tfrac{1}{2}(1 - \cos 2\omega_c t) + \tfrac{1}{2}\sin (\omega_c t + \omega_c t) + \tfrac{1}{2}\sin (\omega_c t - \omega_c t)$$

$$I = -\tfrac{1}{2} + \tfrac{1}{2}\cos 2\omega_c t + \tfrac{1}{2}\sin 2\omega_c t + \tfrac{1}{2}\sin 0$$

(filtered out)   (equals 0)

$$= -\tfrac{1}{2} \text{ V dc} \ (\text{logic } 0)$$

Again, the receive QPSK signal ($-\sin \omega_c t + \cos \omega_c t$) is one of the inputs to the Q product detector. The other input is the recovered carrier shifted $90°$ in phase ($\cos \omega_c t$). The output of the Q product detector is

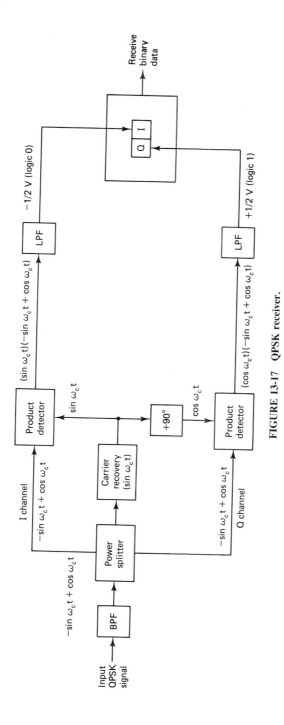

FIGURE 13-17 QPSK receiver.

$$Q = \underbrace{(-\sin \omega_c t + \cos \omega_c t)}_{\text{QPSK input signal}} \underbrace{(\cos \omega_c t)}_{\text{carrier}}$$

$$= \cos^2 \omega_c t - (\sin \omega_c t)(\cos \omega_c t)$$

$$= \tfrac{1}{2}(1 + \cos 2\omega_c t) - \tfrac{1}{2} \sin (\omega_c t + \omega_c t) - \tfrac{1}{2} \sin (\omega_c t - \omega_c t)$$

$$\overset{\text{(filtered out)}}{} \qquad \overset{\text{(equals 0)}}{}$$

$$Q = \tfrac{1}{2} + \tfrac{1}{2} \cos 2\omega_c t - \tfrac{1}{2} \sin 2\omega_c t - \tfrac{1}{2} \sin 0$$

$$= \text{½ V dc (logic 1)}$$

The demodulated I and Q bits (1 and 0, respectively) correspond to the constellation diagram and truth table for the QPSK modulator shown in Fig. 13-14.

Offset QPSK

Offset QPSK (OQPSK) is a modified form of QPSK where the bit waveforms on the I and Q channels are offset or shifted in phase from each other by one-half of a bit time.

Figure 13-18 shows a simplified block diagram, the bit sequence alignment, and the constellation diagram for a OQPSK modulator. Because changes in the I channel occur at the midpoints of the Q-channel bits, and vice versa, there is never more than a single bit change in the dibit code, and therefore there is never more than a 90° shift in the output phase. In conventional QPSK, a change in the input dibit from 00 to 11 or 01 to 10 causes a corresponding 180° shift in the output phase. Therefore, an advantage of OQPSK is the limited phase shift that must be imparted during modulation. A disadvantage of OQPSK is that changes in the output phase occur at twice the data rate in either the I or Q channels. Consequently, with OQPSK the baud and minimum bandwidth are twice that of conventional QPSK for a given transmission bit rate. OQPSK is sometimes called OKQPSK (*offset-keyed PSK*).

EIGHT-PHASE PSK

Eight-phase PSK (8PSK) is an *M*-ary encoding technique where $M = 8$. With an 8PSK modulator, there are eight possible output phases. To encode eight different phases, the incoming bits are considered in groups of 3 bits, called *tribits* ($2^3 = 8$).

8PSK Transmitter

A block diagram of an 8PSK modulator is shown in Figure 13-19. The incoming serial bit stream enters the bit splitter, where it is converted to a parallel, three-channel output (the I or in-phase channel, the Q or in-quadrature channel, and the C or control channel). Consequently, the bit rate in each of the three channels is $F_b/3$. The bits in the I and C channels enter the I-channel 2-to-4-level converter, and the bits in the Q

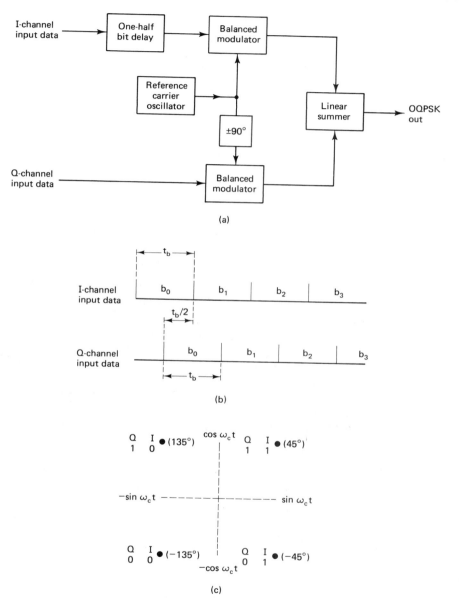

FIGURE 13-18   Offset keyed PSK (OQPSK): (a) block diagram; (b) bit alignment;
(c) constellation diagram.

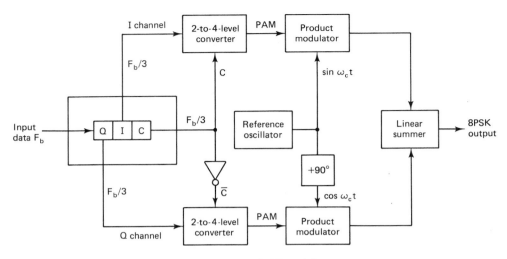

FIGURE 13-19 8PSK modulator.

and $\overline{C}$ channels enter the Q-channel 2-to-4-level converter. Essentially, the 2-to-4-level converters are parallel-input *digital-to-analog converters* (DACs). With 2 input bits, four output voltages are possible. The algorithm for the DACs is quite simple. The I or Q bit determines the polarity of the output analog signal (logic 1 = +V and logic 0 = −V) while the C or $\overline{C}$ bit determines the magnitude (logic 1 = 1.307 V and logic 0 = 0.541 V). Consequently, with two magnitudes and two polarities, four different output conditions are possible.

Figure 13-20 shows the truth table and corresponding output conditions for the 2-to-4-level converters. Because the C and $\overline{C}$ bits can never be the same logic state, the outputs from the I and Q 2-to-4-level converters can never have the same magnitude, although they can have the same polarity. The output of a 2-to-4-level converter is an *M*-ary, *pulse-amplitude-modulated* (PAM) signal where $M = 4$.

I	C	Output
0	0	−0.541 V
0	1	−1.307 V
1	0	+0.541 V
1	1	+1.307 V

Q	$\overline{C}$	Output
0	1	−1.307 V
0	0	−0.541 V
1	1	+1.307 V
1	0	+0.541 V

```
                    +1.307 V
                 ___
                |   +0.541 V
        _____|_____ 0 V
       |  −0.541 V
    ___|
   |  −1.307 V
```

(a)   (b)   (c)

FIGURE 13-20   I- and Q-channel 2-to-4-level converters: (a) I-channel truth table;
(b) Q-channel truth table; (c) PAM levels.

EXAMPLE 13-5

For a tribit input of Q = 0, I = 0, and C = 0 (000), determine the output phase for the 8PSK modulator shown in Figure 13-19.

Solution   The inputs to the I-channel 2-to-4-level converter are $I = 0$ and $C = 0$. From Figure 13-20 the output is -0.541 V. The inputs to the Q-channel 2-to-4-level converter are $Q = 0$ and $\overline{C} = 1$. Again from Figure 13-20, the output is -1.307 V.

Thus the two inputs to the I-channel product modulator are -0.541 and $\sin \omega_c t$. The output is

$$I = (-0.541)(\sin \omega_c t) = -0.541 \sin \omega_c t$$

The two inputs to the Q-channel product modulator are -1.307 V and $\cos \omega_c t$. The output is

$$Q = (-1.307)(\cos \omega_c t) = -1.307 \cos \omega_c t$$

The outputs of the I- and Q-channel product modulators are combined in the linear summer and produce a modulated output of

$$\text{summer output} = -0.541 \sin \omega_c t - 1.307 \cos \omega_c t$$

$$= 1.41 \sin \omega_c t - 112.5°$$

For the remaining tribit codes (001, 010, 011, 100, 101, 110, and 111), the procedure is the same. The results are shown in Figure 13-21.

From Figure 13-21 it can be seen that the angular separation between any two adjacent phasors is 45°, half what it is with QPSK. Therefore, an 8PSK signal can undergo almost a $\pm 22.5°$ phase shift during transmission and still retain its integrity. Also, each phasor is of equal magnitude; the tribit condition (actual information) is again contained only in the phase of the signal. The PAM levels of 1.307 and 0.541 are relative values. Any levels may be used as long as their ratio is 0.541/1.307 and their arc tangent is equal to 22.5°. For example, if their values were doubled to 2.614 and 1.082, the resulting phase angles would not change, although the magnitude of the phasor would increase proportionally.

It should also be noted that the tribit code between any two adjacent phases changes by only one bit. This type of code is called the *Gray code* or, sometimes, the *maximum distance code*. This code is used to reduce the number of transmission errors. If a signal were to undergo a phase shift during transmission, it would most likely be shifted to an adjacent phasor. Using the Gray code results in only a single bit being received in error.

Figure 13-22 shows the output phase versus time relationship of an 8PSK modulator.

Bandwidth Considerations of 8PSK

With 8PSK, since the data are divided into three channels, the bit rate in the I, Q, or C channel is equal to one-third of the binary input data rate ($F_b/3$). (The bit splitter stretches the I, Q, and C bits to three times their input bit length.) Because the I, Q, and C bits are outputted simultaneously and in parallel, the 2-to-4-level converters also see a change in their inputs (and consequently their outputs) at a rate equal to $F_b/3$.

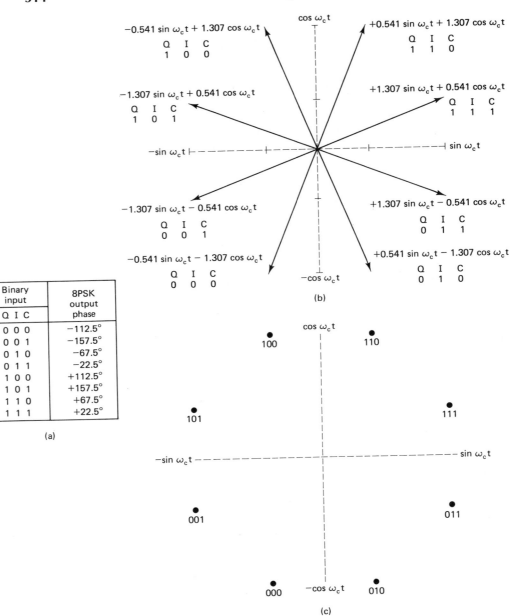

Binary input	8PSK output phase
Q I C	
0 0 0	−112.5°
0 0 1	−157.5°
0 1 0	−67.5°
0 1 1	−22.5°
1 0 0	+112.5°
1 0 1	+157.5°
1 1 0	+67.5°
1 1 1	+22.5°

(a)

(b)

(c)

FIGURE 13-21   8PSK modulator: (a) truth table; (b) phasor diagram; (c) constellation diagram.

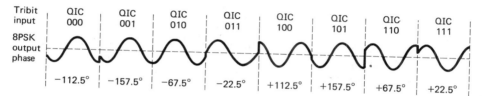

FIGURE 13-22 Output phase versus time relationship for an 8PSK modulator.

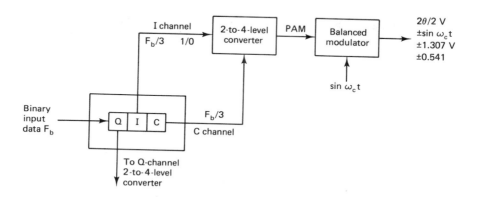

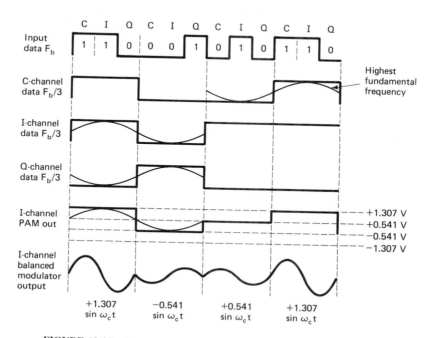

FIGURE 13-23 Bandwidth considerations of an 8PSK modulator.

Figure 13-23 shows the bit timing relationship between the binary input data; the I-, Q-, and C-channel data; and the I and Q PAM signals. It can be seen that the highest fundamental frequency in the I, Q or C channel is equal to one-sixth of the bit rate of the binary input (one cycle in the I, Q, or C channel takes the same amount of time as six input bits). Also, the highest fundamental frequency in either PAM signal is equal to one-sixth of the binary input bit rate.

With an 8PSK modulator, there is one change in phase at the output for every 3 data input bits. Consequently, the baud for 8PSK equals $F_b/3$, the same as the minimum bandwidth. Again, the balanced modulators are product modulators; their outputs are the product of the carrier and the PAM signal. Mathematically, the output of the balanced modulators is

$$\theta = (X \sin \omega_a t)(\sin \omega_c t)$$

where

$$\underbrace{\omega_a t = 2\pi \frac{F_b t}{6}}_{\text{modulating signal}} \quad \text{and} \quad \underbrace{\omega_c t = 2\pi F_c t}_{\text{carrier}}$$

and

$$X = \pm 1.307 \quad \text{or} \quad \pm 0.541$$

Thus

$$\theta = \left(X \sin 2\pi \frac{F_b t}{6}\right)(\sin 2\pi F_c t)$$

$$= \frac{X}{2} \cos 2\pi \left(F_c - \frac{F_b}{6}\right) t - \frac{X}{2} \cos 2\pi \left(F_c + \frac{F_b}{6}\right) t$$

The output frequency spectrum extends from $F_c + F_b/6$ to $F_c - F_b/6$ and the minimum bandwidth (F_N) is

$$\left(F_c + \frac{F_b}{6}\right) - \left(F_c - \frac{F_b}{6}\right) = \frac{2F_b}{6} = \frac{F_b}{3}$$

EXAMPLE 13-6

For an 8PSK modulator with an input data rate (F_b) equal to 10 Mbps and a carrier frequency of 70 MHz, determine the minimum double-sided Nyquist bandwidth (F_N) and the baud. Also, compare the results with those achieved with the BPSK and QPSK modulators in Examples 13-2 and 13-4. Use the 8PSK block diagram shown in Figure 13-19 as the modulator model.

Solution The bit rate in the I, Q, and C channels is equal to one-third of the input bit rate, or

$$F_{bC} = F_{bQ} = F_{bI} = \frac{10 \text{ Mbps}}{3} = 3.33 \text{ Mbps}$$

Therefore, the fastest rate of change and highest fundamental frequency presented to either balanced modulator is

$$F_a = \frac{F_{bC}}{2} \quad \text{or} \quad \frac{F_{bQ}}{2} \quad \text{or} \quad \frac{F_{bI}}{2} = \frac{3.33 \text{ Mbps}}{2} = 1.667 \text{ Mbps}$$

The output wave from the balance modulators is

$$(\sin 2\pi F_a t)(\sin 2\pi F_c t)$$

$$\tfrac{1}{2}\cos 2\pi(F_c - F_a)t - \tfrac{1}{2}\cos 2\pi(F_c + F_a)t$$

$$\tfrac{1}{2}\cos 2\pi[(70 - 1.667) \text{ MHz}]t - \tfrac{1}{2}\cos 2\pi[(70 + 1.667) \text{ MHz}]t$$

$$\tfrac{1}{2}\cos 2\pi(68.333 \text{ MHz})t - \tfrac{1}{2}\cos 2\pi(71.667 \text{ MHz})t$$

The minimum Nyquist bandwidth is

$$F_N = (71.667 - 68.333) \text{ MHz} = 3.333 \text{ MHz}$$

Again, the baud equals the bandwidth; thus

$$\text{baud} = 3.333 \text{ megabaud}$$

The output spectrum is as follows:

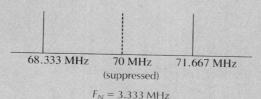

It can be seen that for the same input bit rate the minimum bandwidth required to pass the output of an 8PSK modulator is equal to one-third that of the BPSK modulator in Example 13-2 and 50% less than that required for the QPSK modulator in Example 13-4. Also, in each case the baud has been reduced by the same proportions.

8PSK Receiver

Figure 13-24 shows a block diagram of an 8PSK receiver. The power splitter directs the input 8PSK signal to the I and Q product detectors and the carrier recovery circuit. The carrier recovery circuit reproduces the original reference oscillator signal. The incoming 8PSK signal is mixed with the recovered carrier in the I product detector and with a quadrature carrier in the Q product detector. The outputs of the product detectors are 4-level PAM signals that are fed to the 4-to-2-level *analog-to-digital converters* (ADCs). The outputs from the I-channel 4-to-2-level converter are the I and C bits, while the outputs from the Q-channel 4-to-2-level converter are the Q and $\overline{C}$ bits. The parallel-to-serial logic circuit converts the I/C and Q/$\overline{C}$ bit pairs to serial I, Q, and C output data streams.

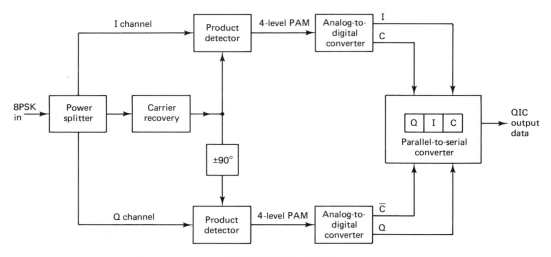

FIGURE 13-24   8PSK receiver.

SIXTEEN-PHASE PSK

Sixteen-phase PSK (16PSK) is an *M*-ary encoding technique where $M = 16$; there are 16 different output phases possible. A 16PSK modulator acts on the incoming data in groups of 4 bits ($2^4 = 16$), called *quadbits*. The output phase does not change until 4 bits have been inputted into the modulator. Therefore, the output rate of change (baud) and the minimum bandwidth are equal to one-fourth of the incoming bit rate ($F_b/4$). The truth table and constellation diagram for a 16PSK transmitter are shown in Figure 13-25.

With 16PSK, the angular separation between adjacent output phases is only 22.5°. Therefore, a 16PSK signal can undergo almost a $\pm 11.25°$ phase shift during transmission and still retain its integrity. Because of this, 16PSK is highly susceptible to phase impairments introduced in the transmission medium and is therefore seldom used.

QUADRATURE AMPLITUDE MODULATION

Quadrature amplitude modulation (QAM) is a form of digital modulation where the digital information is contained in both the amplitude and phase of the transmitted carrier.

EIGHT QAM

Eight QAM (8QAM) is an *M*-ary encoding technique where $M = 8$. Unlike 8PSK, the output signal from an 8QAM modulator is not a constant-amplitude signal.

Bit code	Phase	Bit code	Phase
0000	11.25°	1000	191.25°
0001	33.75°	1001	213.75°
0010	56.25°	1010	236.25°
0011	78.75°	1011	258.75°
0100	101.25°	1100	281.25°
0101	123.75°	1101	303.75°
0110	146.25°	1110	326.25°
0111	168.75°	1111	348.75°

(a)

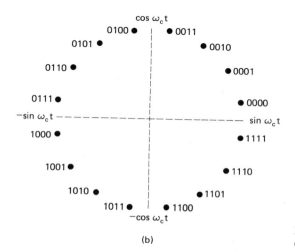

(b)

FIGURE 13-25   16PSK: (a) truth table; (b) constellation diagram.

8QAM Transmitter

Figure 13-26 shows the block diagram of an 8QAM transmitter. As you can see, the only difference between the 8QAM transmitter and the 8PSK transmitter shown in Figure 13-19 is the omission of the inverter between the C channel and the Q product modulator. As with 8PSK, the incoming data are divided into groups of three (tribits): the I, Q, and C channels, each with a bit rate equal to one-third of the incoming data rate. Again, the I and Q bits determine the polarity of the PAM signal at the output of the 2-to-4-level converters, and the C channel determines the magnitude. Because the C bit is fed uninverted to both the I- and Q-channel 2-to-4-level converters, the magnitudes of the I and Q PAM signals are always equal. Their polarities depend on the logic condition of the I and Q bits and therefore may be different. Figure 13-27 shows the truth table for the I- and Q-channel 2-to-4-level converters; they are the same.

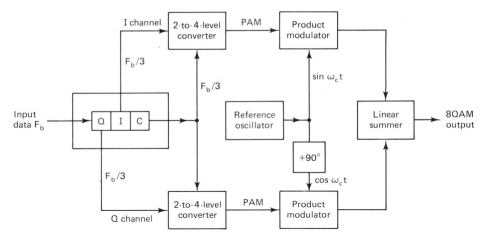

FIGURE 13-26   8QAM transmitter block diagram.

EXAMPLE 13-7

For a tribit input of Q = 0, I = 0, and C = 0 (000), determine the output amplitude and phase for the 8QAM modulator shown in Figure 13-26.

Solution   The inputs to the I-channel 2-to-4-level converter are I = 0 and C = 0. From Figure 13-27 the output is −0.541 V. The inputs to the Q-channel 2-to-4-level converter are Q = 0 and C = 0. Again from Figure 13-27, the output is −0.541 V.

Thus the two inputs to the I-channel product modulator are −0.541 and $\sin \omega_c t$. The output is

$$I = (-0.541)(\sin \omega_c t) = -0.541 \sin \omega_c t$$

The two inputs to the Q-channel product modulator are −0.541 and $\cos \omega_c t$. The output is

$$Q = (-0.541)(\cos \omega_c t) = -0.541 \cos \omega_c t$$

The outputs from the I- and Q-channel product modulators are combined in the linear summer and produce a modulated output of

$$\text{summer output} = -0.541 \sin \omega_c t - 0.541 \cos \omega_c t$$

$$= 0.765 \sin \omega_c t - 135°$$

For the remaining tribit codes (001, 010, 011, 100, 101, 110, and 111), the procedure is the same. The results are shown in Figure 13-28.

I/Q	C	Output
0	0	−0.541
0	1	−1.307 V
1	0	+0.541
1	1	+1.307 V

FIGURE 13-27   Truth table for the I- and Q-channel 2-to-4-level converters.

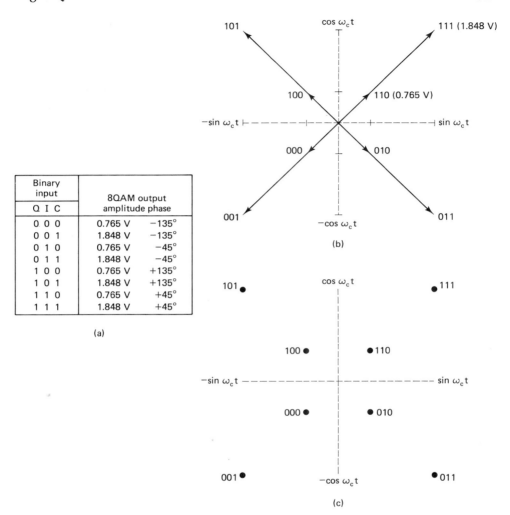

| Binary input | | 8QAM output | |
Q	I	C	amplitude	phase
0	0	0	0.765 V	−135°
0	0	1	1.848 V	−135°
0	1	0	0.765 V	−45°
0	1	1	1.848 V	−45°
1	0	0	0.765 V	+135°
1	0	1	1.848 V	+135°
1	1	0	0.765 V	+45°
1	1	1	1.848 V	+45°

(a)

FIGURE 13-28 **8QAM modulator: (a) truth table; (b) phasor diagram; (c) constellation diagram.**

Figure 13-29 shows the output phase versus time relationship for an 8QAM modulator. Note that there are two output amplitudes and only four phases are possible.

Bandwidth Considerations of 8QAM

In 8QAM, the bit rate in the I and Q channels is one-third of the input binary rate, the same as in 8PSK. As a result, the highest fundamental modulating frequency and the

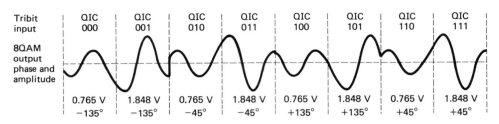

FIGURE 13-29   Output phase and amplitude versus time relationship for 8QAM.

fastest output rate of change in 8QAM are the same as with 8PSK. Therefore, the minimum bandwidth required for 8QAM is $F_b/3$, the same as in 8PSK.

8QAM Receiver

An 8QAM receiver is almost identical to the 8PSK receiver shown in Figure 13-24. The differences are the PAM levels at the output of the product detectors and the binary signals at the output of the analog-to-digital converters. Because there are two transmit amplitudes possible with 8QAM that are different from those achievable with 8PSK, the four demodulated PAM levels in 8QAM are different from those in 8PSK. Therefore, the conversion factor for the analog-to-digital converters must also be different. Also, with 8QAM the binary output signals from the I-channel analog-to-digital converter are the I and C bits, and the binary output signals from the Q-channel analog-to-digital converter are the Q and C bits.

SIXTEEN QAM

Like 16PSK, 16QAM is an *M*-ary system where $M = 16$. The input data are acted on in groups of four ($2^4 = 16$). As with 8QAM, both the phase and amplitude of the transmit carrier are varied.

16QAM Transmitter

The block diagram for a 16QAM transmitter is shown in Figure 13-30. The input binary data are divided into four channels: the I, $\bar{I}$, Q, and $\bar{Q}$. The bit rate in each channel is equal to one-fourth of the input bit rate ($f_b/4$). Four bits are serially clocked into the bit splitter; then they are outputted simultaneously and in parallel with the I, $\bar{I}$, Q, and $\bar{Q}$ channels. The I and Q bits determine the polarity at the output of the 2-to-4-level converters (a logic 1 = positive and a logic 0 = negative). The $\bar{I}$ and $\bar{Q}$ bits determine the magnitude (a logic 1 = 0.821 V and a logic 0 = 0.22 V). Consequently, the 2-to-4-level converters generate a 4-level PAM signal. Two polarities and two magnitudes are possible at the output of each 2-to-4-level converter. They are ±0.22 V and ±0.821 V. The PAM signals modulate the in-phase and quadrature carriers in the product

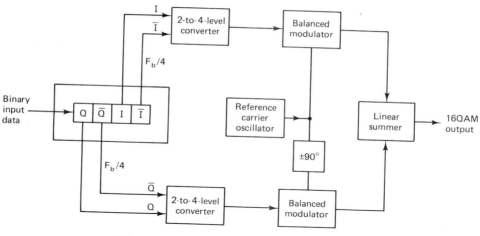

FIGURE 13-30   16QAM transmitter block diagram.

I	Ī	Output
0	0	−0.22 V
0	1	−0.821 V
1	0	+0.22 V
1	1	+0.821 V

(a)

Q	Q̄	Output
0	0	−0.22 V
0	1	−0.821 V
1	0	+0.22 V
1	1	+0.821 V

(b)

**FIGURE 13-31   Truth tables for the I-
and Q-channel 2-to-4-level converters:
(a) I channel; (b) Q channel.**

modulators. Four outputs are possible for each product modulator. For the I product modulator they are $+0.821 \sin \omega_c t$, $-0.821 \sin \omega_c t$, $+0.22 \sin \omega_c t$, and $-0.22 \sin \omega_c t$. For the Q product modulator they are $+0.821 \cos \omega_c t$, $+0.22 \cos \omega_c t$, $-0.821 \cos \omega_c t$, and $-0.22 \cos \omega_c t$. The linear summer combines the outputs from the I- and Q-channel product modulators and produces the 16 output conditions necessary for 16 QAM. Figure 13-31 shows the truth table for the I- and Q-channel 2-to-4-level converters.

EXAMPLE 13-8

For a quadbit input of $I = 0$, $\overline{I} = 0$, $Q = 0$, and $\overline{Q} = 0$ (0000), determine the output amplitude and phase for the 16QAM modulator shown in Figure 13-30.

Solution   The inputs to the I-channel 2-to-4-level converter are $I = 0$ and $\overline{I} = 0$. From Figure 13-31 the output is -0.22 V. The inputs to the Q-channel 2-to-4-level converter are $Q = 0$ and $\overline{Q} = 0$. Again from Figure 13-31, the output is -0.22 V.

Thus the two inputs to the I-channel product modulator are -0.22 V and $\sin \omega_c t$. The output is

$$I = (-0.22)(\sin \omega_c t) = -0.22 \sin \omega_c t$$

The two inputs to the Q-channel product modulator are -0.22 V and $\cos \omega_c t$. The output is

$$Q = (-0.22)(\cos \omega_c t) = -0.22 \cos \omega_c t$$

The outputs from the I- and Q-channel product modulators are combined in the linear summer and produce a modulated output of

$$\text{summer output} = -0.22 \sin \omega_c t - 0.22 \cos \omega_c t$$

$$= 0.311 \sin \omega_c t - 135°$$

For the remaining quadbit codes the procedure is the same. The results are shown in Figure 13-32.

Binary input				16QAM output	
Q	$\overline{Q}$	I	$\overline{I}$		
0	0	0	0	0.311 V	−135°
0	0	0	1	0.850 V	−165°
0	0	1	0	0.311 V	−45°
0	0	1	1	0.850 V	−15°
0	1	0	0	0.850 V	−105°
0	1	0	1	1.161 V	−135°
0	1	1	0	0.850 V	−75°
0	1	1	1	1.161 V	−45°
1	0	0	0	0.311 V	135°
1	0	0	1	0.850 V	175°
1	0	1	0	0.850 V	45°
1	0	1	1	0.850 V	15°
1	1	0	0	0.850 V	105°
1	1	0	1	1.161 V	135°
1	1	1	0	0.850 V	75°
1	1	1	1	1.161 V	45°

(a)

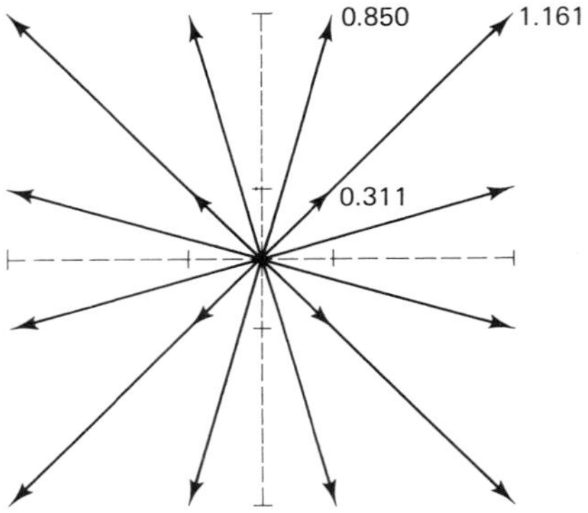

(b)

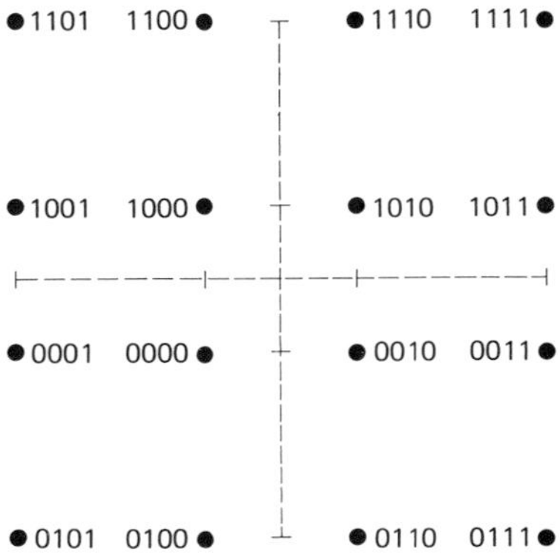

(c)

FIGURE 13-32   16QAM modulator: (a) truth table; (b) phasor diagram; (c) constellation diagram.

Bandwidth Considerations of 16QAM

With 16QAM, since the input data are divided into four channels, the bit rate in the I, $\bar{I}$, Q, or $\bar{Q}$ channel is equal to one-fourth of the binary input data rate ($F_b/4$). (The bit splitter stretches the I, $\bar{I}$, Q, and $\bar{Q}$ bits to four times their input bit length.) Also, because the I, $\bar{I}$, Q, and $\bar{Q}$ bits are outputted simultaneously and in parallel, the 2-to-4-level converters see a change in their inputs and outputs at a rate equal to one-fourth of the input data rate.

Figure 13-33 shows the bit timing relationship between the binary input data;

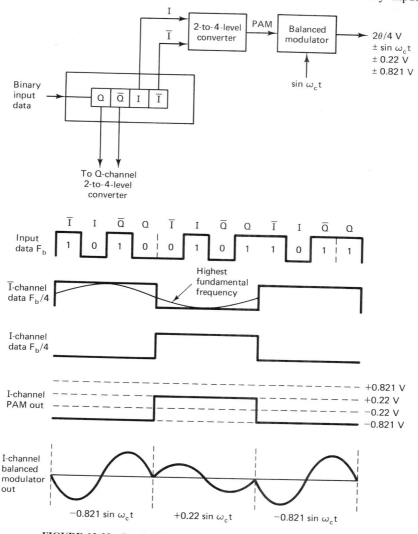

FIGURE 13-33 Bandwidth considerations of a 16QAM modulator.

the I, $\overline{\text{I}}$, Q, and $\overline{\text{Q}}$ channel data; and the I PAM signals. It can be seen that the highest fundamental frequency in the I, $\overline{\text{I}}$, Q, or $\overline{\text{Q}}$ channel is equal to one-eighth of the bit rate of the binary input data (one cycle in the I, $\overline{\text{I}}$, Q, or $\overline{\text{Q}}$ channel takes the same amount of time as 8 input bits). Also, the highest fundamental frequency of either PAM signal is equal to one-eighth of the binary input bit rate.

With a 16QAM modulator, there is one change in the output signal (either its phase, amplitude, or both) for every 4 input data bits. Consequently, the baud equals $F_b/4$, the same as the minimum bandwidth.

Again, the balanced modulators are product modulators and their outputs can be represented mathematically as

$$\theta = (X \sin \omega_a t)(\sin \omega_c t)$$

where

$$\underbrace{\omega_a t = 2\pi \frac{F_b t}{8}}_{\text{modulating signal}} \quad \text{and} \quad \underbrace{\omega_c t = 2\pi F_c t}_{\text{carrier}}$$

and

$$X = \pm 0.22 \quad \text{or} \quad \pm 0.821$$

Thus

$$\theta = \left(X \sin 2\pi \frac{F_b t}{8} \right) (\sin 2\pi F_c t)$$

$$= \frac{X}{2} \cos 2\pi \left(F_c - \frac{F_b}{8} \right) t - \frac{X}{2} \cos 2\pi \left(F_c + \frac{F_b}{8} \right) t$$

The output frequency spectrum extends from $F_c + F_b/8$ to $F_c - F_b/8$ and the minimum bandwidth (F_N) is

$$\left(F_c + \frac{F_b}{8} \right) - \left(F_c - \frac{F_b}{8} \right) = \frac{2F_b}{8} = \frac{F_b}{4}$$

EXAMPLE 13-9

For a 16QAM modulator with an input data rate (F_b) equal to 10 Mbps and a carrier frequency of 70 MHz, determine the minimum double-sided Nyquist frequency (F_N) and the baud. Also, compare the results with those achieved with the BPSK, QPSK, and 8PSK modulators in Examples 13-2, 13-4, and 13-6. Use the 16QAM block diagram shown in Figure 13-26 as the modulator model.

Solution The bit rate in the I, $\overline{\text{I}}$, Q, and $\overline{\text{Q}}$ channels is equal to one-fourth of the input bit rate or

$$F_{bI} = F_{b\overline{I}} = F_{bQ} = F_{b\overline{Q}} = \frac{F_b}{4} = \frac{10 \text{ Mbps}}{4} = 2.5 \text{ Mbps}$$

Therefore, the fastest rate of change and highest fundamental frequency presented to either balanced modulator is

$$F_a = \frac{F_{bI}}{2} \quad \text{or} \quad \frac{F_{b\overline{I}}}{2} \quad \text{or} \quad \frac{F_{bQ}}{2} \quad \text{or} \quad \frac{F_{b\overline{Q}}}{2} = \frac{2.5 \text{ Mbps}}{2} = 1.25 \text{ Mhz}$$

The output wave from the balanced modulator is

$$(\sin 2\pi F_a t)(\sin 2\pi F_c t)$$

$$\tfrac{1}{2} \cos 2\pi (F_c - F_a)t - \tfrac{1}{2} \cos 2\pi (F_c + F_a)t$$

$$\tfrac{1}{2} \cos 2\pi [(70 - 1.25) \text{ MHz}]t - \tfrac{1}{2} \cos 2\pi [(70 + 1.25) \text{ MHz}]t$$

$$\tfrac{1}{2} \cos 2\pi (68.75 \text{ MHz})t - \tfrac{1}{2} \cos 2\pi (71.25 \text{ MHz})t$$

The minimum Nyquist bandwidth is

$$F_N = (71.25 - 68.75) \text{ MHz} = 2.5 \text{ MHz}$$

The baud equals the bandwidth; thus

$$\text{baud} = 2.5 \text{ megabaud}$$

The output spectrum is as follows:

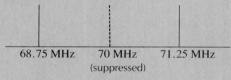

68.75 MHz   70 MHz   71.25 MHz

(suppressed)

$$F_N = 2.5 \text{ MHz}$$

For the same input bit rate, the minimum bandwidth required to pass the output of a 16QAM modulator is equal to one-fourth that of the BPSK moduals, one-half that of QPSK, and 25% less than with 8PSK. For each modulation technique, the baud is also reduced by the same proportions.

BANDWIDTH EFFICIENCY

Bandwidth efficiency (or *information density* as it is sometimes called) is often used to compare the performance of one digital modulation technique to another. In essence, it is the ratio of the transmission bit rate to the minimum bandwidth required for a particular modulation scheme. Bandwidth efficiency is generally normalized to a 1-Hz bandwidth and thus indicates the number of bits that can be propagated through a medium for each hertz of bandwidth. Mathematically, bandwidth efficiency is

$$\text{BW efficiency} = \frac{\text{transmission rate (bps)}}{\text{minimum bandwidth (Hz)}} \tag{13-4}$$

$$= \frac{\text{bits/second}}{\text{hertz}} = \frac{\text{bits/second}}{\text{cycles/second}} = \frac{\text{bits}}{\text{cycle}}$$

EXAMPLE 13-10

Determine the bandwidth efficiencies for the following modulation schemes: BPSK, QPSK, 8PSK, and 16QAM.

Solution Recall from Examples 13-2, 13-4, 13-6, and 13-9 the minimum bandwidths required to propagate a 10-Mbps transmission rate with the following modulation schemes:

Modulation scheme	Minimum bandwidth (MHz)
BPSK	10
QPSK	5
8PSK	3.33
16QAM	2.5

Substituting into Equation 13-4, the bandwidth efficiencies are determined as follows:

$$\text{BPSK:} \quad \text{BW efficiency} = \frac{10 \text{ Mbps}}{10 \text{ MHz}} = \frac{1 \text{ bps}}{\text{Hz}} = \frac{1 \text{ bit}}{\text{cycle}}$$

$$\text{QPSK:} \quad \text{BW efficiency} = \frac{10 \text{ Mbps}}{5 \text{ MHz}} = \frac{2 \text{ bps}}{\text{Hz}} = \frac{2 \text{ bits}}{\text{cycle}}$$

$$\text{8PSK:} \quad \text{BW efficiency} = \frac{10 \text{ Mbps}}{3.33 \text{ MHz}} = \frac{3 \text{ bps}}{\text{Hz}} = \frac{3 \text{ bits}}{\text{cycle}}$$

$$\text{16QAM:} \quad \text{BW efficiency} = \frac{10 \text{ Mbps}}{2.5 \text{ MHz}} = \frac{4 \text{ bps}}{\text{Hz}} = \frac{4 \text{ bits}}{\text{cycle}}$$

The results indicate that BPSK is the least efficient and 16QAM is the most efficient. 16QAM requires one-fourth as much bandwidth as BPSK for the same bit rate.

PSK AND QAM SUMMARY

The various forms of FSK, PSK, and QAM are summarized in Table 13-2.

TABLE 13-2　DIGITAL MODULATION SUMMARY

Modulation	Encoding	Bandwidth (Hz)	Baud	Bandwidth efficiency (bps/Hz)
FSK	Single bit	$\geq F_b$	F_b	≤ 1
BPSK	Single bit	F_b	F_b	1
QPSK	Dibit	$F_b/2$	$F_b/2$	2
8PSK	Tribit	$F_b/3$	$F_b/3$	3
8QAM	Tribit	$F_b/3$	$F_b/3$	3
16PSK	Quadbit	$F_b/4$	$F_b/4$	4
16QAM	Quadbit	$F_b/4$	$F_b/4$	4

CARRIER RECOVERY

Carrier recovery is the process of extracting a phase-coherent reference carrier from a received carrier waveform. This is sometimes called *phase referencing*. In the phase modulation techniques described thus far, the binary data were encoded as a precise phase of the transmitted carrier. (This is referred to as *absolute phase encoding*.) Depending on the encoding method, the angular separation between adjacent phasors varied between 15 and 180°. To correctly demodulate the data, a phase-coherent carrier was recovered and compared with the received signal in a product detector. To determine the absolute phase of the received signal, it is necessary to reproduce a carrier at the receiver that is phase coherent with the transmit reference oscillator. This is the function of the carrier recovery circuit.

With PSK and QAM, the carrier is suppressed in the balanced modulators and is therefore not transmitted. Consequently, at the receiver the carrier cannot simply be tracked with a standard phase-locked loop. A more sophisticated method is used.

One common method of achieving carrier recovery for BPSK is the *squaring loop*. Figure 13-34 shows the block diagram of a squaring loop. The received BPSK waveform is filtered and then squared. The filtering reduces the spectral width of the received noise. The squaring circuit removes the modulation and generates the second harmonic of the carrier frequency. This harmonic is phase tracked by the PLL. The VCO output frequency from the PLL is then divided by 2 and used as the phase reference for the product detectors.

With BPSK, only two output phases are possible: $+\sin \omega_c t$ and $-\sin \omega_c t$. Mathematically, the operation of the squaring circuit can be described as follows.

For a receive signal of $+\sin \omega_c t$ the output of the squaring circuit is

$$\text{output} = (+\sin \omega_c t)(+\sin \omega_c t) = +\sin^2 \omega_c t$$

$$= \tfrac{1}{2}(1 - \cos 2\omega_c t) = \tfrac{1}{2} = \tfrac{1}{2}\cos 2\omega_c t$$

(filtered out)

For a received signal of $-\sin \omega_c t$ the output of the squaring circuit is

$$\text{output} = (-\sin \omega_c t)(-\sin \omega_c t) = +\sin^2 \omega_c t$$

$$= \tfrac{1}{2}(1 - \cos 2\omega_c t) = \tfrac{1}{2} - \tfrac{1}{2}\cos 2\omega_c t$$

(filtered out)

It can be seen that in both cases the output from the squaring circuit contained a dc voltage $(+\tfrac{1}{2}$ V$)$ and a signal at twice the carrier frequency $(\cos 2\omega_c t)$. The dc voltage is removed by filtering, leaving only $\cos 2\omega_c t$.

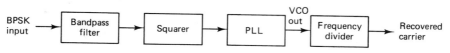

FIGURE 13-34 Squaring loop carrier recovery circuit for a BPSK receiver.

A more elaborate carrier recovery circuit is the *Costas* or *quadrature loop*, which combines carrier recovery with noise suppression and can therefore accurately recover a carrier from a poorer-quality received signal than can a conventional squaring loop.

Carrier recovery circuits for higher-than-binary encoding techniques are similar to BPSK except that circuits which raise the receive signal to the fourth, eighth, and higher powers are used.

DIFFERENTIAL PHASE SHIFT KEYING

Differential phase shift keying (DPSK) is an alternative form of digital modulation where the binary input information is contained in the difference between two successive signaling elements rather than the absolute phase. With DPSK it is not necessary to recover a phase-coherent carrier. Instead, a received signaling element is delayed by one signaling element time slot and then compared to the next received signaling element. The difference in the phase of the two signaling elements determines the logic condition of the data.

DIFFERENTIAL BPSK

DBPSK Transmitter

Figure 13-35a shows a simplified block diagram of a *differential binary phase shift keying* (DBPSK) transmitter. An incoming information bit is XNORed with the preceding bit prior to entering the BPSK modulator (balanced modulator). For the first data bit,

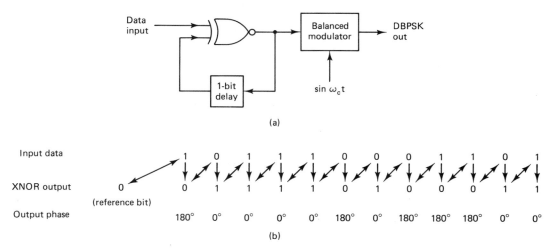

FIGURE 13-35   DBPSK modulator: (a) block diagram; (b) timing diagram.

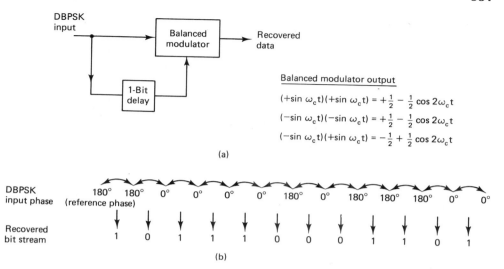

FIGURE 13-36   **DBPSK demodulator: (a) block diagram; (b) timing sequence.**

there is no preceding bit with which to compare it. Therefore, an initial reference bit is assumed. Figure 13-35b shows the relationship between the input data, the XNOR output data, and the phase at the output of the balanced modulator. If the initial reference bit is assumed a logic 1, the output from the XNOR circuit is simply the complement of that shown.

In Figure 13-35b the first data bit is XNORed with the reference bit. If they are the same, the XNOR output is a logic 1; if they are different, the XNOR output is a logic 0. The balanced modulator operates the same as a conventional BPSK modulator; a logic 1 produces $+\sin \omega_c t$ at the output and a logic 0 produces $-\sin \omega_c t$ at the output.

DBPSK Receiver

Figure 13-36 shows the block diagram and timing sequence for a DBPSK receiver. The received signal is delayed by one bit time, then compared with the next signaling element in the balanced modulator. If they are the same, a logic 1 (+ voltage) is generated. If they are different, a logic 0 (− voltage) is generated. If the reference phase is incorrectly assumed, only the first demodulated bit is in error. Differential encoding can be implemented with higher-than-binary digital modulation schemes, although the differential algorithms are much more complicated than for DBPSK.

The primary advantage of DPSK is the simplicity with which it can be implemented. With DPSK, no carrier recovery circuit is needed. A disadvantage of DPSK is that it requires between 1 and 3 dB more signal-to-noise ratio to achieve the same bit error rate as that of absolute PSK.

CLOCK RECOVERY

As with any digital system, digital radio requires precise timing or clock synchronization between the transmit and the receive circuitry. Because of this, it is necessary to regenerate clocks at the receiver that are synchronous with those at the transmitter.

Figure 13-37a shows a simple circuit that is commonly used to recover clocking information from the received data. The recovered data are delayed by one-half a bit time and then compared with the original data in an XOR circuit. The frequency of the clock that is recovered with this method is equal to the received data rate (F_b). Figure 13-37b shows the relationship between the data and the recovered clock timing. From Figure 13-37b it can be seen that as long as the receive data contain a substantial number of transitions (1/0 sequences), the recovered clock is maintained. If the receive data were to undergo an extended period of successive 1's or 0's, the recovered clock would be lost. To prevent this from occurring, the data are scrambled at the transmit end and descrambled at the receive end.

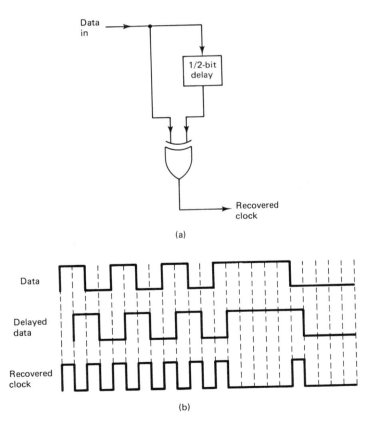

FIGURE 13-37   (a) Clock recovery circuit; (b) timing diagram.

PROBABILITY OF ERROR AND BIT ERROR RATE

Probability of error $P(e)$ and *bit error rate* (BER) are often used interchangeably, although they do have slightly different meanings. $P(e)$ is a theoretical (mathematical) expectation of the error rate for a given system. BER is an empirical (historical) record of a system's actual error performance. For example, if a system has a $P(e)$ of 10^{-5}, this means that mathematically, you can expect one bit error in every 100,000 bits transmitted ($1/10^{-5} = 1/100,000$). If a system has a BER of 10^{-5}, this means that in the past there was one bit error for every 100,000 bits transmitted.

Probability of error is a function of the receiver *carrier-to-noise ratio*. Depending on the *M*-ary used and the desired $P(e)$, the minimum carrier-to-noise ratio varies. In general, the minimum carrier-to-noise ratio required for a QAM system is less than that required for a comparable PSK system (see Table 13-3). Also, the higher the level of encoding used, the higher the minimum carrier-to-noise ratio. Another parameter often used for comparing digital system performances is the energy of the bit-to-noise ratio (E_b/N_0). E_b/N_0 is explained in Chapter 20, where several examples are shown for determining the minimum carrier-to-noise ratio for a given *M*-ary system and desired $P(e)$.

TABLE 13-3 PERFORMANCE COMPARISON OF VARIOUS DIGITAL MODULATION SCHEMES (BER = 10^{-6})

Modulation technique	C/N ratio (dB)	E_b/N_0 ratio (dB)
BPSK	13.6	10.6
QPSK	13.6	10.6
8QAM	13.6	10.6
8PSK	18.8	14
16PSK	24.3	18.3
16QAM	20.5	14.5
32QAM	24.4	17.4
64QAM	26.6	18.8

APPLICATIONS FOR DIGITAL MODULATION

A digitally modulated transceiver (*trans*mitter-re*ceiver*) that uses FSK, PSK, or QAM has many applications. They are used in digitally modulated microwave radio and satellite systems (Chapter 20) with carrier frequencies from tens of megahertz to several gigahertz, and they are also used for voice band data modems (Chapter 14) with carrier frequencies between 300 and 3000 Hz.

QUESTIONS

13-1. Explain *digital transmission* and *digital radio*.

13-2. Define *information capacity*.

13-3. What are the three most predominant modulation schemes used in digital radio systems?

13-4. Explain the relationship between bits per second and baud for an FSK system.

13-5. Define the following terms for FSK modulation: frequency deviation, modulation index, and deviation ratio.

13-6. Explain the relationship between (a) the minimum bandwidth required for an FSK system and the bit rate, and (b) the mark and space frequencies.

13-7. What is the difference between standard FSK and MSK? What is the advantage of MSK?

13-8. Define *PSK*.

13-9. Explain the relationship between bits per second and baud for a BPSK system.

13-10. What is a constellation diagram, and how is it used with PSK?

13-11. Explain the relationship between the minimum bandwidth required for a BPSK system and the bit rate.

13-12. Explain *M-ary*.

13-13. Explain the relationship between bits per second and baud for a QPSK system.

13-14. Explain the significance of the I and Q channels in a QPSK modulator.

13-15. Define *dibit*.

13-16. Explain the relationship between the minimum bandwidth required for a QPSK system and the bit rate.

13-17. What is a coherent demodulator?

13-18. What advantage does OQPSK have over conventional QPSK? What is a disadvantage of OQPSK?

13-19. Explain the relationship between bits per second and baud for an 8PSK system.

13-20. Define *tribit*.

13-21. Explain the relationship between the minimum bandwidth required for an 8PSK system and the bit rate.

13-22. Explain the relationship between bits per second and baud for a 16PSK system.

13-23. Define *quadbit*.

13-24. Define *QAM*.

13-25. Explain the relationship between the minimum bandwidth required for a 16QAM system and the bit rate.

13-26. What is the difference between PSK and QAM?

13-27. Define *bandwidth efficiency*.

13-28. Define *carrier recovery*.

13-29. Explain the differences between absolute PSK and differential PSK.

13-30. What is the purpose of a clock recovery circuit? When is it used?

13-31. What is the difference between probability of error and bit error rate?

PROBLEMS

31-1. For an FSK modulator with space, rest, and mark frequencies of 40, 50, and 60 MHz, respectively, and an input bit rate of 10 Mbps, determine the output baud and minimum bandwidth. Sketch the output spectrum.

31-2. Determine the minimum bandwidth and baud for a BPSK modulator with a carrier frequency of 40 MHz and an input bit rate of 500 kbps. Sketch the output spectrum.

13-3. For the QPSK modulator shown in Figure 13-13, change the +90° phase-shift network to −90° and sketch the new constellation diagram.

13-4. For the QPSK demodulator shown in Figure 13-17, determine the I and Q bits for an input signal of $\sin \omega_c t - \cos \omega_c t$.

13-5. For an 8PSK modulator with an input data rate (F_b) equal to 20 Mbps and a carrier frequency of 100 MHz, determine the minimum double-sided Nyquist bandwidth (F_N) and the baud. Sketch the output spectrum.

13-6. For the 8PSK modulator shown in Figure 13-19, change the reference oscillator to $\cos \omega_c t$ and sketch the new constellation diagram.

13-7. For a 16QAM modulator with an input bit rate (F_b) equal to 20 Mbps and a carrier frequency of 100 MHz, determine the minimum double-sided Nyquist bandwidth (F_N) and the baud. Sketch the output spectrum.

13-8. For the 16QAM modulator shown in Figure 13-26, change the reference oscillator to $\omega_c t$ and determine the output expressions for the following I, $\overline{I}$, Q, and $\overline{Q}$ input conditions: 0000, 1111, 1010, and 0101.

13-9. Determine the bandwidth efficiency for the following modulators.
   (a) QPSK, $F_b = 10$ Mbps
   (b) 8PSK, $F_b = 21$ Mbps
   (c) 16QAM, $F_b = 20$ Mbps

13-10. For the DBPSK modulator shown in Figure 13-35, determine the output phase sequence for the following input bit sequence: 00110011010101 (assume that the reference bit = 1).

Chapter 14

DATA COMMUNICATIONS

INTRODUCTION

Data communications can be defined as the transmission of digital information (usually in binary form) from a source to a destination. The original source data are in digital form and the received data are in digital form, although the data can be transmitted in analog or digital form. The source information can be binary-coded alpha/numeric characters such as ASCII or EBCDIC, microprocessor op-codes, control words, user addresses, program data, or data base information.

A data communications network can be as simple as two personal computers connected together through the public telephone network, or it can comprise a complex network of one or more mainframe computers and hundreds of remote terminals. Data communications networks are used to connect automatic teller machines (ATMs) to bank computers or they can be used to interface computer terminals (CTs) or keyboard displays (KDs) directly to application programs in mainframe computers. Data communications networks are used for airline and hotel reservation systems and for mass media and news networks such as the Associated Press (AP) or United Press International (UPI). The list of applications for data communications networks goes on almost indefinitely.

HISTORY OF DATA COMMUNICATIONS

It is highly likely that data communications began long before recorded time in the form of smoke signals or tom-tom drums, although it is improbable that these signals

536

were binary coded. If we limit the scope of data communications to methods that use electrical signals to transmit binary-coded information, then data communications began in 1837 with the invention of the *telegraph* and the development of the *Morse code* by Samuel F. B. Morse. With telegraph, dots and dashes (analogous to binary 1's and 0's) are transmitted across a wire using electromechanical induction. Various combinations of these dots and dashes were used to represent binary codes for letters, numbers, and punctuation. Actually, the first telegraph was invented in England by Sir Charles Wheatstone and Sir William Cooke, but their contraption required six different wires for a single telegraph line. In 1840, Morse secured an American patent for the telegraph and in 1844 the first telegraph line was established between Baltimore and Washington, D.C. In 1849, the first slow-speed telegraph printer was invented, but it was not until 1860 that high-speed (15 bps) printers were available. In 1850, the Western Union Telegraph Company was formed in Rochester, New York, for the purpose of carrying coded messages from one person to another.

In 1874, Emile Baudot invented a telegraph *multiplexer*, which allowed signals from up to six different telegraph machines to be transmitted simultaneously over a single wire. The telephone was invented in 1876 by Alexander Graham Bell and, consequently, very little new evolved in telegraph until 1899, when Marconi succeeded in sending radio telegraph messages. Telegraph was the only means of sending information across large spans of water until 1920, when the first commercial radio stations were installed.

Bell Laboratories developed the first special-purpose computer in 1940 using electromechanical relays. The first general-purpose computer was an automatic sequence-controlled calculator developed jointly by Harvard University and International Business Machines Corporation (IBM). The UNIVAC computer, built in 1951 by Remington Rand Corporation (now Sperry Rand), was the first mass-produced electronic computer. Since 1951, the number of mainframe computers, small business computers, personal computers, and computer terminals has increased exponentially, creating a situation where more and more people have the need to exchange digital information with each other. Consequently, the need for data communications has also increased exponentially.

Until 1968, the AT&T operating tariff allowed only equipment furnished by AT&T to be connected to AT&T lines. In 1968, a landmark Supreme Court decision, the Carterfone decision, allowed non-Bell companies to interconnect to the vast AT&T communications network. This decision started the *interconnect industry*, which has led to competitive data communications offerings by a large number of independent companies.

DATA COMMUNICATIONS CIRCUITS

Figure 14-1 shows a simplified block diagram of a data communications circuit. There is a source of digital information, a transmission medium, and a destination. Both the source and destination equipment are digital; they process information in the form of binary pulses. The transmission medium may be a digital or an analog facility and

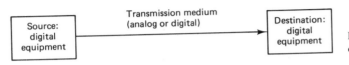

FIGURE 14-1   Data communications circuit: simplified block diagram.

could comprise one or more of the following: metallic wire pair, coaxial cable, microwave radio, satellite radio, or an optical fiber.

Data Communications Circuit Configurations and Topologies

Configurations.   Data communications circuits can be generally categorized as either two-point or multipoint. A *two-point* configuration involves only two locations or stations, whereas a *multipoint* configuration involves three or more stations. A two-point circuit can involve the transfer of information between a mainframe computer and a remote computer terminal, two mainframe computers, or two remote computer terminals. A multipoint circuit is generally used to interconnect a single mainframe computer (*host*) to many remote computer terminals, although any combination of three or more computers or computer terminals constitutes a multipoint circuit.

Topologies.   The topology or architecture of a data communications circuit identifies how the various locations within the network are interconnected. The most common topologies used are the *point to point*, the *star*, the *bus* or *multidrop*, the *ring* or *loop*, and the *mesh*. These are all multipoint configurations except the point to point. Figure 14-2 shows the various circuit configurations and topologies used for data communications networks.

Transmission Modes

Essentially, there are four modes of transmission for data communications circuits: *simplex*, *half duplex*, *full duplex*, and *full/full duplex*.

Simplex.   With simplex operation, data transmission is unidirectional; information can be sent only in one direction. Simplex lines are also called *receive-only*, *transmit-only*, or *one-way-only* lines.

Half duplex (HDX).   In the half-duplex mode, data transmission is possible in both directions, but not at the same time. Half-duplex lines are also called two-way alternate lines.

Full duplex (FDX).   In the full-duplex mode, transmissions are possible in both directions simultaneously, but they must be between the same two stations. Full-duplex lines are also called two-way-simultaneous or simply *duplex* lines.

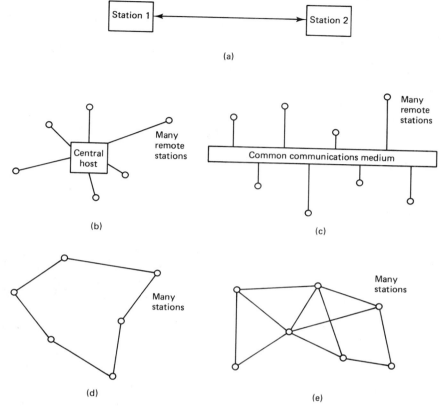

FIGURE 14-2 Data network topologies: (a) point to point; (b) star; (c) bus or multidrop; (d) ring or loop; (e) mesh.

Full/full duplex (F/FDX). In the F/FDX mode, transmission is possible in both directions at the same time but not between the same two stations (i.e., one station is transmitting to a second station and receiving from a third station at the same time). F/FDX is possible only on multipoint circuits.

Two-Wire versus Four-Wire Operation

Two-wire, as the name implies, involves a transmission medium that either uses two wires (a signal and a reference lead) or a configuration that is equivalent to having only two wires. With two-wire operation, simplex, full-, or half-duplex transmission is possible. For full-duplex operation, the signals propagating in opposite directions must occupy different bandwidths; otherwise, they will mix linearly and interfere with each other.

Four-wire, as the name implies, involves a transmission medium that uses four wires (two are used for signals that are propagating in opposite directions and two are used for reference leads) or a configuration that is equivalent to having four wires. With four-wire operation, the signals propagating in opposite directions are physically separated and therefore can occupy the same bandwidths without interfering with each other. Four-wire operation provides more isolation and is preferred over two-wire, although four-wire requires twice as many wires and, consequently, twice the cost.

A transmitter and its associated receiver are equivalent to a two-wire circuit. A transmitter and a receiver for both directions of propagation are equivalent to a four-wire circuit. With full-duplex transmission over a two-wire line, the available bandwidth must be divided in half, thus reducing the information capacity in either direction to one-half of the half-duplex value. Consequently, full-duplex operation over two-wire lines requires twice as much time to transfer the same amount of information.

DATA COMMUNICATIONS CODES

Data communications codes are used for encoding alpha/numeric characters and symbols (punctuation, etc.) and are consequently often called *character sets*, *character languages*, or *character codes*. Essentially, three types of characters are used in data communications codes: *data link control* characters, which are used to facilitate the orderly flow of data from the source to the destination; *graphic control* characters, which involve the syntax or presentation of the data at the receive terminal; and *alpha/numeric* characters, which are used to represent the various symbols used for letters, numbers, and punctuation in the English language.

The first data communications code that saw widespread usage was the Morse code. The Morse code used three unequal-length symbols (dot, dash, and space) to encode alpha/numeric characters, punctuation marks, and an interrogation word.

The Morse code is inadequate for use in modern digital computer equipment because all characters do not have the same number of symbols or take the same length of time to send, and each Morse code operator transmits code at a different rate. Also, with Morse code, there is an insufficient selection of graphic and data link control characters to facilitate the transmission and presentation of the data typically used in contemporary computer applications.

The three most common character sets presently used for character encoding are the Baudot code, the American Standard Code for Information Interchange (ASCII), and the Extended Binary-Coded Decimal Interchange Code (EBCDIC).

Baudot Code

The *Baudot code* (sometimes called the *Telex code*) was the first fixed-length character code. The Baudot code was developed by a French postal engineer, Thomas Murray, in 1875 and named after Emile Baudot, an early pioneer in telegraph printing. The Baudot code is a 5-bit character code that is used primarily for low-speed teletype equipment such as the TWX/Telex system. With a 5-bit code there are only 2^5 or 32 codes possible, which is insufficient to represent the 26 letters of the alphabet, the 10 digits, and the various punctuation marks and control characters. Therefore, the Baudot code uses *figure* shift and *letter* shift characters to expand its capabilities to 58 characters. The latest version of the Baudot code is recommended by the CCITT as the International Alphabet No. 2. The Baudot code is still used by Western Union Company for their TWX and Telex teletype systems. The AP and UPI news services also use the Baudot code for sending news information around the world. The most recent version of the Baudot code is shown in Table 14-1.

TABLE 14-1 BAUDOT CODE

Character shift		Binary code				
Letter	*Figure*	*Bit:* 4	3	2	1	0
A	—	1	1	0	0	0
B	?	1	0	0	1	1
C	:	0	1	1	1	0
D	$	1	0	0	1	0
E	3	1	0	0	0	0
F	!	1	0	1	1	0
G	&	0	1	0	1	1
H	#	0	0	1	0	1
I	8	0	1	1	0	0
J	'	1	1	0	1	0
K	(	1	1	1	1	0
L	)	0	1	0	0	1
M	.	0	0	1	1	1
N	,	0	0	1	1	0
O	9	0	0	0	1	1
P	0	0	1	1	0	1
Q	1	1	1	1	0	1
R	4	0	1	0	1	0
S	bel	1	0	1	0	0
T	5	0	0	0	0	1
U	7	1	1	1	0	0
V	;	0	1	1	1	1
W	2	1	1	0	0	1
X	/	1	0	1	1	1
Y	6	1	0	1	0	1
Z	"	1	0	0	0	1
Figure shift		1	1	1	1	1
Letter shift		1	1	0	1	1
Space		0	0	1	0	0
Line feed (LF)		0	1	0	0	0
Blank (null)		0	0	0	0	0

TABLE 14-2   ASCII-77 CODE—ODD PARITY

	Binary code								Hex		Binary code								Hex
Bit:	7	6	5	4	3	2	1	0		Bit:	7	6	5	4	3	2	1	0	
NUL	1	0	0	0	0	0	0	0	00	@	0	1	0	0	0	0	0	0	40
SOH	0	0	0	0	0	0	0	1	01	A	1	1	0	0	0	0	0	1	41
STX	0	0	0	0	0	0	1	0	02	B	1	1	0	0	0	0	1	0	42
ETX	1	0	0	0	0	0	1	1	03	C	0	1	0	0	0	0	1	1	43
EOT	0	0	0	0	0	1	0	0	04	D	1	1	0	0	0	1	0	0	44
ENQ	1	0	0	0	0	1	0	1	05	E	0	1	0	0	0	1	0	1	45
ACK	1	0	0	0	0	1	1	0	06	F	0	1	0	0	0	1	1	0	46
BEL	0	0	0	0	0	1	1	1	07	G	1	1	0	0	0	1	1	1	47
BS	0	0	0	0	1	0	0	0	08	H	1	1	0	0	1	0	0	0	48
HT	1	0	0	0	1	0	0	1	09	I	0	1	0	0	1	0	0	1	49
NL	1	0	0	0	1	0	1	0	0A	J	0	1	0	0	1	0	1	0	4A
VT	0	0	0	0	1	0	1	1	0B	K	1	1	0	0	1	0	1	1	4B
FF	1	0	0	0	1	1	0	0	0C	L	0	1	0	0	1	1	0	0	4C
CR	0	0	0	0	1	1	0	1	0D	M	1	1	0	0	1	1	0	1	4D
SO	0	0	0	0	1	1	1	0	0E	N	1	1	0	0	1	1	1	0	4E
SI	1	0	0	0	1	1	1	1	0F	O	0	1	0	0	1	1	1	1	4F
DLE	0	0	0	1	0	0	0	0	10	P	1	1	0	1	0	0	0	0	50
DC1	0	0	0	1	0	0	0	1	11	Q	0	1	0	1	0	0	0	1	51
DC2	1	0	0	1	0	0	1	0	12	R	0	1	0	1	0	0	1	0	52
DC3	0	0	0	1	0	0	1	1	13	S	1	1	0	1	0	0	1	1	53
DC4	1	0	0	1	0	1	0	0	14	T	0	1	0	1	0	1	0	0	54
NAK	0	0	0	1	0	1	0	1	15	U	1	1	0	1	0	1	0	1	55
SYN	0	0	0	1	0	1	1	0	16	V	1	1	0	1	0	1	1	0	56
ETB	1	0	0	1	0	1	1	1	17	W	0	1	0	1	0	1	1	1	57
CAN	1	0	0	1	1	0	0	0	18	X	0	1	0	1	1	0	0	0	58
EM	0	0	0	1	1	0	0	1	19	Y	1	1	0	1	1	0	0	1	59
SUB	0	0	0	1	1	0	1	0	1A	Z	1	1	0	1	1	0	1	0	5A
ESC	1	0	0	1	1	0	1	1	1B	[	0	1	0	1	1	0	1	1	5B
FS	0	0	0	1	1	1	0	0	1C	\	1	1	0	1	1	1	0	0	5C
GS	1	0	0	1	1	1	0	1	1D	]	0	1	0	1	1	1	0	1	5D
RS	1	0	0	1	1	1	1	0	1E	∧	0	1	0	1	1	1	1	0	5E
US	0	0	0	1	1	1	1	1	1F	—	1	1	0	1	1	1	1	1	5F
SP	0	0	1	0	0	0	0	0	20	`	1	1	1	0	0	0	0	0	60
!	1	0	1	0	0	0	0	1	21	a	0	1	1	0	0	0	0	1	61
"	1	0	1	0	0	0	1	0	22	b	0	1	1	0	0	0	1	0	62
#	0	0	1	0	0	0	1	1	23	c	1	1	1	0	0	0	1	1	63
$	1	0	1	0	0	1	0	0	24	d	1	1	1	0	0	1	0	0	64
%	0	0	1	0	0	1	0	1	25	e	1	1	1	0	0	1	0	1	65
&	0	0	1	0	0	1	1	0	26	f	1	1	1	0	0	1	1	0	66
'	1	0	1	0	0	1	1	1	27	g	0	1	1	0	0	1	1	1	67
(	1	0	1	0	1	0	0	0	28	h	0	1	1	0	1	0	0	0	68
)	0	0	1	0	1	0	0	1	29	i	1	1	1	0	1	0	0	1	69
*	0	0	1	0	1	0	1	0	2A	j	1	1	1	0	1	0	1	0	6A
+	1	0	1	0	1	0	1	1	2B	k	0	1	1	0	1	0	1	1	6B
,	0	0	1	0	1	1	0	0	2C	l	1	1	1	0	1	1	0	0	6C
-	1	0	1	0	1	1	0	1	2D	m	0	1	1	0	1	1	0	1	6D
.	1	0	1	0	1	1	1	0	2E	n	0	1	1	0	1	1	1	0	6E

TABLE 14-2 *(continued)*

	Binary code								Hex		Binary code								Hex
Bit:	7	6	5	4	3	2	1	0		Bit:	7	6	5	4	3	2	1	0	
/	0	0	1	0	1	1	1	1	2F	o	1	1	1	0	1	1	1	1	6F
0	0	1	0	1	1	0	0	0	30	p	0	1	1	1	0	0	0	0	70
1	0	0	1	1	0	0	0	1	31	q	1	1	1	1	0	0	0	1	71
2	0	0	1	1	0	0	1	0	32	r	1	1	1	1	0	0	1	0	72
3	1	0	1	1	0	0	1	1	33	s	0	1	1	1	0	0	1	1	73
4	0	0	1	1	0	1	0	0	34	t	1	1	1	1	0	1	0	0	74
5	1	0	1	1	0	1	0	1	35	u	0	1	1	1	0	1	0	1	75
6	1	0	1	1	0	1	1	0	36	v	0	1	1	1	0	1	1	0	76
7	0	0	1	1	0	1	1	1	37	w	1	1	1	1	0	1	1	1	77
8	0	0	1	1	1	0	0	0	38	x	1	1	1	1	1	0	0	0	78
9	1	0	1	1	1	0	0	1	39	y	0	1	1	1	1	0	0	1	79
:	1	0	1	1	1	0	1	0	3A	z	0	1	1	1	1	0	1	0	7A
;	0	0	1	1	1	0	1	1	3B	{	1	1	1	1	1	0	1	1	7B
<	1	0	1	1	1	1	0	0	3C	|	0	1	1	1	1	1	0	0	7C
=	0	0	1	1	1	1	0	1	3D	}	1	1	1	1	1	1	0	1	7D
>	0	0	1	1	1	1	1	0	3E	~	1	1	1	1	1	1	1	0	7E
?	1	0	1	1	1	1	1	1	3F	DEL	0	1	1	1	1	1	1	1	7F

NUL = null
SOH = start of heading
STX = start of text
ETX = end of text
EOT = end of transmission
ENQ = enquiry
ACK = acknowledge
BEL = bell
BS = back space
HT = horizontal tab
NL = new line
VT = vertical tab
FF = form feed
CR = carriage return
SO = shift-out
SI = shift-in
DLE = data link escape

DC1 = device control 1
DC2 = device control 2
DC3 = device control 3
DC4 = device control 4
NAK = negative acknowledge
SYN = synchronous
ETB = end of transmission block
CAN = cancel
SUB = substitute
ESC = escape
FS = field separator
GS = group separator
RS = record separator
US = unit separator
SP = space
DEL = delete

ASCII Code

In 1963, in an effort to standardize data communications codes, the United States adopted the Bell System model 33 teletype code as the United States of America Standard Code for Information Interchange (USASCII), better known simply as ASCII-63. Since its adoption, ASCII has generically progressed through the 1965, 1967, and 1977 versions, with the 1977 version being recommended by the CCITT as the International

Alphabet No. 5. ASCII is a 7-bit character set which has 2^7 or 128 codes. With ASCII, the least significant bit (LSB) is designated b_0 and the most significant bit (MSB) is designated b_6. b_7 is not part of the ASCII code but is generally reserved for the parity bit, which is explained later in this chapter. Actually, with any character set, all bits are equally significant because the code does not represent a weighted binary number. It is common with character codes to refer to bits by their order; b_0 is the zero-order bit, b_1 is the first-order bit, b_7 is the seventh-order bit, and so on. With serial transmission, the bit transmitted first is called the LSB. With ASCII, the low-order bit (b_0) is the LSB and is transmitted first. ASCII is probably the code most often used today. The 1977 version of the ASCII code is shown in Table 14-2.

EBCDIC Code

EBCDIC is an 8-bit character code developed by IBM and used extensively in IBM and IBM-compatible equipment. With 8 bits, 2^8 or 256 codes are possible, making EBCDIC the most powerful character set. Note that with EBCDIC the LSB is designated b_7 and the MSB is designated b_0. Therefore, with EBCDIC, the high-order bit (b_7) is transmitted first and the low-order bit (b_0) is transmitted last. The EBCDIC code is shown in Table 14-3.

TABLE 14-3   EBCDIC CODE

	Binary code								Hex		Binary code								Hex
Bit:	0	1	2	3	4	5	6	7		Bit:	0	1	2	3	4	5	6	7	
NUL	0	0	0	0	0	0	0	0	00		1	0	0	0	0	0	0	0	80
SOH	0	0	0	0	0	0	0	1	01	a	1	0	0	0	0	0	0	1	81
STX	0	0	0	0	0	0	1	0	02	b	1	0	0	0	0	0	1	0	82
ETX	0	0	0	0	0	0	1	1	03	c	1	0	0	0	0	0	1	1	83
	0	0	0	0	0	1	0	0	04	d	1	0	0	0	0	1	0	0	84
PT	0	0	0	0	0	1	0	1	05	e	1	0	0	0	0	1	0	1	85
	0	0	0	0	0	1	1	0	06	f	1	0	0	0	0	1	1	0	86
	0	0	0	0	0	1	1	1	07	g	1	0	0	0	0	1	1	1	87
	0	0	0	0	1	0	0	0	08	h	1	0	0	0	1	0	0	0	88
	0	0	0	0	1	0	0	1	09	i	1	0	0	0	1	0	0	1	89
	0	0	0	0	1	0	1	0	0A		1	0	0	0	1	0	1	0	8A
	0	0	0	0	1	0	1	1	0B		1	0	0	0	1	0	1	1	8B
FF	0	0	0	0	1	1	0	0	0C		1	0	0	0	1	1	0	0	8C
	0	0	0	0	1	1	0	1	0D		1	0	0	0	1	1	0	1	8D
	0	0	0	0	1	1	1	0	0E		1	0	0	0	1	1	1	0	8E
	0	0	0	0	1	1	1	1	0F		1	0	0	0	1	1	1	1	8F
DLE	0	0	0	1	0	0	0	0	10		1	0	0	1	0	0	0	0	90
SBA	0	0	0	1	0	0	0	1	11	j	1	0	0	1	0	0	0	1	91
EUA	0	0	0	1	0	0	1	0	12	k	1	0	0	1	0	0	1	0	92
IC	0	0	0	1	0	0	1	1	13	l	1	0	0	1	0	0	1	1	93
	0	0	0	1	0	1	0	0	14	m	1	0	0	1	0	1	0	0	94
NL	0	0	0	1	0	1	0	1	15	n	1	0	0	1	0	1	0	1	95

TABLE 14-3 (*continued*)

Bit:	0	1	2	3	4	5	6	7	Hex	Bit:	0	1	2	3	4	5	6	7	Hex
	0	0	0	1	0	1	1	0	16	o	1	0	0	1	0	1	1	0	96
	0	0	0	1	0	1	1	1	17	p	1	0	0	1	0	1	1	1	97
	0	0	0	1	1	0	0	0	18	q	1	0	0	1	1	0	0	0	98
EM	0	0	0	1	1	0	0	1	19	r	1	0	0	1	1	0	0	1	99
	0	0	0	1	1	0	1	0	1A		1	0	0	1	1	0	1	0	9A
	0	0	0	1	1	0	1	1	1B		1	0	0	1	1	0	1	1	9B
DUP	0	0	0	1	1	1	0	0	1C		1	0	0	1	1	1	0	0	9C
SF	0	0	0	1	1	1	0	1	1D		1	0	0	1	1	1	0	1	9D
FM	0	0	0	1	1	1	1	0	1E		1	0	0	1	1	1	1	0	9E
ITB	0	0	0	1	1	1	1	1	1F		1	0	0	1	1	1	1	1	9F
	0	0	1	0	0	0	0	0	20		1	0	1	0	0	0	0	0	A0
	0	0	1	0	0	0	0	1	21	~	1	0	1	0	0	0	0	1	A1
	0	0	1	0	0	0	1	0	22	s	1	0	1	0	0	0	1	0	A2
	0	0	1	0	0	0	1	1	23	t	1	0	1	0	0	0	1	1	A3
	0	0	1	0	0	1	0	0	24	u	1	0	1	0	0	1	0	0	A4
	0	0	1	0	0	1	0	1	25	v	1	0	1	0	0	1	0	1	A5
ETB	0	0	1	0	0	1	1	0	26	w	1	0	1	0	0	1	1	0	A6
ESC	0	0	1	0	0	1	1	1	27	x	1	0	1	0	0	1	1	1	A7
	0	0	1	0	1	0	0	0	28	y	1	0	1	0	1	0	0	0	A8
	0	0	1	0	1	0	0	1	29	z	1	0	1	0	1	0	0	1	A9
	0	0	1	0	1	0	1	0	2A		1	0	1	0	1	0	1	0	AA
	0	0	1	0	1	0	1	1	2B		1	0	1	0	1	0	1	1	AB
	0	0	1	0	1	1	0	0	2C		1	0	1	0	1	1	0	0	AC
ENQ	0	0	1	0	1	1	0	1	2D		1	0	1	0	1	1	0	1	AD
	0	0	1	0	1	1	1	0	2E		1	0	1	0	1	1	1	0	AE
	0	0	1	0	1	1	1	1	2F		1	0	1	0	1	1	1	1	AF
	0	0	1	1	0	0	0	0	30		1	0	1	1	0	0	0	0	B0
	0	0	1	1	0	0	0	1	31		1	0	1	1	0	0	0	1	B1
SYN	0	0	1	1	0	0	1	0	32		1	0	1	1	0	0	1	0	B2
	0	0	1	1	0	0	1	1	33		1	0	1	1	0	0	1	1	B3
	0	0	1	1	0	1	0	0	34		1	0	1	1	0	1	0	0	B4
	0	0	1	1	0	1	0	1	35		1	0	1	1	0	1	0	1	B5
	0	0	1	1	0	1	1	0	36		1	0	1	1	0	1	1	0	B6
EOT	0	0	1	1	0	1	1	1	37		1	0	1	1	0	1	1	1	B7
	0	0	1	1	1	0	0	0	38		1	0	1	1	1	0	0	0	B8
	0	0	1	1	1	0	0	1	39		1	0	1	1	1	0	0	1	B9
	0	0	1	1	1	0	1	0	3A		1	0	1	1	1	0	1	0	BA
	0	0	1	1	1	0	1	1	3B		1	0	1	1	1	0	1	1	BB
RA	0	0	1	1	1	1	0	0	3C		1	0	1	1	1	1	0	0	BC
NAK	0	0	1	1	1	1	0	1	3D		1	0	1	1	1	1	0	1	BD
	0	0	1	1	1	1	1	0	3E		1	0	1	1	1	1	1	0	BE
SUB	0	0	1	1	1	1	1	1	3F		1	0	1	1	1	1	1	1	BF
SP	0	1	0	0	0	0	0	0	40	{	1	1	0	0	0	0	0	0	C0
	0	1	0	0	0	0	0	1	41	A	1	1	0	0	0	0	0	1	C1
	0	1	0	0	0	0	1	0	42	B	1	1	0	0	0	0	1	0	C2
	0	1	0	0	0	0	1	1	43	C	1	1	0	0	0	0	1	1	C3
	0	1	0	0	0	1	0	0	44	D	1	1	0	0	0	1	0	0	C4

TABLE 14-3 (*continued*)

	Binary code 0	1	2	3	4	5	6	7	Hex	Bit	Binary code 0	1	2	3	4	5	6	7	Hex
	0	1	0	0	0	1	0	1	45	E	1	1	0	0	0	1	0	1	C5
	0	1	0	0	0	1	1	0	46	F	1	1	0	0	0	1	1	0	C6
	0	1	0	0	0	1	1	1	47	G	1	1	0	0	0	1	1	1	C7
	0	1	0	0	1	0	0	0	48	H	1	1	0	0	1	0	0	0	C8
	0	1	0	0	1	0	0	1	49	I	1	1	0	0	1	0	0	1	C9
¢	0	1	0	0	1	0	1	0	4A		1	1	0	0	1	0	1	0	CA
.	0	1	0	0	1	0	1	1	4B		1	1	0	0	1	0	1	1	CB
<	0	1	0	0	1	1	0	0	4C		1	1	0	0	1	1	0	0	CC
(	0	1	0	0	1	1	0	1	4D		1	1	0	0	1	1	0	1	CD
+	0	1	0	0	1	1	1	0	4E		1	1	0	0	1	1	1	0	CE
¦	0	1	0	0	1	1	1	1	4F		1	1	0	0	1	1	1	1	CF
&	0	1	0	1	0	0	0	0	50	}	1	1	0	1	0	0	0	0	D0
	0	1	0	1	0	0	0	1	51	J	1	1	0	1	0	0	0	1	D1
	0	1	0	1	0	0	1	0	52	K	1	1	0	1	0	0	1	0	D2
	0	1	0	1	0	0	1	1	53	L	1	1	0	1	0	0	1	1	D3
	0	1	0	1	0	1	0	0	54	M	1	1	0	1	0	1	0	0	D4
	0	1	0	1	0	1	0	1	55	N	1	1	0	1	0	1	0	1	D5
	0	1	0	1	0	1	1	0	56	O	1	1	0	1	0	1	1	0	D6
	0	1	0	1	0	1	1	1	57	P	1	1	0	1	0	1	1	1	D7
	0	1	0	1	1	0	0	0	58	Q	1	1	0	1	1	0	0	0	D8
	0	1	0	1	1	0	0	1	59	R	1	1	0	1	1	0	0	1	D9
!	0	1	0	1	1	0	1	0	5A		1	1	0	1	1	0	1	0	DA
$	0	1	0	1	1	0	1	1	5B		1	1	0	1	1	0	1	1	DB
*	0	1	0	1	1	1	0	0	5C		1	1	0	1	1	1	0	0	DC
)	0	1	0	1	1	1	0	1	5D		1	1	0	1	1	1	0	1	DD
;	0	1	0	1	1	1	1	0	5E		1	1	0	1	1	1	1	0	DE
¬	0	1	0	1	1	1	1	1	5F		1	1	0	1	1	1	1	1	DF
-	0	1	1	0	0	0	0	0	60	\	1	1	1	0	0	0	0	0	E0
/	0	1	1	0	0	0	0	1	61		1	1	1	0	0	0	0	1	E1
	0	1	1	0	0	0	1	0	62	S	1	1	1	0	0	0	1	0	E2
	0	1	1	0	0	0	1	1	63	T	1	1	1	0	0	0	1	1	E3
	0	1	1	0	0	1	0	0	64	U	1	1	1	0	0	1	0	0	E4
	0	1	1	0	0	1	0	1	65	V	1	1	1	0	0	1	0	1	E5
	0	1	1	0	0	1	1	0	66	W	1	1	1	0	0	1	1	0	E6
	0	1	1	0	0	1	1	1	67	X	1	1	1	0	0	1	1	1	E7
	0	1	1	0	1	0	0	0	68	Y	1	1	1	0	1	0	0	0	E8
	0	1	1	0	1	0	0	1	69	Z	1	1	1	0	1	0	0	1	E9
	0	1	1	0	1	0	1	0	6A		1	1	1	0	1	0	1	0	EA
,	0	1	1	0	1	0	1	1	6B		1	1	1	0	1	0	1	1	EB
%	0	1	1	0	1	1	0	0	6C		1	1	1	0	1	1	0	0	EC
	0	1	1	0	1	1	0	1	6D		1	1	1	0	1	1	0	1	ED
>	0	1	1	0	1	1	1	0	6E		1	1	1	0	1	1	1	0	EE
?	0	1	1	0	1	1	1	1	6F		1	1	1	0	1	1	1	1	EF
	0	1	1	1	0	0	0	0	70	0	1	1	1	1	0	0	0	0	F0
	0	1	1	1	0	0	0	1	71	1	1	1	1	1	0	0	0	1	F1
	0	1	1	1	0	0	1	0	72	2	1	1	1	1	0	0	1	0	F2
	0	1	1	1	0	0	1	1	73	3	1	1	1	1	0	0	1	1	F3

TABLE 14-3 (*continued*)

				Binary code					Hex	Bit:					Binary code					Hex
Bit:	0	1	2	3	4	5	6	7			0	1	2	3	4	5	6	7		
	0	1	1	1	0	1	0	0	74	4	1	1	1	1	0	1	0	0	F4
	0	1	1	1	0	1	0	1	75	5	1	1	1	1	0	1	0	1	F5
	0	1	1	1	0	1	1	0	76	6	1	1	1	1	0	1	1	0	F6
	0	1	1	1	0	1	1	1	77	7	1	1	1	1	0	1	1	1	F7
	0	1	1	1	1	0	0	0	78	8	1	1	1	1	1	0	0	0	F8
▲	0	1	1	1	1	0	0	1	79	9	1	1	1	1	1	0	0	1	F9
:	0	1	1	1	1	0	1	0	7A		1	1	1	1	1	0	1	0	FA
#	0	1	1	1	1	0	1	1	7B		1	1	1	1	1	0	1	1	FB
@	0	1	1	1	1	1	0	0	7C		1	1	1	1	1	1	0	0	FC
▲	0	1	1	1	1	1	0	1	7D		1	1	1	1	1	1	0	1	FD
=	0	1	1	1	1	1	1	0	7E		1	1	1	1	1	1	1	0	FE
"	0	1	1	1	1	1	1	1	7F		1	1	1	1	1	1	1	1	FF

DLE = data link escape

DUP = duplicate
EM = end of medium
ENQ = enquiry
EOT = end of transmission
ESC = escape
ETB = end of transmission block
ETX = end of text
EUA = erase unprotected to address
FF = form feed
FM = field mark
IC = insert cursor

ITB = end of intermediate transmission block
NUL = null
PT = program tab
RA = repeat to address
SBA = set buffer address
SF = start field
SOH = start of heading
SP = space
STX = start of text
SUB = substitute
SYN = synchronous
NAK = negative acknowledge

ERROR CONTROL

A data communications circuit can be as short as a few feet or as long as several thousand miles, and the transmission medium can be as simple as a piece of wire or as complex as a microwave, satellite, or fiber optic system. Therefore, due to the nonideal transmission characteristics that are associated with any communications system, it is inevitable that errors will occur and that it is necessary to develop and implement procedures for error control. Error control can be divided into two general categories: error detection and error correction.

Error Detection

Error detection is simply the process of monitoring the received data and determining when a transmission error has occurred. Error detection techniques do not identify which bit (or bits) is in error, only that an error has occurred. The purpose of error detection is not to prevent errors from occurring but to prevent undetected errors from occurring.

How a system reacts to transmission errors is system dependent and varies considerably. The most common error detection techniques used for data communications circuits are: redundancy, exact-count encoding, parity, vertical and longitudinal redundancy checking, and cyclic redundancy checking.

Redundancy. *Redundancy* involves transmitting each character twice. If the same character is not received twice in succession, a transmission error has occurred. The same concept can be used for messages. If the same sequence of characters is not received twice in succession, in exactly the same order, a transmission error has occurred.

Exact-count encoding. With *exact-count encoding*, the number of 1's in each character is the same. An example of an exact-count encoding scheme is the ARQ code shown in Table 14-4. With the ARQ code, each character has three 1's in it, and therefore a simple count of the number of 1's received can determine if a transmission error has occurred.

Parity. *Parity* is probably the simplest error detection scheme used for data communications systems and is used with both vertical and horizontal redundancy checking. With parity, a single bit (called a *parity bit*) is added to each character to force the total number of 1's in the character, including the parity bit, to be either an odd number (odd parity) or an even number (even parity). For example, the ASCII code for the letter "C" is 43 hex or P1000011 binary, with the P bit representing the parity bit. There are three 1's in the code, not counting the parity bit. If odd parity is used, the P bit is made a 0, keeping the total number of 1's at three, an odd number. If even parity is used, the P bit is made a 1 and the total number of 1's is four, an even number.

Taking a closer look at parity, it can be seen that the parity bit is independent of the number of 0's in the code and unaffected by pairs of 1's. For the letter "C," if all the 0 bits are dropped, the code is P1——11. For odd parity, the P bit is still a 0 and for even parity, the P bit is still a 1. If pairs of 1's are also excluded, the code is either P1——, P——1, or P——1—. Again, for odd parity the P bit is a 0, and for even parity the P bit is a 1.

The definition of parity is *equivalence* of *equality*. A logic gate that will determine when all its inputs are equal is the XOR gate. With an XOR gate, if all the inputs are equal (either all 0's or all 1's), the output is a 0. If all inputs are not equal, the output is a 1. Figure 14-3 shows two circuits that are commonly used to generate a parity bit. Essentially, both circuits go through a comparison process eliminating 0's and pairs of 1's. The circuit shown in Figure 14-3a uses *sequential (serial)* comparison, while the circuit shown in Figure 14-3b uses *combinational (parallel)* comparison. With the sequential parity generator b_0 is XORed with b_1, the result is XORed with b_2, and so on. The result of the last XOR operation is compared with a *bias* bit. If even parity is desired, the bias bit is made a logic 0. If odd parity is desired, the bias bit is made a logic 1. The output of the circuit is the parity bit, which is appended to the

TABLE 14-4 ARQ EXACT-COUNT CODE

	Binary code							Character	
Bit:	1	2	3	4	5	6	7	Letter	Figure
	0	0	0	1	1	1	0	Letter shift	
	0	1	0	0	1	1	0	Figure shift	
	0	0	1	1	0	1	0	A	—
	0	0	1	1	0	0	1	B	?
	1	0	0	1	1	0	0	C	:
	0	0	1	1	1	0	0	D	(WRU)
	0	1	1	1	0	0	0	E	3
	0	0	1	0	0	1	1	F	%
	1	1	0	0	0	0	1	G	(a
	1	0	1	0	0	1	0	H	£
	1	1	1	0	0	0	0	I	8
	0	1	0	0	0	1	1	J	(bell)
	0	0	0	1	0	1	1	K	(
	1	1	0	0	0	1	0	L	)
	1	0	1	0	0	0	1	M	.
	1	0	1	0	1	0	0	N	,
	1	0	0	0	1	1	0	O	9
	1	0	0	1	0	1	0	P	0
	0	0	0	1	1	0	1	Q	1
	1	1	0	0	1	0	0	R	4
	0	1	0	1	0	1	0	S	'
	1	0	0	0	1	0	1	T	5
	0	1	1	0	0	1	0	U	7
	1	0	0	1	0	0	1	V	=
	0	1	0	0	1	0	1	W	2
	0	0	1	0	1	1	0	X	/
	0	0	1	0	1	0	1	Y	6
	0	1	1	0	0	0	1	Z	+
	0	0	0	0	1	1	1		(blank)
	1	1	0	1	0	0	0		(space)
	1	0	1	1	0	0	0		(line feed)
	1	0	0	0	0	1	1		(carriage return)

character code. With the parallel parity generator, comparisons are made in layers or levels. Pairs of bits (b_0 and b_1, b_2 and b_3, etc.) are XORed. The results of the first-level XOR gates are then XORed together. The process continues until only one bit is left, which is XORed with the bias bit. Again, if even parity is desired, the bias bit is made a logic 0 and if odd parity is desired, the bias bit is made a logic 1.

The circuits shown in Figure 14-3 can also be used for the parity checker in the receiver. A parity checker uses the same procedure as a parity generator except that the logic condition of the final comparison is used to determine if a parity violation has occurred (for odd parity a 1 indicates an error and a 0 indicates no error; for even parity, a 1 indicates an error and a 0 indicates no error).

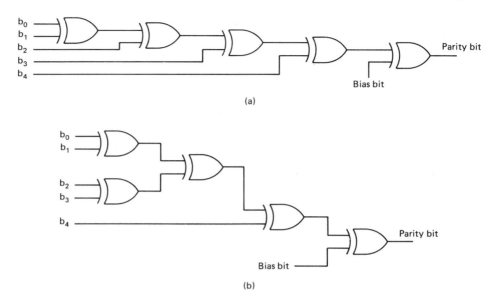

FIGURE 14-3   Parity generators: (a) serial; (b) parallel. 1, odd parity; 2, even parity.

The primary advantage of parity is its simplicity. The disadvantage is that when an even number of bits are received in error, the parity checker will not detect it (i.e., if the logic conditions of 2 bits are changed, the parity remains the same). Consequently, parity, over a long period of time, will detect only 50% of the transmission errors (this assumes an equal probability that an even or an odd number of bits could be in error).

Vertical and horizontal redundancy checking.   *Vertical redundancy checking* (VRC) is an error detection scheme that uses parity to determine if a transmission error has occurred within a character. Therefore, VRC is sometimes called *character parity*. With VRC, each character has a parity bit added to it prior to transmission. It may use even or odd parity. The example shown under the topic "parity" involving the ASCII character "C" is an example of how VRC is used.

Horizontal or longitudinal redundancy checking (HRC or LRC) is an error detection scheme that uses parity to determine if a transmission error has occurred in a message and is therefore sometimes called *message parity*. With LRC, each bit position has a parity bit. In other words, b_0 from each character in the message is XORed with b_0 from all of the other characters in the message. Similarly, b_1, b_2, and so on, are XORed with their respective bits from all the other characters in the message. Essentially, LRC is the result of XORing the "characters" that make up a message, whereas VRC is the XORing of the bits within a single character. With LRC, only even parity is used.

The LRC bit sequence is computed in the transmitter prior to sending the data, then transmitted as though it were the last character of the message. At the receiver, the LRC is recomputed from the data and the recomputed LRC is compared with the LRC transmitted with the message. If they are the same, it is assumed that no transmission errors have occurred. If they are different, a transmission error must have occurred.

Example 14-1 shows how VRC and LRC are determined.

EXAMPLE 14-1

Determine the VRC and LRC for the following ASCII-encoded message: THE CAT. Use odd parity for VRC and even parity for LRC.

Character		T	H	E	sp	C	A	T	LRC
Hex		*54*	*48*	*45*	*20*	*43*	*41*	*54*	*2F*
LSB	b_0	0	0	1	0	1	1	0	1
	b_1	0	0	0	0	1	0	0	1
	b_2	1	0	1	0	0	0	1	1
ASCII code	b_3	0	1	0	0	0	0	0	1
	b_4	1	0	0	0	0	0	1	0
	b_5	0	0	0	1	0	0	0	1
MSB	b_6	1	1	1	0	1	1	1	0
VRC	b_7	0	1	0	0	0	1	0	0

The LRC is 2FH or 00101111 binary. In ASCII, this is the character /.

The VRC bit for each character is computed in the vertical direction, and the LRC bits are computed in the horizontal direction. This is the same scheme that was used with the early teletype paper tapes and keypunch cards and has subsequently been carried over to present-day data communications applications.

The group of characters that make up the message (i.e., THE CAT) is often called a *block* of data. Therefore, the bit sequence for the LRC is often called a *block check character* (BCC) or a *block check sequence* (BCS). BCS is more appropriate because the LRC has no function as a character (i.e., it is not an alpha/numeric, graphic, or data link control character); the LRC is simply a *sequence of bits* used for error detection.

Historically, LRC detects between 95 and 98% of all transmission errors. LRC will not detect transmission errors when an even number of characters have an error in the same bit position. For example, if b_4 in two different characters is in error, the LRC is still valid even though multiple transmission errors have occurred.

If VRC and LRC are used simultaneously, the only time an error would go undetected is when an even number of bits in an even number of characters were in error and the same bit positions in each character are in error, which is highly unlikely to happen. VRC does not identify which bit is in error in a character, and LRC does not

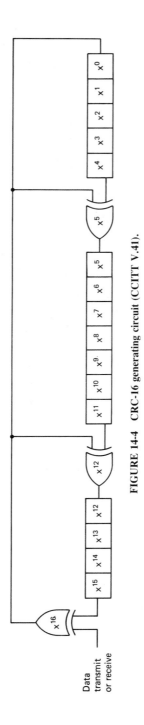

FIGURE 14-4 CRC-16 generating circuit (CCITT V.41).

552

identify which character has an error in it. However, for single bit errors, VRC used together with LRC will identify which bit is in error. Otherwise, VRC and LRC only identify that an error has occurred.

Cyclic redundancy checking. Probably the most reliable scheme for error detection is *cyclic redundancy checking* (CRC). With CRC, approximately 99.95% of all transmission errors are detected. CRC is generally used with 8-bit codes such as EBCDIC or 7-bit codes when parity is not used.

In the United States, the most common CRC code is CRC-16, which is identical to the international standard, CCITT's V.41. With CRC-16, 16 bits are used for the BCS. Essentially, the CRC character is the remainder of a division process. A data message polynomial $G(x)$ is divided by a generator polynomial function $P(x)$, the quotient is discarded, and the remainder is truncated to 16 bits and added to the message as the BCS. With CRC generation, the division is not accomplished with a standard arithmetic division process. Instead of using straight subtraction, the remainder is derived from an XOR operation. At the receiver, the data stream and the BCS are divided by the same generating function $P(x)$. If no transmission errors have occurred, the remainder will be zero.

The generating polynomial for CRC-16 is

$$P(x) = x^{16} + x^{12} + x^5 + x^0$$

where

$$x^0 = 1.$$

The number of bits in the CRC code is equal to the highest exponent of the generating polynomial. The exponents identify the bit positions that contain a 1. Therefore, b_{16}, b_{12}, b_5, and b_0 are 1's and all of the other bit positions are 0's.

Figure 14-4 shows the block diagram for a circuit that will generate a CRC-16 BSC for the CCITT V.41 standard. Note that for each bit position of the generating polynomial that is a 1 there is an XOR gate.

EXAMPLE 14-2

Determine the BSC for the following data and CRC generating polynomials:

$$\text{data } G(x) = x^7 + x^5 + x^4 + x^2 + x^1 + x^0 \quad \text{or} \quad 10110111$$

$$\text{CRC } P(x) = x^5 + x^4 + x^1 + x^0 \quad \text{or} \quad 110011$$

Solution First $G(x)$ is multiplied by the number of bits in the CRC generating polynomial, 5.

$$x^5(x^7 + x^5 + x^4 + x^2 + x^1 + x^0) = x^{12} + x^{10} + x^9 + x^7 + x^6 + x^5$$

$$= 1011011100000$$

```
                              11010111
               110011|1011011100000
                      110011
                       111101
                       110011
                         111010
                         110011
                          100100
                          110011
                           101110
                           110011
                            111010
                            110011
                             01001 = CRC
```

The CRC is appended to the data to give the following transmitted data stream:

```
              G(x)         CRC
           10110111    01001
```

At the receiver, the transmitted data are again divided by $P(x)$.

```
                              11010111
               110011|1011011101001
                      110011
                       111101
                       110011
                         111010
                         110011
                          100110
                          110011
                           101010
                           110011
                            110011
                            110011
                             000000  remainder = 0
                                     no error occurred
```

Error Correction

Essentially, there are three methods of error correction: symbol substitution, retransmission, and forward error correction.

Symbol substitution. *Symbol substitution* was designed to be used in a human environment: when there is a human being at the receive terminal to analyze the received data and make decisions on its integrity. With symbol substitution, if a character is received in error, rather than revert to a higher level of error correction or display the incorrect character, a unique character that is undefined by the character code, such as a reverse question mark (ʕ), is substituted for the bad character. If the character in error cannot be discerned by the operator, retransmission is called for (i.e., symbol substitution is a form of selective retransmission). For example, if the message "Name" had an error in the first character, it would be displayed as "ʕ ame." An operator can discern the correct message by inspection and retransmission is unnecessary. However, if the message "$ʕ,000.00" were received, an operator could not determine the correct character, and retransmission is required.

Retransmission. *Retransmission*, as the name implies, is when a message is received in error and the receive terminal automatically calls for retransmission of the entire message. Retransmission is often called ARQ, which is an old radio communications term that means *automatic request for retransmission*. ARQ is probably the most reliable method of error correction, although it is not always the most efficient. Impairments on transmission media occur in bursts. If short messages are used, the likelihood that an impairment will occur during a transmission is small. However, short messages require more acknowledgments and line turnarounds than do long messages. Acknowledgments and line turnarounds for error control are forms of *overhead* (characters other than data that must be transmitted). With long messages, less turnaround time is needed, although the likelihood that a transmission error will occur is higher than for short messages. It can be shown statistically that message blocks between 256 and 512 characters are of optimum size when using ARQ for error correction.

Forward error correction. *Forward error correction* (FEC) is the only error correction scheme that actually detects and corrects transmission errors at the receive end without calling for retransmission.

With FEC, bits are added to the message prior to transmission. A popular error-correcting code is the *Hamming code*, developed by R. W. Hamming at Bell Laboratories. The number of bits in the Hamming code is dependent on the number of bits in the data character. The number of Hamming bits that must be added to a character is determined from the following expression:

$$2^n \geq m + n + 1 \tag{14-1}$$

where

n = number of Hamming bits
m = number of bits in the data character

EXAMPLE 14-3

For a 12-bit data string of 101100010010, determine the number of Hamming bits required, arbitrarily place the Hamming bits into the data string, determine the condition of each Hamming bit, assume an arbitrary single-bit transmission error, and prove that the Hamming code will detect the error.

Solution   Substituting into Equation 14-1, the number of Hamming bits is

$$2^n \geq m + n + 1$$

for $n = 4$:

$$2^4 = 16 \geq m + n + 1 = 12 + 4 + 1 = 17$$

$16 < 17$; therefore, 4 Hamming bits are insufficient.
For $n = 5$:

$$2^5 = 32 \geq m + n + 1 = 12 + 5 + 1 = 18$$

$32 > 18$; therefore, 5 Hamming bits are sufficient to meet the criterion of Equation 2-1. Therefore, a total of $12 + 5 = 17$ bits make up the data stream.

Arbitrarily place 5 Hamming bits into the data stream:

```
17 16 15 14 13 12 11 10 9 8 7 6 5 4 3 2 1
 H  1  0  1  H  1  0  0 H H 0 1 0 H 0 1 0
```

To determine the logic condition of the Hamming bits, express all bit positions that contain a 1 as a 5-bit binary number and XOR them together.

Bit position	Binary number
2	00010
6	00110
XOR	00100
12	01100
XOR	01000
14	01110
XOR	00110
16	10000
XOR	10110 = Hamming code

$$b_{17} = 1, \quad b_{13} = 0, \quad b_9 = 1, \quad b_8 = 1, \quad b_4 = 0$$

The 17-bit encoded data stream becomes

```
H       H     H H      H
1 1 0 1 0 1 0 0 1 1 0 1 0 0 0 1 0
```

Assume that during transmission, an error occurs in bit position 14. The received data stream is

```
1 1 0 0 0 1 0 0 1 1 0 1 0 0 0 1 0
```

At the receiver to determine the bit position in error, extract the Hamming bits and XOR them with the binary code for each data bit position that contains a 1.

Bit position	Binary number
Hamming code	10110
2	10110
XOR	10100
6	00110
XOR	10010
12	01100
XOR	11110
16	10000
XOR	01110 = binary 14

Bit position 14 was received in error. To fix the error, simply complement bit 14.

The Hamming code described here will detect only single-bit errors. It cannot be used to identify multiple-bit errors or errors in the Hamming bits themselves. The Hamming code, like all FEC codes, requires the addition of bits to the data, consequently lengthening the transmitted message. The purpose of FEC codes is to reduce or eliminate the wasted time of retransmissions. However, the addition of the FEC bits to each message wastes transmission time in itself. Obviously, a trade-off is made between ARQ and FEC and system requirements determine which method is best suited to a particular system.

SYNCHRONIZATION

Synchronize means to coincide or agree in time. In data communications, there are four types of synchronization that must be achieved: bit or clock synchronization, modem or carrier synchronization, character synchronization, and message synchronization. The clock and carrier recovery circuits discussed in Chapter 13 accomplish bit and carrier synchronization, and message synchronization is discussed in Chapter 15.

Character Synchronization

Clock synchronization ensures that the transmitter and receiver agree on a precise time slot for the occurrence of a bit. When a continuous string of data is received, it is necessary to identify which bits belong to which characters and which bit is the least significant data bit, the parity bit, and the stop bit. In essence, this is character synchronization: identifying the beginning and the end of a character code. In data communications circuits, there are two formats used to achieve character synchronization: asynchronous and synchronous.

Asynchronous data format.   With *asynchronous data*, each character is framed between a *start* and a *stop* bit. Figure 14-5 shows the format used to frame a character for asynchronous data transmission. The first bit transmitted is the start bit and is always a logic 0. The character code bits are transmitted next beginning with the LSB and continuing through the MSB. The parity bit (if used) is transmitted directly after the MSB of the character. The last bit transmitted is the stop bit, which is always a logic 1. There can be either 1, 1.5, or 2 stop bits.

A logic 0 is used for the start bit because an idle condition (no data transmission) on a data communications circuit is identified by the transmission of continuous 1's (these are often called *idle line 1's*). Therefore, the start bit of the first character is identified by a high-to-low transition in the received data, and the bit that immediately follows the start bit is the LSB of the character code. All stop bits are logic 1's, which guarantees a high-to-low transition at the beginning of each character. After the start bit is detected, the data and parity bits are clocked into the receiver. If data are transmitted in real time (i.e., as an operator types data into their computer terminal), the number of idle line 1's between each character will vary. During this *dead time*, the receiver will simply wait for the occurrence of another start bit before clocking in the next character.

Stop bit (1, 1.5, 2)	Parity bit			Data bits (5–7)					Start bit
1	1	1/0	b_6 MSB	b_5	b_4	b_3	b_2	b_1	b_0 LSB	0 LSB

FIGURE 14-5   Asynchronous data format.

EXAMPLE 14-4

For the following string of asynchronous ASCII-encoded data, identify each character (assume even parity and 2 stop bits).

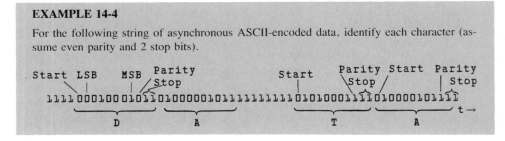

Synchronous data format.   With *synchronous data*, rather than frame each character independently with start and stop bits, a unique synchronizing character called a SYN character is transmitted at the beginning of each message. For example, with ASCII code, the SYN character is 16H. The receiver disregards incoming data until it receives the SYN character, then it clocks in the next 7 bits and interprets them as a character. The character that is used to signify the end of a transmission varies with the type of protocol used and what kind of transmission it is. Message-terminating characters are discussed in Chapter 15.

With asynchronous data, it is not necessary that the transmit and receive clocks be continuously synchronized. It is only necessary that they operate at approximately the same rate and be synchronized at the beginning of each character. This was the purpose of the start bit, to establish a time reference for character synchronization. With synchronous data, the transmit and receive clocks must be synchronized because character synchronization occurs only once at the beginning of the message.

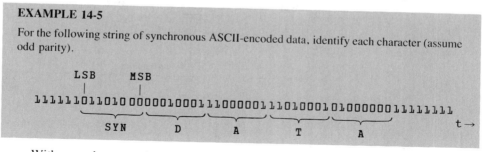

EXAMPLE 14-5

For the following string of synchronous ASCII-encoded data, identify each character (assume odd parity).

With asynchronous data, each character has 2 or 3 bits added to each character (1 start and 1 or 2 stop bits). These bits are additional overhead and thus reduce the efficiency of the transmission (i.e., the ratio of information bits to total transmitted bits). Synchronous data have two SYN characters (16 bits of overhead) added to each message. Therefore, asynchronous data are more efficient for short messages, and synchronous data are more efficient for long messages.

DATA COMMUNICATIONS HARDWARE

Figure 14-6 shows the block diagram of a multipoint data communications circuit that uses a bus topology. This arrangement is one of the most common configurations used for data communications circuits. At one station there is a mainframe computer and at each of the other two stations there is a *cluster* of computer terminals. The hardware and associated circuitry that connect the host computer to the remote computer terminals is called a *data communications link*. The station with the mainframe is called the *host* or *primary* and the other stations are called *secondaries* or simply *remotes*. An arrangement such as this is called a *centralized network*; there is one centrally located station (the host) with the responsiblity of ensuring an orderly flow of data between the remote stations and itself. Data flow is controlled by an applications program which is stored at the primary station.

At the primary station there is a mainframe computer, a *line control unit* (LCU), and a *data modem* (a data modem is commonly referred to simply as a *modem*). At each secondary station there is a modem, an LCU, and terminal equipment, such as computer terminals, printers, and so on. The mainframe is the host of the network and is where the applications program is stored for each circuit it serves. For simplicity, Figure 14-6 shows only one circuit served by the primary, although there can be many different circuits served by one mainframe computer. The primary station has the capability of storing, processing, or retransmitting the data it receives from the secondary stations. The primary also stores software for data base management.

The LCU at the primary station is more complicated than the LCUs at the secondary

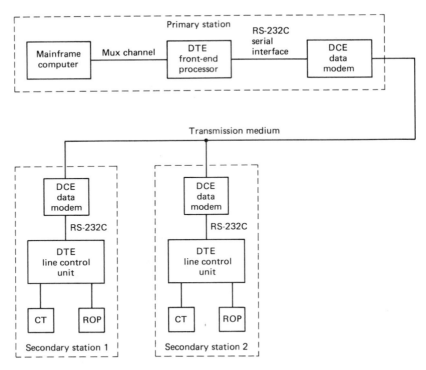

FIGURE 14-6   Multipoint data communications circuit block diagram.

stations. The LCU at the primary station directs data traffic to and from many different circuits, which could all have different characteristics (i.e., different bit rates, character codes, data formats, etc.). The LCU at a secondary station directs data traffic between one data link and a few terminal devices which all operate at the same speed and use the same character code. Generally speaking, if the LCU has software associated with it, it is called a *front-end processor* (FEP). The LCU at the primary station is usually an FEP.

Line Control Unit

The LCU has several important functions. The LCU at the primary station serves as an interface between the host computer and the circuits that it serves. Each circuit served is connected to a different port on the LCU. The LCU directs the flow of input and output data between the different data communications links and their respective applications program. The LCU performs parallel-to-serial and serial-to-parallel conversion of data. The mux interface channel between the mainframe computer and the LCU transfers data in parallel. Data transfer between the modem and the LCU is done serially. The LCU also houses the circuitry that performs error detection and correction. Also, data link control (DLC) characters are inserted and deleted in the LCU. Data link control characters are explained in Chapter 15.

The LCU operates on the data when it is in digital form and is therefore called *data terminal equipment* (DTE). Essentially, any piece of equipment between the mainframe computer and the modem or the station equipment and its modem is classified as data terminal equipment. The modem is called *data communications equipment* (DCE) because it interfaces the digital DTE to the analog transmission line.

Within the LCU, there is a single integrated circuit that performs several of the LCU's functions. This circuit is called a UART when asynchronous transmission is used and a USRT when synchronous transmission is used.

Universal asynchronous receiver/transmitter (UART). The UART is used for asynchronous transmission of data between the DTE and the DCE. Asynchronous transmission means that an asynchronous data format is used and there is no clocking information transferred between the DTE and the DCE. The primary functions of the UART are:

1. To perform serial-to-parallel and parallel-to-serial conversion of data
2. To perform error detection by inserting and checking parity bits
3. To insert and detect start and stop bits

Functionally, the UART is divided into two sections: the transmitter and the receiver. Figure 14-7a shows a simplified block diagram of a UART transmitter.

Prior to transferring data in either direction, a *control* word must be programmed into the UART control register to indicate the nature of the data, such as the number of data bits; if parity is used, and if so, whether it is even or odd; and the number of stop bits. Essentially, the start bit is the only bit that is not optional; there is always only one start bit and it must be a logic 0. Figure 14-7b shows how to program the control word for the various functions. In the UART, the control word is used to set up the data-, parity-, and stop-bit steering logic circuit.

UART transmitter. The operation of the UART transmitter section is really quite simple. The UART sends a transmit buffer empty (TBMT) signal to the DTE to indicate that it is ready to receive data. When the DTE senses an active condition on TBMT, it sends a parallel data character to the transmit data lines (TD0–TD7) and strobes them into the transmit buffer register with the transmit data strobe signal ($\overline{\text{TDS}}$). The contents of the transmit buffer register are transferred to the transit shift register when the transmit-end-of-character (TEOC) signal goes active (the TEOC signal simply tells the buffer register when the shift register is empty and available to receive data). The data pass through the steering logic circuit, where they pick up the appropriate start, stop, and parity bits. After data have been loaded into the transmit shift register, they are serially outputted on the transmit serial output (TSO) pin with a bit rate equal to the transmit clock (TCP) frequency. While the data in the transmit shift register are sequentially clocked out, the DTE loads the next character into the buffer register. The process

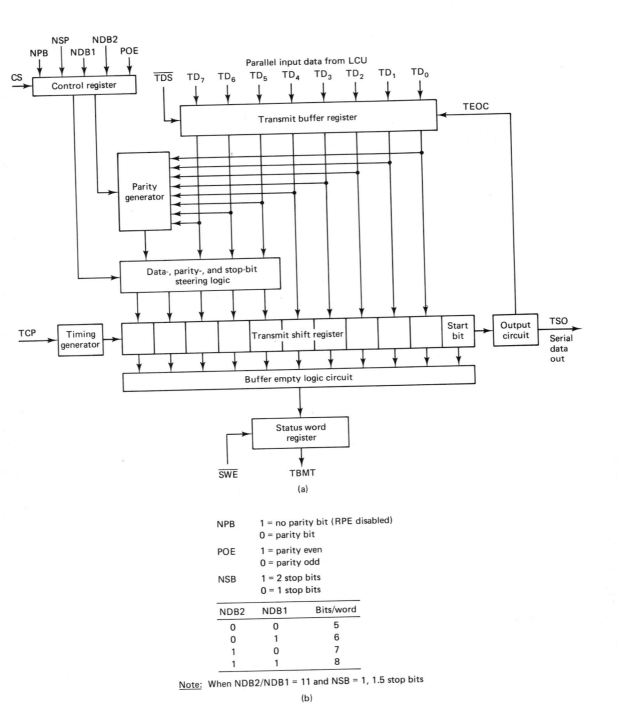

FIGURE 14-7   UART transmitter: (a) simplified block diagram; (b) control word.

continues until the DTE has transferred all its data. The preceding sequence is shown in Figure 14-8.

UART receiver. A simplified block diagram of a UART receiver is shown in Figure 14-9. The number of stop bits, data bits, and the parity-bit information for the UART receiver are determined by the same control word that is used by the transmitter (i.e., the type of parity, the number of stop bits, and the number of data bits used for the UART receiver must be the same as that used for the UART transmitter).

The UART receiver ignores idle line 1's. When a valid start bit is detected by the start bit verification circuit, the data character is serially clocked into the receive shift register. If parity is used, the parity bit is checked in the parity check circuit. After one complete data character is loaded into the shift register, the character is transferred in parallel into the buffer register and the receive data available (RDA) flag is set in the status word register. To read the status register, the DTE activates status word enable ($\overline{SWE}$) and if it is active, reads the character from the buffer register by placing an active condition on the receive data enable (RDE) pin. After reading the

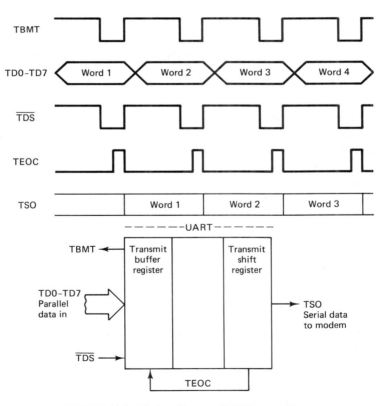

FIGURE 14-8 Timing diagram: UART transmitter.

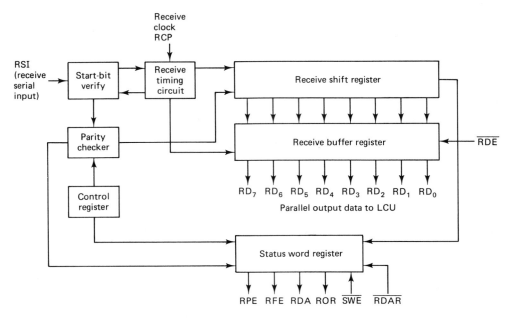

FIGURE 14-9   Simplified block diagram of a UART receiver.

data, the DTE places an active signal on the receive data available reset ($\overline{\text{RDAR}}$) pin, which resets the RDA pin. Meanwhile, the next character is received and clocked into the receive shift register and the process repeats itself until all the data have been received. The preceding sequence is shown in Figure 14-10.

The status word register is also used for diagnostic information. The receive parity error (RPE) flag is set when a received character has a parity error in it. The receive framing error (RFE) flag is set when a character is received without any or an improper number of stop bits. The receive overrun (ROR) flag is set when a character in the buffer register is written over with another character (i.e., the DTE failed to service an active conditon on RDA before the next character was received by the shift register).

The receive clock for the UART (RCP) is 16 times higher than the receive data rate. This allows the start-bit verification circuit to determine if a high-to-low transition in the received data is actually a valid start bit and not simply a negative-going noise spike. Figure 14-11 shows how this is accomplished. The incoming idle line 1's (continuous high condition) are sampled at a rate 16 times the actual bit rate. This assures that a high-to-low transition is detected within p of a bit time after it occurs. Once a low is detected, the verification circuit counts off seven clock pulses, then resamples the data. If it is still low, it is assumed that a valid start bit has been detected. If it has reverted to the high condition, it is assumed that the high-to-low transition was simply a noise pulse and is therefore ignored. Once a valid start bit has been detected and verified, the verification circuit samples the incoming data once every

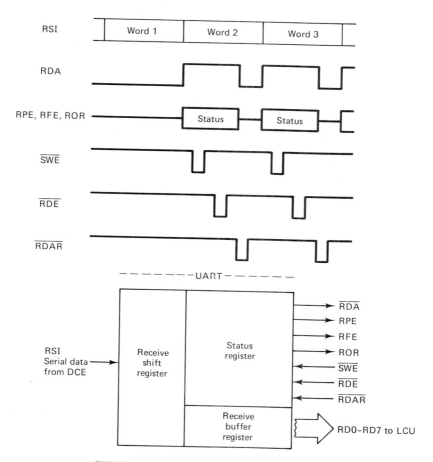

FIGURE 14-10   Timing diagram: UART receiver.

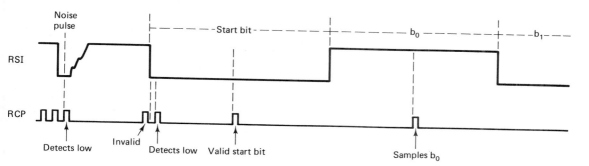

FIGURE 14-11   Start-bit verification.

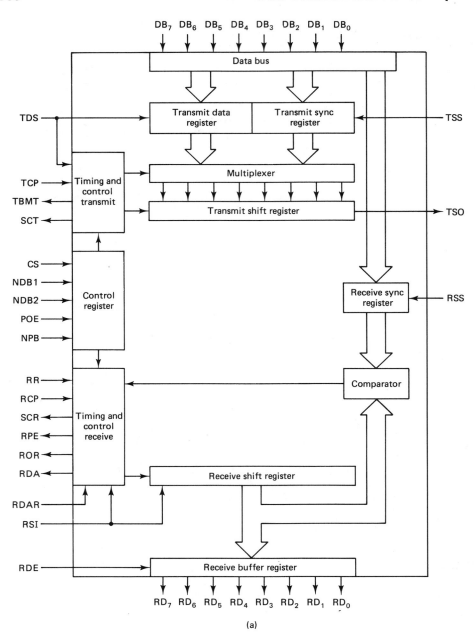

FIGURE 14-12   USRT transceiver: (a) block diagram; (b) control word.

NPB   1 = no parity bit (RPE disabled)
        0 = parity bit

POE   1 = parity even
        0 = parity odd

NDB2	NDB1	Bits/word
0	0	5
0	1	6
1	0	7
1	1	8

(b)

FIGURE 14-12   (*cont'd*)

16 clock cycles, which is equal to the data rate. Sampling at 16 times the bit rate also establishes the sample time to within $\frac{1}{16}$ of a bit time from the center of a bit.

Universal synchronous receiver/transmitter (USRT). The USRT is used for synchronous data transmission between the DTE and the DCE. Synchronous transmission means that there is clocking information transferred between the USRT and the modem and each transmission begins with a unique SYN character. The primary functions of the USRT are:

1. To perform serial-to-parallel and parallel-to-serial conversion of data
2. To perform error detection by inserting and checking parity bits
3. To insert and detect SYN characters

The block diagram of the USRT is shown in Figure 14-12a. The USRT operates very similarly to the UART, and therefore only the differences are explained. With the USRT, start and stop bits are not allowed. Instead, unique SYN characters are loaded into the transmit and receive SYN registers prior to transferring data. The programming information for the control word is shown in Figure 14-12b.

USRT transmitter. The transmit clock signal (TCP) is set at the desired bit rate and the desired SYN character is loaded from the parallel input pins (DB1–DB8) into the transmit SYN register by pulsing transmit SYN strobe (TSS). Data are loaded into the transmit data register from DB1–DB8 by pulsing the transmit data strobe (TDS). The next character transmitted is extracted from the transmit data register provided that the TDS pulse occurs during the presently transmitted character. If TDS is not pulsed, the next transmitted character is extracted from the transmit SYN register and the SYN character transmitted (SCT) signal is set. The transmit buffer empty (TBMT) signal is used to request the next character from the DTE. The serial output data appears on the transmit serial output (TSO) pin.

USRT receiver. The receive clock signal (RCP) is set at the desired bit rate and the desired SYN character is loaded into the receive SYN register from DB1–DB8 by pulsing receive SYN strobe (RSS). On a high-to-low transition of the receiver rest input (RR), the receiver is placed in the search (bit phase) mode. In the search mode, serially received data are examined on a bit-by-bit basis until a SYN character is found. After each bit is clocked into the receive shift register, its contents are compared to the contents of the receive SYN register. If they are identical, a SYN character has been found and the SYN character receive (SCR) output is set. This character is transferred into the receive buffer register and the receiver is placed into the character mode. In the character mode, receive data are examined on a character-by-character basis and receiver flags for receive data available (RDA), receiver overrun (ROR), receive parity error (RPE), and SYN character received are provided to the status word register. Parallel receive data are outputted to the DTE on RB1–RB8.

SERIAL INTERFACES

To ensure an orderly flow of data between the line control unit and the modem, a *serial interface* is placed between them. This interface coordinates the flow of data, control signals, and timing information between the DTE and the DCE.

Before serial interfaces were standardized, every company that manufactured data communications equipment used a different interface configuration. More specifically, the cabling arrangement between the DTE and the DCE, the type and size of the connectors used, and the voltage levels varied considerably from vender to vender. To interconnect equipment manufactured by different companies, special level converters, cables, and connectors had to be built. The Electronic Industries Association (EIA), in an effort to standardize interface equipment between the data terminal equipment and data communications equipment, agreed on a set of standards which are called the RS-232C specifications. The RS-232C specifications identify the mechanical, electrical, and functional description for the interface between the DTE and the DCE. The RS-232C interface is similar to the combined CCITT standards V.28 (electrical specifications) and V.24 (functional description) and is designed for serial transmission of data up to 20,000 bps for a distance of approximately 50 ft. The EIA has recently adopted a new set of standards called the RS-449A, which when used in conjunction with the RS-422A or RS-423A standard, can operate at data rates up to 10 Mbps and span distances up to 1200 m.

RS-232C Interface

The RS-232C interface specifies a 25-wire cable with a DB25P/DB25S-compatible connector. Figure 14-13 shows the electrical characteristics of the RS-232C interface. The terminal load capacitance of the cable is specified as 2500 pF, which includes cable capacitance. The impedance at the terminating end must be between 3000 and 7000 Ω, and the output impedance is specified as greater than 300 Ω. With these electrical specifications and for a maximum bit rate of 20,000 bps, the nominal maximum length of the RS-232C interface is approximately 50 ft.

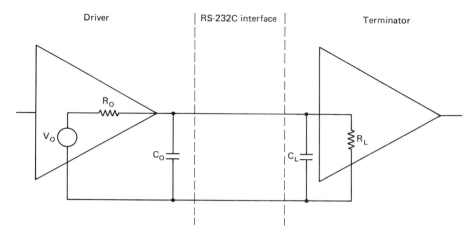

FIGURE 14-13   RS-232C electrical specifications.

Although the RS-232C interface is simply a cable and two connectors, the standard also specifies limitations on the voltage levels that the DTE and DCE can output onto or receive from the cable. In both the DTE and DCE, there are circuits that convert the internal logic level to RS-232C values. For example, a DTE uses TTL logic and is interfaced to a DCE which uses ECL logic; they are not compatible. Voltage-leveling circuits convert the internal voltage values of the DTE and DCE to RS-232C values. If both the DCE and DCE output and input RS-232C levels, they are electrically compatible regardless of which logic family they use internally. A leveler is called a *driver* if it outputs a signal voltage to the cable and a *terminator* if it accepts a signal voltage from the cable. Table 14-5 lists the voltage limits for both drivers and terminators. Note that the data lines use negative logic and the control lines use positive logic.

From Table 14-5 it can be seen that the limits for a driver are more inclusive

TABLE 14-5   RS232C VOLTAGE SPECIFICATIONS (V DC)

	Data pins	
	Logic 1	*Logic 0*
Driver	−5 to −15	+5 to +15
Terminator	−3 to −25	+3 to +25
	Control pins	
	Enable "on"	*Disable "off"*
Driver	+5 to +15	−5 to −15
Terminator	+3 to +25	−3 to −25

than those for a terminator. The driver can output any voltage between +5 and +15 or −5 and −15 V dc, and a terminator will accept any voltage between +3 and +25 and −3 and −25 V dc. The difference in the voltage levels between a driver and a terminator is called *noise margin*. The noise margin reduces the susceptibility of the interface to noise transients on the cable. Typical voltages used for data and control signals are ±7 V dc and ±10 V dc.

The pins on the RS-232C interface cable are functionally categorized as either ground, data, control (handshaking), or timing pins. All the pins are unidirectional (signals are propagated only from the DTE to the DCE, or vice versa). Table 14-6 lists the 25 pins of the RS-232C interface, their designations, and the direction of signal propagation (i.e., either toward the DTE or toward the DCE). The RS-232C specifications designate the ground, data, control, and timing pins as A, B, C, and D, respectively. These are nondescriptive designations. It is more practical and useful to use acronyms to designate the pins that reflect the pin functions. Table 14-6 lists the CCITT and EIA designations and the nomenclature more commonly used by industry in the United States.

EIA RS-232C pin functions.   Twenty of the 25 pins of the RS-232C interface are designated for specific purposes or functions. Pins 9, 10, 11, 18, and 25 are unassigned; pins 1 and 7 are grounds; pins 2, 3, 14, and 16 are data pins; pins 15, 17, and 24 are timing pins; and all the other assigned pins are reserved for control or handshaking signals. There are two full-duplex data channels available with the RS-232C interface; one channel is for primary data (actual information) and the second channel is for secondary data (diagnostic information and handshaking signals). The functions of the 20 assigned pins are summarized below.

Pin 1—protective ground. This pin is frame ground and is used for protection against electrical shock. Pin 1 should be connected to the third-wire ground of the ac electrical system at one end of the cable (either at the DTE or the DCE, but not at both ends).

Pin 2—transmit data (TD). Serial data on the primary channel from the DTE to the DCE are transmitted on this pin. TD is enabled by an active condition on the CS pin.

Pin 3—received data (RD). Serial data on the primary channel are transferred from the DCE to the DTE on this pin. RD is enabled by an active condition on the RLSD pin.

Pin 4—request to send (RS). The DTE bids for the primary communications channel from the DCE on this pin. An active condition on RS turns on the modem's analog carrier. The analog carrier is modulated by a unique bit pattern called a training sequence which is used to initialize the communications channel and synchronize the receive modem. RS cannot go active unless pin 6 (DSR) is active.

TABLE 14-6   EIA RS-232C PIN DESIGNATIONS

Pin number	EIA nomenclature	Common acronym	Direction
1	Protective ground (AA)	GND	None
2	Transmitted data (BA)	TD, SD	DTE to DCE
3	Received data (BB)	RD	DCE to DTE
4	Request to send (CA)	RS, RTS	DTE to DCE
5	Clear to send (CB)	CS, CTS	DCE to DTE
6	Data set ready (CC)	DSR, MR	DCE to DTE
7	Signal ground (AB)	GND	None
8	Received line signal detect (CF)	RLSD	DCE to DTE
9	Unassigned		
10	Unassigned		
11	Unassigned		
12	Secondary received line signal detect (SCF)	SRLSD	DCE to DTE
13	Secondary clear to send (SCB)	SCS	DCE to DTE
14	Secondary transmitted data (SBA)	STD	DTE to DCE
15	Transmission signal element timing (DB)	SCT	DCE to DTE
16	Secondary received data (SBB)	SRD	DCE to DTE
17	Receiver signal element timing (DD)	SCR	DCE to DTE
18	Unassigned		
19	Secondary request to send (SCA)	SRS	DTE to DCE
20	Data terminal ready (CD)	DTR	DTE to DCE
21	Signal quality detector (CG)	SQD	DCE to DTE
22	Ring indicator (CE)	RI	DCE to DTE
23	Data signal rate selector (CH)	DSRS	DTE to DCE
24	Transmit signal element timing (DA)	SCTE	DTE to DCE
25	Unassigned		

Pin 5—clear to send (CS). This signal is a handshake from the DCE to the DTE in response to an active condition on request to send. CS enables the TD pin.

Pin 6—data set ready (DSR). On this pin the DCE indicates the availability of the communications channel. DSR is active as long as the DCE is connected to the communications channel (i.e., the modem or the communications channel is not being tested or is not in the voice mode).

Pin 7—signal ground. This pin is the signal reference for all the data, control and timing pins. Usually, this pin is strapped to frame ground (pin 1).

Pin 8—receive line signal detect (RLSD). The DCE uses this pin to signal the DTE when the DCE is receiving an analog carrier on the primary data channel. RSLD enables the RD pin.

Pin 9. Unassigned.

Pin 10. Unassigned.

Pin 11. Unassigned.

Pin 12—secondary receive line signal detect (SRLSD). This pin is active when the DCE is receiving an analog carrier on the secondary channel. SRLSD enables the SRD pin.

Pin 13—secondary clear to send (SCS). This pin is used by the DCE to send a handshake to the DTE in response to an active condition on the secondary request to send pin. SCS enables the STD pin.

Pin 14—secondary transmit data (STD). Diagnostic data are transferred from the DTE to the DCE on this pin. STD is enabled by an active condition on the SCS pin.

Pin 15—transmission signal element timing (SCT). Transmit clocking signals are sent from the DCE to the DTE on this pin.

Pin 16—secondary received data (SRD). Diagnostic data are transferred from the DCE to the DTE on this pin. SRD is enabled by an active condition on the SCS pin.

Pin 17—receive signal element timing (SCR). Receive clocking signals are sent from the DCE to the DTE on this pin. The clock frequency is equal to the bit rate of the primary data channel.

Pin 18. Unassigned.

Pin 19—secondary request to send (SRS). The DTE bids for the secondary communications channel from the DCE on this pin.

Pin 20—data terminal ready (DTR). The DTE sends information to the DCE on this pin concerning the availability of the data terminal equipment (i.e., access to the mainframe at the primary station or status of the computer terminal at the secondary station). DTR is used primarily with dial-up data communications circuits to handshake with RI.

Pin 21—signal quality detector (SQD). The DCE sends signals to the DTE on this pin that reflect the quality of the received analog carrier.

Pin 22—ring indicator (RI). This pin is used with dial-up lines for the DCE to signal the DTE that there is an incoming call.

Pin 23—data signal rate selector (DSRS). The DTE uses this pin to select the transmission bit rate (clock frequency) of the DCE.

Pin 24—transmit signal element timing (SCTE). Transmit clocking signals are sent from the DTE to the DCE on this pin when the master clock oscillator is located in the DTE.

Pin 25. Unassigned.

Pins 1 through 8 are used with both asynchronous and synchronous modems. Pins 15, 17, and 24 are used for only synchronous modems. Pins 12, 13, 14, 16, and 19 are used only when the DCE is equipped with a secondary channel. Pins 19 and 22 are used exclusively for dial-up telephone connections.

The basic operation of the RS-232C interface is shown in Figure 14-14 and described as follows. When the DTE has primary data to send, it enables request to send ($t = 0$ ms). After a predetermined time delay (50 ms), CS goes active. During the RS/CS delay the modem is outputting an analog carrier that is modulated by a unique bit

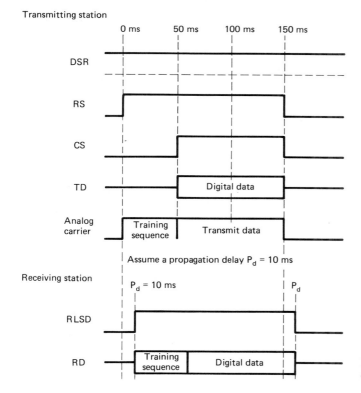

FIGURE 14-14 Timing diagram: basic operation of the RS-232C interface.

pattern called a *training sequence*. The training sequence is used to initialize the communications line and synchronize the carrier and clock recovery circuits in the receive modem. After the RS/CS delay, TD is enabled and the DTE begins to transmit data. After the receive DTE detects an analog carrier, RD is enabled. When the transmission is complete (t = 150 ms), RS goes low turning off the analog carrier and shutting off CS. For a more detailed explanation, timing diagrams, and illustrative examples, see V. Alisouskas and W. Tomasi, *Digital and Data Communications* (Englewood Cliffs, N.J.: Prentice-Hall, 1985).

RS-449A Interface

Contemporary data rates have exceeded the capabilities of the RS-232C interface. Therefore, it was necessary to adopt and implement a new standard that allows higher bit rates to be transmitted for longer distances. The RS-232C has a maximum bit rate of 20,000 bps and a maximum distance of approximately 50 ft. Consequently, the EIA

TABLE 14-7   EIA RS-449A PRIMARY CHANNEL PIN DESIGNATIONS

Pin number	Mnemonic	Circuit name
1	None	Shield
2	SI	Signaling rate indicator
3,21	None	Spare
4,22	SD	Send data
5,23	ST	Send timing
6,24	RD	Receive data
7,25	RS	Request to send
8,26	RT	Receive timing
9,27	CS	Clear to send
10	LL	Local loopback
11,29	DM	Data mode
12,30	TR	Terminal ready
13,31	RR	Receiver ready
14	RL	Remote loopback
15	IC	Incoming call
16	SF/SR	Select frequency/signaling rate
17,23	TT	Terminal timing
18	TM	Test mode
19	SG	Signal ground
20	RC	Receive common
28	IS	Terminal in service
32	SS	Select standby
33	SQ	Signal quality
34	NS	New signal
36	SB	Standby indicator
37	SC	Send common

**TABLE 14-8   EIA RS-449A SECONDARY
DIAGNOSTIC CHANNEL PIN DESIGNATIONS**

Pin number	Mnemonic	Circuit name
1	None	Shield
2	SRR	Secondary receiver ready
3	SSD	Secondary send data
4	SRD	Secondary receive data
5	SG	Signal ground
6	RC	Receive common
7	SRS	Secondary request to send
8	SCS	Secondary clear to send
9	SC	Send common

has adopted a new standard: the RS-449A interface. The RS-449A is essentially an updated version of the RS-232C except that the RS-449A outlines only the mechanical and functional specifications of the cable and connectors.

The RS-449A specifies two cables: one with 37 wires that is used for serial data transmission and one with 9 wires that is used for secondary diagnostic information. Table 14-7 lists the 37 pins of the RS-449A primary cable and their designations, and Table 14-8 lists the 9 pins of the diagnostic cable and their designations. Note that the acronyms used with the RS-449A are more descriptive than those recommended by the EIA for the RS-232C. The functions specified by the RS-449A are very similar to the RS-232C. The major difference between the two standards is the separation of the primary data and secondary diagnostic channels onto two cables.

The electrical specifications used with the RS-449A are specified by either the RS-422A or the RS-423A standard. The RS-422A standard specifies a balanced interface cable that will operate at bit rates up to 10 Mbps and span distances up to 1200 m. This does not mean that 10 Mbps can be transmitted 1200 m. At 10 Mbps the maximum distance is 15 m, and 90 kbps is the maximum bit rate that can be transmitted 1200 m. The RS-423A standard specifies an unbalanced interface cable that will operate at a maximum line speed of 100 kbps and span a maximum distance of 90 m.

Figure 14-15 shows the *balanced* digital interface circuit for the RS-422A, and Figure 14-16 shows the *unbalanced* digital interface circuit for the RS-423A.

TRANSMISSION MEDIA AND DATA MODEMS

In its simplest form, data communication is the transmittal of digital information between two DTEs. The DTEs may be separated by a few feet or several thousand miles. At the present time, there is an insufficient number of transmission media to carry digital information from source to destination in digital form. Therefore, the most convenient

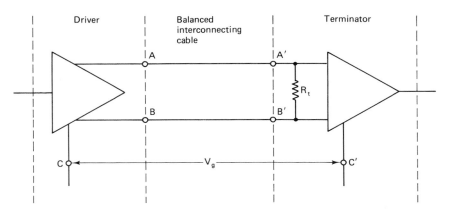

FIGURE 14-15   RS-422A interface circuit. R_t, **Optional cable termination resistance;** V_g, **ground potential difference; A, B, driver interface points; A', B', terminator interface points; C, driver circuit ground; C', terminator circuit ground; A-B, balanced driver output; A'-B', balanced terminator input.**

alternative is to use the existing public telephone network (PTN) as the transmission media for data communications circuits. Unfortunately, the PTN was designed (and most of it constructed) long before the advent of large-scale data communications. The PTN was intended to be used for transferring voice telephone communications signals, not digital data. Therefore, to use the PTN for data communications, the data must be converted to a form more suitable for transmission over analog carrier systems.

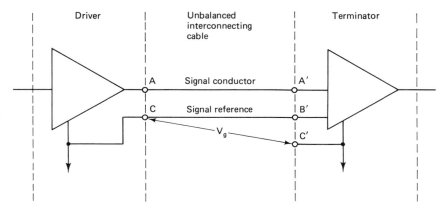

FIGURE 14-16   RS-423A interface circuit. A, C, Driver interface; A', B', terminator interface; V_g, **ground potential difference; C, driver circuit ground; C', terminator circuit ground.**

Transmission Media

As stated previously, the public telephone network is a convenient alternative to constructing alternate digital facilities (at a tremendous cost) for carrying only digital data. The public telephone network comprises over 2000 local telephone companies and several long-distance common carriers such as Microwave Communications Incorporated (MCI), GTE Sprint, and the American Telephone and Telegraph Company (AT&T). Local telephone companies provide voice and data services for relatively small geographic areas, whereas long-distance common carriers provide voice and data services for relatively large geographic areas.

Essentially, there are two types of circuits available from the public telephone network: *direct distance dialing* (DDD) and *private line*. The DDD network is commonly called the *dial-up network*. Anyone who has a telephone number subscribes to the DDD network. With the DDD network, data links are established and disconnected in the same manner as normal voice calls are established and disconnected—with a standard telephone or some kind of an automatic dial/answer machine. Data links that are established through the DDD network use *common usage* equipment and facilities. Common usage means that a subscriber uses the equipment and transmission medium for the duration of the call, then they are relinquished to the network for other subscribers to use. With private line circuits, a subscriber has a permanent dedicated communications link 24 hours a day.

Figure 14-17 shows a simplified block diagram of a telephone communications link. Each subscriber has a dedicated cable facility between his station and the nearest telephone office called a *local loop*. The local loop is used by the subscriber to access

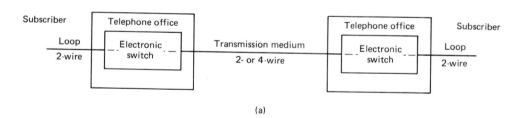

(a)

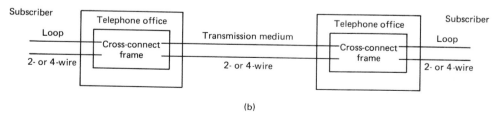

(b)

FIGURE 14-17 Telephone communications link: (a) direct distance dialing; (b) dedicated private line.

the PTN. The facilities used to interconnect telephone offices are called *trunk* circuits and can be a metallic cable, a digital carrier system, a microwave radio, a fiber optic link, or a satellite radio system, depending on the distance between the two offices. For temporary connections using the DDD network, telephone offices are interconnected through sophisticated electronic switching systems (ESS) and use intricate switching arrangements. With private line circuits, data links are permanently hardwired through telephone offices without going through a switch. Dial-up data links are preferred when there are a large number of subscribers in a network or if there is a small volume of data traffic. Private line circuits are preferred for limited-access networks when there is a large volume of data throughput.

The quality of a dial-up circuit is guaranteed to meet the minimum requirements for a *voice band* (VB) communications circuit. With a private line circuit, the communications link can be improved by adding amplifiers and equalizers to the circuit. This is called *conditioning* the line. A voice-grade circuit using the PTN has an ideal passband from 0 to 4 kHz, although the usable passband is limited to approximately 300 to 3000 Hz. The minimum-quality circuit available using the PTN is called a basic voice grade (VG) circuit. The quality of a dial-up circuit is guaranteed to meet basic requirements and can be as good as a private line circuit. However, with the DDD network, the transmission characteristics of the data link vary from call to call, while in a private line circuit they remain relatively constant. With the DDD network, *contention* can be a problem; each subscriber must contend for a connection through the network with every other subscriber in the network. With private line circuits, there is no contention because each circuit has only one subscriber. Consequently, there are several advantages that private line circuits have over dial-up networks: increased availability, more consistent performance, greater reliability, and lower costs for moderate to high volumes of data. Dial-up circuits are limited to two-wire operation, whereas private line circuits can operate either two- or four-wire.

Data Modems

The primary purpose of the data modem is to interface the digital terminal equipment to an analog communications channel. The data modem is also called a DCE, a *dataset*, a *dataphone*, or simply a *modem*. At the transmit end, the modem converts digital pulses from the serial interface to analog signals, and at the receive end, the modem converts analog signals to digital pulses.

Modems are generally classified as either asynchronous or synchronous and use either FSK, PSK, or QAM modulation. With synchronous modems, clocking information is recovered in the receive modem; with asynchronous modems, it is not. Asynchronous modems use FSK modulation and are restricted to low-speed applications (below 2000 bps). Synchronous modems use PSK and QAM modulation and are used for medium-speed (2400 to 4800 bps) and high-speed (9600 bps) applications.

Asynchronous modems. Asynchronous modems are used primarily for low-speed dial-up circuits. There are several standard modem designs commonly used for asynchronous data transmission. For half-duplex operation using the two-wire DDD network or full-duplex operation with four-wire private line circuit, the Bell System 202T/S or equivalent is a popular modem. The 202T is a four-wire, full-duplex modem and the 202S is a two-wire, half-duplex modem. The 202T/S uses FSK modulation. The mark and space frequencies are 1200 and 2200 Hz, respectively. With a 1200-bps data rate, the modulation index for the 202T/S is 0.83 and requires approximately 2400 Hz of bandwidth. The output spectrum for the 202T/S modem is shown in Figure 14-18.

To operate full duplex with a two-wire dial-up circuit, it is necessary to divide the usable bandwidth of a voice band circuit in half, creating two equal-capacity data channels. A popular modem that does this is the Bell System 103 or equivalent. The 103 modem is capable of full-duplex operation over a two-wire line at bit rates up to 300 bps. With the 103 modem, there are two data channels each with separate mark and space frequencies. One channel is the *low-band channel* and occupies a passband from 300 to 1650 Hz. The second channel is the *high-band channel* and occupies a passband from 1650 to 3000 Hz. The mark and space frequencies for the low-band channel are 1270 and 1070 Hz, respectively. The mark and space frequencies for the high-band channel are 2225 and 2025 Hz, respectively. For a bit rate of 300 bps, the modulation index for the 103 modem is 0.67. The output spectrum for the 103 modem is shown in Figure 14-19. The high- and low-band data channels occupy different frequency bands and can therefore use the same two-wire facility without interfering with each other. This is called *frequency-division multiplexing* and is explained in detail in Chapter 18.

The low-band channel is commonly called the *originate channel* and the high-band channel is called the *answer channel*. It is standard procedure on a dial-up circuit for the station that originates the call to transmit on the low-band frequencies and receive on the high-band frequencies, and the station that answers the call to transmit on the high-band frequencies and receive on the low-band frequencies.

Synchronous modems. Synchronous modems are used for medium- and high-speed data transmission and use either PSK or QAM modulation. With synchronous modems the transmit clock, together with the data, digitally modulate an analog carrier.

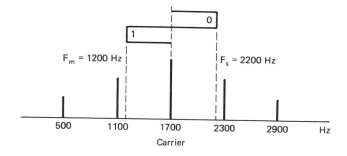

FIGURE 14-18 Output spectrum for a 202T/S modem. Carrier frequency = 1700 Hz, input data = 1200 bps alternating 1/0 pattern, modulation index = 0.83.

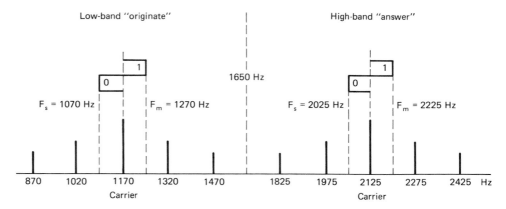

FIGURE 14-19   Output spectrum for a 103 modem. Carrier frequency: low band = 1170, high band = 2125; input data = 300 bps alternating 1/0 sequence; modulation index = 0.67.

The modulated carrier is transmitted to the receive modem, where a coherent carrier is recovered and used to demodulate the data. The transmit clock is recovered from the data and used to clock the received data into the DTE. Because of the clock and carrier recovery circuits, a synchronous modem is more complicated and thus more expensive than its asynchronous counterpart.

PSK modulation is used for medium-speed (2400 to 4800 bps) synchronous modems. More specifically, QPSK is used with 2400-bps modems and 8PSK is used with 4800-bps modems. QPSK has a bandwidth efficiency of 2 bps/Hz; therefore, the baud rate and minimum bandwidth for a 2400-bps synchronous modem are 1200 baud and 1200 Hz. The standard 2400-bps synchronous modem is the Bell System 201C or equivalent. The 201C uses a 1600-Hz carrier and has an output spectrum that extends from 1000 to 2200 Hz. 8PSK has a bandwidth efficiency of 3 bps/Hz; therefore, the baud rate and minimum bandwidth for 4800-bps synchronous modems are 1600 baud and 1600 Hz. The standard 4800-bps synchronous modem is the Bell System 208A or equivalent. The 208A also uses a 1600-Hz carrier but has an output spectrum that extends from 800 to 2400 Hz. Both the 201C and 208A are full-duplex modems designed to be used with four-wire private line circuits. The 201C and 208A can operate over two-wire dial-up circuits but only in the simplex mode. There are half-duplex two-wire versions of both models: the 201B and 208B.

High-speed synchronous modems operate at 9600 bps and use 16QAM modulation. 16QAM has a bandwidth efficiency of 4 bps/Hz; therefore, the baud rate and minimum bandwidth for 9600-bps synchronous modems are 2400 baud and 2400 Hz. The standard 9600-bps modem is the Bell System 209A or equivalent. The 209A uses a 1650-Hz carrier and has an output spectrum that extends from 450 to 2850 Hz. The Bell System 209A is a four-wire synchronous modem designed to be used on full-duplex private line circuits. The 209B is the two-wire version designed for half-duplex dial-up circuits.

Normally, an asynchronous data format is used with asynchronous modems and a synchronous data format is used with synchronous modems. However, asynchronous data are occasionally used with synchronous modems; this is called *isochronous transmission*. Synchronous data are never used with asynchronous modems.

Table 14-9 summarizes the standard Bell System modems.

TABLE 14-9   MODEM SUMMARY

Bell designation	Line facility	Operating mode	Synchronization	Type of modulation	Maximum data rate (bps)
103	Dial-up	FDX	Asynchronous	FSK	300
113A	Dial-up	Simplex	Asynchronous	FSK	300
113B	Dial-up	Simplex	Asynchronous	FSK	300
201B	Dial-up	HDX/FDX	Synchronous	QPSK	2400
201C	Private	HDX/FDX	Synchronous	QPSK	2400
202S	Dial-up	HDX	Asynchronous	FSK	1200
202T	Private	HDX/FDX	Asynchronous	FSK	1200 (basic) 1800 (Cl conditioning)
208A	Private	HDX/FDX	Synchronous	8PSK	4800
208B	Dial-up	HDX	Synchronous	8PSK	4800
209A	Private	HDX/FDX	Synchronous	16QAM	9600
209B	Dial-up	HDX	Synchronous	16QAM	9600

QUESTIONS

14-1. Define *data communications*.

14-2. What was the significance of the Carterfone decision?

14-3. Explain the difference between a two-point and a multipoint circuit.

14-4. What is a data communications topology?

14-5. Define the four transmission modes for data communications circuits.

14-6. Which of the four transmission modes can be used only with multipoint circuits?

14-7. Explain the differences between two-wire and four-wire circuits.

14-8. What is a data communications code? What are some of the other names for data communications codes?

14-9. What are the three types of characters used in data communications codes?

14-10. Which data communications code is the most powerful? Why?

14-11. What are the two general categories of error control? What is the difference between them?

14-12. Explain the following error detection techniques: redundancy, exact-count encoding, parity, vertical redundancy checking, longitudinal redundancy checking, and cyclic redundancy checking.

14-13. Which error detection technique is the simplest?

14-14. Which error detection technique is the most reliable?

14-15. Explain the following error correction techniques: symbol substitution, retransmission, and forward error correction.

14-16. Which error correction technique is designed to be used in a human environment?

14-17. Which error correction technique is the most reliable?

14-18. Define *character synchronization*.

14-19. Describe the asynchronous data format.

14-20. Describe the synchronous data format.

14-21. Which data format is best suited to long messages? Why?

14-22. What is a cluster?

14-23. Describe the functions of a control unit.

14-24. What is the purpose of the data modem?

14-25. What are the primary functions of the UART?

14-26. What is the maximum number of bits that can make up a single character with a UART?

14-27. What do the status signals RPE, RFE, and ROR indicate?

14-28. Why does the receive clock for a UART operate 16 times faster than the receive bit rate?

14-29. What are the major differences between a UART and a USRT?

14-30. What is the purpose of the serial interface?

14-31. What is the most prominent serial interface in the United States?

14-32. Why did the EIA establish the RS-232C interface?

14-33. What is the nominal maximum length for the RS-232C interface?

14-34. What are the four general classifications of pins on the RS-232C interface?

14-35. What is the maximum positive voltage that a driver will output?

14-36. Which classification of pins uses negative logic?

14-37. What is the primary difference between the RS-449A interface and the RS-232C interface?

14-38. Higher bit rates are possible with a (balanced, unbalanced) interface cable.

14-39. Who provides the most commonly used transmission medium for data communications circuits? Why?

14-40. Explain the differences between DDD circuits and private line circuits.

14-41. Define the following terms: local loop, trunk, common usage, and dial switch.

14-42. What is a DCE?

14-43. What is the primary difference between a synchronous and an asynchronous modem?

14-44. What is necessary for full-duplex operation using a two-wire circuit?

14-45. What do *originate* and *answer mode* mean?

14-46. What modulation scheme is used for low-speed applications? For medium-speed applications? For high-speed applications?

14-47. Why are synchronous modems required for medium- and high-speed applications?

PROBLEMS

14-1. Determine the LRC and VCR for the following message (use even parity for LRC and odd parity for VRC).

```
P S S                                                   E B P
A Y Y T D A T A sp C O M M U N I C A T I O N S T C A
D N N X                                                 X S D
```

14-2. Determine the BCS for the following data- and CRC-generating polynomials:

$$G(x) = x^7 + x^4 + x^2 + x^0 = 1\ 0\ 0\ 1\ 0\ 1\ 0\ 1$$

$$P(x) = x^5 + x^4 + x^1 + x^0 = 1\ 1\ 0\ 0\ 1\ 1$$

14-3. How many Hamming bits are required for a single ASCII character?

14-4. Determine the Hamming bits for the ASCII character "B." Insert the Hamming bits into every other location starting at the left.

Chapter 15

DATA COMMUNICATIONS PROTOCOLS

INTRODUCTION

In essence, a *protocol* is a set of customs or regulations dealing with formality or precedence such as diplomatic or military protocol. A data communications protocol is a set of rules governing the orderly exchange of data information. As stated previously, the function of a line control unit is to control the flow of data between the applications program and the remote terminals. Therefore, there must be a set of rules that govern how an LCU reacts to or initiates different types of transmissions. This set of rules is called a *data link protocol*. Essentially, a data link protocol is a set of procedures, including precise character sequences, that ensure an orderly exchange of data between two LCUs.

In a data communications circuit, the station that is presently transmitting is called the *master* and the receiving station is called the *slave*. In a centralized network, the primary station controls when each secondary station can transmit. When a secondary station is transmitting, it is the master and the primary station is now the slave. The role of master is temporary and which station is master is delegated by the primary. Initially, the primary is master. The primary station solicits each secondary station, in turn, by *polling* it. A poll is an invitation from the primary to a secondary to transmit a message. Secondaries cannot poll a primary. When a primary polls a secondary, the primary is initiating a *line turnaround*; the polled secondary has been designated the master and must respond. If the primary *selects* a secondary, the secondary is identified as a receiver. A selection is an interrogation by the primary of a secondary to determine the secondary's status (i.e., ready to receive or not ready to receive a message). Secondary stations cannot select the primary. Transmissions from the primary go to all the secondaries; it is up to the secondary stations to individually decode each transmission and determine if it is intended for them. When a secondary transmits, it sends only to the primary.

Data link protocols are generally categorized as either asynchronous or synchronous. As a rule, asynchronous protocols use an asynchronous data format and asynchronous modems, whereas synchronous protocols use a synchronous data format and synchronous modems.

Asynchronous Protocols

Two of the most commonly used asynchronous data protocols are the Bell System's *selective calling system* (8A1/8B1) and IBM's *asynchronous data link protocol* (83B). In essence, these two protocols are the same set of procedures.

Asynchronous protocols are *character oriented*. That is, unique data link control characters such as end of transmission (EOT) and start of text (STX), no matter where they occur in a transmission, warrant the same action or perform the same function. For example, the end-of-transmission character used with ASCII is 04H. No matter when 04H is received by a secondary, the LCU is cleared and placed in the line monitor mode. Consequently, care must be taken to ensure that the bit sequences for data link control characters do not occur within a message unless they are intended to perform their designated data link functions. Vertical redundancy checking (parity) is the only type of error detection used with asynchronous protocols, and symbol substitution and ARQ (retransmission) are used for error correction. With asynchronous protocols, each secondary station is generally limited to a single terminal/printer pair. This station arrangement is called a *stand alone*. With the stand-alone configuration, all messages transmitted from or received on the terminal CRT are also written on the printer. Thus the printer simply generates a hard copy of all transmissions.

In addition to the line monitoring mode, a remote station can be in any one of three operating modes: *transmit*, *receive*, and *local*. A secondary station is in the transmit mode whenever it has been designated master. In the transmit mode, the secondary can send formatted messages or acknowledgments. A secondary is in the receive mode whenever it has been selected by the primary. In the receive mode, the secondary can receive formatted messages from the primary. For a terminal operator to enter information into his or her computer terminal, the terminal must be in the local mode. A terminal can be placed in the local mode through software sent from the primary or the operator can do it manually from the keyboard.

The polling sequence for most asynchronous protocols is quite simple and usually encompasses sending one or two data link control characters, then a *station polling address*. A typical polling sequence is

```
E  D
O  C  A
T  3
```

The EOT character is the *clearing* character and always precedes the polling sequence. EOT places all the secondaries in the line monitor mode. When in the line monitor mode, a secondary station listens to the line for its polling or selection address. When DC3 immediately follows EOT, it indicates that the next character is a station polling address. For this example, the station polling address is the single ASCII character "A." Station "A" has been designated the master and must respond with either a formatted message or an acknowledgement. There are two acknowledgment sequences that may be transmitted in response to a poll. They are listed below together with their functions.

Acknowledgement	Function
A \C K	No message to transmit, ready to receive
\\	No message to transmit, not ready to receive

The selection sequence, which is very similar to the polling sequence, is

$$\begin{matrix} E \\ O & X & Y \\ T \end{matrix}$$

Again, the EOT character is transmitted first to ensure that all the secondary stations are in the line monitor mode. Following the EOT is a two-character selection address "XY." Station XY has been selected by the primary and designated as a receiver. Once selected, a secondary station must respond with one of three acknowledgment sequences indicating its status. They are listed below together with their functions.

Acknowledgment	Function
A \C K	Ready to receive
\\	Not ready to receive, terminal in local, or printer out of paper
**	Not ready to receive, have a formatted message to transmit

More than one station can be selected simultaneously with *group* or *broadcast* addresses. Group addresses are used when the primary desires to select more than one but not all of the remote stations. There is a single broadcast address that is used to select simultaneously all the remote stations. With asynchronous protocols, acknowledgment procedures for group and broadcast selections are somewhat involved and for this reason are seldom used.

Messages transmitted from the primary and secondary use exactly the same data format. The format is as follows:

```
S                         E
T   message data          O
X                         T
```

The preceding format is used by the secondary to transmit data to the primary in response to a poll. The STX and EOT characters frame the message. STX precedes the data and indicates that the message begins with the character that immediately follows it. The EOT character signals the end of the message and relinquishes the role of master to the primary. The same format is used when the primary transmits a message except that the STX and EOT characters have an additional function. The STX is a *blinding* character. Upon receipt of the STX character, all previously unselected stations are "blinded," which means that they ignore all transmissions except EOT. Consequently, the subsequent message transmitted by the primary is received only by the previously selected station. The unselected secondaries remain blinded until they receive an EOT character, at which time they will return to the line monitor mode and again listen to the line for their polling or selection addresses. STX and EOT are not part of the message; they are data link control characters and are inserted and deleted by the LCU.

Sometimes it is necessary or desirable to transmit coded data in addition to the message that are used only for data link management, such as date, time of message, message number, message priority, routing information, and so on. This bookkeeping information is not part of the message; it is overhead and is transmitted as *heading* information. To identify the heading, the message begins with a start-of-heading character (SOH). SOH is transmitted first, followed by the heading information, STX, then the message. The entire sequence is terminated with an EOT character. When a heading is included, STX terminates the heading and also indicates the beginning of the message. The format for transmitting heading information together with message data is

```
S               S               E
O   heading     T   message data   O
H               X               T
```

Synchronous Protocols

With synchronous protocols, a secondary station can have more than a single terminal/printer pair. The group of devices is commonly called a *cluster*. A single LCU can serve a cluster with as many as 50 devices (terminals and printers). Synchronous protocols can be either character or bit oriented. The most commonly used character-oriented synchronous protocol is IBM's 3270 binary synchronous communications (BSC or bisync), and the most popular bit-oriented protocol (BPO) is IBM's synchronous data link communications (SDLC).

IBM's bisync protocol.   With bisync, each transmission is preceded by a unique SYN character: 16H for ASCII and 32H for EBCDIC. The SYN character places the receive USRT in the character or byte mode and prepares it to receive data in 8-bit groupings. With bisync, SYN characters are always transmitted in pairs (hence the name "bisync"). Therefore, if 8 successive bits are received in the middle of a message that are equivalent to a SYN character, they are ignored. For example, the characters "A" and "b" have the following hex and binary codes:

$$A = 41H = 0\ 1\ 0\ 0\ 0\ 0\ 0\ 1$$
$$b = 62H = 0\ 1\ 1\ 0\ 0\ 0\ 1\ 0$$

If the ASCII characters A and b occur successively during a message or heading, the following bit sequence occurs:

```
      A (41H)        b (62H)
 ╭───────────────╮╭───────────────╮
 0 1 0 0 0 0 0 1 0 1 1 0 0 0 1 0
           ╰───────────────╯
              SYN (16H)
```

As you can see, it appears that a SYN character has been transmitted when actually it has not. To avoid this situation, SYN characters are always transmitted in pairs, and consequently, if only one is received, it is ignored. The likelihood of two false SYN characters occurring one immediately after the other is remote.

With synchronous protocols, the concepts of polling, selecting, and acknowledging are identical to those used with asynchronous protocols except, with bisync, group, and broadcast selections are not allowed. There are two polling formats used with bisync: general and specific. The format for a general poll is

```
P S S E P S S S S   E P
A Y Y O A Y Y P P " " N A
D N N T D N N A A     Q D
```

The PAD character at the beginning of the sequence is called a *leading* pad and is either a 55H or an AAH (01010101 or 10101010 binary). As you can see, a leading pad is simply a string of alternating 1's and 0's. The purpose of the leading pad is to ensure that transitions occur in the data prior to the actual message. The transitions are needed for clock recovery in the receive modem to maintain bit synchronization. Next, there are two SYN characters to establish character synchronization. The EOT character is again used as a clearing character and places all the secondary stations into the line monitor mode. The PAD character immediately following the second SYN character is simply a string of successive logic 1's that is used for a time fill, giving each of the secondary stations time to clear. The number of 1's transmitted during this time fill may not be a multiple of 8 bits. Consequently, the two SYN characters are repeated to reestablish character synchronization. The SPA is not an ASCII or EBCDIC character. The letters SPA stand for *station polling address*. Each secondary station has a unique SPA. Two SPAs are transmitted for the purpose of error detection (redundancy). A secondary will not respond to a poll unless its SPA appears twice. The two quotation marks signify that the poll is for any device at that station that is in the send mode. If two or more devices are in the send mode when a general poll is received, the LCU determines which device's message is transmitted. The enquiry (ENQ) character is sometimes called a *format* or *line turnaround character* because it completes the polling format and initiates a line turnaround (i.e., the secondary station identified by the SPA is designated master and must respond).

The PAD character at the end of the polling sequence is called a *trailing* pad and is simply a 7FH (DEL or delete character). The purpose of the trailing pad is to ensure that the RLSD signal in the receive modem is held active long enough for the entire received message to be demodulated. If the carrier were shut off immediately at the end of the message, RLSD would go inactive and disable the receive data pin. If the last character of the message were not completely demodulated, the end of it would be cut off.

The format for a specific poll is

```
P S S E P S S S     E P
A Y Y O A Y Y P P D D N A
D N N T D N N A A A A Q D
```

The character sequence for a specific poll is similar to that of a general poll except that two DAs (*device addresses*) are substituted for the two quotation marks. With a specific poll, both the station and device addresses are included. Therefore, a specific poll is an invitation to transmit to a specific device at a given station. Again, two DAs are transmitted for redundancy error detection.

The character sequence for a selection is

```
P S S E P S S S     E P
A Y Y O A Y Y S S D D N A
D N N T D N N A A A A Q D
```

The sequence for a selection is similar to that of a specific poll except that two SSA characters are substituted for the two SPAs. SSA stands for "station select address." All selections are specific; they are for a specific device (device DA). Table 15-1 lists the SPAs, SSAs, and DAs for a network that can have a maximum of 32 stations and the LCU at each station can serve a 32-device cluster.

EXAMPLE 15-1

Determine the character sequences for (a) a general poll for station 8, (b) a specific poll for device 6 at station 8, and (c) a selection of device 6 at station 8.

TABLE 15-1   STATION AND DEVICE ADDRESSES

Station or device number	SPA	SSA	DA	Station or device number	SPA	SSA	DA
0	sp	-	sp	16	&	Ø	&
1	A	./	A	17	J	1	J
2	B	S	B	18	K	2	K
3	C	T	C	19	L	3	L
4	D	U	D	20	M	4	M
5	E	V	E	21	N	5	N
6	F	W	F	22	O	6	O
7	G	X	G	23	P	7	P
8	H	Y	H	24	Q	8	Q
9	I	Z	I	25	R	9	R
10	[	!	[	26	]	:	]
11	.	,	.	27	$	#	$
12	<	%	<	28	*	@	*
13	(	—	(	29	)	'	)
14	+	>	+	30	;	=	;
15	!	?	!	31	∧	"	∧

Solution   (a) From Table 15-1 the SPA for station 8 is H; therefore, the sequence for a general poll is

```
P S S E P S S        E P
A Y Y O A Y Y H H ⌐ ⌐ N A
D N N T D N N        Q D
```

    (b) From Table 15-1 the DA for device 6 is F; therefore, the sequence for a specific poll is

```
P S S E P S S        E P
A Y Y O A Y Y H H F F N A
D N N T D N N        Q D
```

    (c) From Table 15-1 the SSA for station 8 is Y; therefore, the sequence for a selection is

```
P S S E P S S        E P
A Y Y O A Y Y Y Y F F N A
D N N T D N N        Q D
```

    With bisync, there are only two ways in which a secondary can respond to a poll: with a formatted message or with a *handshake*. A handshake is simply a response from the secondary that indicates it has no formatted messages to transmit (i.e., a handshake is a negative acknowledgment to a poll). The character sequence for a hand-shake is

```
P S S E P
A Y Y O A
D N N T D
```

A secondary can respond to a selection with either a positive or a negative acknowl-edgement. A positive acknowledgment to a selection indicates that the device selected is ready to receive. The character sequence for a positive acknowledgment is

```
P S S D   P
A Y Y L θ A
D N N E   D
```

A negative acknowledgment to a selection indicates that the device selected is not ready to receive. A negative acknowledgment is called a *reverse interrupt* (RVI). The character sequence for an RVI is

```
P S S D   P
A Y Y L < A
D N N E   D
```

With bisync, formatted messages are sent from a secondary to the primary in response to a poll and sent from the primary to a secondary after the secondary has been selected. Formated messages use the following format:

```
P S S S         S          E B P
A Y Y O heading T message  T C A
D N N H         X          X C D
```

Note: If CRC-16 is used for error detection, there are two block check characters.

Longitudinal redundancy checking (LRC) is used for error detection with ASCII-coded messages, and cyclic redundancy checking (CRC) is used for EBCDIC. The BCC is computed beginning with the first character after SOH and continues through and includes ETX. (If there is no heading, the BCC is computed beginning with the first character after STX.) With synchronous protocols, data are transmitted in blocks. Blocks of data are generally limited to 256 characters. ETX is used to terminate the last block of a message. ETB is used for multiple block messages to terminate all message blocks except the last one. The last block of a message is always terminated with ETX. All BCCs must be acknowledged by the receiving station. A positive acknowl-edgment indicates that the BCC was good and a negative acknowledgment means that the BCC was bad. A negative acknowledgement is an automatic request for retransmis-sion. The character sequences for positive and negative acknowledgments are as follows:

Positive acknowledgment:

```
P S S D   P        P S S D   P
A Y Y L θ A   or   A Y Y L 1 A
D N N E   D        D N N E   D

 even-numbered      odd-numbered
    blocks             blocks
```

Negative acknowledgment:

```
P S S N P
A Y Y A A
D N N K D
```

Examples of dialogue using bisync protocol

```
P S S E P S S       E P
A Y Y O A Y Y A A " " N A ──────────→
D N N T D N N       Q D
```

Primary station sends a general poll for station 1.

```
         P S S E P
◄─────── A Y Y O A
         D N N T D
```

Station 1 responds with a negative acknowledgment—no messages to transmit.

```
P S S E P S S       E P
A Y Y O A Y Y B B " " N A ──────────→
D N N T D N N       Q D
```

Primary station sends a general poll for station 2.

```
      P S S          S         E B P
◄──── A Y Y O heading T message T C A
      D Y Y H         X block 1 B C D
```

Station 2 responds with the first block of a multiblock message.

```
P S S D   P
A Y Y L 1 A ──────→
D N N E   D
```

Primary sends a positive acknowledgment indicating that block 1 was received without any errors—because block 1 is an odd-numbered block, DLE 1 is used.

```
      P S S          E B P
◄──── A Y Y T message T C A
      D N N X block 2 X C D
```

Station 2 sends the second and final block of the message—note that there is no heading at the beginning of the second block—a heading is transmitted only with the first block of a message.

```
                    P S S N P
              ——————A Y Y A A——————————————————————▶
                    D N N K D
```

Primary sends a negative acknowledgment to station 2 indicating that block 2 was received with an error and must be retransmitted.

```
              P S S S              E B P
        ◀—————A Y Y T   message    T C A——————
              D N N X   block 2    X C D
```

Station 2 resends block 2.

```
                    P S S D  P
              ——————A Y Y L θ A——————————————————▶
                    D N N E  D
```

Primary sends a positive acknowledgment to station 2 indicating that block 2 was received without any errors—because block 2 is an even-numbered block, DLE θ is used.

```
                    P S S E P
              ◀—————A Y Y O A——————————————————————
                    D N N T D
```

Secondary responds with a handshake—a secondary sends a handshake whenever it is its turn to transmit but it has nothing to say.

```
              P S S E P S S       E P
        ——————A Y Y O A Y Y T T E E N A——————————▶
              D N N T D N N       Q D
```

Primary selects station 3, device 5.

```
                    P S S D  P
              ◀—————A Y Y L θ A——————————————————
                    D N N E  D
```

Station 3 sends a positive acknowledgment to the selection; device 5 is ready to receive.

```
              P S S S             S           E B P
        ——————A Y Y O   heading   T  message  T C A——————▶
              D N N H             X  block 1  X C D
```

Primary sends a single block message to station 3.

```
        P  S S  D   P
←────── A  Y Y  L 1 A ──────
        D  N N  E   D
```

Station 3 responds with a positive acknowledgment indicating the block
of data was received without any errors.

Transparency.   It is possible that a device that is attached to one of the ports
of a station LCU is not a computer terminal or a printer. For example, a microprocessor-
controlled monitor system that is used to monitor environmental conditions (temperature,
humidity, etc.) or a security alarm system. If so, the data transferred between it and
the applications program are not ASCII- or EBCDIC-encoded characters; they are microprocessor op-codes or binary-encoded data. Consequently, it is possible that an 8-bit
sequence could occur in the message that is equivalent to a data link control character.
For example, if the binary code 00000011 (03H) occurred in a message, the LCU
would misinterpret it as the ASCII code for ETX. Consequently, the receive LCU
would prematurely terminate the message and interpret the next 8-bit sequence as a
BCC. To prevent this from occurring, the LCU is made *transparent* to the data. With
bisync, a *data link escape* character (DLE) is used to achieve transparency. To place
an LCU in the transparent mode, STX is preceded by a DLE (i.e., the LCU simply
transfers the data to the selected device without searching through the message for data
link control characters). To come out of the transparent mode, DLE ETX is transmitted.
To transmit a DLE as part of the text, it must be preceded by DLE (i.e., DLE DLE).
Actually, there are only five characters that it is necessary to precede with DLE:

1. *DLE STX*: places the receive LCU into the transparent mode.
2. *DLE ETX*: used to terminate the last block of transparent text and take the LCU
   out of the transparent mode.
3. *DLE ETB*: used to terminate blocks of transparent text other than the final block.
4. *DLE ITB*: used to terminate blocks of transparent text other than the final block
   when ITB is used for a block terminating character.
5. *DLE SYN*: used only with transparent messages that are more than 1 s long.
   With bisync, two SYN characters are inserted in the text every 1 s to ensure that
   the receive LCU does not lose character synchronization. In a multipoint circuit
   with a polling environment, it is highly unlikely that any blocks of data would
   exceed 1 s in duration. SYN character insertion is used almost exclusively for
   two-point circuits.

Synchronous Data Link Communications

Synchronous data link communications (SDLC) is a synchronous *bit-oriented* protocol
developed by IBM. A bit-oriented protocol (BOP) is a discipline for serial-by-bit information transfer over a data communication channel. With a BOP, data link control informa-

tion is transferred and interpreted on a bit-by-bit basis rather than with unique data link control characters. SDLC can transfer data either simplex, half-duplex, or full duplex. With a BOP, there is a single control field that performs essentially all the data link control functions. The character language used with SDLC is EBCDIC and data are transferred in groups called *frames*. Frames are generally limited to 256 characters in length. There are two types of stations in SDLC: primary stations and secondary stations. The *primary station* controls data exchange on the communications channel and issues *commands*. The *secondary station* receives commands and returns *responses* to the primary.

There are three transmission states with SDLC: transient, idle, and active. The *transient state* exists before and after the initial transmission and after each line turnaround. An *idle state* is presumed after 15 or more consecutive 1's have been received. The *active state* exists whenever either the primary or a secondary station is transmitting information or control signals.

Figure 15-1 shows the frame format used with SDLC. The frames sent from the primary and the frames sent from a secondary use exactly the same format. There are five fields used with SDLC: the flag field, the address field, the control field, the text or information field, and the frame check field.

Information field. All information transmitted in an SDLC frame must be in the information field (I field), and the number of bits in the I field must be a multiple of 8. An I field is not allowed with all SDLC frames. The types of frames that allow an I field are discussed later.

Flag field. There are two flag fields per frame: the beginning flag and the ending flag. The flags are used for the *delimiting sequence* and to achieve character synchronization. The delimiting sequence sets the limits of the frame (i.e., when the frame begins and when it ends). The flag is used with SDLC in the same manner that SYN characters are used with bisync, to achieve character synchronization. The sequence for a flag is

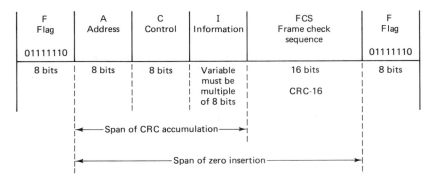

FIGURE 15-1 SDLC frame format.

7EH, 01111110 binary, or the EBCDIC character "=." There are several variations of how flags are used. They are:

1. One beginning and one ending flag for each frame.

```
        beginning flag                           ending flag
    . . . 01111110 address control text FCC 01111110 . . .
```

2. The ending flag from one frame can be used for the beginning flag for the next frame.

```
                              ←─────────────────── frame N + 1 ──────→
    ────── frame N ──────→
    . . . text FCC 01111110 address control text FCC 01111110 . . .
                        /    \
                ending flag  beginning flag
                  frame N      frame N + 1
```

3. The last zero of an ending flag is also the first zero of the beginning flag of the next frame.

```
                              ←─────────────── frame N + 1 ──────
                                     shared 0
                  frame N ──────→   /
    . . . text FCC 0111111O1111110 address control FCC . . .
                        /       \
                ending flag   beginning flag
                  frame N       frame N + 1
```

4. Flags are transmitted in lieu of idle line 1's.

```
    0111111011111101111110111110 address control text . . .
        /    /    /
    idle line flags   beginning flag
```

Address field.   The address field has 8 bits; thus 256 addresses are possible with SDLC. The address 00H (00000000) is called the *null* or *void address* and is never assigned to a secondary. The null address is used for network testing. The address FFH (11111111) is the *broadcast address* and is common to all secondaries. The remaining 254 addresses can be used as *unique* station addresses or as *group* addresses. In frames sent from the primary, the address field contains the address of the destination station (a secondary). In frames sent from a secondary, the address field contains the address of that secondary. Therefore, the address is always that of a secondary. The primary station has no address because all transmissions from secondary stations go to the primary.

Control field. The control field is an 8-bit field that identifies the type of frame it is. The control field is used for polling, confirming previously received information frames, and several other data link management functions. There are three frame formats used with SDLC: information, supervisory, and unnumbered.

Information frame. With an information frame there must be an information field. Information frames are used for transmitting sequenced information. The bit pattern for the control field of an information frame is

$$
\begin{array}{lcccccccc}
\text{Bit:} & b_0 & b_1 & b_2 & b_3 & b_4 & b_5 & b_6 & b_7 \\
\text{Function:} & \longleftarrow & \text{nr} & \longrightarrow & \underset{\overline{\text{P or }\overline{\text{F}}}}{\text{P or F}} & \longleftarrow & \text{ns} & \longrightarrow & 0
\end{array}
$$

An information frame is identified by a 0 in the least significant bit position (b_7 with EBCDIC code). Bits b_4, b_5, and b_6 are used for numbering transmitted frames (ns = number sent). With 3 bits, the binary numbers 000 through 111 (0–7) can be represented. The first frame transmitted is designated frame 000, the second frame 001, and so on up to frame 111 (the eighth frame); then the count cycles back to 000 and repeats.

Bits b_0, b_1, and b_2 are used to confirm correctly received information frames (nr = number received) and to automatically request retransmission of incorrectly received information frames. The nr is the number of the next frame that the transmitting station expects to receive, or the number of the next frame that the receiving station will transmit. The nr confirms received frames through nr-1. Frame nr-1 is the last frame received without a transmission error. Any transmitted I frame not confirmed must be retransmitted. Together, the ns and nr bits are used for error correction (ARQ). The primary must keep track of an ns and nr for each secondary. Each secondary must keep track of only its ns and nr. After all frames have been confirmed, the primary's ns must agree with the secondary's nr, and vice versa. For the example shown next, the primary and secondary stations begin with their ns and nr counters reset to 000. The primary sends three numbered information frames (ns = 0, 1, and 2). At the same time the primary sends nr = 0 because the next frame it expects to receive is frame 0, which is the secondary's present ns. The secondary responds with two information frames (ns = 0 and 1). The secondary received all three frames from the primary without any errors, so the nr transmitted in the secondary's control field is 3 (which is the number of the next frame that the primary will send). The primary now sends information frames 3 and 4 with an nr = 2. The nr = 2 confirms the correct reception of frames 0 and 1. The secondary responds with frames ns = 2, 3, and 4 with an nr = 4. The nr = 4 confirms reception of only frame 3 from the primary (nr-1). Consequently, the primary must retransmit frame 4. Frame 4 is retransmitted together with four additional frames (ns = 5, 6, 7, and 0). The primary's nr = 5, which confirms frames 2, 3, and 4 from the secondary. Finally, the secondary sends information frame 5 with an nr = 1. The nr = 1 confirms frames 4, 5, 6, 7 and 0 from the primary. At this point, all the frames transmitted have been confirmed except frame 5 from the secondary.

Primary's ns:	0 1 2		3 4		4 5 6 7 0
Primary's nr:	0 0 0		2 2		5 5 5 5 5
Secondary's ns:		0 1		2 3 4	5
Secondary's nr:		3 3		4 4 4	1

With SDLC, a station can never send more than seven numbered frames without receiving a confirmation. For example, if the primary sent eight frames (ns = 0, 1, 2, 3, 4, 5, 6, and 7) and the secondary responded with an nr = 0, it is ambiguous which frames are being confirmed. Does nr = 0 mean that all eight frames were received correctly, or that frame 0 had an error in it and all eight frames must be retransmitted? (With SDLC, all previously transmitted frames beginning with frame nr-1 must be retransmitted.)

Bit b_3 is the *poll* (P) or *not-a-poll* ($\overline{P}$) bit when sent from the primary and the *final* (F) or *not-a-final* ($\overline{F}$) bit when sent by a secondary. In a frame sent by the primary, if the primary desires to poll the secondary, the P bit is set (1). If the primary does not wish to poll the secondary, the P bit is reset (0). A secondary cannot transmit unless it receives a frame addressed to it with the P bit set. In a frame sent from a secondary, if it is the last (final) frame of the message, the F bit is set (1). If it is not the final frame, the F bit is reset (0). With I frames, the primary can select a secondary station, send formatted information, confirm previously received I frames, and poll with a single transmission.

EXAMPLE 15-2

Determine the bit pattern for the control field of a frame sent from the primary to a secondary station for the following conditions: primary is sending information frame 3, it is a poll, and the primary is confirming the correct reception of frames 2, 3, and 4 from the secondary.

Solution

$b_7 = 0$ because it is an information frame.

b_4, b_5, and b_6 are 011 (binary 3 for ns =3).

$b_3 = 1$, it is a polling frame.

b_0, b_1, and b_2 are 101 (binary 5 for nr = 5).

control field = B6H.

$$b_0 \; b_1 \; b_2 \; b_3 \; b_4 \; b_5 \; b_6 \; b_7$$
$$1 \quad 0 \quad 1 \quad 1 \quad 0 \quad 1 \quad 1 \quad 0$$

Supervisory frame.   An information field is not allowed with a supervisory frame. Consequently, supervisory frames cannot be used to transfer information; they are used to assist in the transfer of information. Supervisory frames are used to confirm previously received information frames, convey ready or busy conditions, and to report frame numbering errors. The bit pattern for the control field of a supervisory frame is

Bit: b_0 b_1 b_2 b_3 b_4 b_5 b_6 b_7

Function: $-$ $-$ nr $-$ $-$ P or F   X   X   0   1
                              $\overline{\text{P or F}}$

A supervisory frame is identified by a 01 in bit positions b_6 and b_7, respectively, of the control field. With the supervisory format, bit b_3 is again the poll/not-a-poll or final/not-a-final bit and b_0, b_1, and b_2 are the nr bits. However, with a supervisory format, b_5 and b_6 are used to indicate either the receive status of the station transmitting the frame or to request transmission or retransmission of sequenced information frames. With two bits, there are four combinations possible. The four combinations and their functions are as follows:

b_4	b_5	Receiver status
0	0	Ready to receive (RR)
0	1	Ready not to receive (RNR)
1	0	Reject (REJ)
1	1	Not used with SDLC

When the primary sends a supervisory frame with the P bit set and a status of ready to receive, it is equivalent to a general poll with bisync. Supervisory frames are used by the primary for polling and for confirming previously received information frames when there is no information to send. A secondary uses the supervisory format for confirming previously received information frames and for reporting its receive status to the primary. If a secondary sends a supervisory frame with RNR status, the primary cannot send it numbered information frames until that status is cleared. RNR is cleared when a secondary sends an information frame with the F bit = 1 or a RR or REJ frame with the F bit = 0. The REJ command/response is used to confirm information frames through nr-1 and to request retransmission of numbered information frames beginning with the frame number identified in the REJ frame. An information field is prohibited with a supervisory frame and the REJ command/response is used only with full-duplex operation.

EXAMPLE 15-3

Determine the bit pattern for the control field of a frame sent from a secondary station to the primary for the following conditions: the secondary is ready to receive, it is the final frame, and the secondary station is confirming frames 3, 4, and 5.

Solution

b_6 and b_7 = 01 because it is a supervisory frame.

b_4 and b_5 = 00 (ready to receive).

b_3 = 1 (it is the final frame).

b_0, b_1, and b_2 = 110 (binary 6 for nr = 6).

control field = D1H

$$b_0 \ b_1 \ b_2 \ b_3 \ b_4 \ b_5 \ b_6 \ b_7$$
$$1 \ \ 1 \ \ 0 \ \ 1 \ \ 0 \ \ 0 \ \ 0 \ \ 1$$

Unnumbered frame. An unnumbered frame is identified by making bits b_6 and b_7 in the control field 11. The bit pattern for the control field of an unnumbered frame is

Bit:	b_0 b_1 b_2	b_3	b_4 b_5 b_6 b_7
Function:	X X X	P or F $\overline{\text{P}}$ or $\overline{\text{F}}$	X X 1 1

With an unnumbered frame, bit b_3 is again either the P/$\overline{\text{P}}$ or F/$\overline{\text{F}}$ bit. Bits b_0, b_1, b_2, b_4, and b_5 are used for various unnumbered commands and responses. With 5 bits available, 32 unnumbered commands/responses are possible. The control field in an unnumbered frame sent by the primary is a command. The control field in an unnumbered frame sent by a secondary is a response. With unnumbered frames, there are no ns or nr bits. Therefore, numbered information frames cannot be sent or confirmed with the unnumbered format. Unnumbered frames are used to send network control and status information. Two examples of control functions are (1) placing secondary stations on-line and off-line and (2) LCU initialization. Table 15-2 lists several of the more commonly used unnumbered commands and responses. An information field is prohibited with all the unnumbered commands/responses except UI, FRMR, CFGR, TEST and XID.

A secondary station must be in one of three modes: the initialization mode, the normal response mode, or the normal disconnect mode. The procedures for the *initialization mode* are system specified and vary considerably. A secondary in the *normal response mode* cannot initiate unsolicited transmissions; it can transmit only in response to a frame received with the P bit set. When in the *normal disconnect mode*, a secondary is off-line. In this mode, a secondary can receive only a TEST, XID, CFGR, SNRM, or SIM command from the primary and can respond only if the P bit is set.

TABLE 15-2   UNNUMBERED COMMANDS AND RESPONSES

b_0	Binary configuration	b_7	Acronym	Command	Response	I Field prohibited	Resets ns and nr
000	P/F	0011	UI	Yes	Yes	No	No
000	F	0111	RIM	No	Yes	Yes	No
000	P	0111	SIM	Yes	No	Yes	Yes
100	P	0011	SNRM	Yes	No	Yes	Yes
000	F	1111	DM	No	Yes	Yes	No
010	P	0011	DISC	Yes	No	Yes	No
011	F	0011	UA	No	Yes	Yes	No
100	F	0111	FRMR	No	Yes	No	No
111	F	1111	BCN	No	Yes	Yes	No
110	P/F	0111	CFGR	Yes	Yes	No	No
010	F	0011	RD	No	Yes	Yes	No
101	P/F	1111	XID	Yes	Yes	No	No
001	P	0011	UP	Yes	No	Yes	No
111	P/F	0011	TEST	Yes	Yes	No	No

The unnumbered commands and responses are summarized below.

Unnumbered information (UI). UI is a command/response that is used to send unnumbered information. Unnumbered information transmitted in the I field is not confirmed.

Set initialization mode (SIM). SIM is a command that places the secondary station into the initialization mode. The initialization procedure is system specified and varies from a simple self-test of the station controller to executing a complete IPL (initial program logic) program. SIM resets the ns and nr counters at the primary and secondary stations. A secondary is expected to respond to a SIM command with a UA response.

Request initialization mode (RIM). RIM is a response sent by a secondary station to request the primary to send an SIM command.

Set normal response mode (SNRM). SNRM is a command that places a secondary station in the normal response mode (NRM). A secondary station cannot send or receive numbered information frames unless it is in the normal response mode. Essentially, SNRM places a secondary station on-line. SNRM resets the ns and nr counters at the primary and secondary stations. UA is the normal response to an SNRM command. Unsolicited responses are not allowed when the secondary is in the NRM. A secondary remains in the NRM until it receives a DISC or SIM command.

Disconnect mode (DM). DM is a response that is sent from a secondary station if the primary attempts to send numbered information frames to it when the secondary is in the normal disconnect mode.

Request disconnect (RD). RD is a response sent when a secondary wishes to be placed in the disconnect mode.

Disconnect (DISC). DISC is a command that places a secondary station in the normal disconnect mode (NDM). A secondary cannot send or receive numbered information frames when it is in the normal disconnect mode. When in the normal disconnect mode, a secondary can receive only an SIM or SNRM command and can transmit only a DM response. The expected response to a DISC command is UA.

Unnumbered acknowledgment (UA). UA is an affirmative response that indicates compliance to a SIM, SNRM, or DISC command. UA is also used to acknowledge unnumbered information frames.

Frame reject (FRMR). FRMR is for reporting procedual errors. The FRMR sequence is a response transmitted when the secondary has received an invalid frame from the primary. A received frame may be invalid for any one of the following reasons:

1. The control field contains an invalid or unassigned command.
2. The amount of data in the information field exceeds the buffer space at the secondary.
3. An information field is received in a frame that does not allow information.

4. The nr received is incongruous with the secondary's ns. For example, if the secondary transmitted ns frames 2, 3, and 4 and then the primary responded with an nr of 7.

A secondary cannot release itself from the FRMR condition, nor does it act on the frame that caused the condition. The secondary repeats the FRMR response until it receives one of the following *mode-setting* commands: SNRM, DISC, or SIM. The information field for a FRMR response always contains three bytes (24 bits) and has the following format:

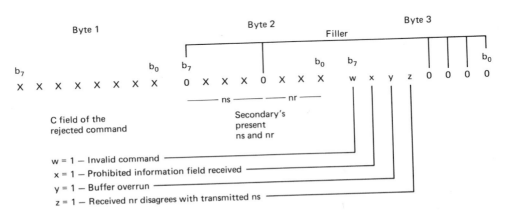

TEST. TEST is a command that can be sent in any mode to solicit a TEST response. If an information field is included with the command, the secondary returns it with the response. The TEST command/response is exchanged for link testing purposes.

Exchange station identification (XID). As a command, XID solicits the identification of the secondary station. An information field can be included in the frame to convey the identification data of either the primary or secondary station. For dial-up circuits, it is often necessary that the secondary station identify itself before the primary will exchange information frames with it, although XID is not restricted to only dial-up data circuits.

Frame check sequence field. The FCS field contains the error detection mechanism for SDLC. The FCS is equivalent to the BCC used with bisync. SDLC uses CRC-16 and the following generating polynomial: $x^{16} + x^{12} + x^5 + x^1$.

SDLC Loop Operation

An SDLC *loop* is operated in the half-duplex mode. The primary difference between the loop and bus configurations is that in a loop, all transmissions travel in the same direction on the communications channel. In a loop configuration, only one station transmits at a time. The primary transmits first, then each secondary station responds sequentially. In an SDLC loop, the transmit port of the primary station controller is connected to one or more secondary stations in a serial fashion; then the loop is terminated back at the receive port of the primary. Figure 15-2 shows an SDLC loop configuration.

In an SDLC loop, the primary transmits frames that are addressed to any or all of the secondary stations. Each frame transmitted by the primary contains an address of the secondary station to which that frame is directed. Each secondary station, in turn, decodes the address field of every frame, then serves as a repeater for all stations that are down-loop from it. If a secondary detects a frame with its address, it accepts

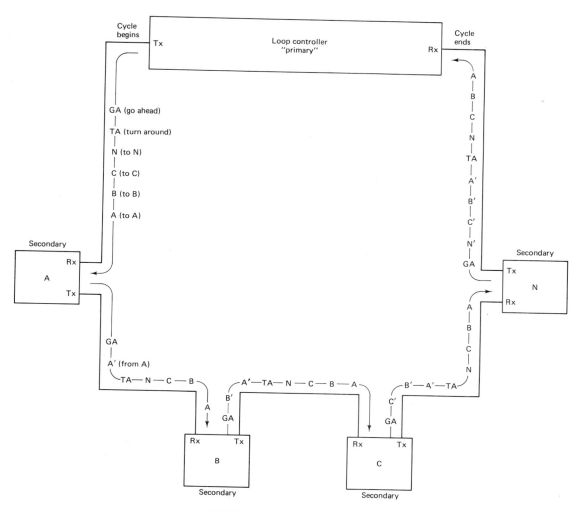

FIGURE 15-2   SDLC loop configuration.

the frame, then passes it on to the next down-loop station. All frames transmitted by the primary are returned to the primary. When the primary has completed transmitting, it follows the last flag with eight consecutive 0's. A flag followed by eight consecutive 0's is called a *turnaround* sequence which signals the end of the primary's transmission. Immediately following the turnaround sequence, the primary transmits continuous 1's, which generates a *go-ahead* sequence (01111111). A secondary cannot transmit until it has received a frame addressed to it with the P bit set, a turnaround sequence, and then a go-ahead sequence. Once the primary has begun transmitting 1's, it goes into the receive mode.

The first down-loop secondary station that has received a frame addressed to it with the P bit set, changes the seventh 1 bit in the go-ahead sequence to a 0, thus creating a flag. That flag becomes the beginning flag of the secondary's response frame or frames. After the secondary has transmitted its last frame, it again becomes a repeater for the idle line 1's from the primary. These idle line 1's again become the go-ahead sequence for the next secondary station. The next down-loop station that has received a frame addressed to it with the P bit set detects the turnaround sequence, any frames transmitted from up-loop secondaries, and then the go-ahead sequence. Each secondary station inserts its response frames immediately after the last repeated frame. The cycle is completed when the primary receives its own turnaround sequence, a series of response frames, and then the go-ahead sequence.

Configure command/response.   The configure command/response (CFGR) is an unnumbered command/response that is used only in a loop configuration. CFGR contains a one-byte *function descriptor* (essentially a subcommand) in the information field. A CFGR command is acknowledged with a CFGR response. If the low-order bit of the function descriptor is set, a specified function is initiated. If it is reset, the specified function is cleared. There are six subcommands that can appear in the configure command's function field.

1. *Clear—00000000.* A clear subcommand causes all previously set functions to be cleared by the secondary. The secondary's response to a clear subcommand is another clear subcommand, 00000000.

2. *Beacon test (BCN)—0000000X.* The beacon test causes the secondary receiving it to turn on or turn off its carrier. If the X bit is set, the secondary suppresses transmission of the carrier. If the X bit is reset, the secondary resumes transmission of the carrier. The beacon test is used to isolate an open-loop problem. Also, whenever a secondary detects the loss of a receive carrier, it automatically begins to transmit its beacon response. The secondary will continue transmiting the beacon until the loop resumes normal status.

3. *Monitor mode—0000010X.* The monitor command causes the addressed secondary to place itself into a monitor (receive only) mode. Once in the monitor mode, a secondary cannot transmit until it receives a monitor mode clear (00000100) or a clear (00000000) subcommand.

4. *Wrap—0000100X*. The wrap command causes the secondary station to loop its transmissions directly to its receiver input. The wrap command places the secondary effectively off-line for the duration of the test. A secondary station does not send the results of a wrap test to the primary.

5. *Self-test—0000101X*. The self-test subcommand causes the addressed secondary to initiate a series of internal diagnostic tests. When the tests are completed, the secondary will respond. If the P bit in the configure command is set, the secondary will respond following completion of the self-test at its earliest opportunity. If the P bit is reset, the secondary will respond following completion of the test to the next poll-type frame it receives. All other transmissions are ignored by the secondary while it is performing the self-tests. The secondary indicates the results of the self-test by setting or resetting the low-order bit (X) of its self-test response. A 1 indicates that the tests were unsuccessful, and a 0 indicates that they were successful.

6. *Modified link test—0000110X*. If the modified link test function is set (X bit set), the secondary station will respond to a TEST command with a TEST response that has an information field containing the first byte of the TEST command information field repeated *n* times. The number *n* is system implementation dependent. If the X bit is reset, the secondary station will respond to a TEST command, with or without an information field, with a TEST response with a zero-length information field. The modified link test is an optional subcommand and is only used to provide an alternative form of link test to that previously described for the TEST command.

Transparency

The transparency mechanism used with SDLC is called *zero-bit insertion* or *zero stuffing*. The flag bit sequence (01111110) can occur in a frame where this pattern is not intended to be a flag. For example, any time that 7EH occurs in the address, control, information, or FCS field it would be interpreted as a flag and disrupt character synchronization. Therefore, 7EH must be prohibited from occurring except when it is intended to be a flag. To prevent a 7EH sequence from occurring, a zero is automatically inserted after any occurrence of five consecutive 1's except in a designated flag sequence (i.e., flags are not zero inserted). When five consecutive 1's are received and the next bit is a 0, the 0 is deleted or removed. If the next bit is a 1, it must be a valid flag. An example of zero insertion/deletion is shown below.

Original frame bits at the transmit station:

```
01111110   01101111   11010011   1110001100110101   01111110
  flag      address    control          FCS            flag
```

After zero insertion but prior to transmission:

```
01111110   01101111   101010011   111000011001101012   01111110
  flag     address     control          FCS             flag
                          \
                     inserted zeros
```

After zero deletion at the receive end:

```
01111110   01101111   11010011   111000110011010101   01111110
  flag     address     control          FCS             flag
```

Message Abort

Message abort is used to prematurely terminate a frame. Generally, this is only done to accommodate high-priority messages such as emergency link recovery procedures, and so on. A message abort is any occurrence of 7 to 14 consecutive 1's. Zeros are not inserted in an abort sequence. A message abort terminates an existing frame and immediately begins the higher-priority frame. If more than 14 consecutive 1's occur in succession, it is considered an idle line condition. Therefore, 15 or more successive 1's place the circuit into the idle state.

Invert-on-Zero Encoding

A binary synchronous transmission such as SDLC is time synchronized to enable identification of sequential binary digits. Synchronous data communications assumes that bit or clock synchronization is provided by either the DCE or the DTE. With synchronous transmissions, a receiver samples incoming data at the same rate that they were transmitted. Although minor variations in timing can exist, synchronous modems provide received data clock recovery and dynamically adjusted sample timing to keep sample times midway between bits. For a DTE or a DCE to recover the clock, it is necessary that transitions occur in the data. *Invert-on-zero coding* is an encoding scheme that guarantees at least one transition in the data for every 7 bits transmitted. Invert-on-zero coding is also called NRZI (*nonreturn-to-zero inverted*).

With NRZI encoding, the data are encoded in the transmitter, then decoded in the receiver. Figure 15-3 shows an example of NRZI encoding. 1's are unaffected by the NRZI encoder. However, 0's invert the encoded transmission level. Consequently, consecutive 0's generate an alternating high/low sequence. With SDLC, there can never be more than six 1's in succession (a flag). Therefore, a high-to-low transition is guaranteed to occur at least once for every 7 bits transmitted except during a message abort or an idle line condition. In a NRZI decoder, whenever a high/low transition occurs in the received data, a 0 is generated. The absence of a transition simply generates a 1. In Figure 15-3, a high level is assumed prior to encoding the incoming data.

NRZI encoding was intended to be used with asynchronous modems which do not have clock recovery capabilities. Consequently, the DTE must provide time synchro-

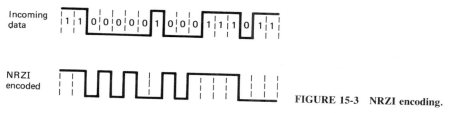

FIGURE 15-3   NRZI encoding.

nization which is aided by using NRZI-encoded data. Synchronous modems have built in scramblers and descramblers which ensure that transitions occur in the data, and thus NRZI encoding is unnecessary. The NRZI encoder/decoder is placed between the DTE and the DCE.

High-Level Data Link Control

In 1975, the International Standards Organization (ISO) defined several sets of sub-standards that, when combined, are called *high-level data link control* (HDLC). Since HDLC is a superset of SDLC, only the added capabilities are explained.

HDLC comprises three standards (subdivisions) that, when combined, outline the frame structure, control standards, and class of operation for a bit-oriented data link control (DLC).

ISO 3309–1976(E).   This standard defines the frame structure, delimiting, sequence, and transparency mechanism used with HDLC. These are essentially the same as with SDLC except that HDLC has extended addressing capabilities and checks the FCS in a slightly different manner. The delimiting sequence used with HDLC is identical to SDLC: a 01111110 sequence.

HDLC can use either the *basic* 8-bit address field or an *extended* addressing format. With extended addressing the address field may be extended recursively. If b_0 in the address byte is a logic 1, the 7 remaining bits are the secondary's address (the ISO defines the low-order bit as b_0, whereas SDLC designates the high-order bit as b_0). If b_0 is a logic 0, the next byte is also part of the address. If b_0 of the second byte is a 0, a third address byte follows, and so on, until an address byte with a logic 1 for the low-order bit is encountered. Essentially, there are 7 bits available in each address byte for address encoding. An example of a three-byte extended addressing scheme is shown below, b_0 in the first two bytes of the address field are 0's, indicating that additional address bytes follow and b_0 in the third address byte is a logic 1, which terminates the address field.

```
                 b₀ = 0       b₀ = 0       b₀ = 1
                   |            |            |
                   |            |            |
   01111110    0XXXXXXX     0XXXXXXX     1XXXXXXX   . . .

     flag          three-byte address field       control field, etc.
```

HDLC uses CRC-16 with a generating polynomial specified by CCITT V.41 as the FCS. At the transmit station, the CRC is computed such that if it is included in the FCS computation at the receive end, the remainder for an errorless transmission is always F0BBH.

ISO 4335–1979(E).   This standard defines the elements of procedure for HDLC. The control field, information field, and supervisory format have increased capabilities over SDLC.

Control field. With HDLC, the control field can be extended to 16 bits. Seven bits are for the ns and 7 bits are for the nr. Therefore, with the extended control format, there can be a maximum of 127 outstanding (unconfirmed) frames at any given time.

Information field. HDLC permits any number of bits in the information field of an information command or response (SDLC is limited to 8-bit bytes). With HDLC any number of bits may be used for a character in the I field as long as all characters have the same number of bits.

Supervisory format. With HDLC, the supervisory format includes a fourth status condition: selective reject (SREJ). SREJ is identified by a 11 in bit position b_4 and b_5 of a supervisory control field. With a SREJ, a single frame can be rejected. A SREJ calls for the retransmission of only the frame identified by nr, whereas a REJ calls for the retransmission of all frames beginning with nr. For example, the primary sends I frames ns = 2, 3, 4, and 5. Frame 3 was received in error. A REJ would call for a retransmission of frames 3, 4, and 5; a SREJ would call for the retransmission of only frame 3. SREJ can be used to call for the retransmission of any number of frames except that only one is identified at a time.

Operational modes. HDLC has two operational modes not specified in SDLC: asynchronous response mode and asynchronous disconnect mode.

1. *Asynchronous response mode (ARM).* With the ARM, secondary stations are allowed to send unsolicited responses. To transmit, a secondary does not need to have received a frame from the primary with the P bit set. However, if a secondary receives a frame with the P bit set, it must respond with a frame with the F bit set.

2. *Asynchronous disconnect mode (ADM).* An ADM is identical to the normal disconnect mode except that the secondary can initiate a DM or RIM response at any time.

ISO 7809–1985(E).   This standard combines previous standards 6159(E) (unbalanced) and 6256(E) (balanced) and outlines the class of operation necessary to establish the link-level protocol.

Unbalanced operation. This class of operation is logically equivalent to a multipoint private line circuit with a polling environment. There is a single primary station responsible for central control of the network. Data transmission may be either half- or full-duplex.

Balanced operation. This class of operation is logically equivalent to a two-point private line circuit. Each station has equal data link responsibilities, and channel access is through contention using the asynchronous response mode. Data transmission may be half- or full-duplex.

PUBLIC DATA NETWORK

A *public data network* (PDN) is a switched data communications network similar to the public telephone network except that a PDN is designed for transferring data only. Public data networks combine the concepts of both *value-added networks* (VANs) and *packet-switching networks*.

Value-Added Network

A value-added network "*adds value*" to the services or facilities provided by a common carrier to provide new types of communication services. Examples of added values are error control, enhanced connection reliability, dynamic routing, failure protection, logical multiplexing, and data format conversions. A VAN comprises an organization that leases communications lines from common carriers such as AT&T and MCI and adds new types of communications services to those lines. Examples of value-added networks are GTE Telnet, DATAPAC, TRANSPAC, and Tymnet Inc.

Packet-Switching Network

Packet switching involves dividing data messages into small bundles of information and transmitting them through communications networks to their intended destinations using computer-controlled switches. Three common switching techniques are used with public data networks: *circuit switching*, *message switching*, and *packet switching*.

Circuit switching. Circuit switching is used for making a standard telephone call on the public telephone network. The call is established, information is transferred, and then the call is disconnected. The time required to establish the call is called the *setup* time. Once the call has been established, the circuits interconnected by the network switches are allocated to a single user for the duration of the call. After a call has been established, information is transferred in *real time*. When a call is terminated, the circuits and switches are once again available for another user. Because there are a limited number of circuits and switching paths available, *blocking* can occur. Blocking is when a call cannot be completed because there are no facilities or switching paths available between the source and destination locations. When circuit switching is used for data transfer, the terminal equipment at the source and destination must be compatible; they must use compatible modems and the same bit rate, character set, and protocol.

A circuit switch is a *transparent* switch. The switch is transparent to the data; it does nothing more than interconnect the source and destination terminal equipment. A circuit switch adds no value to the circuit.

Message switching.   Message switching is a form of *store-and-forward* network. Data, including source and destination identification codes, are transmitted into the network and stored in a switch. Each switch within the network has message storage capabilities. The network transfers the data from switch to switch when it is convenient to do so. Consequently, data are not transferred in real time; there can be a delay at each switch. With message switching, blocking cannot occur. However, the delay time from message transmission to reception varies from call to call and can be quite long (possibly as long as 24 hours). With message switching, once the information has entered the network, it is converted to a more suitable format for transmission through the network. At the receive end, the data are converted to a format compatible with the receiving data terminal equipment. Therefore, with message switching, the source and destination data terminal equipment do not need to be compatible. Message switching is more efficient than circuit switching because data that enter the network during busy times can be held and transmitted later when the load has decreased.

A message switch is a *transactional* switch because it does more than simply transfer the data from the source to the destination. A message switch can store data or change its format and bit rate, then convert the data back to their original form or an entirely different form at the receive end. Message switching multiplexes data from different sources onto a common facility.

Packet switching.   With packet switching, data are divided into smaller segments called *packets* prior to transmission through the network. Because a packet can be held in memory at a switch for a short period of time, packet switching is sometimes called a *hold-and-forward* network. With packet switching, a message is divided into packets and each packet can take a different path through the network. Consequently, all packets do not necessarily arrive at the receive end at the same time or in the same order in which they were transmitted. Because packets are small, the hold time is generally quite short and message transfer is near real time and blocking cannot occur. However, packet-switching networks require complex and expensive switching arrangements and complicated protocols. A packet switch is also a transactional switch. Circuit, message, and packet switching techniques are summarized in Table 15-3.

CCITT X.1 International User Class of Service

The CCITT X.1 standard divides the various classes of service into three basic modes of transmission for a public data network. The three modes are: *start/stop*, *synchronous*, and *packet*.

TABLE 15-3   SWITCHING TECHNIQUE SUMMARY

Circuit switching	Message switching	Packet switching
Dedicated transmission path	No dedicated transmission path	No dedicated transmission path
Continuous transmission of data	Transmission of messages	Transmission of packets
Operates in real time	Not real time	Near real time
Messages not stored	Messages stored	Messages held for short time
Path established for entire message	Route established for each message	Route established for each packet
Call setup delay	Message transmission delay	Packet transmission delay
Busy signal if called party busy	No busy signal	No busy signal
Blocking may occur	Blocking cannot occur	Blocking cannot occur
User responsible for message-loss protection	Network responsible for lost messages	Network may be responsible for each packet but not for entire message
No speed or code conversion	Speed and code conversion	Speed and code conversion
Fixed bandwidth transmission (i.e., fixed information capacity)	Dynamic use of bandwidth	Dynamic use of bandwidth
No overhead bits after initial setup delay	Overhead bits in each message	Overhead bits in each packet

Start/stop mode.   With the start/stop mode, data are transferred from the source to the network and from the network to the destination in an asynchronous data format (i.e., each character is framed within a start and stop bit). Call control signaling is done in International Alphabet No. 5 (ASCII-77). Two common protocols used for start/stop transmission are IBM's 83B protocol and AT&T's 8A1/B1 selective calling arrangement.

Synchronous mode.   With the synchronous mode, data are transferred from the source to the network and from the network to the destination in a synchronous data format (i.e., each message is preceded by a unique synchronizing character). Call control signaling is identical to that used with private line data circuits and common protocols used for synchronous transmission are IBM's 3270 bisync, Burrough's BASIC, and UNIVAC's UNISCOPE.

Packet mode.   With the packet mode, data are transferred from the source to the network and from the network to the destination in a frame format. The ISO HDLC frame format is the standard data link protocol used with the packet mode. Within the network, data are divided into smaller packets and transferred in accordance with the CCITT X.25 user to network interface protocol.

Figure 15-4 illustrates a typical layout for a public data network showing each of the three modes of operation. The packet assembler/disassembler (PAD) interfaces user data to X.25 format when the user's data are in either the asynchronous or synchronous mode of operation. A PAD is unnecessary when the user is operating in the packet

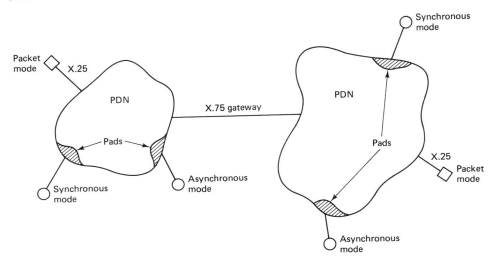

FIGURE 15-4   **Public data network.**

mode. X.75 is recommended by the CCITT for the *gateway* protocol. A gateway is used to interface two public data networks.

ISO PROTOCOL HIERARCHY

The ISO international protocol hierarchy is shown in Figure 15-5. This hierarchy was developed to facilitate the intercommunications of data processing equipment by separating network responsibilities into seven *levels* or *layers*. The basic concept of layering responsibilities is that each layer adds value to services provided by the sets of lower layers. In this way, the highest level is offered the full set of services needed to run a distributed data application.

From Figure 15-5 it can be seen that each level of the hierarchy adds overhead to the data. In fact, if all seven levels are addressed, less than 15% of the transmitted message is source information. The basic services provided by each layer of the hierarchy are summarized below.

1. *Physical layer.* The physical layer is the lowest level of the hierarchy and specifies the physical, electrical, functional, and procedural standards for accessing the data communications network. The specifications outlined by the physical layer are similar to those specified by the EIA RS-232C interface standard.

2. *Data link layer.* The data link layer is responsible for communications between primary and secondary nodes within the network. The data link layer provides a means to activate, maintain, and deactivate the data link. The data link layer provides the final framing of the information envelope, facilitates the orderly flow of data between nodes, and allows for error detection and correction.

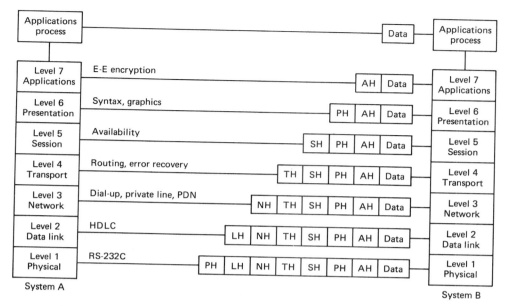

FIGURE 15-5 ISO international protocol hierarchy. AH, Applications header; PH, presentation header; SH, session header; TH, transport header; NH, network header; LH, link header; PH, physical header.

3. *Network layer.* The network layer determines which network configuration (dial-up, leased, or packet) is most appropriate for the function provided by the network.

4. *Transport layer.* The transport layer controls the end-to-end integrity of the message, which includes message routing, segmenting, and error recovery. The transport layer acts as the interface between the network and session layers.

5. *Session layer.* The session layer is responsible for network availability (i.e., buffer storage and processor capacity). Session responsibilities includes network log-on and log-off procedures and user authentication. A session is the temporary condition that exists when data are actually in the process of being transferred and does not include procedures such as call establishment, setup, or disconnect procedures. The session layer determines the type of dialogue available (i.e., simplex, half-duplex, or full-duplex).

6. *Presentation layer.* The presentation layer is concerned with syntax or representation. Presentation functions include data formatting, encoding, the encryption/decryption of messages, dialogue procedures, synchronization, interruption, and termination. The presentation layer performs code and character set translation and determines the display mechanism for messages.

7. *Application layer.* The application layer is analogous to the general manager of the network. The application layer controls the sequence of activities within an applica-

tion and also the sequence of events between the computer application and the user of another application. The application layer communicates directly with the user's application program.

CCITT X.25 USER-TO-NETWORK INTERFACE PROTOCOL

In 1976, the CCITT designated the X.25 user interface as the international standard for packet network access. Keep in mind that X.25 is strictly a *user-to-network* interface and addresses only the physical, data link, and network layers in the ISO seven layer model. X.25 uses existing standards whenever possible. For example, X.25 specifies X.21, X.26, and X.27 standards as the physical interface, which correspond to EIA RS-232C, RS-423A, and RS-422A standards, respectively. X.25 defines HDLC as the international standard for the data link layer and the American National Standards Institute (ANSI) 3.66 *advanced data communications control procedures* (ADCCP) as the U.S. standard. ANSI 3.66 and ISO HDLC specify exactly the same set of data link control procedures. However, ANSI 3.66 and HDLC were designed for private line data circuits with a polling environment. Consequently, the addressing and control procedures outlined by them are not appropriate for packet data networks. ANSI 3.66 and HDLC were selected for the data link layer because of their frame format, delimiting sequence, transparency mechanism, and error detection method.

The network layer of X.25 specifies three switching services offered in a switched data network: permanent virtual circuit, virtual call, and datagram.

Permanent Virtual Circuit

A *permanent virtual circuit* (PVC) is logically equivalent to a two-point dedicated private line circuit except slower. A PVC is slower because a hardwired connection is not provided. Each time a connection is requested, the appropriate switches and circuits must be established through the network to provide the interconnection. A PVC establishes a connection between two predetermined subscribers of the network on demand, but not permanently. With a PVC, a source and destination address are unnecessary because the two users are fixed.

Virtual Call

A *virtual call* (VC) is logically equivalent to making a telephone call through the DDD network. A VC is a one-to-many arrangement. Any VC subscriber can access any other VC subscriber through a network of switches and communication channels. Virtual calls are temporary connections that use common usage equipment and circuits. The source must provide its address and the address of the destination before a VC can be completed.

Datagram

A *datagram* (DG) is, at best, vaguely defined by X.25 and, until it is completely outlined, has very limited usefulness. With a DG, users send small packets of data into the network. The network does not acknowledge packets nor does it guarantee successful transmission. However, if a message will fit into a single packet, a DG is somewhat reliable. This is called a *single-packet-per-segment* protocol.

X.25 Packet Format

A virtual call is the most efficient service offered for a packet network. There are two packet formats used with virtual calls: a call request packet and a data transfer packet.

Call request packet. Figure 15-6 shows the field format for a call request packet. The delimiting sequence is 01111110 (an HDLC flag), and the error detection/correction mechanism is CRC-16 with ARQ. The link address field and the control field have little use and are therefore seldom used with packet networks. The rest of the fields are defined in sequence.

Format identifier. The format identifier identifies whether the packet is a new call request or a previously established call. The format identifier also identifies the packet numbering sequence (either 0–7 or 0–127).

Logical channel identifier (LCI). The LCI is a 12-bit binary number that identifies the source and destination users for a given virtual call. After a source user has gained access to the network and has identified the destination user, they are assigned an LCI. In subsequent packets, the source and destination addresses are unnecessary; only the LCI is needed. When two users disconnect, the LCI is relinquished and can be reassigned two new users. There are 4096 LCIs available. Therefore, there may be as many as 4096 virtual calls established at any given time.

Packet type. This field is used to identify the function and the content of the packet (i.e., new request, call clear, call reset, etc.).

Calling address length. This 4-bit field gives the number of digits (in binary) that appear in the calling address field. With 4 bits, up to 15 digits can be specified.

Called address length. This field is the same as the calling address field except that it identifies the number of digits that appear in the called address field.

Called address. This field contains the destination address. Up to 15 BCD digits can be assigned to a destination user.

Calling address. This field is the same as the called address field except that it contains up to 15 BCD digits that can be assigned to a source user.

Flag	Link address field	Link control field	Format identifier	Logical channel identifier	Packet type	Calling address length	Called address length	Called address	Calling address	0	Facilities field length	Facilities field	Protocol ID	User data	Frame check sequence	Flag
8	8	8	4	12	8	4	4	To 60	To 60	2	6	To 512	32	To 96	16	8

Bits:

FIGURE 15-6   Call request packet format.

Facilities length field. This field identifies (in binary) the number of 8-bit octets present in the facilities field.

Facilities field. This field contains up to 512 bits of optional network facility information, such as reverse billing information, closed user groups, and whether it is a simplex transmit or simplex receive connection.

Protocol identifier. This 32-bit field is reserved for the subscriber to insert user-level protocol functions such as log-on procedures and user identification practices.

User data field. Up to 96 bits of user data can be transmitted with a call request packet. These are unnumbered data which are not confirmed. This field is generally used for user passwords.

Data transfer packet. Figure 15-7 shows the field format for a data transfer packet. A data transfer packet is similar to a call request packet except that a data transfer packet has considerably less overhead and can accommodate a much larger user data field. The data transfer packet contains a send and receive packet sequence field that were not included with the call request format.

The flag, link address, link control, format identifier, LCI, and FCS fields are identical to those used with the call request packet. The send and receive packet sequence fields are described as follows.

Send packet sequence field. This field is used in the same manner that the ns and nr sequences are used with SDLC and HDLC. P(s) is analogous to ns, and P(r) is analogous to nr. Each successive data transfer packet is assigned the next P(s) number in sequence. The P(s) can be a 3- or a 7-bit binary number and thus number packets from either 0–7 or 0–127. The numbering sequence is identified in the format identifier. The send packet field always contains 8 bits and the unused bits are reset.

Receive packet sequence field. P(r) is used to confirm received packets and call for retransmission of packets received in error (ARQ). The I field in a data transfer packet can have considerably more source information than an I field in a call request packet.

Flag	Link address field	Link control field	Format identifier	Logical channel identifier	Send packet sequence number P(s)	0	Receive packet sequence number P(r)	0	User data	Frame check sequence	Flag
Bits: 8	8	8	4	12	3/7	5/1	3/7	5/1	To 1024	16	8

FIGURE 15-7 Data transfer packet format.

LOCAL AREA NETWORKS

A *local area network* (LAN) is a data communications network that is designed to provide two-way communications between a large variety of data communications terminal equipment within a relatively small geographic area. LANs are privately owned and operated and are used to interconnect data terminal equipment in the same building or building complex.

Local Area Network System Considerations

Topology.   The topology or physical architecture of a LAN identifies how the stations are interconnected. The most common configurations used with LANs are the star, bus, ring, and mesh topologies.

Connecting medium.   Presently, all LANs use coaxial cable as the transmission medium, although in the near future it is likely that fiber optic cables will also be used. Fiber cables can operate at higher bit rates and, consequently, have a larger capacity to transfer information than coaxial cables. LANs that use a coaxial cable are limited to an overall length of approximately 1500 m. Fiber links are expected to far exceed this distance.

Transmission format.   There are two basic approaches to transmission format for LANs: *baseband* and *broadband*. Baseband transmission uses the connecting medium as a single-channel device. Only one station can transmit at a time and all stations must transmit and receive the same types of signals (encoding schemes and bit rates). Essentially, a baseband format time division multiplexes signals onto the transmission medium. Broadband transmission uses the connecting medium as a multichannel device. Each channel occupies a different frequency band (i.e., frequency-division multiplexing). Consequently, each channel can contain different encoding schemes and operate at different bit rates. A broadband network permits voice, digital data, and video to be transmitted simultaneously over the same transmission medium. However, broadband systems require RF modems, amplifiers, and more complicated transceivers than baseband systems. For this reason, baseband systems are more prevalent. Table 15-4 summarizes baseband and broadband transmission techniques.

Channel Accessing

Channel accessing describes the mechanism used by a station to gain access to a local area network. There are essentially two methods used for channel accessing with LANs: carrier sense, multiple access with collision detection (CSMA/CD) and token passing.

Carrier sense, multiple access with collision detection.   With CSMA/CD, a station monitors (listens to) the line to determine if the line is busy. If a station has a message to transmit but the line is busy, it waits for an idle condition before it transmits its message. If two stations begin transmitting at the same time, a *collision* occurs. When this happens, both stations cease transmitting (*back off*) and each station waits a random period of time before attempting a retransmission. The random delay time for each station is different and therefore allows for prioritizing the stations on the network.

TABLE 15-4   TRANSMISSION FORMAT SUMMARY

Baseband	Broadband
Characteristics	
Digital signaling	Analog signaling (requires RF modem)
Entire bandwidth used by signal	FDM possible (i.e., multiple data channels)
Bidirectional	Unidirectional
Bus topology	Bus topology
Maximum length approximately 1500 meters	Maximum length up to tens of kilometers
Advantages	
Less expensive	High capacity
Simpler technology	Multiple traffic types
Easy and quick to install	More flexible circuit configurations
	Larger area covered
Disadvantages	
Single channel	Modem required
Limited capacity	Complex installation and maintenance
Grounding problems	Double propagation delay
Limited distance	

With CSMA/CD, stations must contend for the network. A station is not guaranteed access to the network. To detect the occurrence of a collision, a station must be capable of transmitting and receiving simultaneously. CSMA/CD is used by most baseband LANs in the bus configuration. Ethernet is a popular local area network that uses baseband transmission with CSMA/CD. The transmission rate with Ethernet is 10 Mbps over a coaxial cable. Collision detection is accomplished by monitoring the line for phase violations in a Manchester-encoded (biphase) digital encoding scheme.

Token passing.   Token passing is a channel-accessing arrangement that is best suited for a ring topology with either a baseband or a broadband network. With token passing, an electrical *token* (*code*) is circulated around the ring from station to station. Each station, in turn, acquires the token. In order to transmit, a station must first possess the token; then the station removes the token and places its message on the line. After a station transmits, it passes the token on to the next sequential station. With token passing, each station has equal access to the transmission medium. The Cambridge ring is a popular local area network that uses baseband transmission with token passing. The transmission rate with a Cambridge ring is 10 Mbps. Table 15-5 lists several local area networks and some of their characteristics.

TABLE 15-5   LOCAL AREA NETWORK SUMMARY

Ethernet:	Developed by Xerox Corporation in conjunction with Digital Equipment Corporation and Intel Corporation; baseband system using CSMA/CD; 10 Mbps
Wangnet:	Developed by Wang Computer Corporation; broadband system using CSMA/CD
Localnet:	Developed by Sytek Corporation; broadband system using CSMA/CD
Domain:	Developed by Apollo Computer Corporation; broadband network using token passing
Cambridge ring:	Developed by the University of Cambridge; baseband system using CSMA/CD; 10 Mbps

In 1980, the IEEE local area network committee was established to standardize the means of connecting digital computer equipment and peripherals with the local area network environment. In 1983, the committee established IEEE standards 802.3 (CSMA/CD) and 802.4 (token passing) for a bus topology. IEEE standard 802.5, which is still pending approval, addresses token passing with a ring topology.

QUESTIONS

15-1. Define *data communications protocol*.

15-2. What is a master station? A slave station?

15-3. Define *polling* and *selecting*.

15-4. What is the difference between a synchronous and an asynchronous protocol?

15-5. What is the difference between a character-oriented protocol and a bit-oriented protocol?

15-6. Define the three operating modes used with data communications circuits.

15-7. What is the function of the clearing character?

15-8. What is a unique address? A group address? A broadcast address?

15-9. What does a negative acknowledgment to a poll indicate?

15-10. What is the purpose of a heading?

15-11. Why is IBM's 3270 synchronous protocol called bisync?

15-12. Why are SYN characters always transmitted in pairs?

15-13. What is an SPA? An SSA? A DA?

15-14. What is the purpose of a leading pad? A trailing pad?

15-15. What is the difference between a general poll and a specific poll?

15-16. What is a handshake?

15-17. (Primary, secondary) stations transmit polls.

15-18. What does a negative acknowledgment to a poll indicate?

15-19. What is a positive acknowledgment to a poll?

15-20. What is the difference between ETX, ETB, and ITB?

15-21. What character is used to terminate a heading and begin a block of text?

15-22. What is transparency? When is it necessary? Why?

15-23. What is the difference between a command and a response with SDLC?

15-24. What are the three transmission states used with SDLC? Explain them.

15-25. What are the five fields used with an SDLC frame? Briefly explain each.

15-26. What is the delimiting sequence used with SDLC?

15-27. What is the null address in SDLC? When is it used?

15-28. What are the three frame formats used with SDLC? Explain what each format is used for.

15-29. How is an information frame identified in SDLC? A supervisory frame? An unnumbered frame?

15-30. What are the purposes of the nr and ns sequences in SDLC?

15-31. With SDLC, when is the P bit set? The F bit?

15-32. What is the maximum number of unconfirmed frames that can be outstanding at any one time with SDLC? Why?

15-33. With SDLC, which frame formats can have an information field?

15-34. With SLDC, which frame formats can be used to confirm previously received frames?

15-35. What command/response is used for reporting procedual errors with SDLC?

15-36. Explain the three modes in SDLC that a secondary station can be in.

15-37. When is the configure command/response used with SDLC?

15-38. What is a go-ahead sequence? A turnaround sequence?

15-39. What is the transparency mechanism used with SDLC?

15-40. What is a message abort? When is it transmitted?

15-41. Explain invert-on-zero encoding. Why is it used?

15-42. What supervisory condition exists with HDLC that is not included with SDLC?

15-43. What is the delimiting sequence used with HDLC? The transparency mechanism?

15-44. Explain extended addressing as it is used with HDLC.

15-45. What is the difference between the basic control format and the extended control format with HDLC?

15-46. What is the difference in the information fields used with SDLC and HDLC?

15-47. What operational modes are included with HDLC that are not included with SDLC?

15-48. What is a public data network?

15-49. Describe a value-added network.

15-50. Explain the differences in circuit-, message-, and packet-switching techniques.

15-51. What is blocking? With which switching techniques is blocking possible?

15-52. What is a transparent switch? A transactional switch?

15-53. What is a packet?

15-54. What is the difference between a store-and-forward and a hold-and-forward network?

15-55. Explain the three modes of transmission for public data networks.

15-56. What is the user-to-network protocol designated by CCITT?

15-57. What is the user-to-network protocol designated by ANSI?

15-58. Which layers of the ISO protocol hierarchy are addressed by X.25?

15-59. Explain the following terms: permanent virtual circuit, virtual call, and datagram.

15-60. Why was HDLC selected as the link-level protocol for X.25?

15-61. Briefly explain the fields that make up an X.25 call request packet.

15-62. Describe a local area network.

15-63. What is the connecting medium used with local area networks?

15-64. Explain the two transmission formats used with local area networks.

15-65. Explain CSMA/CD.

15-66. Explain token passing.

PROBLEMS

15-1. Determine the hex code for the control field in an SDLC frame for the following conditions: information frame, poll, transmitting frame 4, and confirming reception of frames 2, 3, and 4.

15-2 Determine the hex code for the control field in an SDLC frame for the following conditions: supervisory frame, ready to receive, final, confirming reception of frames 6, 7, and 0.

15-3. Insert 0's into the following SDLC data stream.

11100100001111111100111110100111101011111111100101

15-4. Delete 0's from the following SDLC data stream.

01011111010001101111011101101011110101110001111100

15-5. Sketch the NRZI levels for the following data stream (start with a high condition).

1 0 0 1 1 1 0 0 1 0 1 0

Chapter 16

DIGITAL
TRANSMISSION

INTRODUCTION

As stated previously, digital transmission is the transmittal of digital pulses between two points in a communications system. The original source information may already be in digital form or it may be analog signals that must be converted to digital pulses prior to transmission and converted back to analog form at the receive end. With digital transmission systems, a physical facility such as a metallic wire pair, a coaxial cable, or a fiber optic link is required to interconnect the two points in the system. The pulses are contained in and propagate down the facility.

Advantages of Digital Transmission

1. The primary advantage of digital transmission is noise immunity. Analog signals are more susceptible than digital pulses to undesired amplitude, frequency, and phase variations. This is because with digital transmission, it is not necessary to evaluate these parameters as precisely as with analog transmission. Instead, the received pulses are evaluated during a sample interval, and a simple determination is made whether the pulse is above or below a certain threshold.

2. Digital pulses are better suited to processing and multiplexing than analog signals. Digital pulses can be stored easily, whereas analog signals cannot. Also, the transmission rate of a digital system can easily be changed to adapt to different environments and to interface with different types of equipment. Multiplexing is explained in detail in Chapter 17.

Disadvantages of Digital Transmission

1. The transmission of digitally encoded analog signals requires more bandwidth than simply transmitting the analog signal.
2. Analog signals must be converted to digital codes prior to transmission and converted back to analog at the receiver.
3. Digital transmission requires precise time synchronization between transmitter and receiver clocks.
4. Digital transmission systems are incompatible with existing analog facilities.

PULSE MODULATION

Pulse modulation includes many different methods of transferring pulses from a source to a destination. The four predominant methods are *pulse width modulation* (PWM), *pulse position modulation* (PPM), *pulse amplitude modulation* (PAM), and *pulse code modulation* (PCM). The four most common methods of pulse modulation are summarized below and shown in Figure 16-1.

1. *PWM*. This method is sometimes called pulse duration modulation (PDM) or pulse length modulation (PLM). The pulse width (active portion of the duty cycle) is proportional to the amplitude of the analog signal.
2. *PPM*. The position of a constant-width pulse within a prescribed time slot is varied according to the amplitude of the analog signal.
3. *PAM*. The amplitude of a constant-width, constant-position pulse is varied according to the amplitude of the analog signal.

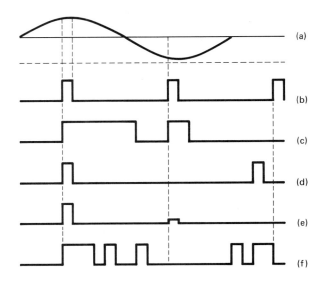

(a)

(b)

(c)

(d)

(e)

(f)

FIGURE 16-1   Pulse modulation: (a) analog signal; (b) sample pulse; (c) PWM; (d) PPM; (e) PAM; (f) PCM.

4. *PCM*. The analog signal is sampled and converted to a fixed-length, serial binary number for transmission. The binary number varies according to the amplitude of the analog signal.

PAM is used as an intermediate form of modulation with PSK, QAM, and PCM, although it is seldom used by itself. PWM and PPM are used in special-purpose communications systems (usually for the military) but are seldom used for commercial systems. PCM is by far the most prevalent method of pulse transmission and consequently, will be the topic of discussion for the remainder of this chapter.

PULSE CODE MODULATION

Pulse code modulation (PCM) is the only one of the pulse modulation techniques previously mentioned that is a digital transmission system. With PCM, the pulses are of fixed length and fixed amplitude. PCM is a binary system; a pulse or lack of a pulse within a prescribed time slot represents either a logic 1 or a logic 0 condition. With PWM, PPM, or PAM; a single pulse does not represent a single binary digit (bit).

Figure 16-2 shows a simplified block diagram of a single-channel, *simplex (one-way-only)* PCM system. The bandpass filter limits the input analog signal to the standard voice band frequency range 300 to 3000 Hz. The *sample-and-hold* circuit periodically samples the analog input and converts those samples to a multilevel PAM signal. The *analog-to-digital converter* (ADC) converts the PAM samples to a serial binary data stream for transmission. The transmission medium is generally a metallic wire pair.

At the receive end, the *digital-to-analog converter* (DAC) converts the serial binary data stream to a multilevel PAM signal. The sample-and-hold circuit and low-pass filter convert the PAM signal back to its original analog form. An integrated circuit

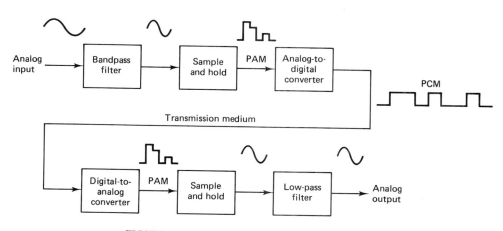

FIGURE 16-2 Simplified PCM system block diagram.

that performs the PCM encoding and decoding is called a *codec* (*code*r/*deco*der). The codec is explained in detail in Chapter 17.

Sample-and-Hold Circuit

The purpose of the sample-and-hold circuit is to sample periodically the continually changing analog input signal and convert the sample to a series of constant-amplitude PAM levels. For the ADC to accurately convert a signal to a digital code, the signal must be relatively constant. If not, before the ADC can complete the conversion, the input would change. Therefore, the ADC would continually be attempting to follow the analog changes and never stabilize on any PCM code.

Figure 16-3 shows the schematic diagram of a sample-and-hold circuit. The FET acts like a simple switch. When turned "on," it provides a low-impedance path to deposit the analog sample across capacitor C1. The time that Q1 is "on" is called the *aperture* or *acquisition time*. Essentially, C1 is the hold circuit. When Q1 is "off," the capacitor does not have a complete path to discharge through and therefore stores the sampled voltage. The *storage time* of the capacitor is also called the A/D *conversion time* because it is during this time that the ADC converts the sample voltage to a digital code. The acquisition time should be very short. This assures that a minimum change occurs in the analog signal while it is being deposited across C1. If the input to the ADC is changing while it is performing the conversion, distortion results. This distortion is called *aperture distortion*. Thus, by having a short aperture time and keeping the input to the ADC relatively constant, the sample-and-hold circuit reduces aperture distortion. If the analog signal is sampled for a short period of time and the sample voltage is held at a constant amplitude during the A/D conversion time, this is called *flat-top sampling*. If the sample time is made longer and the analog-to-digital conversion takes place with a changing analog signal, this is called *natural sampling*. Natural sampling introduces more aperture distortion than flat-top sampling and requires a faster A/D converter.

Figure 16-4 shows the input analog signal, the sampling pulse, and the waveform developed across C1. It is important that the output impedance of voltage follower Z1

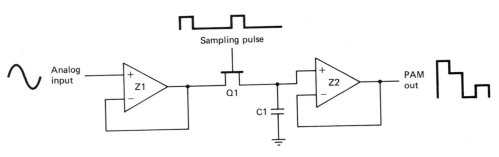

FIGURE 16-3   Sample-and-hold circuit.

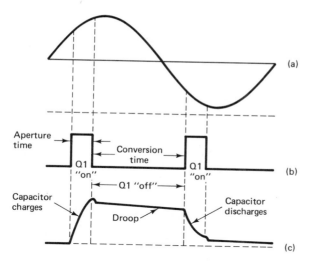

Aperture time

Conversion time

Q1 "on"

Q1 "on"

Q1 "off"

Capacitor charges

Droop

Capacitor discharges

(a)

(b)

(c)

FIGURE 16-4   Sample-and-hold wave-forms: (a) analog in; (b) sample pulse; (c) capacitor voltage.

and the "on" resistance of Q1 be as small as possible. This assures that the *RC* charging time constant of the capacitor is kept very short, allowing the capacitor to charge or discharge rapidly during the short acquisition time. The rapid drop in the capacitor voltage immediately following each sample pulse is due to the redistribution of the charge across C1. The interelectrode capacitance between the gate and drain of the FET is placed in series with C1 when the FET is "off," thus acting like a capacitive voltage-divider network. Also, note the gradual discharge across the capacitor during the conversion time. This is called *droop* and is caused by the capacitor discharging through its own leakage resistance and the input impedance of voltage follower Z2. Therefore, it is important that the input impedance of Z2 and the leakage resistance of C1 be as high as possible. Essentially, voltage followers Z1 and Z2 isolate the sample-and-hold circuit (Q1 and C1) from the input and output circuitry.

EXAMPLE 16-1

For the sample-and-hold circuit shown in Figure 16-3, determine the largest-value capacitor that can be used. Use an output impedance for Z1 of 10 Ω, an "on" resistance for Q1 of 10 Ω, an acquisition time of 10 μs, a maximum peak-to-peak input voltage of 10 V, a maximum output current from Z1 of 10 mA, and an accuracy of 1%.

Solution   The expression for the current through a capacitor is

$$i = C\frac{dv}{dt}$$

Rearranging and solving for *C* yields

$$C = i\frac{dt}{dv}$$

where

C = maximum capacitance
i = maximum output current from Z1, 10 mA
dv = maximum change in voltage across C1, which equals 10 V
dt = charge time, which equals the aperture time, 10 µs

Therefore,

$$C_{max} = \frac{(10 \text{ mA}) (10 \text{ µs})}{10 \text{ V}} = 10 \text{ nF}$$

The charge time constant for C when Q1 is "on" is

$$\tau = RC$$

where

τ = one charge time constant
R = output impedance of Z1 plus the "on" resistance of Q1
C = capacitance value of C1

Rearranging and solving for C gives us

$$C_{max} = \frac{\tau}{R}$$

The charge time of capacitor C1 is also dependent on the accuracy desired from the device. The percent accuracy and its required RC time constant are summarized as follows:

Accuracy (%)	Charge time
10	3τ
1	4τ
0.1	7τ
0.01	9τ

For an accuracy of 1%,

$$C = \frac{10 \text{ µs}}{4 (20)} = 125 \text{ nF}$$

To satisfy the output current limitations of Z1, a maximum capacitance of 10 nF was required. To satisfy the accuracy requirements, 125 nF was required. To satisfy both requirements, the smaller-value capacitor must be used. Therefore, C1 can be no larger than 10 nF.

Sampling Rate

The Nyquist sampling theorem establishes the *minimum sampling rate* (F_s) that can be used for a given PCM system. For a sample to be reproduced accurately at the receiver, each cycle of the analog input signal (F_a) must be sampled at least twice. Consequently, the minimum sampling rate is equal to twice the highest audio input frequency. If F_s is less than two times F_a, distortion will result. This distortion is called *aliasing* or *foldover distortion*. Mathematically, the minimum Nyquist sample rate is

$$F_s \geq 2F_a \qquad\qquad (16\text{-}1)$$

where

F_s = minimum Nyquist sample rate

F_a = highest frequency to be sampled

Essentially, a sample-and-hold circuit is an AM modulator. The switch is a non-linear device that has two inputs: the sampling pulse and the input analog signal. Consequently, *nonlinear mixing (heterodyning)* occurs between these two signals. Figure 16-5a shows the frequency-domain representation of the output spectrum from a sample-and-hold circuit. The output includes the two original inputs (the audio and the fundamental frequency of the sampling pulse), their sum and difference frequencies ($F_s \pm F_a$), all the harmonics of F_s and F_a ($2F_s$, $2F_a$, $3F_s$, $3F_a$, etc.), and their associated sidebands ($2F_s \pm F_a$, $3F_s \pm F_a$, etc.).

Because the sampling pulse is a repetitive waveform, it is made up of a series of harmonically related sine waves. Each of these sine waves is amplitude modulated by the analog signal and produces sum and difference frequencies symetrical around each of the harmonics of F_s. Each sum and difference frequency generated is separated from its respective center frequency by F_a. As long as F_s is at least twice F_a, none of the side frequencies from one harmonic will spill into the sidebands of another harmonic and aliasing does not occur. Figure 16-5b shows the results when an analog input frequency greater than $F_s/2$ modulates F_s. The side frequencies from one harmonic foldover into the sideband of another harmonic. The frequency that folds over is an alias of the input signal (hence the names ''aliasing'' or ''foldover distortion''). If an alias side frequency from the first harmonic folds over into the input audio spectrum, it cannot be removed through filtering or any other technique.

EXAMPLE 16-2

For a PCM system with a maximum audio input frequency of 4 kHz, determine the minimum sample rate and the alias frequency produced if a 5-kHz audio signal were allowed to enter the sample-and-hold circuit.

Solution Using Nyquist's sampling theorem (Equation 16-1), we have

$$F_s \geq 2F_a \qquad \text{therefore, } F_s \geq 8 \text{ kHz}$$

If a 5-kHz audio frequency entered the sample-and-hold circuit, the output spectrum shown in Figure 16-16 is produced. It can be seen that the 5-kHz signal produces an alias frequency of 3 kHz that folds over into the original audio spectrum.

The input bandpass filter shown in Figure 16-2 is called an *antialiasing* or *antifold-over filter*. Its upper cutoff frequency is chosen such that no frequency greater than

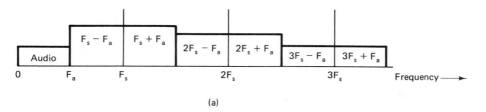

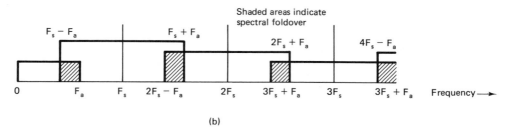

(b)

FIGURE 16-5   Output spectrum for a sample-and-hold circuit: (a) no aliasing; (b) aliasing distortion.

one-half of the sampling rate is allowed to enter the sample-and-hold circuit, thus eliminating the possibility of foldover distortion occurring.

PCM CODES

With PCM, the analog input signal is sampled, then converted to a serial binary code. The binary code is transmitted to the receiver, where it is converted back to the original analog signal. The binary codes used for PCM are n-bit codes, where n may be any positive whole number greater than 1. The codes currently used for PCM are *sign-magnitude codes*, where the *most significant bit* (MSB) is the sign bit and the remaining bits are used for magnitude. Table 16-1 shows an n-bit PCM code where n equals 3. The most significant bit is used to represent the sign of the sample (logic 1 = positive and logic 0 = negative). The two remaining bits represent the magnitude. With 2 magnitude bits, there are four codes possible for positive numbers and four possible for negative numbers. Consequently, there is a total of eight possible codes ($2^3 = 8$).

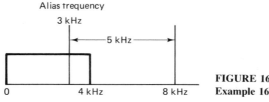

FIGURE 16-6   Output spectrum for Example 16-2.

TABLE 16-1   3-BIT PCM CODE

Sign	Magnitude		Level	Decimal
1	1	1		+3
1	1	0		+2
1	0	1		+1
1	0	0⎫		⎰ +0
0	0	0⎭		⎱ −0
0	0	1		−1
0	1	0		−2
0	1	1		−3

Folded Binary Code

The PCM code shown in Table 16-1 is called a *folded binary code*. Except for the sign bit, the codes on the bottom half of the table are a mirror image of the codes on the top half. (If the negative codes were folded over on top of the positive codes, they would match perfectly.) Also, with folded binary there are two codes assigned to zero volts: 100 (+0) and 000 (−0). For this example, the magnitude of the minimum step size is 1 V. Therefore, the maximum voltage that may be encoded with this scheme is +3 V (111) or −3 V (011). If the magnitude of a sample exceeds the highest quantization interval, *overload distortion* (also called *peak limiting*) occurs. Assigning PCM codes to absolute magnitudes is called *quantizing*. The magnitude of the minimum step size is called *resolution*, which is equal in magnitude to the voltage of the least significant bit (V_{lsb} or the magnitude of the minimum step size of the DAC). The resolution is the minimum voltage other than 0 V that can be decoded by the DAC at the receiver. The smaller the magnitude of the minimum step size, the better (smaller) the resolution and the more accurately the quantization interval will resemble the actual analog sample.

In Table 16-1, each 3-bit code has a range of input voltages that will be converted to that code. For example, any voltage between +0.5 and +1.5 will be converted to the code 101. Any voltage between +1.5 and +2.5 will be encoded as 110. Each code has a *quantization range* equal to + or − one-half the resolution except the codes for +0 V and −0 V. The 0-V codes each have an input range equal to only one-half the resolution, but because there are two 0-V codes, the range for 0 V is also + or − one-half the resolution. Consequently, the maximum input voltage to the system is equal to the voltage of the highest magnitude code plus one-half of the voltage of the least significant bit.

Figure 16-7 shows an analog input signal, the sampling pulse, the corresponding PAM signal, and the PCM code. The analog signal is sampled three times. The first sample occurs at time t_1 when the analog voltage is +2 V. The PCM code that corresponds to sample 1 is 110. Sample 2 occurs at time t_2 when the analog voltage is −1 V. The corresponding PCM code is 001. To determine the PCM code for a particular sample,

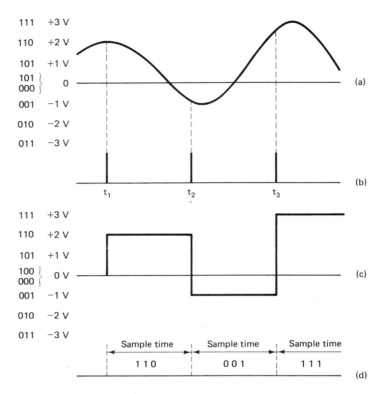

FIGURE 16-7   (a) Analog input signal; (b) sample pulse; (c) PAM signal; (d) PCM code.

simply divide the voltage of the sample by the resolution, convert it to an *n*-bit binary code, and add the sign bit to it. For sample 1, the sign bit is 1, indicating a positive voltage. The magnitude code (10) corresponds to a binary 2. Two times 1 V equals 2 V, the magnitude of the sample.

Sample 3 occurs at time t_3. The voltage at this time is +2.6 V. The folded PCM code for +2.6 V is 2.6/1 = 2.6. There is no code for this magnitude. If successive approximation ADCs are used, the magnitude of the sample is rounded off to the nearest valid code (111 or +3 V for this example). This results in an error when the code is converted back to analog by the DAC at the receive end. This error is called *quantization error* (Qe). The quantization error is equivalent to additive noise (it alters the signal amplitude). Like noise, the quantization error may add to or subtract from the actual signal. Consequently, quantization error is also called *quantization noise* (Qn) and its maximum magnitude is one-half the voltage of the minimum step size ($V_{lsb}/2$). For this example, Qe = 1 V/2 or 0.5 V.

Dynamic Range

The number of PCM bits transmitted per sample is determined by several variables, which include maximum allowable input amplitude, resolution, and dynamic range. *Dynamic range* (DR) is the ratio of the largest possible magnitude to the smallest possible magnitude that can be decoded by the DAC. Mathematically, dynamic range is

$$DR = \frac{V_{max}}{V_{min}}$$

(16-2a)

where V_{min} is equal to the resolution and V_{max} is the maximum voltage that can be decoded by the DACs. Thus

$$DR = \frac{V_{max}}{\text{resolution}}$$

For the system shown in Table 16-1,

$$DR = \frac{3 \text{ V}}{1 \text{ V}} = 3$$

It is common to represent dynamic range in decibels; therefore,

$$DR = 20 \log \frac{V_{max}}{V_{min}} = 20 \log \frac{3}{1} = 9.54 \text{ dB}$$

(16-2b)

A dynamic range of 3 indicates that the ratio of the largest to the smallest decoded receive signal is 3.

If the resolution is halved (to 0.5 V), V_{max} is also reduced by a factor of two (to 1.5 V) and the dynamic range remains constant.

$$DR = \frac{1.5}{0.5} = 3$$

If the resolution is doubled (to 2 V), V_{max} also doubles (to 6 V) and the dynamic range again remains constant.

$$DR = \frac{6}{2} = 3$$

Therefore, dynamic range is independent of resolution for a given PCM code.

The number of bits used for a PCM code depends on the dynamic range. With a 2-bit PCM code, the minimum decodable magnitude has a binary code of 01. The maximum magnitude is 11. The ratio of the maximum binary code to the minimum binary code is 3, the same as the dynamic range. Because the minimum binary code is always 1, DR is simply the maximum binary number for a system. Consequently, to determine the number of bits required for a PCM code the following mathematical relationship is used:

$$2^n - 1 \geq DR$$

and for a minimum value of n,

$$2^n - 1 = DR \qquad (16\text{-}3a)$$

where

$$n = \text{number of PCM bits}$$
$$DR = \text{absolute value of DR}$$

Why $2^n - 1$? One PCM code is used for 0 V, which is not considered for dynamic range. Therefore,

$$2^n = DR + 1 \qquad (16\text{-}3b)$$

To solve for n, convert to logs:

$$\log 2^n = \log (DR + 1) \qquad (16\text{-}3c)$$

$$n \log 2 = \log (DR + 1)$$

$$n = \frac{\log (3 + 1)}{\log 2} = \frac{0.602}{0.301} = 2$$

For a dynamic range of 3, a PCM code with 2 bits is required.

EXAMPLE 16-3

A PCM system has the following parameters: a maximum analog input frequency of 4 kHz, a maximum decoded voltage at the receiver of $\pm 2.55 V_p$, and a minimum dynamic range of 46 dB. Determine the following: minimum sample rate, minimum number of bits used in the PCM code, resolution, and quantization error.

Solution   Substituting into Equation 16-1, the minimum sample rate is

$$F_s = 2F_a = 2 (4 \text{ kHz}) = 8 \text{ kHz}$$

To determine the absolute value of the dynamic range, substitute into Equation 16-2b:

$$46 \text{ dB} = 20 \log \frac{V_{max}}{V_{min}}$$

$$2.3 = \log \frac{V_{max}}{V_{min}}$$

$$10^{2.3} = \frac{V_{max}}{V_{min}} = DR$$

$$199.5 = DR$$

Substitute into Equation 16-4c and solve for n:

$$n = \frac{\log (199.5 + 1)}{\log 2} = 7.63$$

The closest whole number greater than 7.63 is 8; therefore, 8 bits must be used for the magnitude.

Because the input amplitude range is $\pm 2.55 V_p$, one additional bit, the sign bit, is required. Therefore, the total number of PCM bits is 9 and the total number of PCM codes is 2^9 or 512. (There are 255 positive codes, 255 negative codes, and 2 zero codes.)

To determine the actual dynamic range, substitute into Equation 16-3c:

$$DR = 20 \log 255 = 48.13 \text{ dB}$$

To determine the resolution, divide the maximum $+$ or $-$ magnitude by the number of positive or negative nonzero PCM codes.

$$\text{resolution} = \frac{V_{max}}{2^n - 1}$$

$$= \frac{2.55}{2^8 - 1} = \frac{2.55}{256 - 1} = 0.01 \text{ V}$$

The maximum quantization error is

$$Qe = \frac{\text{resolution}}{2} = \frac{0.01}{2} = 0.005 \text{ V}$$

Coding Efficiency

Coding efficiency is a numerical indication of how efficiently a PCM code is utilized. Coding efficiency is the ratio of the minimum number of bits required to achieve a certain dynamic range to the actual number of PCM bits used. Mathematically, coding efficiency is

$$\text{coding efficiency} = \frac{\text{minimum number of bits}}{\text{actual number of bits}} \times 100 \qquad (16\text{-}4)$$

The coding efficiency for Example 16-3 is

$$\text{coding efficiency} = \frac{7.63}{8} \times 100 = 95.38\%$$

Signal-to-Quantization Noise Ratio

The 3-bit PCM coding scheme described in the preceding section is a linear code. That is, the magnitude change between any two successive codes is uniform. Consequently, the magnitude of their quantization error is also equal. The maximum quantization noise is the voltage of the least significant bit divided by 2. Therefore, the worst possible *signal-to-quantization noise ratio* (SQR) occurs when the input signal is at its minimum amplitude (101 or 001). Mathematically, the worst-case SQR is

$$SQR = \frac{\text{minimum voltage}}{\text{quantization noise}} = \frac{V_{lsb}}{V_{lsb}/2} = 2$$

For a maximum amplitude input signal of 3 V (either 111 or 011), the maximum quantization noise is also the voltage of the least significant bit divided by 2. Therefore, the SQR for a maximum input signal condition is

$$SQR = \frac{\text{maximum voltage}}{\text{quantization noise}} = \frac{V_{max}}{V_{lsb}/2} = \frac{3}{0.5} = 6$$

From the preceding example it can be seen that even though the magnitude of error remains constant throughout the entire PCM code, the percentage of error does not; it decreases as the magnitude or the input signal increases. As a result, the SQR is not constant.

The preceding expression for SQR is for voltage and presumes the maximum quantization error and a constant-amplitude analog signal; therefore, it is of little practical use and is shown only for comparison purposes. In reality and as shown in Figure 16-7, the difference between the PAM waveform and the analog input waveform varies in magnitude. Therefore, the signal-to-quantization noise ratio is not constant. Generally, the quantization error or distortion caused by digitizing an analog sample is expressed as an average noise power-to-average signal power ratio. For linear PCM codes (all quantization intervals have equal magnitudes), the signal-to-quantizing noise ratio (also called *signal-to-distortion ratio* or *signal-to-noise ratio*) is determined as follows:

$$SQR \text{ (dB)} = 10 \log \frac{v^2/R}{(q^2/12)/R}$$

where

$$R = \text{resistance}$$
$$v = \text{rms signal voltage}$$

$$q = \text{quantization interval (step size)}$$

$$\frac{v^2}{R} = \text{rms signal power}$$

$$\frac{q^2/12}{R} = \text{average rms quantization noise power}$$

If the resistances are assumed to be equal,

$$SQR \text{ (dB)} = 10 \log \frac{v^2}{q^2/12} \qquad (16\text{-}5a)$$

$$= 10.8 + 20 \log \frac{v}{q} \qquad (16\text{-}5b)$$

Linear versus Nonlinear PCM Codes

Early PCM systems used *linear codes* (i.e., the magnitude change between any two successive steps is uniform). With linear encoding, the accuracy (resolution) for the higher-magnitude analog signals is the same as for the lower-magnitude signals, and the SQR for the lower-magnitude signals is less than for the higher-magnitude signals. With voice transmission, low-amplitude signals are more likely to occur than large-amplitude signals. Therefore, if there were more codes for the lower magnitudes, it would increase the accuracy where the accuracy is needed. As a result, there would be fewer codes available for the higher amplitudes, which would increase the quantization error for the larger-amplitude signals (thus decreasing the SQR). Such a coding technique is called *nonlinear* or *nonuniform encoding*. With nonlinear encoding, the step size increases with the magnitude of the input signal. Figure 16-8 shows the step outputs from a linear and a nonlinear ADC.

Note, with nonlinear encoding, there are more codes at the bottom of the scale than there are at the top, thus increasing the accuracy for the smaller signals. Also note that the distance between successive codes is greater for the higher-amplitude signals, thus increasing the quantization error and reducing the SQR. Also, because the ratio of V_{max} to V_{min} is increased with nonlinear encoding, the dynamic range is larger than with a uniform code. It is evident that nonlinear encoding is a compromise; SQR is sacrificed for the high-amplitude signals to achieve more accuracy for the low-amplitude signals and to achieve a larger dynamic range.

It is difficult to fabricate nonlinear ADCs; consequently, alternative methods of achieving the same results have been devised.

Idle Channel Noise

During times when there is no analog input signal, the only input to the PAM sampler is random, thermal noise. This noise is called *idle channel noise* and is converted to a PAM sample just as if it were a signal. Consequently, even input noise is quantized by the ADC. Figure 16-9 shows a way to reduce idle channel noise by a method called *midtread quantization*. With midtread quantizing, the first quantization interval

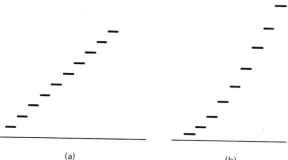

(a) (b)

FIGURE 16-8 (a) Linear versus
(b) nonlinear encoding.

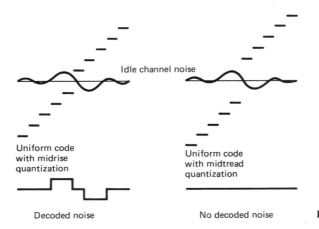

FIGURE 16-9   Idle channel noise.

is made larger in amplitude then the rest of the steps. Consequently, input noise can be quite large and still be quantized as a positive or negative zero code. As a result, the noise is suppressed during the encoding process.

In the PCM codes described thus far, the lowest-magnitude positive and negative codes have the same voltage range as all the other codes (+ or − one-half the resolution). This is called *midrise quantization*. Figure 16-9 contrasts the idle channel noise transmitted with a midrise PCM code to the idle channel noise transmitted when midtread quantization is used. The advantage of midtread quantization is less idle channel noise. The disadvantage is a larger possible magnitude for Qe in the lowest quantization interval.

With a folded binary PCM code, residual noise that fluctuates slightly above and below 0 V is converted to either a + or − zero PCM code and is consequently eliminated. In systems that do not use the two 0-V assignments, the residual noise could cause the PCM encoder to alternate between the zero code and the minimum + or − code. Consequently, the decoder would reproduce the encoded noise. With a folded binary code, most of the residual noise is inherently eliminated by the encoder.

Companding

Companding is the process of *compressing*, then *expanding*. With companded systems, the higher-amplitude analog signals are compressed (amplified less than the lower-amplitude signals) prior to transmission, then expanded (amplified more than the smaller-amplitude signals) at the receiver.

Figure 16-10 illustrates the process of companding. An input signal with a dynamic range of 120 dB is compressed to 60 dB for transmission, then expanded to 120 dB at the receiver. With PCM, companding may be accomplished through analog or digital techniques. Early PCM systems used analog companding, whereas more modern systems use digital companding.

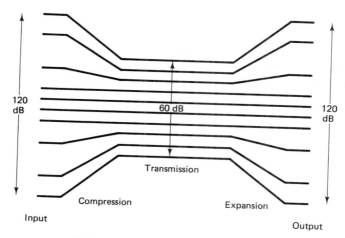

FIGURE 16-10 **Basic companding process.**

Analog Companding

Historically, analog compression was implemented using specially designed diodes inserted in the analog signal path in the PCM transmitter prior to the sample-and-hold circuit. Analog expansion was also implemented with diodes that were placed just after the receive low-pass filter. Figure 16-11 shows the basic process of analog companding. In the transmitter, the analog signal is compressed, sampled, then converted to a linear PCM code. In the receiver, the PCM code is converted to a PAM signal, filtered, then expanded back to its original input amplitude characteristics.

Different signal distributions require different companding characteristics. For instance, voice signals require relatively constant SQR performance over a wide dynamic range, which means that the distortion must be proportional to signal amplitude for

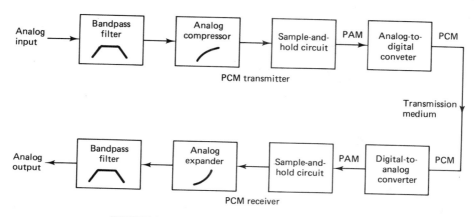

FIGURE 16-11 **PCM system with analog companding.**

any input signal level. This requires a logarithmic compression ratio. A truly logarithmic assignment code requires an infinite dynamic range and an infinite number of PCM codes, which is impossible. There are two methods of analog companding currently being used that closely approximate a logarithmic function and are often called *log-PCM* codes. They are μ-*law* and *A-law companding.*

μ-Law companding.   In the United States and Japan, μ-law companding is used. The compression characteristic for μ-law is

$$V_{out} = \frac{V_{max} \times \ln (1 + \mu V_{in}/V_{max})}{\ln (1 + \mu)} \qquad (16\text{-}6)$$

where

V_{max} = maximum uncompressed analog input amplitude
V_{in} = amplitude of the input signal at a particular instant of time
μ = parameter used to define the amount of compression
V_{out} = compressed output amplitude

Figure 16-12 shows the compression for several values of μ. Note that the higher the μ, the more compression. Also note that for a μ = 0, the curve is linear (no compression).

The parameter μ determines the range of signal power in which the SQR is relatively constant. Voice transmission requires a minimum dynamic range of 40 dB and a 7-bit PCM code. For a relatively constant SQR and a 40-dB dynamic range, μ = 100 or larger is required. The early Bell System digital transmission systems used a 7-bit PCM code with μ = 100. The most recent digital transmission systems use 8-bit PCM codes and μ = 255.

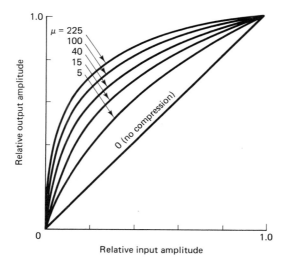

FIGURE 16-12   μ-Law compression characteristics.

EXAMPLE 16-4

For a compressor with $\mu = 255$, determine the gain for the following values of V_{in}: V_{max}, 0.75 V_{max}, 0.5 V_{max}, and 0.25 V_{max}.

Solution Substituting into Equation 16-6, the following gains are achieved for various input magnitudes:

V_{in}	Gain
V_{max}	1
$0.75V_{max}$	1.26
$0.5V_{max}$	1.75
$0.25V_{max}$	3

It can be seen that as the input signal amplitude increases, the gain decreases or is compressed.

A-Law companding. In Europe, the CCITT has established A-law companding to be used to approximate true logarithmic companding. For an intended dynamic range, A-law companding has a slightly flatter SQR than μ-law. A-law companding, however, is inferior to μ-law in terms of small-signal quality (idle channel noise). The compression characteristic for A-law companding is

$$V_{out} = V_{max} \frac{AV_{in}/V_{max}}{1 + \ln A} \qquad 0 \le \frac{V_{in}}{V_{max}} \le \frac{1}{A} \qquad (16\text{-}7a)$$

$$= V_{max} \frac{1 + \ln (AV_{in}/V_{max})}{1 + \ln A} \qquad \frac{1}{A} \le \frac{V_{in}}{V_{max}} \le 1 \qquad (16\text{-}7b)$$

Digital Companding

Digital companding involves compression at the transmit end after the input sample has been converted to a linear PCM code and expansion at the receive end prior to PCM decoding. Figure 16-13 shows the block diagram of a digitally companded PCM system.

With digital companding, the analog signal is first sampled and converted to a linear code, then the linear code is digitally compressed. At the receive end, the compressed PCM code is received, expanded, then decoded. The most recent digitally compressed PCM systems use a 12-bit linear code and an 8-bit compressed code. This companding process closely resembles a $\mu = 255$ analog compression curve by approximating the curve with a set of eight straight line *segments* (segments 0 through 7). The slope of each successive segment is exactly one-half that of the previous segment. Figure 16-14 shows the 12-bit-to-8-bit digital compression curve for positive values only. The curve for negative values is identical except the inverse. Although there are 16 segments (eight positive and eight negative) this scheme is often called *13-segment*

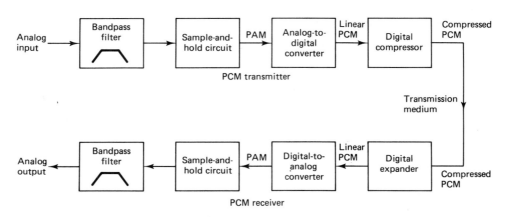

FIGURE 16-13   Digitally companded PCM system.

compression. This is because the curve for segments $+0$, $+1$, -0, and -1 is a straight line with a constant slope and is often considered as one segment.

    The digital companding algorithm for a 12-bit-linear-to-8-bit-compressed code is actually quite simple. The 8-bit compressed code is comprised of a sign bit, a 3-bit

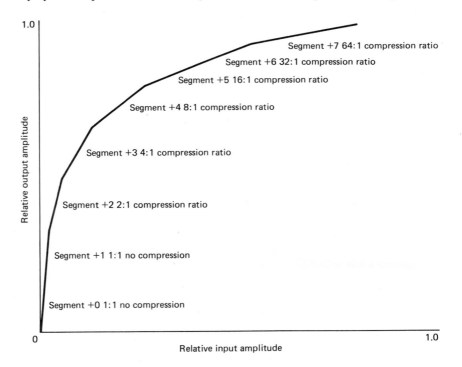

FIGURE 16-14   µ255 compression characteristics (positive values only).

segment identifier, and a 4-bit magnitude code which identifies the *quantization interval* within the specified segment (see Figure 16-15a).

In the μ255 encoding table shown in Figure 16-15b, the bit positions designated with an X are truncated during compression and are consequently lost. Bits designated A, B, C, and D are transmitted as is. The sign bit (s) is also transmitted as is. Note that for segments 0 and 1, the original 12 bits are duplicated exactly at the output of the decoder (Figure 16-15c), whereas for segment 7, only the most significant 6 bits are recovered. With 11 magnitude bits, there are 2048 possible codes. There are 16 codes in segment 0 and in segment 1. In segment 2, there are 32 codes; segment 3 has 64. Each successive segment beginning with segment 3 has twice as many codes as the previous segment. In each of the eight segments, only sixteen 12-bit codes can be recovered. Consequently, in segments 0 and 1, there is no compression (of the 16 possible codes, all 16 can be recovered). In segment 2, there is a compression ratio of 2:1 (32 possible transmit codes and 16 possible recovered codes). In segment 3, there is a 4:1 compression ratio (64 possible transmit codes and 16 possible recovered codes). The compression ratio doubles with each successive segment. The compression ratio in segment 7 is 1024:16 or 64:1.

The compression process is as follows. The analog signal is sampled and converted to a linear 12-bit sign-magnitude code. The sign bit is transferred directly to the 8-bit code. The segment is determined by counting the number of leading 0's in the 11-bit magnitude portion of the code beginning with the MSB. Subtract the number of leading 0's (not to exceed 7) from 7. The result is the segment number, which is converted to

Sign bit 1 = + 0 = −	3-Bit segment identifier	4-Bit quantization interval A B C D
	000 to 111	0000 to 1111

(a)

Segment	12-Bit linear code	8-Bit compressed code	8-Bit compressed code	12-Bit recovered code	Segment
0	s0000000ABCD	s000ABCD	s000ABCD	s0000000ABCD	0
1	s0000001ABCD	s001ABCD	s001ABCD	s0000001ABCD	1
2	s000001ABCDX	s010ABCD	s010ABCD	s000001ABCD1	2
3	s00001ABCDXX	s011ABCD	s011ABCD	s00001ABCD10	3
4	s0001ABCDXXX	s100ABCD	s100ABCD	s0001ABCD100	4
5	s001ABCDXXXX	s101ABCD	s101ABCD	s001ABCD1000	5
6	s01ABCDXXXXX	s110ABCD	s110ABCD	s01ABCD10000	6
7	s1ABCDXXXXXX	s111ABCD	s111ABCD	s1ABCD100000	7

(b)   (c)

FIGURE 16-15   **12-bit to 8-bit digital companding: (a) 8-bit μ255 compressed code format; (b) μ255 encoding table; (c) μ255 decoding table.**

a 3-bit binary number and substituted into the 8-bit code as the segment identifier. The four magnitude bits (A, B, C, and D) are the quantization interval and are substituted into the least significant 4 bits of the 8-bit compressed code.

Essentially, segments 2 through 7 are subdivided into smaller subsegments. Each segment has 16 subsegments, which correspond to the 16 conditions possible for the bits A, B, C, and D (0000–1111). In segment 2 there are two codes per subsegment. In segment 3 there are four. The number of codes per subsegment doubles with each subsequent segment. Consequently, in segment 7, each subsegment has 64 codes. Figure 16-16 shows the breakdown of segments versus subsegments for segments 2, 5, and 7. Note that in each subsegment, all 12-bit codes, once compressed and expanded, yield a single 12-bit code. This is shown in Figure 16-16.

From Figures 16-15 and 16-16, it can be seen that the most significant of the truncated bits is reinserted at the decoder as a 1. The remaining truncated bits are reinserted as 0's. This ensures that the maximum magnitude of error introduced by the compression and expansion process is minimized. Essentially, the decoder guesses what the truncated bits were prior to encoding. The most logical guess is halfway between the minimum- and maximum-magnitude codes. For example, in segment 5, the 5 least significant bits are truncated during compression. At the receiver, the decoder must determine what those bits were. The possibilities are any code between 00000 and 11111. The logical guess is 10000, approximately half the maximum magnitude. Consequently, the maximum compression error is slightly more than one-half the magnitude of that segment.

EXAMPLE 16-5

For a resolution of 0.01 V and analog sample voltages of (a) 0.05 V, (b) 0.32 V, and (c) 10.23 V, determine the 12-bit linear code, the 8-bit compressed code, and the recovered 12-bit code.

Solution   (a) To determine the 12-bit linear code for 0.05 V, simply divide the sample voltage by the resolution and convert the result to a 12-bit sign-magnitude binary number.

12-bit linear code:

$$\frac{0.05 \text{ V}}{0.01 \text{ V}} = 5 = \begin{array}{l} 1\ 0\ 0\ 0\ 0\ 0\ 0\ 0\ 0\ 1\ 0\ 1 \\ \text{s ------magnitude-----} \\ \text{(11-bit binary number)} \end{array}$$

8-bit compressed code:

```
        1   0  0  0  0  0  0  0   0  1  0  1
        s        (? - ? = 0 or 000)    A  B  C  D
        1             0  0  0       0  1  0  1
        ↑
    sign bit         unit            quanti-
     (+)           identifier         zation
                  (segment 0)        interval
```

Segment	12-Bit linear code		12-Bit expanded code	Subsegment
7	s11111111111 .... s11111000000	} 64:1	s11111100000	15
7	s11110111111 .... s11110000000	} 64:1	s11110100000	14
7	s11101111111 .... s11101000000	} 64:1	s11101100000	13
7	s11100111111 .... s11100000000	} 64:1	s11100100000	12
7	s11011111111 .... s11011000000	} 64:1	s11011100000	11
7	s11010111111 .... s11010000000	} 64:1	s11010100000	10
7	s11001111111 .... s11001000000	} 64:1	s11001100000	9
7	s11000111111 .... s11000000000	} 64:1	s11000100000	8
7	s10111111111 .... s10111000000	} 64:1	s10111100000	7
7	s10110111111 .... s10110000000	} 64:1	s10110100000	6
7	s10101111111 .... s10101000000	} 64:1	s10101100000	5
7	s10100111111 .... s10100000000	} 64:1	s10100100000	4
7	s10011111111 .... s10011000000	} 64:1	s10011100000	3
7	s10010111111 .... s10010000000	} 64:1	s10010100000	2
7	s10001111111 .... s10010000000	} 64:1	s10001100000	1
7	s10001111111 .... s10000000000	} 64:1	s10000100000	0

s1ABCD------

(a)

FIGURE 16-16   12-bit segments divided into subsegments: (a) segment 7; (b) segment 5; (c) segment 2.

Segment	12-Bit linear code		12-Bit expanded code	Subsegment
5	s00111111111 .... s00111110000	} 16:1	s00111111000	15
5	s00111101111 .... s00111100000	} 16:1	s00111101000	14
5	s00111011111 .... s00111010000	} 16:1	s00111011000	13
5	s00111001111 .... s00111000000	} 16:1	s00111001000	12
5	s00110111111 .... s00110110000	} 16:1	s00110111000	11
5	s00110101111 .... s00110100000	} 16:1	s00110101000	10
5	s00110011111 .... s00110010000	} 16:1	s00110011000	9
5	s00110001111 .... s00110000000	} 16:1	s00110001000	8
5	s00101111111 .... s00101110000	} 16:1	s00101111000	7
5	s00101101111 .... s00101100000	} 16:1	s00101101000	6
5	s00101011111 .... s00101010000	} 16:1	s00101011000	5
5	s00101001111 .... s00101000000	} 16:1	s00101001000	4
5	s00100111111 .... s00100110000	} 16:1	s00100111000	3
5	s00100101111 .... s00100100000	} 16:1	s00100101000	2
5	s00100011111 .... s00100010000	} 16:1	s00100011000	1
5	s00100001111 .... s00100000000	} 16:1	s00100001000	0
	s001ABCD----			

(b)

FIGURE 16-16 (*continued*)

Segment	12-Bit linear code			12-Bit expanded code	Subsegment
2	s00000111111 .... s00000111110	}	2:1	s00000111111	15
2	s00000111101 .... s00000111100	}	2:1	s00000111101	14
2	s00000111011 .... s00000111010	}	2:1	s00000111011	13
2	s00000111001 .... s00000111000	}	2:1	s00000111001	12
2	s00000110111 .... s00000110110	}	2:1	s00000110111	11
2	s00000110101 .... s00000110100	}	2:1	s00000110101	10
2	s00000110011 .... s00000110010	}	2:1	s00000110011	9
2	s00000110001 .... s00000110000	}	2:1	s00000110001	8
2	s00000101111 .... s00000101110	}	2:1	s00000101111	7
2	s00000101101 .... s00000101100	}	2:1	s00000101101	6
2	s00000101011 .... s00000101010	}	2:1	s00000101011	5
2	s00000101001 .... s00000101000	}	2:1	s00000101001	4
2	s00000100111 .... s00000100110	}	2:1	s00000100111	3
2	s00000100101 .... s00000100100	}	2:1	s00000100101	2
2	s00000100011 .... s00000100010	}	2:1	s00000100011	1
2	s00000100001 .... s00000100000	}	2:1	s00000100001	0
	s000001ABCD-				

(c)

FIGURE 16-16 (*continued*)

12-bit recovered code:

```
  1            0  0  0                 0   1   0   1
  s      (7 - 0 = 7 leading 0's)      A   B   C   D
  1     0   0  0  0  0   0   0         0   1   0   1
  ↑
 sign bit      segment identifier       quantization
            determines the               interval
            number of leading
                 0's
```

As you can see, the recovered 12-bit code is exactly the same as the original 12-bit linear code. This is true for all codes in segment 0 and 1. Consequently, there is no compression error in these two segments.

(b) For the 0.32-V sample:

12-bit linear code:

$$\frac{0.32\ V}{0.01\ V} = 32 = \begin{array}{l} 1\ 0\ 0\ 0\ 0\ 0\ 1\ 0\ 0\ 0\ 0\ 0 \\ s\ ------magnitude----- \end{array}$$

8-bit compressed code:

```
  1      0  0  0  0  0  1     0   0   0   0
  s      (7 - 5 = 2 or 010)   A   B   C   D   X
  1         0  1  0           0   0   0   0    ↑
 (+)         (segment 2)                  truncated
```

12-bit recovered code:

```
  1            0  1  0             0   0   0   0
  s      (7 - 2 = 5 leading 0's)   A   B   C   D   X
  1     0  0  0  0  0  1           0   0   0   0   1
                      ↑                           ↑
                  inserted                    inserted
```

Note the two inserted 1's in the decoded 12-bit code. The least significant bit is determined from the decoding table in Figure 16-15. The stuffed 1 in bit position 6 was dropped during the 12-bit-to-8-bit conversion. Transmission of this bit is redundant because if it were not a 1, the sample would not be in segment 3. Consequently, in all segments except 0, a 1 is automatically inserted after the reinserted zeros. For this sample, there is an error in the received voltage equal to the resolution, 0.01 V. In segment 2, for every two 12-bit codes possible, there is only one recovered 12-bit code. Thus a coding compression of 2:1 is realized.

(c) To determine the codes for 10.23 V, the process is the same:

12-bit linear code:

```
    1   1   1111    111111
    ↑
    s    ABCD   truncated
```

8-bit compressed code:

```
1    111    1111
↑
s  segment  ABCD
```

12-bit recovered code:

```
1  1  1111    100000
↑  ↑  ‿‿‿‿    ‿‿‿‿‿‿
s  |  ABCD  inserted
inserted
```

The difference in the original 12-bit linear code and the recovered 12-bit code is

```
  111111111111
- 111111100000
  000000011111 = 31(0.01 V) = 0.31 V
```

Percentage Error

For comparison purposes, the following formula is used for computing the *percentage of error* introduced by digital compression:

$$\% \text{ error} = \frac{|\text{Tx voltage} - \text{Rx voltage}|}{\text{Rx voltage}} \times 100 \qquad (16\text{-}8)$$

EXAMPLE 16-6

The maximum percentage of error will occur for the smallest number in the lowest subsegment within any given segment. Because there is no compression error in segments 0 and 1, for segment 3 the maximum % error is computed as follows:

```
Transmit 12-bit code:  s00001000000
Receive 12-bit code:   s00001000010
Magnitude of error:     00000000010
```

$$\% \text{ error} = \frac{|1000000 - 1000010|}{1000010} \times 100$$

$$= \frac{|64 - 66|}{66} \times 100 = 3.03\%$$

For segment 7:

```
Transmit 12-bit code:  s10000000000
Receive 12-bit code:   s10000100000
Magnitude of error:     00000100000
```

$$\% \text{ error} = \frac{|10000000000 - 10000100000|}{10000100000} \times 100$$

$$= \frac{|1024 - 1056|}{1056} \times 100 = 3.03\%$$

Although the magnitude of error is higher for segment 7, the percentage of error is the same. The maximum percentage of error is the same for segments 3 through 7, and consequently, the SQR degradation is the same for each segment.

Although there are several ways in which the 12-bit-to-8-bit compression and the 8-bit-to-12-bit expansion can be accomplished with hardware, the simplest and most economical is with a look-up table in ROM (read-only memory).

Essentially every function performed by a PCM encoder and decoder is now accomplished with a single integrated-circuit chip called a *codec*. Most of the more recently developed codecs include an antialiasing (bandpass) filter, a sample-and-hold circuit, and an analog-to-digital converter in the transmit section and a digital-to-analog converter, a sample-and-hold circuit, and a bandpass filter in the receive section. The operation of a codec is explained in detail in Chapter 17.

DELTA MODULATION PCM

Delta modulation uses a single-bit PCM code to achieve digital transmission of analog signals. With conventional PCM, each code is a binary representation of both the sign and magnitude of a particular sample. Therefore, multiple-bit codes are required to represent the many values that the sample can be. With delta modulation, rather than transmit a coded representation of the sample, only a single bit is transmitted which simply indicates whether that sample is larger or smaller than the previous sample. The algorithm for a delta modulation system is quite simple. If the current sample is smaller than the previous sample, a logic 0 is transmitted. If the current sample is larger than the previous sample, a logic 1 is transmitted.

Delta Modulation Transmitter

Figure 16-17 shows a block diagram of a delta modulation transmitter. The input analog is sampled and converted to a PAM signal which is compared to the output of the DAC. The output of the DAC is a voltage equal to the magnitude of the previous sample, which was stored in the up-down counter as a binary number. The up-down counter is incremented or decremented depending on whether the previous sample is larger or smaller than the current sample. The up-down counter is clocked at a rate equal to the sample rate. Therefore, the up-down counter is updated after each comparison.

Figure 16-18 shows the ideal operation of a delta modulation encoder. Initially, the up-down counter is zeroed and the DAC is outputting 0 V. The first sample is taken, converted to a PAM signal, and compared to zero volts. The output of the comparator is a logic 1 condition (+V), indicating that the current sample is larger in amplitude than the previous sample. On the next clock pulse, the up-down counter is incremented to a count of 1. The DAC now outputs a voltage equal to the magnitude

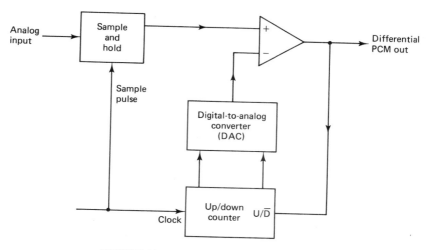

FIGURE 16-17 Delta modulation transmitter.

of the minimum step size (resolution). The steps change value at a rate equal to the clock frequency (sample rate). Consequently, with the input signal shown, the up-down counter follows the input analog signal up until the output of the DAC exceeds the analog sample; then the up-down counter will begin counting down until the output of the DAC drops below the sample amplitude. In the idealized situation (shown in Figure 16-18), the DAC output follows the input signal. Each time the up-down counter is incremented, a logic 1 is transmitted, and each time the up-down counter is decremented, a logic 0 is transmitted.

Delta Modulation Receiver

Figure 16-19 shows the block diagram of a delta modulation receiver. As you can see, the receiver is almost identical to the transmitter except for the comparator. As the logic 1's and 0's are received, the up-down counter is incremented or decremented accordingly. Consequently, the output of the DAC in the decoder is identical to the output of the DAC in the transmitter.

With delta modulation, each sample requires the transmission of only one bit; therefore, the bit rates associated with delta modulation are lower than conventional

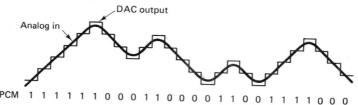

FIGURE 16-18 Ideal operation of a delta modulation encoder.

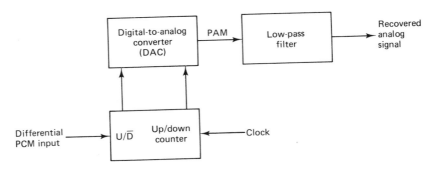

FIGURE 16-19   Delta modulation receiver.

PCM systems. However, there are two problems associated with delta modulation that do not occur with conventional PCM: slope overload and granular noise.

   Slope overload.   Figure 16-20 shows what happens when the analog input signal changes at a faster rate than the DAC can keep up with. The slope of the analog signal is greater than the delta modulator can maintain. This is called *slope overload*. Increasing the clock frequency reduces the probability of slope overload occurring. Another way is to increase the magnitude of the minimum step size.

   Granular noise.   Figure 16-21 contrasts the original and reconstructed signals associated with a delta modulation system. It can be seen that when the original analog input signal has a relatively constant amplitude, the reconstructed signal has variations that were not present in the original signal. This is called *granular noise*. Granular noise in delta modulation is analogous to quantization noise in conventional PCM.

   Granular noise can be reduced by decreasing the step size. Therefore, to reduce the granular noise, a small resolution is needed, and to reduce the possibility of slope overload occurring, a large resolution is required. Obviously, a compromise is necessary.

   Granular noise is more prevalent in analog signals that have gradual slopes and whose amplitudes vary only a small amount. Slope overload is more prevalent in analog signals that have steep slopes or whose amplitudes vary rapidly.

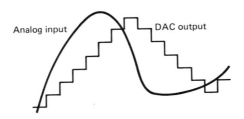

FIGURE 16-20   Slope overload distortion.

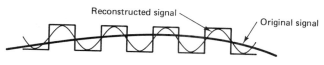

FIGURE 16-21   Granular noise.

ADAPTIVE DELTA MODULATION PCM

Adaptive delta modulation is a delta modulation system where the step size of the DAC is automatically varied depending on the amplitude characteristics of the analog input signal. Figure 16-22 shows how an adaptive delta modulator works. When the output of the transmitter is a string of consecutive 1's or 0's, this indicates that the slope of the DAC output is less than the slope of the analog signal in either the positive or negative direction. Essentially, the DAC has lost track of exactly where the analog samples are and the possibility of slope overload occurring is high. With an adaptive delta modulator, after a predetermined number of consecutive 1's or 0's, the step size is automatically increased. After the next sample, if the DAC output amplitude is still below the sample amplitude, the next step is increased even further until eventually the DAC catches up with the analog signal. When an alternating sequence of 1's and 0's is occurring, this indicates that the possibility of granular noise occurring is high. Consequently, the DAC will automatically revert to its minimum step size and thus reduce the magnitude of the noise error.

A common algorithm for an adaptive delta modulator is when three consecutive 1's or 0's occur, the step size of the DAC is increased or decreased by a factor of 1.5. Various other algorithms may be used for adaptive delta modulators, depending on particular system requirements.

DIFFERENTIAL PULSE CODE MODULATION

In a typical PCM-encoded speech waveform, there are often successive samples taken in which there is little difference between the amplitudes of the two samples. This

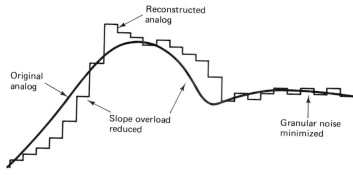

FIGURE 16-22   Adaptive delta modulation.

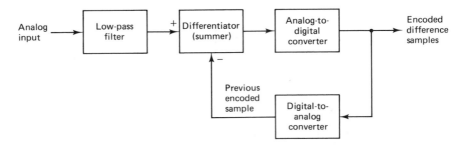

FIGURE 16-23 DPCM transmitter.

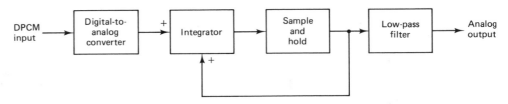

FIGURE 16-24 DPCM receiver.

facilitates transmitting several identical PCM codes, which is redundant. Differential pulse code modulation (DPCM) is designed specifically to take advantage of the sample-to-sample redundancies in typical speech waveforms. With DPCM, the difference in the amplitude of two successive samples is transmitted rather than the actual sample. Since the range of sample differences is typically less than the range of individual samples, fewer bits are required for DPCM than conventional PCM.

    Figure 16-23 shows a simplified block diagram of a DPCM transmitter. The analog input signal is bandlimited to one-half of the sample rate, then compared to the preceding DPCM signal in the differentiator. The output of the differentiator is the difference between the two signals. The difference is PCM encoded and transmitted. The A/D converter operates the same as in a conventional PCM system except that it typically uses fewer bits per sample.

    Figure 16-24 shows a simplified block diagram of a DPCM receiver. Each received sample is converted back to analog, stored, and then summed with the next sample received. In the receiver shown in Figure 16-24 the integration is performed on the analog signals, although it could also be performed digitally.

QUESTIONS

16-1. Contrast the advantages and disadvantages of digital transmission.

16-2. What are the four most common methods of pulse modulation?

16-3. Which method listed in Question 16.2 is the only form of pulse modulation that is a digital transmission system? Explain?

16-4. What is the purpose of the sample-and-hold circuit?

16-5. Define *aperture* and *acquisition time*.

16-6. What is the difference between natural and flat-top sampling?

16-7. Define *droop*. What causes it?

16-8. What is the Nyquist sampling rate?

16-9. Define and state the causes of foldover distortion.

16-10. Explain the difference between a magnitude-only code and a sign-magnitude code.

16-11. Explain overload distortion.

16-12. Explain quantizing.

16-13. What is quantization range? Quantization error?

16-14. Define *dynamic range*.

16-15. Explain the relationship between dynamic range, resolution, and the number of bits in a PCM code.

16-16. Explain coding efficiency.

16-17. What is SQR? What is the relationship between SQR, resolution, dynamic range, and the number of bits in a PCM code?

16-18. Contrast linear and nonlinear PCM codes.

16-19. Explain idle channel noise.

16-20. Contrast midtread and midrise quantization.

16-21. Define *companding*.

16-22. What does the parameter μ determine?

16-23. Briefly explain the process of digital companding.

16-24. What is the effect of digital compression on SQR, resolution, quantization interval, and quantization noise?

16-25. Contrast delta modulatin PCM and standard PCM.

16-26. Define *slope overload* and *granular noise*.

16-27. What is the difference between adaptive delta modulation and conventional delta modulation?

16-28. Contrast differential and conventional PCM.

PROBLEMS

16-1. Determine the Nyquist sample rate for a maximum analog input frequency of 4 kHz; 10 kHz.

16-2. For the sample-and-hold circuit shown in Figure 16-3, determine the largest-value capacitor that can be used. Use the following parameters: an output impedance for Z1 = 20 Ω, an "on" resistance of Q1 of 20 Ω, an acquisition time of 10 μs, a maximum output current from Z1 of 20 mA, and an accuracy of 1%.

16-3. For a sample rate of 20 kHz, determine the maximum analog input frequency.

16-4. Determine the alias frequency for a 4-kHz sample rate and an analog input frequency of 1.5 kHz.

16-5. Determine the dynamic range for a 10-bit sign-magnitude PCM code.

16-6. Determine the minimum number of bits required in a PCM code for a dynamic range of 80 dB. What is the coding efficiency?

16-7. For a resolution of 0.04 V, determine the voltages for the following linear 7-bit sign-magnitude PCM codes:

   (a) 0110101
   (b) 0000011
   (c) 1000001
   (d) 0111111
   (e) 1000000

16-8. Determine the SQR for a 2-V rms signal and a quantization noise magnitude of 0.2 V.

16-9. Determine the resolution and quantization noise for an 8-bit linear sign-magnitude PCM code for a maximum decoded voltage of $1.27V_p$.

16-10. A 12-bit linear PCM code is digitally compressed into 8 bits. The resolution = 0.03 V. Determine the following for an analog input voltage of 1.465 V.

   (a) 12-Bit linear PCM code.
   (b) 8-Bit compressed code.
   (c) Decoded 12-bit code.
   (d) Decoded voltage.
   (e) Percentage error.

16-11. For a 12-bit linear PCM code with a resolution of 0.02 V, determine the voltage range that would be converted to the following PCM codes.

   (a) 1 0 0 0 0 0 0 0 0 0 0 1
   (b) 0 0 0 0 0 0 0 0 0 0 0 0
   (c) 1 1 0 0 0 0 0 0 0 0 0 0
   (d) 0 1 0 0 0 0 0 0 0 0 0 0
   (e) 1 0 0 1 0 0 0 0 0 0 0 1
   (f) 1 0 1 0 1 0 1 0 1 0 1 0

16-12. For each of the following 12-bit linear PCM codes, determine their input voltage range and the 8-bit compressed code to which they would be converted.

   (a) 1 0 0 0 0 0 0 1 0 0 0
   (b) 1 0 0 0 0 0 0 1 0 0 1
   (c) 1 0 0 0 0 0 1 0 0 0 0
   (d) 0 0 0 0 0 1 0 0 0 0 0
   (e) 0 1 0 0 0 0 0 0 0 0 0
   (f) 0 1 0 0 0 1 0 0 0 0 0

Chapter 17

DIGITAL MULTIPLEXING

INTRODUCTION

Multiplexing is the transmission of information (either voice or data) from more than one source to more than one destination on the same transmission medium (*facility*). Transmissions occur on the same facility but not necessarily at the same time. The transmission medium may be a metallic wire pair, a coaxial cable, a microwave radio, a satellite radio, or a fiber optic cable. There are several ways in which multiplexing can be achieved, although the two most common methods are *frequency-division multiplexing* (FDM) and *time-division multiplexing* (TDM). Frequency-division multiplexing is discussed in Chapter 18.

TIME-DIVISION MULTIPLEXING

With TDM, transmissions from multiple sources occur on the same facility but not at the same time. Transmissions from various sources are *interleaved* in the time domain. The most common type of modulation used with TDM systems is PCM. With a PCM-TDM system, several voice band channels are sampled, converted to PCM codes, then time-division multiplexed onto a single metallic cable pair.

Figure 17-1a shows a simplified block diagram of a two-channel PCM-TDM carrier system. Each channel is alternately sampled and converted to a PCM code. While the PCM code for channel 1 is being transmitted, channel 2 is sampled and converted to a PCM code. While the PCM code from channel 2 is being transmitted, the next sample is taken from channel 1 and converted to a PCM code. This process continues and

657

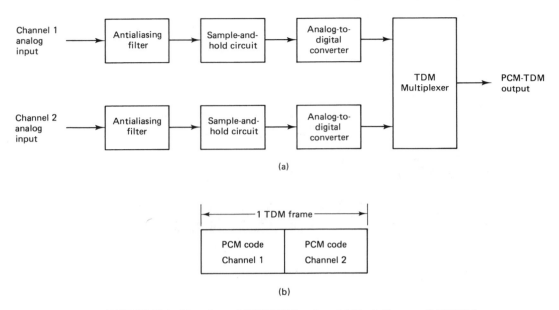

FIGURE 17-1   Two-channel PCM-TDM system: (a) block diagram; (b) TDM frame.

samples are taken alternately from each channel, converted to PCM codes, and transmitted. The multiplexer is simply a switch with two inputs and one output. Channel 1 and channel 2 are alternately selected and connected to the multiplexer output. The time it takes to transmit one sample from each channel is called the *frame time*.

The PCM code for each channel occupies a fixed time slot (epoch) within the total TDM frame. With a two-channel system, the time allocated for each channel is equal to one-half of the total frame time. A sample from each channel is taken once during each frame. Therefore, the total frame time is equal to the reciprocal of the sample rate $(1/F_s)$. Figure 17-1b shows the TDM frame allocation for a two-channel system.

T1 DIGITAL CARRIER SYSTEM

Figure 17-2 shows the block diagram of the Bell System T1 digital carrier system. This system is the North American telephone standard. A T1 carrier time-division multiplexes 24 PCM encoded samples for transmission over a single metallic wire pair. Again, the multiplexer is simply a switch except that now it has 24 inputs and one output. The 24 voice band channels are sequentially selected and connected to the multiplexer output. Each voice band channel occupies a 300- to 3000-Hz bandwidth.

Simply time-division multiplexing 24 voice band channels does not in itself constitute a T1 carrier. At this point, the output of the multiplexer is simply a multiplexed

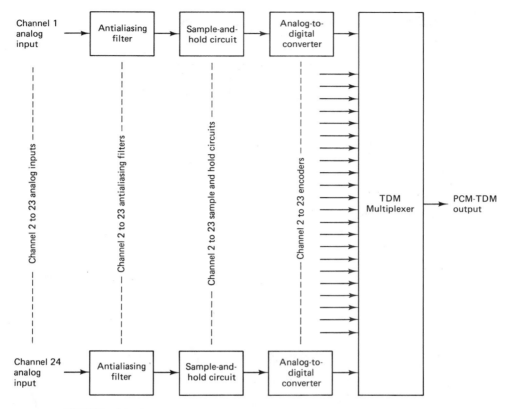

FIGURE 17-2 Bell system T1 PCM-TDM digital carrier system block diagram.

digital signal (DS-1). It does not actually become a T1 carrier until it is line encoded and placed on special conditioned wire pairs called *T1 lines*. This is explained in more detail later in this chapter under the heading "North American Digital Hierarchy."

With the Bell System T1 carrier system, D-type (digital) channel banks perform the sampling, encoding, and multiplexing of 24 voice band channels. Each channel contains an 8-bit PCM code and is sampled 8000 times per second. (Each channel is sampled at the same rate but not at the same time; see Figure 17-3.) Therefore, a 64-kbps PCM encoded sample is transmitted for each voice band channel during each frame.

$$\frac{8 \text{ bits}}{\text{sample}} \times \frac{8000 \text{ samples}}{\text{second}} = 64 \text{ kbps}$$

Within each frame an additional bit called a *framing bit* is added. The framing bit occurs at an 8000-bps rate and is recovered in the receiver circuitry and used to

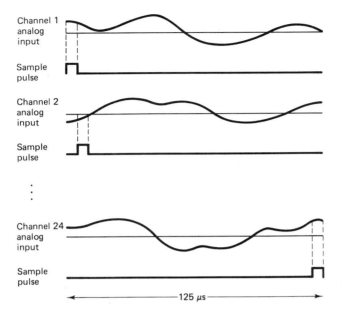

FIGURE 17-3   T1 sampling sequence.

maintain frame and sample synchronization between the TDM transmitter and receiver. As a result, each TDM frame contains 193 bits.

$$\frac{8 \text{ bits}}{\text{channel}} \times \frac{24 \text{ channels}}{\text{frame}} = \frac{192 \text{ bits}}{\text{frame}} + \frac{1 \text{ framing bit}}{\text{frame}} = \frac{193 \text{ bits}}{\text{frame}}$$

As a result, the line speed (bps) for the T1 carrier is

$$\text{line speed} = \frac{193 \text{ bits}}{\text{frame}} \times \frac{8000 \text{ frames}}{\text{second}} = 1.544 \text{ Mbps}$$

D-Type Channel Banks

The early T1 carrier systems were equipped with D1A channel banks which use a 7-bit magnitude-only PCM code with analog companding and $\mu = 100$. A later version of the D1 channel bank (D1D) used an 8-bit sign-magnitude PCM code. With D1A channel banks an eighth bit (the s bit) is added to each PCM code for the purpose of *signaling* (supervision: on-hook, off-hook, dial pulsing, etc.). Consequently, the signaling rate for D1 channel banks is 8 kbps. Also, with D1 channel banks, the framing bit sequence is simply an alternating 1/0 pattern. Figure 17-4 shows the frame and sample alignment for the T1 carrier system using D1A channel banks.

       Generically, the T1 carrier system has progressed through the D2, D3, and D4 channel banks, which use a digitally companded, 8-bit sign-magnitude compressed PCM code with $\mu = 255$. In the D1 channel bank, the compression and expansion characteris-

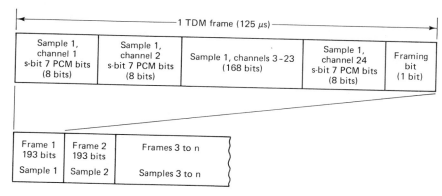

FIGURE 17-4 **T1 carrier system frame and sample alignment using D1 channel banks.**

tics were implemented in circuitry separate from the encoder and decoder. The D2, D3, and D4 channel banks incorporate the companding functions directly in the encoders and decoders. Although the D2 and D3 channel banks are functionally similar, the D3 channel banks were the first to incorporate integrated circuitry into their design, which reduced their size and power consumption. D4 channel banks were the first to incorporate separate customized LSI integrated circuits (codecs) for each voice band channel. With D1, D2, and D3 channel banks, common equipment performs the encoding and decoding functions. Consequently, a single equipment malfunction constitutes a total system failure.

D1A channel banks use a magnitude-only code; consequently, an error in the most significant bit (MSB) of a channel sample always produces a decoded error equal to one-half the total quantization range (one-half of V_{max}). Because D1D, D2, D3, and D4 channel banks use a sign-magnitude code, an error in the MSB (sign bit) causes a decoded error equal to twice the sample magnitude (from $+V$ to $-V$, or vice versa). The worst-case error is equal to twice the total quantization range. However, maximum amplitude samples occur rarely and most errors with D1D, D2, D3, and D4 coding are less than one-half the coding range. On the average, the error performance with a sign-magnitude code is less than with a magnitude-only code.

Superframe Format

The 8-kbps signaling rate used with D1 channel banks is excessive for voice transmission. Therefore, with D2 and D3 channel banks, a signaling bit is substituted only into the least significant bit (LSB) of every sixth frame. Therefore, five out of every six frames have 8-bit resolution, while one out of every six frames (the signaling frame) has only 7-bit resolution. Consequently, the signaling rate on each channel is 1.333 kbps (8000 bps/6) and the effective number of bits per sample is actually $7\frac{5}{6}$ bits and not 8.

Because only every sixth frame includes a signaling bit, it is necessary that all the frames are numbered so that the receiver knows when to extract the signaling information. Also, because the signaling is accomplished with a 2-bit binary word, it is necessary to identify the MSB and LSB of the signaling word. Consequently, the *superframe* format shown in Figure 17-5 was devised. Within each superframe, there are 12 consecutively numbered frames (1–12). The signaling bits are substituted in frames 6 and 12; the MSB into frame 6 and the LSB into frame 12. Frames 1–6 are called the A-highway with frame 6 designated as the A-channel signaling frame. Frames 7–12 are called the B-highway with frame 12 designated as the B-channel signaling frame. Therefore, in addition to identifying the signaling frames, the sixth and twelfth frames must be positively identified.

To identify frames 6 and 12, a different framing bit sequence is used for the odd- and even-numbered frames. The odd frames (frames 1, 3, 5, 7, 9, and 11) have an alternating 1/0 pattern, and the even frames (frames 2, 4, 6, 8, 10, and 12) have a 0 0 1 1 1 0 repetitive pattern. As a result, the combined bit pattern for the framing bits is a 1 0 0 0 1 1 0 1 1 1 0 0 repetitive pattern. The odd-numbered frames are used for frame and sample synchronization, while the even-numbered frames are used to identify the A and B channel signaling frames (6 and 12). Frame 6 is identified by a 0/1 transition in the framing bit between frames 4 and 6. Frame 12 is identified by a 1/0 transition in the framing bit between frames 10 and 12.

Figure 17-6 shows the frame, sample, and signaling alignment for the T1 carrier system using D2 or D3 channel banks.

In addition to *multiframe alignment* bits and PCM sample bits, certain time slots are used to indicate alarm conditions. For example, in the case of a transmit power supply failure, a common equipment failure, or loss of multiframe alignment; the second bit in each channel is made a 0 until the alarm condition has cleared. Also, the framing bit in frame 12 is complemented whenever multiframe alignment is lost (this is assumed whenever frame alignment is lost). In addition, there are special framing conditions that must be avoided in order to maintain clock and bit synchronization at the receive demultiplexing equipment. These special conditions are explained later in this chapter.

D4 Channel Bank

D4 channel banks time-division multiplex 48 voice band channels and operate at a transmission rate of 3.152 Mbps. This is slightly more than twice the line speed for 24-channel D1, D2, or D3 channel banks. This is because with D4 channel banks, rather than transmit a single framing bit with each frame, a 10-bit frame synchronization pattern is used. Consequently, the total number of bits in a D4 (DS-1C) TDM frame is

$$\frac{8 \text{ bits}}{\text{channel}} \times \frac{48 \text{ channels}}{\text{frame}} = \frac{384 \text{ bits}}{\text{frame}} + \frac{10 \text{ syn bits}}{\text{frame}} = \frac{394 \text{ bits}}{\text{frame}}$$

and the line speed is

$$\text{line speed} = \frac{394 \text{ bits}}{\text{frame}} \times \frac{8000 \text{ frames}}{\text{second}} = 3.152 \text{ Mbps}$$

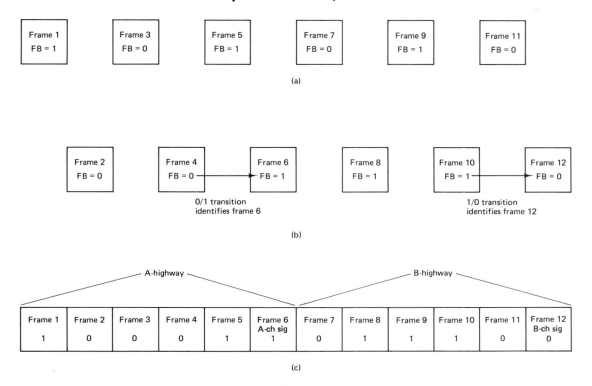

FIGURE 17-5 Framing bit sequence for the T1 superframe format: (a) frame synchronizing bits (odd-numbered frames); (b) signaling frame alignment bits (even-numbered frames); (c) composite frame alignment.

The framing for the DS-1 (T1) system or the framing pattern for the DS-1C (T1C) time-division-multiplexed carrier systems are added to the multiplexed digital signal at the output of the multiplexer. Figure 17-7 shows the framing bit circuitry for the 24-channel T1 carrier system using either D1, D2, or D3 channel banks (DS-1). Note that the bit rate at the output of the TDM multiplexer is 1.536 Mbps and the bit rate at the output of the 193-bit shift register is 1.544 Mbps. The difference (8 kbps) is because of the added framing bit.

CCITT TIME-DIVISION-MULTIPLEXED CARRIER SYSTEM

Figure 17-8 shows the frame alignment for the CCITT (Comité Consultatif International Téléphonique et Télégraphique) European standard PCM-TDM system. With the CCITT system, an 125-µs frame is divided into 32 equal time slots. Time slot 0 is used for a frame alignment pattern and for an alarm channel. Time slot 17 is used for a common

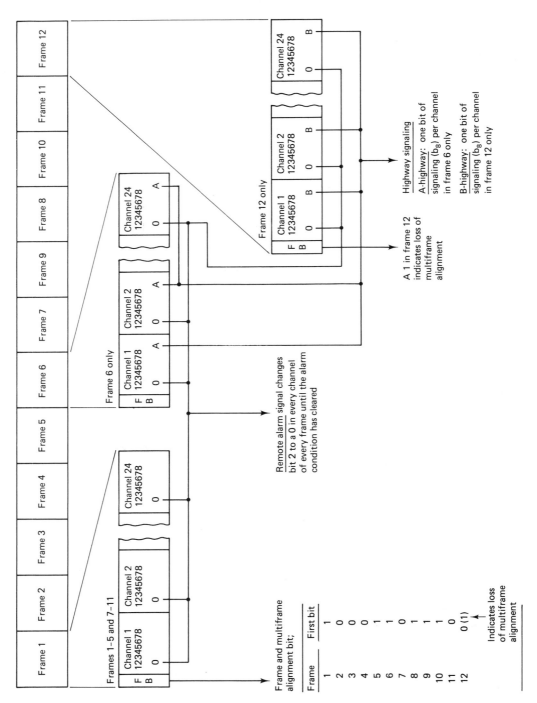

FIGURE 17-6 T1 carrier frame, sample, and signaling alignment for D2 and D3 channel banks.

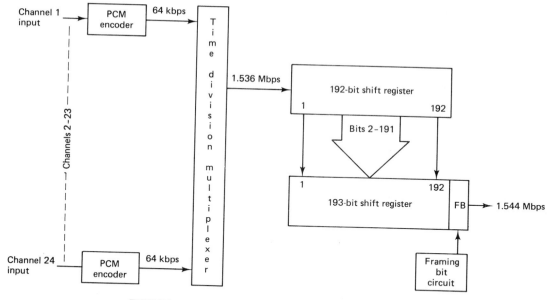

FIGURE 17-7 Framing bit circuitry for the DS-1 T1 carrier system.

signaling channel. The signaling for all the voice band channels is accomplished on the common signaling channel. Consequently, there are 30 voice band channels time-division multiplexed into each CCITT frame.

With the CCITT standard, each time slot has 8 bits. Consequently, the total number of bits per frame is

$$\frac{8 \text{ bits}}{\text{time slot}} \times \frac{32 \text{ time slots}}{\text{frame}} = \frac{256 \text{ bits}}{\text{frame}}$$

and the line speed is

$$\text{line speed} = \frac{256 \text{ bits}}{\text{frame}} \times \frac{8000 \text{ frames}}{\text{second}} = 2.048 \text{ Mbps}$$

CODECS

A *codec* is a large-scale-integration (LSI) chip designed for use in the telecommunications industry for *private branch exchanges* (PBXs), central office switches, digital handsets, voice store-and-forward systems, and digital echo suppressors. Essentially, the codec is applicable for any purpose that requires the digitizing of analog signals, such as in a PCM-TDM carrier system.

"Codec" is a generic term that refers to the *co*ding and *dec*oding functions per-

Time slot 0	Time slot 1	Time slots 2–16	Time slot 17	Time slots 18–30	Time slot 31
Framing and alarm channel	Voice channel 1	Voice channels 2–15	Common signaling channel	Voice channels 16–29	Voice channel 30
8 bits	8 bits	112 bits	8 bits	112 bits	8 bits

(a)

Time slot 17

	Bits	
Frame	1234	5678
0	0000	xyxx
1	ch 1	ch 16
2	ch 2	ch 17
3	ch 3	ch 18
4	ch 4	ch 19
5	ch 5	ch 20
6	ch 6	ch 21
7	ch 7	ch 22
8	ch 8	ch 23
9	ch 9	ch 24
10	ch 10	ch 25
11	ch 11	ch 26
12	ch 12	ch 27
13	ch 13	ch 28
14	ch 14	ch 29
15	ch 15	ch 30

16 frames equal one multiframe; 500 multiframes are transmitted each second

x = spare
y = loss of multiframe alignment if a 1

4 bits per channel are transmitted once every 16 frames, resulting in a 500-bps signaling rate for each channel

(b)

FIGURE 17-8   CCITT TDM frame alignment and common signaling channel alignment: (a) CCITT TDM frame (125 μs, 256 bits, 2.048 Mbps); (b) common signaling channel.

formed by a device that converts analog signals to digital codes and digital codes to analog signals. Recently developed codecs are called *combo* chips because they combine codec and filter functions in the same LSI package. The input/output filter performs the following functions: bandlimiting, noise rejection, antialiasing, and reconstruction of analog audio waveforms after decoding. The codec performs the following functions: analog sampling, encoding/decoding (analog-to-digital and digital-to-analog conversions), and digital companding.

2913/14 COMBO CHIP

The 2913/14 is a combo chip that can provide the analog-to-digital and the digital-to-analog conversions and the transmit and receive filtering necessary to interface a full-duplex (four-wire) voice telephone circuit to the PCM highway of a TDM carrier system. Essentially, the 2913/14 combo chip replaces the older 2910A/11A codec and 2912A filter chip. The 2913 (20-pin package)/2914 (24-pin package) combo chip is manufactured with HMOS technology. There is a newer CHMOS version (29C13/14) that is functionally identical to the HMOS version except that it comes in a 28-pin package and has three low-power modes of operation. In addition, the 2916/17 and 29C16/17 are 16-pin limited-feature versions of the 2913/14 and 29C13/14. The following discussion is limited to the 2914 combo chip, although extrapolation to the other versions is quite simple.

Table 17-1 lists several of the combo chips available and their prominent features. Table 17-2 lists the pin names for the 2914 and gives a brief description of each of their functions. Figure 17-9 shows the block diagram of a 2914 combo chip.

General Operation

The following major functions are provided by the 2914 combo chip:

1. Bandpass filtering of the analog signals prior to encoding and after decoding
2. Encoding and decoding of voice and call progress signals
3. Encoding and decoding of signaling and supervision information
4. Digital companding

TABLE 17-1   FEATURES OF SEVERAL CODEC/FILTER COMBO CHIPS

2916 (16-pin)	2917 (16-pin)	2913 (20-pin)	2914 (24-pin)
μ-law companding only	A-law companding only	μ/A-law companding	μ/A-law companding
Master clock 2.048 MHz only	Master clock 2.048 MHz only	Master clock 1.536 MHz, 1.544 MHz, or 2.048 MHz	Master clock 1.536 MHz, 1.544 MHz, or 2.048 MHz
Fixed data rate	Fixed data rate	Fixed data rate	Fixed data rate
Variable data rate 64 kbps–2.048 Mbps	Variable data rate 64 kbps–4.048 Mbps	Variable data rate 64 kbps–4.096 Mbps	Variable data rate 64 kbps–4.096 Mbps
78-dB dynamic	78-dB dynamic range	78-dB dynamic range	78-dB dynamic range
ATT D3/4 compatible	ATT D3/4 compatible	ATT D3/4 compatible	ATT D3/4 compatible
Single-ended input	Single-ended input	Differential input	Differential input
Single-ended output	Single-ended output	Differential output	Differential output
Gain adjust transmit only	Gain adjust transmit only	Gain adjust transmit and receive	Gain adjust transmit and receive
Synchronous clocks	Synchronous clocks	Synchronous clocks	Synchronous clocks Asynchronous clocks
———	———	———	Analog loopback
———	———	———	Signaling

TABLE 17-2 2914 COMBO CHIP

Symbol	Name	Function
VBB	Power (-5 V)	Negative supply voltage.
PWRO+	Receive power amplifier output	Noninverting output of the receive power amplifier. This output can drive transformer hybrids or high-impedance loads directly in either a differential or single-ended mode.
PWRO−	Receive power amplifier output	Inverting output of the receive power amplifier. Functionally, PWRO− is identical and complementary to PWRO+.
GSR	Receive gain control	Input to the gain-setting network on the receive power amplifier. Transmission level can be adjusted over a 12-dB range depending on the voltage at GSR.
$\overline{\text{PDN}}$	Power-down select	When $\overline{\text{PDN}}$ is a TTL high, the 2914 is active. When $\overline{\text{PDN}}$ is low, the 2914 is powered down.
CLKSEL	Master clock frequency select	Input that must be pinstrapped to reflect the master clock frequency at CLKX, CLKR. CLKSEL = VBB . . . . 2.048 MHz CLKSEL = GRDD . . . . 1.544 MHz CLKSEL = VCC . . . . 1.536 MHz
LOOP	Analog loopback	When this pin is a TTL high, the analog output (PWRO+) is internally connected to the analog input (VFXI+), GSR is internally connected to PWRO−, and VFXI− is internally connected to GSX.
SIGR	Receive signaling bit output	Signaling bit output from the receiver. In the fixed data rate mode only, SIGR outputs the logic state of the eighth bit of the PCM word in the most recent signaling frame.
DCLKR	Receive variable data rate	Selects either the fixed or variable data rate mode of operation. When DCLKR is tied to VBB, the fixed data rate mode is selected. When DCLKR is not connected to VBB, the 2914 operates in the variable data rate mode and will accept TTL input levels from 64 kHz to 4096 MHz.
DR	Receive PCM highway input	PCM data are clocked in on this lead on eight consecutive negative transitions of the receive data rate clock; CLKR in the fixed data rate mode and DCLKR in the variable data rate mode.
FSR	Receive frame synchronization clock	8-kHz frame synchronization clock input/time slot enable for the receive channel. Also in the fixed data rate mode, this lead designates the signaling and nonsignaling frames. In the variable data rate mode this signal must remain active high for the entire length

TABLE 17-2 . **(continued)**

Symbol	Name	Function
		of the PCM word (8 PCM bits). The receive channel goes into the standby mode whenever this input is TTL low for 300 ms.
GRDD	Digital ground	Digital ground for all internal logic circuits. This pin is not internally tied to GRDA.
CLKR	Receive master clock	Receive master clock and data rate clock in the fixed data rate mode, master clock only in the variable data rate mode.
CLKX	Transmit master clock	Transmit master clock and data rate clock in the fixed data rate mode, master clock only in the variable data rate mode.
FSX	Transmit frame synchronization clock	8-kHz frame synchronization clock input/time slot enable for the transmit channel. Operates independently but in an analogous manner to FSR.
DX	Transmit PCM output	PCM data are clocked out on this lead on eight consecutive positive transitions of the transmit data rate clock; CLKX in the fixed data rate mode and DCLKX in the variable data rate mode.
$\overline{\text{TSX}}$/ DCLKX	Times-slot strobe/ buffer enable Transmit variable data rate	Transmit channel timeslot strobe (output) or data rate clock (input) for the transmit channel. In the fixed data rate mode, this pin is an open drain output designed to be used as an enable signal for a three-state buffer. In variable data rate mode, this pin is the transmit data rate clock input which can operate at data rates between 64 kbps and 4096 kbps.
SIGX/ ASEL	Transmit signaling input μ- or A-law select	A dual-purpose pin. When connected to VBB, A-law companding is selected. When it is not connected to VBB, this pin is a TTL-level input for signaling bits. This input is substituted into the least significant bit position of the PCM word during every signaling frame.
GRDA	Analog ground	Analog ground return for all internal voice circuits. Not internally connected to GRDD.
VFXI+	Noninverting analog input	Noninverting analog input to uncommitted transmit operational amplifier.
VFXI−	Inverting analog input	Inverting analog input to uncommitted transmit operational amplifier.
GSX	Transmit gain control	Output terminal of on-chip uncommitted operational amplifier. Internally, this is the voice signal input to the transmit BPF.
VCC	Power (+5 V)	Positive supply voltage.

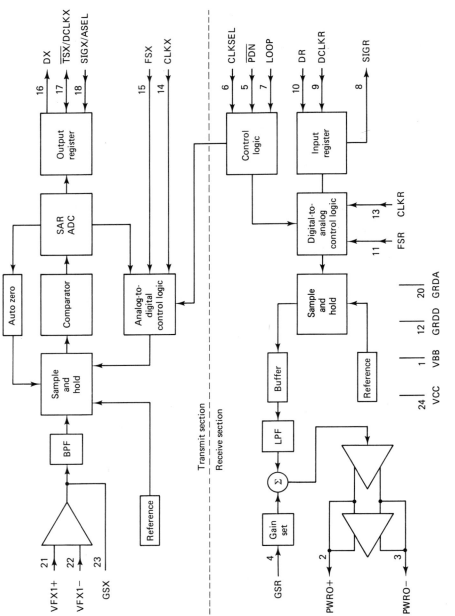

FIGURE 17-9 Block diagram of the 2914 combo chip.

System Reliability Features

The 2914 combo chip is powered up by pulsing the *transmit frame synchronization input* (FSX) and/or the *receive frame synchronization input* ($\overline{\text{FSR}}$), while a TTL high (inactive condition) is applied to the *power down select pin* ($\overline{\text{PDN}}$) and all clocks and power supplies are connected. The 2914 has an internal reset on all power-ups (when VBB or VCC are applied or temporarily interrupted). This ensures the validity of the digital output and thereby maintains the integrity of the PCM highway.

On the transmit channel, PCM *data output* (DX) and *transmit timeslot strobe* ($\overline{\text{TSX}}$) are held in a high-impedance state for approximately four frames (500 µs) after power-up. After this delay DX, $\overline{\text{TSX}}$, and signaling are functional and will occur in the proper time slots. Due to the auto-zeroing circuit on the transmit channel, the analog circuit requires approximately 60 ms to reach equilibrium. Therefore, signaling information such as on/off hook detection is available almost immediately while analog input signals are not available until after the 60-ms delay.

On the receive channel, the *signaling bit output pin* SIGR is also held low (inactive) for approximately 500 µs after power-up and remains inactive until updated by reception of a signaling frame.

$\overline{\text{TSX}}$ and DX are placed in the high-impedance state and SIGR is held low for approximately 20 µs after an interruption of the *master clock* (CLKX). Such an interruption could be caused by some kind of fault condition.

Power-Down and Standby Modes

To minimize power consumption, two power-down modes are provided in which most 2914 functions are disabled. Only the power-down, clock, and frame synchronization buffers are enabled in these modes.

The power-down is enabled by placing an external TTL low signal on $\overline{\text{PDN}}$. In this mode power consumption is reduced to an average of 5 mW.

The standby mode for the transmit and receive channels is separately controlled by removing FSX and/or FSR.

Fixed-Data-Rate Mode

In the *fixed-data-rate mode*, the master *transmit* and *receive clocks* (CLKX and CLKR) perform the following functions:

1. Provide the master clock for the on-board switched capacitor filters
2. Provide the clock for the analog-to-digital and digital-to-analog converters
3. Determine the input and output data rates between the codec and the PCM highway

Therefore, in the fixed-data-rate mode, the transmit and receive data rates must be either 1.536, 1.544, or 2.048 Mbps, the same as the master clock rate.

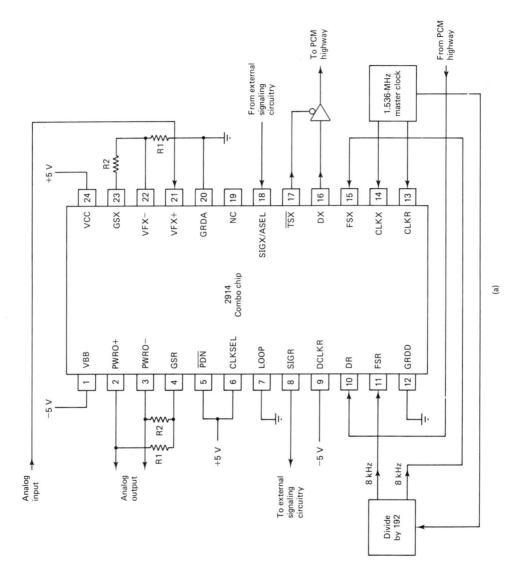

FIGURE 17-10 Single-channel PCM system using the 2914 combo chip in the fixed-data-rate mode: (a) block diagram; (b) timing sequence.

(a)

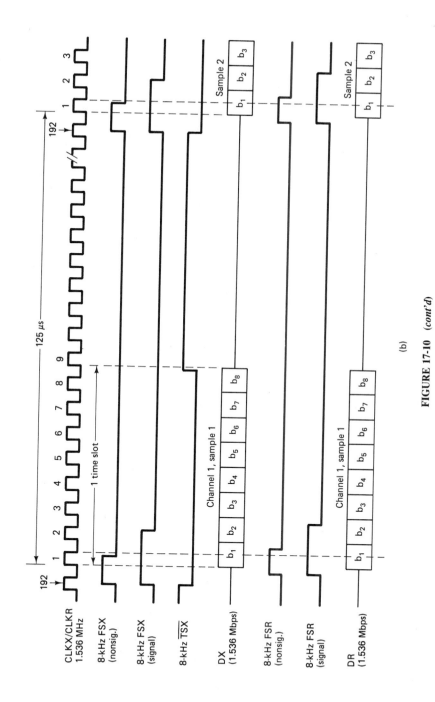

FIGURE 17-10 (cont'd)

673

Transmit and receive frame synchronizing pulses (FSX and FSR) are 8-kHz inputs which set the transmit and receive sampling rates and distinguish between *signaling* and *nonsignaling* frames. $\overline{TSX}$ is a *time-slot strobe buffer enable* output which is used to gate the PCM word onto the PCM highway when an external buffer is used to drive the line. $\overline{TSX}$ is also used as an external gating pulse for a time-division multiplexer (see Figure 17-10).

Data are transmitted to the PCM highway from DX on the first eight positive transitions of CLKX following the rising edge of FSX. On the receive channel, data are received from the PCM highway from DR on the first eight falling edges of CLKR after the occurrence of FSR. Therefore, the occurrence of FSX and FSR must be synchronized between codecs in a multiple-channel system to ensure that only one codec is transmitting to or receiving from the PCM highway at any given time.

Figure 17-10 shows the block diagram and timing sequence for a single-channel PCM system using the 2914 combo chip in the fixed-data-rate mode and operating with a master clock frequency of 1.536 MHz. In the fixed-data-rate mode, data are inputted and outputted in short bursts. (This mode of operation is sometimes called the *burst mode*.) With only a single channel, the PCM highway is active only $\frac{1}{24}$ of the total frame time. Additional channels can be added to the system provided that their transmissions are synchronized so that they do not occur at the same time as transmissions from any other channel.

From Figure 17-10 the following observations can be made:

1. The input and output bit rates from the codec are equal to the master clock frequency, 1.536 Mbps.
2. The codec inputs and outputs 64,000 PCM bits per second.
3. The data output (DX) and data input (DR) are active only $\frac{1}{24}$ of the total frame time (125 μs).

To add channels to the system shown in Figure 17-10, the occurrence of the FSX, FSR, and $\overline{TSX}$ signals for each additional channel must be synchronized so that they follow a timely sequence and do not allow more than one codec to transmit or receive at the same time. Figure 17-11 shows the block diagram and timing sequence for a 24-channel PCM-TDM system operating with a master clock frequency of 1.536 MHz.

Variable-Data-Rate Mode

The *variable-data-rate mode* allows for a flexible data input and output clock frequency. It provides the ability to vary the frequency of the transmit and receive bit clocks. In the variable data rate mode, a master clock frequency of 1.536, 1.544, or 2.048 MHz is still required for proper operation of the on-board bandpass filters and the analog-to-digital and digital-to-analog converters. However, in the variable-data-rate mode, DCLKR and DCLKX become the data clocks for the receive and transmit PCM highways, respectively. When FSX is high, data are transmitted onto the PCM highway on the next eight consecutive positive transitions of DCLKX. Similarly, while FSR is high, data from the PCM highway are clocked into the codec on the next eight consecutive negative transitions of DCLKR. This mode of operation is sometimes called the *shift register mode*.

On the transmit channel, the last transmitted PCM word is repeated in all remaining time slots in the 125-μs frame as long as DCLKX is pulsed and FSX is held active high. This feature allows the PCM word to be transmitted to the PCM highway more than once per frame. Signaling is not allowed in the variable-data-rate mode because this mode provides no means to specify a signaling frame.

Figure 17-12 shows the block diagram and timing sequence for a two-channel PCM-TDM system using the 2914 combo chip in the variable-data-rate mode with a master clock frequency of 1.536 MHz, a sample rate of 8 kHz, and a transmit and receive data rate of 128 kbps.

With a sample rate of 8 kHz, the frame time is 125 μs. Therefore, one 8-bit PCM word from each channel is transmitted and/or received during each 125-μs frame. For 16 bits to occur in 125 μs, a 128-kHz transmit and receive data clock is required.

$$\frac{1 \text{ channel}}{8 \text{ bits}} \times \frac{1 \text{ frame}}{2 \text{ channels}} \times \frac{125 \text{ μs}}{\text{frame}} = \frac{125 \text{ μs}}{16 \text{ bits}} = \frac{7.8125 \text{ μs}}{\text{bit}}$$

$$\text{bit rate} = \frac{1}{t_b} = \frac{1}{7.8125 \text{ μs}} = 128 \text{ kbps}$$

The transmit and receive enable signals (FSX and FSR) for each codec are active for one-half of the total frame time. Consequently, 8-kHz, 50% duty cycle transmit and receive data enable signals (FXS and FXR) are fed directly to one codec and fed to the other codec 180° out of phase (inverted), thereby enabling only one codec at a time.

To expand to a four-channel system, simply increase the transmit and receive data clock rates to 256 kHz and change the enable signals to an 8-kHz, 25% duty cycle pulse.

Supervisory Signaling

With the 2914 combo chip, *supervisory signaling* can be used only in the fixed-data-rate mode. A transmit signaling frame is identified by making the FSX and FSR pulses twice their normal width. During a transmit signaling frame, the signal present on input SIGX is substituted into the least significant bit position (b_1) of the encoded PCM word. At the receive end, the signaling bit is extracted from the PCM word prior to decoding and placed on output SIGR until updated by reception of another signaling frame.

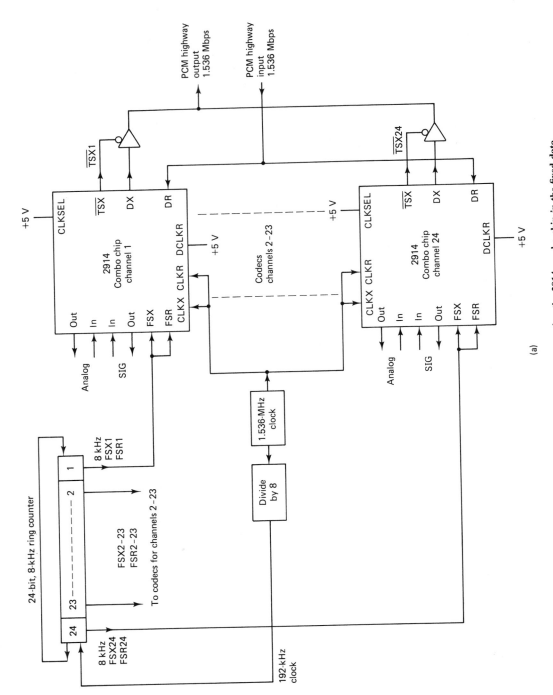

FIGURE 17-11 24-channel PCM-TDM system using the 2914 combo chip in the fixed-data-rate mode and operating with a master clock frequency of 1.536 MHz: (a) block diagram; (b) timing diagram.

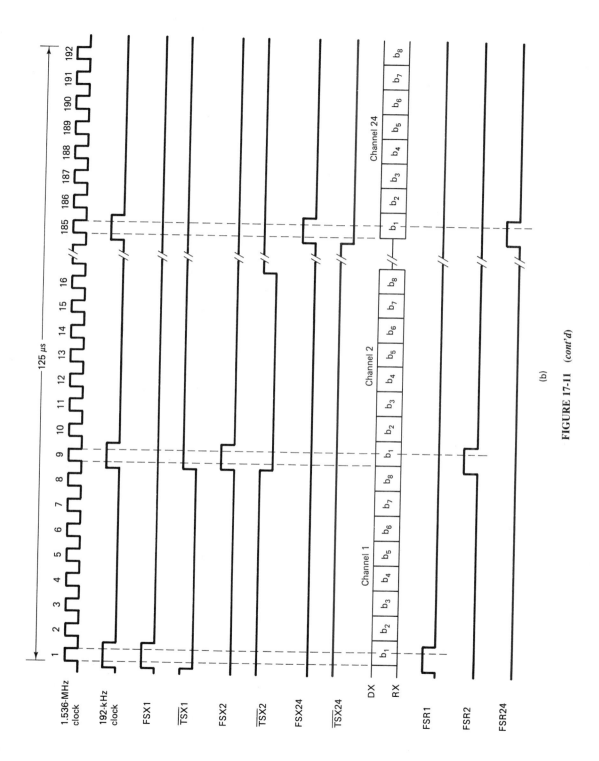

FIGURE 17-11 (cont'd)

(b)

677

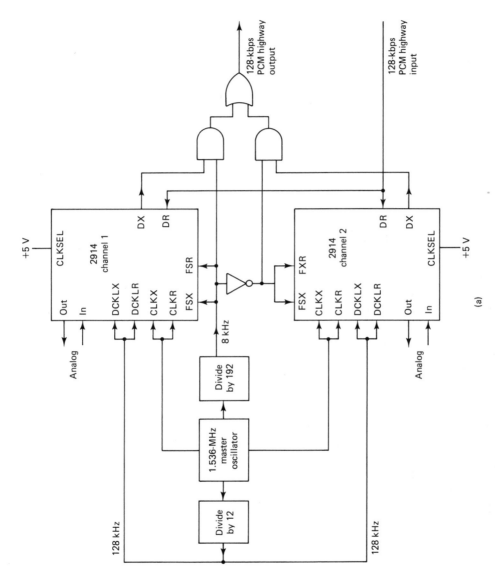

FIGURE 17-12 Two-channel PCM-TDM system using the 2914 combo chip in the variable-data-rate mode with a master clock frequency of 1.536 MHz: (a) block diagram; (b) timing diagram.

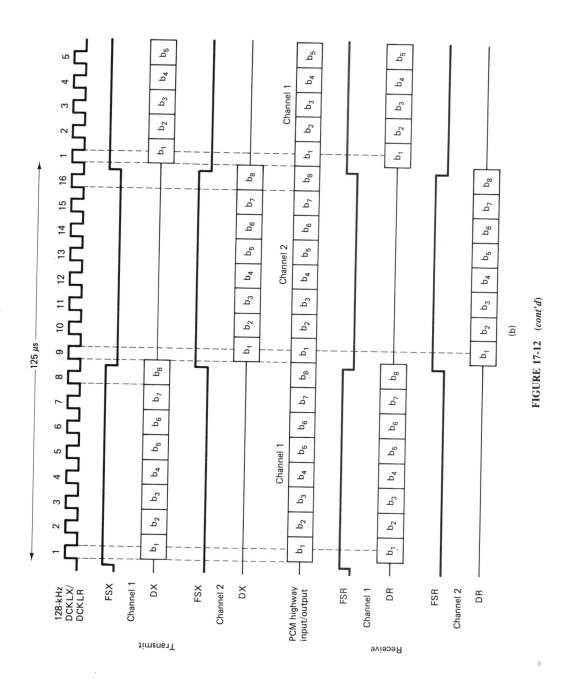

FIGURE 17-12 *(cont'd)*

679

Asynchronous Operation

Asynchronous operation is when the master transmit and receive clocks are derived from separate independent sources. The 2914 combo chip can be operated in either the synchronous or asynchronous mode. The 2914 has separate digital-to-analog converters and voltage references in the transmit and receive channels, which allows them to be operated completely independent of each other. With either synchronous or asynchronous operation, the master clock, data clock, and time-slot strobe must be synchronized at the beginning of each frame. In the variable data rate mode, CLKX and DCLKX must be synchronized once per frame but may be different frequencies.

Transmit Filter Gain

The analog input to the transmit section of the 2914 is equipped with an uncommitted operational amplifier that can operate in the single-ended or differential mode. Figure 17-13 shows a circuit configuration commonly used to provide input gain. To operate with unity gain, simply strap VFXI− to GSX and apply the analog input to VFXI+.

Receive Output Power Amplifier

The 2914 is equipped with an internal balanced output amplifier that may be used as two separate single ended outputs or as a single differential output. Figure 17-14 shows the gain setting configuration for the output amplifier operating in the differential mode. To operate with a single ended output and unity gain, simply pin strap PWRO− to GSR and take the output from PWRO+.

NORTH AMERICAN DIGITAL HIERARCHY

Multiplexing signals in digital form lends itself easily to interconnecting digital transmission facilities with different transmission bit rates. Figure 17-15 shows the American

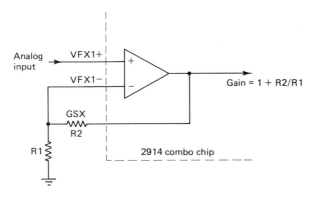

FIGURE 17-13 Transmit filter gain circuit.

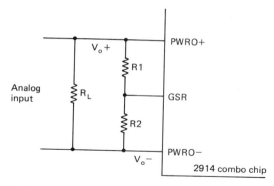

FIGURE 17-14   Receive output power amplifier. PWRO+ and PWRO− are low-impedance complementary outputs. The voltages at the nodes are V_o+ at PWRO+ and $V_o−$ at PWRO−. R1 and R2 comprise a gain-setting resistor network with the center tap connected to the GSR input. A value greater than 10 kΩ for R1 and a value less than 100 kΩ for R2 is recommended because (1) the parallel combination of R1 + R2 and RL set the total load impedance to the analog sink, and (2) the total capacitance at the GSR input and the parallel combination of R1 and R2 define a time constant that has to be minimized to avoid inaccuracies. VA represents the maximum available digital miliwatt output response (VA = 3.006 V rms).

$$V_o = -A(VA)$$

where

$$A = \frac{1 + R1/R2}{4 + R1/R2}$$

For design purposes, a useful form is R1/R2 as a function of A.

$$R1/R2 = \frac{4A - 1}{1 - A}$$

Telephone and Telegraph Company's (AT&T) North American Digital Hierarchy for multiplexing digital signals with the same bit rates into a single pulse stream suitable for transmission on the next higher level of the hierarchy. To upgrade from one level in the hierarchy to the next higher level, special devices called *muldems* (*mul*tiplexers/*dem*ultiplexers) are used. Muldems can handle bit-rate conversions in both directions. The muldem designations (M12, M23, etc.) identify the input and output digital signals associated with that muldem. For instance, an M12 muldem is a multiplexer/demultiplexer that interfaces DS-1 and DS-2 *digital signals*. An M23 muldem interfaces DS-2 and DS-3 signals. DS-1 signals may be further multiplexed or line encoded and placed on specially conditioned lines called T1 lines. DS-2, DS-3, and DS-4 signals may be placed on T2, T3, and T4M lines, respectively.

Digital signals are routed at central locations called *digital cross-connects*. A digital cross-connect (DSX) provides a convenient place to make hardwire interconnects and to perform routine maintenance and troubleshooting. Each type of digital signal (DS-1,

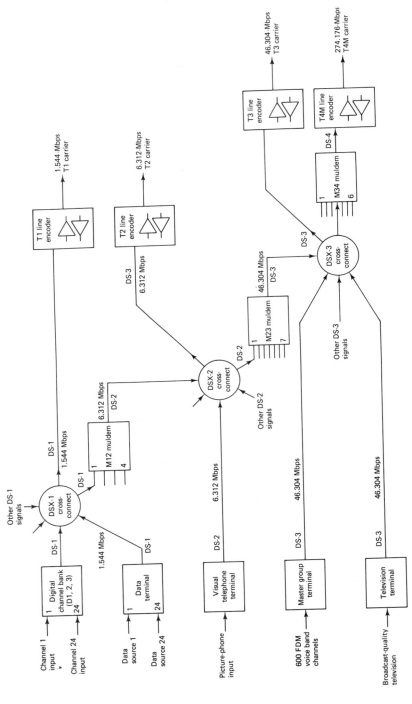

FIGURE 17-15 North American Digital Hierarchy.

DS-2, etc.) has its own digital switch (DSX-1 DSX-2, etc.). The output from a digital switch may be upgraded to the next higher level or line encoded and placed on their respective T lines (T1, T2, etc.).

Table 17-3 lists the digital signals, their bit rates, channel capacities, and services offered for the line types included in the North American Digital Hierarchy.

When the bandwidth of the signals to be transmitted is such that after digital conversion it occupies the entire capacity of a digital transmission line, a single-channel terminal is provided. Examples of such single-channel terminals are picturephone, mastergroup, and commercial television terminals.

Mastergroup and Commercial Television Terminals

Figure 17-16 shows the block diagram of a mastergroup and commercial television terminal. The mastergroup terminal receives voice band channels that have already been frequency-division multiplexed (a topic covered in Chapter 18) without requiring that each voice band channel be demultiplexed to voice frequencies. The signal processor provides frequency shifting for the mastergroup signal (shifts it from a 564- to 3084-kHz bandwidth to a 0- to 2520-kHz bandwidth) and dc restoration for the television signal. By shifting the mastergroup band, it is possible to sample at a 5.1-MHz rate. Sampling of the commercial television signal is at twice that rate or 10.2 MHz.

To meet the transmission requirements, a 9-bit PCM code is used to digitize each sample of the mastergroup or television signal. The digital output from the terminal is therefore approximately 46 Mbps for the mastergroup and twice that much (92 Mbps) for the television signal.

The digital terminal shown in Figure 17-16 has three specific functions: it converts the parallel data from the output of the encoder to serial data, it inserts frame synchronizing bits, and it converts the serial binary signal to a form more suitable for transmission. In addition, for the commercial television terminal, the 92-Mbps digital signal must be split into two 46-Mbps digital signals because there is no 92-Mbps line speed in the digital hierarchy.

TABLE 17-3 SUMMARY OF THE NORTH AMERICAN DIGITAL HIERARCHY

Line type	Digital signal	Bit rate (Mbps)	Channel capacities	Services offered
T1	DS-1	1.544	24	Voice band telephone
T1C	DS-1C	3.152	48	Voice band telephone
T2	DS-2	6.312	96	Voice band telephone and picture-phone
T3	DS-3	44.736	672	Voice band telephone, picturephone, and broadcast-quality television
T4M	DS-4	274.176	4032	same as T3 except more capacity

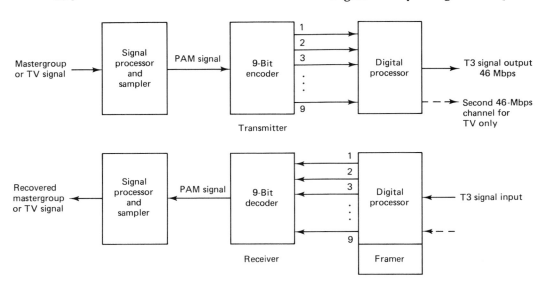

FIGURE 17-16   **Block diagram of a mastergroup or commercial television digital terminal.**

Picturephone Terminal

Essentially, *picturephone* is a low-quality video transmission for use between nondedicated subscribers. For economic reasons it is desirable to encode a picturephone signal into the T2 capacity of 6.312 Mbps, which is substantially less than that for commercial network broadcast signals. This substantially reduces the cost and makes the service affordable. At the same time, this permits the transmission of adequate detail and contrast resolution to satisfy the average picturephone subscriber. Picturephone service is ideally suited to a differential PCM code. Differential PCM is similar to conventional PCM except that the exact magnitude of a sample is not transmitted. Instead, only the difference between that sample and the previous sample is encoded and transmitted. To encode the difference between samples requires substantially fewer bits than encoding the actual sample.

Data Terminal

The portion of communications traffic that involves data (signals other than voice) is increasing exponentially. Also, in most cases, the data rates generated by each individual subscriber are substantially less than the data rate capacities of digital lines. Therefore, it seems only logical that terminals be designed that transmit data signals from several sources over the same digital line.

    Data signals could be sampled directly; however, this would require excessively high sample rates resulting in excessively high transmission bit rates, especially for

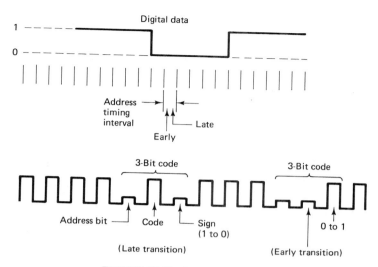

FIGURE 17-17 Data coding format.

sequences of data with few or no transitions. A more efficient method is one that codes the transition times. Such a method is shown in Figure 17-17. With the coding format shown, a 3-bit code is used to identify when transitions occur in the data and whether that transition is from a 1 to a 0, or vice versa. The first bit of the code is called the address bit. When this bit is a logic 1 this indicates that no transition occurred, a logic 0 indicates that a transition did occur. The second bit indicates whether the transition occurred during the first half (0) or during the second half (1) of the sample interval. The third bit indicates the sign or direction of the transition; a 1 for this bit indicates a 0-to-1 transition and a 0 indicates a 1-to-0 transition. Consequently, when there are no transitions in the data, a signal of all 1's is transmitted. Transmission of only the address bit would be sufficient; however, the sign bit provides a degree of error protection and limits error propagation (when one error leads to a second error, etc.). The efficiency of this format is approximately 33%; there are 3 code bits for each data bit. The advantage of using a coded format rather than the original data is that coded data are more efficiently substituted for voice in analog systems. To transmit a 250-kbps data signal, the same bandwidth is required to transmit 60 voice channels with analog multiplexing. With this coded format, a 50-kbps data signal displaces three 64-kbps PCM encoded channels, and a 250-kbps data stream displaces only 12 voice band channels.

LINE ENCODING

Line encoding involves converting standard logic levels (TTL, CMOS, etc.) to a form more suitable to telephone line transmission. Essentially, there are four primary factors that must be considered when selecting a line-encoding format:

1. Timing (clock) recovery
2. Transmission bandwidth
3. Ease of detection and decoding
4. Error detection

Transmission Voltages

Transmission voltages or levels can be categorized as either *unipolar* (UP) or *bipolar* (BP). Unipolar transmission of binary data involves the transmission of only a single nonzero voltage level (e.g., $+V$ for a logic 1 and 0 V or ground for a logic 0). In bipolar transmission, two nonzero voltage levels are involved (e.g., $+V$ for a logic 1 and $-V$ for a logic 0).

Duty Cycle

The *duty cycle* of a binary pulse can also be used to categorize the type of transmission. If the binary pulse is maintained for the entire bit time, this is called *nonreturn-to-zero* (NRZ). If the active time of the binary pulse is less than 100% of the bit time, this is called *return-to-zero* (RZ).

Unipolar and bipolar transmission voltages and return-to-zero and nonreturn-to-zero encoding can be combined in several ways to achieve a particular line encoding scheme. Figure 17-18 shows five line-encoding possibilities.

In Figure 17-18a, there is only one nonzero voltage level ($+V$ = logic 1); a zero voltage simply implies a binary 0. Also, each logic 1 maintains the positive voltage for the entire bit time (100% duty cycle). Consequently, Figure 17-18a represents a unipolar nonreturn-to-zero signal (UPNRZ). In Figure 17-18b, there are two nonzero voltages ($+V$ = logic 1 and $-V$ = logic 0) and a 100% duty cycle is used. Figure 17-18b represents a bipolar nonreturn-to-zero signal (BPNRZ). In Figure 17-18c, only one nonzero voltage is used but each pulse is active for only 50% of the bit time. Consequently, Figure 17-18c represents a unipolar return-to-zero signal (UPRZ). In Figure 17-18d, there are two nonzero voltages ($+V$ = logic 1 and $-V$ = logic 0). Also, each pulse is active only 50% of the total bit time. Consequently, Figure 17-18d represents a bipolar return-to-zero (BPRZ) signal. In Figure 17-18e, there are again two nonzero voltage levels ($-V$ and $+V$), but here both polarities represent a logic 1 and 0 V represents a logic 0. This method of encoding is called *alternate mark inversion* (AMI). With AMI transmissions, each successive logic 1 is inverted in polarity from the previous logic 1. Because return-to-zero is used, this encoding technique is called *bipolar-return-to-zero alternate mark inversion* (BPRZ-AMI).

The method of line encoding used determines the minimum bandwidth required for transmission, how easily a clock may be extracted from it, how easily it may be decoded, and whether it offers a convenient means of detecting errors.

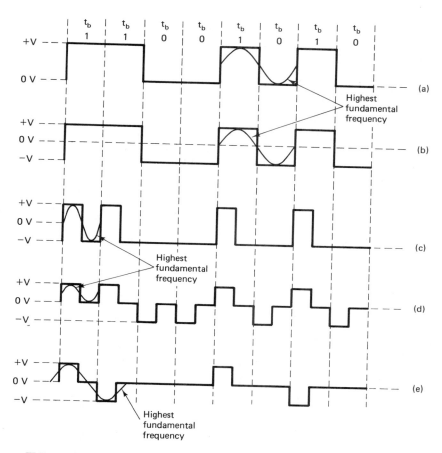

FIGURE 17-18 Line-encoding formats: (a) UPNRZ; (b) BPNRZ; (c) UPRZ; (d) BPRZ; (e) BPRZ-AMI.

Bandwidth Considerations

To determine the minimum bandwidth required to propagate a line-encoded signal, you must determine the highest fundamental frequency associated with it (see Figure 17-18). The highest fundamental frequency is determined from the worst-case (fastest transition) binary bit sequence. With UPNRZ, the worst-case condition is an alternating 1/0 sequence; the highest fundamental frequency takes the time of 2 bits and is therefore equal to one-half the bit rate. With BPNRZ, again the worst-case condition is an alternating 1/0 sequence and the highest fundamental frequency is one-half of the bit rate. With UPRZ, the worst-case condition is two successive 1's. The minimum bandwidth is therefore equal to the bit rate. With BPRZ, the worst-case condition is either successive

1's or 0's and the minimum bandwidth is again equal to the bit rate. With BPRZ-AMI, the worst-case condition is two or more consecutive 1's, and the minimum bandwidth is equal to one-half of the bit rate.

Clock Recovery

To recover and maintain clocking information from received data, there must be a sufficient number of transitions in the data signal. With UPNRZ and BPNRZ, a long string of consecutive 1's or 0's generates a data signal void of transitions and is therefore inadequate for clock synchronization. With UPRZ and BPRZ-AMI, a long string of 0's also generates a data signal void of transitions. With BPRZ, a transition occurs in each bit position regardless of whether the bit is a 1 or a 0. In the clock recovery circuit, the data are simply full-wave rectified to produce a data-independent clock equal to the receive bit rate. Therefore, BPRZ encoding is best suited for clock recovery. If long sequences of 0's are prevented from occurring, BPRZ-AMI encoding is sufficient to ensure clock synchronization.

Error Detection

With UPNRZ, BPNRZ, UPRZ, and BPRZ transmissions, there is no way to determine if the incoming data have errors. With BPRZ-AMI transmissions, an error in any bit will cause a bipolar violation (the reception of two or more consecutive 1's with the same polarity). Therefore, BPRZ-AMI has a built-in error detection mechanism.

Ease of Detection and Decoding

Because unipolar transmission involves the transmission of only one polarity voltage, there is a dc average voltage associated with the signal equal to $+V/2$. Assuming an equal probability of 1's and 0's occurring, bipolar transmissions have an average dc component of 0 V. A dc component is undesirable because it biases the input to a conventional threshold detector (a biased comparator) and could cause a misinterpretation of the logic condition of the received pulses. Therefore, bipolar transmission is better suited to data detection.

Table 17-4 summarizes the minimum bandwidth, average dc voltage, clock recovery, and error detection capabilities of the line-encoding formats shown in Figure 17-18. From Table 17-4 it can be seen that BPRZ-AMI encoding has the best overall characteristics and is therefore the most common method used.

T CARRIERS

T carriers involve the transmission of PCM-encoded time-division-multiplexed digital signals. In addition, T carriers utilize special line-encoded signals and metallic cables that have been conditioned to meet the relatively high bandwidths required for high-speed digital transmissions. Digital signals deteriorate as they are propagated along a

TABLE 17-4 LINE-ENCODING SUMMARY

Encoding format	Minimum BW	Average DC	Clock recovery	Error detection
UPNRZ	$F_b/2^a$	$+V/2$	Poor	No
BPNRZ	$F_b/2^a$	$0 V^a$	Poor	No
UPRZ	F_b	$+V/2$	Good	No
BPRZ	F_b	$0 V^a$	Best^a	No
BPRZ-AMI	$F_b/2^a$	$0 V^a$	Good	Yes^a

^{a} Denotes best performance or quality.

cable due to power loss in the metallic conductors and the low-pass filtering inherent in parallel wire transmission lines. Consequently, *regenerative repeaters* must be placed at periodic intervals. The distance between repeaters is dependent on the transmission bit rate and the line-encoding technique used.

Figure 17-19 shows the block diagram of a regenerative repeater. Essentially, there are three functional blocks: an amplifier-equalizer, a timing circuit, and the regenerator. The amplifier-equalizer shapes the incoming digital signal and raises their power level so that a pulse/no pulse decision can be made by the regenerator circuit. The timing circuit recovers the clocking information from the received data and provides the proper timing information to the regenerator so that decisions can be made at the optimum time that minimizes the chance of an error occurring. Spacing of the repeaters is designed to maintain an adequate signal-to-noise ratio for error-free performance. The signal-to-noise ratio (*S/N*) at the output of a regenerator is exactly what it was at the output of the transmit terminal or at the output of the previous regenerator (i.e., the *S/N* does not deteriorate as a digital signal propagates through a regenerator; in fact, a regenerator reconstructs the original pulses and produces the original *S/N* ratio).

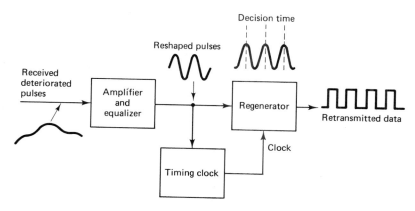

FIGURE 17-19 Regenerative repeater block diagram.

T1 and T1C Carrier Systems

The T1 carrier system utilizes PCM and TDM techniques to provide short-haul transmission of 24 voice band signals. The lengths of T1 carrier systems range from about 5 to 50 miles. T1 carriers use BPRZ-AMI encoding with regenerative repeaters placed every 6000 ft; 6000 ft was chosen because telephone company manholes are located at approximately 6000-ft intervals and these same manholes are used for placement of the repeaters, facilitating convenient installation, maintenance, and repair. The transmission medium for T1 carriers is either a 19- or 22-gauge wire pair.

Because T1 carriers use BPRZ-AMI encoding, they are susceptible to losing synchronization on a long string of consecutive 0's. With a folded binary PCM code, the possibility of generating a long string of consecutive 0's is high (whenever a channel is idle it generates a ± 0-V code which is either seven or eight consecutive 0's). If two or more adjacent voice channels are idle, there is a high probability that a long string of consecutive 0's will be transmitted. To reduce this possibility, the PCM code is inverted prior to transmission and inverted again at the receiver prior to decoding. Consequently, the only time a long string of consecutive 0's is transmitted is when two or more adjacent voice band channels each encode the maximum positive sample voltage, which is unlikely to happen.

With T1 and T1C carrier systems, provisions are taken to prevent more than 14 consecutive 0's from occurring. The transmissions from each frame are monitored for the presence of either 15 consecutive 0's or any one PCM sample (8 bits) without at least one nonzero bit. If either of these conditions occurs, a 1 is substituted into the appropriate bit position. The worst-case conditions are as follows:

```
                   MSB     LSB MSB     LSB
Original           1000 0000   0000 0001    14 consecutive 0's
DS-1 signal                                 (no substitution)


                   MSB     LSB MSB     LSB
Original           1000 0000   0000 0000    15 consecutive 0's
DS-1 signal


Substituted        1000 0000   0000 0010
DS-1 signal                         ↑
                                Substituted
                                bit
```

A 1 is substituted into the second least significant bit. This introduces an encoding error equal to twice the amplitude resolution. This bit is selected rather than the least significant bit because, with the superframe format, during every sixth frame the LSB is the signaling bit and to alter it would alter the signaling word.

```
                   MSB      LSB MSB      LSB MSB      LSB
     Original      1010 1000    0000 0000   0000 0001
     DS-1 signal
```

```
Substituted    1010  1000    0000 0010    0000 0001
DS-1 signal                         ↑
                              Substituted
                                  bit
```

The process shown is used for T1 and T1C carrier systems. Also, if at any time 32 consecutive 0's are received, it is assumed that the system is not generating pulses and is therefore out of service; this is because the occurrence of 32 consecutive 0's is highly unlikely.

T2 Carrier System

The T2 carrier utilizes PCM to time-division multiplex 96 voice band channels into a single 6.312-Mbps data signal for transmission up to 500 miles over a special LOCAP cable. A T2 carrier is also used to carry a single picturephone signal. T2 carriers use BPRZ-AMI encoding. However, because of the higher transmission rate, clock synchronization becomes more critical. A sequence of six consecutive 0's could be sufficient to cause loss of clock synchronization. Therefore, T2 carrier systems use an alternative method of ensuring that ample transitions occur in the data. This method is called *binary six zero substitution* (B6ZS).

With B6ZS, whenever six consecutive 0's occur, one of the following codes is substituted in its place: $0 - + 0 + -$ or $0 + - 0 - +$. The $+$ and $-$ represent positive and negative logic 1s. A zero simply indicates a logic 0 condition. The 6-bit code substituted for the six 0's is selected to purposely cause a bipolar violation. If the violation is caught at the receiver and the B6ZS code is detected, the original six 0's can be substituted back into the data signal. The substituted patterns cause a bipolar violation in the second and fifth bits of the substituted pattern. If DS-2 signals are multiplexed to form DS-3 signals, the B6ZS code must be detected and stripped from the DS-2 signal prior to DS-3 multiplexing. An example of B6ZS is as follows:

```
                    MSB       LSB MSB       LSB MSB
Original           +000   −0+0   000−   0000      000+. . .
data signal                                ⌣⎯⌣
                                          6 0's

                                      Substituted
                                        pattern
                                         ⌢⎯⌢
Encoded            +000   −0+0   000−   0−+0   +−0+
data

                        Bipolar
                        violations
```

T3 Carrier System

A T3 carrier time-division multiplexes 672, PCM-encoded voice channels for transmission over a single metallic cable. The transmission rate for T3 signals is 44.736 Mbps. The encoding technique used with T3 carriers is *binary three zero substitution* (B3ZS). Substitutions are made for any occurrence of three consecutive 0's. There are four substitution patterns used: $00-$, $-0-$, $00+$, and $+0+$. The pattern chosen should cause a bipolar error in the third substitute bit. An example is as follows:

```
              MSB      LSB  MSB      LSB  MSB      LSB
  Original   0+00 00-0    +000 0-00    +-00 00+0
  data
              3 0's        3 0's        3 0's

                        Substituted pattern

  Encoded    0+00 +0-0   + -0- 0-00    + -00 -0+0
  data
                         Bipolar
                         violations
```

T4M Carrier System

A T4M carrier time-division multiplexes 4032 PCM-encoded voice band channels for transmission over a single coaxial cable up to 500 miles. The transmission rate is sufficiently high that substitute patterns are impractical. Instead, T4M-carriers transmit scrambled unipolar NRZ digital signals where the scrambling and descrambling functions are performed in the subscriber's terminal equipment.

FRAME SYNCHRONIZATION

With TDM systems it is imperative that a frame is identified and that individual time slots (samples) within the frame are also identified. To acquire frame synchronization, there is a certain amount of overhead that must be added to the transmission. There are five methods commonly used to establish frame synchronization: added digit framing, robbed digit framing, added channel framing, statistical framing, and unique line signal framing.

Added Digit Framing

T1 carriers using D1, D2, or D3 channel banks use *added digit framing*. There is a special *framing digit* (framing pulse) added to each frame. Consequently, for an 8-kHz sample rate (125-μs frame), there are 8000 digits added per second. With T1 carriers, an alternating 1/0 frame synchronizing pattern is used.

To acquire frame synchronization, the receive terminal searches through the incoming data until it finds the alternating 1/0 sequence used for the framing bit pattern. This encompasses testing a bit, counting off 193 bits, then testing again for the opposite condition. This process continues until an alternating 1/0 sequence is found. Initial frame synchronization is dependent on the total frame time, the number of bits per frame, and the period of each bit. Searching through all possible bit positions requires N tests, where N is the number of bit positions in the frame. On the average, the receiving terminal dwells at a false framing position for two frame periods during a search; therefore, the maximum average synchronization time is

$$\text{synchronization time} = 2NT = 2N^2 t$$

where

T = frame period or Nt
N = number of bits per frame
t = bit time

For the T1 carrier, $N = 193$, $T = 125$ μs, and $t = 0.648$ μs; therefore, a maximum of 74,498 bits must be tested and the maximum average synchronization time is 48.25 ms.

Robbed Digit Framing

When a short frame time is used, added digit framing is very inefficient. This occurs in single-channel PCM systems such as those used in television terminals. An alternative solution is to replace the least significant bit of every nth frame with a framing bit. The parameter n is chosen as a compromise between reframe time and signal impairment. For $n = 10$, the SQR is impaired by only 1 dB. *Robbed digit framing* does not interrupt transmission, but instead, periodically replaces information bits with forced data errors to maintain clock synchronization.

Added Channel Framing

Essentially, *added channel framing* is the same as added digit framing except that digits are added in groups or words instead of as individual bits. The CCITT multiplexing scheme previously discussed uses added channel framing. One of the 32 time slots in each frame is dedicated to a unique synchronizing sequence. The average frame synchronization time for added channel framing is

$$\text{synchronization time (bits)} = \frac{N^2}{2(2^L - 1)}$$

where

N = number of bits per frame
L = number of bits in the frame code

For the CCITT 32-channel system, $N = 256$ and $L = 8$. Therefore, the average number of bits needed to acquire frame synchronization is 128.5. At 2.048 Mbps, the synchronization time is approximately 62.7 μs.

Statistical Framing

With *statistical framing*, it is not necessary to either rob or add digits. With the Gray code, the second bit is a 1 in the central half of the code range and 0 at the extremes. Therefore, a signal that has a centrally peaked amplitude distribution generates a high probability of a 1 in the second digit. A mastergroup signal has such a distribution. With a mastergroup encoder, the probability that the second bit will be a 1 is 95%. For any other bit it is less than 50%. Therefore, the second bit can be used for a framing bit.

Unique Line Code Framing

With *unique line code framing*, the framing bit is different from the information bits. It is either made higher or lower in amplitude or of a different time duration. The earliest PCM/TDM systems used unique line code framing. D1 channel banks used framing pulses that were twice the amplitude of normal data bits. With unique line code framing, added digit or added word framing can be used with it or data bits can be used to simultaneously convey information and carry synchronizing signals. The

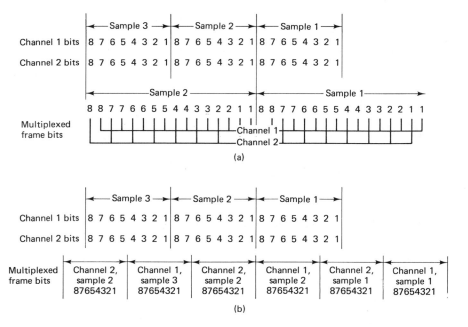

FIGURE 17-20 Interleaving: (a) bit; (b) word.

advantage of unique line code framing is synchronization is immediate and automatic. The disadvantage is the additional processing requirements required to generate and recognize the unique framing bit.

BIT INTERLEAVING VERSUS WORD INTERLEAVING

When time-division multiplexing two or more PCM systems, it is necessary to interleave the transmissions from the various terminals in the time domain. Figure 17-20 shows two methods of interleaving PCM transmissions: *bit interleaving* and *word interleaving*.

T1 carrier systems use word interleaving; 8-bit samples from each channel are interleaved into a single 24-channel TDM frame. Higher-speed TDM systems and delta modulation systems use bit interleaving. The decision as to which type of interleaving to use is usually determined by the nature of the signals to be multiplexed.

QUESTIONS

17-1. Define *multiplexing*.

17-2. Describe time-division multiplexing.

17-3. Describe the Bell System T1 carrier system.

17-4. What is the purpose of the signaling bit?

17-5. What is frame synchronization? How is it achieved in a PCM/TDM system?

17-6. Describe the superframe format. Why is it used?

17-7. What is a codec? A combo chip?

17-8. What is a fixed-data-rate mode?

17-9. What is a variable-data-rate mode?

17-10. What is a DSX? What is it used for?

17-11. Explain *line encoding*.

17-12. Briefly explain unipolar and bipolar transmission.

17-13. Briefly explain return-to-zero and nonreturn-to-zero transmission.

17-14. Contrast the bandwidth considerations of return-to-zero and nonreturn-to-zero transmission.

17-15. Contrast the clock recovery capabilities with return-to-zero and nonreturn-to-zero transmission.

17-16. Contrast the error detection and decoding capabilities of return-to-zero and nonreturn-to-zero transmission.

17-17. What is a regenerative repeater?

17-18. Explain B6ZS and B3ZS. When or why would you use one rather than the other?

17-19. Briefly explain the following framing techniques: added digit framing, robbed digit framing, added channel framing, statistical framing, and unique line code framing.

17-20. Contrast bit and word interleaving.

PROBLEMS

17-1. A PCM/TDM system multiplexes 24 voice band channels. Each sample is encoded into 7 bits and a framing bit is added to each frame. The sampling rate is 9000 samples/second. BPRZ-AMI encoding is the line format. Determine:
 (a) Line speed in bits per second.
 (b) Minimum Nyquist bandwidth.

17-2. A PCM/TDM system multiplexes 32 voice band channels each with a bandwidth of 0 to 4 kHz. Each sample is encoded with an 8-bit PCM code. UPNRZ encoding is used. Determine:
 (a) Minimum sample rate.
 (b) Line speed in bits per second.
 (c) Minimum Nyquist bandwidth.

17-3. For the following bit sequence, draw the timing diagram for UPRZ, UPNRZ, BPRZ, BPNRZ, and BPRZ-AMI encoding:

   bit stream: 1 1 1 0 0 1 0 1 0 1 1 0 0

17-4. Encode the following BPRZ-AMI data stream with B6ZS and B3ZS.

   + - 0 0 0 0 + - + 0 - 0 0 0 0 + - 0 0 + - + 0

Chapter 18

FREQUENCY-DIVISION MULTIPLEXING

INTRODUCTION

In *frequency-division multiplexing* (FDM), multiple sources that originally occupied the same frequency spectrum are each converted to a different frequency band and transmitted simultaneously over a single transmission medium. Thus many relatively narrowband channels can be transmitted over a single wideband transmission system.

FDM is an analog multiplexing scheme; the information entering an FDM system is analog and it remains analog throughout transmission. An example of FDM is the AM commercial broadcast band, which occupies a frequency spectrum from 535 to 1605 kHz. Each station carries an intelligence signal with a bandwidth of 0 to 5 kHz. If the audio from each station were transmitted with their original frequency spectrum, it would be impossible to separate one station from another. Instead, each station amplitude modulates a different carrier frequency and produces a 10-kHz double-sideband signal. Because adjacent stations' carrier frequencies are separated by 10 kHz, the total commercial AM band is divided into 107 10-kHz frequency slots stacked next to each other in the frequency domain. To receive a particular station, a receiver is simply tuned to the frequency band associated with that station's transmissions. Figure 18-1 shows how commercial AM broadcast station signals are frequency-division multiplexed and transmitted over a single transmission medium (free space).

There are many other applications for FDM such as commercial FM and television broadcasting and high-volume telecommunications systems. Within any of the commercial broadcast bands, each station's transmissions are independent of all the other stations' transmissions. Consequently, the multiplexing (stacking) process is accomplished without any synchronization between stations. With a high-volume telephone communication

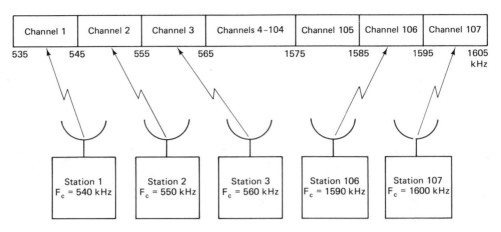

FIGURE 18-1 **Frequency-division-multiplexing commercial AM broadcast band stations.**

system, many voice band telephone channels may originate from a common source and terminate in a common destination. The source and destination terminal equipment is most likely a high-capacity *electronic switching system* (ESS). Because of the possibility of a large number of narrowband channels originating and terminating at the same location, all multiplexing and demultiplexing operations must be synchronized.

AT&T's FDM HIERARCHY

Although AT&T is no longer the only long-distance common carrier in the United States, they still provide a vast majority of the long-distance services and if for no other reason than their overwhelming size, have essentially become the standards organization for the telephone industry in North America.

AT&T's nationwide communications network is subdivided into two classifications: *short haul* (short distance) and *long haul* (long distance). The T1 carrier explained in Chapter 17 is an example of a short-haul communications system.

Long-Haul Communications with FDM

Figure 18-2 shows AT&T's North American FDM Hierarchy for long-haul communications. Only a transmit terminal is shown, although a complete set of inverse functions must be performed at the receiving terminal.

Message Channel

The *message channel* is the basic building block of the FDM hierarchy. The basic message channel was originally intended for voice transmission, although it now includes

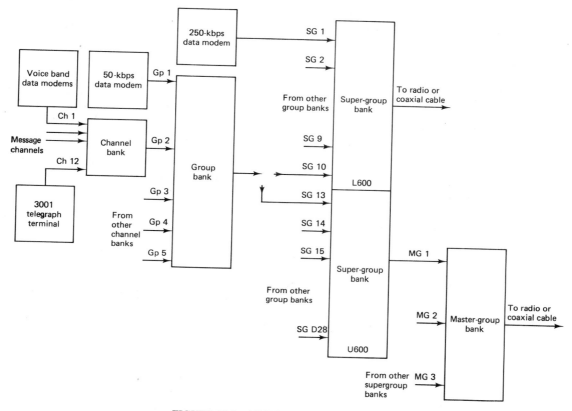

FIGURE 18-2 AT&T's long-haul FDM hierarchy.

any transmissions that utilize voice band frequencies (0 to 4 kHz) such as voice band data circuits. The basic voice band (VB) circuit is called a 3002 channel and is actually bandlimited to a 300- to 3000-Hz band, although for practical considerations, it is considered a 4-kHz channel. The basic 3002 channel can be subdivided into 24 narrower 3001 (telegraph) channels that have been frequency-division multiplexed to form a single 3002 channel.

Basic Group

A *group* is the next higher level in the FDM hierarchy above the basic message channel and is, consequently, the first multiplexing step for the message channels. A basic group is comprised of 12 voice band channels stacked on top of each other in the frequency domain. The 12-channel modulating block is called an *A-type* (analog) channel bank. The 12-channel *group* output of the A-type channel bank is the standard building block for most long-haul *broadband* communications systems. Additions and deletions in total system capacity are accomplished with a minimum of one group (12 VB channels). The A-type channel bank has generically progressed from the early A1 channel bank to the most recent A6 channel bank.

Basic Supergroup

The next higher level in the FDM hierarchy shown in Figure 18-2 is the combination of five groups into a *supergroup*. The multiplexing of five groups is accomplished in a group bank. A single supergroup can carry information from 60 VB channels or handle high-speed data up to 250 kbps.

Basic Mastergroup

The next higher level in the FDM hierarchy is the basic *mastergroup*. A mastergroup is comprised of 10 supergroups (10 supergroups of five groups each = 600 VB channels). Supergroups are combined in supergroup banks to form mastergroups. There are two categories of mastergroups (U600 and L600) which occupy different frequency bands. The type of mastergroup used depends on the system capacity and whether the transmission medium is a coaxial cable or a microwave radio.

Larger Groupings

Master groups can be further multiplexed in mastergroup banks to form *jumbogroups*, *multi-jumbogroups*, and *superjumbogroups*. A basic FDM/FM microwave radio channel carries three mastergroups (1800 VB channels), a jumbogroup has 3600 VB channels, and a superjumbogroup has three jumbogroups (10,800 VB channels).

COMPOSITE BASEBAND SIGNAL

Baseband describes the modulating signal (intelligence) in a communications system. A single message channel is baseband. A group, supergroup, or mastergroup is also baseband. The composite baseband signal is the total intelligence signal prior to modulation of the final carrier. In Figure 18-2 the output of a channel bank is baseband. Also, the output of a group or supergroup bank is baseband. The final output of the FDM multiplexer is the *composite* (total) baseband. The formation of the composite baseband signal can include channel, group, supergroup, and mastergroup banks, depending on the capacity of the system.

Formation of a Group

Figure 18-3a shows how a group is formed with an A-type channel bank. Each voice band channel is bandlimited with an antialiasing filter prior to modulating the channel carrier. FDM uses single-sideband suppressed carrier (SSBSC) modulation. The combination of the balanced modulator and the bandpass filter make up the SSBSC modulator. A balanced modulator is a double-sideband suppressed carrier modulator and the bandpass filter is tuned to the difference between the carrier and the input voice band frequencies (LSB). The ideal input frequency range for a single voice band channel is 0 to 4 kHz. The carrier frequencies for the channel banks are determined from the following expression:

$$F_c = 112 - 4n \text{ kHz}$$

where n is the channel number. Table 18-1 lists the carrier frequencies for channels 1 through 12. Therefore, for channel 1, a 0- to 4-kHz band of frequencies modulates a 108-kHz carrier. Mathematically, the output of the channel 1 bandpass filter is

$$F_{\text{out}} = F_c - F_i$$

where

$$F_c = \text{channel carrier frequency } (112 - 4n \text{ kHz})$$
$$F_i = \text{channel frequency spectrum } (0 \text{ to } 4 \text{ kHz})$$

For channel 1:

$$F_{\text{out}} = 108 \text{ kHz} - (0 \text{ to } 4 \text{ kHz}) = 104 \text{ to } 108 \text{ kHz}$$

For channel 2:

$$F_{\text{out}} = 104 \text{ kHz} - (0 \text{ to } 4 \text{ kHz}) = 100 \text{ to } 104 \text{ kHz}$$

For channel 12:

$$F_{\text{out}} = 64 \text{ kHz} - (0 \text{ to } 4 \text{ kHz}) = 60 \text{ to } 64 \text{ kHz}$$

TABLE 18-1   CHANNEL CARRIER FREQUENCIES

Channel	Carrier frequency (kHz)
1	108
2	104
3	100
4	96
5	92
6	88
7	84
8	80
9	76
10	72
11	68
12	64

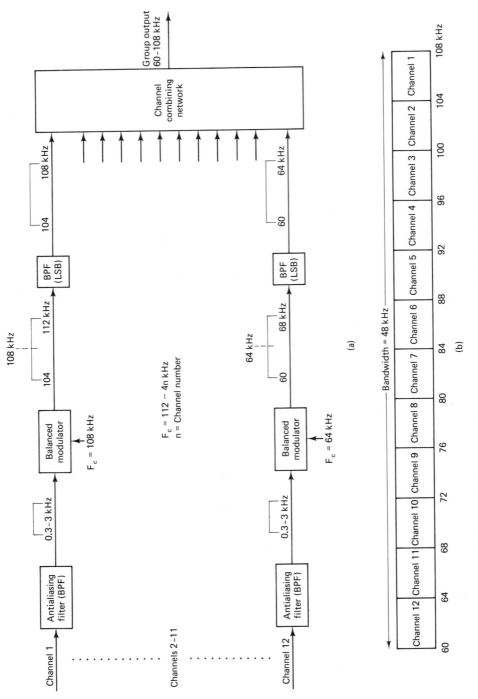

FIGURE 18-3 Formation of a group: (a) A-type channel bank block diagram; (b) output spectrum.

The outputs from the 12 A-type channel modulators are summed in the *linear combiner* to produce the total group spectrum shown in Figure 18-3b (60 to 108 kHz). Note that the total group bandwidth is equal to 48 kHz (12 channels × 4 kHz).

Formation of a Supergroup

Figure 18-4a shows how a supergroup is formed with a group bank and combining a network. Five groups are combined to form a supergroup. The frequency spectrum for each group is 60 to 108 kHz. Each group is mixed with a different group carrier frequency in a balanced modulator then bandlimited with a bandpass filter tuned to the difference frequency band (LSB) to produce a SSBSC signal. The group carrier frequencies are derived from the following expression:

$$F_c = 372 + 48n \text{ kHz}$$

where n is the group number. Table 18-2 lists the carrier frequencies for groups 1 through 5. For group 1, a 60- to 108-kHz group signal modulates a 420-kHz group carrier frequency. Mathematically, the output of the group 1 bandpass filter is

$$F_{out} = F_c - F_i$$

where

F_c = group carrier frequency (372 + 48n kHz)
F_i = group frequency spectrum (60 to 108 kHz)

TABLE 18-2 GROUP CARRIER FREQUENCIES

Group	Carrier frequency (kHz)
1	420
2	468
3	516
4	564
5	612

For group 1:

$$F_{out} = 420 \text{ kHz} - (60 \text{ to } 108 \text{ kHz}) = 312 \text{ to } 360 \text{ kHz}$$

For group 2:

$$F_{out} = 468 \text{ kHz} - (60 \text{ to } 108 \text{ kHz}) = 360 \text{ to } 408 \text{ kHz}$$

For group 5:

$$F_{out} = 612 \text{ kHz} - (60 \text{ to } 108 \text{ kHz}) = 504 \text{ to } 552 \text{ kHz}$$

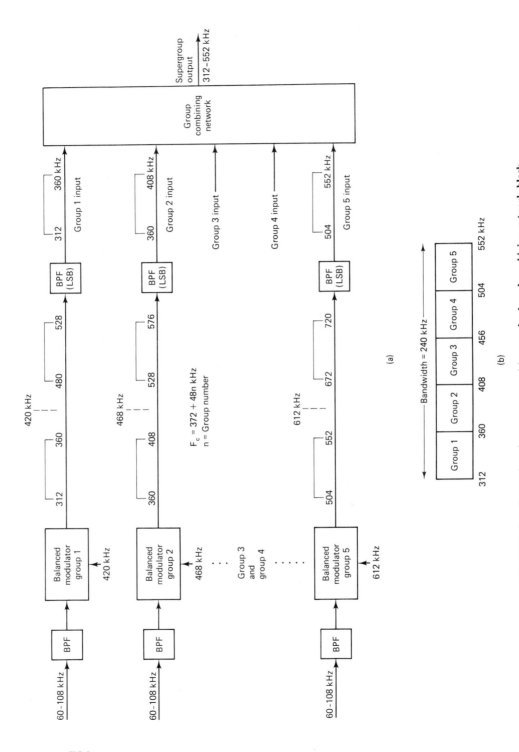

FIGURE 18-4 Formation of a supergroup: (a) group bank and combining network block diagram; (b) output spectrum.

The outputs from the five group modulators are summed in the linear combiner to produce the total supergroup spectrum shown in Figure 18-4b (312 to 552 kHz). Note that the total supergroup bandwidth is equal to 240 kHz (60 channels × 4 kHz).

Formation of a Mastergroup

There are two types of mastergroups: L600 and U600 type. The L600 mastergroup is used for low-capacity microwave systems, while the U600 mastergroup may be further multiplexed and used for higher-capacity microwave radio systems.

U600 Mastergroup. Figure 18-5a shows how a U600 mastergroup is formed with a supergroup bank and combining network. Ten supergroups are combined to form a mastergroup. The frequency spectrum for each supergroup is 312 to 552 kHz. Each supergroup is mixed with a different supergroup carrier frequency in a balanced modulator. The output is then bandlimited to the difference frequency band (LSB) to form a SSBSC signal. The 10 supergroup carrier frequencies are listed in Table 18-3. For supergroup 13, a 312- to 552-kHz supergroup band of frequencies modulates a 1116-kHz carrier frequency. Mathematically, the output from the supergroup 13 bandpass filter is

$$F_{\text{out}} = F_c - F_i$$

where

F_c = supergroup carrier frequency
F_i = supergroup frequency
spectrum (312 to 552 kHz)

For supergroup 13:

$$F_{\text{out}} = 1116 \text{ kHz} - (312 \text{ to } 552 \text{ kHz}) = 564 \text{ to } 804 \text{ kHz}$$

For supergroup 14:

$$F_{\text{out}} = 1364 \text{ kHz} - (312 \text{ to } 552 \text{ kHz}) = 812 \text{ to } 1052 \text{ kHz}$$

For supergroup D28:

$$F_{\text{out}} = 3396 \text{ kHz} - (312 \text{ to } 552 \text{ kHz}) = 2844 \text{ to } 3084 \text{ kHz}$$

The outputs from the 10 supergroup modulators are summed in the linear summer to produce the total mastergroup spectrum shown in Figure 18-4b (564 to 3084 kHz). Note that between any two adjacent supergroups there is a void band of frequencies that is not included within any supergroup band. These voids are called *guard bands*. The guard bands are necessary because the demultiplexing process is accomplished through filtering and down-converting. Without the guard bands, it would be difficult to separate one supergroup from an adjacent supergroup. The guard bands reduce the *quality factor* (Q) required to perform the necessary filtering. The guard band is 8 kHz

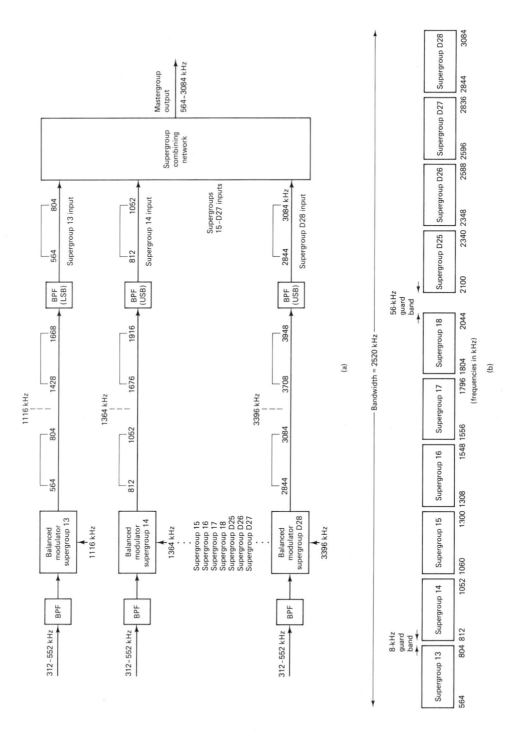

FIGURE 18-5 Formation of a U600 mastergroup: (a) supergroup bank and combining network block diagram; (b) output spectrum.

706

TABLE 18-3 SUPERGROUP CARRIER FREQUENCIES FOR A U600 MASTERGROUP

Supergroup	Carrier frequency (kHz)
13	1116
14	1364
15	1612
16	1860
17	2108
18	2356
D25	2652
D26	2900
D27	3148
D28	3396

between all supergroups except 18 and D25, where it is 56 kHz. Consequently, the bandwidth of a U600 mastergroup is 2520 kHz (564 to 3084 kHz), which is greater than is necessary to stack 600 voice band channels (600 × 4 kHz = 2400 kHz).

Guard bands were not necessary between adjacent groups because the group frequencies are sufficiently low and it is relatively easy to build bandpass filters to separate one group from another.

In the channel bank, the antialiasing filter at the channel input passes a 0.3-to 3-kHz band. The separation between adjacent channel carrier frequencies is 4 kHz. Therefore, there is a 1300-Hz guard band between adjacent channels. This is shown in Figure 18-6.

L600 Mastergroup. With an L600 mastergroup, 10 supergroups are combined as with the U600 mastergroup except that the supergroup carrier frequencies are lower. Table 18-4 lists the supergroup carrier frequencies for a L600 mastergroup. With an L600 mastergroup, the composite baseband spectrum occupies a lower-frequency band than the U-type mastergroup (Figure 18-7). An L600 mastergroup is not further multiplexed. Therefore, the maximum channel capacity for a microwave or coaxial cable system using a single L600 mastergroup is 600 voice band channels.

Formation of a Radio Channel

A *radio channel* comprise either a single L600 mastergroup or up to three U600 mastergroups (1800 voice band channels). Figure 18-8 shows how an 1800-channel composite FDM baseband signal is formed for transmission over a single microwave radio channel. Mastergroup 1 is transmitted directly as is, while mastergroups 2 and 3 undergo an additional multiplexing step. The three mastergroups are summed in a mastergroup

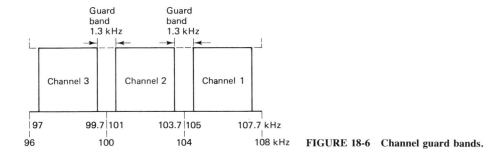

FIGURE 18-6   Channel guard bands.

combining network to produce the output spectrum shown in Figure 18-8b. Note the 80-kHz guard band between adjacent mastergroups.

The system shown in Figure 18-8 can be increased from 1800 voice band channels to 1860 by adding an additional supergroup (supergroup 12) directly to mastergroup 1. The addition 312- to 552-kHz supergroup band extends the output spectrum to 312 to 8284 kHz.

Frequency Translation in FDM

Essentially, FDM is the process of transposing or translating a given input frequency band to some higher frequency band, where it is combined with other translated signals. With the L1860 FDM system shown in Figure 18-2, a single voice band channel may undergo as many as four frequency translations before it is transmitted. When troubleshooting an FDM system, it is necessary that the frequency band of a given channel be known at the various levels of multiplexing.

**TABLE 18-4   SUPERGROUP
CARRIER FREQUENCIES FOR
A L600 MASTERGROUP**

Supergroup	Carrier frequency (kHz)
1	612
2	Direct
3	1116
4	1364
5	1612
6	1860
7	2108
8	2356
9	1860
10	3100

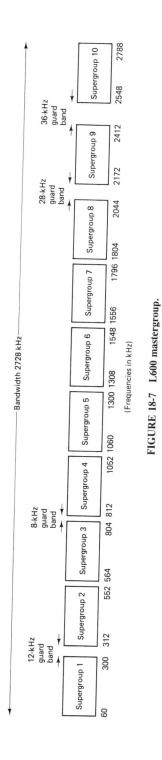

FIGURE 18-7 L600 mastergroup.

(Frequencies in kHz)

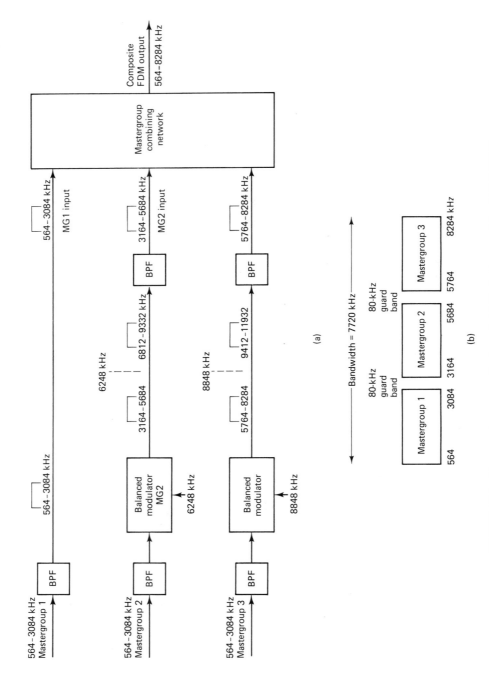

FIGURE 18-8 Three-mastergroup radio channel: (a) block diagram; (b) output spectrum.

EXAMPLE 18-1

For a single-voice-band channel:

(a) Determine its frequency band at the output of the channel, group, supergroup, and mastergroup combiners when it is assigned to channel 4, group 2, supergroup 16, and mastergroup 2.

(b) Determine the frequency that a 1-kHz tone on the same channel would translate to.

Solution (a) For an ideal bandwidth of 0 to 4 kHz, the frequency band at the channel, group, supergroup, and master group combiners is determined as follows:

$$\text{channel bank out} = 96 \text{ kHz} - (0 \text{ to } 4 \text{ kHz}) = 92 \text{ to } 96 \text{ kHz}$$

$$\text{GP bank out} = 468 \text{ kHz} - (92 \text{ to } 96 \text{ kHz}) = 372 \text{ to } 376 \text{ kHz}$$

$$\text{SG bank out} = 1860 \text{ kHz} - (372 \text{ to } 376 \text{ kHz}) = 1484 \text{ to } 1488 \text{ kHz}$$

$$\text{MG bank out} = 6248 \text{ kHz} - (1484 \text{ to } 1488 \text{ kHz}) = 4760 \text{ to } 4764 \text{ kHz}$$

(b) For a 1-kHz test tone,

$$\text{channel bank out} = 96 \text{ kHz} - 1 \text{ kHz} = 95 \text{ kHz}$$

$$\text{GP bank out} = 468 \text{ kHz} - 95 \text{ kHz} = 373 \text{ kHz}$$

$$\text{SG bank out} = 1860 \text{ kHz} - 373 \text{ kHz} = 1487 \text{ kHz}$$

$$\text{MG bank out} = 6248 \text{ kHz} - 1487 \text{ kH} = 4761 \text{ kHz}$$

L CARRIERS

L carrier systems transmit frequency-division-multiplexed voice band signals over a coaxial cable for distances up to 4000 miles. L carriers have generically progressed from the early L1 and L3 systems to the high-capacity L4, L5, and L6 systems. L carrier systems combine many coaxial cables into a single tube and carry dozens of mastergroups and literally thousands of two-way simultaneous voice transmissions. In the near future, L cables are destined to be replaced by even higher-capacity fiber-optic systems.

Carrier Synchronization

With FDM, the receive channel, group, supergroup, and mastergroup carrier frequencies must be synchronized to the transmit carrier frequencies. If they are not synchronized, the recovered voice band signals will be offset in frequency from their original spectrum by the difference in the two carrier frequencies. FDM uses single-sideband suppressed carrier transmission. The carriers are suppressed in the balanced modulators at the transmit terminal and therefore cannot be recovered in the receive terminal directly from the composite baseband signal. Consequently, a carrier pilot frequency is transmitted together with the baseband signal for the purpose of carrier synchronization.

The channel, group, supergroup, and mastergroup carrier frequencies are all integral multiples of 4 kHz. Therefore, if the carrier frequencies at the transmit and receive terminals are derived from a single 4-kHz master oscillator, all of the transmit and receive carriers will be synchronized.

In an FDM communications system, one station is designated as the *master station*. Every other station in the system is a slave. That is, there is a single 4-kHz master oscillator from which all carrier frequencies in the system are derived. The 4-kHz master oscillator is multiplied to either a 64-, 312-, or 552-kHz pilot frequency, combined with the composite baseband signal, and transmitted to each slave station. The slave stations detect the pilot, divide it down to a 4-kHz base frequency, and synchronize their 4-kHz slave oscillators to it. Each slave station then multiplies the synchronous 4-kHz signal to produce synchronous channel, group, supergroup, and mastergroup carrier frequencies. If the master 4-kHz oscillator drifts in frequency, each slave station's 4-kHz oscillator tracks the frequency shift and the system remains synchronous.

D-Type Supergroups

With the U600 mastergroup, supergroups 25 through 28 are preceded by the letter ''D.'' These supergroup carrier frequencies are derived in a slightly different fashion than the other carrier frequencies. Except for the D-supergroups, all carrier frequencies are generated through integer multiplication of the 4-kHz base frequency. The D-supergroups are generated through both multiplication and heterodyning. The supergroup 15- to 18-carrier frequencies are mixed with a 1040-kHz harmonic and the sum frequencies become the D25 to D28 carrier frequencies.

This method of deriving the higher supergroup carrier frequencies has no effect on the frequency offset introduced by frequency drift of the 4-kHz base oscillator. However, this technique reduces the amount of *phase jitter* (incidental phase modulation generally caused by ac ripple present in dc power supplies) in the higher supergroup carrier frequencies. When two frequencies with phase jitter are combined in a mixer, the total phase jitter in the sum frequency is less than the algebraic sum of the phase jitter of the two signals. Consequently, using D-type supergroup carriers has no effect on the frequency shift but reduces the magnitude of the total phase jitter in the higher supergroups. Before this technique was introduced, the voice band channels located in the higher supergroups could not be used for voice band data transmission when PSK modulation was used because the phase jitter caused excessive transmission errors.

Amplitude Regulation

Figure 18-9a shows the gain characteristics for an ideal transmission medium. For the ideal situation (Figure 18-9a), the gain for all baseband frequencies is the same. In a more practical situation, the gain is not the same for all frequencies (Figure 18-9b). Therefore, the demultiplexed voice band channels do not have the same amplitude characteristics as the original voice band signals had. This is called *amplitude distortion*. To reduce amplitude distortion, filters with the opposite characteristics as those introduced

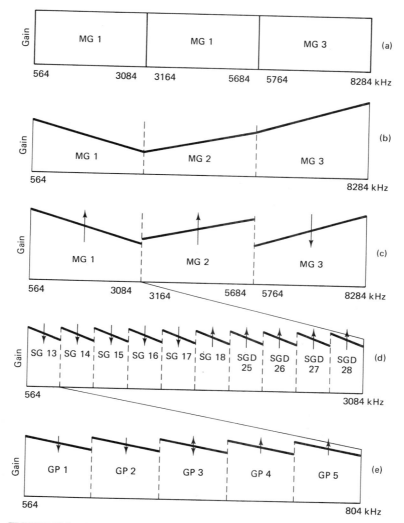

FIGURE 18-9 Gain characteristics: (a) ideal gain versus frequency characteristics; (b) amplitude distortion; (c) mastergroup regulation; (d) supergroup regulation; (e) group regulation.

in the transmission medium can be added to the system, thus canceling the distortion. This is impractical because every transmission system has different characteristics and a special filter would have to be designed and built for each system.

Automatic gain devices (regulators) are used in the receiver demultiplexing equipment to compensate for amplitude distortion introduced in the transmission medium. Amplitude regulation is accomplished in several stages. First, the amplitude of each mastergroup is adjusted or regulated (mastergroup regulation; Figure 18-9c), then each supergroup within each mastergroup is regulated (supergroup regulation; Figure 18-9d). The last stage of regulation is performed at a group level (group regulation; Figure 18-9e).

Regulation is performed by monitoring the power level of a mastergroup, supergroup, or group *pilot*, then regulating the entire frequency band associated with it, depending on the pilot level. Pilots are monitored rather than the actual signal level because the signal levels vary depending on how many channels are in use at a given time. A pilot is a continuous signal with a constant power level.

Figure 18-10 shows how the regulation pilots are nested within the composite baseband signal. Each group has a 104.08-kHz pilot added to it in the channel combining network. Consequently, each supergroup has five group pilots. The group 1 pilot is also the supergroup pilot. Thus each mastergroup has 50 group pilots, of which 10 are also supergroup pilots. A separate 2840-kHz mastergroup pilot is added to each mastergroup in the supergroup combining network, making a total of 51 pilots per mastergroup.

Figure 18-11 is a partial block diagram for an FDM demultiplexer that shows how the pilots are monitored and used to separately regulate the mastergroups, supergroups, and groups automatically.

HYBRID DATA

With *hybrid* data it is possible to combine digitally encoded signals with FDM signals and transmit them as one composite baseband signal. There are four primary types of hybrid data: data under voice (DUV), data above voice (DAV), data above video (DAVID), and data in voice (DIV).

Data under Voice

Figure 18-12a shows the block diagram of AT&T's 1.544-Mbps *data under FDM voice* system. With the L1800 FDM system explained earlier in this chapter, the 0 to 564-kHz frequency spectrum is void of baseband signals. With FM transmission, the lower baseband frequencies realize the highest signal-to-noise ratios. Consequently, the best portion of the baseband spectrum was unused. DUV is a means of utilizing this spectrum for the transmission of digitally encoded signals. A T1 carrier system can be converted to a quasi-analog signal and then frequency-division multiplexed onto the lower portion of the FDM spectrum.

In Figure 18-12a, the *elastic store* removes timing jitter from the incoming data stream. The data are then *scrambled* to suppress the discrete high-power spectral components. The advantage of scrambling is the randomized data output spectrum is continuous and has a predictable effect on the FDM radio system. In other words, the data present a load to the system equivalent to adding additional FDM voice channels. The serial seven-level *partial response* encoder (correlative coder) compresses the data bandwidth and allows a 1.544-Mbps signal to be transmitted in a bandwidth less than 400 kHz. The low-pass filter performs the final spectral shaping of the digital information and

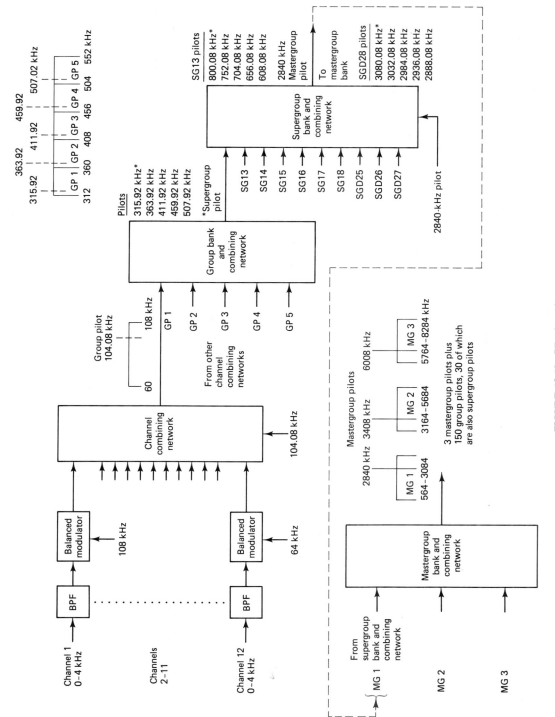

FIGURE 18-10 Pilot insertion.

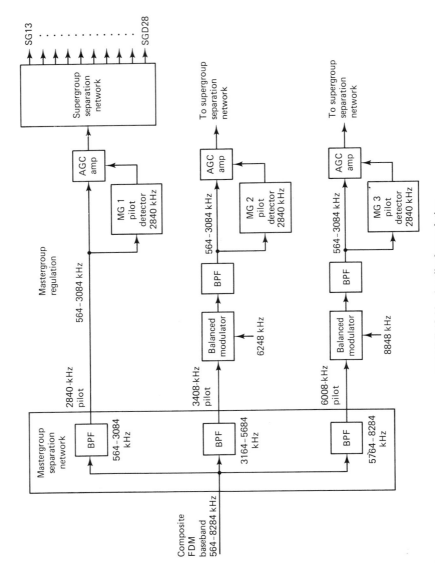

FIGURE 18-11 Amplitude regulation.

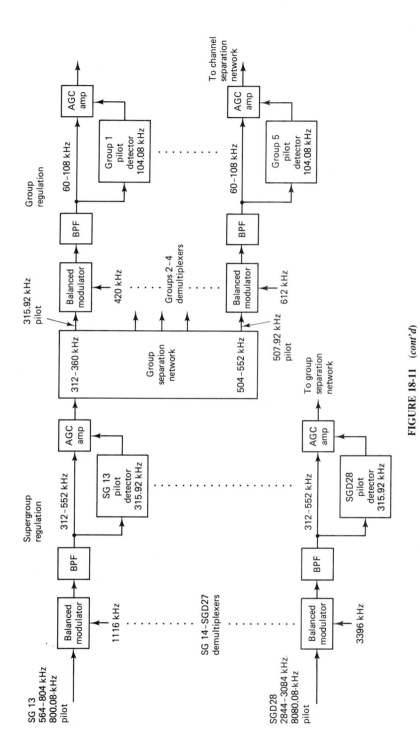

FIGURE 18-11 *(cont'd)*

717

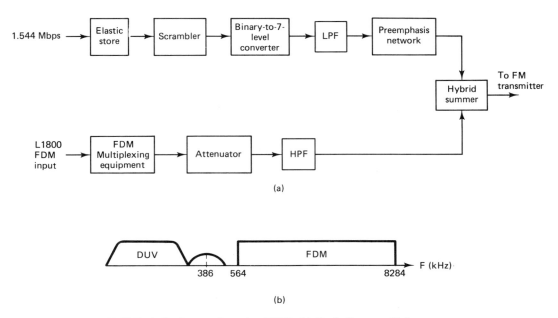

(a)

(b)

FIGURE 18-12   Data under voice (DUV): (a) block diagram; (b) frequency spectrum.

suppresses the spectral power above 386 kHz. This prevents the DUV information from interfering with the 386-kHz pilot control tone. The DUV signal is preemphasized and combined with the L1800 baseband signal. The output spectrum is shown in Figure 18-12b.

AT&T uses DUV for *digital data service* (DDS). DDS is intended to provide a communications medium for the transfer of digital data from station to station without the use of a data modem. DDS circuits are guaranteed to average 99.5% error-free seconds at 56 kbps.

Data above Voice

Figure 18-13 shows the block diagram and frequency spectrum for a *data above voice* system. The advantage of DAV is for FDM systems that extend into the low end of the baseband spectrum; the low-frequency baseband does not have to be vacated for data transmission. With DAV, data PSK modulates a carrier which is then up-converted to a frequency above the FDM message. With DAV, up to 3.152 Mbps can be cost-effectively transmitted using existing FDM/FM microwave systems (Chapter 19).

Data above Video

Essentially, *data above video* is the same as DAV except that the lower baseband spectrum is a *vestigal sideband* video signal rather than a composite FDM signal. Figure 18-14 shows the frequency spectrum for a DAVID system.

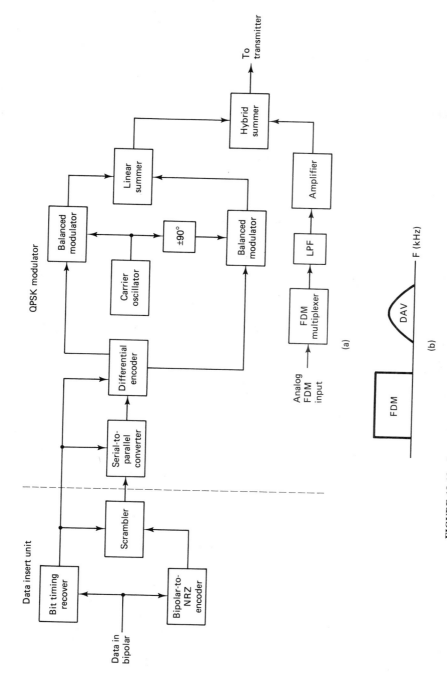

FIGURE 18-13 Data above voice (DAV): (a) block diagram; (b) frequency spectrum.

FIGURE 18-14   Data above
video (DAVID).

Data in Voice

Data in voice, developed by Fujitsu of Japan, uses an eight-level PAM-VSB modulation technique with steep filtering. It uses a highly compressed partial response encoding technique which gives it a high bandwidth efficiency of nearly 5 bps/Hz (1.544-Mbps data are transmitted in a 344-kHz bandwidth).

QUESTIONS

18-1. Describe frequency-division multiplexing.

18-2. Describe a message channel.

18-3. Describe the formation of a group, a supergroup, and a mastergroup.

18-4. Define *baseband* and *composite baseband*.

18-5. Describe the modulators used in FDM multiplexers.

18-6. Describe the difference between an L600 and a U600 mastergroup.

18-7. What is a guard band? When is a guard band used?

18-8. Are FDM-multiplexed communications systems synchronous? Explain.

18-9. Why are D-type supergroups used?

18-10. What are the two types of pilots used with FDM systems, and what is the purpose of each?

18-11. What are the four primary types of hybrid data networks?

18-12. What is the difference between a DUV and a DAV network?

18-13. At what level is the 104.08-kHz pilot inserted?

PROBLEMS

18-1. Calculate the 12 channel carrier frequencies for the U600 FDM system.

18-2. Calculate the five group carrier frequencies for the U600 FDM system.

18-3. Calculate the frequency range for a single channel at the output of the channel, group, supergroup, and mastergroup combining networks for channel 3, group 4, supergroup 15, mastergroup 2.

18-4. Determine the frequency that a 1-kHz test tone will translate to at the output of the channel, group, supergroup, and mastergroup combining networks for channel 5, group 5, supergroup 27, mastergroup 3.

18-5. Determine the frequency at the output of the mastergroup combining network for a group pilot of 104.08 kHz on group 2, supergroup 13, mastergroup 2.

18-6. Calculate the frequency range for group 4, supergroup 18, mastergroup 1 at the output of the mastergroup combining network.

18-7. Calculate the frequency range for supergroup 15, mastergroup 2 at the output of the master-group combining network.

Chapter 19

MICROWAVE COMMUNICATIONS AND SYSTEM GAIN

INTRODUCTION

Presently, terrestrial (earth) *microwave radio relay systems* provide less than half of the total message circuit mileage in the United States. However, at one time microwave systems carried the bulk of long-distance communications for the public telephone network, military and governmental agencies, and specialized private communications networks. There are many different types of microwave systems that operate over distances varing from 15 to 4000 miles in length. *Intrastate* or *feeder service* systems are generally categorized as *short haul* because they are used for relatively short distances. *Long-haul* radio systems are those used for relatively long distances, such as interstate and backbone route applications. Microwave system capacities range from less than 12 voice band channels to more than 22,000. Early microwave systems carried frequency-division-multiplexed voice band circuits and used conventional, noncoherent frequency modulation techniques. More recently developed microwave systems carry pulse-code-modulated time-division-multiplexed voice band circuits and use more modern digital modulation techniques, such as phase shift keying and quadrature amplitude modulation. This chapter deals primarily with conventional FDM/FM microwave systems, and Chapter 20 deals with the more modern PCM/PSK techniques.

SIMPLIFIED MICROWAVE SYSTEM

A simplified block diagram of a microwave radio system is shown in Figure 19-1. The *baseband* is the composite signal that modulates the FM carrier and may comprise one or more of the following:

722

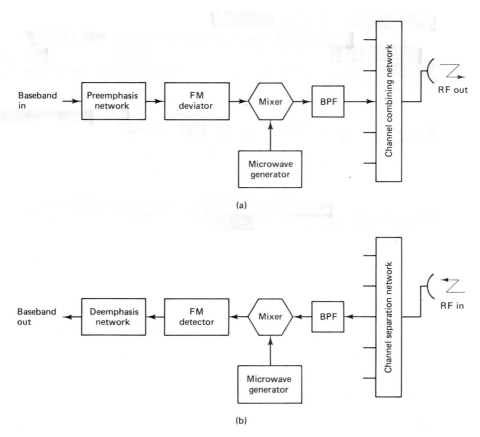

FIGURE 19-1 **Simplified block diagram of a microwave system: (a) transmitter; (b) receiver.**

1. Frequency-division-multiplexed voice band channels
2. Time-division-multiplexed voice band channels
3. Broadcast-quality composite video or picturephone

Microwave Transmitter

In the *microwave transmitter* (Figure 19-1a), a *preemphasis* network precedes the FM deviator. The preemphasis network provides an artificial boost in amplitude to the higher baseband frequencies. This allows the lower baseband frequencies to frequency modulate the IF carrier and the higher baseband frequencies to phase modulate it. This scheme assures a more uniform signal-to-noise ratio throughout the entire baseband spectrum. An FM deviator provides the modulation of the IF carrier which eventually becomes the main microwave carrier. Typically, IF carrier frequencies are between 60 and 80

MHz, with 70 MHz the most common. *Low-index* frequency modulation is used in the FM deviator. Typically, modulation indices are kept between 0.5 and 1. This produces a *narrowband* FM signal at the output of the deviator. Consequently, the IF bandwidth resembles conventional AM and is approximately equal to twice the highest baseband frequency.

The IF and its associated sidebands are up-converted to the microwave region by the AM mixer, microwave oscillator, and bandpass filter. Mixing, rather than multiplying, is used to translate the IF frequencies to RF frequencies because the modulation index is unchanged by the heterodyning process. Multiplying the IF carrier would also multiply the frequency deviation and the modulation index, thus increasing the bandwidth. Typically, frequencies above 1000 MHz (1 GHz) are considered microwave frequencies. Presently, there are microwave systems operating with carrier frequencies up to approximately 18 GHz. The most common microwave frequencies currently being used are the 2-, 4-, 6-, 12-, and 14-GHz bands. The channel combining network provides a means of connecting more than one microwave transmitter to a single transmission line feeding the antenna.

Microwave Receiver

In the receiver (Figure 19-1b), the channel separation network provides the isolation and filtering necessary to separate individual microwave channels and direct them to their respective receivers. The bandpass filter, AM mixer, and microwave oscillator down-convert the RF microwave frequencies to IF frequencies and pass them on to the FM demodulator. The FM demodulator is a conventional, *noncoherent* FM detector (i.e., a discriminator or a ratio detector). At the output of the FM detector, a deemphasis network restores the baseband signal to its original amplitude versus frequency characteristics.

MICROWAVE REPEATERS

The permissible distance between a microwave transmitter and its associated microwave receiver depends on several system variables, such as transmitter output power, receiver noise threshold, terrain, atmospheric conditions, system capacity, reliability objectives, and performance expectations. Typically, this distance is between 15 and 40 miles. Longhaul microwave systems span distances considerably longer than this. Consequently, a single-hop microwave system, such as the one shown in Figure 19-1, is inadequate for most practical system applications. With systems that are longer than 40 miles or when geographical obstructions, such as a mountain, block the transmission path, *repeaters* are needed. A microwave repeater is a receiver and a transmitter placed back to back or in tandem with the system. A block diagram of a microwave repeater is shown in Figure 19-2. The repeater station receives a signal, amplifies and reshapes it, then retransmits the signal to the next repeater or terminal station downline from it.

Basically, there are two types of microwave repeaters: *baseband* and *IF* (Figure

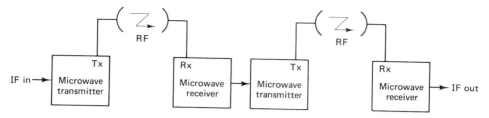

FIGURE 19-2   Microwave repeater.

19-3). IF repeaters are also called *heterodyne* repeaters. With an IF repeater (Figure 19-3a), the received RF carrier is down-converted to an IF frequency, amplified, reshaped, up-converted to an RF frequency, and then retransmitted. The signal is never demodulated beyond IF. Consequently, the baseband intelligence is unmodified by the repeater. With a baseband repeater (Figure 19-3b), the received RF carrier is down-converted to an IF frequency, amplified, filtered, and then further demodulated to baseband. The baseband signal, which is typically frequency-division-multiplexed voice band channels, is further demodulated to a mastergroup, supergroup, group, or even channel level. This allows the baseband signal to be reconfigured to meet the routing needs of the overall communications network. Once the baseband signal has been reconfigured, it FM modulates an IF carrier which is up-converted to an RF carrier and then retransmitted.

Figure 19-3c shows another baseband repeater configuration. The repeater demodulates the RF to baseband, amplifies and reshapes it, then modulates the FM carrier. With this technique, the baseband is not reconfigured. Essentially, this configuration accomplishes the same thing that an IF repeater accomplishes. The difference is that in a baseband configuration, the amplifier and equalizer act on baseband frequencies rather than IF frequencies. The baseband frequencies are generally less than 9 MHz, whereas the IF frequencies are in the range 60 to 80 MHz. Consequently, the filters and amplifiers necessary for baseband repeaters are simpler to design and less expensive than the ones required for IF repeaters. The disadvantage of a baseband configuration is the addition of the FM terminal equipment.

DIVERSITY

Microwave systems use *line-of-sight* transmission. There must be a direct, line-of-sight signal path between the transmit and the receive antennas. Consequently, if that signal path undergoes a severe degradation, a service interruption will occur. *Diversity* suggests that there is more than one transmission path or method of transmission available between a transmitter and a receiver. In a microwave system, the purpose of using diversity is to increase the reliability of the system by increasing its availability. When there is more than one transmission path or method of transmission available, the system can select the path or method that produces the highest-quality received signal. Generally,

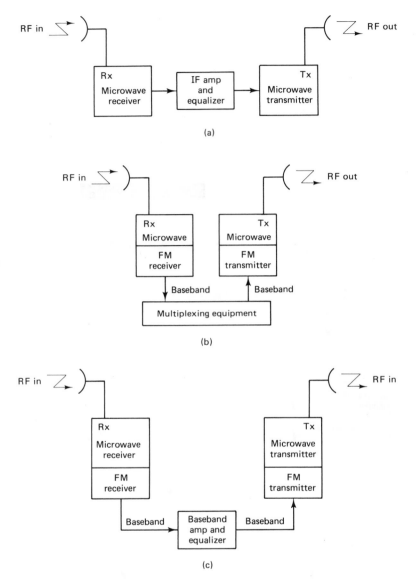

FIGURE 19-3   Microwave repeaters: (a) IF; (b) and (c) baseband.

the highest quality is determined by evaluating the carrier-to-noise (C/N) ratio at the receiver input or by simply measuring the received carrier power. Although there are many ways of achieving diversity, the most common methods used are *frequency*, *space*, and *polarization*.

Frequency Diversity

Frequency diversity is simply modulating two different RF carrier frequencies with the same IF intelligence, then transmitting both RF signals to a given destination. At the destination, both carriers are demodulated, and the one that yields the better-quality IF signal is selected. Figure 19-4 shows a single-channel frequency-diversity microwave system.

In Figure 19-4a, the IF input signal is fed to a power splitter, which directs it to microwave transmitters A and B. The RF outputs from the two transmitters are combined in the channel-combining network and fed to the transmit antenna. At the receive end (Figure 19-4b), the channel separator directs the A and B RF carriers to their respective microwave receivers, where they are down-converted to IF. The quality detector circuit determines which channel, A or B, is the higher quality and directs that channel through the IF switch to be further demodulated to baseband. Many of the temporary, adverse atmospheric conditions that degrade an RF signal are frequency selective; they may degrade one frequency more than another. Therefore, over a given period of time, the IF switch may switch back and forth from receiver A to receiver B, and vice versa many times.

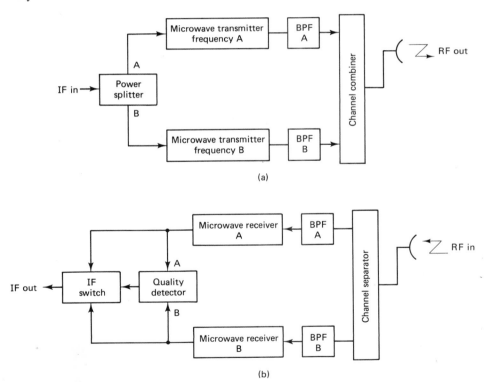

FIGURE 19-4   Frequency diversity microwave system: (a) transmitter; (b) receiver.

Space Diversity

With space diversity, the output of a transmitter is fed to two or more antennas that are physically separated by an appreciable number of wavelengths. Similarly, at the receiving end, there may be more than one antenna providing the input signal to the receiver. If multiple receiving antennas are used, they must also be separated by an appreciable number of wavelengths. Figure 19-5 shows a single-channel space-diversity microwave system.

When space diversity is used, it is important that the electrical distance from a transmitter to each of its antennas and to a receiver from each of its antennas is an equal multiple of wavelengths long. This is to ensure that when two or more signals of the same frequency arrive at the input to a receiver, they are in phase and additive. If received out of phase, they will cancel and, consequently, result in less received signal power than if simply one antenna system were used. Adverse atmospheric conditions are often isolated to a very small geographical area. With space diversity, there is more than one transmission path between a transmitter and a receiver. When adverse atmospheric conditions exist in one of the paths, it is unlikely that the alternate path is experiencing the same degradation. Consequently, the probability of receiving an accept-

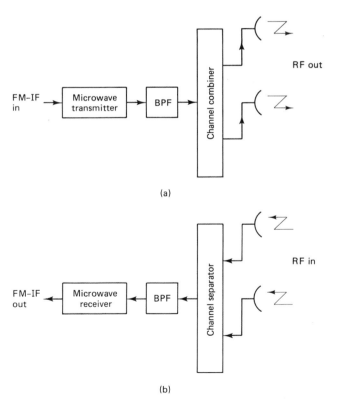

(a)

(b)

FIGURE 19-5 Space-diversity microwave system: (a) transmitter; (b) receiver.

able signal is higher when space diversity is used than when no diversity is used. An alternate method of space diversity uses a single transmitting antenna and two receiving antennas separated vertically. Depending on the atmospheric conditions at a particular time, one of the receiving antennas should be receiving an adequate signal. Again, there are two transmission paths that are unlikely to be affected simultaneously by fading.

Polarization Diversity

With polarization diversity, a single RF carrier is propagated with two different electromagnetic polarizations (vertical and horizontal). Electromagnetic waves of different polarizations do not necessarily experience the same transmission impairments. Polarization diversity is generally used in conjunction with space diversity. One transmit/receive antenna pair is vertically polarized and the other is horizontally polarized. It is also possible to use frequency, space, and polarization diversity simultaneously.

PROTECTION SWITCHING

Radio path losses vary with atmospheric conditions. Over a period of time, the atmospheric conditions between transmitting and receiving antenna can vary significantly, causing a corresponding reduction in the received signal strength of 20, 30, 40, or more dB. This reduction in signal strength is referred to as a *radio fade*. *Automatic gain control circuits*, built into radio receivers, can compensate for fades of 25 to 40 dB, depending on the system design. However, when fades in excess of 40 dB occur, this is equivalent to a total loss of the received signal. When this happens, service continuity is lost. To avoid a service interruption during periods of deep fades or equipment failures, alternate facilities are temporarily made available in what is called a *protection switching* arrangement. Essentially, there are two types of protection switching arrangements: *hot standby* and *diversity*. With hot standby protection, each working radio channel has a dedicated backup or spare channel. With diversity protection, a single backup channel is made available to as many as 11 working channels. Hot standby systems offer 100% protection for each working radio channel. A diversity system offers 100% protection only to the first working channel that fails. If two radio channels fail at the same time, a service interruption will occur.

Hot Standby

Figure 19-6a shows a single-channel hot standby protection switching arrangement. At the transmitting end, the IF goes into a *head-end bridge*, which splits the signal power and directs it to the working and the spare (standby) microwave channels simultaneously. Consequently, both the working and standby channels are carrying the same baseband information. At the receiving end, the IF switch passes the IF signal from the working channel to the FM terminal equipment. The IF switch continuously monitors the received signal power on the working channel and if it fails, switches to the standby channel. When the IF signal on the working channel is restored, the IF switch resumes its normal position.

Diversity

Figure 19-6b shows a diversity protection switching arrangement. This system has two working channels (channel 1 and channel 2), one spare channel, and an *auxiliary* channel. The IF switch at the receive end continuously monitors the receive signal strength of both working channels. If either one should fail, the IF switch detects a loss of carrier and sends back to the transmitting station IF switch a VF (*voice frequency*) tone-encoded signal that directs it to switch the IF signal from the failed channel onto the spare microwave channel. When the failed channel is restored, the IF switches resume their normal positions. The auxiliary channel simply provides a transmission path between the two IF switches. Typically, the auxiliary channel is a low-capacity low-power microwave radio that is designed to be used for a maintenance channel only.

Reliability

The number of repeater stations between protection switches depends on the *reliability objectives* of the system. Typically, there are between two and six repeaters between switching stations.

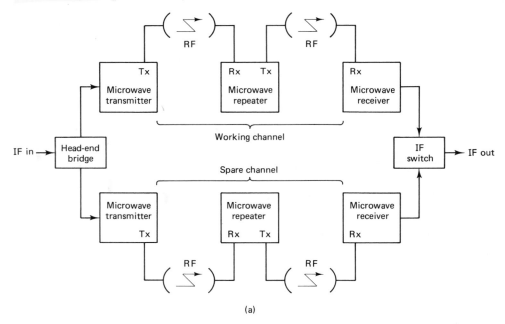

(a)

FIGURE 19-6   Microwave protection switching arrangements: (a) hot standby; (b) diversity.

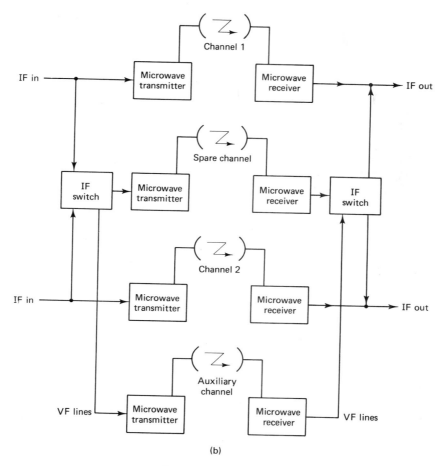

FIGURE 19-6 (*cont'd*)

As you can see, diversity systems and protection switching arrangements are quite similar. The primary difference between the two is that diversity systems are permanent arrangements and are intended only to compensate for temporary, abnormal atmospheric conditions between only two selected stations in a system. Protection switching arrangements, on the other hand, compensate for both radio fades and equipment failures and may include from six to eight repeater stations between switches. Protection channels may also be used as temporary communication facilities, while routine maintenance is performed on a regular working channel. With a protection switching arrangement, all signal paths and radio equipment are protected. Diversity is used selectively, that is, only between stations that historically experience severe fading a high percentage of the time.

A statistical study of outage time (i.e., service interruptions) caused by radio fades, equipment failures, and maintenance is important in the design of a microwave radio system. From such a study, engineering decisions can be made on the type of diversity system and protection switching arrangement best suited for a particular application.

MICROWAVE RADIO STATIONS

Basically, there are two types of microwave stations: terminals and repeaters. *Terminal stations* are points in the system where baseband signals are either originated or terminated. *Repeater stations* are points in a system where baseband signals may be reconfigured or where RF carriers are simply "repeated" or amplified.

Terminal Station

Essentially, a terminal station consists of four major sections: the baseband, wire line entrance link (WLEL), FM-IF, and RF sections. Figure 19-7 shows the block diagram of the baseband, WLEL, and FM-IF sections. As mentioned previously, the baseband may be one of several different types of signals. For our example, frequency-division-multiplexed voice band channels are used.

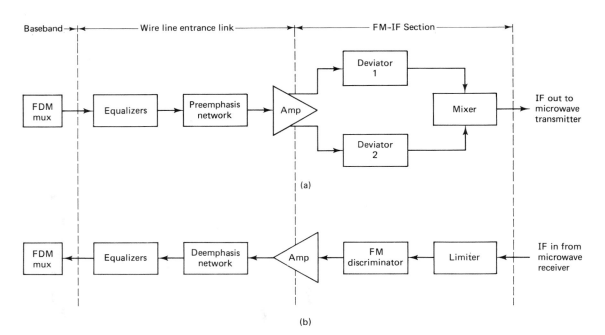

FIGURE 19-7   Microwave terminal station, baseband, wire line entrance link, and FM-IF: (a) transmitter; (b) receiver.

Wire line entrance link (WLEL). Very often in large communications networks such as the American Telephone and Telegraph Company (AT&T), the building that houses the radio station is quite large. Consequently, it is desirable that similar equipment be physically placed at a common location (i.e., all FDM equipment in the same room). This simplifies alarm systems, providing dc power to the equipment, maintenance, and other general cabling requirements. Dissimilar equipment may be separated by a considerable distance. For example, the distance between the FDM multiplexing equipment and the FM-IF section is typically several hundred feet and in some cases several miles. For this reason a WLEL is required. A WLEL serves as the interface between the multiplex terminal equipment and the FM-IF equipment. A WLEL generally consists of an amplifier and an equalizer (which together compensate for cable transmission losses) and a level-shaping device commonly called pre- and deemphasis networks.

IF section. The FM terminal equipment shown in Figure 19-7 generates a frequency-modulated IF carrier. This is accomplished by mixing the outputs of two deviated oscillators that differ in frequency by the desired IF carrier. The oscillators are deviated in phase opposition, which reduces the magnitude of phase deviation required of a single deviator by a factor of 2. This technique also reduces deviation linearity requirements for the oscillators and provides for the partial cancellation of unwanted modulation products. Again, the receiver is a conventional noncoherent FM detector.

RF section. A block diagram of the RF section of a microwave terminal station is shown in Figure 19-8. The IF signal enters the transmitter (Figure 19-8a) through a protection switch. The IF and compression amplifiers help keep the IF signal power constant and at approximately the required input level to the transmit modulator (*transmod*). A transmod is a balanced modulator that when used in conjunction with a microwave generator, power amplifier, and bandpass filter, up-converts the IF carrier to an RF carrier and amplifies the RF to the desired output power. Power amplifiers for microwave radios must be capable of amplifying very high frequencies and pass a very wide bandwidth signal. *Klystron tubes*, *traveling-wave tubes* (TWTs), and *IMPATT* (*i*mpact/*a*valanche and *T*ransit *T*ime) diodes are several of the devices currently being used in microwave power amplifiers. Because high-gain antennas are used and the distance between microwave stations is relatively short, it is not necessary to develop a high output power from the transmitter output amplifiers. Typical gains for microwave antennas range from 40 to 80 dB, and typical transmitter output powers are between 0.5 and 10 W.

A *microwave generator* provides the RF carrier input to the up-converter. It is called a microwave generator rather than an oscillator because it is difficult to construct a stable circuit that will oscillate in the gigahertz range. Instead, a crystal-controlled oscillator operating in the range 5 to 25 MHz is used to provide a base frequency that is multiplied up to the desired RF carrier frequency.

An *isolator* is a unidirectional device often made from a ferrite material. The isolator is used in conjunction with a channel-combining network to prevent the output of one transmitter from interfering with the output of another transmitter.

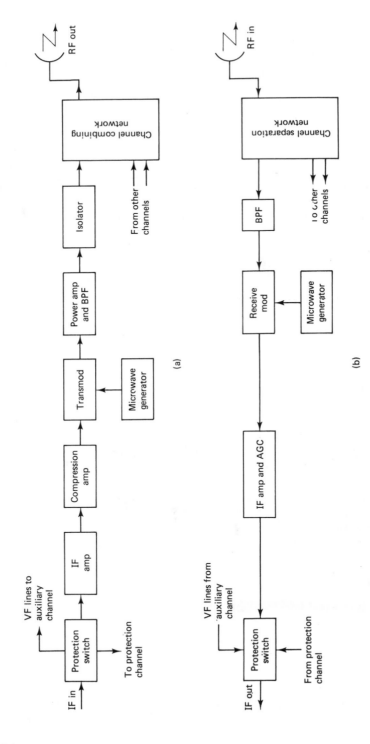

FIGURE 19-8 Microwave terminal station: (a) transmitter; (b) receiver.

The RF receiver (Figure 19-8b) is essentially the same as the transmitter except that it works in the opposite direction. However, one difference is the presence of an IF amplifier in the receiver. This IF amplifier has an *automatic gain control* (AGC) circuit. Also, very often, there are no RF amplifiers in the receiver. Typically, a very sensitive, low-noise-balanced demodulator is used for the receive demodulator (receive mod). This eliminates the need for an RF amplifier and improves the overall signal-to-noise ratio. When RF amplifiers are required, high-quality, *low-noise amplifiers* (LNAs) are used. Examples of commonly used LNAs are tunnel diodes and parametric amplifiers.

Repeater Station

Figure 19-9 shows the block diagram of a microwave IF repeater station. The received RF signal enters the receiver through the channel separation network and bandpass filter. The receive mod down-converts the RF carrier to IF. The IF AMP/AGC and equalizer circuits amplify and reshape the IF. The equalizer compensates for *gain versus frequency nonlinearities* and *envelope delay distortion* introduced in the system. Again, the transmod up-converts the IF to RF for retransmission. However, in a repeater station, the method used to generate the RF microwave carrier frequencies is slightly different from the method used in a terminal station. In the IF repeater, only one microwave generator is required to supply both the transmod and the receive mod with an RF carrier signal. The microwave generator, shift oscillator, and shift modulator allow the repeater to receive one RF carrier frequency, down-convert it to IF, and then up-convert the IF to a different RF carrier frequency (Figure 19-10a). It is possible for station C to receive the transmissions from both station A and station B simultaneously (this is called *multihop interference*). This can occur only when three stations are placed in a geographical straight line in the system. To prevent this from occurring, the allocated bandwidth for the system is divided in half, creating a low-frequency and a high-frequency band. Each station, in turn, alternates from a low-band to a high-band transmit carrier frequency (Figure 19-10b). If a transmission from station A is received by station C, it will be rejected in the channel separation network and cause no interference. This is called a high/low microwave repeater system. The rules are simple: If a repeater station receives a low-band RF carrier, it retransmits a high-band RF carrier, and vice versa. The only time that multiple carriers of the same frequency can be received is when a transmission from one station is received from another station that is three hops away. This is unlikely to happen.

Another reason for using a high/low-frequency scheme is to prevent the power that "leaks" out the back and sides of a transmit antenna from interferring with the signal entering the input of a neaby receive antenna. This is called *ringaround*. All antennas, no matter how high their gain or how directive their radiation pattern, radiate a small percentage of their power out the back and sides; giving a finite *front-to-back* ratio for the antenna. Although the front-to-back ratio of a typical microwave antenna is quite high, the relatively small amount of power that is radiated out the back of the antenna may be quite substantial compared to a normal received carrier power in the

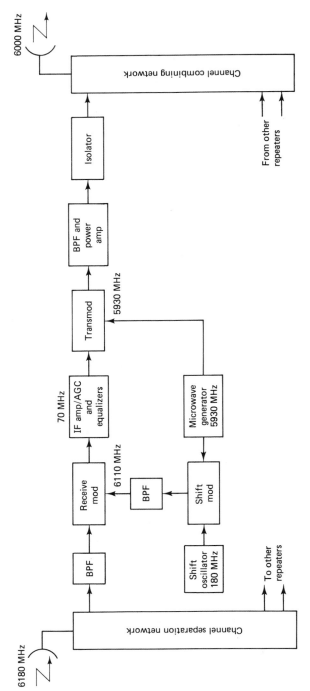

FIGURE 19-9   Microwave IF repeater station.

736

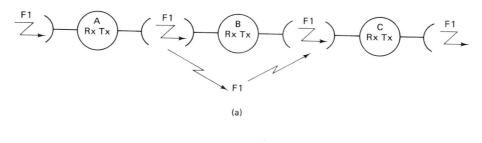

(a)

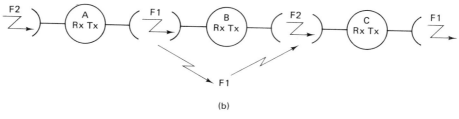

(b)

FIGURE 19-10   (a) Multihop interference and (b) high/low microwave system.

system. If the transmit and receive carrier frequencies are different, filters in the receiver separation network will prevent ringaround from occurring.

A high/low microwave repeater station (Figure 19-10b) needs two microwave carrier supplies for the down- and up-converting process. Rather than use two microwave generators, a single generator together with a shift oscillator, a shift modulator, and a bandpass filter can generate the two required signals. One output from the microwave generator is fed directly into the transmod and another output (from the same microwave generator) is mixed with the shift oscillator signal in the shift modulator to produce a second microwave carrier frequency. The second microwave carrier frequency is offset from the first by the shift oscillator frequency. The second microwave carrier frequency is fed into the receive modulator.

EXAMPLE 19-1

In Figure 19-9 the received RF carrier frequency is 6180 MHz, and the transmitted RF carrier frequency is 6000 MHz. With a 70-MHz IF frequency, a 5930-MHz microwave generator frequency, and a 180-MHz shift oscillator frequency, the output filter of the shift mod must be tuned to 6110 MHz. This is the sum of the microwave generator and the shift oscillator frequencies (5930 MHz + 180 MHz = 6110 MHz).

This process does not reduce the number of oscillators required, but it is simpler and cheaper to build one microwave generator and one relatively low-frequency shift oscillator than to build two microwave generators. The obvious disadvantage of the high/low scheme is that the number of channels available in a given bandwidth is cut in half.

Figure 19-11 shows a high/low-frequency plan with eight channels (four high-band and four low-band). Each channel occupies a 29.7-MHz bandwidth. The west terminal transmits the low-band frequencies and receives the high-band frequencies. Channel 1 and 3 (Figure 19-11a) are designated as *V channels*. This means that they are propagated with vertical polarization. Channels 2 and 4 are designated as H or horizontally polarized channels. This is not a polarization diversity system. Channels 1 through 4 are totally independent of each other; they carry different baseband information. The transmission of *orthogonally* polarized carriers (90° out of phase) further enhances the isolation between the transmit and receive signals. In the west-to-east direction, the repeater receives the low-band and transmits the high-band frequencies. After channel 1 is received and down-converted to IF, it is up-converted to a different RF frequency and polarization for retransmission. The low-band channel 1 corresponds to the high-band channel 11, channel 2 to channel 12, and so on. The east-to-west direction (Figure 19-11b) propagates the high- and low-band carriers in the sequence opposite to the west-to-east system. The polarizations are also reversed. If some of the power from channel 1 of the west terminal were to propagate directly to the east terminal receiver, it has a different frequency and polarization than channel 11's transmissions. Consequently, it would not interfere with the reception of channel 11 (no multihop interference). Also, note that none of the transmit or receive channels at the repeater station has both the same frequency and polarization. Consequently, the interference from the transmitters to the receivers due to ringaround is insignificant.

SYSTEM GAIN

In its simplest form, *system gain* is the difference between the nominal output power of a transmitter and the minimum input power to a receiver. System gain must be greater than or equal to the sum of all the gains and losses incurred by a signal as it propagates from a transmitter to a receiver. In essence, it represents the net loss of a radio system. System gain is used to predict the reliability of a system for given system parameters. Mathematically, system gain is

$$G_s = P_t - C_{min}$$

where

G_s = system gain (dB)
P_t = transmitter output power (dBm)
C_{min} = minimum receiver input power for a given quality objective (dBm)

and where

$$P_t - C_{min} \geq \text{losses} + \text{gains}$$

Gains:

A_t = transmit antenna gain (dB) relative to an isotropic radiator

A_r = receive antenna gain (dB) relative to an isotropic radiator

Losses:

L_p = free-space path loss between antennas (dB)

L_f = waveguide feeder loss (dB) between the distribution network (channel combining network or channel separation network) and its respective antenna (see Table 19-1)

L_b = total coupling or branching loss (dB) in the circulators, filters, and distribution network between the output of a transmitter or the input to a receiver and its respective waveguide feed (see Table 19-1)

FM = fade margin for a given reliability objective

Mathematically, system gain is

$$G_s = P_t - C_{min} \geq FM + L_p + L_f + L_b - A_t - A_r \qquad (19\text{-}1)$$

where all values are expressed in dB or dBm. Because system gain is indicative of a net loss, the losses are represented with positive dB values and the gains are represented with negative dB values. Figure 19-12 shows an overall microwave system diagram and indicates where the respective losses and gains are incurred.

TABLE 19-1   SYSTEM GAIN PARAMETERS

Frequency (GHz)	Feeder loss, L_f		Branching loss (dB) Diversity:		Antenna gain, A_t or A_r	
	Type	Loss (dB/100 m)	Frequency	Space	Size (m)	Gain (dB)
1.8	Air-filled coaxial cable	5.4	5	2	1.2	25.2
					2.4	31.2
					3.0	33.2
					3.7	34.7
7.4	EWP 64 eliptical waveguide	4.7	3	2	1.5	38.8
					2.4	43.1
					3.0	44.8
					3.7	46.5
8.0	EWP 69 eliptical waveguide	6.5	3	2	2.4	43.8
					3.0	45.6
					3.7	47.3
					4.8	49.8

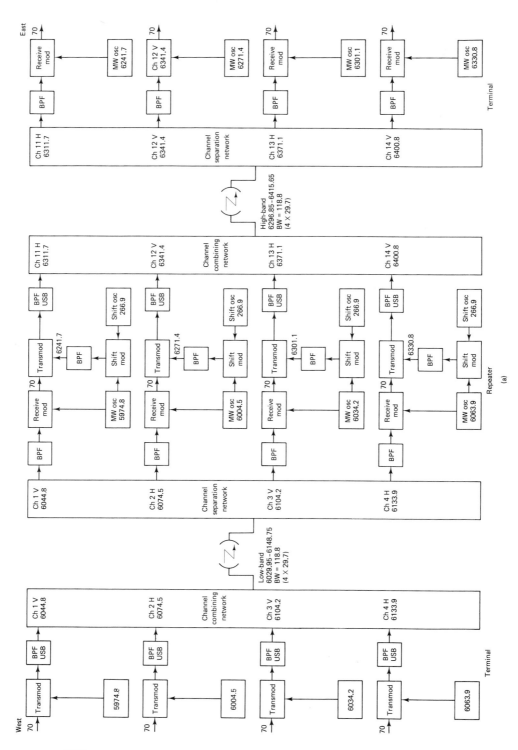

FIGURE 19-11 Eight-channel high/low frequency plan: (a) west to east; (b) east to west. All frequencies in megahertz.

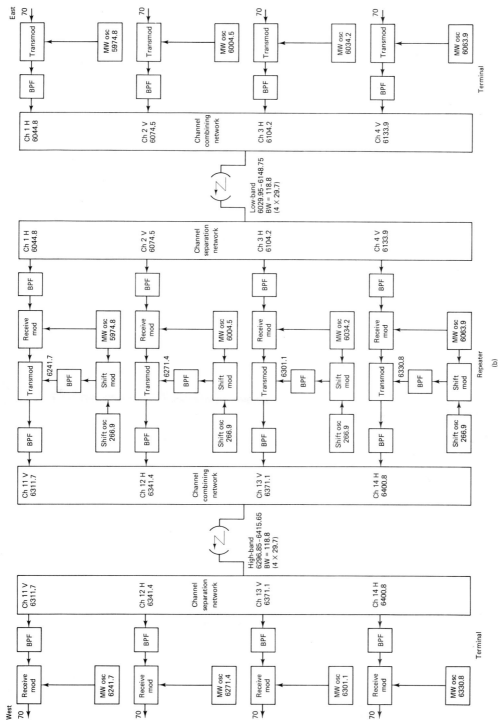

FIGURE 19-11 (*cont'd*)

Tomasi 7.11 cont.

741

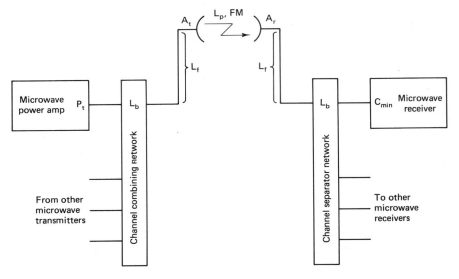

FIGURE 19-12   System gains and losses.

Free-Space Path Loss

Free-space path loss is defined as the loss incurred by an electromagnetic wave as it propagates in a straight line through a vacuum with no absorption or reflection of energy from nearby objects. The expression for free-space path loss is given as

$$L_p = \left(\frac{4\pi D}{\lambda}\right)^2 = \left(\frac{4\pi FD}{C}\right)^2$$

where

L_p = free-space path loss
D = distance
F = frequency
λ = wavelength
C = velocity of light in free space (3×10^8 m/s)

Converting to dB yields

$$L_p(\text{dB}) = 20 \log \frac{4\pi FD}{C} = 20 \log \frac{4\pi}{C} + 20 \log F + 20 \log D$$

When the frequency is given in MHz and the distance in km,

$$L_p(\text{dB}) = 20 \log \frac{4\pi(10)^6(10)^3}{3 \times 10^8} + 20 \log F \text{ (MHz)} + 20 \log D \text{ (km)}$$

$$= 32.4 + 20 \log F \text{ (MHz)} + 20 \log D \text{ (km)}$$

(19-2)

When the frequency is given in GHz and the distance in km,

$$L_p \text{(dB)} = 92.4 + 20 \log F \text{ (GHz)} + 20 \log D \text{ (km)} \qquad (19\text{-}3)$$

Similar conversions can be made using distance in miles, frequency in kHz, and so on.

EXAMPLE 19-2

For a carrier frequency of 6 GHz and a distance of 50 km, determine the free-space path loss.

Solution

$$L_p \text{(dB)} = 32.4 + 20 \log 6000 + 20 \log 50$$
$$= 32.4 + 75.6 + 34$$
$$= 142 \text{ dB}$$

or

$$L_p \text{(dB)} = 92.4 + 20 \log 6 + 20 \log 50$$
$$= 92.4 + 15.6 + 34$$
$$= 142 \text{ dB}$$

Fade Margin

Essentially, *fade margin* is a "fudge factor" included in the system gain equation that considers the nonideal and less predictable characteristics of radio-wave propagation, such as *multipath propagation (multipath loss)* and *terrain sensitivity*. These characteristics cause temporary, abnormal atmospheric conditions that alter the free-space path loss and are usually detrimental to the overall system performance. Fade margin also considers system reliability objectives. Thus fade margin is included in the system gain equation as a loss.

Solving the Barnett–Vignant reliability equations for a specified annual system availability for an unprotected, nondiversity system yields the following expression:

$$FM = \underbrace{30 \log D}_{\substack{\text{multipath} \\ \text{effect}}} + \underbrace{10 \log (6ABF)}_{\substack{\text{terrain} \\ \text{sensitivity}}} - \underbrace{10 \log (1 - R)}_{\substack{\text{reliability} \\ \text{objectives}}} - \underbrace{70}_{\text{constant}} \qquad (19\text{-}4)$$

where

FM = fade margin (dB)
D = distance (km)
F = frequency (GHz)
R = reliability expressed as a decimal (i.e., 99.99% = 0.9999 reliability)
$1 - R$ = reliability objective for a one-way 400-km route
A = roughness factor

= 4 over water or a very smooth terrain

= 1 over an average terrain

= 0.25 over a very rough, mountainous terrain

B = factor to convert a worst-month probability to an annual probability

= 1 to convert an annual availability to a worst-month basis

= 0.5 for hot humid areas

= 0.25 for average inland areas

= 0.125 for very dry or mountainous areas

EXAMPLE 19-3

Consider a space-diversity microwave radio system operating at an RF carrier frequency of 1.8 GHz. Each station has a 2.4-m-diameter parabolic antenna that is fed by 100 m of air-filled coaxial cable. The terrain is smooth and the area has a humid climate. The distance between stations is 40 km. A reliability objective of 99.99% is desired. Determine the system gain.

Solution   Substituting into Equation 19-4, we find that the fade margin is

$$\text{FM} = 30 \log 40 + 10 \log (6) (4) (0.5) (1.8) - 10 \log (1 - 0.9999) - 70$$

$$= 48.06 + 13.34 - (-40) - 70$$

$$= 48.06 + 13.34 + 40 - 70$$

$$= 31.4 \text{ dB}$$

Substituting into Equation 19-3, we obtain path loss

$$L_p = 92.4 + 20 \log 1.8 + 20 \log 40$$

$$= 92.4 + 5.11 + 32.04$$

$$= 129.55 \text{ dB}$$

From Table 19-1,

$$L_b = 4 \text{ dB } (2 + 2 = 4)$$

$$L_f = 10.8 \text{ dB } (100 \text{ m} + 100 \text{ m} = 200 \text{ m})$$

$$A_t = A_r = 31.2 \text{ dB}$$

Substituting into Equation 19-1 gives us system gain

$$G_s = 31.4 + 129.55 + 10.8 + 4 - 31.2 - 31.2 = 113.35 \text{ dB}$$

The results indicate that for this sytem to perform at 99.99% reliability with the given terrain, distribution networks, transmission lines, and antennas, the transmitter output power must be at least 113.35 dB more than the minimum receive signal level.

Receiver Threshold

Carrier-to-noise (*C/N*) is probably the most important parameter considered when evaluating the performance of a microwave communications system. The minimum wideband carrier power (C_{min}) at the input to a receiver that will produce a usable baseband output is called the receiver *threshold* or, sometimes, receiver *sensitivity*. The receiver threshold is dependent on the wideband noise power present at the input of a receiver, the noise introduced within the receiver, and the noise sensitivity of the baseband detector. Before C_{min} can be calculated, the input noise power must be determined. The input noise power is expressed mathematically as

$$N = KTB$$

where

N = noise power
K = Boltzmann's constant (1.38×10^{-23} J/K)
T = equivalent noise temperature of the receiver (K) (room temperature = 290 K)
B = noise bandwidth (Hz)

Expressed in dBm,

$$N \text{ (dBm)} = 10 \log \frac{KTB}{0.001} = 10 \log \frac{KT}{0.001} + 10 \log B$$

For a 1-Hz bandwidth at room temperature,

$$N = 10 \log \frac{(1.38 \times 10^{-23})(290)}{0.001} + 10 \log 1$$

$$= -174 \text{ dBm}$$

Thus

$$N \text{ (dBm)} = -174 \text{ dBm} + 10 \log B \qquad (19\text{-}5)$$

EXAMPLE 19-4

For an equivalent noise bandwidth of 10 MHz, determine the noise power.

Solution Substituting into Equation 19-5 yields

$$N = -174 \text{ dBm} + 10 \log (10 \times 10^{6})$$

$$= -174 \text{ dBm} + 70 \text{ dB} = -104 \text{ dBm}$$

If the minimum C/N requirement for a receiver with a 10-MHz noise bandwidth is 24 dB, the minimum receive carrier power is

$$C_{min} = \frac{C}{N} \text{ (dB)} + N \text{ (dB)}$$

$$= 24 \text{ dB} + (-104 \text{ dBm}) = -80 \text{ dBm}$$

For a system gain of 113.35 dB, it would require a minimum transmit carrier power (P_t) of

$$P_t = G_s + C_{min}$$

$$= 113.35 \text{ dB} + (-80 \text{ dBm}) = 33.35 \text{ dBm}$$

This indicates that a minimum transmit power of 33.35 dBm (2.16 W) is required to achieve a carrier-to-noise ratio of 24 dB with a system gain of 113.35 dB and a bandwidth of 10 MHz.

Carrier-to-Noise versus Signal-to-Noise

Carrier-to-noise (*C/N*) is the ratio of the wideband "carrier" (actually, not just the carrier, but rather the carrier and its associated sidebands) to the wideband noise power (the noise bandwidth of the receiver). *C/N* can be determined at an RF or an IF point in the receiver. Essentially, *C/N* is a *predetection* (before the FM demodulator) signal-to-noise ratio. Signal-to-noise (*S/N*) is a *postdetection* (after the FM demodulator) ratio. At a baseband point in the receiver, a single voice band channel can be separated from the rest of the baseband and measured independently. At an RF or IF point in the receiver, it is impossible to separate a single voice band channel from the composite FM signal. For example, a typical bandwidth for a single microwave channel is 30 MHz. The bandwidth of a voice band channel is 4 kHz. *C/N* is the ratio of the power of the composite RF signal to the total noise power in the 30-MHz bandwidth. *S/N* is the ratio of the power of a single voice band channel to the noise power in a 4-kHz bandwidth.

Noise Figure

In its simplest form, *noise figure* (F) is the signal-to-noise ratio of an ideal noiseless device divided by the *S/N* ratio at the output of an amplifier or a receiver. In a more practical sense, noise figure is defined as the ratio of the *S/N* ratio at the input to a device divided by the *S/N* ratio at the output. Mathematically, noise figure is

$$F = \frac{(S/N)_{in}}{(S/N)_{out}} \qquad \text{and} \qquad F \text{ (dB)} = 10 \log \frac{(S/N)_{in}}{(S/N)_{out}}$$

Thus noise figure is a ratio of ratios. The noise figure of a totally noiseless device is 1 or 0 dB. Remember, the noise present at the input to an amplifier is amplified by the same gain as the signal. Consequently, only noise added within the amplifier can decrease the signal-to-noise ratio at the output and increase the noise figure. (Keep in mind, the higher the noise figure, the worse the *S/N* ratio at the output.)

Essentially, noise figure indicates the relative increase of the noise power to the increase in signal power. A noise figure of 10 means that the device added sufficient noise to reduce the *S/N* ratio by a factor of 10, or the noise power increased tenfold in respect to the increase in signal power.

When two or more amplifiers or devices are cascaded together (Figure 19-13), the total noise figure (NF) is an accumulation of the individual noise figures. Mathematically, the total noise figure is

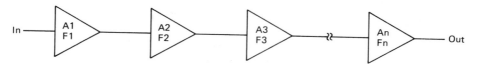

FIGURE 19-13   Total noise figure.

$$NF = F_1 + \frac{F_2 - 1}{A_1} + \frac{F_3 - 1}{A_1 A_2} + \frac{F_4 - 1}{A_1 A_2 A_3} \quad \text{etc.} \tag{19-6}$$

where

NF = total noise figure
F_1 = noise figure of amplifier 1
F_2 = noise figure of amplifier 2
F_3 = noise figure of amplifier 3
A_1 = gain of amplifier 1
A_2 = gain of amplifier 2

Note: Noise figures and gains are expressed as absolute values rather than dB.

It can be seen that the noise figure of the first amplifier (F1) contributes the most toward the overall noise figure. The noise introduced in the first stage is amplified by each of the succeeding amplifiers. Therefore, when compared to the noise introduced in the first stage, the noise added by each succeeding amplifier is effectively reduced by a factor equal to the product of the gains of the preceding amplifiers.

When precise noise calculations (0.1 dB or less) are necessary, it is generally more convenient to express noise figure in terms of noise temperature or equivalent noise temperature rather than as an absolute power (Chapter 20). Because noise power (N) is proportional to temperature, the noise present at the input to a device can be expressed as a function of the device's environmental temperature (T) and its equivalent noise temperature (T_e). To convert noise figure to a term dependent on temperature only, refer to Figure 19-14.

Let

N_d = noise power added by a single amplifier

Then

$$N_d = KT_e B$$

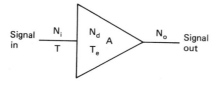

FIGURE 19-14   Noise figure as a function of temperature.

where T_e is the equivalent noise temperature. Let

N_o = total output noise power of an amplifier
N_i = total input noise power of an amplifier
A = gain of an amplifier

Therefore,

N_o may be expressed as
$$N_o = AN_i + AN_d$$

and

$$N_o = AKTB + AKT_eB$$

Simplifying yields

$$N_o = AKB\,(T + T_e)$$

and the overall noise figure (NF) equals

$$\text{NF} = \frac{(S/N)_{\text{in}}}{(S/N)_{\text{out}}} = \frac{S/N_i}{AS/N_o} = \frac{N_o}{AN_i} = \frac{AKB\,(T + T_e)}{AKTB} \qquad (19\text{-}7)$$

$$= \frac{T + T_e}{T} = 1 + \frac{T_e}{T}$$

EXAMPLE 19-5

In Figure 19-13, let $F_1 = F_2 = F_3 = 3$ dB and $A_1 = A_2 = A_3 = 10$ dB. Solve for the total noise figure.

Solution   Substituting into Equation 19-6 (*note*: All gains and noise figures have been converted to absolute values) yields

$$\text{NF} = F_1 + \frac{F_2 - 1}{A_1} + \frac{F_3 - 1}{A_1 A_2}$$

$$= 2 + \frac{2 - 1}{10} + \frac{2 - 1}{100}$$

$$= 2.11 \text{ or } 10 \log 2.11 = 3.24 \text{ dB}$$

An overall noise figure of 3.24 dB indicates that the *S/N* ratio at the output of A3 is 3.24 dB less than the *S/N* ratio at the input to A1.

The noise figure of a receiver must be considered when determining C_{min}. The noise figure is included in the system gain equation as an equivalent loss. (Essentially, a gain in the total noise power is equivalent to a corresponding loss in the signal power.)

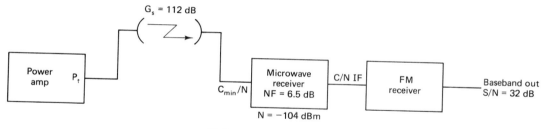

FIGURE 19-15   System gain example.

EXAMPLE 19-6

Refer to Figure 19-15. For a system gain of 112 dB, a total noise figure of 6.5 dB, an input noise power of −104 dBm, and a minimum $(S/N)_{out}$ of the FM demodulator of 32 dB, determine the minimum receive carrier power and the minimum transmit power.

Solution   To achieve a S/N ratio of 32 dB out of the FM demodulator, an input C/N of 15 dB is required (17 dB of improvement due to FM quieting). Solving for the receiver input carrier-to-noise ratio gives

$$\frac{C_{min}}{N} = \frac{C}{N} + NF = 15\ dB + 6.5\ dB = 21.5\ dB$$

Thus

$$C_{min} = \frac{C_{min}}{N} + N$$

$$= 21.5\ dB + (-104\ dBm) = -82.5\ dBm$$

$$P_t = G_s + C_{min}$$

$$= 112\ dB + (-82.5\ dBm) = 29.5\ dBm$$

EXAMPLE 19-7

For the system shown in Figure 19-16, determine the following: G_s, C_{min}/N, C_{min}, N, G_s, and P_t.

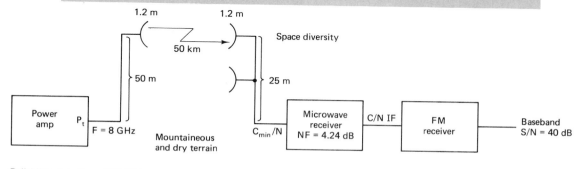

Reliability objective = 99.999%
Bandwidth = 6.3 MHz

FIGURE 19-16   System gain example.

Solution The minimum C/N at the input to the FM receiver is 23 dB.

$$\frac{C_{min}}{N} = \frac{C}{N} + NF$$

$$= 23 \text{ dB} + 4.24 \text{ dB} = 27.24 \text{ dB}$$

Substituting into Equation 19-5 yields

$$N = -174 \text{ dBm} + 10 \log B$$

$$= -174 \text{ dBm} + 68 \text{ dB} = -106 \text{ dBm}$$

$$C_{min} = \frac{C_{min}}{N} + N$$

$$= 27.24 \text{ dB} + (-106 \text{ dBm}) = -78.76 \text{ dBm}$$

Substituting into Equation 19-4 gives us

$$FM = 30 \log 50 + 10 \log [(6)(0.25)(0.125)(8)]$$

$$- 10 \log (1 - 0.99999) - 70$$

$$= 32.76 \text{ dB}$$

Substituting into Equation 19-3, we have

$$L_p = 92.4 \text{ dB} + 20 \log 8 + 20 \log 50$$

$$= 92.4 \text{ dB} + 18.06 \text{ dB} + 33.98 \text{ dB} = 144.44 \text{ dB}$$

From Table 19-1,

$$L_b = 4 \text{ dB}$$

$$L_f = 0.75 (6.5 \text{ dB}) = 4.875 \text{ dB}$$

$$A_t = A_r = 37.8 \text{ dB}$$

Note: The gain of an antenna increases or decreases proportional to the square of its diameter (i.e., if its diameter changes by a factor of 2, its gain changes by a factor of 4 or 6 dB). Substituting into Equation 19-1 yields

$$G_s = 32.76 + 144.44 + 4.875 + 4 - 37.8 - 37.8 = 110.475 \text{ dB}$$

$$P_t = G_s + C_{min}$$

$$= 110.475 \text{ dB} + (-78.76 \text{ dBm}) = 31.715 \text{ dBm}$$

QUESTIONS

19-1. What constitutes a short-haul microwave system? A long-haul microwave system?

19-2. Describe the baseband signal for a microwave system.

19-3. Why do FDM/FM microwave systems use low-index FM?

19-4. Describe a microwave repeater. Contrast baseband and IF repeaters.

19-5. Define *diversity*. Describe the three most commonly used diversity schemes.

19-6. Describe a protection switching arrangement. Contrast the two types of protection switching arrangements.

19-7. Briefly describe the four major sections of a microwave terminal station.

19-8. Define *ringaround*.

19-9. Briefly describe a high/low microwave system.

19-10. Define *system gain*.

19-11. Define the following terms: free-space path loss, branching loss, and feeder loss.

19-12. Define *fade margin*. Describe multipath losses, terrain, sensitivity, and reliability objectives and how they effect fade margin.

19-13. Define *receiver threshold*.

19-14. Contrast carrier-to-noise ratio and signal-to-noise ratio.

19-15. Define *noise figure*.

PROBLEMS

19-1. Calculate the noise power at the input to a receiver that has a radio carrier frequency of 4 GHz and a bandwidth of 30 MHz (assume room temperature).

19-2. Determine the path loss for a 3.4-GHz signal propagating 20,000 m.

19-3. Determine the fade margin for a 60-km microwave hop. The RF carrier frequency is 6 GHz, the terrain is very smooth and dry, and the reliability objective is 99.95%.

19-4. Determine the noise power for a 20-MHz bandwidth at the input to a receiver with an input noise temperature of 290°C.

19-5. For a system gain of 120 dB, a minimum input C/N of 30 dB, and an input noise power of -115 dBm, determine the minimum transmit power (P_t).

19-6. Determine the amount of loss contributed to a reliability objective of 99.98%.

19-7. Determine the terrain sensitivity loss for a 4-GHz carrier that is propagating over a very dry, mountainous area.

19-8. A frequency-diversity microwave system operates at an RF carrier frequency of 7.4 GHz. The IF is a low-index frequency-modulated subcarrier. The baseband signal is the 1800-channel FDM system described in Chapter 6 (564 to 8284 kHz). The antennas are 4.8-m-diameter parabolic dishes. The feeder lengths are 150 m at one station and 50 m at the other station. The reliability objective is 99.999%. The system propagates over an average terrain that has a very dry climate. The distance between stations is 50 km. The minimum carrier-to-noise ratio at the receiver input is 30 dB. Determine the following: fade margin, antenna gain, free-space path loss, total branching and feeder losses, receiver input noise power, C_{min}, minimum transmit power, and system gain.

19-9. Determine the overall noise figure for a receiver that has two RF amplifiers each with a noise figure of 6 dB and a gain of 10 dB, a mixer down-converter with a noise figure of 10 dB, and a conversion gain of -6 dB, and 40 dB of IF gain with a noise figure of 6 dB.

19-10. A microwave receiver has a total input noise power of -102 dBm and an overall noise figure of 4 dB. For a minimum C/N ratio of 20 dB at the input to the FM detector, determine the minimum receive carrier power.

Chapter 20

SATELLITE COMMUNICATIONS

INTRODUCTION

In the early 1960s, the American Telephone and Telegraph Company (AT&T) released studies indicating that a few powerful satellites of advanced design could handle more traffic than the entire AT&T long-distance communications network. The cost of these satellites was estimated to be only a fraction of the cost of equivalent terrestrial microwave facilities. Unfortunately, because AT&T was a utility, government regulations prevented them from developing the satellite systems. Smaller and much less lucrative corporations were left to develop the satellite systems, and AT&T continued to invest billions of dollars each year in conventional terrestrial microwave systems. Because of this, early developments in satellite technology were slow in coming.

Throughout the years the prices of most goods and services have increased substantially; however, satellite communications services have become more affordable each year. In most instances, satellite systems offer more flexibility than submarine cables, buried underground cables, line-of-sight microwave radio, tropospheric scatter radio, or optical fiber systems.

Essentially, a satellite is a radio repeater in the sky (*transponder*). A satellite system consists of a transponder, a ground-based station to control its operation, and a user network of earth stations that provide the facilities for transmission and reception of communications traffic through the satellite system. Satellite transmissions are categorized as either *bus* or *payload*. The bus includes control mechanisms that support the payload operation. The payload is the actual user information that is conveyed through the system. Although in recent years new data services and television broadcasting are more and more in demand, the transmission of conventional speech telephone signals (in analog or digital form) is still the bulk of the satellite payload.

752

HISTORY OF SATELLITES

The simplest type of satellite is a *passive reflector*, a device that simply "bounces" a signal from one place to another. The moon is a natural satellite of the earth and, consequently, in the late 1940s and early 1950s, became the first satellite transponder. In 1954, the U.S. Navy successfully transmitted the first messages over this earth-to-moon-to-earth relay. In 1956, a relay service was established between Washington, D.C., and Hawaii and, until 1962, offered reliable long-distance communications. Service was limited only by the availability of the moon.

In 1957, Russia launched *Sputnik I*, the first *active* earth satellite. An active satellite is capable of receiving, amplifying, and retransmitting information to and from earth stations. *Sputnik I* transmitted telemetry information for 21 days. Later in the same year, the United States launched *Explorer I*, which transmitted telemetry information for nearly 5 months.

In 1958, NASA launched *Score*, a 150-pound conical-shaped projectory. With an on-board tape recording, *Score* rebroadcasted President Eisenhower's 1958 Christmas message. *Score* was the first artificial satellite used for relaying terrestrial communications. *Score* was a *delayed repeater satellite*; it received transmissions from earth stations, stored them on magnetic tape, and rebroadcasted them to ground stations farther along its orbit.

In 1960, NASA in conjunction with Bell Telephone Laboratories and the Jet Propulsion Laboratory launched *Echo*, a 100-ft-diameter plastic balloon with an aluminum coating. *Echo* passively reflected radio signals from a large earth antenna. *Echo* was simple and reliable but required extremely high power transmitters at the earth stations. The first transatlantic transmission using a satellite transponder was accomplished using *Echo*. Also in 1960, the Department of Defense launched *Courier*. *Courier* transmitted 3 W of power and lasted only 17 days.

In 1962, AT&T launched *Telstar I*, the first satellite to receive and transmit simultaneously. The electronic equipment in *Telstar I* was damaged by radiation from the newly discovered Van Allen belts and, consequently, lasted only a few weeks. *Telstar II* was electronically identical to *Telstar I*, but it was made more radiation resistant. *Telstar II* was successfully launched in 1963. It was used for telephone, television, facsimile, and data transmissions. The first successful transatlantic transmission of video was accomplished with *Telstar II*.

Early satellites were both of the passive and active type. Again, a passive satellite is one that simply reflects a signal back to earth; there are no gain devices on board to amplify or repeat the signal. An active satellite is one that electronically repeats a signal back to earth (i.e., receives, amplifies, and retransmits the signal). An advantage of passive satellites is that they do not require sophisticated electronic equipment on board, although they are not necessarily void of power. Some passive satellites require a *radio beacon transmitter* for tracking and ranging purposes. A beacon is a continuously transmitted unmodulated carrier that an earth station can lock onto and use to align its antennas or to determine the exact location of the satellite. A disadvantage of passive satellites is their inefficient use of transmitted power. With *Echo*, for example, only 1 part in every 10^{18} of the earth station transmitted power was actually returned to the earth station receiving antenna.

ORBITAL SATELLITES

The satellites mentioned thus far are of the *orbital* or *nonsynchronous* type. That is, they rotate around the earth in a low-altitude elliptical or circular pattern with an angular velocity greater than (*prograde*) or less than (*retrograde*) that of Earth. Consequently, they are continuously gaining or falling back on Earth and do not remain stationary to any particular point on Earth. Thus they have to be used when available, which may be as short a period of time as 15 minutes per orbit. Another disadvantage of orbital satellites is the need for complicated and expensive tracking equipment at the earth stations. Each Earth station must locate the satellite as it comes into view on each orbit and then lock its antenna onto the satellite and track it as it passes overhead. A major advantage of orbital satellites is that propulsion rockets are not required on board the satellites to keep them in their respective orbits.

One of the more interesting orbital satellite systems is the Soviet *Molniya* system. It is presently the only nonsynchronous-orbit commercial satellite system in use. *Molniya* uses a highly elliptical orbit with *apogee* at about 40,000 km and *perigee* at about 1000 km (see Figure 20-1). The apogee is the maximum distance from earth a satellite orbit reaches; the perigee is the minimum distance. With the *Molniya* system, the apogee is reached while over the northern hemisphere and the perigee while over the southern hemisphere. The size of the ellipse was chosen to make its period exactly one-half of a sidereal day (the time it takes the earth to rotate back to the same constellation). Because of its unique orbital pattern, the *Molniya* satellite is synchronous with the rotation of the earth. During its 12-h orbit, it spends about 11 h over the north hemisphere.

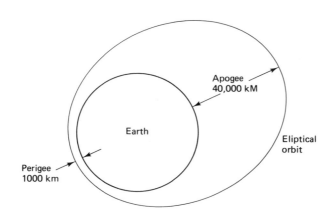

FIGURE 20-1 Soviet *Molniya* satellite orbit.

GEOSTATIONARY SATELLITES

Geostationary or *geosynchronous* satellites are satellites that orbit in a circular pattern with an angular velocity equal to that of Earth. Consequently, they remain in a fixed position in respect to a given point on Earth. An obvious advantage is they are available to all the earth stations within their *shadow* 100% of the time. The shadow of a satellite includes all earth stations that have a line-of-sight path to it and lie within the radiation pattern of the satellite's antennas. An obvious disadvantage is they require sophisticated

TABLE 20-1   CURRENT SATELLITE COMMUNICATIONS SYSTEMS

	Characteristic System				
	Westar	*Intelsat V*	*SBS*	*Fleet-satcom*	*Anik D*
Operator	Western Union Telegraph	Intelsat	Satellite Business Systems	U.S. Dept. of Defense	Telsat Canada
Frequency band	C	C and Ku	Ku	UHF, X	C, Ku
Coverage	Consus	Global, zonal, spot	Consus	Global	Canada, northern U.S.
Number of transponders	12	21	10	12	24
Transponder BW (MHz)	36	36–77	43	0.005–0.5	36
EIRP (dBw)	33	23.5–29	40–43.7	26–28	36
Multiple Access	FDMA, TDMA	FDMA, TDMA, reuse	TDMA	FDMA	FDMA
Modulation	FM, QPSK	FDM/FM, QPSK	QPSK	FM, QPSK	FDM, FM, FM/TVD, SCPC
Service	Fixed tele, TTY	Fixed tele, TVD	Fixed tele, TVD	Mobile military	Fixed tele

C-band: 3.4–6.425 GHz
Ku-band: 10.95–14.5 GHz
X-band: 7.25–8.4 GHz

TTY   teletype
TVD   TV distribution
FDMA frequency-division multiple access
TDMA time-division multiple access
Consus continental United States

and heavy propulsion devices on board to keep them in a fixed orbit. The orbital time of a geosynchronous satellite is 24 h, the same as Earth.

Syncom I, launched in February 1963, was the first attempt to place a geosynchronous satellite into orbit. *Syncom I* was lost during orbit injection. *Syncom II* and *Syncom III* were successfully launched in February 1963 and August 1964, respectively. The *Syncom III* satellite was used to broadcast the 1964 Olympic Games from Tokyo. The *Syncom* projects demonstrated the feasibility of using geosynchronous satellites.

Since the *Syncom* projects, a number of nations and private corporations have successfully launched satellites that are currently being used to provide national as well as regional and international global communications. There are more than 80 satellite communications systems operating in the world today. They provide worldwide fixed common-carrier telephone and data circuits; point-to-point cable television (CATV); network television distribution; music broadcasting; mobile telephone service; and private networks for corporations, governmental agencies, and military applications. A commercial global satellite network known as *Intelsat* (International Telecommunications Satellite Organization) is owned and operated by a consortium of more than 100 countries. *Intelsat* is managed by the designated communications entities in their respective countries. The *Intelsat* network provides high-quality, reliable service to its member countries. Table 20-1 is a partial list of current international and domestic satellite systems and their primary payload.

ORBITAL PATTERNS

Once projected, a satellite remains in orbit because the centrifugal force caused by its rotation around the earth is counterbalanced by the earth's gravitational pull. The closer to earth the satellite rotates, the greater the gravitational pull and the greater the velocity required to keep it from being pulled to earth. Low-altitude satellites that orbit close to Earth (100 to 300 miles in height) travel at approximately 17,500 miles per hour. At this speed, it takes approximately $1\frac{1}{2}$ h to rotate around the entire earth. Consequently, the time that the satellite is in line of sight of a particular earth station is only $\frac{1}{4}$ h or less per orbit. Medium-altitude satellites (6000 to 12,000 miles in height) have a rotation period of 5 to 12 h and remain in line of sight of a particular earth station for 2 to 4 h per orbit. High-altitude, geosynchronous satellites (19,000 to 25,000 miles in height) travel at approximately 6879 miles per hour and have a rotation period of 24 h, exactly the same as the earth. Consequently, they remain in a *fixed* position in respect to a given earth station and have a 24-h availability time. Figure 20-2 shows a low-, medium-, and high-altitude satellite orbit. It can be seen that three equally spaced, high-altitude geosynchronous satellites rotating around the earth above the equator can cover the entire earth except for the unpopulated areas of the north and south poles.

Figure 20-3 shows the three paths that a satellite may take as it rotates around the earth. When the satellite rotates in an orbit above the equator, it is called an *equatorial*

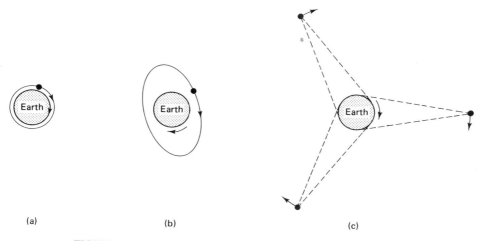

FIGURE 20-2 **Satellite orbits: (a) low altitude (circular orbit, 100–300 mi); (b) medium altitude (elliptical orbit, 6000–12,000 mi); (c) high altitude (geosynchronous orbit, 19,000–25,000 mi).**

orbit. When the satellite rotates in an orbit that takes it over the north and south poles, it is called a *polar orbit.* Any other orbital path is called an *inclined orbit.*

It is interesting to note that 100% of the earth's surface can be covered with a single satellite in a polar orbit. The satellite is rotating around the earth in a longitudinal orbit while the earth is rotating on a latitudinal axis. Consequently, the satellite's radiation pattern is a diagonal spiral around the earth which somewhat resembles a barber pole. As a result, every location on earth lies within the radiation pattern of the satellite twice each day.

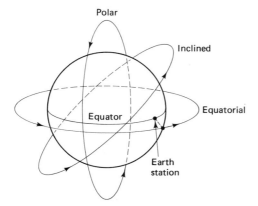

FIGURE 20-3 **Satellite orbits.**

SUMMARY

Advantages of Geosynchronous Orbits

1. The satellite remains almost stationary in respect to a given earth station. Consequently, expensive tracking equipment is not required at the earth stations.
2. There is no need to switch from one satellite to another as they orbit overhead. Consequently, there are no breaks in transmission because of the switching times.
3. High-altitude geosynchronous satellites can cover a much larger area of the earth than their low-altitude orbital counterparts.
4. The effects of Doppler shift are negligible.

Disadvantages of Geosynchronous Orbits

1. The higher altitudes of geosynchronous satellites introduce much longer propagation times. The round-trip propagation delay between two earth stations through a geosynchronous satellite is 500 to 600 ms.
2. Geosynchronous satellites require higher transmit powers and more sensitive receivers because of the longer distances and greater path losses.
3. High-precision spacemanship is required to place a geosynchronous satellite into orbit and to keep it there. Also, propulsion engines are required on board the satellites to keep them in their respective orbits.

LOOK ANGLES

To orient an antenna toward a satellite, it is necessary to know the *elevation angle* and *azimuth* (Figure 20-4). These are called the *look angles*.

Angle of Elevation

The angle of elevation is the angle formed between the plane of a wave radiated from an earth station antenna and the horizon, or the angle subtended at the earth station antenna between the satellite and the earth's horizon. The smaller the angle of elevation, the greater the distance a propagated wave must pass through the earth's atmosphere. As with any wave propagated through the earth's atmosphere, it suffers absorption and may also be severely contaminated by noise. Consequently, if the angle of elevation is too small and the distance the wave is within the earth's atmosphere is too long, the wave may deteriorate to a degree that it provides inadequate transmission. Generally, 5° is considered as the minimum acceptable angle of elevation. Figure 20-5 shows how the angle of elevation affects the signal strength of a propagated wave due to normal atmospheric absorption, absorption due to thick fog, and absorption due to a

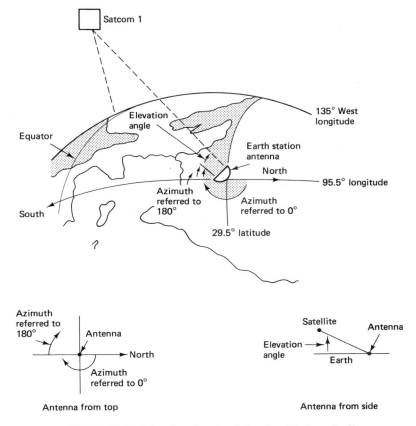

FIGURE 20-4   Azimuth and angle of elevation "look angles."

heavy rain. It can be seen that the 14/12-GHz band (Figure 20-5b) is more severely affected than the 6/4-GHz band (Figure 20-5a). This is due to the smaller wavelengths associated with the higher frequencies. Also, at elevation angles less than 5°, the attenuation increases rapidly.

Azimuth

Azimuth is defined as the horizontal pointing angle of an antenna. It is measured in a clockwise direction in degrees from true north. The angle of elevation and the azimuth both depend on the latitude of the earth station and the longitude of both the earth station and the orbiting satellite. For a geosynchronous satellite in an equatorial orbit, the procedure is as follows: From a good map, determine the longitude and latitude of the earth station. From Table 20-2, determine the longitude of the satellite of interest. Calculate the difference, in degrees (ΔL), between the longitude of the satellite and the

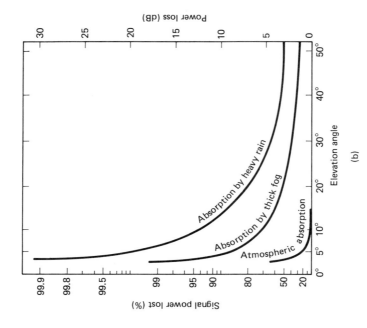

(a)

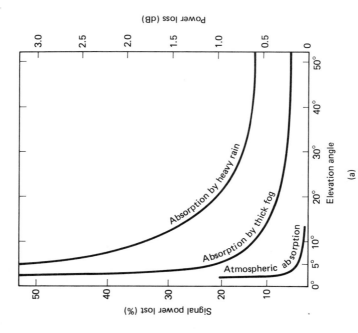

(b)

FIGURE 20-5 Attenuation due to atmospheric absorption: (a) 6/4-GHz band; (b) 14/12-GHz band.

760

TABLE 20-2   LONGITUDINAL POSITION OF SEVERAL CURRENT SYNCHRONOUS SATELLITES PARKED IN AN EQUATORIAL ARC[a]

Satellite	Longitude (°W)
Anik 1	104
Anik 2	109
Anik 3	114
Westar I	99
Westar II	123.5
Westar III	91
Satcom 1	135
Satcom 2	119
Comstar D2	95
Palapa 1	277
Palapa 2	283

[a] 0° latitude.

longitude of the earth station. Then, from Figure 20-6, determine the azimuth and elevation angle for the antenna. Figure 20-6 is for a geosynchronous satellite in an equatorial orbit.

EXAMPLE 20-1

An earth station is located at Houston, Texas, which has a longitude of 95.5°W and a latitude of 29.5°N. The satellite of interest is RCA's *Satcom 1*, which has a longitude of 135°W. Determine the azimuth and elevation angle for the earth station antenna.

Solution First determine the difference between the longitude of the earth station and the satellite.

$$\Delta L = 135° - 95.5° = 39.5°$$

Locate the intersection of ΔL and the latitude of the earth station on Figure 20-6. From the figure the angle of elevation is approximately 35·, and the azimuth is approximately 59° west of south.

ORBITAL SPACING AND FREQUENCY ALLOCATION

Geosynchronous satellites must share a limited space and frequency spectrum within a given arc of a geostationary orbit. Satellites operating at or near the same frequency must be sufficiently separated in space to avoid interfering with each other (Figure 20-7). There is a realistic limit to the number of satellite structures that can be stationed

Azimuth angle referenced to 180° (degrees)

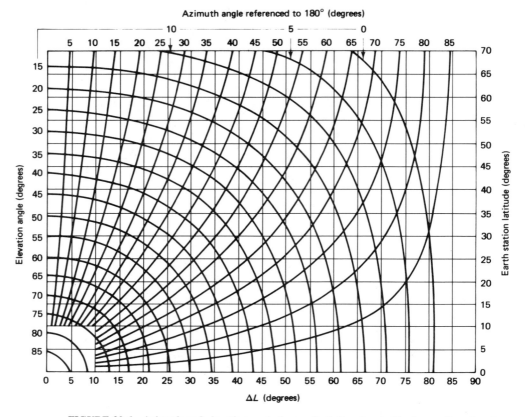

FIGURE 20-6 Azimuth and elevation angle for earth stations located in the northern hemisphere (referred to 180°).

(*parked*) within a given area in space. The required *spatial separation* is dependent on the following variables:

1. Beamwidths and sidelobe radiation of both the earth station and satellite antennas
2. RF carrier frequency
3. Encoding or modulation technique used
4. Acceptable limits of interference
5. Transmit carrier power

Generally, 3 to 6° of spatial separation is required depending on the variables stated above.

The most common carrier frequencies used for satellite communications are the 6/4- and 14/12-GHz bands. The first number is the up-link (earth station-to-transponder) frequency, and the second number is the down-link (transponder-to-earth station) fre-

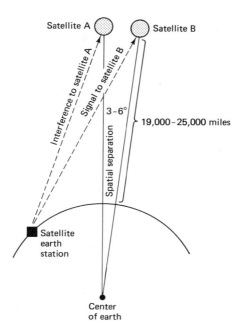

FIGURE 20-7 Spatial separation of satellites in geosynchronous orbit.

quency. Different up-link and down-link frequencies are used to prevent ringaround from occurring (Chapter 19). The higher the carrier frequency, the smaller the diameter required of an antenna for a given gain. Most domestic satellites use the 6/4-GHz band. Unfortunately, this band is also used extensively for terrestrial microwave systems. Care must be taken when designing a satellite network to avoid interference from or interference with established microwave links.

Certain positions in the geosynchronous orbit are in higher demand than the others. For example, the mid-Atlantic position which is used to interconnect North America and Europe is in exceptionally high demand. The mid-Pacific position is another.

The frequencies allocated by WARC (World Administrative Radio Conference) are summarized in Figure 20-8. Table 20-3 shows the bandwidths available for various services in the United States. These services include *fixed-point* (between earth stations located at fixed geographical points on earth), *broadcast* (wide-area coverage), *mobile* (ground-to-aircraft, ships, or land vehicles), and *intersatellite* (satellite-to-satellite cross-links).

RADIATION PATTERNS: FOOTPRINTS

The area of the earth covered by a satellite depends on the location of the satellite in its geosynchronous orbit, its carrier frequency, and the gain of its antennas. Satellite engineers select the antenna and carrier frequency for a particular spacecraft to concentrate the limited transmitted power on a specific area of the earth's surface. The geographical

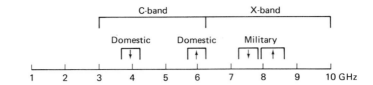

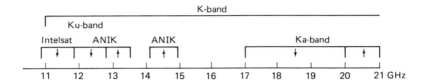

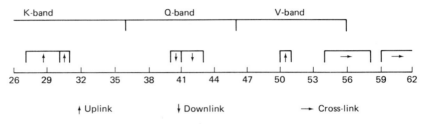

↑ Uplink   ↓ Downlink   → Cross-link

FIGURE 20-8   **WARC satellite frequency assignments.**

TABLE 20-3   RF SATELLITE BANDWIDTHS AVAILABLE IN THE UNITED STATES

Band	Frequency band (GHz) Up-link	Down-link	Bandwidth (MHz)
C	5.9–6.4	3.7–4.2	500
X	7.9–8.4	7.25–7.75	500
Ku	14–14.5	11.7–12.2	500
Ka	27–30	17–20	—
	30–31	20–21	—
V	50–51	40–41	1000
Q	—	41–43	2000
V	54–58		3900
(ISL)	59–64		5000

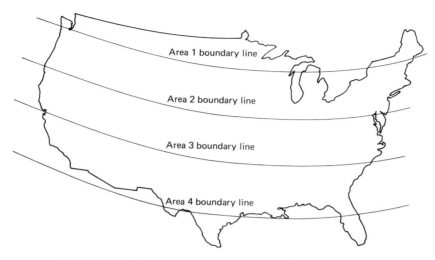

FIGURE 20-9 Satellite antenna radiation patterns ("footprints").

representation of a satellite antenna's radiation pattern is called a *footprint* (Figure 20-9). The contour lines represent limits of equal receive power density.

The radiation pattern from a satellite antenna may be categorized as either *spot*, *zonal*, or *earth* (Figure 20-10). The radiation patterns of earth coverage antennas have a beamwidth of approximately 17° and include coverage of approximately one-third of the earth's surface. Zonal coverage includes an area less than one-third of the earth's surface. Spot beams concentrate the radiated power in a very small geographic area.

Reuse

When an allocated frequency band is filled, additional capacity can be achieved by *reuse* of the frequency spectrum. By increasing the size of an antenna (i.e., increasing the antenna gain) the beamwidth of the antenna is also reduced. Thus different beams of the same frequency can be directed to different geographical areas of the earth. This is called frequency reuse. Another method of frequency reuse is to use dual polarization. Different information signals can be transmitted to different earth station receivers using the same band of frequencies simply by orienting their electromagnetic polarizations in an orthogonal manner (90° out of phase). Dual polarization is less effective because the earth's atmosphere has a tendency to reorient or repolarize an electromagnetic wave as it passes through. Reuse is simply another way to increase the capacity of a limited bandwidth.

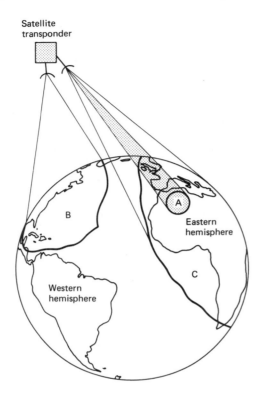

FIGURE 20-10   Beams: A, spot; B, zonal; C, earth.

SATELLITE SYSTEM LINK MODELS

Essentially, a satellite system consists of three basic sections: the uplink, the satellite transponder, and the downlink.

Uplink Model

The primary component within the *uplink* section of a satellite system is the earth station transmitter. A typical earth station transmitter consists of an IF modulator, an IF-to-RF microwave up-converter, a high-power amplifier (HPA), and some means of bandlimiting the final output spectrum (i.e., an output bandpass filter). Figure 20-11 shows the block diagram of a satellite earth station transmitter. The IF modulator converts the input baseband signals to either an FM, a PSK, or a QAM modulated intermediate frequency. The up-converter (mixer and bandpass filter) converts the IF to an appropriate RF carrier frequency. The HPA provides adequate input sensitivity and output power to propagate the signal to the satellite transponder. HPAs commonly used are klystons and traveling-wave tubes.

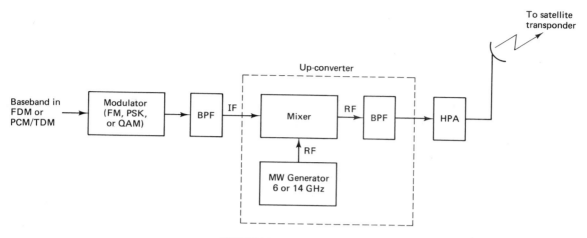

FIGURE 20-11   Satellite uplink model.

Transponder

A typical *satellite transponder* consists of an input bandlimiting device (BPF), an input *low-noise amplifier* (LNA), a *frequency translator*, a low-level power amplifier, and an output bandpass filter. Figure 20-12 shows a simplified block diagram of a satellite transponder. This transponder is an RF-to-RF repeater. Other transponder configurations are IF and baseband repeaters similar to those used in microwave repeaters. In Figure 20-12, the input BPF limits the total noise applied to the input of the LNA. (A common device used as an LNA is a tunnel diode.) The output of the LNA is fed to a frequency translator (a shift oscillator and a BPF) which converts the high-band uplink frequency to the low-band downlink frequency. The low-level power amplifier, which is commonly

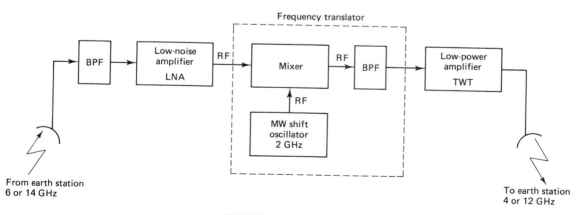

FIGURE 20-12   Satellite transponder.

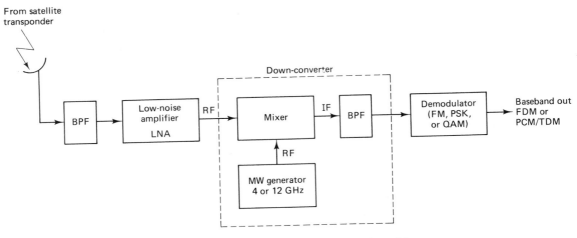

FIGURE 20-13   Satellite downlink model.

a traveling-wave tube, amplifies the RF signal for transmission through the downlink to the earth station receivers. Each RF satellite channel requires a separate transponder.

Downlink Model

An earth station receiver includes an input BPF, an LNA, and an RF-to-IF down-converter. Figure 20-13 shows a block diagram of a typical earth station receiver. Again, the BPF limits the input noise power to the LNA. The LNA is a highly sensitive, low-noise device such as a tunnel diode amplifier or a parametric amplifier. The RF-to-IF down-converter is a mixer/bandpass filter combination which converts the received RF signal to an IF frequency.

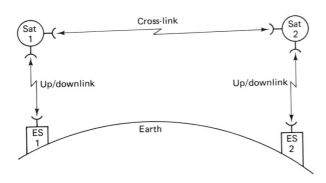

FIGURE 20-14   Intersatellite link.

Cross-Links

Occasionally, there is an application where it is necessary to communicate between satellites. This is done using *satellite cross-links* or *intersatellite links* (ISLs), shown in Figure 20-14. A disadvantage of using an ISL is that both the transmitter and receiver are *spacebound*. Consequently, both the transmitter's output power and the receiver's input sensitivity are limited.

SATELLITE SYSTEM PARAMETERS

Transmit Power and Bit Energy

High-power amplifiers used in earth station transmitters and the traveling-wave tubes typically used in satellite transponders are *nonlinear devices*; their gain (output power versus input power) is dependent on input signal level. A typical input/output power characteristic curve is shown in Figure 20-15. It can be seen that as the input power is reduced by 5 dB, the output power is reduced by only 2 dB. There is an obvious *power compression*. To reduce the amount of intermodulation distortion caused by the nonlinear amplification of the HPA, the input power must be reduced (*backed off*) by several dB. This allows the HPA to operate in a more *linear* region. The amount the input level is backed off is equivalent to a loss and is appropriately called *back-off loss* (L_{bo}).

   To operate as efficiently as possible, a power amplifier should be operated as close as possible to saturation. The *saturated output power* is designated P_o (sat) or simply P_t. The output power of a typical satellite earth station transmitter is much

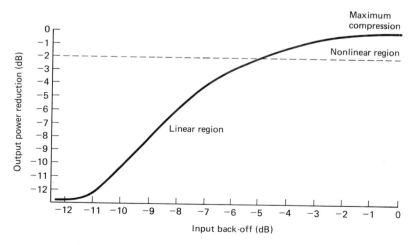

FIGURE 20-15   HPA input/output characteristic curve.

higher than the output power from a terrestrial microwave power amplifier. Consequently, when dealing with satellite systems, P_t is generally expressed in dBW (decibels in respect to 1 W) rather than in dBm (decibels in respect to 1 mW).

Most modern satellite systems use either phase shift keying (PSK) or quadrature amplitude modulation (QAM) rather than conventional frequency modulation (FM). With PSK and QAM, the input baseband is generally a PCM-encoded, time-division-multiplexed signal which is digital in nature. Also, with PSK and QAM, several bits may be encoded in a single transmit signaling element (baud). Consequently, a parameter more meaningful than carrier power is *energy per bit* (E_b). Mathematically, E_b is

$$E_b = P_t T_b \qquad (20\text{-}1a)$$

where

E_b = energy of a single bit (J/bit)
P_t = total carrier power (W)
T_b = time of a single bit (s) ,

or because $T_b = 1/F_b$, where F_b is the bit rate in bits per second.

$$E_b = \frac{P_t}{F_b} \qquad (20\text{-}1b)$$

EXAMPLE 20-2

For a total transmit power (P_t) of 1000 W, determine the energy per bit (E_b) for a transmission rate of 50 Mbps.

Solution

$$T_b = \frac{1}{F_b} = \frac{1}{50 \times 10^6 \text{ bps}} = 0.02 \times 10^{-6} \text{ s}$$

(It appears that the units for T_b should be s/bit but the per bit is implied in the definition of T_b, time of bit.)

Substituting into Equation 20-1a yields

$$E_b = 1000 \text{ J/s } (0.02 \times 10^{-6} \text{ s/bit}) = 20 \text{ } \mu\text{J}$$

(Again the units appear to be J/bit, but the per bit is implied in the definition of E_b, energy per bit.)

$$E_b = \frac{1000 \text{ J/s}}{50 \times 10^6 \text{ bps}} = 20 \text{ } \mu\text{J}$$

Expressed as a log,

$$E_b = 10 \log(20 \times 10^{-6}) = -47 \text{ dBJ}$$

It is common to express P_t in dBW and E_b in dBW/bps. Thus

$$P_t = 10 \log 1000 = 30 \text{ dBW}$$

$$E_b = P_t - 10 \log F_b$$

$$= P_t - 10 \log(50 \times 10^6)$$

$$= 30 \text{ dBW} - 77 \text{ dB} = -47 \text{ dBW/bps}$$

or simply -47 dBW.

Effective Isotropic Radiated Power

Effective isotropic radiated power (EIRP) is defined as an equivalent transmit power and is expressed mathematically as

$$\text{EIRP} = P_r A_t$$

where

EIRP = effective isotropic radiated power (W)
P_r = total power radiated from an antenna (W)
A_t = transmit antenna gain (W/W or a unitless ratio)

Expressed as a log,

$$\text{EIRP (dBW)} = P_r\text{(dBW)} + A_t\text{(dB)}$$

In respect to the transmitter output,

$$P_r = P_t - L_{bo} - L_{bf}$$

Thus

$$\text{EIRP} = P_t - L_{bo} - L_{bf} + A_t \qquad (20\text{-}2)$$

where

P_t = actual power output of the transmitter (dBW)
L_{bo} = back-off losses of HPA (dB)
L_{bf} = total branching and feeder loss (dB)
A_t = transmit antenna gain (dB)

EXAMPLE 20-3

For an earth station transmitter with an output power of 40 dBW (10,000 W), a back-off loss of 3 dB, a total branching and feeder loss of 3 dB, and a transmit antenna gain of 40 dB, determine the EIRP.

Solution Substituting into Equation 20-2 yields

$$\text{EIRP} = P_t - L_{bo} - L_{bf} + A_t$$

$$= 40 \text{ dBW} - 3 \text{ dB} - 3\text{dB} + 40 \text{ dB} = 74 \text{ dBW}$$

Equivalent Noise Temperature

With terrestrial microwave systems, the noise introduced in a receiver or a component within a receiver was commonly specified by the parameter noise figure. In satellite communications systems, it is often necessary to differentiate or measure noise in increments as small as a tenth or a hundredth of a decibel. Noise figure, in its standard form, is inadequate for such precise calculations. Consequently, it is common to use *environmental temperature* (T) and *equivalent noise temperature* (T_e) when evaluating the performance of a satellite system. In Chapter 19 total noise power was expressed mathematically as

$$N = KTB$$

Rearranging and solving for T gives us

$$T = \frac{N}{KB}$$

where

N = total noise power (W)
K = Boltzmann's constant (J/K)
B = bandwidth (Hz)
T = temperature of the environment (K)

Again from Chapter 19 (Equation 19-7),

$$NF = 1 + \frac{T_e}{T}$$

where

T_e = equivalent noise temperature (K)
NF = noise figure (absolute value)
T = temperature of the environment (K)

Rearranging Equation 19-7, we have

$$T_e = T(NF - 1)$$

Typically, equivalent noise temperatures of the receivers used in satellite transponders are about 1000 K. For earth station receivers T_e values are between 20 and 1000 K. Equivalent noise temperature is generally more useful when expressed logarithmically with the unit of dBK, as follows:

$$T_e \text{ (dBK)} = 10 \log T_e$$

For an equivalent noise temperature of 100 K, T_e (dBK) is

$$T_e \text{ (dBK)} = 10 \log 100 \text{ or } 20 \text{ dBK}$$

Equivalent noise temperature is a hypothetical value that can be calculated but cannot be measured. Equivalent noise temperature is often used rather than noise figure because it is a more accurate method of expressing the noise contributed by a device or a receiver when evaluating its performance. Essentially, equivalent noise temperature (T_e) is the noise present at the input to a device or amplifier plus the noise added internally by that device. This allows us to analyze the noise characteristics of a device by simply evaluating an equivalent input noise temperature. As you will see in subsequent discussions, T_e is a very useful parameter when evaluating the performance of a satellite system.

EXAMPLE 20-4

Convert noise figures of 4 and 4.01 to equivalent noise temperatures. Use 300 K for the environmental temperature.

Solution Substituting into Equation 20-7 yields

$$T_e = T(\text{NF} - 1)$$

For NF = 4:

$$T_e = 300(4 - 1) = 900 \text{ K}$$

For NF = 4.01:

$$T_e = 300(4.01 - 1) = 903 \text{ K}$$

It can be seen that the 3° difference in the equivalent temperatures is 300 times as large as the difference between the two noise figures. Consequently, equivalent noise temperature is a more accurate way of comparing the noise performances of two receivers or devices.

Noise Density

Simply stated, *noise density* (N_o) is the total noise power normalized to a 1-Hz bandwidth, or the noise power present in a 1-Hz bandwidth. Mathematically, noise density is

$$N_o = \frac{N}{B} \quad \text{or} \quad KT_e \tag{20-3a}$$

where

N_o = noise density (W/Hz) (N_o is generally expressed as simply watts; the per hertz is implied in the definition of N_o)
N = total noise power (W)
B = bandwidth (Hz)
K = Boltzmann's constant (J/K)
T_e = equivalent noise temperature (K)

Expressed as a log,

$$N_o \text{ (dBW/Hz)} = 10 \log N - 10 \log B \tag{20-3b}$$

$$= 10 \log K + 10 \log T_e \tag{20-3c}$$

EXAMPLE 20-5

For an equivalent noise bandwidth of 10 MHz and a total noise power of 0.0276 pW, determine the noise density and equivalent noise temperature.

Solution   Substituting into Equation 20-3a, we have

$$N_o = \frac{N}{B} = \frac{276 \times 10^{-16}\,\text{W}}{10 \times 10^6\,\text{Hz}} = 276 \times 10^{-23}\,\frac{\text{W}}{\text{Hz}}$$

or simply, 276×10^{-23} W.

$$N_o = 10 \log (276 \times 10^{-23}) = -205.6\,\text{dBW/Hz}$$

or simply -205.6 dBW. Substituting into Equation 20-3b gives us

$$N_o = N\,(\text{dBW}) - B\,(\text{dB/Hz})$$

$$= -135.6\,\text{dBW} - 70\,(\text{dB/Hz}) = -205.6\,\text{dBW}$$

Rearranging Equation 20-3a and solving for equivalent noise temperature yields

$$T_e = \frac{N_o}{K}$$

$$= \frac{276 \times 10^{-23}\,\text{J/cycle}}{1.38 \times 10^{-23}\,\text{J/K}} = 200°\,\text{K/cycle}$$

$$= 10 \log 200 = 23\,\text{dBK}$$

$$= N_o\,(\text{dBW}) - 10 \log K$$

$$= -205.6\,\text{dBW} - (-228.6\,\text{dBWK}) = 23\,\text{dBK}$$

Carrier-to-Noise Density Ratio

C/N_o is the average wideband carrier power-to-noise density ratio. The *wideband carrier power* is the combined power of the carrier and its associated sidebands. The noise is the thermal noise present in a normalized 1-Hz bandwidth. The carrier-to-noise density ratio may also be written as a function of noise temperature. Mathematically, C/N_o is

$$\frac{C}{N_o} = \frac{C}{KT_e} \tag{20-4a}$$

Expressed as a log,

$$\frac{C}{N_o}\,(\text{dB}) = C\,(\text{dBW}) - N_o\,(\text{dBW}) \tag{20-4b}$$

Energy of Bit-to-Noise Density Ratio

E_b/N_o is one of the most important and most often used parameters when evaluating a digital radio system. The E_b/N_o ratio is a convenient way to compare digital systems that use different transmission rates, modulation schemes, or encoding techniques. Mathematically, E_b/N_o is

$$\frac{E_b}{N_o} = \frac{C/F_b}{N/B} = \frac{CB}{NF_b}$$ (20-5)

E_b/N_o is a convenient term used for digital system calculations and performance comparisons, but in the real world, it is more convenient to measure the wideband carrier power-to-noise density ratio and convert it to E_b/N_o. Rearranging Equation 20-5 yields the following expression:

$$\frac{E_b}{N_o} = \frac{C}{N} \times \frac{B}{F_b}$$

The E_b/N_o ratio is the product of the carrier-to-noise ratio (C/N) and the noise bandwidth-to-bit ratio (B/F_b). Expressed as a log,

$$\frac{E_b}{N_o}\,(\text{dB}) = \frac{C}{N}\,(\text{dB}) + \frac{B}{F_b}\,(\text{dB})$$ (20-6)

The energy per bit (E_b) will remain constant as long as the total wideband carrier power (C) and the transmission rate (bps) remain unchanged. Also, the noise density (N_o) will remain constant as long as the noise temperature remains constant. The following conclusion can be made: For a given carrier power, bit rate, and noise temperature, the E_b/N_o ratio will remain constant regardless of the encoding technique, modulation scheme, or bandwidth used.

Figure 20-16 graphically illustrates the relationship between an expected probability of error $P(e)$ and the minimum C/N ratio required to achieve the $P(e)$. The C/N specified is for the minimum double-sided Nyquist bandwidth. Figure 20-17 graphically illustrates the relationship between an expected $P(e)$ and the minimum E_b/N_o ratio required to achieve that $P(e)$.

A $P(e)$ of 10^{-5} $(1/10^5)$ indicates a probability that 1 bit will be in error for every 100,000 bits transmitted. $P(e)$ is analogous to the bit error rate (BER).

EXAMPLE 20-6

A coherent binary phase-shift-keyed (BPSK) transmitter operates at a bit rate of 20 Mbps. For a probability of error $P(e)$ of 10^{-4}:
    (a) Determine the minimum theoretical C/N and E_b/N_o ratios for a receiver bandwidth equal to the minimum double-sided Nyquist bandwidth.
    (b) Determine the C/N if the noise is measured at a point prior to the bandpass filter, where the bandwidth is equal to twice the Nyquist bandwidth.
    (c) Determine the C/N if the noise is measured at a point prior to the bandpass filter where the bandwidth is equal to three times the Nyquist bandwidth.

Solution   (a) With BPSK, the minimum bandwidth is equal to the bit rate, 20 MHz. From Figure 20-16, the minimum C/N is 8.8 dB. Substituting into Equation 20-6 gives us

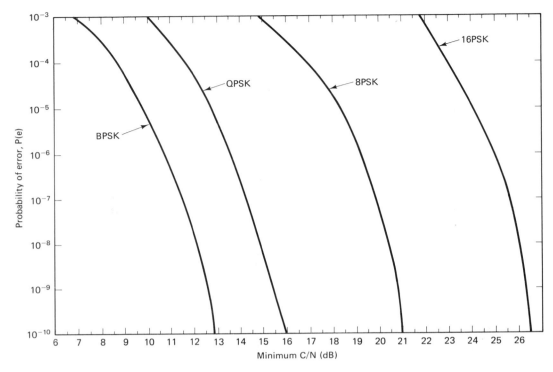

FIGURE 20-16 Probability of error $P(e)$ versus C/N for various digital modulation schemes. (Bandwidth equals minimum double-sided Nyquist bandwidth.)

$$\frac{E_b}{N_o}\,(\text{dB}) = \frac{C}{N}\,(\text{dB}) + \frac{B}{F_b}\,(\text{dB})$$

$$= 8.8\,\text{dB} + 10 \log \frac{20 \times 10^6}{20 \times 10^6}$$

$$= 8.8\,\text{dB} + 0\,\text{dB} = 8.8\,\text{db}$$

Note: The minimum E_b/N_o equals the minimum C/N when the receiver noise bandwidth equals the minimum Nyquist bandwidth. The minimum E_b/N_o of 8.8 can be verified from Figure 20-17.

What effect does increasing the noise bandwidth have on the minimum C/N and E_b/N_o ratios? The wideband carrier power is totally independent of the noise bandwidth. Similarly, an increase in the bandwidth causes a corresponding increase in the noise power. Consequently, a decrease in C/N is realized that is directly proportional to the increase in the noise bandwidth. E_b is dependent on the wideband carrier power and the bit rate only. Therefore, E_b is unaffected by an increase in the noise bandwidth. N_o is the noise power normalized to a 1-Hz bandwidth and, consequently, is also unaffected by an increase in the noise bandwidth.

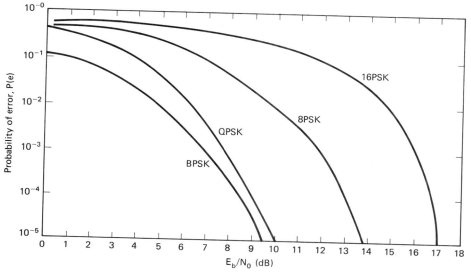

FIGURE 20-17 Probability of error $P(e)$ versus E_b/N_0 ratio for various digital modulation schemes.

(b) Since E_b/N_o is independent of bandwidth, measuring the C/N at a point in the receiver where the bandwidth is equal to twice the minimum Nyquist bandwidth has absolutely no effect on E_b/N_o. Therefore, E_b/N_o becomes the constant in Equation 20-6 and is used to solve for the new value of C/N. Rearranging Equation 20-6 and using the calculated E_b/N_o ratio, we have

$$\frac{C}{N}\,(dB) = \frac{E_b}{N_o}\,(dB) - \frac{B}{F_b}\,(dB)$$

$$= 8.8\,dB - 10\log\frac{40 \times 10^6}{20 \times 10^6}$$

$$= 8.8\,dB - 10\log 2$$

$$= 8.8\,dB - 3\,dB = 5.8\,dB$$

(c) Measuring the C/N ratio at a point in the receiver where the bandwidth equals three times the minimum bandwidth yields the following results for C/N.

$$\frac{C}{N} = \frac{E_b}{N_o} - 10\log\frac{60 \times 10^6}{20 \times 10^6}$$

$$= 8.8\,dB - 10\log 3$$

$$= 4.03\,dB$$

The C/N ratios of 8.8, 5.8, and 4.03 dB indicate the C/N ratios that would be measured at the three specified points in the receiver to achieve the desired minimum E_b/N_0 and $P(e)$.

Because E_b/N_0 cannot be directly measured to determine the E_b/N_0 ratio, the wideband carrier-to-noise ratio is measured and then substituted into Equation 20-6. Consequently, to accurately determine the E_b/N_0 ratio, the noise bandwidth of the receiver must be known.

EXAMPLE 20-7

A coherent 8PSK transmitter operates at a bit rate of 90 Mbps. For a probability of error of 10^{-5}:

    (a) Determine the minimum theoretical C/N and E_b/N_o ratios for a receiver bandwidth equal to the minimum double-sided Nyquist bandwidth.

    (b) Determine the C/N if the noise is measured at a point prior to the bandpass filter where the bandwidth is equal to twice the Nyquist bandwidth.

    (c) Determine the C/N if the noise is measured at a point prior to the bandpass filter where the bandwidth is equal to three times the Nyquist bandwidth.

Solution   (a) 8PSK has a bandwidth efficiency of 3 bps/Hz and, consequently, requires a minimum bandwidth of one-third the bit rate or 30 MHz. From Figure 20-16, the minimum C/N is 18.5 dB. Substituting into Equation 20-6, we obtain

$$\frac{E_b}{N_o}(\text{dB}) = 18.5 \text{ dB} + 10 \log \frac{30 \text{ MHz}}{90 \text{ Mbps}}$$

$$= 18.5 \text{ dB} + (-4.8 \text{ dB}) = 13.7 \text{ db}$$

(b) Rearranging Equation 20-6 and substituting in E_b/N_o yields

$$\frac{C}{N}(\text{dB}) = 13.7 \text{ dB} - 10 \log \frac{60 \text{ MHz}}{90 \text{ Mbps}}$$

$$= 13.7 \text{ dB} - (-1.77 \text{ dB}) = 15.47 \text{ dB}$$

(c) Again, rearranging Equation 20-6 and substituting in E_b/N_o gives us

$$\frac{C}{N}(\text{dB}) = 13.7 \text{ (dB)} - 10 \log \frac{90 \text{ MHz}}{90 \text{ Mbps}}$$

$$= 13.7 \text{ dB} - 0 \text{ dB} = 13.7 \text{ dB}$$

It should be evident from Examples 20-6 and 20-7 that the E_b/N_o and C/N ratios are equal only when the noise bandwidth is equal to the bit rate. Also, as the bandwidth at the point of measurement increases, the C/N decreases.

When the modulation scheme, bit rate, bandwidth, and C/N ratios of two digital radio systems are different; it is often difficult to determine which system has the lower probability of error. Because E_b/N_o is independent of bit rate, bandwidth, and modulation scheme; it is a convenient common denominator to use for comparing the probability of error performance of two digital radio systems.

EXAMPLE 20-8

Compare the performance characteristics of the two digital systems listed below, and determine which system has the lower probability of error.

	QPSK	8PSK
Bit rate	40 Mbps	60 Mbps
Bandwidth	1.5 × minimum	2 × minimum
C/N	10.75 dB	13.76 dB

Solution Substituting into Equation 20-6 for the QPSK system gives us

$$\frac{E_b}{N_o}(dB) = \frac{C}{N}(dB) + 10\log\frac{B}{F_b}$$

$$= 10.75\ dB + 10\log\frac{1.5\times20\ MHz}{40\ Mbps}$$

$$= 10.75\ dB + (-1.25\ dB)$$

$$= 9.5\ dB$$

From Figure 20-17, the $P(e)$ is 10^{-4}.

Substituting into Equation 20-6 for the 8PSK system gives us

$$\frac{E_b}{N_o}(dB) = 13.76\ dB + 10\log\frac{2\times20\ MHz}{60\ Mbps}$$

$$= 13.76\ dB + (-1.76\ dB)$$

$$= 12\ dB$$

From Figure 20-17, the $P(e)$ is 10^{-3}

Although the QPSK system has a lower C/N and E_b/N_o ratio, the $P(e)$ of the QPSK system is 10 times lower (better) than the 8PSK system.

Gain-to-Equivalent Noise Temperature Ratio

Essentially, *gain-to-equivalent noise temperature ratio* (G/T_e) is a figure of merit used to represent the quality of a satellite or an earth station receiver. The G/T_e of a receiver is the ratio of the receive antenna gain to the equivalent noise temperature (T_e) of the receiver. Because of the extremely small receive carrier powers typically experienced with satellite systems, very often an LNA is physically located at the feedpoint of the antenna. When this is the case, G/T_e is a ratio of the gain of the receiving antenna plus the gain of the LNA to the equivalent noise temperature. Mathematically, gain-to-equivalent noise temperature ratio is

$$\frac{G}{T_e} = \frac{A_r + A(LNA)}{T_e} \tag{20-7}$$

Expressed in logs, we have

$$\frac{G}{T_e}(dBK^{-1}) = A_r(dB) + A(LNA)(dB) - T_e(dBK^{-1}) \tag{20-8}$$

G/T_e is a very useful parameter for determining the E_b/N_0 and C/N ratios at the satellite transponder and earth station receivers. G/T_e is essentially the only parameter required at a satellite or an earth station receiver when completing a link budget.

EXAMPLE 20-9

For a satellite transponder with a receiver antenna gain of 22 dB, an LNA gain of 10 dB, and an equivalent noise temperature of 22 dBK^{-1}; determine the G/T_e figure of merit.

Solution   Substituting into Equation 20-8 yields

$$\frac{G}{T_e} \text{(dBK}^{-1}) = 22 \text{ dB} + 10 \text{ dB} - 22 \text{ dBK}^{-1}$$

$$= 10 \text{ dBK}^{-1}$$

SATELLITE SYSTEM LINK EQUATIONS

The error performance of a digital satellite system is quite predictable. Figure 20-18 shows a simplified block diagram of a digital satellite system and identifies the various gains and losses that may affect the system performance. When evaluating the performance of a digital satellite system, the uplink and downlink parameters are first considered

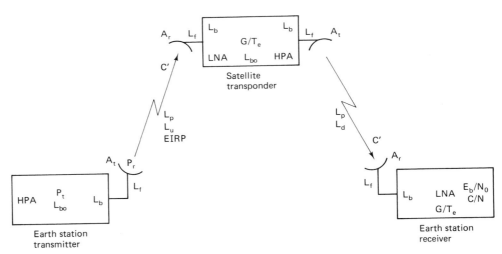

FIGURE 20-18   Overall satellite system showing the gains and losses incurred in both the uplink and downlink sections. HPA, High-power amplifier; P_t, HPA output power; L_{bo}, back-off loss; L_f, feeder loss; L_b, branching loss; A_t, transmit antenna gain; P_r, total radiated power $= P_t - L_{bo} - L_b - L_f$; EIRP, effective isotropic radiated power $= P_r A_t$; L_u, additional uplink losses due to atmosphere; L_p, path loss; A_r, receive antenna gain; G/T_e, gain-to-equivalent noise ratio; L_d, additional downlink losses due to atmosphere; LNA, low-noise amplifier; C/T_e, carrier-to-equivalent noise ratio; C/N_0, carrier-to-noise density ratio; E_b/N_0, energy of bit-to-noise density ratio; C/N, carrier-to-noise ratio.

separately, then the overall performance is determined by combining them in the appropriate manner. Keep in mind, a digital microwave or satellite radio simply means the original and demodulated baseband signals are digital in nature. The RF portion of the radio is analog; that is, FSK, PSK, QAM, or some other higher-level modulation riding on an analog microwave carrier.

LINK EQUATIONS

The following *link equations* are used to separately analyze the uplink and the downlink sections of a single radio-frequency carrier satellite system. These equations consider only the ideal gains and losses and effects of thermal noise associated with the earth station transmitter, earth station receiver, and the satellite transponder. The nonideal aspects of the system are discussed later in this chapter.

Uplink Equation

$$\frac{C}{N_0} = \frac{A_t P_r (L_p L_u) A_r}{K T_e} = \frac{A_t P_r (L_p L_u)}{K} \times \frac{G}{T_e}$$

where L_d and L_u are the additional uplink and downlink atmospheric losses, respectively. The uplink and downlink signals must pass through the earth's atmosphere, where they are partially absorbed by the moisture, oxygen, and particulates in the air. Depending on the elevation angle, the distance the RF signal travels through the atmosphere varies from one earth station to another. Because L_u and L_d represent losses, they are decimal values less than 1. G/T_e is the receiving antenna gain divided by the equivalent input noise temperature.

Expressed as a log,

$$\frac{C}{N_0} = \underbrace{10 \log A_t P_r}_{\substack{\text{EIRP} \\ \text{earth} \\ \text{station}}} - \underbrace{20 \log \left(\frac{4\pi D}{\lambda}\right)}_{\substack{\text{free-space} \\ \text{path loss}}} + \underbrace{10 \log \left(\frac{G}{T_e}\right)}_{\substack{\text{satellite} \\ G/T_e}} - \underbrace{10 \log L_u}_{\substack{\text{additional} \\ \text{atmospheric} \\ \text{losses}}} - \underbrace{10 \log K}_{\substack{\text{Boltzmann's} \\ \text{constant}}}$$

$$= \text{EIRP (dBW)} - L_p \text{(dB)} + \frac{G}{T_e} \text{(dBK}^{-1}) - L_u \text{(dB)} - K \text{(dBWK)}$$

Downlink Equation

$$\frac{C}{N_0} = \frac{A_t P_r (L_p L_d) A_r}{K T_e} = \frac{A_t A_r (L_p L_d)}{K} \times \frac{G}{T_e}$$

Expressed as a log

$$\frac{C}{N_{\mathrm{o}}} = \underbrace{10 \log A_t P_r}_{\substack{\text{EIRP} \\ \text{satellite}}} - \underbrace{20 \log \left(\frac{4\pi D}{\lambda}\right)}_{\substack{\text{free-space} \\ \text{path loss}}} + \underbrace{10 \log \frac{G}{T_e}}_{\substack{\text{+ earth station} \\ G/T_e}} - \underbrace{10 \log L_d}_{\substack{\text{additional} \\ \text{atmospheric} \\ \text{losses}}} - \underbrace{10 \log K}_{\substack{\text{Boltzmann's} \\ \text{constant}}}$$

$$= \text{EIRP (dBW)} - L_p\,(\text{dB}) + \frac{G}{T_e}\,(\text{dBK}^{-1}) - L_d\,(\text{dB}) - K\,(\text{dBWK})$$

LINK BUDGET

Table 20-4 lists the system parameters for three typical satellite communication systems. The systems and their parameters are not necessarily for an existing or future system; they are hypothetical examples only. The system parameters are used to construct a *link budget*. A link budget identifies the system parameters and is used to determine the projected C/N and E_b/N_0 ratios at both the satellite and earth station receivers for a given modulation scheme and desired $P(e)$.

EXAMPLE 20-10

Complete the link budget for a satellite system with the following parameters.

Uplink

1. Earth station transmitter output power at saturation, 2000 W	33 dBW
2. Earth station back-off loss	3 dB
3. Earth station branching and feeder losses	4 dB
4. Earth station transmit antenna gain (from Figure 20-19, 15 m at 14 GHz)	64 dB
5. Additional uplink atmospheric losses	0.6 dB
6. Free-space path loss (from Figure 20-20, at 14 Ghz)	206.5 db
7. Satellite receiver G/T_e ratio	-5.3 dBK^{-1}
8. Satellite branching and feeder losses	0 dB
9. Bit rate	120 Mbps
10. Modulation scheme	8 PSK

Downlink

1. Satellite transmitter output power at saturation 10 W	10 dBW
2. Satellite back-off loss	0.1 dB
3. Satellite branching and feeder losses	0.5 dB
4. Satellite transmit antenna gain (from Figure 20-19, 0.37 m at 12 GHz)	30.8 dB

TABLE 20-4   SYSTEM PARAMETERS FOR THREE HYPOTHETICAL SATELLITE SYSTEMS

	System A: 6/4 GHz, earth coverage QPSK modulation, 60 Mbps	System B: 14/12 GHz, earth coverage 8PSK modulation, 90 Mbps	System C: 14/12 GHz, earth coverage 8PSK modulation, 120 Mbps
Uplink			
Transmitter output power (saturation, dBW)	35	25	33
Earth station back-off loss (dB)	2	2	3
Earth station branching and feeder loss (dB)	3	3	4
Additional atmospheric (dB)	0.6	0.4	0.6
Earth station antenna gain (dB)	55	46	64
Free-space path loss (dB)	200	208	206.5
Satellite receive antenna gain (dB)	20	45	23.7
Satellite branching and feeder loss (dB)	1	1	0
Satellite equivalent noise temperature (K)	1000	800	800
Satellite G/T_e (dBK^{-1})	-10	16	-5.3
Downlink			
Transmitter output power (saturation, dBW)	18	20	30.8
Satellite back-off loss (dB)	0.5	0.2	0.1
Satellite branching and feeder loss (dB)	1	1	0.5
Additional atmospheric loss (dB)	0.8	1.4	0.4
Satellite antenna gain (dB)	16	44	10
Free-space path loss (dB)	197	206	205.6
Earth station receive antenna gain (dB)	51	44	62
Earth station branching and feeder loss (dB)	3	3	0
Earth station equivalent noise temperature (K)	250	1000	270
Earth station G/T_e (dBK^{-1})	27	14	37.7

5. Additional downlink atmospheric losses	0.4 dB
6. Free-space path loss (from Figure 20-20, at 12 GHz)	205.6 dB
7. Earth station receive antenna gain (15 m, 12 GHz)	62 dB
8. Earth station branching and feeder losses	0 dB
9. Earth station equivalent noise temperature	270 K
10. Earth station G/T_e ratio	37.7 dBK^{-1}

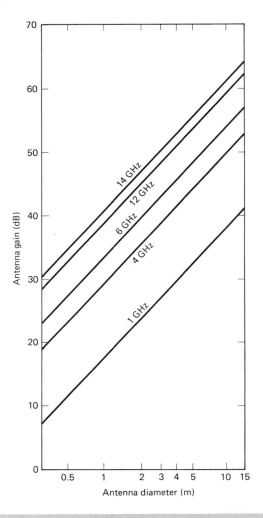

FIGURE 20-19   Antenna gain based on the gain equation for a parabolic antenna:

$$A \text{ (dB)} = 10 \log \eta \, (\pi \, D/\lambda)^2$$

where D is the antenna diameter, $\lambda =$ the wavelength, and $\eta =$ the antenna efficiency. Here $\eta = 0.55$. To correct for a 100% efficient antenna, add 2.66 dB to the value.

11. Bit rate	120 Mbps
12. Modulation scheme	8 PSK

Solution   *Uplink budget*: Expressed as a log,

$$\text{EIRP (earth station)} = P_t + A_t - L_{\text{bo}} - L_{bf}$$

$$= 33 \text{ dBW} + 64 \text{ dB} - 3 \text{ dB} - 4 \text{ dB} = 90 \text{ dBW}$$

Carrier power density at the satellite antenna:

$$C' = \text{EIRP (earth station)} - L_p - L_u$$

$$= 90 \text{ dBw} - 206.5 \text{ dB} - 0.6 \text{ dB} = -117.1 \text{ dBW}$$

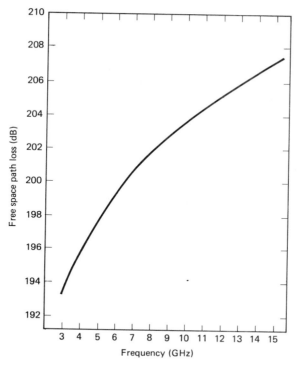

Elevation angle correction:	
Angle	+dB
90°	0
45°	0.44
0°	1.33

FIGURE 20-20 Free-space path loss (L_p) determined from

$$L_p = 183.5 = 20 \log F \text{ (GHz)}$$

Elevation angle = 90°, distance = 35,930 km.

C/N_0 at the satellite:

$$\frac{C}{N_0} = \frac{C}{KT_e} = \frac{C}{T_e} \times \frac{1}{K} \qquad \text{where } \frac{C}{T_e} = C' \times \frac{G}{T_e}$$

Thus

$$\frac{C}{N_0} = C' \times \frac{G}{T_e} \times \frac{1}{K}$$

Expressed as a log,

$$\frac{C}{N_0} \text{(dB)} = C'\text{(dBW)} + \frac{G}{T_e}\text{(dBK}^{-1}) - 10 \log (1.38 \times 10^{-23})$$

$$\frac{C}{N_0} = -117.1 \text{ dBW} + (-5.3 \text{ dBK}^{-1}) - (-228.6 \text{ dBWK}) = 106.2 \text{ dB}$$

Thus

$$\frac{E_b}{N_0}(dB) = \frac{C/F_b}{N_0}(dB) = \frac{C}{N_0}(dB) - 10 \log F_b$$

$$\frac{E_b}{N_0} = 106.2 \text{ dB} - 10 (\log 120 \times 10^6) = 25.4 \text{ dB}$$

and for a minimum bandwidth system,

$$\frac{C}{N} = \frac{E_b}{N_0} - \frac{B}{F_b} = 25.4 - 10 \log \frac{40 \times 10^6}{120 \times 10^6} = 30.2 \text{ dB}$$

Downlink budget: Expressed as a log,

EIRP (satellite transponder) $= P_t + A_t - L_{bo} - L_{bf}$

$$= 10 \text{ dBW} + 30.8 \text{ dB} - 0.1 \text{ dB} - 0.5 \text{ dB}$$

$$= 40.2 \text{ dBW}$$

Carrier power density at earth station antenna:

$$C' = \text{EIRP (dBW)} - L_p \text{ (dB)} - L_d \text{ (dB)}$$

$$= 40.2 \text{ dBW} - 205.6 \text{ dB} - 0.4 \text{ dB} = -165.8 \text{ dBW}$$

C/N_0 at the earth station receiver:

$$\frac{C}{N_0} = \frac{C}{KT_e} = \frac{C}{T_e} \times \frac{1}{K} \qquad \text{where } \frac{C}{T_e} = C' \times \frac{G}{T_e}$$

Thus

$$\frac{C}{N_0} = C' \times \frac{G}{T_e} \times \frac{1}{K}$$

Expressed as a log,

$$\frac{C}{N_0}(dB) = C' \text{ (dBW)} + \frac{G}{T_e}(dBK^{-1}) - 10 \log(1.38 \times 10^{-23})$$

$$= -165.8 \text{ dBW} + (37.7 \text{ dBK}^{-1}) - (-228.6 \text{ dBWK}) = 100.5 \text{ dB}$$

An alternative method of solving for C/N_0 is

$$\frac{C}{N_0}(dB) = C' \text{ (dBW)} + A_r \text{ (dB)} - T_e (dBk^{-1}) - K \text{ (dBWK)}$$

$$= -165.8 \text{ dBW} + 62 \text{ dB} - 10 \log 270 - (-228.6 \text{ dBWK})$$

$$\frac{C}{N_0} = -165.8 \text{ dBW} + 62 \text{ dB} - 24.3 \text{ dBK}^{-1} + 228.6 \text{ dBWK} = 100.5 \text{ dB}$$

$$\frac{E_b}{N_0}(dB) = \frac{C}{N_0}(dB) - 10 \log F_b$$

$$= 100.5 \text{ dB} - 10 \log (120 \times 10^6)$$

$$= 100.5 \text{ dB} - 80.8 \text{ dB} = 19.7 \text{ dB}$$

and for a minimum bandwidth system,

$$\frac{C}{N} = \frac{E_b}{N_o} - \frac{B}{F_b} = 19.7 - 10 \log \frac{40 \times 10^6}{120 \times 10^6} = 24.5 \text{ dB}$$

With careful analysis and a little algebra, it can be shown that the overall energy of bit-to-noise density ratio (E_b/N_o), which includes the combined effects of the uplink

TABLE 20-5   LINK BUDGET FOR EXAMPLE 20-10

Uplink

1. Earth station transmitter output power at saturation, 2000 W	33 dBW
2. Earth station back-off loss	3 dB
3. Earth station branching and feeder losses	4 dB
4. Earth station transmit antenna gain	64 dB
5. Earth station EIRP	90 dBW
6. Additional uplink atmospheric losses	0.6 dB
7. Free-space path loss	206.5 dB
8. Carrier power density at satellite	-117.1 dBW
9. Satellite branching and feeder losses	0 dB
10. Satellite G/T_e ratio	-5.3 dBK^{-1}
11. Satellite C/T_e ratio	-122.4 dBWK^{-1}
12. Satellite C/N_0 ratio	106.2 dB
13. Satellite C/N ratio	30.2 dB
14. Satellite E_b/N_0 ratio	25.4 dB
15. Bit rate	120 Mbps
16. Modulation scheme	8 PSK

Downlink

1. Satellite transmitter output power at saturation, 10 W	10 dBW
2. Satellite back-off loss	0.1 dB
3. Satellite branching and feeder losses	0.5 dB
4. Satellite transmit antenna gain	30.8 dB
5. Satellite EIRP	40.2 dBW
6. Additional downlink atmospheric losses	0.4 dB
7. Free-space path loss	205.6 dB
8. Earth station receive antenna gain	62 dB
9. Earth station equivalent noise temperature	270 K
10. Earth station branching and feeder losses	0 dB
11. Earth station G/T_e ratio	37.7 dBK^{-1}
12. Carrier power density at earth station	-165.8 dBW
13. Earth station C/T_e ratio	-128.1 dBWK^{-1}
14. Earth station C/N_0 ratio	100.5 dB
15. Earth station C/N ratio	24.5 dB
16. Earth station E_b/N_0 ratio	19.7 dB
17. Bit rate	120 Mbps
18. Modulation scheme	8 PSK

ratio $(E_b/N_o)_u$ and the downlink ratio $(E_b/N_o)_d$, is a standard product over the sum relationship and is expressed mathematically as

$$\frac{E_b}{N_o} \text{(overall)} = \frac{(E_b/N_o)_u (E_b/N_o)_d}{(E_b/N_o)_u + (E_b/N_o)_d} \tag{20-9}$$

where all E_b/N_o ratios are in absolute values. For Example 20-10, the overall E_b/N_o ratio is

$$\frac{E_b}{N_o} \text{(overall)} = \frac{(346.7)(93.3)}{346.7 + 93.3} = 73.5$$

$$= 10 \log 73.5 = 18.7 \text{ dB}$$

As with all product-over-sum relationships, the smaller of the two numbers dominates. If one number is substantially smaller than the other, the overall result is approximately equal to the smaller of the two numbers.

The system parameters used for Example 20-10 were taken from system C in Table 20-4. A complete link budget for the system is shown in Table 20-5.

NONIDEAL SYSTEM PARAMETERS

Additional *nonideal parameters* include the following impairments: AM/AM conversion and AM/PM conversion, which result from nonlinear amplification in HPAs and limiters; *pointing error*, which occurs when the earth station and satellite antennas are not exactly aligned; *phase jitter*, which results from imperfect carrier recovery in receivers; *nonideal filtering*, due to the imperfections introduced in bandpass filters; *timing error*, due to imperfect clock recovery in receivers; and *frequency translation errors* introduced in the satellite transponders. The degradation caused by the preceding impairments effectively reduces the E_b/N_o ratios determined in the link budget calculations. Consequently, they have to be included in the link budget as equivalent losses. An in-depth coverage of the nonideal parameters is beyond the intent of this text.

QUESTIONS

20-1. Briefly describe a satellite.

20-2. What is a passive satellite? An active satellite?

20-3. Contrast nonsynchronous and synchronous satellites.

20-4. Define *prograde* and *retrograde*.

20-5. Define *apogee* and *perigee*.

20-6. Briefly explain the characteristics of low-, medium-, and high-altitude satellite orbits.

20-7. Explain equatorial, polar, and inclined orbits.

20-8. Contrast the advantages and disadvantages of geosynchronous satellites.

20-9. Define *look angles, angle of elevation,* and *azimuth.*

20-10. Define *satellite spatial separation* and list its restrictions.

20-11. Describe a "footprint."

20-12. Describe spot, zonal, and earth coverage radiation patterns.

20-13. Explain *reuse.*

20-14. Briefly describe the functional characteristics of an uplink, a transponder, and a downlink model for a satellite system.

20-15. Define *back-off loss* and its relationship to saturated and transmit power.

20-16. Define *bit energy.*

20-17. Define *effective isotropic radiated power.*

20-18. Define *equivalent noise temperature.*

20-19. Define *noise density.*

20-20. Define *carrier-to-noise density ratio* and *energy of bit-to-noise density ratio.*

20-21. Define *gain-to-equivalent noise temperature ratio.*

20-22. Describe what a satellite link budget is and how it is used.

PROBLEMS

20-1. An earth station is located at Houston, Texas, which has a longitude of 99.5° and a latitude of 29.5° north. The satellite of interest is Satcom 2. Determine the look angles for the earth station antenna.

20-2. A satellite system operates at 14-GHz uplink and 11-GHz downlink and has a projected $P(e)$ of 10^{-7}. The modulation scheme is 8PSK, and the system will carry 120 Mbps. The equivalent noise temperature of the receiver is 400 K, and the receiver noise bandwidth is equal to the minimum Nyquist frequency. Determine the following parameters: minimum theoretical C/N ratio, minimum theoretical E_b/N_o ratio, noise density, total receiver input noise, minimum receive carrier power, and the minimum energy per bit at the receiver input.

20-3. A satellite system operates at 6-GHz uplink and 4-GHz downlink and has a projected $P(e)$ of 10^{-6}. The modulation scheme is QPSK and the system will carry 100 Mbps. The equivalent receiver noise temperature is 290 K, and the receiver noise bandwidth is equal to the minimum Nyquist frequency. Determine the following:
   (a) The C/N ratio that would be measured at a point in the receiver prior to the BPF where the bandwidth is equal to $1\frac{1}{2}$ times the minimum Nyquist frequency.
   (b) The C/N ratio that would be measured at a point in the receiver prior to the BPF where the bandwidth is equal to 3 times the minimum Nyquist frequency.

20-4. Which system has the best projected BER?
   (a) 8QAM, $C/N = 15$ dB, $B = 2F_N$, $F_b = 60$ Mbps.
   (b) QPSK, $C/N = 16$ dB, $B = F_N$, $F_b = 40$ Mbps.

20-5. An earth station satellite transmitter has an HPA with a rated saturated output power of 10,000 W. The back-off ratio is 6 dB, the branching loss is 2 dB, the feeder loss is 4 dB, and the antenna gain is 40 dB. Determine the actual radiated power and the EIRP.

20-6. Determine the total noise power for a receiver with an input bandwidth of 20 MHz and an equivalent noise temperature of 600 K.

20-7. Determine the noise density for Problem 8-6.

20-8. Determine the minimum C/N ratio required to achieve a $P(e)$ of 10^{-5} for an 8PSK receiver with a bandwidth equal to F_N.

20-9. Determine the energy of bit-to-noise density ratio when the receiver input carrier power is -100 dBW, the receiver input noise temperature is 290 K, and a 60-Mbps transmission rate is used.

20-10. Determine the carrier-to-noise density ratio for a receiver with a -70-dBW input carrier power, an equivalent noise temperature of 180 K, and a bandwidth of 20 MHz.

20-11. Determine the minimum C/N ratio for an 8PSK system when the transmission rate is 60 Mbps, the minimum energy of bit-to-noise density ratio is 15 dB, and the receiver bandwidth is equal to the minimum Nyquist frequency.

20-12. For an earth station receiver with an equivalent input temperature of 200 K, a noise bandwidth of 20 MHz, a receive antenna gain of 50 dB, and a carrier frequency of 12 GHz, determine the following: G/T_e, N_o, and N.

20-13. For a satellite with an uplink E_b/N_o of 14 dB and a downlink E_b/N_o of 18 dB, determine the overall E_b/N_o ratio.

20-14. Complete the following link budget:

Uplink Parameters

1. Earth station transmitter output power at saturation, 1 kW
2. Earth station back-off loss, 3 dB
3. Earth station total branching and feeder losses, 3 dB
4. Earth station transmit antenna gain for a 10-m parabolic dish at 14 GHz
5. Free-space path loss for 14 GHz
6. Additional uplink losses due to the earth's atmosphere, 0.8 dB
7. Satellite transponder G/Te, -4.6 dBk
8. Transmission bit rate, 90 Mbps, 8 PSK

Downlink Parameters

1. Satellite transmitter output power at saturation, 10 W
2. Satellite station transmit antenna gain for a 0.5-m parabolic dish at 12 GHz
3. Satellite modulation back-off loss, 0.8 dB
4. Free-space path loss for 12 GHz
5. Additional downlink losses due to earth's atmosphere, 0.6 dB
6. Earth station receive antenna gain for a 10-m parabolic dish at 12 GHz
7. Earth station equivalent noise temperature, 200 K
8. Earth station branching and feeder losses, 0 dB
9. Transmission bit rate, 90 Mbps, 8 PSK

Chapter 21

SATELLITE MULTIPLE-ACCESS ARRANGEMENTS

FDM/FM SATELLITE SYSTEMS

Figure 21-1a shows a single-link (two earth stations) *fixed-frequency* FDM/FM system using a single satellite transponder. With earth coverage antennas and for full-duplex operation, each link requires two RF satellite channels (i.e., four RF carrier frequencies, two uplink and two downlink). In Figure 21-1a, earth station 1 transmits on a high-band carrier (F11, F12, F13, etc.) and receives on a low-band carrier (F1, F2, F3, etc.). To avoid interfering with earth station 1, earth station 2 must transmit and receive on different RF carrier frequencies. The RF carrier frequencies are fixed and the satellite transponder is simply an RF-to-RF repeater that provides the uplink/downlink frequency translation. This arrangement is economically impractical and extremely inefficient as well. Additional earth stations can communicate through different transponders within the same satellite structure (Figure 21-1b). Each additional link requires four more RF carrier frequencies. It is unlikely that any two-point link would require the capacity available in an entire RF satellite channel. Consequently, most of the available bandwidth is wasted. Also, with this arrangement, each earth station can communicate with only one other earth station. The RF satellite channels are fixed between any two earth stations; thus the voice band channels from each earth station are dedicated to a single destination.

In a system where three or more earth stations wish to communicate with each other, fixed-frequency or *dedicated channel* systems such as those shown in Figure 21-1 are inadequate; a method of *multiple accessing* is required. That is, each earth station using the satellite system has a means of communicating with each of the other earth stations in the system through a common satellite transponder. Multiple accessing

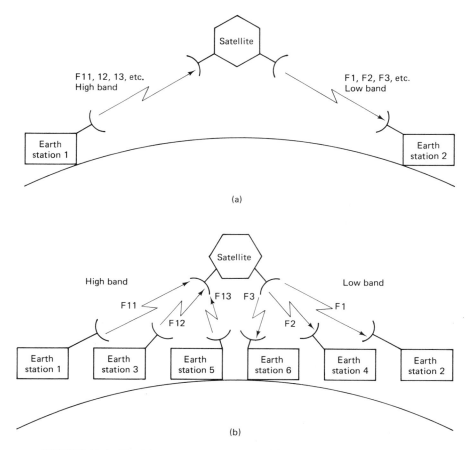

FIGURE 21-1   Fixed-frequency earth station satellite system: (a) single link; (b) multiple link.

is sometimes called *multiple destination* because the transmissions from each earth station are received by all the other earth stations in the system. The voice band channels between any two earth stations may be *preassigned* (*dedicated*) or *demand-assigned* (*switched*). When preassignment is used, a given number of the available voice band channels from each earth station are assigned a dedicated destination. With demand assignment, voice band channels are assigned on an as-needed basis. Demand assignment provides more versatility and more efficient use of the available frequency spectrum. On the other hand, demand assignment requires a control mechanism that is common to all the earth stations to keep track of channel routing and the availability of each voice band channel.

Remember, in an FDM/FM satellite system, each RF channel requires a separate transponder. Also, with FDM/FM transmissions, it is impossible to differentiate (separate) multiple transmissions that occupy the same bandwidth. Fixed-frequency systems may be used in a multiple-access configuration by switching the RF carriers at the satellite, reconfiguring the baseband signals with multiplexing/demultiplexing equipment on board the satellite, or by using multiple spot beam antennas (reuse). All three of these methods require relatively complicated, expensive, and heavy hardware on the spacecraft.

MULTIPLE ACCESSING

Figure 21-2 shows the three most commonly used multiple accessing arrangements: frequency-division multiple accessing (FDMA), time-division multiple accessing (TDMA), and code-division multiple accessing (CDMA). With FDMA, each earth sta-

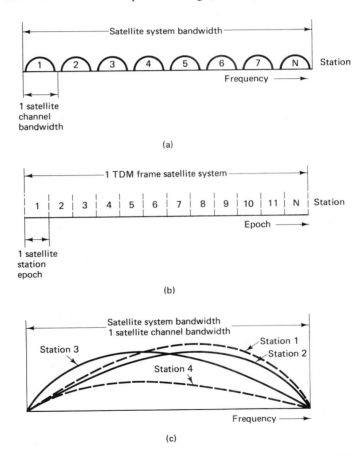

FIGURE 21-2 Multiple-accessing arrangements: (a) FDMA; (b) TDMA; (c) CDMA.

tion's transmissions are assigned specific uplink and downlink frequency bands within an allotted satellite channel bandwidth; they may be preassigned or demand assigned. Consequently, transmissions from different earth stations are separated in the frequency domain. With TDMA, each earth station transmits a short burst of information during a specific time slot (*epoch*) within a TDMA frame. The bursts must be synchronized so that each station's *burst* arrives at the satellite at a different time. Consequently, transmissions from different earth stations are separated in the time domain. With CDMA, all earth stations transmit within the same frequency band and, for all practical purposes, have no limitation on when they may transmit or on which carrier frequency. Carrier separation is accomplished with *envelope encryption/decryption* techniques.

Frequency-Division Multiple Access

Frequency-division multiple access (FDMA) is a method of multiple accessing where a given RF channel bandwidth is divided into smaller frequency bands called *subdivisions*. Each subdivision is used to carry one voice band channel. A control mechanism is used to ensure that no two earth stations transmit on the same subdivision at the same time. Essentially, the control mechanism designates a receive station for each of the subdivisions. In demand-assignment systems, the control mechanism is also used to establish or terminate the voice band links between the source and destination earth stations. Consequently, any of the subdivisions may be used by any of the participating earth stations at any given time. Typically, each subdivision is used to carry a single 4-kHz voice band channel, but occasionally, groups, supergroups, or even mastergroups are assigned a larger subdivision.

    SPADE system.   The first FDMA demand-assignment system for satellites was developed by COMSAT for use on the INTELSAT IV satellite. This system was called *SPADE* (single-channel-per-carrier PCM multiple-access demand assignment equipment). Figures 21-3 and 21-4 show the block diagram and IF frequency assignments for SPADE, respectively.

    With SPADE, 800 PCM-encoded voice band channels separately QPSK modulate an IF carrier frequency (hence the name *single carrier per channel*, SCPC). Each 4-kHz voice band channel is sampled at an 8-kHz rate and converted to an 8-bit PCM code. This produces a 64-kbps PCM code for each voice band channel. The PCM code from each voice band channel QPSK modulates a different IF carrier frequency. With QPSK, the minimum required bandwidth is equal to one-half the input bit rate. Consequently, the output of each QPSK modulator requires a minimum bandwidth of 32 kHz. Each channel is allocated a 45-kHz bandwidth, which allows for a 13-kHz guard band between each frequency-division-multiplexed channel. The IF carrier frequencies begin at 52.0225 MHz (low-band channel 1) and increase in 45-kHz steps to 87.9775 MHz (high-band channel 400). The entire 36-MHz band (52 to 88 MHz) is divided in half, producing two 400-channel bands (a low-band and a high-band). For full-duplex operation, four hundred 45-kHz channels are used for one direction of transmission and 400 are used for the opposite direction. Also, channels 1, 2, and 400 from each

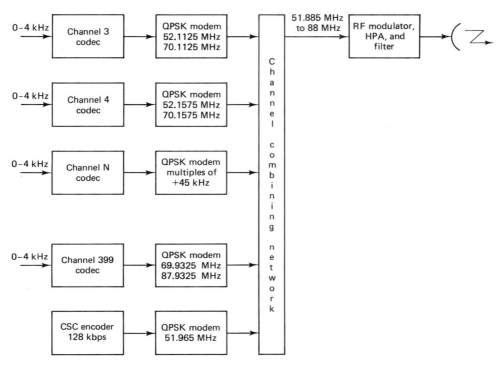

FIGURE 21-3   FDMA, SPADE earth station transmitter.

band are left permanently vacant. This reduces the number of usable full-duplex voice band channels to 397. The 6-GHz C-band extends from 5.725 to 6.425 GHz (700 MHz). This allows for approximately nineteen 36-MHz RF channels per system. Each RF channel has a capacity of 397 full-duplex voice band channels.

Each IF channel (Figure 21-4) has a 160-kHz *common signaling channel* (CSC). The CSC is a time-division-multiplexed transmission that is frequency-division multiplexed onto the IF spectrum below the QPSK-encoded voice band channels. Figure 21-5 shows the TDM frame structure for the CSC. The total frame time is 50 ms, which is subdivided into fifty 1-ms epochs. Each earth station transmits on the CSC channel only during its preassigned 1-ms time slot. The CSC signal is a 128-bit binary code. To transmit a 128-bit code in 1 ms, a transmission rate of 128 kbps is required. The CSC code is used for establishing and disconnecting voice band links between two earth station users when demand-assignment channel allocation is used.

EXAMPLE 21-1

For the system shown in Figure 21-6, a user earth station in New York wishes to establish a voice band link between itself and London. New York randomly selects an idle voice band channel. It then transmits a binary-coded message to London on the CSC channel

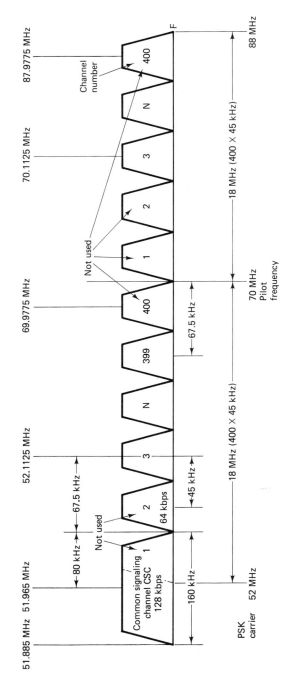

FIGURE 21-4 Carrier frequency assignments for the Intelsat single channel-per-carrier *P*CM multiple *a*ccess *d*emand assignment *e*quipment (SPADE).

796

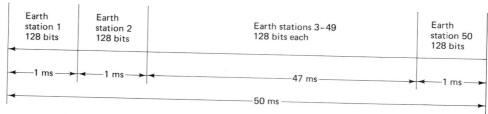

128 bits/1 ms × 1000 ms/1 s = 128 kbps or 6400 bits/frame × 1 frame/50 ms = 128 kbps

FIGURE 21-5   FDMA, SPADE common signaling channel (CSC).

during its respective time slot, requesting that a link be established on the randomly selected channel. London responds on the CSC channel during its time slot with a binary code, either confirming or denying the establishment of the voice band link. The link is disconnected in a similar manner when the users are finished.

The CSC channel occupies a 160-kHz bandwidth, which includes the 45 kHz for low-band channel 1. Consequently, the CSC channel extends from 51.885 MHz to 52.045 MHz. The 128-kbps CSC binary code QPSK modulates a 51.965-MHz carrier. The minimum bandwidth required for the CSC channel is 64 kHz; this results in a 48-kHz guard band on either side of the CSC signal.

With FDMA, each earth station may transmit simultaneously within the same 36-MHz RF spectrum, but on different voice band channels. Consequently, simultaneous transmissions of voice band channels from all earth stations within the satellite network are interleaved in the frequency domain in the satellite transponder. Transmissions of CSC signals are interleaved in the time domain.

An obvious disadvantage of FDMA is that carriers from multiple earth stations

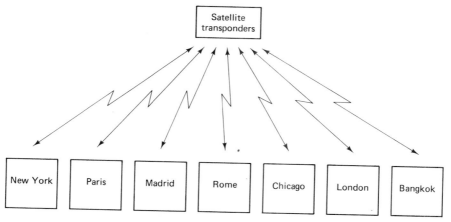

FIGURE 21-6   Diagram of the system for Example 21-11.

may be present in a satellite transponder at the same time. This results in cross-modulation distortion between the various earth station transmissions. This is alleviated somewhat by shutting off the IF subcarriers on all unused 45-kHz voice band channels. Because balanced modulators are used in the generation of QPSK, carrier suppression is inherent. This also reduces the power load on a system and increases its capacity by reducing the idle channel power.

Time-Division Multiple Access

Time-division multiple access (TDMA) is the predominant multiple-access method used today. It provides the most efficient method of transmitting digitally modulated carriers (PSK). TDMA is a method of time-division multiplexing digitally modulated carriers between participating earth stations within a satellite network through a common satellite transponder. With TDMA, each earth station transmits a short *burst* of a digitally modulated carrier during a precise time slot (epoch) within a TDMA frame. Each station's burst is synchronized so that it arrives at the satellite transponder at a different time. Consequently, only one earth station's carrier is present in the transponder at any given time, thus avoiding a collision with another station's carrier. The transponder is an RF-to-RF repeater that simply receives the earth station transmissions, amplifies them, and then retransmits them in a down-link beam which is received by all the participating earth stations. Each earth station receives the bursts from all other earth stations and must select from them the traffic destined only for itself.

Figure 21-7 shows a basic TDMA frame. Transmissions from all earth stations are synchronized to a *reference burst*. Figure 21-7 shows the reference burst as a separate

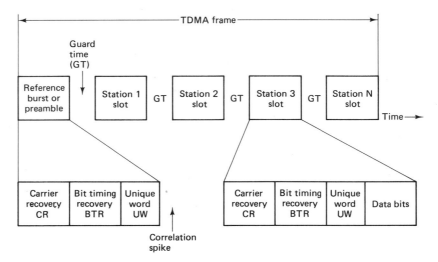

FIGURE 21-7 Basic time-division-multiple accessing (TDMA) frame.

transmission, but it may be the *preamble* which precedes a reference station's transmission of data. Also, there may be more than one synchronizing reference burst.

The reference burst contains a *carrier recovery sequence* (CRS) from which all receiving stations recover a frequency and phase coherent carrier for PSK demodulation. Also included in the reference burst is a binary sequence for *bit timing recovery* (BTR, i.e., clock recovery). At the end of each reference burst, a *unique word* (UW) is transmitted. The UW sequence is used to establish a precise time reference that each of the earth stations use to synchronize the transmission of its burst. The UW is typically a string of successive binary 1's terminated with a binary 0. Each earth station receiver demodulates and integrates the UW sequence. Figure 21-8 shows the result of the integration process. The integrator and threshold detector are designed so that the threshold voltage is reached precisely when the last bit of the UW sequence is integrated. This generates a *correlation spike* at the output of the threshold detector at the exact time the UW sequence ends.

Each earth station synchronizes the transmission of its carrier to the occurrence of the UW correlation spike. Each station waits a different length of time before it begins transmitting. Consequently, no two stations will transmit carrier at the same time. Note the *guard time* (GT) between transmissions from successive stations. This is analogous to a guard band in a frequency-division-multiplexed system. Each station precedes the transmission of data with a *preamble*. The preamble is logically equivalent to the reference burst. Because each station's transmissions must be received by all

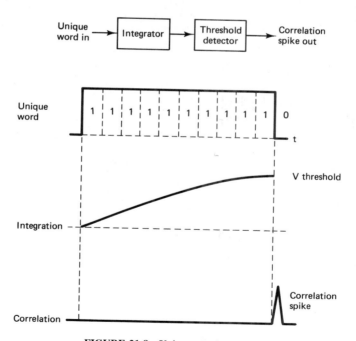

FIGURE 21-8   Unique word correlator.

other earth stations, all stations must recover carrier and clocking information prior to demodulating the data. If demand assignment is used, a common signaling channel must also be included in the preamble.

CEPT primary multiplex frame.   Figures 21-9 and 21-10 show the block diagram and timing sequence for the CEPT primary multiplex frame respectively (CEPT— Conference of European Postal and Telecommunications Administrations; the CEPT sets many of the European telecommunications standards). This is a commonly used TDMA frame format for digital satellite systems.

Essentially, TDMA is a *store-and-forward* system. Earth stations can transmit only during their specified time slot, although the incoming voice band signals are continuous. Consequently, it is necessary to sample and store the voice band signals prior to transmission. The CEPT frame is made up of 8-bit PCM encoded samples from 16 independent voice band channels. Each channel has a separate codec that samples the incoming voice signals at a 16-kHz rate and converts those samples to an 8-bit binary code. This results in 128 kbps transmitted at a 2.0408-MHz rate from each

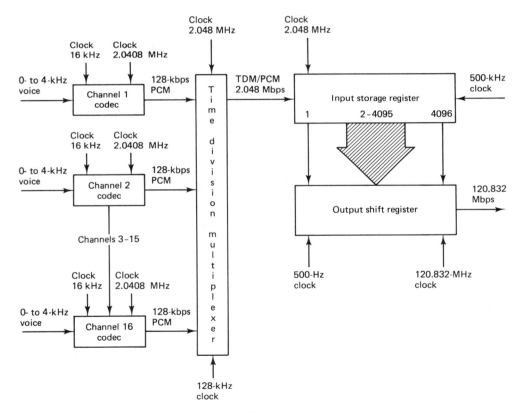

FIGURE 21-9   TDMA, CEPT primary multiplex frame transmitter.

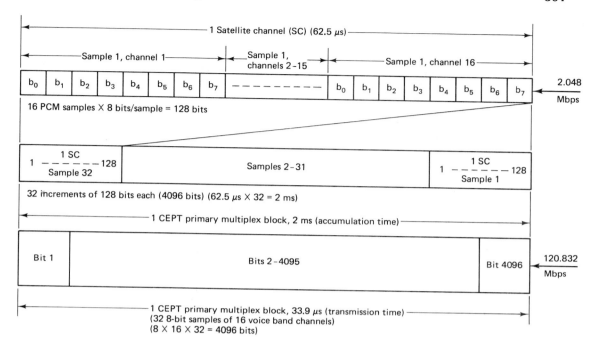

FIGURE 21-10   TDMA, CEPT primary multiplex frame.

voice channel codec. The sixteen 128-kbps transmissions are time-division multiplexed into a subframe that contains one 8-bit sample from each of the 16 channels (128 bits). It requires only 62.5 μs to accumulate the 128 bits (2.048-Mbps transmission rate). The CEPT multiplex format specifies a 2-ms frame time. Consequently, each earth station can transmit only once every 2 ms and therefore must store the PCM-encoded samples. The 128 bits accumulated during the first sample of each voice band channel are stored in a holding register while a second sample is taken from each channel and converted into another 128-bit *subframe*. This 128-bit sequence is stored in the holding register behind the first 128 bits. The process continues for 32 subframes (32 × 62.5 μs = 2 ms). After 2 ms, thirty-two 8-bit samples have been taken from each of 16 voice band channels for a total of 4096 bits (32 × 8 × 16 = 4096). At this time, the 4096 bits are transferred to an output shift register for transmission. Because the total TDMA frame is 2 ms long and during this 2-ms period each of the participating earth stations must transmit at different times, the individual transmissions from each station must occur in a significantly shorter time period. In the CEPT frame, a transmission rate of 120.832 Mbps is used. This rate is the fifty-ninth multiple of 2.048 Mbps. Consequently, the actual transmission of the 4096 accumulated bits takes approximately 33.9 μs. At the earth station receivers, the 4096 bits are stored in a holding register and shifted out to their PCM decoders at a 2.048-Mbps rate. Because all the clock rates (500 Hz, 16 kHz, 128 kHz, 2.048 MHz, and 120.832 MHz) are synchronized, the PCM codes are accumulated, stored, transmitted, received, and then decoded in perfect synchronization. To the users, the voice transmission is a continuous process.

There are several advantages of TDMA over FDMA. The first, and probably the most significant, is that with TDMA only the carrier from one earth station is present in the satellite transponder at any given time, thus reducing intermodulation distortion. Second, with FDMA, each earth station must be capable of transmitting and receiving on a multitude of carrier frequencies to achieve multiple accessing capabilities. Third, TDMA is much better suited to the transmission of digital information than FDMA. Digital signals are naturally acclimated to storage, rate conversions, and time-domain processing than their analog counterparts.

The primary disadvantage of TDMA as compared to FDMA is that in TDMA precise synchronization is required. Each earth station's transmissions must occur during an exact time slot. Also, bit and frame timing must be achieved and maintained with TDMA.

Code-Division Multiple Access (Spread-Spectrum Multiple Accessing)

With FDMA, earth stations are limited to a specific bandwidth within a satellite channel or system but have no restriction on when they can transmit. With TDMA, earth station's transmissions are restricted to a precise time slot but have no restriction on what frequency or bandwidth they may use within a specified satellite system or channel allocation. With *code-division multiple access* (CDMA), there are no restrictions on time or bandwidth. Each earth station transmitter may transmit whenever it wishes and can use any or all of the bandwidth allocated a particular satellite system or channel. Because there is no limitation on the bandwidth, CDMA is sometimes referred to as *spread-spectrum multiple access*; transmissions can spread throughout the entire allocated bandwidth spectrum. Transmissions are separated through envelope encryption/decryption techniques. That is, each earth station's transmissions are encoded with a unique binary word called a *chip code*. Each station has a unique chip code. To receive a particular earth station's transmission, a receive station must know the chip code for that station.

Figure 21-11 shows the block diagram of a CDMA encoder and decoder. In the encoder (Figure 21-11a), the input data (which may be PCM-encoded voice band signals of raw digital data) is multiplied by a unique chip code. The product code PSK modulates an IF carrier which is up-converted to RF for transmission. At the receiver (Figure 21-11b), the RF is down-converted to IF. From the IF, a coherent PSK carrier is recovered. Also, the chip code is acquired and used to synchronize the receive station's code generator. Keep in mind, the receiving station knows the chip code but must generate a chip code that is synchronous in time with the receive code. The recovered synchronous chip code multiplies the recovered PSK carrier and generates a PSK modulated signal that contains the PSK carrier plus the chip code. The received IF signal that contains the chip code, the PSK carrier, and the data information is compared to the received IF signal in the *correlator*. The function of the correlator is to compare the two signals and recover the original data. Essentially, the correlator subtracts the recovered PSK

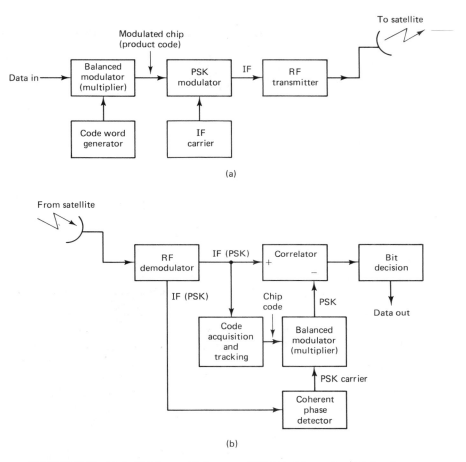

FIGURE 21-11 Code-division multiple access (CDMA): (a) encoder; (b) decoder.

carrier + chip code from the received PSK carrier + chip code + data. The resultant is the data.

The correlation is accomplished on the analog signals. Figure 21-12 shows how the encoding and decoding is accomplished. Figure 21-12a shows the correlation of the correctly received chip code. A +1 indicates an in-phase carrier and a −1 indicates an out-of-phase carrier. The chip code is multiplied by the data (either +1 or −1). The product is either an in-phase code or one that is 180° out of phase with the chip code. In the receiver, the recovered synchronous chip code is compared in the correlator to the received signaling elements. If the phases are the same, a +1 is produced; if they are 180° out of phase, a −1 is produced. It can be seen that if all the recovered chips correlate favorably with the incoming chip code, the output of the correlator will be a +6 (which is the case when a logic 1 is received). If all the code chips correlate

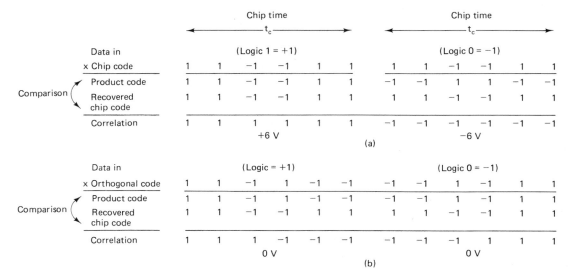

FIGURE 21-12 CDMA code/data alignment: (a) correct code; (b) orthogonal code.

180° out of phase, a −6 is generated (which is the case when a logic 0 is received). The bit decision circuit is simply a threshold detector. Depending on whether a +6 or a −6 is generated, the threshold detector will output a logic 1 or a logic 0, respectively.

As the name implies, the correlator looks for a correlation (similarity) between the incoming coded signal and the recovered chip code. When a correlation occurs, the bit decision circuit generates the corresponding logic condition.

With CDMA, all earth stations within the system may transmit on the same frequency at the same time. Consequently, an earth station receiver may be receiving coded PSK signals simultaneously from more than one transmitter. When this is the case, the job of the correlator becomes considerably more difficult. The correlator must compare the recovered chip code with the entire received spectrum and separate from it only the chip code from the desired earth station transmitter. Consequently, the chip code from one earth station must not correlate with the chip codes from any of the other earth stations.

Figure 21-12b shows how such a coding scheme is achieved. If half of the chips within a code were made the same and half were made exactly the opposite, the resultant would be zero cross correlation between chip codes. Such a code is called an *orthogonal code*. In Figure 21-12b it can be seen that when the orthogonal code is compared with the original chip code, there is no correlation (i.e., the sum of the comparison is zero). Consequently, the orthogonal code, although received simultaneously with the desired chip code, had absolutely no effect on the correlation process. For this example, the orthogonal code is received in exact time synchronization with the desired chip code; this is not always the case. For systems that do not have time synchronous transmissions, codes must be developed where there is no correlation between one station's code and any phase of another station's code. For more than two participating earth stations, this is impossible to do. A code set has been developed called the *Gold code*. With the Gold code, there is a minimum correlation between different chips' codes. For a reasonable number of users, it is impossible to achieve perfect orthogonal codes. You can design only for a minimum *cross correlation* between chips.

One of the advantages of CDMA was that the entire bandwith of a satellite channel or system may be used for each transmission from every earth station. For our example, the chip rate was six times the original bit rate. Consequently, the actual transmission rate of information was one-sixth of the PSK modulation rate, and the bandwidth required is six times that required to simply transmit the original data as binary. Because of the coding inefficiency resulting from transmitting chips for bits, the advantage of more bandwidth is partially offset and is thus less of an advantage. Also, if the transmission of chips from the various earth stations must be synchronized, precise timing is required for the system to work. Therefore, the disadvantage of requiring time synchronization in TDMA systems is also present with CDMA. In short, CDMA is not all that it is cracked up to be. The only significant advantage of CDMA is immunity to interference (jamming), which makes CDMA ideally suited for military applications.

FREQUENCY HOPPING

Frequency hopping is a form of CDMA where a digital code is used to continually change the frequency of the carrier. With frequency hopping, the total available bandwidth is partitioned into smaller frequency bands and the total transmission time is subdivided into smaller time slots. The idea is to transmit within a limited frequency band for only a short period of time, then switch to another frequency band, and so on. This process continues indefinitely. The frequency hopping pattern is determined by a binary code. Each station uses a different code sequence. A typical *hopping pattern (frequency-time matrix)* is shown in Figure 21-13.

With frequency hopping, each earth station within a CDMA network is assigned a different frequency hopping pattern. Each transmitter switches (hops) from one frequency band to the next according to their assigned pattern. With frequency hopping, each station uses the entire RF spectrum but never occupies more than a small portion of that spectrum at any one time.

FSK is the modulation scheme most commonly used with frequency hopping. When it is a given station's turn to transmit, it sends one of the two frequencies (either mark or space) for the particular band in which it is transmitting. The number of stations in a given frequency hopping system is limited by the number of unique hopping patterns that can be generated.

CHANNEL CAPACITY

Essentially, there are two methods used to interface terrestrial voice band channels with satellite channels: digital noninterpolated interfaces (DNI) and digital speech interpolated interfaces (DSI).

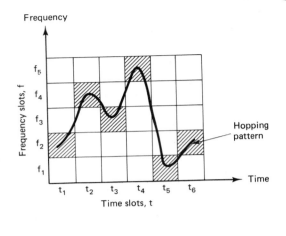

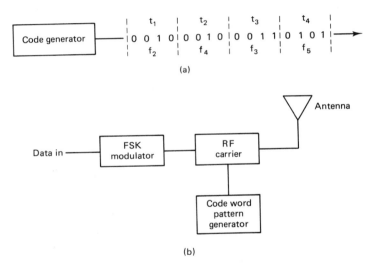

FIGURE 21-13 **Frequency hopping: (a) frequency time-hopping matrix; (b) frequency hopping transmitter.**

Digital Noninterpolated Interfaces

A *digital noninterpolated interface* assigns an individual terrestrial channel (TC) to a particular satellite channel (SC) for the duration of the call. A DNI system can carry no more traffic than the number of satellite channels it has. Once a TC has been assigned an SC, the SC is unavailable to the other TCs for the duration of the call. DNI is a form of preassignment; each TC has a permanent dedicated SC.

Digital Speech Interpolated Interfaces

A *digital speech interpolated interface* assigns a terrestrial channel to a satellite channel only when speech energy is present on the TC. DSI interfaces have *speech detectors* that are similar to *echo suppressors*; they sense speech energy, then seize an SC. Whenever a speech detector senses energy on a TC, the TC is assigned to an SC. The SC assigned is randomly selected from the idle SCs. On a given TC, each time speech energy is detected, the TC could be assigned to a different SC. Therefore, a single TC can use several SCs for a single call. For demultiplexing purposes, the TC/SC assignment information must be conveyed to the receive terminal. This is done on a common signaling channel similar to the one used on the SPADE system. DSI is a form of demand assignment; SCs are randomly assigned on an as-needed basis.

With DSI it is apparent that there is a *channel compression*; there can be more TCs assigned than there are SCs. Generally, a TC:SC ratio of 2:1 is used. For a full-duplex (two-way simultaneous) communication circuit, there is speech in each direction 40% of the time, and for 20% of the time the circuit is idle in both directions. Therefore, a DSI gain slightly more than 2 is realized. The DSI gain is affected by a phenomenon called *competitive clipping*. Competitive clipping is when speech energy is detected on a TC and there is no SC to assign it to. During the *wait* time, speech information is lost. Competitive clipping is not noticed by a subscriber if its duration is less than 50 ms.

To further enhance the channel capacity, a technique called *bit stealing* is used. With bit stealing, channels can be added to fully loaded systems by stealing bits from the in-use channels. Generally, an overload channel is generated by stealing the least significant bit from seven other satellite channels. Bit stealing results in eight channels with 7-bit resolution for the time that the *overload channel* is in use. Consequently, bit stealing results in a lower SQR than normal.

Time-Assignment Speech Interpolation

Time-assignment speech interpolation (TASI) is a form of analog channel compression that has been used for suboceanic cables for many years. TASI is very similar to DSI except that the signals interpolated are analog rather than digital. TASI also uses a 2:1 compression ratio. TASI was also the first means used to scramble voice for military security. TASI is similar to a packet data network; the voice message is chopped up into smaller segments comprised of sounds or portions of sounds. The sounds are sent through the network as separate bundles of energy, then put back together at the receive end to reform the original voice message.

QUESTIONS

21-1. Discuss the drawbacks of using FDM/FM modulation for satellite multiple-accessing systems.

21-2. Contrast *preassignment* and *demand assignment*.

21-3. What are the three most common multiple-accessing arrangements used with satellite systems?

21-4. Briefly describe the multiple-accessing arrangements listed in Question 21.3.

21-5. Briefly describe the operation of Comsat's *Spade* system.

21-6. What is meant by *single carrier per channel*?

21-7. What is a common signaling channel, and how is it used?

21-8. Describe what a reference burst is for TDMA and explain the following terms: preamble, carrier recovery sequence, bit timing recovery, unique word, and correlation spike.

21-9. Describe guard time.

21-10. Briefly describe the operation of the CEPT primary multiplex frame.

21-11. What is a store-and-forward system?

21-12. What is the primary advantage of TDMA as compared to FDMA?

21-13. What is the primary advantage of FDMA as compared to TDMA?

21-14. Briefly describe the operation of a CDMA multiple-accessing system.

21-15. Describe a chip code.

21-16. Describe what is meant by an orthogonal code.

21-17. Describe cross correlation.

21-18. What are the advantages of CDMA as compared to TDMA and FDMA?

21-19. What are the disadvantages of CDMA?

21-20. What is a Gold code?

21-21. Describe frequency hopping.

21-22. What is a frequency-time matrix?

21-23. Describe digital noninterpolated interfaces.

21-24. Describe digital speech interpolated interfaces.

21-25. What is channel compression, and how is it accomplished with a DSI system?

21-26. Describe competitive clipping.

21-27. What is meant by *bit stealing*?

21-28. Describe time-assignment speech interpolation.

PROBLEMS

21-1. How many satellite transponders are required to interlink six earth stations with FDM/FM modulation?

21-2. For the *Spade* system, what are the carrier frequencies for channel 7? What are the allocated passbands for channel 7? What are the actual passband frequencies (excluding guard bands) required?

21-3. If a 512-bit preamble precedes each CEPT station's transmission, what is the maximum number of earth stations that can be linked together with a single satellite transponder?

21-4. Determine an orthogonal code for the following chip code (101010). Prove that your selection will not produce any cross correlation for an in-phase comparison. Determine the cross correlation for each out-of-phase condition that is possible.

Chapter 22

FIBER OPTIC
COMMUNICATIONS

INTRODUCTION

During the past 10 years, the electronic communications industry has experienced many remarkable and dramatic changes. A phenomenal increase in voice, data, and video communications has caused a corresponding increase in the demand for more economical and larger capacity communications systems. This has caused a technical revolution in the electronic communications industry. Terrestrial microwave systems have long since reached their capacity, and satellite systems can provide, at best, only a temporary relief to the ever-increasing demand. It is obvious that economical communications systems that can handle large capacities and provide high-quality service are needed.

Communications systems that use light as the carrier of information have recently received a great deal of attention. As we shall see later in this chapter, propagating light waves through the earth's atmosphere is difficult and impractical. Consequently, systems that use glass or plastic fiber cables to "contain" a light wave and guide it from a source to a destination are presently being investigated at several prominent research and development laboratories. Communications systems that carry information through a *guided fiber cable* are called *fiber optic* systems.

The *information-carrying capacity* of a communications system is directly proportional to its bandwidth; the wider the bandwidth, the greater its information-carrying capacity. For comparison purposes, it is common to express the bandwidth of a system as a percentage of its carrier frequency. For instance, a VHF radio system operating at 100 MHz has a bandwidth equal to 10 MHz (i.e., 10% of the carrier frequency). A microwave radio system operating at 6 GHz with a bandwidth equal to 10% of its carrier frequency would have a bandwidth equal to 600 MHz. Thus the higher the carrier frequency, the wider the bandwidth possible and consequently, the greater the information-carrying capacity. Light frequencies used in fiber optic systems are between 10^{14} and 10^{15} Hz (100,000 to 1,000,000 GHz). Ten percent of 1,000,000 GHz is 100,000 GHz. To meet today's communications needs or the needs of the foreseeable future, 100,000 GHz is an excessive bandwidth. However, it does illustrate the capabilities of fiber optic systems.

HISTORY OF FIBER OPTICS

In 1880, Alexander Graham Bell experimented with an apparatus he called a *photophone*. The photophone was a device constructed from mirrors and selenium detectors that transmitted sound waves over a beam of light. The photophone was awkward, unreliable, and had no real practical application. Actually, visual light was a primary means of communicating long before electronic communications came about. Smoke signals and mirrors were used ages ago to convey short, simple messages. Bell's contraption, however, was the first attempt at using a beam of light for carrying information.

Transmission of light waves for any useful distance through the earth's atmosphere is impractical because water vapor, oxygen, and particulates in the air absorb and attenuate the ultrahigh light frequencies. Consequently, the only practical type of optical communications system is one that uses a fiber guide. In 1930, J. L. Baird, an English scientist, and C. W. Hansell, a scientist from the United States, were granted patents for scanning and transmitting television images through uncoated fiber cables. A few years later a German scientist named H. Lamm successfully transmitted images through a single glass fiber. At that time, most people considered fiber optics more of a toy or a laboratory stunt and consequently, it was not until the early 1950s that any substantial breakthrough was made in the field of fiber optics.

In 1951, A. C. S. van Heel of Holland and H. H. Hopkins and N. S. Kapany of England experimented with light transmission through *bundles* of fibers. Their studies led to the development of the *flexible fiberscope*, which is used extensively in the medical field. It was Kapany who coined the term "fiber optics" in 1956.

The *laser* (*l*ight *a*mplification by *s*timulated *e*mission of *r*adiation) was invented in 1960. The laser's relatively high output power, high frequency of operation, and capability of carrying an extremely wide bandwidth signal make it ideally suited for high-capacity communications systems. The invention of the laser greatly accelerated research efforts in fiber optic communications, although it was not until 1967 that K. C. Kao and G. A. Bockham of the Standard Telecommunications Laboratory in England proposed a new communications medium using *cladded* fiber cables.

The fiber cables available in the 1960s were extremely *lossy* (more than 1000 dB/km), which limited optical transmissions to short distances. In 1970, Kapron, Keck, and Maurer of Corning Glass Works in Corning, New York, developed an optical fiber with losses less than 20 dB/km. That was the "big" breakthrough needed to permit practical fiber optics communications systems. Since 1970, fiber optics technology has grown exponentially. Recently, Bell Laboratories successfully transmitted 1 billion bps through a fiber cable for 75 miles without a regenerator. AT&T has projected they will have a transatlantic fiber cable installed and operational by 1988.

In the late 1970s and early 1980s, the refinement of optical cables and the development of high-quality, affordable light sources and detectors have opened the door to the development of high-quality, high-capacity, and efficient fiber optics communications systems.

FIBER OPTIC VERSUS METALLIC CABLE FACILITIES

Communications through glass or plastic fiber cables has several overwhelming advantages over communications over conventional *metallic* or *coaxial* cable facilities.

Advantages of Fiber Systems

1. Fiber systems have a greater capacity due to the inherently larger bandwidths available with optical frequencies. Metallic cables exhibit capacitance between and inductance along their conductors. These properties cause them to act like low-pass filters which limit their bandwidths.

2. Fiber systems are immune to crosstalk between cables caused by *magnetic induction*. Glass or plastic fibers are nonconductors of electricity and therefore do not have a magnetic field associated with them. In metallic cables, the primary cause of crosstalk is magnetic induction between conductors located near each other.

3. Fiber cables are immune to *static* interference caused by lightning, electric motors, fluorescent lights, and other electrical noise sources. This immunity is also attributable to the fact that optical fibers are nonconductors of electricity. Also, fiber cables do not radiate energy and therefore cannot cause interference with other communications systems. This characteristic makes fiber systems ideally suited to military applications, where the effects of nuclear weapons (EMP—electromagnetic pulse interference) has a devastating effect on conventional communications systems.

4. Fiber cables are more resistive to environmental extremes. They operate over a larger temperature variation than their metallic counterparts, and fiber cables are affected less by corrosive liquids and gases.

5. Fiber cables are safer and easier to install and maintain. Because glass and plastic fibers are nonconductors, there are no electrical currents or voltages associated with them. Fibers can be used around volatile liquids and gases without worrying about their causing explosions or fires. Fibers are smaller and much more lightweight than their metallic counterparts. Consequently, they are easier to work with. Also, fiber cables require less storage space and are cheaper to transport.

6. Fiber cables are more secure than their copper counterparts. It is virtually impossible to tap into a fiber cable without the user knowing about it. This is another quality attractive for military applications.

7. Although it has not yet been proven, it is projected that fiber systems will last longer than metallic facilities. This assumption is based on the higher tolerances that fiber cables have to changes in the environment.

8. The long-term cost of a fiber optic system is projected to be less than that of its metallic counterpart.

Disadvantages of Fiber Systems

At the present time, there are few disadvantages of fiber systems. The only significant disadvantage is the higher initial cost of installing a fiber system, although in the future it is believed that the cost of installing a fiber system will be reduced dramatically. Another disadvantage of fiber systems is the fact that they are unproven; there are no systems that have been in operation for an extended period of time.

ELECTROMAGNETIC SPECTRUM

The total electromagnetic frequency spectrum is shown in Figure 22-1. It can be seen that the frequency spectrum extends from the *subsonic* frequencies (a few hertz) to *cosmic rays* (10^{22} Hz). The frequencies used for fiber optic systems extend from approximately 10^{14} to 10^{15} Hz (infrared to ultraviolet). This frequency subspectrum is called *visible light*, although it extends above and below the actual sensitivity of the human eye.

When dealing with ultrahigh-frequency electromagnetic waves, such as light, it is common to use units of *wavelength* rather than frequency. The wavelengths associated with light frequencies are shown in Figure 22-2. There are two common units for wavelength: *nanometer* (nm) and *angstrom* (Å). One nanometer is 10^{-9} m, and 1 angstrom is 10^{-10} m. Therefore, 1 nanometer is equal to 10 angstroms.

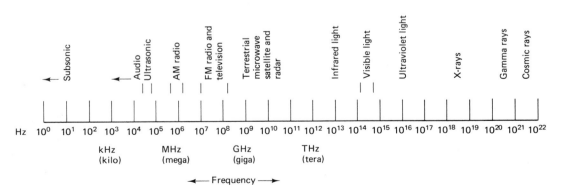

FIGURE 22-1   Electromagnetic frequency spectrum.

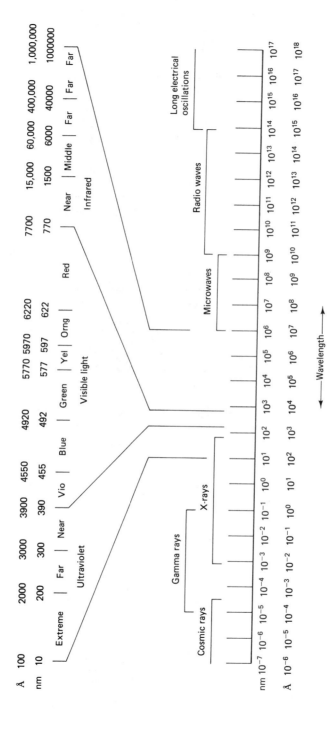

FIGURE 22-2 Electromagnetic wavelength spectrum.

FIBER OPTIC COMMUNICATIONS SYSTEM

Figure 22-3 shows a simplified block diagram of a fiber optic communications link. The three primary building blocks of the link are the *transmitter*, the *receiver*, and the *fiber guide*. The transmitter consists of an analog or digital interface, a voltage-to-current converter, a light source, and a source-to-fiber light coupler. The fiber guide is either an ultra-pure glass or plastic cable. The receiver includes a fiber-to-light detector coupling device, a photo detector, a current-to-voltage converter, an amplifier, and an analog or digital interface.

In a fiber optic transmitter, the light source can be modulated by a digital or an analog signal. For analog modulation, the input interface matches impedances and limits the input signal amplitude. For digital modulation, the original source may already be in digital form or, if in analog form, it must be converted to a digital pulse stream. For the latter case, an analog-to-digital converter must be included in the interface.

The voltage-to-current converter serves as an electrical interface between the input circuitry and the light source. The light source is either a light-emitting diode (LED) or an injection laser diode (ILD). The amount of light emitted by either an LED or an ILD is proportional to the amount of drive current. Thus the voltage-to-current converter converts an input signal voltage to a current which is used to drive the light source.

The source-to-fiber coupler is a mechanical interface. Its function is to couple the light emitted by the source into the optical fiber cable. The optical fiber consists of a glass or plastic fiber core, a cladding, and a protective jacket. The fiber-to-light detector coupling device is also a mechanical coupler. Its function is to couple as much light as possible from the fiber cable into the light detector.

The light detector is very often either a PIN (*positive-intrinsic-negative*) diode or an APD (*avalanche photodiode*). Both the APD and the PIN diode convert light energy to current. Consequently, a current-to-voltage converter is required. The current-to-voltage converter transforms changes in detector current to changes in output signal voltage.

The analog or digital interface at the receiver output is also an electrical interface. If analog modulation is used, the interface matches impedances and signal levels to the

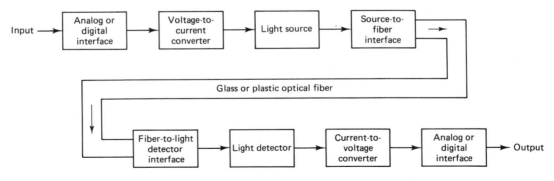

FIGURE 22-3   Fiber optic communications link.

output circuitry. If digital modulation is used, the interface must include a digital-to-analog converter.

OPTICAL FIBERS

Fiber Types

Essentially, there are three varieties of optical fibers available today. All three varieties are constructed of either glass, plastic, or a combination of glass and plastic. The three varieties are:

1. Plastic core and cladding
2. Glass core with plastic cladding (often called PCS fiber, plastic-clad silica)
3. Glass core and glass cladding (often called SCS, silica-clad silica)

Presently, Bell Laboratories is investigating the possibility of using a fourth variety that uses a *nonsilicate* substance, *zinc chloride*. Preliminary experiments have indicated that fibers made of this substance will be as much as 1000 times as efficient as glass, their silica-based counterpart.

Plastic fibers have several advantages over glass fibers. First, plastic fibers are more flexible and, consequently, more rugged than glass. They are easy to install, can better withstand stress, are less expensive, and weigh approximately 60% less than glass. The disadvantage of plastic fibers is their high attenuation characteristic; they do not propagate light as efficiently as glass. Consequently, plastic fibers are limited to relatively short runs, such as within a single building or a building complex.

Fibers with glass cores exhibit low attenuation characteristics. However, PCS fibers are slightly better than SCS fibers. Also, PCS fibers are less affected by radiation and are therefore more attractive to military applications. SCS fibers have the best propagation characteristics and they are easier to terminate than PCS fibers. Unfortunately, SCS cables are the least rugged, and they are more susceptible to increases in attenuation when exposed to radiation.

The selection of a fiber for a given application is a function of specific system requirements. There are always trade-offs based on the economics and logistics of a particular application.

Fiber Construction

There are many different cable designs available today. Figure 22-4 shows examples of several fiber optic cable configurations. Depending on the configuration, the cable may include a *core*, a *cladding*, a *protective tube*, *buffers*, *strength members*, and one or more *protective jackets*.

With the *loose* tube construction (shown in Figure 22-4a) each fiber is contained

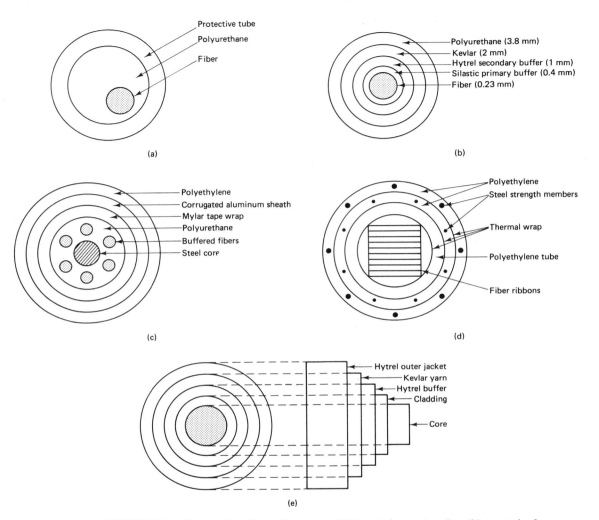

FIGURE 22-4 Fiber optic cable configurations: (A) loose tube construction; (b) constrained fiber; (c) multiple strands; (d) telephone cable; (e) plastic-clad silica cable.

in a protective tube. Inside the protective tube, a polyurethane compound encapsules the fiber and prevents the intrusion of water.

Figure 22-4b shows the construction of a *constrained* fiber cable. Surrounding the fiber cable are a primary and a secondary buffer. The buffer jackets provide protection for the fiber from external mechanical influences which could cause fiber breakage or excessive optical attenuation. Kelvar is a yarn-type material that increases the tensile strength of the cable. Again, an outer protective tube is filled with polyurethane, which prevents moisture from coming into contact with the fiber core.

Figure 22-4c shows a *multiple-strand* configuration. To increase the tensile strength, a steel central member and a layer of Mylar tape wrap are included in the package. Figure 22-4d shows a *ribbon* configuration, which is frequently seen in telephone systems using fiber optics. Figure 22-4e shows both the end and side views of a plastic-clad silica cable.

The type of cable construction used depends on the performance requirements of the system and both the economic and environmental constraints.

LIGHT PROPAGATION

The Physics of Light

Although the performance of optical fibers can be analyzed completely by application of Maxwell's equations, this is necessarily complex. For most practical applications, Maxwell's equations may be substituted by the application of *geometric ray tracing*, which will yield a sufficiently detailed analysis.

Velocity of Propagation

Electromagnetic energy, such as light, travels at approximately 300,000,000 m/s (186,000 miles per second) in free space. Also, the velocity of propagation is the same for all light frequencies in free space. However, it has been demonstrated that in materials more dense than free space, the velocity is reduced. When the velocity of an electromagnetic wave is reduced as it passes from one medium to another medium of a denser material, the light ray is *refracted* (bent) toward the normal. Also, in materials more dense than free space, all light frequencies do not propagate at the same velocity.

Refraction

Figure 22-5a shows how a light ray is refracted as it passes from a material of a given density into a less dense material. (Actually, the light ray is not bent, but rather, it changes direction at the interface.) Figure 22-5b shows how sunlight, which contains all light frequencies, is affected as it passes through a material more dense than free space. Refraction occurs at both air/glass interfaces. The violet wavelengths are refracted the most, and the red wavelengths are refracted the least. The spectral separation of white light in this manner is called *prismatic refraction*. It is this phenomenon that causes rainbows; water droplets in the atmosphere act like small prisms that split the white sunlight into the various wavelengths, creating a visible spectrum of color.

Refractive Index

The amount of bending or refraction that occurs at the interface of two materials of different densities is quite predictable and depends on the *refractive index* (also called

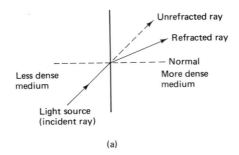

(a)

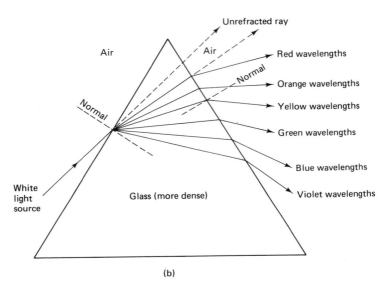

(b)

FIGURE 22-5   Refraction of light: (A) light refraction; (b) prismatic refraction.

index of refraction) of the two materials. The refractive index is simply the ratio of the velocity of propagation of a light ray in free space to the velocity of propagation of a light ray in a given material. Mathematically, the refractive index is

$$n = \frac{c}{v}$$

where

        c = speed of light in free space
        v = speed of light in a given material

Although the refractive index is also a function of frequency, the variation in most applications is insignificant and therefore omitted from this discussion. The indexes of refraction of several common materials are given in Table 22-1.

TABLE 22-1   TYPICAL INDEXES OF REFRACTION

Medium	Index of refraction[a]
Vacuum	1.0
Air	1.0003 (1.0)
Water	1.33
Ethyl alcohol	1.36
Fused quartz	1.46
Glass fiber	1.5–1.9
Diamond	2.0–2.42
Silicon	3.4
Gallium-arsenide	3.6

[a] Index of refraction is based on a wavelength of light emitted from a sodium flame (5890 Å).

How a light ray reacts when it meets the interface of two transmissive materials that have different indexes of refraction can be explained with *Snell's law*. Snell's law simply states:

$$n_1 \sin \theta_1 = n_2 \sin \theta_2 \qquad (22\text{-}1)$$

where

   n_1 = refractive index of material 1
   n_2 = refractive index of material 2
   θ_1 = angle of incidence
   θ_2 = angle of refraction

A refractive index model for Snell's law is shown in Figure 22-6. At the interface, the incident ray may be refracted toward the normal or away from it, depending on whether n_1 is less than or greater than n_2.

Figure 22-7 shows how a light ray is refracted as it travels from a more dense (higher refractive index) material into a less dense (lower refractive index) material. It can be seen that the light ray changes direction at the interface, and the angle of refraction is greater than the angle of incidence. Consequently, when a light ray enters a less dense material, the ray bends away from the normal. The normal is simply a line drawn perpendicular to the interface at the point where the incident ray strikes the interface. Similarly, when a light ray enters a more dense material, the ray bends toward the normal.

EXAMPLE 22-1

In Figure 22-7, let medium 1 be glass and medium 2 be ethyl alcohol. For an angle of incidence of 30°, determine the angle of refraction.

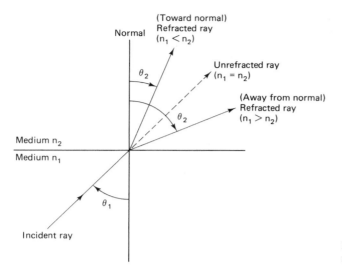

FIGURE 22-6   Refractive
model for Snell's law.

Solution   From Table 22-1,

$$n_1 \text{ (glass)} = 1.5$$

$$n_2 \text{ (ethyl alcohol)} = 1.36$$

Rearranging Equation 22-1 and substituting for n_1, n_2, and θ_1 gives us

$$\frac{n_1}{n_2} \sin \theta_1 = \sin \theta_2$$

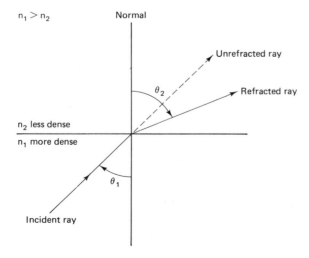

FIGURE 22-7   Light ray refracted
away from the normal.

$$\frac{1.5}{1.36} \sin 30 = 0.5514 = \sin \theta_2$$

$$\theta_2 = \sin^{-1} 0.5514 = 33.47°$$

The result indicates that the light ray refracted (bent) or changed direction by 3.47° at the interface. Because the light was traveling from a more dense material into a less dense material, the ray bent away from the normal.

Critical Angle

Figure 22-8 shows a condition in which an *incident ray* is at an angle such that the angle of refraction is 90° and the refracted ray is along the interface. (It is important to note that the light ray is traveling from a medium of higher refractive index to a medium with a lower refractive index.) Again, using Snell's law,

$$\sin \theta_1 = \frac{n_2}{n_1} \sin \theta_2$$

With $\theta_2 = 90°$,

$$\sin \theta_1 = \frac{n_2}{n_1} (1) \qquad \text{or} \qquad \sin \theta_1 = \frac{n_2}{n_1}$$

and

$$\sin^{-1} \frac{n_2}{n_1} = \theta_1 = \theta_c \tag{22-2}$$

where θ_c is the critical angle.

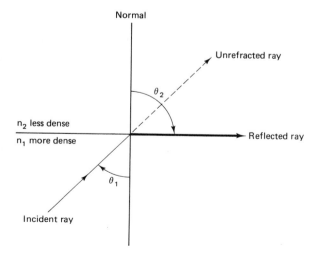

FIGURE 22-8   Critical angle reflection.

The *critical angle* is defined as the minimum angle of incidence at which a light ray may strike the interface of two media and result in an angle of refraction of 90° or greater. (This definition pertains only when the light ray is traveling from a more dense medium into a less dense medium.) If the angle of refraction is 90° or greater, the light ray is not allowed to penetrate the less dense material. Consequently, total reflection takes place at the interface, and the angle of reflection is equal to the angle of incidence. Figure 22-9 shows a comparison of the angle of refraction and the angle of reflection when the angle of incidence is less than or more than the critical angle.

PROPAGATION OF LIGHT THROUGH AN OPTICAL FIBER

Light can be propagated down an optical fiber cable by either reflection or refraction. How the light is propagated depends on the *mode of propagation* and the *index profile* of the fiber.

Mode of Propagation

In fiber optics terminology, the word *mode* simply means path. If there is only one path for light to take down the cable, it is called *single mode*. If there is more than one path, it is called *multimode*. Figure 22-10 shows single and multimode propagation of light down an optical fiber.

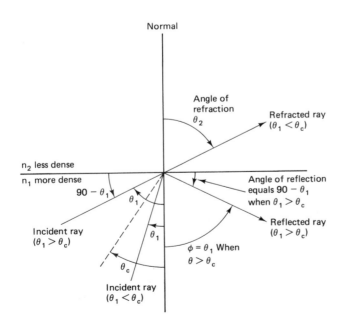

FIGURE 22-9 Angle of reflection and refraction.

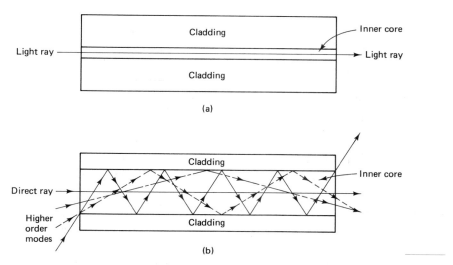

FIGURE 22-10   Modes of propagation: (a) single mode; (b) multimode.

Index Profile

The index profile of an optical fiber is a graphical representation of the refractive index of the core. The refractive index is plotted on the horizontal axis and the radial distance from the core axis is plotted on the vertical axis. Figure 22-11 shows the core index profiles of three types of fiber cables.

There are two basic types of index profiles: step and graded. A *step-index fiber* has a central core with a uniform refractive index. The core is surrounded by an outside cladding with a uniform refractive index less than that of the central core. From Figure 22-11 it can be seen that in a step-index fiber there is an abrupt change in the refractive index at the core/cladding interface. In a *graded-index fiber* there is no cladding, and the refractive index of the core is nonuniform; it is highest at the center and decreases gradually toward the outer edge.

OPTICAL FIBER CONFIGURATIONS

Essentially, there are three types of optical fiber configurations: single-mode step-index, multimode step-index, and multimode graded-index.

Single-Mode Step-Index Fiber

A *single-mode step-index fiber* has a central core that is sufficiently small so that there is essentially only one path that light may take as it propagates down the cable. This type of fiber is shown in Figure 22-12. In the simplest form of single-mode step-index

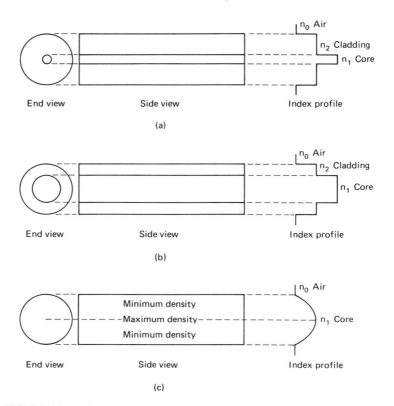

FIGURE 22-11   Core index profiles: (a) single-mode step index; (b) multimode step index; (c) multimode graded index.

fiber, the outside cladding is simply air (Figure 22-12a). The refractive index of the glass core (n_1) is approximately 1.5, and the refractive index of the air cladding (n_0) is 1. The large difference in the refractive indexes results in a small critical angle (approximately 42°) at the glass/air interface. Consequently, the fiber will accept light from a wide aperture. This makes it relatively easy to couple light from a source into the cable. However, this type of fiber is typically very weak and of limited practical use.

A more practical type of single-mode step-index fiber is one that has a cladding other than air (Figure 22-12b). The refractive index of the cladding (n_2) is slightly less than that of the central core (n_1) and is uniform throughout the cladding. This type of cable is physically stronger than the air-clad fiber, but the critical angle is also much higher (approximately 77°). This results in a small acceptance angle and a narrow source-to-fiber aperture, making it much more difficult to couple light into the fiber from a light source.

With both types of single-mode step-index fibers, light is propagated down the fiber through reflection. Light rays that enter the fiber propagate straight down the core or, perhaps, are reflected once. Consequently, all light rays follow approximately the

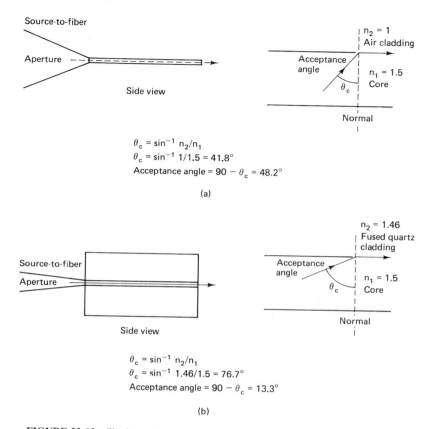

$\theta_c = \sin^{-1} n_2/n_1$
$\theta_c = \sin^{-1} 1/1.5 = 41.8°$
Acceptance angle = $90 - \theta_c = 48.2°$

(a)

$\theta_c = \sin^{-1} n_2/n_1$
$\theta_c = \sin^{-1} 1.46/1.5 = 76.7°$
Acceptance angle = $90 - \theta_c = 13.3°$

(b)

FIGURE 22-12 **Single-mode step-index fibers: (a) air cladding; (b) glass cladding.**

same path down the cable and take approximately the same amount of time to travel the length of the cable. This is one overwhelming advantage of single-mode step-index fibers and will be explained in more detail later.

Multimode Step-Index Fiber

A *multimode step-index fiber* is shown in Figure 22-13. It is similar to the single-mode configuration except that the center core is much larger. This type of fiber has a larger light-to-fiber aperture and, consequently, allows more light to enter the cable. The light rays that strike the core/cladding interface at an angle greater than the critical angle (ray A) are propagated down the core in a zigzag fashion, continuously reflecting off the interface boundary. Light rays that strike the core/cladding interface at an angle less than the critical angle (ray B) enter the cladding and are lost. It can be seen that there are many paths that a light ray may follow as it propagates down the fiber. As a

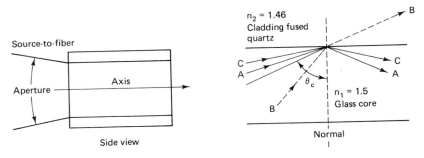

FIGURE 22-13   Multimode step-index fiber.

result, all light rays do not follow the same path and, consequently, do not take the same amount of time to travel the length of the fiber.

Multimode Graded-Index Fiber

A *multimode graded-index fiber* is shown in Figure 22-14. A multimode graded-index fiber is characterized by a central core that has a refractive index that is nonuniform; it is maximum at the center and decreases gradually toward the outer edge. Light is propagated down this type of fiber through refraction. As a light ray propagates diagonally across the core, it is continually intersecting a less-dense-to-more dense interface. Consequently, the light rays are constantly being refracted, which results in a continuous bending of the light rays. Light enters the fiber at many different angles. As they propagate down the fiber, the light rays that travel in the outermost area of the fiber travel a greater distance than the rays traveling near the center. Because the refractive index decreases with distance from the center and the velocity is inversely proportional to the refractive index, the light rays traveling farthest from the center propagate at a higher velocity. Consequently, they take approximately the same amount of time to travel the length of the fiber.

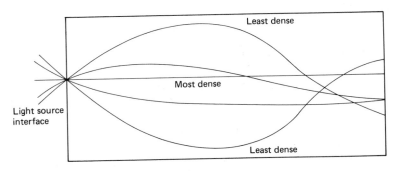

FIGURE 22-14   Multimode graded-index fiber.

COMPARISON OF THE THREE TYPES OF OPTICAL FIBERS

Single-Mode Step-Index Fiber

Advantages

1. There is minimum dispersion. Because all rays propagating down the fiber take approximately the same path, they take approximately the same amount of time to travel down the cable. Consequently, a pulse of light entering the cable can be reproduced at the receiving end very accurately.
2. Because of the high accuracy in reproducing transmitted pulses at the receive end, larger bandwidths and higher information transmission rates are possible with single-mode step-index fibers than with the other types of fibers.

Disadvantages

1. Because the central core is very small, it is difficult to couple light into and out of this type of fiber. The source-to-fiber aperture is the smallest of all the fiber types.
2. Again, because of the small central core, a highly directive light source such as a laser is required to couple light into a single-mode step-index fiber.
3. Single-mode step-index fibers are expensive and difficult to manufacture.

Multimode Step-Index Fiber

Advantages

1. Multimode step-index fibers are inexpensive and simple to manufacture.
2. It is easy to couple light into and out of multimode step-index fibers; they have a relatively large source-to-fiber aperture.

Disadvantages

1. Light rays take many different paths down the fiber, which results in large differences in their propagation times. Because of this, rays traveling down this type of fiber have a tendency to spread out. Consequently, a pulse of light propagating down a multimode step-index fiber is distorted more than with the other types of fibers.
2. The bandwidth and rate of information transfer possible with this type of cable are less than the other types.

Multimode Graded-Index Fiber

Essentially, there are no outstanding advantages or disadvantages of this type of fiber. Multimode graded-index fibers are easier to couple light into and out of than single-mode step-index fibers but more difficult than multimode step-index fibers. Distortion due to multiple propagation paths is greater than in single-mode step-index fibers but less than in multimode step-index fibers. Graded-index fibers are easier to manufacture than single-mode step-index fibers but more difficult than multimode step-index fibers. The multimode graded-index fiber is considered an intermediate fiber compared to the other types.

ACCEPTANCE ANGLE AND ACCEPTANCE CONE

In previous discussions, the *source-to-fiber aperture* was mentioned several times, and the *critical* and *acceptance* angles at the point where a light ray strikes the core/cladding interface were explained. The following discussion deals with the light-gathering ability of the fiber, the ability to couple light from the source into the fiber cable.

Figure 22-15 shows the source end of a fiber cable. When light rays enter the fiber, they strike the air/glass interface at normal A. The refractive index of air is 1 and the refractive index of the glass core is 1.5. Consequently, the light entering at the air/glass interface propagates from a less dense medium into a more dense medium. Under these conditions and according to Snell's law, the light rays will refract toward the normal. This causes the light rays to change direction and propagate diagonally down the core at an angle (θ_c) which is different than the external angle of incidence at the air/glass interface (θ_{in}). In order for a ray of light to propagate down the cable,

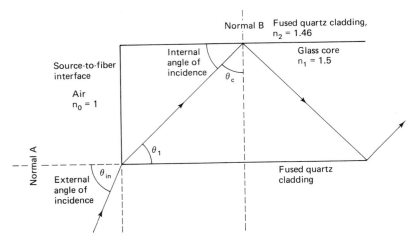

FIGURE 22-15   Ray propagation into and down an optical fiber cable.

it must strike the internal core/cladding interface at an angle that is greater than the critical angle (θ_c).

Applying Snell's law to the external angle of incidence yields the following expression:

$$n_0 \sin \theta_{in} = n_1 \sin \theta_1 \qquad (22\text{-}3)$$

and

$$\theta_1 = 90 - \theta_c$$

Thus

$$\sin \theta_1 = \sin (90 - \theta_c) = \cos \theta_c \qquad (22\text{-}4)$$

Substituting Equation 22-4 into Equation 22-3 yields the following expression:

$$n_0 \sin \theta_{in} = n_1 \cos \theta_c$$

Rearranging and solving for $\sin \theta_{in}$ gives us

$$\sin \theta_{in} = \frac{n_1}{n_0} \cos \theta_c \qquad (10\text{-}5)$$

Figure 22-16 shows the geometric relationship of Equation 22-5.

From Figure 22-16 and using the Pythagorean theorem, we obtain

$$\cos \theta_c = \frac{\sqrt{n_1^2 - n_2^2}}{n_1} \qquad (22\text{-}6)$$

Substituting Equation 22-6 into Equation 22-5 yields

$$\sin \theta_{in} = \frac{n_1}{n_0} \frac{\sqrt{n_1^2 - n_2^2}}{n_1}$$

Reducing the equation gives

$$\sin \theta_{in} = \frac{\sqrt{n_1^2 - n_2^2}}{n_0} \qquad (22\text{-}7)$$

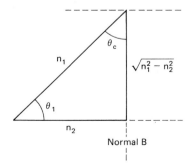

Normal B

FIGURE 22-16 Geometric relationship of Equation 10-5.

and

$$\theta_{in} = \sin^{-1} \frac{\sqrt{n_1^2 - n_2^2}}{n_0} \qquad (22\text{-}8)$$

Because light rays generally enter the fiber from an air medium, n_0 equals 1. This simplifies Equation 22-8 to

$$\theta_{in(max)} = \sin^{-1} \sqrt{n_1^2 - n_2^2} \qquad (22\text{-}9)$$

θ_{in} is called the *acceptance angle* or *acceptance cone* half-angle. It defines the maximum angle in which external light rays may strike the air/fiber interface and still propagate down the fiber with a response that is no greater than 10 dB down from the peak value. Rotating the acceptance angle around the fiber axis describes the acceptance cone of the fiber. This is shown in Figure 22-17.

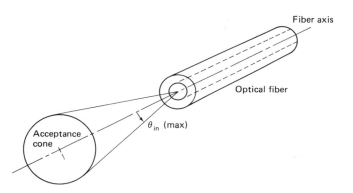

FIGURE 22-17 **Acceptance cone of a fiber cable.**

Numerical Aperture

Numerical aperture (NA) is a figure of merit that is used to measure the light-gathering or light-collecting ability of an optical fiber. The larger the magnitude of NA, the greater the amount of light accepted by the fiber from the external light source. For a step-index fiber, numerical aperture is mathematically defined as the sin of the acceptance half-angle. Thus

$$NA = \sin \theta_{in}$$

and

$$NA = \sqrt{n_1^2 - n_2^2} \qquad (22\text{-}10)$$

Also,

$$\sin^{-1} NA = \theta_{in}$$

For a graded index, NA is simply the sin of the critical angle:

$$NA = \sin \theta_c$$

EXAMPLE 22-2

For this example refer to Figure 22-15. For a multimode step-index fiber with a glass core ($n_1 = 1.5$) and a fused quartz cladding ($n_2 = 1.46$), determine the critical angle (θ_c), acceptance angle (θ_{in}), and numerical aperture. The source-to-fiber media is air.

Solution Substituting into Equation 22-2, we have

$$\theta_c = \sin^{-1} \frac{n_2}{n_1} = \sin^{-1} \frac{1.46}{1.5} = 76.7°$$

Substituting into Equation 22-9 yields

$$\theta_{in} = \sin^{-1} \sqrt{n_1^2 - n_2^2} = \sin^{-1} \sqrt{1.5^2 - 1.46^2}$$
$$= 20.2°$$

Substituting into Equation 22-10 gives us

$$NA = \sin \theta_{in} = \sin 20.2$$
$$= 0.344$$

LOSSES IN OPTICAL FIBER CABLES

Transmission losses in optical fiber cables are one of the most important characteristics of the fiber. Losses in the fiber result in a reduction in the light power and thus reduce the system bandwidth, information transmission rate, efficiency, and overall system capacity. The predominant fiber losses are as follows:

1. Absorption losses
2. Material or Rayleigh scattering losses
3. Chromatic or wavelength dispersion
4. Radiation losses
5. Modal dispersion
6. Coupling losses

Absorption Losses

Absorption loss in optical fibers is analogous to power dissipation in copper cables; impurities in the fiber absorb the light and convert it to heat. The ultrapure glass used to manufacture optical fibers is approximately 99.9999% pure. Still, absorption losses between 1 and 1000 dB/km are typical. Essentially, there are three factors that contribute to the absorption losses in optical fibers: ultraviolet absorption, infrared absorption, and ion resonance absorption.

Ultraviolet absorption.   Ultraviolet absorption is caused by valence electrons in the silica material from which fibers are manufactured. Light *ionizes* the valence electrons into conduction. The ionization is equivalent to a loss in the total light field and, consequently, contributes to the transmission losses of the fiber.

Infrared absorption.   Infrared absorption is a result of *photons* of light that are absorbed by the atoms of the glass core molecules. The absorbed photons are converted to random mechanical vibrations typical of heating.

Ion resonance absorption.   Ion resonance absorption is caused by OH^- ions in the material. The source of the OH^- ions is water molecules that have been trapped in the glass during the manufacturing process. Ion absorption is also caused by iron, copper, and chromium molecules.

Figure 22-18 shows typical losses in optical fiber cables due to ultraviolet, infrared, and ion resonance absorption.

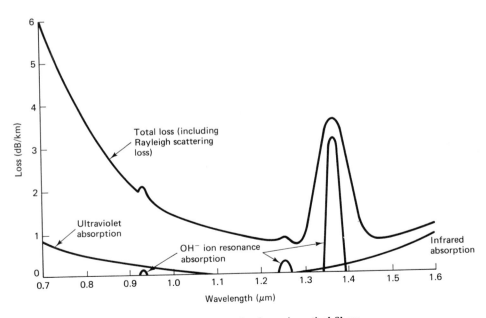

FIGURE 22-18   **Absorption losses in optical fibers.**

Material or Rayleigh Scattering Losses

During the manufacturing process, glass is extruded (drawn into long fibers of very small diameter). During this process, the glass is in a plastic state (not liquid and not solid). The tension applied to the glass during this process causes the cooling glass to develop submicroscopic irregularities that are permanently formed in the fiber. When light rays that are propagating down a fiber strike one of these impurities, they are *diffracted*. Diffraction causes the light to disperse or spread out in many directions. Some of the diffracted light continues down the fiber and some of it escapes through the cladding. The light rays that escape represent a loss in light power. This is called *Rayleigh scattering loss*. Figure 22-19 graphically shows the relationship between wavelength and Rayleigh scattering loss.

Chromatic or Wavelength Dispersion

As stated previously, the refractive index of a material is wavelength dependent. Light-emitting diodes (LEDs) emit light that contains a combination of wavelengths. Each wavelength within the composite light signal travels at a different velocity. Consequently, light rays that are simultaneously emitted from an LED and propagated down an optical fiber do not arrive at the far end of the fiber at the same time. This results in a distorted

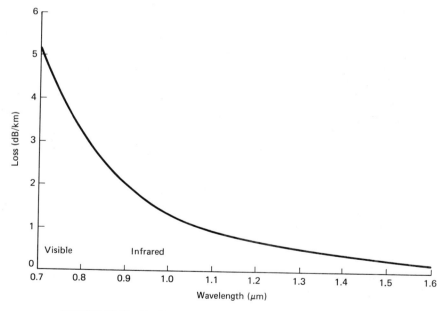

FIGURE 22-19   Rayleigh scattering loss as a function of wavelength.

receive signal and is called *chromatic distortion*. Chromatic distortion can be eliminated by using a monochromatic source such as an injection laser diode (ILD).

Radiation Losses

Radiation losses are caused by small bends and kinks in the fiber. Essentially, there are two types of bends: microbends and constant-radius bends. *Microbending* occurs as a result of differences in the thermal contraction rates between the core and cladding material. A microbend represents a discontinuity in the fiber where Rayleigh scattering can occur. *Constant-radius bends* occur when fibers are bent during handling or installation.

Modal Dispersion

Modal dispersion, or *pulse spreading*, is caused by the difference in the propagation times of light rays that take different paths down a fiber. Obviously, modal dispersion can occur only in multimode fibers. It can be reduced considerably by using graded-index fibers and almost entirely eliminated by using single-mode step-index fibers.

Modal dispersion can cause a pulse of light energy to spread out as it propagates down a fiber. If the pulse spreading is sufficiently severe, one pulse may fall back on top of the next pulse (this is an example of intersymbol interference). In a multimode step-index fiber, a light ray that propagates straight down the axis of the fiber takes the least amount of time to travel the length of the fiber. A light ray that strikes the core/cladding interface at the critical angle will undergo the largest number of internal reflections and, consequently, take the longest time to travel the length of the fiber.

Figure 22-20 shows three rays of light propagating down a multimode step-index fiber. The lowest-order mode (ray 1) travels in a path parallel to the axis of the fiber. The middle-order mode (ray 2) bounces several times at the interface before traveling the length of the fiber. The highest-order mode (ray 3) makes many trips back and forth across the fiber as it propagates the entire length. It can be seen that ray 3 travels a considerably longer distance than ray 1 as it propagates down the fiber. Consequently,

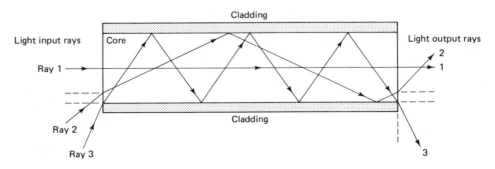

FIGURE 22-20   Light propagation down a multimode step-index fiber.

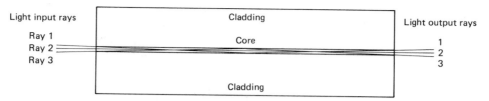

FIGURE 22-21   Light propagation down a single-mode step-index fiber.

if the three rays of light were emitted into the fiber at the same time and represented a pulse of light energy, the three rays would reach the far end of the fiber at different times and result in a spreading out of the light energy in respect to time. This is called modal dispersion and results in a stretched pulse which is also reduced in amplitude at the output of the fiber. All three rays of light propagate through the same material at the same velocity, but ray 3 must travel a longer distance and, consequently, takes a longer period of time to propagate down the fiber.

Figure 22-21 shows light rays propagating down a single-mode step-index fiber. Because the radial dimension of the fiber is sufficiently small, there is only a single path for each of the rays to follow as they propagate down the length of the fiber. Consequently, each ray of light travels the same distance in a given period of time and the light rays have exactly the same time relationship at the far end of the fiber as they had when they entered the cable. The result is no *modal dispersion* or *pulse stretching*.

Figure 22-22 shows light propagating down a multimode graded-index fiber. Three rays are shown traveling in three different modes. Each ray travels a different path but they all take approximately the same amount of time to propagate the length of fiber. This is because the refractive index of the fiber decreases with distance from the center, and the velocity at which a ray travels is inversely proportional to the refractive index. Consequently, the farther rays 2 and 3 travel from the center of the fiber, the faster they propagate.

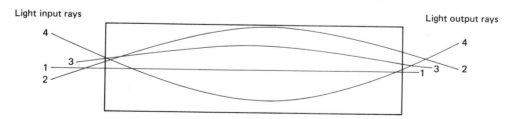

FIGURE 22-22   Light propagation down a multimode graded-index fiber.

Figure 22-23 shows the relative time/energy relationship of a pulse of light as it propagates down a fiber cable. It can be seen that as the pulse propagates down the fiber, the light rays that make up the pulse spread out in time, which causes a corresponding reduction in the pulse amplitude and stretching of the pulse width. It can also be

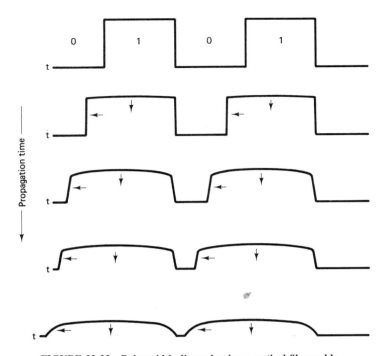

FIGURE 22-23   Pulse-width dispersion in an optical fiber cable.

seen that as light energy from one pulse falls back in time, it will interfere with the next pulse. This is called *pulse spreading* or *pulse-width dispersion* and causes errors in digital transmission.

Figure 22-24a shows a unipolar return-to-zero (UPRZ) digital transmission. With UPRZ transmission (assuming a very narrow pulse) if light energy from pulse A were to fall back (*spread*) one bit time (T_b), it would interfere with pulse B and change what was a logic 0 to a logic 1. Figure 22-24b shows a unipolar nonreturn-to-zero (UPNRZ) digital transmission where each pulse is equal to the bit time. With UPNRZ transmission, if energy from pulse A were to fall back one-half of a bit time, it would interfere with pulse B. Consequently, UPRZ transmissions can tolerate twice as much delay or spread as UPNRZ transmissions.

The difference between the absolute delay times of the fastest and slowest rays of light propagating down a fiber is called the *pulse-spreading constant* (Δt) and is generally expressed in nanoseconds per kilometer (ns/km). The total pulse spread (ΔT) is then equal to the pulse spreading constant (Δt) times the total fiber length (L). Mathematically, ΔT is

$$\Delta T \text{ (ns)} = \Delta t \left(\frac{\text{ns}}{\text{km}}\right) \times L \text{ (km)} \qquad (22\text{-}11)$$

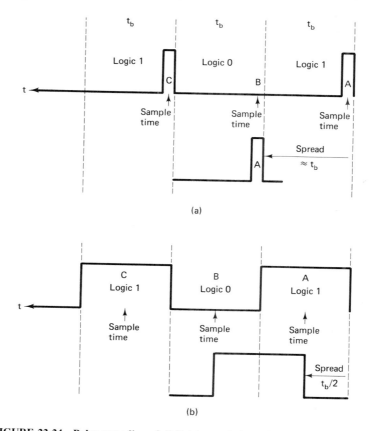

FIGURE 22-24 Pulse spreading of digital transmissions: (a) UPRZ; (b) UPNRZ.

For UPRZ transmissions, the maximum data transmission rate in bits per second (bps) is expressed as

$$F_b(\text{bps}) = \frac{1}{\Delta t \times L} \tag{22-12}$$

and for UPNRZ transmissions, the maximum transmission rate is

$$F_b(\text{bps}) = \frac{1}{2\,\Delta t \times L} \tag{22-13}$$

EXAMPLE 22-3

For an optical fiber 10 km long with a pulse-spreading constant of 5 ns/km, determine the maximum digital transmission rates for (a) return-to-zero and (b) nonreturn-to-zero transmissions.

Solution   (a) Substituting into Equation 22-12 yields

$$F_b = \frac{1}{5 \text{ ns/km} \times 10 \text{ km}} = 20 \text{ Mbps}$$

(b) Substituting into Equation 22-13 yields

$$F_b = \frac{1}{(2 \times 5 \text{ ns/km}) \times 10 \text{ km}} = 10 \text{ Mbps}$$

The results indicate that the digital transmission rate possible for this optical fiber is twice as high (20 Mbps versus 10 Mbps) for UPRZ as for UPNRZ transmission.

Coupling Losses

In fiber cables coupling losses can occur at any of the following three types of optical junctions: light source-to-fiber connections, fiber-to-fiber connections, and fiber-to-photo-detector connections. Junction losses are most often caused by one of the following alignment problems: lateral misalignment, gap misalignment, angular misalignment, and imperfect surface finishes. These impairments are shown in Figure 22-25.

Lateral misalignment.   This is shown in Figure 22-25a and is the lateral or axial displacement between two pieces of adjoining fiber cables. The amount of loss can be from a couple of tenths of a decibel to several decibels. This loss is generally negligible if the fiber axes are aligned to within 5% of the smaller fiber's diameter.

Gap misalignment.   This is shown in Figure 22-25b and is sometimes called *end separation*. When *splices* are made in optical fibers, the fibers should actually touch. The farther apart the fibers are, the greater the loss of light. If two fibers are joined with a connector, the ends should not touch. This is because the two ends rubbing against each other in the connector could cause damage to either or both fibers.

Angular misalignment.   This is shown in Figure 22-25c and is sometimes called *angular displacement*. If the angular displacement is less than 2°, the loss will be less than 0.5 dB.

Imperfect surface finish.   This is shown in Figure 22-25d. The ends of the two adjoining fibers should be highly polished and fit together squarely. If the fiber ends are less than 3· off from perpendicular, the losses will be less than 0.5 dB.

LIGHT SOURCES

Essentially, there are two devices commonly used to generate light for fiber optic communications systems: light-emitting diodes (LEDs) and injection laser diodes (ILDs). Both

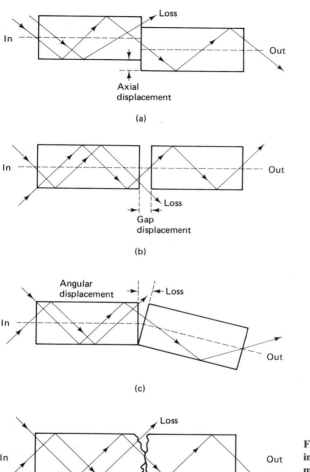

(a)

(b)

(c)

(d)

FIGURE 22-25 Fiber alignment impairments: **(a)** lateral misalignment; **(b)** gap displacement; **(c)** angular misalignment; **(d)** surface finish.

devices have advantages and disadvantages and selection of one device over the other is determined by system requirements.

Light-Emitting Diodes

Essentially, a *light-emitting diode* (LED) is simply a P-N junction diode. It is usually made from a semiconductor material such as aluminum-gallium-arsenide (AlGaAs) or gallium-arsenide-phosphide (GaAsP). LEDs emit light by spontaneous emission; light is emitted as a result of the recombination of electrons and holes. When forward biased, minority carriers are injected across the *p-n* junction. Once across the junction, these

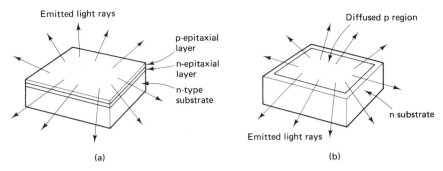

FIGURE 22-26 **Homojunction LED structures: (a) silicon-doped gallium arsenide;**
(b) planar diffused.

minority carriers recombine with majority carriers and give up energy in the form of light. This process is essentially the same as in a conventional diode except that in LEDs certain semiconductor materials and dopants are chosen such that the process is radiative; a photon is produced. A photon is a quantum of electromagnetic wave energy. Photons are particles that travel at the speed of light but at rest have no mass. In conventional semiconductor diodes (germanium and silicon, for example), the process is primarily nonradiative and no photons are generated. The energy gap of the material used to construct an LED determines whether the light emitted by it is invisible or visible and of what color.

The simplest LED structures are homojunction, epitaxially grown, or single-diffused devices and are shown in Figure 22-26. *Epitaxially grown LEDs* are generally constructed of silicon-doped gallium-arsenide (Figure 22-26a). A typical wavelength of light emitted from this construction is 940 nm, and a typical output power is approximately 3 mW at 100 mA of forward current. *Planar diffused* (*homojunction*) *LEDs* (Figure 22-26b) output approximately 500 μW at a wavelength of 900 nm. The primary disadvantage of homojunction LEDs is the nondirectionality of their light emission, which makes them a poor choice as a light source for fiber optic systems.

The *planar heterojunction LED* (Figure 22-27) is quite similar to the epitaxially grown LED except that the geometry is designed such that the forward current is concen-

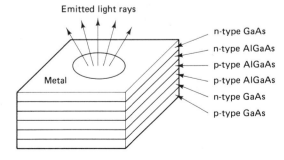

FIGURE 22-27 **Planar heterojunction**
LED.

trated to a very small area of the active layer. Because of this the planar heterojunction LED has several advantages over the homojunction type. They are:

1. The increase in current density generates a more brilliant light spot.
2. The smaller emitting area makes it easier to couple its emitted light into a fiber.
3. The small effective area has a smaller capacitance, which allows the planar heterojunction LED to be used at higher speeds.

Burrus etched-well surface-emitting LED.   For the more practical applications, such as telecommunications, data rates in excess of 100 Mbps are required. For these applications, the etched-well LED was developed. Burrus and Dawson of Bell laboratories developed the etched-well LED. It is a surface-emitting LED and is shown in Figure 22-28. The Burrus etched-well LED emits light in many directions. The etched well helps concentrate the emitted light to a very small area. Also, domed lenses can be placed over the emitting surface to direct the light into a smaller area. These devices are more efficient than the standard surface emitters and they allow more power to be coupled into the optical fiber, but they are also more difficult and expensive to manufacture.

Edge-emitting LED.   The edge-emitting LED, which was developed by RCA, is shown in Figure 22-29. These LEDs emit a more directional light pattern than do the surface-emitting LEDs. The construction is similar to the planar and Burrus diodes except that the emitting surface is a stripe rather than a confined circular area. The light is emitted from an active stripe and forms an elliptical beam. Surface-emitting LEDs are more commonly used than edge emitters because they emit more light. However, the coupling losses with surface emitters are greater and they have narrower bandwidths.

The *radiant* light power emitted from an LED is a linear function of the forward current passing through the device (Figure 22-30). It can also be seen that the optical output power of an LED is, in part, a function of the operating temperature.

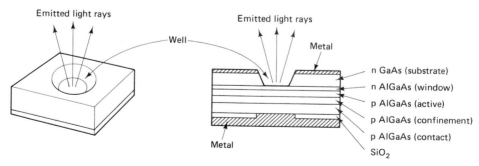

FIGURE 22-28   Burrus etched-well surface-emitting LED.

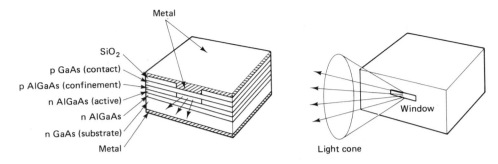

FIGURE 22-29   Edge-emitting LED.

Injection Laser Diode

The word *laser* is an acronym for *l*ight *a*mplification by *s*timulated *e*mission of *r*adiation. Lasers are constructed from many different materials, including gases, liquids, and solids, although the type of laser used most often for fiber optic communications is the semiconductor laser.

The *injection laser diode* (ILD) is similar to the LED. In fact, below a certain threshold current, an ILD acts like an LED. Above the threshold current, an ILD oscillates; lasing occurs. As current passes through a forward-biased *p-n* junction diode, light is emitted by spontaneous emission at a frequency determined by the energy gap of the semiconductor material. When a particular current level is reached, the number of minority carriers and photons produced on either side of the *p-n* junction reaches a level where they begin to collide with already excited minority carriers. This causes an

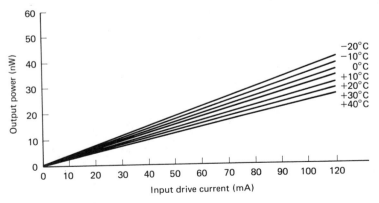

FIGURE 22-30   Output power versus forward current and operating temperature for an LED.

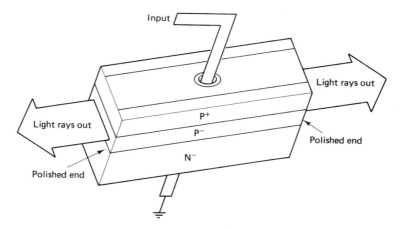

FIGURE 22-31 Injection laser diode construction.

increase in the ionization energy level and makes the carriers unstable. When this happens, a typical carrier recombines with an opposite type of carrier at an energy level that is above its normal before-collision value. In the process, two photons are created; one is stimulated by another. Essentially, a gain in the number of photons is realized. For this to happen, a large forward current that can provide many carriers (holes and electrons) is required.

The construction of an ILD is similar to that of an LED (Figure 22-31) except that the ends are highly polished. The mirror-like ends trap the photons in the active region and, as they reflect back and forth, stimulate free electrons to recombine with holes at a higher-than-normal energy level. This process is called *lasing*.

The radiant output light power of a typical ILD is shown in Figure 22-32. It can be seen that very little output power is realized until the threshold current is reached;

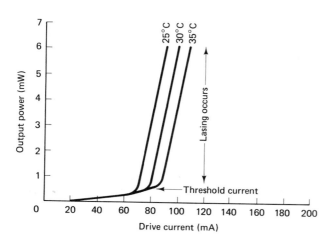

FIGURE 22-32 Output power versus forward current and temperature for an ILD.

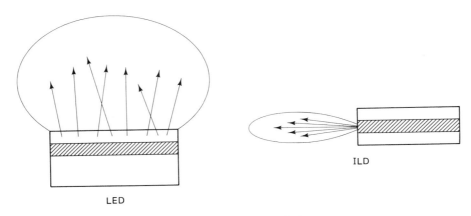

FIGURE 22-33   LED and ILD radiation patterns.

then lasing occurs. After lasing begins, the optical output power increases dramatically, with small increases in drive current. It can also be seen that the magnitude of the optical output power of the ILD is more dependent on operating temperature than is the LED.

Figure 22-33 shows the light radiation patterns typical of an LED and an ILD. Because light is radiated out the end of an ILD in a narrow concentrated beam, it has a more direct radiation pattern.

Advantages of ILDs

1. Because ILDs have a more direct radiation pattern, it is easier to couple their light into an optical fiber. This reduces the coupling losses and allows smaller fibers to be used.
2. The radiant output power from an ILD is greater than that for an LED. A typical output power for an ILD is 5 mW (7 dBm) and 0.5 mW (−3 dBm) for LEDs. This allows ILDs to provide a higher drive power and to be used for systems that operate over longer distances.
3. ILDs can be used at higher bit rates than can LEDs.
4. ILDs generate monochromatic light, which reduces chromatic or wavelength dispersion.

Disadvantages of ILDs

1. ILDs are typically on the order of 10 times more expensive than LEDs.
2. Because ILDs operate at higher powers, they typically have a much shorter lifetime than LEDs.
3. ILDs are more temperature dependent than LEDs.

LIGHT DETECTORS

There are two devices that are commonly used to detect light energy in fiber optic communications receivers; PIN (positive-intrinsic-negative) diodes and APD (avalanche photodiodes).

PIN Diodes

A *PIN diode* is a *depletion-layer photodiode* and is probably the most common device used as a light detector in fiber optic communications systems. Figure 22-34 shows the basic construction of a PIN diode. A very lightly doped (almost pure or intrinsic) layer of *n*-type semiconductor material is sandwiched between the junction of the two heavily doped *n*- and *p*-type contact areas. Light enters the device through a very small window and falls on the carrier-void intrinsic material. The intrinsic material is made thick enough so that most of the photons that enter the device are absorbed by this layer. Essentially, the PIN photodiode operates just the opposite of an LED. Most of the photons are absorbed by electrons in the valence band of the intrinsic material. When the photons are absorbed, they add sufficient energy to generate carriers in the depletion region and allow current to flow through the device.

Photoelectric effect. Light entering through the window of a PIN diode is absorbed by the intrinsic material and adds enough energy to cause electrons to move from the valence band into the conduction band. The increase in the number of electrons that move into the conduction band is matched by an increase in the number of holes in the valence band. To cause current to flow in a photodiode, sufficient light must be absorbed to give valence electrons enough energy to jump the energy gap. The energy gap for silicon is 1.12 eV (electrode volts). Mathematically, the operation is as follows.

For silicon, the energy gap (E_g) equals 1.12 eV:

$$1 \text{ eV} = 1.6 \times 10^{-19} \text{ J}$$

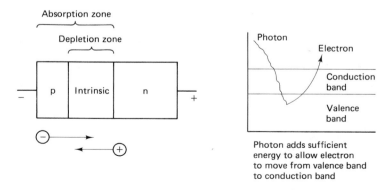

FIGURE 22-34 PIN photodiode construction.

Thus the energy gap for silicon is

$$E_g = (1.12 \text{ eV}) \left(1.6 \times 10^{-19} \frac{\text{J}}{\text{eV}} \right) = 1.792 \times 10^{-19} \text{J}$$

and

$$\text{energy } (E) = hf$$

where

$$h = \text{Planck's constant} = 6.6256 \times 10^{-34} \text{ J/Hz}$$
$$f = \text{frequency (Hz)}$$

Rearranging and solving for f yields

$$f = \frac{E}{h}$$

For a silicon photodiode,

$$f = \frac{1.792 \times 10^{19} \text{ J}}{6.6256 \times 10^{-34} \text{ J/Hz}}$$

$$= 2.705 \times 10^{14} \text{ Hz}$$

Converting to wavelength yields

$$\lambda = \frac{c}{f} = \frac{3 \times 10^8 \text{ m/s}}{2.705 \times 10^{14} \text{ Hz}} = 1109 \text{ nm/cycle}$$

Consequently, light wavelengths of 1109 nm or shorter, or light frequencies of 2.705 $\times$ 10^{14} Hz or higher, are required to generate enough electrons to jump the energy gap of a silicon photodiode.

Avalanche Photodiodes

Figure 22-35 shows the basic construction of an *avalanche photodiode* (APD). An APD is a *pipn* structure. Light enters the diode and is absorbed by the thin, heavily doped *n*-layer. This causes a high electric field intensity to be developed across the *i-p-n* junction. The high reverse-biased field intensity causes impact ionization to occur near the breakdown voltage of the junction. During impact ionization, a carrier can gain

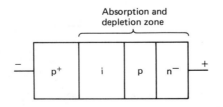

FIGURE 22-35   Avalanche photodiode construction.

sufficient energy to ionize other bound electrons. These ionized carriers, in turn, cause more ionizations to occur. The process continues like an avalanche and is, effectively, equivalent to an internal gain or carrier multiplication. Consequently, APDs are more sensitive than PIN diodes and require less additional amplification. The disadvantages of APDs are relatively long transit times and additional internally generated noise due to the avalanche multiplication factor.

Characteristics of Light Detectors

The most important characteristics of light detectors are:

Responsivity. This is a measure of the conversion efficiency of a photodetector. It is the ratio of the output current of a photodiode to the input optical power and has the unit of amperes/watt. Responsivity is generally given for a particular wavelength or frequency.

Dark current. This is the leakage current that flows through a photodiode with no light input. Dark current is caused by thermally generated carriers in the diode.

Transit time. This is the time it takes a light-induced carrier to travel across the depletion area. This parameter determines the maximum bit rate possible with a particular photodiode.

Spectral response. This parameter determines the range or system length that can be achieved for a given wavelength. Generally, relative spectral response is graphed as a function of wavelength or frequency. Figure 22-36 is an illustrative example of a spectral response curve. It can be seen that this particular photodiode more efficiently absorbs energy in the range 800 to 820 nm.

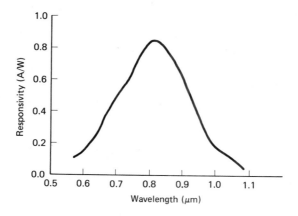

FIGURE 22-36 Spectral response curve.

QUESTIONS

22-1. Define a fiber optic system.

22-2. What is the relationship between information capacity and bandwidth.

22-3. What development in 1951 was a substantial breakthrough in the field of fiber optics? In 1960? In 1970?

22-4. Contrast the advantages and disadvantages of fiber optic cables and metallic cables.

22-5. Outline the primary building blocks of a fiber optic system.

22-6. Contrast glass and plastic fiber cables.

22-7. Briefly describe the construction of a fiber optic cable.

22-8. Define the following terms: velocity of propagation, refraction, and refractive index.

22-9. State Snell's law for refraction and outline its significance in fiber optic cables.

22-10. Define *critical angle*.

22-11. Describe what is meant by mode of operation; by index profile.

22-12. Describe a step-index fiber cable; a graded-index cable.

22-13. Contrast the advantages and disadvantages of step-index, graded-index, single-mode propagation, and multimode propagation.

22-14. Why is single-mode propagation impossible with graded-index fibers?

22-15. Describe the source-to-fiber aperture.

22-16. What are the acceptance angle and the acceptance cone for a fiber cable?

22-17. Define *numerical aperture*.

22-18. List and briefly describe the losses associated with fiber cables.

22-19. What is *pulse spreading*?

22-20. Define *pulse spreading constant*.

22-21. List and briefly describe the various coupling losses.

22-22. Briefly describe the operation of a light-emitting diode.

22-23. What are the two primary types of LEDs?

22-24. Briefly describe the operation of an injection laser diode.

22-25. What is lasing?

22-26. Contrast the advantages and disadvantages of ILDs and LEDs.

22-27. Briefly describe the function of a photodiode.

22-28. Describe the photoelectric effect.

22-29. Explain the difference between a PIN diode and an APD.

22-30. List and describe the primary characteristics of light detectors.

PROBLEMS

22-1. Determine the wavelengths in nanometers and angstroms for the following light frequencies.

    (a) 3.45×10^{14} Hz

    (b) 3.62×10^{14} Hz

    (c) 3.21×10^{14} Hz

22-2. Determine the light frequency for the following wavelengths.
   (a) 670 nm
   (b) 7800 Å
   (c) 710 nm

22-3. For a glass ($n = 1.5$)/quartz ($n = 1.38$) interface and an angle of incidence of 35°, determine the angle of refraction.

22-4. Determine the critical angle for the fiber described in Problem 22-3.

22-5. Determine the acceptance angle for the cable described in Problem 22-3.

22-6. Determine the numerical aperture for the cable described in Problem 22-3.

22-7. Determine the maximum bit rate for RZ and NRZ encoding for the following pulse-spreading constants and cable lengths.
   (a) $\Delta t = 10$ ns/m, $L = 100$ m
   (b) $\Delta t = 20$ ns/m, $L - 1000$ m
   (c) $\Delta t = 2000$ ns/km, $L = 2$ km

22-8. Determine the lowest light frequency that can be detected by a photodiode with an energy gap $= 1.2$ eV.

Appendix A

THE SMITH CHART

INTRODUCTION

Mathematical solutions for transmission line impedances are laborious. Therefore, it is common practice to use charts to graphically solve transmission line impedance problems. Equation A-1 is the formula for determining the impedance at a given point on a transmission line.

$$Z = Z_O \left[\frac{Z_L + jZ_O \tan\beta \ S}{Z_O + jZ_L \tan\beta \ S} \right] \tag{A-1}$$

where

Z = line impedance at a given point
Z_L = load impedance
Z_O = line characteristic impedance
βS = distance from the load to the point where the impedance value is to be calculated

There are several charts available on which the properties of transmission lines are graphically presented. However, the most useful graphical representations are those that give the impedance relations that exist along a lossless transmission line for varying load conditions. The *Smith chart* is the most widely used transmission-line calculator of this type. The Smith chart is a special kind of impedance coordinate system that portrays the relationship of impedance at any point along a uniform transmission line to the impedance at any other point on the line.

The Smith chart was developed by Philip H. Smith at Bell Telephone Laboratories

IMPEDANCE OR ADMITTANCE COORDINATES

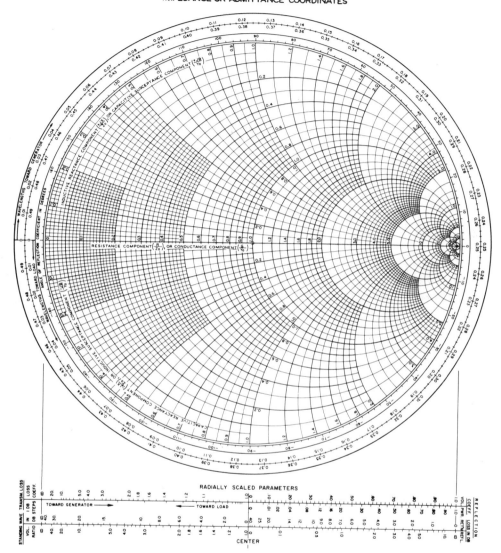

FIGURE A-1 Smith chart, transmission line calculator.

and was originally described in an article entitled "Transmission Line Calculator"(*Electronics*, January, 1939). A Smith chart is shown in Figure A-1. This chart is based on two sets of *orthogonal* circles. One set represents the ratio of the resistive component of the line impedance (*R*) to the characteristic impedance of the line (Z_O), which for a lossless line is also purely resistive. The second set of circles represents the ratio of the reactive component of the line impedance $(\pm jX)$ to the characteristic impedance of the line (Z_O). Parameters plotted on the Smith chart include:

1. Impedance (or admittance) at any point along a transmission line
    a. Reflection coefficient magnitude (Γ)
    b. Reflection coefficient angle in degrees
2. Length of transmission line between any two points in wavelengths
3. Attenuation between any two points
    a. Standing-wave loss coefficient
    b. Reflection loss
4. Voltage or current standing-wave ratio
    a. Standing-wave ratio
    b. Limits of voltage and currrent due to standing waves

SMITH CHART DERIVATION

The characteristic impedance of a transmission line, Z_O, is made up of both *real* and *imaginary* components of either sign (i.e., $Z = \pm R \pm jX$). Figure A-2(a) shows three typical circuit elements, and Figure A-2(b) shows their impedances graphed on a *rectangular* coordinate plane. All values of Z that correspond to passive networks must be plotted on or to the right of the imaginary axis of the Z plane (this is because a negative real component implies that the network is capable of supplying energy). In order to

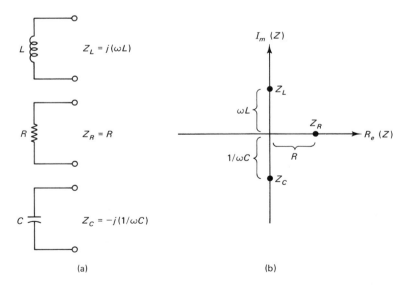

FIGURE A-2 (a) Typical circuit elements; (b) impedances graphed on rectangular coordinate plane. (Note: ω is the angular frequency at which Z is measured.)

display the impedance of all possible passive networks on a rectangular plot, the plot must extend to infinity in three directions $(+R, +jX,$ and $-jX)$. The Smith chart overcomes this limitation by plotting the complex *reflection coefficient*,

$$\Gamma = \frac{z - 1}{z + 1} \tag{A-2}$$

where

$z =$ the impedance normalized to the characteristic impedance (i.e., $z = Z/Z_O$)

Equation A-2 shows that for all passive impedance values, z, the magnitude of Γ is between 0 and 1. Also, since $|\Gamma| \leq 1$, the entire right side of the z plane can be mapped onto a circular area on the Γ plane. The resulting circle has a radius $r = 1$ and a center at $\Gamma = 0$, which corresponds to $z = 1$ or $Z = Z_O$.

Lines of constant $R_e(z)$. Figure A-3(a) shows the rectangular plot of four lines of constant resistance $R_e(z) = 0, 0.5, 1,$ and 2. For example, any impedance with a real part $R_e = 1$ will lie on the $R = 1$ line. Impedances with a positive reactive component (X_L) will fall above the real axis, while impedances with a negative reactive component (X_C) will fall below the real axis. Figure A-3b shows the same four values of R mapped onto the Γ plane. $R_e(z)$ are now circles of $Re(\Gamma)$. However, inductive impedances are still transferred to the area above the horizontal axis, and capacitive impedances are transferred to the area below the horizontal axis. The primary difference between the two graphs is that with the circular plot, the lines no longer extend to infinity. The infinity points all meet on the plane at a distance of 1, to the right of the origin. This implies, of course, that for $z = \infty$ (whether real, inductive, or capacitive), $\Gamma = 1$.

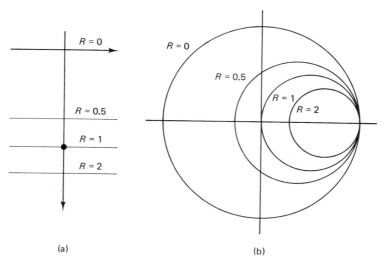

(a)   (b)

FIGURE A-3   **(a) Rectangular plot; (b) Γ plane.**

Lines of constant $I_m(z)$.   Figure A-4(a) shows the rectangular plot of three lines of constant inductance ($+jX = 0.5$, 1, and 2), three lines of constant capacitance ($-jX = -0.5$, -1, and -2), and a line of zero reactance ($jX = 0$). Figure A-4(b) shows the same seven values of jX plotted onto the plane. It can be seen that all values of infinite magnitude again meet at $\Gamma = 1$. The entire rectangular z plane curls to the right and its three axes (which previously extended infinitely) meet at the intersection of the $\Gamma = 1$ circle and the horizontal axis.

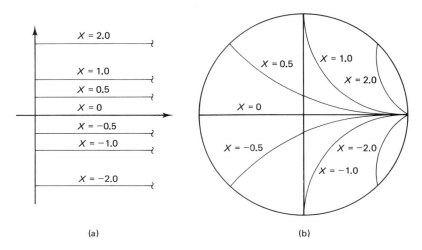

(a)   (b)

FIGURE A-4   (a) Rectangular plot; (b) Γ plane.

Impedance inversion (admittance).   *Admittance* (Y) is the mathematical inverse of Z (i.e., $Y = 1/Z$). Y, or for that matter any complex number, can be found graphically using the Smith chart by simply plotting z on the complex Γ plane, then rotating this point 180° about $\Gamma = 0$. By rotating every point on the chart by 180°, a second set of coordinates (the y coordinates) can be developed that are an inverted mirror image of the original chart. See Figure A-5(a). Occasionally, the admittance coordinates are super-imposed on the same chart as the impedance coordinates. See Figure A-5(b). Using the combination chart, both impedance and admittance values can be read directly by using the proper set of coordinates.

Complex conjugate.   The *complex conjugate* can easily be determined using the Smith chart by simply reversing the sign of the angle of Γ. On the Smith chart, Γ is usually written in polar form and angles become more negative (phase lagging) when

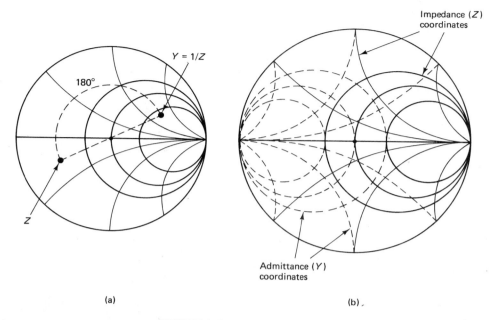

FIGURE A-5 Impedance inversion.

rotated in a clockwise direction around the chart. Hence, 0° is on the right end of the real axis and ±180° is on the left end. For example, let $\Gamma = 0.5\underline{/+150°}$. The complex conjugate, Γ', is $0.5\underline{/-150°}$. In Figure A-6, it is shown that Γ' is found by mirroring Γ about the real axis.

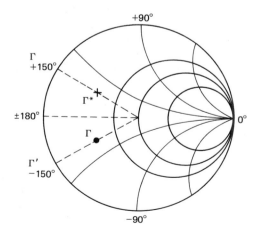

FIGURE A-6 Complex conjugate.

Plotting Impedance, Admittance, and SWR on the Smith Chart

Any impedance Z can be plotted on the Smith chart by simply *normalizing* the impedance value to the characteristic impedance (i.e., $z = Z/Z_O$) and plotting the real and imaginary

IMPEDANCE OR ADMITTANCE COORDINATES

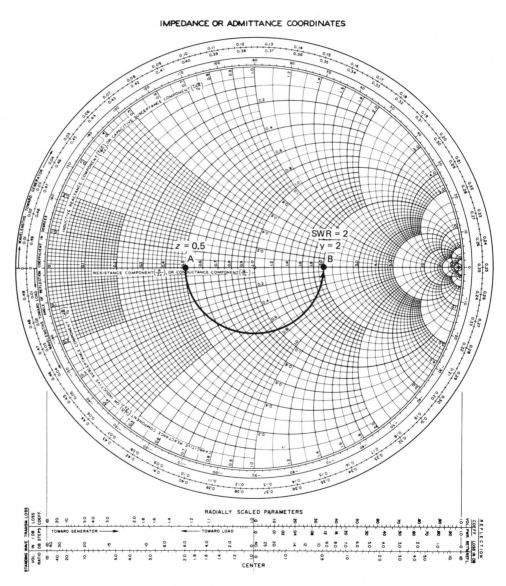

FIGURE A-7　Resistive impedance.

parts. For example, for a characteristic impedance $Z_O = 50$ ohms and an impedance $Z = 25$ ohms resistive, the normalized impedance z is determined as follows:

$$z = \frac{Z}{Z_O} = \frac{25}{50} = 0.5$$

Because z is purely resistive, its plot must fall directly on the horizontal axis ($\pm jX = 0$). $Z = 25$ is plotted on Figure A-7 at point A (i.e., $z = 0.5$). Rotating $180°$ around the chart gives a normalized admittance value $y = 2$ (where $y = Y/Y_O$). y is plotted on Figure A-7 at point B.

As previously stated, a very important characteristic of the Smith chart is that any lossless line can be represented by a circle having its origin at $1 \pm j0$ (i.e., the center of the chart) and radius equal to the distance between the origin and the impedance plot. Therefore, the *standing wave ratio*, *SWR*, corresponding to any particular circle is equal to the value of Z/Z_O at which the circle crosses the horizontal axis on the right side of the chart. Therefore, for this example, SWR $= 0.5$ (i.e., $Z/Z_O = 25/50 = 0.5$). It also should be noted that any impedance or admittance point can be rotated $180°$ by simply drawing a straight line from the point through the center of the chart to where the line intersects the circle on the opposite side.

For a characteristic impedance $Z_O = 50$ and an inductive load $Z = +j25$, the normalized impedance z is determined as follows:

$$z = \frac{Z}{Z_O} = \frac{+jX}{Z_O} = \frac{+j25}{50} = +j0.5$$

Because z is purely inductive, its plot must fall on the $R = 0$ axis which is the outer circle on the chart. $z = +j0.5$ is plotted on Figure A-8 at point A and its admittance $y = -j2$ is graphically found by simply rotating $180°$ around the chart (point B). SWR for this example must lie on the far right end of the horizontal axis which is plotted at point C and corresponds to SWR ∞, which is inevitable for a purely reactive load. SWR is plotted at point C.

For a complex impedance $Z = 25 + j25$, z is determined as follows:

$$z = \frac{25 + j25}{50} = 0.5 + j0.5$$

$$z = 0.707 \;\underline{/45°}$$

therefore

$$Z = 0.707 \;\underline{/45°} \times 50 = 35.35 \;\underline{/45°}$$

and

$$Y = \frac{1}{35.35 \;\underline{/45°}} = 0.02829 \;\underline{/-45°}$$

thus

$$y = \frac{Y}{Y_O} = \frac{0.02829}{0.02} = 1.414$$

and

$$y = 1 - j1$$

IMPEDANCE OR ADMITTANCE COORDINATES

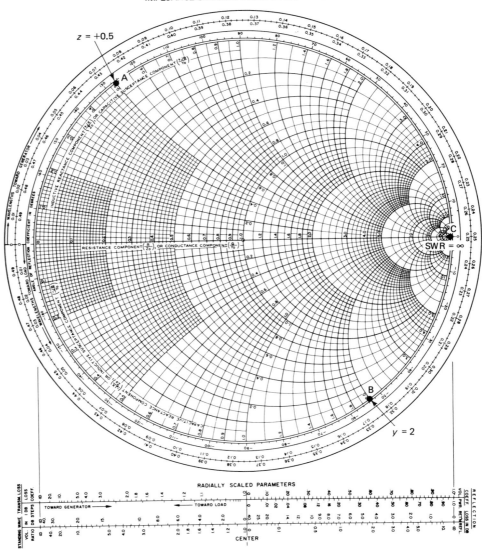

FIGURE A-8   Inductive load.

z is plotted on the Smith chart by locating the point where the $R = 0.5$ arc intersects the $X = 0.5$ arc on the top half of the chart. $z = 0.5 + j0.5$ is plotted on Figure A-9 at point A, and y is plotted at point B $(1 - j1)$. From the chart, SWR is approximately 2.6 (point C).

IMPEDANCE OR ADMITTANCE COORDINATES

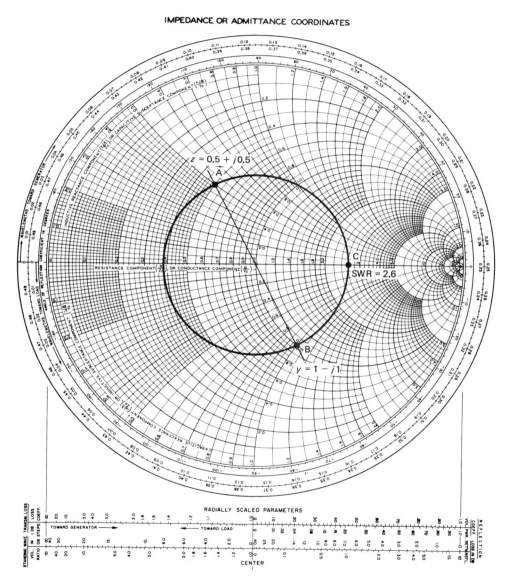

FIGURE A-9   Complex impedance.

Input Impedance and the Smith Chart

The Smith chart can be used to determine the input impedance of a transmission line at any distance from the load. The two outermost scales on the Smith chart indicate distance in *wavelengths* (see Figure A-1). The outside scale gives distance from the load toward the generator and increases in a clockwise direction, and the second scale

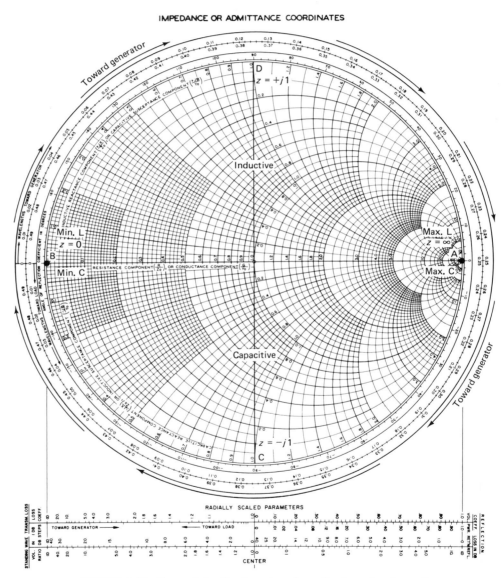

FIGURE A-10   **Transmission line input impedance for shorted and open line.**

gives distance from the source toward the load and increases in a counterclockwise direction. However, neither scale necessarily indicates the position of either the source or the load. One complete revolution (360°) represents a distance of one-half wavelength (0.5 λ), half of a revolution (180°) represents a distance of one-quarter wavelength (0.25 λ), etc.

A transmission line that is terminated in an open circuit has an impedance at the open end which is purely resistive and equal to infinity (Chapter 9). On the Smith chart, this point is plotted on the right end of the $X = 0$ line (point A on Figure A-10). As you move toward the source (generator) the input impedance is found by rotating around the chart in a clockwise direction. It can be seen that input impedance immediately becomes capacitive and maximum. As you rotate farther around the circle (move toward the generator), the capacitance decreases to a normalized value of unity (i.e., $z = -j1$) at a distance of one-eighth wavelength from the load (point C on Figure A-10) and a minimum value just short of one-quarter wavelength. At a distance of one-quarter wavelength, the input impedance is purely resistive and equal to 0 Ω (point B on Figure A-10). As stated in Chapter 9, there is an impedance inversion every one-quarter wavelength on a transmission line. Moving just past one-quarter wavelength, the impedance becomes inductive and minimum, then the inductance increases to a normalized value of unity (i.e., $z = +j1$) at a distance of three-eighths wavelength from the load (point D on Figure A-10) and a maximum value just short of one-half wavelength. At a distance of one-half wavelength, the input impedance is again purely resistive and equal to infinity (return to point A on Figure A-10). The results of the preceding analysis are identical to those achieved with phasor analysis in Chapter 9 and plotted in Figure 9-20.

A similar analysis can be done with a transmission line that is terminated in a short circuit, although the opposite impedance variations are achieved as with an open load. At the load, the input impedance is purely resistive and equal to 0. Therefore, the load is located at point B on Figure A-10 and point A represents a distance one-quarter wavelength from the load. Point D is a distance of one-eighth wavelength from the load and point C a distance of three-eighths wavelength. The results of such an analysis are identical to those achieved with phasors in Chapter 9 and plotted in Figure 9-21.

For a transmission line terminated in a purely resistive load not equal to Z_o, Smith chart analysis is very similar to the process described in the preceding section. For example, for a load impedance $Z_L = 37.5$ Ω resistive and a transmission line characteristic impedance $Z_O = 75$ Ω, the input impedance at various distances from the load is determined as follows:

1. The normalized load impedance z is

$$z = \frac{Z_L}{Z_O} = \frac{37.5}{75} = 0.5$$

2. $z = 0.5$ is plotted on the Smith chart (point A on Figure A-11). A circle is drawn that passes through point A with its center located at the intersection of the $R = 1$ circle and the $x = 0$ arc.

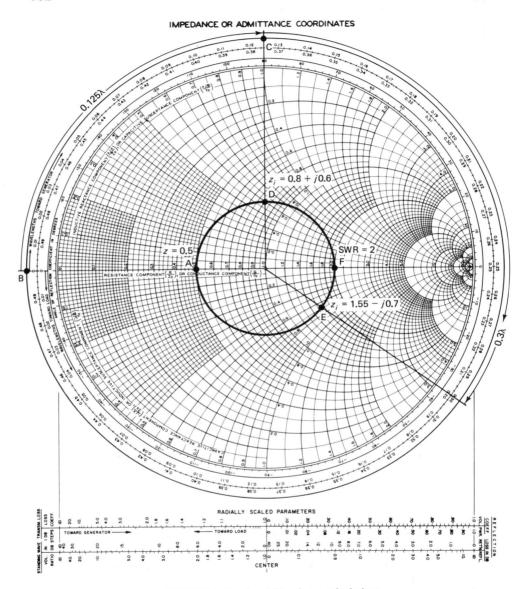

FIGURE A-11   Input impedance calculations.

3. SWR is read directly from the intersection of the $z = 0.5$ circle and the $X = 0$ line on the right side (point F), SWR $= 2$. The impedance circle can be used to describe all impedances along the transmission line. Therefore, the input impedance (Z_i) at a distance of 0.125λ from the load is determined by extending the z circle to the outside of the chart, moving point A to a similar position on the outside scale (point B on Figure A-11), and moving around the scale in a clockwise direction a distance of 0.125 λ.

4. Rotate from point B a distance equal to the length of the transmission line (point C on Figure A-11). Transfer this point to a similar position on the $z = 0.5$ circle (point D on Figure A-11). The normalized input impedance is located at point D ($0.8 + j0.6$). The actual input impedance is found by multiplying the normalized impedance by the characteristic impedance of the line. Therefore, the input impedance Z_i is

$$Z_i = (0.8 + j0.6)75 = 60 + j45$$

Input impedances for other distances from the load are determined in the same way. Simply rotate in a clockwise direction from the initial point a distance equal to the length of the transmission line. At a distance of 0.3 λ from the load, the normalized input impedance is found at point E ($z = 1.55 - j0.7$ and $Z_i = 116.25 - j52.5$). For distances greater than 0.5 λ, simply continue rotating around the circle with each complete rotation accounting for 0.5 λ. A length of 1.125 λ is found by rotating around the circle two complete revolutions and an additional 0.125 λ.

EXAMPLE A-1

Determine the input impedance and SWR for a transmission line 1.25 λ long with a characteristic impedance $Z_O = 50 \ \Omega$ and a load impedance $Z_L = 30 + j40 \ \Omega$.

Solution   The normalized load impedance z is

$$z = \frac{30 + j40}{50} = 0.6 + j0.8$$

z is plotted on Figure A-12 at point A and the impedance circle is drawn. SWR is read off the Smith chart from point B.

$$SWR = 3.4$$

The input impedance 1.3 λ from the load is determined by rotating from point C 1.25 in a clockwise direction. Two complete revolutions account for 1 λ. Therefore, the additional 0.25 λ is simply added to point C.

$$0.12 \ \lambda + 0.25 \ \lambda = 0.37 \ \lambda \qquad \text{point D}$$

Point D is moved to a similar position on the $z = 0.6 + j0.8$ circle (point E) and the input impedance is read directly from the chart.

$$z_i = 0.55 - j0.9$$
$$Z_i = z_i Z_O = 50(0.55 - j0.9) = 27.5 - j45$$

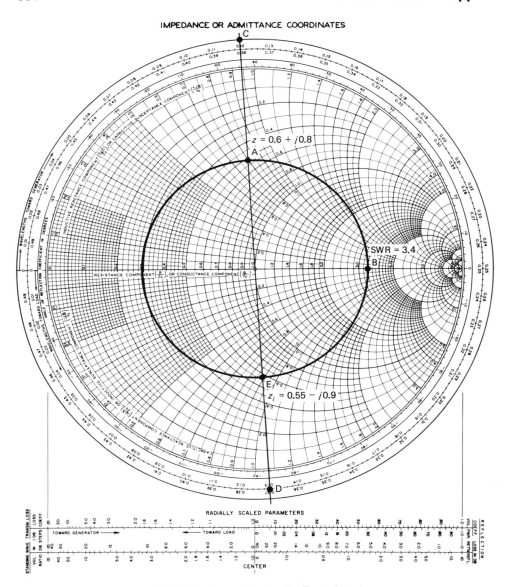

IMPEDANCE OR ADMITTANCE COORDINATES

$z = 0.6 + j0.8$

SWR = 3.4

$z_i = 0.55 - j0.9$

RADIALLY SCALED PARAMETERS

FIGURE A-12 Smith chart for Example A-1.

Quarter-wave Transformer Matching with the Smith Chart

As described in Chapter 9, a length of transmission line acts like a transformer (i.e., there is an impedance inversion every one-quarter wavelength). Therefore, a transmission line with the proper length located the correct distance from the load can be used to match a load to the impedance of the transmission line. The procedure for matching a load to a transmission line with a quarter-wave transformer using the Smith chart is outlined in the following steps.

1. A load $Z_L = 75 + j50 \ \Omega$ can be matched to a 50-Ω source with a quarter-wave transformer. The normalized load impedance z is

$$z = \frac{75 + j50}{50} = 1.5 + j1$$

2. $z = 1.5 + j1$ is plotted on the Smith chart (point A, Figure A-13) and the impedance circle is drawn.

3. Extend point A to the outermost scale (point B). The characteristic impedance of an ideal transmission line is purely resistive. Therefore, if a quarter-wave transformer is located at a distance from the load where the input impedance is purely resistive, the transformer can match the transmission line to the load. There are two points on the impedance circle where the input impedance is purely resistive: either end of the horizontal $X = 0$ line (points C and D on Figure A-13). Therefore, the distance from the load to a point where the input impedance is purely resistive is determined by simply calculating the distance in wavelengths from point B on Figure A-13 to either point C or D, whichever is the shortest. The distance from Point B to point C is

point C	0.250 λ
−point D	−0.192 λ
distance	0.058 λ

If a quarter-wave transformer is placed 0.058 λ from the load, the input impedance is read directly from Figure A-13, $z_i = 2.4$ (point E).

4. Note that 2.4 is also the SWR of the mismatched line and is read directly from the chart.

5. The actual input impedance $Z_i = 50(2.4) = 120 \ \Omega$. The characteristic impedance of the quarter-wave transformer is determined from Equation 9-32.

$$Z_O' = \sqrt{Z_O Z_1} = \sqrt{50 \times 120} = 77.5 \ \Omega$$

Thus, if a quarter-wave length of a 77.5-Ω transmission line is placed 0.058 λ from the load, the line is matched. It should be noted that a quarter-wave transformer does not totally eliminate standing waves on the transmission line. It simply eliminates them from the transformer back to the source. Standing waves are still present on the line between the transformer and the load.

IMPEDANCE OR ADMITTANCE COORDINATES

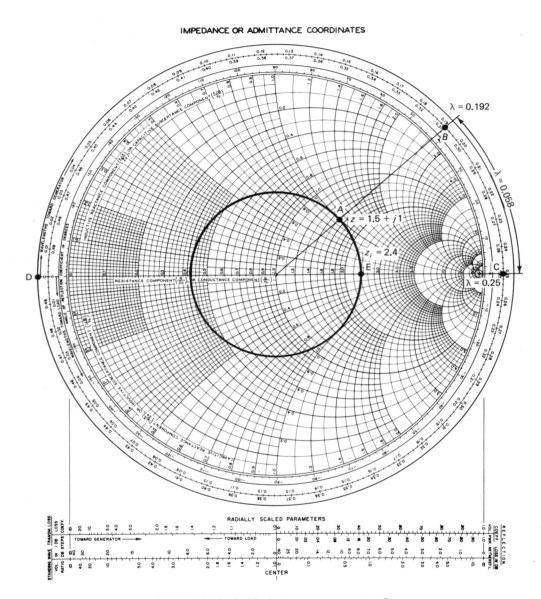

RADIALLY SCALED PARAMETERS

FIGURE A-13   Smith chart, quarter-wave transformer.

EXAMPLE A-2

Determine the SWR, characteristic impedance of a quarter-wave transformer, and the distance the transformer must be placed from the load to match a 75-Ω transmission line to a load $Z_L = 25 - j50$.

Solution The normalized load impedance z is

$$z = \frac{25 - j50}{75} = 0.33 - j0.66$$

z is plotted at point A on Figure A-14 and the corresponding impedance circle is drawn. The SWR is read directly from point F

$$\text{SWR} = 4.6$$

The closest point on the Smith chart where z_i is purely resistive is point D. Therefore, the distance that the quarter-wave transformer must be placed from the load is

point D	0.5 λ
−point B	0.4 λ
distance	0.1 λ

The normalized input impedance is found by moving point D to a similar point on the z circle (point E), $z_i = 2.2$. The actual input impedance is

$$Z_i = 2.2(75) = 165 \ \Omega$$

The characteristic impedance of the quarter-wavelength transformer is again found from Equation 9-32.

$$Z_O' = \sqrt{75 \times 165} = 111.24$$

Stub Matching with the Smith Chart

As described in Chapter 9, shorted and open stubs can be used to cancel the reactive portion of a complex load impedance and thus match the load to the transmission line. Shorted stubs are preferred because open stubs have a greater tendency to radiate.

Matching a complex load, $Z_L = 50 - j100$, to a 75-Ω transmission line using a shorted stub is accomplished quite simply with the aid of a Smith chart. The procedure is outlined in the following steps:

1. The normalized load impedance z is

$$z = \frac{50 - j100}{75} = 0.67 - j1.33$$

2. $z = 0.67 - j1.33$ is plotted on the Smith chart shown in Figure A-15 at point A and the impedance circle is drawn in. Because stubs are shunted across the load (i.e., placed in parallel with the load), admittances are used rather than impedances to simplify the calculations, and the circles and arcs on the Smith chart are now used for conductance and susceptance.

IMPEDANCE OR ADMITTANCE COORDINATES

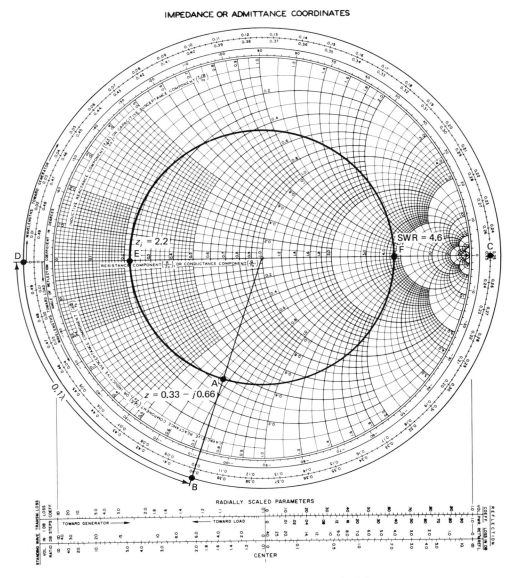

FIGURE A-14 Smith chart for Example A-2.

3. The normalized admittance y is determined from the Smith chart by simply rotating the impedance plot, z, 180°. This is done on the Smith chart by simply drawing a line from point A through the center of the chart to the opposite side of the circle (point B).

4. Rotate the admittance point clockwise to a point on the impedance circle where

IMPEDANCE OR ADMITTANCE COORDINATES

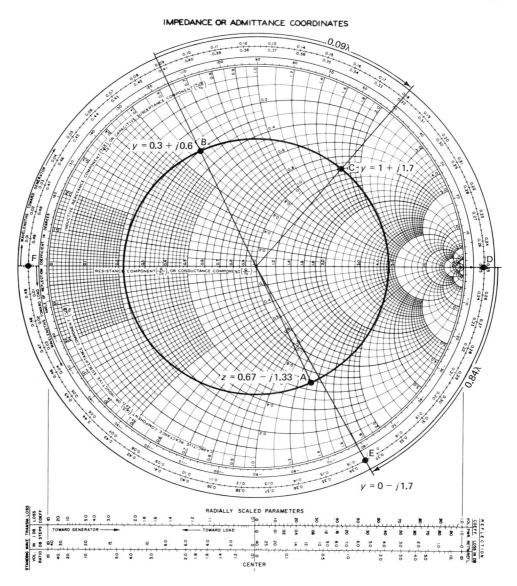

FIGURE A-15 Stub matching, Smith chart.

it intersects the $R = 1$ circle (point C). The real component of the input impedance at this point is equal to the characteristic impedance Z_O, $Z_{in} = R \pm jX$ where $R = Z_O$). At point C, the admittance $y = 1 + j1.7$.

5. The distance from point B to point C is how far from the load the stub must be placed. For this example, the distance is $0.18\,\lambda - 0.09\,\lambda = 0.09\,\lambda$. The stub must have an impedance with a zero resistive component and a susceptance that has the opposite polarity (i.e., $y = 0 - j1.7$).

6. To find the length of the stub with an admittance $y = 0 - j1.7$, move around the outside circle of the Smith chart (the circle where $R = 0$) beginning at point D until an admittance $y = 1.7$ is found (point E). You begin at point D because a shorted

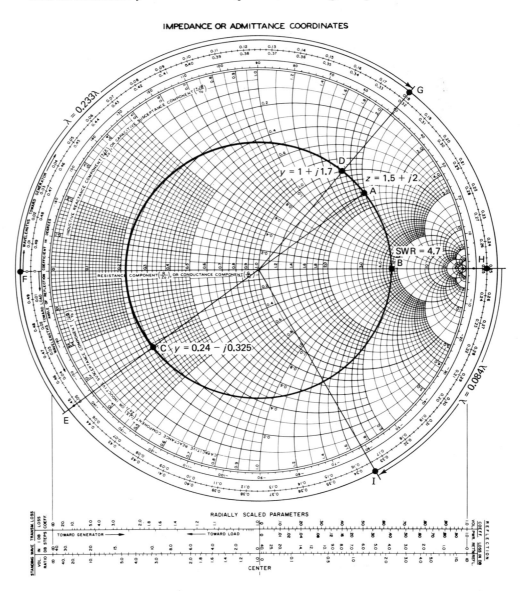

FIGURE A-16 Smith chart for Example A-3.

stub has minimum resistance ($R = 0$) and, consequently, a susceptance B = infinity. Point D is such a point. (If an open stub were used, you would begin your rotation at the opposite side of the $X = 0$ line — point F).

7. The distance from point D to point E is the length of the stub. For this example, the length of the stub is $0.334\ \lambda - 0.25\ \lambda = 0.084\ \lambda$.

EXAMPLE A-3

For a transmission line with a characteristic impedance $Z_O = 300\ \Omega$ and a load with a complex impedance $Z_L = 450 + j600$; determine SWR, the distance a shorted stub must be placed from the load to match the load to the line, and the length of the stub.

Solution The normalized load impedance z is

$$z = \frac{450 + j600}{300} = 1.5 + j2$$

$z = 1.5 + j2$ is plotted on Figure A-16 at point A and the corresponding impedance circle is drawn.

SWR is read directly from the chart at point B.

$$\text{SWR} = 4.7$$

Rotate point A 180° around the impedance circle to determine the normalized admittance.

$$y = 0.24 - j0.325 \text{ (point C)}$$

To determine the distance from the load to the stub, rotate clockwise around the outer scale beginning from point C until the circle intersects the $R = 1$ circle (point D).

$$y = 1 + j1.7$$

The distance from point C to point D is the sum of the distances from point E to point F and point F to point G.

$$
\begin{aligned}
\text{E to F} &= 0.5\ \lambda - 0.447\ \lambda &= 0.053\ \lambda \\
+\,\text{F to G} &= \underline{0.18\ \lambda - 0\ \lambda} &= \underline{0.18\ \ \lambda} \\
& \text{total distance} &= 0.233\ \lambda
\end{aligned}
$$

To determine the length of the shorted stub, calculate the distance from the $y = \infty$ point (point H) to the $y = 0 - j1.7$ point (point I).

$$\text{stub length} = 0.334\ \lambda - 0.25\ \lambda = 0.084\ \lambda$$

SOLUTIONS TO ODD-NUMBERED PROBLEMS

CHAPTER 1

1-1. $V_1 = 10.19V_p$, $V_2 = 0V_p$, $V_3 = 3.4V_p$, $V_4 = 0V$, $V_5 = 2.04V_p$

1-3. %2nd order = 50%, %3rd order = 25%, %THD = 55.9%

1-5. 2nd order intermodulation cross products = 1883, 1890, 1892, 1899, 1901, and 1908 Hz

1-7. (a) $N = 9.3 \times 10^{-15}$ w, $N = -110.3$ dBm
   (b) 2.5 dB decrease
   (c) 3 dB increase

1-9. Duty cycle = 20%

1-11. %2nd IMD = 0.77%

CHAPTER 2

2-1. (a) decrease of 1600 Hz
   (b) decrease of 3200 Hz
   (c) increase of 3200 Hz

2-3. $F_o = 159.2$ kHz

2-5. $C_d = 0.003$ μF

2-7. Nonlinear amplifier with two input frequencies: 1 and 5 kHz. The cross product frequencies are 4 and 6 kHz.

CHAPTER 3

3-1. $m = 0.25$, M $= 25\%$

3-3. (a) $V_c = 25V_p$ (b) $E_m = 15V_p$ (c) $m = 0.6$, (d) $M = 60\%$

3-5. $V_c = 16V_p$, $V_{usb} = V_{lsb} = 3.2V_p$

3-7. (a) $V_{usb} = V_{lsb} = 4V_p$ (b) $V_c = 12V_p$ (c) $E_m = 8V_p$
(d) $m = 0.67$ (e) $M = 67\%$

3-9. (a) $P_{sbt} = 20W_p$ (b) $1020W_p$

3-11. $A_o = 33$, $A_{max} = 51$, $A_{min} = 15$

3-13. (a) $A_{max} = 112$, $A_{min} = 48$
(b) $V_{max} = 0.244V_p$, $V_{min} = 0.096V_p$

3-15. (a) 190 to 210 kHz
(b) 193 and 207 kHz
(c) Maximum bandwidth B $= 20$ kHz
(d)

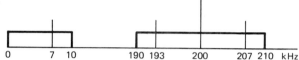

CHAPTER 4

4-1. SF $= 12$, %Selectivity $= 1200\%$

4-3. $T_e = 900°$K

4-5. IF carrier $= 600$ kHz, $IF_{usf} = 604$ kHz, $IF_{lsb} = 596$ kHz

4-7. (a) Image frequency $= 1810$ kHz
(b) IFRR $= 122$

4-9. (a) B $= 20$ kHz
(b) B $= 14276$ Hz

CHAPTER 5

5-1. (a) $F = 185$ kHz
(b) $\Delta F = 90$ kHz
(c) $200–60/6 = 120/6 = 22.\dot{5}$ kHz/V

5-3. (a) $K_v = 12000$ Hz/rad
(b) $\Delta F = 10$ kHz
(c) $V_{out} = 0.67$ V
(d) $V_d = 0.167$ V
(e) $\theta_e = 0.833$ rad
(f) $\Delta F_{max} = 18.84$ kHz

5-5. $\Delta F = 2$ kHz

5-7. $\Delta F_{max} = 31.4$ kHz

5-9. $F = 1.57$ MHz or 1.23 MHz

5-11. Channel 14

5-13. $F_c = 10$ kHz

CHAPTER 6

6-1. (a) $F_{out} = 396$ to 404 kHz

(b) 397.2 kHz and 402.8 kHz

6-3. (a)

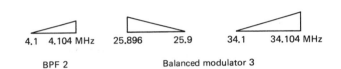

96   104 kHz   100 104 kHz   3.896   3.9   4.1   4.104 MHz

Balanced modulator 1   BPF 1   Balanced modulator 2

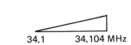

4.1   4.104 MHz   25.896   25.9   34.1   34.104 MHz

BPF 2   Balanced modulator 3

34.1   34.104 MHz

BPF 3

(b) BPF1 = 101.5 kHz, BPF2 = 4.1015 MHz, BPF3 = 34.1015 MHz

6-5. (a)   (b) 28.0022 MHz

28   28.003 MHz

6-7. (a)   (b) 497 kHz

496   500 kHz

6-9. (a)

196   204 kHz   196   200 kHz   196   204 kHz   200   204 kHz

Balanced modulator A   BPF A   Balanced modulator B   BPF B

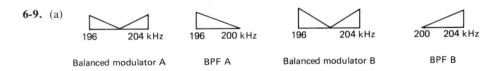

196   204 kHz   196   200 204 kHz

Hybrid   Linear summer

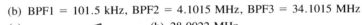
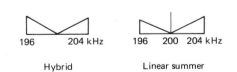

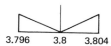  

3.796   3.8   3.804   4.196   4.2   4.204 MHz   4.196   4.2   4.204 MHz

Balanced modulator 3   BPF 3

 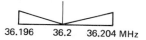

27.796   27.8   27.804   36.196   36.2   36.204 MHz

Balanced modulator 4

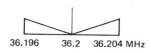

36.196   36.2   36.204 MHz

BPF 4

(b) BPFA = 197.5 kHz
BPFB = 203 kHz
BPF3 = 4.1975, 4.2, and 4.203 MHz
BPF4 = 36.1975, 36.2, and 36.203 MHz

6-11. IF = 10.602 MHz, BF0 = 10.6 MHz

6-13. F_{out} = 202 and 203 kHz

PEP = 5.76w

P_{ave} = 2.88w

CHAPTER 7

7-1. 0.5 kHz/V

7-3. (a) ΔF = 40 kHz (b) CS = 80 kHz (c) m = 20

7-5. %mod = 80%

7-7. (a) 4 sets

(b) and (c)

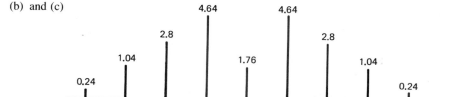

4.64   4.64

2.8   2.8

1.04   1.76   1.04

0.24   0.24

792   794   796   798   800   802   804   806   808

KHz

7-9. $\Delta F = 50$ kHz

7-11. DR = 2, B = 100 kHz

7-13.

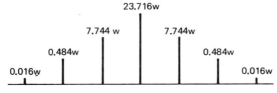

7-15. ~~0.0377~~r /89 rel.

7-17. (a) Combining network out m = 0.0036 r
Power amplifier out m = 7.2 r

(b) Combining network out $\Delta F = 7.2$ Hz
Power amplifier out $\Delta F = 14.4$ kHz

(c) $F_c = 90$ MHz

CHAPTER 8

8-1. 22 dB

8-3. $F_{lo} = 103.45$ MHz, IF = 114.15 MHz

8-5. 20.3335 to 20.4665 MHz

8-7. $V_{out} = 0.8$V

8-9. 13.3 dB

CHAPTER 9

9-1. 300,000 m, 3,000 m, 300 m, 0.3 m

9-3. $Z_o = 261$ Ω

9-5. $Z_o = 111.8$ Ω

9-7. $\Gamma = 0.05$

9-9. SWR = 12

9-11. SWR = 1.5

CHAPTER 10

10-1. $\mathcal{P} = 0.2$ μW/m²

10-3. The power density decreases by a factor of 27 (3^3).

10-5. MUF = 14.14 MHz

10-7. $\mathcal{E} = 0.0058$ V/m

10-9. The power density decreases by a factor of 16 (4^2).

10-11. $\mathcal{D} = 20$ dB

10-13. d = 8.94 miles

10-15. d = 17.89 miles

CHAPTER 11

11-1. (a) $R_r = 25$ ohms

(b) $\eta = 92.6\%$

(c) $P_r = 92.6$ W

11-3. $A_p = 38.13$ dB

11-5. EIRP = 164,000 W

11-7. 0.106 μW/m

11-9. $\mathscr{D} = 16$ dB

11-11. $\eta = 97.9\%$

11-13. $\eta = 98.2\%$

CHAPTER 12

12-1.

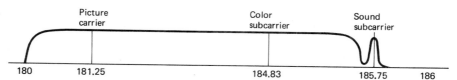

| 180 | 181.25 | | 184.83 | 185.75 | 186 |

12-3. 31.5 horizontal scan lines

12-5. 1.27 ms

12-7. Channel 55 = 721.75 MHz, Channel 66 = 787.75 MHz

12-9. I = 0.248 V, Q = -0.082 V

CHAPTER 13

13-1. 10 megabaud; minimum bandwidth = 40 MHz

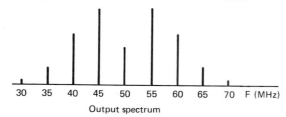

Output spectrum

13-3.

I	Q	Output expression
0	0	$-\sin \omega_c t + \cos \omega_c t$
0	1	$-\sin \omega_c t - \cos \omega_c t$
1	0	$+\sin \omega_c t + \cos \omega_c t$
1	1	$+\sin \omega_c t - \cos \omega_c t$

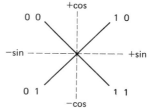

13-5. 6.67 megabaud; minimum bandwidth = 6.67 MHz

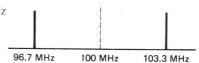

| 96.7 MHz | 100 MHz | 103.3 MHz |

13-7. 5 megabaud; minimum bandwidth = 5 MHz

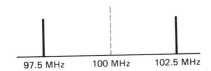

97.5 MHz   100 MHz   102.5 MHz

13-9. (a) 2 bps/Hz   (b) 3 bps/Hz   (c) 4 bps/Hz

CHAPTER 14

14-1.

	D A T A Sp	C O M M U N I C A T I O N S	EXT	BCS
b_0	0 1 0 1 0	1 1 1 1 1 0 1 1 1 0 1 1 0 1	1	0
b_1	0 0 0 0 0	1 1 0 0 0 1 0 1 0 0 0 1 1 1	1	0
b_2	1 0 1 0 0	0 1 1 1 1 1 0 0 0 1 0 1 1 0	0	0
b_3	0 0 0 0 0	0 1 1 1 0 1 1 0 0 0 1 1 1 0	0	0
b_4	0 0 1 0 0	0 0 0 0 1 0 0 0 0 1 0 0 0 1	0	0
b_5	0 0 0 0 1	0 0 0 0 0 0 0 0 0 0 0 0 0 0	0	1
b_6	1 1 1 1 0	1 1 1 1 1 1 1 1 1 1 1 1 1 1	0	0
VRC	1 1 0 1 0	0 0 1 1 1 1 0 0 1 0 0 0 1 1	1	1

14-3. Four Hamming bits

14-5. 10111001 binary or B9H

14-7. 1110001000011111 11110011111 0100111101011111 11111 1001011

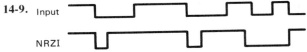

inserted zeros

14-9. Input

NRZI

CHAPTER 16

16-1. 4 kHz, FS = 8 kHz and 10 kHz, FS = 20 kHz

16-3. 10 kHz

16-5. 511 or 54 dB

16-7. −1.16 V, 0.12 V, +0.04 V, −2.52 V, +0.24 V

16-9. Resolution = 0.01 V, Qe = 0.005 V

16-11. (a) +0.01 to +0.03V
     (b) −0.01 to 0 V
     (c) +10.23 to +10.25 V
     (d) −10.23 to −10.25 V
     (e) +5.13 to +5.15 V
     (f) +6.81 to +6.83 V

CHAPTER 17

17-1. (a) 1.521 Mbps
(b) 760.5 kHz

17-3.

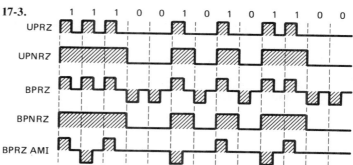

CHAPTER 18

18-1. Ch 1, 108 kHz; Ch 2, 104 kHz; Ch 3, 100 kHz; Ch 4, 96 kHz; Ch 5, 92 kHz; Ch 6, 88 kHz; Ch 7, 84 kHz; Ch 8, 80 kHz; Ch 9, 76 kHz; Ch 10, 72 kHz; Ch 11, 68 kHz; Ch 12, 64 kHz

18-3. Ch = 96 to 100 kHz, GP = 464 to 468 kHz, SG = 1144 to 1148 kHz, MG = 5100 to 5104 kHz

18-5. 5495.92 kHz   **18-7.** 4948 to 5188 kHz

CHAPTER 19

19-1. −99.23 dBm   **19-7.** −1.25 dB

19-3. 28.9 dB   **19-9.** 6.58 dB

19-5. 35 dBm

CHAPTER 20

20-1. Elevation angle = 40·, azimuth = 29· west of south

20-3. (a) 11.74 dB   (b) 19.2 dB

20-5. Radiated power = 28 dBW, EIRP = 68 dBW

20-7. −200.8 dBW   **20-11.** 19.77 dB

20-9. 26.18 dB   **20-13.** 12.5 dB

CHAPTER 21

21-1 15 transponders   **21-3.** 45 stations

CHAPTER 22

22-1. (a) 869 nm, 8690 Å
(b) 828 nm, 8280 Å
(c) 935 nm, 9350 Å

22-3. 38.57°

22-5. 36°

22-7. (a) RZ = 1 Mbps, NRZ = 500 kbps
(b) RZ = 50 kbps, NRZ = 25 kbps
(c) RZ = 250 kbp, NRZ = 125 kbps

INDEX

880

883

D